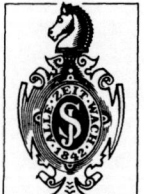

Hiroshi Tomita (Ed.)

Trace Elements in Clinical Medicine

Proceedings of the Second Meeting of the
International Society for Trace Element Research in Humans (ISTERH)
August 28–September 1, 1989, Tokyo

With 170 Figures

Springer-Verlag Tokyo Berlin Heidelberg New York
London Paris Hong Kong

HIROSHI TOMITA M.D.
Professor and Chairman
Department of Otorhinolaryngology
Nihon University School of Medicine
Ohyaguchi, Itabashi-ku
Tokyo, 173 Japan

ISBN-13:978-4-431-68122-9   e-ISBN-13:978-4-431-68120-5
DOI: 10.1007/978-4-431-68120-5

Library of Congress Cataloging-in-Publication Data
International Society for Trace Element Research in Humans. Meeting (2nd: 1989: Tokyo, Japan)
Trace elements in clinical medicine: proceedings of the Second Meetings of the International Society
for Trace Element Research in Humans (ISTERH), August 28 - September 1, 1989, Tokyo/
Hiroshi Tomita (ed.).   p.   cm.   Includes bibliographical references. ISBN-13:978-4-
431-68122-9 (U.S.) 1. Trace element deficiency diseases — Congresses. 2. Trace elements —
Pathophysiology — Congresses. I. Tomita, Hiroshi, 1930-. II. Title. [DNLM: 1. Trace Elements
— congresses. QU 130 I614t 1989] RC627.T7I58 1989 616.3'99–dc20 DNLM/DLC

© Springer-Verlag Tokyo 1990
Softcover reprint of the hardcover 1st edition 1990

The use of registered names, trademarks, etc. in this publication does not imply, even in the
absence of a specific statement, that such names are exempt from the relevant protective
laws and regulations and therefore free for general use.

Product liability: The publisher can give no guarantee for information about drug dosage and
application thereof contained in this book. In every individual case the respective user must
check its accuracy by consulting other pharmaceutical literature.

Preface

The Second Meeting of the International Society for Trace Element Research in Humans (ISTERH) was held in Tokyo from August 28 through September 1, 1989. On August 27, the day preceding the opening of the meeting, a typhoon made a direct attack on Tokyo, welcoming guests from all over the world in a rather violent way. To our great relief, the weather during the week of the meeting turned out to be exceptionally agreeable for that time of year in Tokyo. We were also pleased to see the entire scheduled course of the meeting, including the social activities, carried out smoothly and the contents of the program favorably appraised. The meeting was attended by 518 scientists from 30 countries. Recent unrest following steps toward democratization in the Communist bloc gave rise to some apprehension; therefore, we were particularly pleased to see attendants from China, Eastern Europe, and the Soviet Union. No one could possibly have predicted the drastic change in Eastern Europe that followed, but again, we were relieved to learn (by a subsequent letter) of the safety of an attendant from Rumania, who had been our greatest concern.

A total of 384 papers were contributed to the meeting. The abstracts for all have been published in the *Journal of Trace Elements in Experimental Medicine*, vol 2, No. 2/3 (1989). This proceedings carries 64 subjects introduced in the Special Session. The latest ISTERH meeting concerned the role of essential trace elements in human disease and the preservation of health. Having just finished compiling the proceedings, we realize that scanning the Table of Contents alone will be an intriguing experience for readers. The reader may find a clue to the discovery of the cause of an unknown disease or see the potentials of trace elements as curatives. From a wide range of subjects, we have chosen the following topics which we hope will be of interest:

1. Dynamics of trace elements in human beings suffering from various diseases, particularly neural, cardiovascular, hepatic, gastrointestinal, renal, and genetic disorders and various stresses
2. Symptoms of deficiency of essential trace elements and appropriate treatment
3. Dynamics of trace elements in medical treatment, particularly in clinical surgery, total parenteral nutrition, and dialysis
4. Relations between drugs having chelating effects and neural and renal diseases
5. Diagnosis of marginal zinc deficiency
6. Effects of trace elements on genes and immunity
7. Problems of nutritional requirements of trace elements
8. Various problems on quantification of trace elements

A domino effect such as that which recently swept the political scene in Eastern Europe is not uncommon in the world of science. The impetus created by the meeting resulted in the foundation of a new academic society, namely, the Japan Society for Biomedical Research on Trace Elements. We sincerely hope that the proceedings will further accelerate the move toward research on the relations between

trace elements and human disease. We extend our deepest gratitude to the many colleagues, organizations, and companies that kindly contributed to the success of the Second Meeting of ISTERH, particularly to the backup provided by Nihon University. I also wish to express my personal thanks to Nihon University School of Medicine's Associate Professor S. Takeuchi (Ph.D., Chairman of the Department of Chemistry) and Associate Professor M. Ikeda (M.D., Department of Otolaryngology) who gave me their wholehearted support in the overall preparations and management of the meeting as well as in the compilation of the proceedings.

January 1990   H. TOMITA

Second Meeting of the International Society for Trace Element Research in Humans

Organized by:
Organizing Committee of Second Meeting of the International Society for Trace
    Element Research in Humans

Sponsored by:
Foundation for Advancement of International Sciences

Supported by:
Nihon University
Science Council of Japan
Tokyo Metropolitan Government
Japan Analytical Instruments Manufacturers' Association
Japan Pediatric Society
Japanese Society of Nephrology
Japanese Society of Nutrition and Food Science
Japan Society of Obstetrics and Gynecology
Japan Society of Surgical Metabolism and Nutrition
Pharmaceutical Society of Japan
Society for Promotion of International Otorhinolaryngology (SPIO)
Society of Environment Science, Japan
The Japanese Biochemical Society
The Japan Society for Analytical Chemistry
The Japanese Society for Hygiene
The Japanese Society of Gastroenterology
The Japanese Society of Toxicological Science
Brain Science Foundation
The Foundation for Promotion of Food Science and Technology
The Great Britain-Sasakawa Foundation
Inoue Foundation for Science
Life Science Foundation of Japan
The Mochida Memorial Foundation for Medical and Pharmaceutical Research
The Naito Foundation
Shimadzu Science Foundation
Uehara Memorial Foundation
Yoshida Foundation for Science and Technology
International Union of Nutritional Science
UNESCO Club, Lyon, France

Committee Members

President:
H. Tomita (Tokyo, Japan)

Vice-president:
K. Yamaguchi (Tokyo, Japan)

Secretary General:
K. Nomiyama (Tochigi, Japan)

Treasurer:
A. Okada (Osaka, Japan)

Advisers:
S. Miyake (Tokyo, Japan)   Y. Yamane (Chiba, Japan)
K. Tsuchiya (Kitakyushu, Japan)

Executive Committee Members:
Y. Arakawa (Tokyo, Japan)   I. Matsuda (Kumamoto, Japan)
M. Ikeda (Tokyo, Japan)   K. Saito (Sapporo, Japan)
K. Imamura (Tsukuba, Japan)   K. Soda (Kyoto, Japan)
N. Imura (Tokyo, Japan)   S. Takeuchi (Tokyo, Japan)
S. Kimura (Sendai, Japan)

Scientific Program Committee Members:
J. Aaseth (Elverum, Norway)   Y. Kojima (Sapporo, Japan)
M. Abdulla (Karachi, Pakistan)   F. Marumo (Tokyo, Japan)
A.C. Alfrey (Denver, CO, USA)   T. Morishima (Tokyo, Japan)
R.A. Anderson (Beltsville, MD, USA) R.M. Parr (Vienna, Austria)
M. Anke (Jena, DDR)   A.S. Prasad (Detroit, MI, USA)
I. Bremner (Aberdeen, UK)   J.R.J. Sorenson
G. Chazot (Lyon, France)   (Little Rock, AR, USA)
T.W. Clarkson (Rochester, NY, USA)   S. Suita (Fukuoka, Japan)
D.E. Danford (Bethesda, MD, USA)   F.W. Sunderman, Jr.
D.M. Danks (Melbourne, Australia)   (Farmington, CT, USA)
K.H. Falchuk (Boston, MA, USA)   T. Suzuki (Tokyo, Japan)
H. Haraguchi (Nagoya, Japan)   Y. Takagi (Osaka, Japan)
Y. Itokawa (Kyoto, Japan)   O. Wada (Tokyo, Japan)

Steering Committee of the International Society for Trace Element Research in Humans:
Chairman: A.S. Prasad (Detroit, MI, USA)
Secretary: C. McClain (Lexington, KY, USA)
Former Secretary: M. Abdulla (Karachi, Pakistan)

Managing Subcommittee:
S.M. Ahmed (Kyoto, Japan)   H. Ishikawa (Tokyo, Japan)
S. Endo (Tokyo, Japan)   E. Ishiyama (Tokyo, Japan)
R. Haranaka (Tokyo, Japan)   M. Kawamura (Sendai, Japan)
Y. Hayashi (Tokyo, Japan)   M. Nishimuta (Tokyo, Japan)
T. Ikeda (Tokyo, Japan)   M. Oishi (Tokyo, Japan)

H. Osada (Tokyo, Japan)
T. Sasaki (Tokyo, Japan)
K.T. Suzuki (Tsukuba, Japan)
K. Taguchi (Tokyo, Japan)
C. Tohyama (Tsukuba, Japan)

Y. Tsukamoto (Tokyo, Japan)
H. Uchida (Kanagawa, Japan)
Y. Yamada (Tokyo, Japan)
E. Yokoyama (Tokyo, Japan)

Executive Congress Coordinators

F. Akahori (Kanagawa, Japan)
M. Arima (Tokyo, Japan)
M. Chiba (Tokyo, Japan)
M. Chino (Tokyo, Japan)
H. Dambara (Nagano, Japan)
A. Gemba (Saitama, Japan)
K. Hanada (Aomori, Japan)
A. Harada (Osaka, Japan)
N. Hashizume (Tokyo, Japan)
S. Horiguchi (Osaka, Japan)
N. Ishinishi (Fukuoka, Japan)
K. Kaizu (Kitakyushu, Japan)
K. Kawano (Tokyo, Japan)
K. Kikuchi (Tokyo, Japan)
S. Kojima (Kumamoto, Japan)
T. Matsuo (Hyogo, Japan)
Y. Matsuo (Tokyo, Japan)
M. Mishima (Tokyo, Japan)
N. Murayama (Nagano, Japan)
Y. Nishi (Hiroshima, Japan)
T. Nishima (Tokyo, Japan)
M. Nishimura (Shizuoka, Japan)
M. Nishimuta (Tokyo, Japan)
K. Nogawa (Ishikawa, Japan)
S. Nomoto (Nagano, Japan)
M. Ohsawa (Kanagawa, Japan)
A. Okubo (Tokyo, Japan)

M. Okuni (Tokyo, Japan)
J. Okada (Tokyo, Japan)
S. Okada (Shizuoka, Japan)
K. Okita (Yamaguchi, Japan)
S. Sano (Shiga, Japan)
S. Sato (Morioka, Japan)
H. Satoh (Sendai, Japan)
C. Sekiya (Asahikawa, Japan)
S. Suzuki (Gumma, Japan)
K.T. Suzuki (Tsukuba, Japan)
M. Tachi (Tokyo, Japan)
S. Takagi (Tokyo, Japan)
T. Takasu (Tokyo, Japan)
S. Takita (Tokyo, Japan)
K. Tanaka (Kobe, Japan)
H. Tanaka (Kyoto, Japan)
T. Tanaka (Tokyo, Japan)
R. Tokunaga (Osaka, Japan)
K. Tomokuni (Saga, Japan)
M. Uda (Tokyo, Japan)
K. Waku (Kanagawa, Japan)
Y. Yase (Wakayama, Japan)
M. Yamagishi (Kitakyushu, Japan)
Y. Yamamura (Kanagawa, Japan)
Y. Yamori (Shimane, Japan)
H. Yoshikawa (Gifu, Japan)

Contents

Dialysis and Renal Disorders

Neoplasma

Immunity

Metal-Binding Proteins and Metalloenzymes

New Analytical Techniques for Trace Elements

Environment and Trace Elements

Raulin Award Lecture

Discovery of Human Zinc Deficiency and Marginal Deficiency of Zinc

ANANDA S. PRASAD

Department of Internal Medicine, Division of Hematology-Oncology, Wayne State University School of Medicine, and Harper-Grace Hospitals, Detroit, MI, USA, and Veterans Administration Medical Center, Allen Park, MI, USA

ABSTRACT

The importance of zinc for human health was first documented in 1963. During the past twenty five years, deficiency of zinc in humans due to nutritional factors and several disease states has now been recognized.

Although the clinical, biochemical, and diagnostic apsects of severe and moderate levels of zinc deficiency in humans are well charecterized, the recognition of mild levels of zinc has been difficult. Recently we have established an experimental human model for the study of mild to marginal deficiency of zinc in order to characterize clinical, biochemical and immunological aspects of this level of deficiency. Our studies indicate that measurement of zinc in lymphocytes, granulocytes, and platelets provide sensitive indicators of mild zinc deficiency in humans.

The clinical manifestations of a mild level of deficiency of zinc in humans include decreased serum testosterone level and oligospermia in males, decreased lean body mass, hyperammonemia, neurosensory changes, anergy, decreased serum thymulin activity, and decreased IL-2 activity. The assay of serum for thymulin activity before and after in vitro addition of zinc also provides an additional sensitive criteria for mild deficiency of zinc in our experience.

KEY WORDS

zinc deficiency, zinc, marginal deficiency, thymulin and zinc

INTRODUCTION

In biological systems, Raulin (1869) showed for the first time that zinc was essential for the growth of Aspergillus niger. Zinc was shown to be essential for higher plants in 1926 (Somner and Lipman, 1926). In 1934, Todd, Elvehjem, and Hart (1934) reported that zinc was essential for the rat. In 1955, a disease called parakeratosis in swine was reported by Tucker and Salmon (1955) to be due to a deficiency of zinc. The essentiality of zinc for the growth of the chicken was shown by O'Dell et al (1958).

In animals, the manifestations of zinc deficiency included growth failure, loss of hair, thickening and hyperkeratinization of the epidermis, and testicular atrophy. Deficiency of zinc in breeding hens resulted in decreased hatchability, gross embryonic anomalies characterized by abnormal skeletal development, and weakness in chicks that hatched (Blamberg, 1960).

Although the essentiality of zinc for animals was established, its ubiquity made it seem improbable that zinc deficiency in humans could lead to significant problems in clinical medicine. During the past twenty-five years, however, it has become apparent that deficiency of zinc in humans is quite prevalent.

DISCOVERY OF HUMAN ZINC DEFICIENCY

Studies in Iran

In Shiraz, I met Dr. James A. Halsted who was a Fulbright Professor at Pahlevi University and was primarily involved with Saadi hospital, an equivalent of a charitable city-hospital in the U.S.A. In the fall of 1958, I was invited by Dr. Halsted to discuss a patient with anemia at the medical center grand rounds at the Saadi Hospital. The case was presented to me by the chief resident, Dr. M. Nadimi, a graduate of the Shiraz Medical School.

The patient was a 21 year old male, who looked like a 10 year old boy. In addition to severe growth retardation and anemia he had hypogonadism, hepatosplenomegaly, rough and dry skin, mental lethargy and geophagia. The patient ate only bread from wheat flour and the intake of animal protein was negligible. He consumed nearly one pound of clay daily. Later we discovered that the habit of geophagia (clay eating) was fairly common in the villages around Shiraz. Further studies documented the existence of iron deficiency anemia in our patient. There was no evidence of blood loss. Inasmuch as 10 additional similar cases were brought to the hospital for my care within a short period of time hypopituitarism as an explanation for growth retardation and hypogonadism was considered to be very unlikely.

The anemia of the subjects promptly responded to oral administration of iron. The probable factors responsible for anemia in these patients were: i) the total amount of available iron in the diet was insufficient, ii) excessive sweating probably caused greater iron loss from the skin than would occur in a temperate climate, and finally, iii) geophagia may have further decreased iron absorption as was observed later by Minnich et al (1968). The anemia was corrected by administration of oral iron in every case. Following therapy with orally administered ferrous sulfate (1 g daily) and a nutritious hospital diet containing adequate animal protein, the anemia was corrected, hepatosplenomegaly improved, they grew pubic hair, and their genitalia size increased (Prasad et al, 1961). Liver function tests were unremarkable except for the serum alkaline phosphatase which increased following treatment. Retrospectively one might explain this observation on two bases: (i) ordinary pharmaceutical preparation of iron might have contained appreciable quantities of zinc as a contaminant and (ii) animal protein most likely supplied available zinc, thus inducing the activity of alkaline phosphatase, an established zinc metalloenzyme.

It was difficult to explain all of the clinical features solely on the basis of tissue iron deficiency, inasmuch as growth retardation and testicular atrophy are not seen in iron-deficient experimental animals. The possibility that zinc deficiency may have been present was considered. As noted earlier, zinc deficiency was known to produce retardation of growth and testicular atrophy in animals. Inasmuch as heavy metals may form insoluble complexes with phosphate, we speculated that some factors responsible for decreased availability of iron in these patients with geophagia may also have decreased the availability of zinc. O'Dell and Savage (1960) first observed that phytate (inositol hexaphosphate) which is present in cereal grains, markedly impaired the absorption of zinc. Changes in the activity of alkaline phosphatase following zinc supplementation to deficient animals were also similar to those observed in our subjects fed adequate diet. Thus, in these subjects dwarfism, testicular atrophy, retardation of skeletal maturation, and changes in serum alkaline phosphatase could have been explained on the basis of zinc deficiency. I then moved to Egypt. In Egypt patients similar to the growth-retarded Iranian subjects were encountered. The clinical features were remarkably similar, except for the following: the Iranian patients had more pronounced hepatosplenomegaly, they gave history of geophagia and none had any hookworm infection in contrast to Egyptian subjects who had both schistosomiasis and hookworm infestations, and none gave a history of geophagia.

The dietary history of the Egyptian subjects was similar to that of the Iranians. The consumption of animal protein was negligible. Their diet consisted mainly of bread and beans (vicia fava). These subjects were demonstrated to have a zinc deficiency. This conclusion was based on the following: the zinc concentrations in plasma, red cells, and hair were

decreased and radioactive zinc-65 studies revealed that the plasma zinc
turnover was greater, the 24-hour exchangeable pool was smaller, and the
excretion of zinc-65 in stool and urine was less in the subjects than in the
controls (Prasad et al, 1963).

Further studies in Egypt showed that the rate of growth was greater in
patients who received supplemental zinc as compared to those who received
iron instead or those receiving only an adequate animal protein diet. Pubic
hair appeared in all cases within 7 to 12 weeks after zinc supplementation
was initiated. Genitalia increased to normal size and secondary sexual
characteristics developed within 12 to 24 weeks in all patients receiving
zinc. In contrast, no such changes were observed in a comparable length of
time in the iron-supplemented group or in the group on an animal protein
diet. Thus, the growth retardation and gonadal hypofunction in these
subjects were related to a deficiency of zinc. The anemia was due to iron
deficiency and responded to oral iron treatment. These studies clearly
showed that severe anemia and iron deficiency were not causative factors for
growth retardation and hypogonadism in human subjects.

Chronology of Other Important Observations in Human Zinc Deficiency

In 1956 Vallee et al (1956) reported that the serum zinc levels were
decreased in patients with cirrhosis of the liver and suggested that these
subjects had conditioned deficiency of zinc due to hyperzincuria.
Unfortunately the clinical consequences of this deficiency in their subjects
were not recognized.

The effects of zinc deficiency on growth and gonadal development in
humans were first recognized by Prasad et al (1961, 1963) in the early
sixties and it was also observed that a nutritional deficiency of zinc may
occur in certain populations of the world under most practical conditions.

It is now becoming evident that nutritional, as well as conditioned
deficiency of zinc may complicate many disease states in human subjects. In
1968 MacMahon et al (1968) observed for the first time, zinc deficiency in a
patient with steatorrhea. Several other examples of zinc deficiency in
patients with malabsorption have now been recorded (McClain et al, 1988).

In the U.S.A. Caggiano et al (1969) were the first to report a case of
zinc deficiency in a Puerto Rican subject with dwarfism, hypogonadism,
hypogammaglobulinemia, giardiasis, stronglyloidiasis and schistosomiasis.
This patient responded to zinc therapy so far as growth and development was
concerned.

In 1972, a number of Denver, U.S.A., children from middle class families,
were reported to exhibit evidence of symptomatic nutritional zinc deficiency
(Hambidge et al, 1972). Growth retardation, poor appetite, and impaired
taste acuity were related to zinc deficiency in those children. Zinc
supplementation corrected the above features. Later symptomatic zinc
deficiency in U.S.A. infants was also reported. Indeed it is currently
believed that the risk of sub-optimal zinc nutrition may pose a problem for a
substantial section of U.S.A. population.

Halsted et al (1972) published the results of their study involving a
group of 15 men who were rejected at the Iranian Army Induction Center
because of "malnutrition." Two women, 19 and 20 years old were also
included. A unique feature was that all were 19 or 20 years old. Their
clinical features were similar to those reported by Prasad et al (1961,
1963). They were studied for 6 to 12 months. One group was given a
well-balanced diet containing ample animal protein plus a placebo capsule. A
second group was given the same diet plus a capsule of zinc sulfate
containing 27 mg of zinc. A third group received the diet without additional
supplement for 6 months, followed by the diet plus zinc for another 6 month
period. The two women lived in the house of Dr. Ronaghy and received the
same treatment and observation program.

The development in subjects receiving the diet alone was slow while markedly enhanced in those receiving zinc. The zinc supplemented subjects grew considerably faster than those receiving the well-balanced diet alone. The zinc supplemented subjects showed evidence of early onset of sexual function, as defined by nocturnal emission in males and menarche in females (Halsted et al, 1972).

Zinc deficiency in human populations throughout the world is prevalent, although its incidence is not known. Clinical pictures similar to those reported in zinc-deficient dwarfs have been observed in many countries. It is believed that zinc deficiency should be present in countries where primarily cereal proteins are consumed by the population. One would also expect to see a spectrum of zinc deficiency, ranging from severe cases to marginally deficient examples, in any given population.

In 1973, Barnes and Moynahan (1973) studied a 2-year old girl with severe acrodermatitis enteropathica who was being treated with diiodohydroxyquinoline and a lactose-deficient synthetic diet. The clinical response to this therapy was not satisfactory, and the physicians sought to identify contributing factors. It was noted that the concentration of zinc in the patient's serum was profoundly decreased; therefore, they administered oral zinc sulfate. The skin lesions and gastrointestinal symptoms cleared completely and the patient was discharged from the hospital. When zinc was inadvertently omitted from the child's regimen, she suffered a relapse which promptly responded to oral zinc. In their initial reports the author's attributed zinc deficiency in this patient to the synthetic diet. It became soon clear that zinc might be fundamental to the pathogenesis of this rare inherited disorder and that the clinical improvement reflected improvement in zinc status. This original observation was quickly confirmed in other cases throughout the world. The underlying pathogenesis of the zinc deficiency in these patients is, most likely, due to malabsorption of zinc, the mechanism of which is not well understood at present.

In 1974 a landmark decision to establish recommended dietary allowances for humans for zinc was made by the National Research Council, Food and Nutrition Board of the National Academy of Sciences.

In 1975 Kay and Tasman-Jones (1975) reported the occurrence of severe zinc deficiency in subjects receiving total parenteral nutrition for prolonged periods without zinc. Almost simultaneously, Okada et al (1976) and Arakawa et al (1976) reported similar observations in subjects receiving total parenteral nutrition without zinc. This observation is now well documented in the literature and indeed in the United States zinc is being routinely included in total parenteral fluids for subjects who are likely to receive such therapy for extended periods.

An example of severe parakeratosis in man related to deficiency of zinc was first reported by Klingberg et al in 1976 (1976), in a patient who received penicillamine therapy for Wilson's disease. Zinc supplementation completely reversed the clinical manifestations.

Recent reports suggest that several clinical manifestations in patients with sickle cell disease such as growth retardation, hypogonadism in the males, lack of prompt healing of chronic leg ulcers, abnormal dark adaptation and abnormality in cell mediated immunity are related to a deficiency of zinc (Prasad et al, 1975, 1981, 1984; Warth et al, 1981, Prasad et al, 1988). The exact pathogenesis of zinc deficiency in sickle cell disease is not well understood and further studies are needed. Hyperzincuria due to abnormal renal tubular function has been noted in such subjects and this may be a contributing factor in the pathogenesis of zinc deficiency.

Although the role of zinc in humans has been now defined and its deficiency recognized in several clinical conditions, it is only recently that an experimental human model has been developed which allowed a study of the specific effects of a mild zinc deficient state in man (Prasad et al, 1988, 1978; Abbasi et al, 1980). This model also provides assessment of sensitive parameters which could be utilized clinically for diagnosing marginal zinc deficiency.

CLINICAL SPECTRUM OF HUMAN ZINC DEFICIENCY

During the past two decades, a spectrum of clinical deficiency of zinc in human subjects has been recognized. A severe deficiency of zinc may be life-threatening as has been reported to occur in patients with acrodermatitis enteropathica, following total parenteral nutrition following penicillamine therapy, and acute alcoholism. The clinical manifestations of severely zinc deficient subjects include bullous-pustular dermatitis, diarrhea, alopecia, mental disturbances, and intercurrent infections due to cell mediated immune disorders, and if untreated, the condition becomes fatal.

Growth retardation, male hypogonadism, skin changes, poor appetite, mental lethargy, abnormal dark adaptation and delayed wound healing are some of the manifestations of moderate zinc deficiency in human subjects. Moderate deficiency of zinc due to nutritional factors, malabsorption, sickle cell disease, chronic renal disease and other debilitating diseases have now been well documented in the literature. A beneficial effect of zinc in wound healing was first reported by Pories and Strain in 1966 (1966). This observation remained controversial for several years, however, most studies now provide evidence that zinc supplementation does promote wound healing in zinc deficient patients and that zinc therapy in zinc sufficient subjects is not effective for wound healing.

Abnormalities of taste was first related to a deficiency of zinc in man by Henkin and Bradley (1969). Decreased taste acuity (hypogeusia) has been observed in zinc deficient subjects, such as patients with liver disease, malabsorption syndrome, and chronic uremia, and after burns and administration of penicillamine or histidine. A double blind study, however, failed to show the effectiveness of zinc in treatment of hypogeusia in various diseases. Recently, however, Mahajan et al (1980) reported that zinc was effective in improving taste acuity in subjects with chronic uremia. Their studies were carried out in a double-blind fashion. This may suggest that depletion of zinc may lead to decreased taste acuity, but not all cases of hypogeusia are due to zinc deficiency. The role of zinc in hypogeusia needs to be further delineated.

MARGINAL DEFICIENCY OF ZINC: EXPERIMENTAL HUMAN MODEL

Although the clinical, biochemical, and diagnostic aspects of severe and moderate levels of zinc deficiency in humans are now well recognized, the recognition of mild or marginal deficiency of zinc has been difficult. It was, therefore, considered desirable to develop a human model that would allow us to study the effects of a mild zinc-deficient state in humans and also provide us with sensitive variables that could be used clinically for diagnosing zinc deficiency. We have succeeded in establishing such a model and I will now summarize some of our important results.

Male volunteers between the ages of 20 to 45 years were selected for these studies. The experimental protocol for our studies was reviewed and approved by the Human and Animal Investigation Committee of Wayne State University and informed consents were obtained in each case. Before the study, a thorough history, physical examination and routine laboratory tests including complete blood count, liver function, sequential multiple analyzer-12, and serum electrolytes were performed and found normal. Prior to the study, the volunteers were ambulatory and were encouraged to do daily moderate exercise throughout the study period.

The subjects were given a hospital diet containing animal protein daily for 4 weeks. This diet averaged 12 mg of zinc/day, consistent with the recommended dietary allowance of the National Research Council, National Academy of Sciences. After that, they received a semi-purified soy protein-based experimental diet which supplied 3.0 to 5.0 mg of zinc on a daily basis. The details of methodology for preparation of experimental diet

have been published elsewhere (Rabbani et al, 1987). This regime was
continued for 28 weeks, at the end of which they received 27 mg of zinc
supplement while still consuming experimental diet. The supplementation was
continued for 12 weeks.

Throughout the study, the levels of all nutrients including protein,
amino acids, vitamins, and mineral (both macro and microelements) were kept
constant, meeting the standards set by Recommended Dietary Allowances, except
for zinc, which was varied as outlined above. By this technique, we were
able to induce a specific zinc deficiency in human volunteers.

The peripheral blood cells (lymphocytes, granulocytes, and platelets) for
zinc assay were isolated by a modification of a previously published method
(Prasad et al, 1978). Special care was taken to remove red cells from the
granulocytes, platelets from the granuloctyes and lymphocytes, and trapped
plasma from the platelets. Extreme care was exercised to avoid exogenous
zinc contamination throughout the assay procedure. Zinc was assayed in the
samples by means of an atomic absorption spectrophotometer with a 655 furnace
and 254 Fastac Auto Sampler (Instrumentation Laboratory, Inc., Lexington,
MA).

In our experimental human model studies, we created a negative zinc
balance of approximately one mg/d and we calculated that in a six month
period a total of approximately 180 mg of negative zinc balance was achieved
(Prasad et al, 1978). This is a small fraction of the total body zinc.
Although the body of an adult 70 kg male contains 2300 mg zinc, only 10
percent exchanges with an isotopic dose within one week (Prasad et al, 1963;
Foster et al, 1979; Hambidge et al, 1983). Approximately 28% percent of zinc
resides in the bone, 62% percent in the muscles, 1.8% in the liver, and 0.1%
percent in the plasma pool. In an adult animal model, zinc concentrations of
muscle and bone do not change as a result of mild or marginal zinc
deficiency. It appears that in cases of mild or marginal deficiency of zinc
in humans, one cannot expect a uniform distribution of the deficit over the
entire body pool, but that most likely those compartments with high turnover
rates (liver and peripheral blood cells such as lymphocytes, neutrophils and
platelets) would suffer inproportionate deficit. Thus, if one were to
consider that only 200 to 400 mg of zinc which is represented by liver zinc
and mobile exchangeable pool is the critical pool, a negative balance of 180
mg from this pool may be a considerable fraction.

When the level of zinc deficiency was very mild (5.0 mg of zinc intake
during zinc restricted period), the plasma zinc level remained more or less
within the normal range and it decreased only after 4 to 5 months of zinc
restriction. On the other hand, zinc levels in lymphocytes, granulocytes,
and platelets decreased within 8 to 12 weeks, suggesting that the assay of
cellular zinc may provide a sensitive criteria for diagnosing mild deficiency
of zinc.

In subjects who received 3.0 mg of dietary zinc during zinc-restricted
period, the plasma zinc level remained above 100 ug/dl for 2 months on
zinc-restricted diet. In the third month, a significant decrease in the
plasma zinc level was observed (Prasad et al, 1988). In these volunteers,
the zinc in platelets decreased within 1 months, and zinc levels in
lymphocytes and granulocytes decreased within 2 months following institution
of zinc restricted diet (Prasad et al, 1988). The maximum decline in
cellular zinc levels was observed at the end of 6 months of restricted
dietary zinc intake.

We have previously reported the effects of mild zinc deficiency induced
by dietary means on gonadal functions in male volunteers (Abbasi et al,
1980). These subjects had normal serum androgens, FSH, LH, and sperm count
prior to zinc restriction.

The sperm count declined slightly during zinc restriction and continued
to decline in the early phase of zinc repletion. Oligospermia (total sperm
count per ejaculate of less than 40 million) was observed in four out of five
subjects as a result of dietary zinc restriction.

The baseline serum testosterone decreased significantly during the early phase of zinc repletion and returned to normal levels during the late phase of zinc repletion. There was a slight decline in the maximal rise of serum testosterone after GnRH stimulation during zinc restriction and a more significant decline during the early phase of zinc repletion with recovery to normal level during the late phase of zinc repletion. The changes of serum dihydrotestosterone were similar to those of serum testosterone, but statistically not significant.

The nature of the delayed effect of zinc deficiency on sperm count is not well understood. The developmental progression of spermatogenesis, from the origin of spermatozoa, is a prolonged process. The duration of human spermatogenesis is 74 ± 4.5 days. Therefore, an insult affecting the germinal cells, as induced by zinc deficiency, may not become evident until several months later.

One unexpected finding in our studies in the human experimental model was that the plasma ammonia level increased during the zinc restricted period (Prasad et al, 1978). This was corrected following supplementation with zinc. We have reported similar findings in zinc-deficient rats (Rabbani and Prasad, 1978) and related this observation to abnormalities induced in the activity of ornithine transcarbamoylase activity in the liver, an enzyme known to be important in ammonia utilization and urea synthesis.

In additional experiments, half of our volunteers showed abnormal dark adaptation following zinc restricted diet, which was corrected by supplementation with zinc. Taste test was performed in our volunteers and they showed hypogeusia as a result of zinc restriction. Zinc supplementation corrected this clinical neuro-sensory abnormality (unpublished data).

In 2 subjects who received 3.0 mg of zinc in the diet during zinc restricted period, we determined body composition and we observed that the lean body mass decreased as a result of mild deficiency of zinc. This was also corrected following zinc supplementation (unpublished data).

We have recently assayed activity of serum thymulin in mildly zinc deficient human subjects (Prasad et al, 1988). Thymulin is a thymus specific hormone and it requires the presence of zinc to express its biological activity (Dardenne et al, 1982). Recent data demonstrate that thymulin binds to high affinity receptors on T cells, induces several T cell markers, and promotes T cell function including allogenic cytotoxicity, suppressor functions, and IL-2 production (Pleau et al, 1980).

As a result of mild deficiency of zinc, the activity of thymulin in serum was significantly decreased and was corrected by both in vivo and in vitro zinc supplementation. The in vitro supplementation studies indicated that the inactive thymulin peptide was present in the serum in zinc deficient subjects and was activated by addition of zinc (Prasad et al, 1988). The assay of serum thymulin activity with or without zinc addition in vitro, thus may be utilized as sensitive criteria for the diagnosis of mild zinc deficiency in humans.

An increase in T_{101^-}, sIg- cells, decrease in T4+/T8+ ratio, and decreased IL-2 activity were observed in the experimental human model during the zinc depletion phase, all of which were corrected after repletion with zinc (Prasad et al, 1988). We had previously reported that NK activity was also sensitive to zinc restriction (Tapazoglou, 1985), thus it appears that zinc may play a very important and critical role in the functions of T cells in humans.

Our studies thus far show that a mild or marginal deficiency of zinc in humans is characterized by neuro-sensory changes, oligospermia in males, decreased serum testosterone level, hyper-ammonemia, decreased lean body mass, decreased serum thymulin activity, decreased IL-2 activity, decreased NK activity, and alterations in T cell subpopulations. All the above manifestations are correctable by supplementation with zinc. It is, therefore, important to recognize mild or marginal deficiency of zinc in humans, inasmuch as this is easily correctable.

REFERENCES

Abbasi AA, Prasad AS, Rabbani P, Du Mouchelle E (1980). Experimental zinc
    deficiency in man: Effect on testicular function. 96:544-550.
Arakawa T, Tamura T, Igarashi Y (1976). Zinc deficiency in two infants using
    parenteral alimentation for diarrhea. Am J Clin Nutr 29:197-204.
Barnes PM, Moynahan EJ (1973). Zinc deficiency in acrodermatitis
    enteropathica: multiple dietary intolerance treated with synthetic zinc.
    Proc R Soc Med 66:327-329.
Blamberg DL, Blackwood UB, Supplee WC, Combs GF (1960). Effect of zinc
    deficiency in hens on hatchability and embryonic development. Proc Soc
    Exp Biol Med 104:217-220.
Caggiano V, Schnitzler R, Strauss W, Baker RK, Carter AC, Josephson
    AS,Wallach S (1969). Zinc deficiency in a patient with retarded growth,
    hypogonadism, hypogammaglobulinemia, and chronic infection. Am J Med Sci
    257:305-319.
Dardenne M, Pleau JM, Nabarra P, et al (1982). Contribution of zinc and
    other metals to the biological acitivity of the serum thymic factor.
    Proc Natl Acad Sci USA 79:5370-5373.
Foster DM, Aamodt RL, Henkin RI, Berman M (1979). Zinc metabolism in
    humans. A kinetic model. Am J Physiol 237:R340-R349.
Halsted JA, Ronaghy HA, Adabi P, Haghshenass M, Amirhakemi GH, Barakat RH,
    Reinhold JC (1972). Zinc deficiency in man: The Shiraz experiment. Am
    J Med 53:277-284.
Hambidge KM, Casey CE, Krebs NJ (1985). Zinc. In: Mertz M, ed. Trace
    elements in human and animal nutrition. Academic Press, New York
    5(2):1-137.
Hambidge KM, Hambidge C, Jacobs M, Baum JD (1972). Low levels of zinc in
    hair, anorexia, poor growth, and hypogeusia in children. Pediatr Res
    6:868-874.
Henkin RI, Bradley DF (1969). Regulation of taste acuity by thiols and metal
    ions. Proc Natl Acad Sci USA 62:30-37.
Kay RG, Tasman-Jones C (1975). Zinc deficiency and intravenous feeding.
    Lancet 2:605-606.
Klingberg WG, Prasad AS, Oberleas D (1976). Zinc deficiency following
    penicillamine therapy. In: Prasad AS, (ed). Trace elements in human
    health and disease. Academic Press, New York 1:51-65.
MacMahon RA, Parker ML, McKinnon M (1968). Zinc treatment in
    malabsorption. Med J Aust 2:210-212.
Mahajan SK, Prasad AS, Lambujon J, Abbasi AA, Briggs WA, McDonald FD
    (1980). Improvement of uremic hypogeusia by zinc: A double-blind
    study. Am J Clin Nutr 33:1517-1521.
McClain CJ, Adams L, Shedlofsky S (1988). Zinc and the gastrointestinal
    system. In: Prasad AS, (ed). Essential and toxic trace elements in
    human health and disease. Alan R. Liss, New York 55-73.
Minnich V, Okevogla A, Tarcon Y, Arcasoy A, Cin S, Yorukoglu O, Renda F,
    Demirag B (1968). The effect of clay on iron absorption as a possible
    cause for anemia of Turkish subjects with pica. Am J Clin Nutr 21:78-86.
Okada A, Takagi Y, Itakura T, Satani M, Manabi H, Iida Y, Tanigaki T, Iwasaki
    M, Kasahara N (1976). Skin lesions during intravenous hyperalimentation:
    Zinc deficiency. Surgery 80:629-635.
O'Dell BL, Savage JE (1960). Effect of phytic acid on zinc availability.
    Proc Soc Exp Biol Med 103:304-306.
O'Dell BL, Newberne PM, Savage JE (1958). Significane of dietary zinc for
    the growing chicken. J Nutr 65:503-518.
Pleau JM, Fuentes V, Morgat JL, Bach JF (1980). Specific receptor for the
    serum thymic factor (FTS) in lymphoblastoid cultured cell lines. Proc
    Natl Acad Sci USA 77:2861-2865.
Pories WJ, Strain WH (1966). Zinc and wound healing. In: Prasad AS, (ed).
    Zinc metabolism. Charles C. Thomas, Springfield IL378-394.
Prasad AS, Abbasi AA, Rabbani P, DuMouchelle E (1981). Effect of zinc
    supplementation on serum testosterone level in adult male sickle cell
    anemia subjects. Am J Hematol 19:119-127.
Prasad AS, Cossack ZT (1984). Zinc supplementation and growth in sickle cell
    disease. Ann Intern Med 100:367-371.

Prasad AS, Halsted JA, Nadimi M (1961). Syndrome of iron deficiency anemia, hepatosplenomegaly, hypogonadism, dwarfism and geophagia. Am J Med 31:532-546.

Prasad AS, Meftah S, Abdallah J, Kaplan J, Brewer GH, Bach JF, Dardenne M (1988). Serum thymulin in human zinc deficiency. J Clin Invest 82:1202-1210.

Prasad AS, Miale A, Farid Z, Schulert A, Sandstead HH (1963). Zinc metabolism in patients with the syndrome of iron deficiency anemia, hypogonadism, and dwarfism. J Lab Clin Med 61:547-549.

Prasad AS, Rabbani, Abbasi A, Bowersox F, Fox MRS (1978). Experimental zinc deficiency in humans. Ann Intern Med 89:483-490.

Prasad AS, Schoomaker EB, Ortega J, Brewer GJ, Oberlease D, Oelshlegel FJ (1975). Zinc deficiency in sickle cell disease. Clin Chem 21:582-587.

Rabbani P, Prasad AS (1978). Plasma ammonia and liver ornithinine transcarbamolyase activity in zinc deficient rat. Am J Physiol 235(2):E203-E206.

Rabbani PI, Prasad AS, Tsai R, Harland BF, Fox MRS (1987). Dietary model for production of experimental zinc deficiency in man. Am J Clin Nutr 45:1514-1525.

Raulin J (1869). Etudes chimiques sur la vegetation. Ann Sci Nat. XI Bot. 93-299.

Sommer AL, Lipman CB (1926). Evidence of indispensable nature of zinc and boron for higher green plants. Plant Physiol 1:231-249.

Todd WR, Elvehjem CA, Hart EB (1934). Zinc in the nutrition of the rat. Am J Physiol 107:146-156.

Tapazoglou E, Prasad AS, Hill G, Brewer GJ, Kaplan J (1985). Decreased natural killer cell activity in zinc deficient subjects with sickle cell disease. J Lab Clin Med 105:19-22.

Tucker HF, Salmon WD (1955). Parakeratosis or zinc deficiency disease in pigs. Proc Soc Exp Biol Med 88:613-616.

Vallee BL, Wacker WEC, Bartholomay AF, Robin ED (1965). Zinc metabolism in hepatic dysfunction. I. Serum zinc concentrations in Laennec's cirrhosis and their validation by sequential analysis. N Eng J Med 255:403-408.

Wang H, Prasad AS, DuMouchelle E (1989). Zinc in platelets, lymphocytes, and granulcoytes by flameless atomic absorption spectrophotometry. J Micronutrient Anal 5:181-190.

Warth JA, Prasad AS, Zwas F, Frank RN (1981). Abnormal dark adaptation in sickle cell anemia. J Lab Clin Med 98:189-194.

Supported in part by grants from NIH/NIDDK No. DK-31401 and the Veterans Administration Medical Research Service

Send correspondence to: Ananda S. Prasad, M.D., Ph.D., University Health Center - 5C, 4201 St. Antoine, Detroit, Michigan 48201 USA

Neural Disorders

Zinc in Taste and Smell Disorders

HIROSHI TOMITA

Department of Otorhinolaryngology, Nihon University, School of Medicine, 30-1 Oyaguchi, Itabashi-ku, Tokyo, 173 Japan

ABSTRACT

The study provides the statistical results of clinical tests conducted on 1,500 cases whose main complaints consisted of taste disorder. It was verified, through zinc administration treatment and animal experiments, that a major part of these cases were disorder on the taste bud level, and that they were caused by zinc deficiency. We also obtained evidence that zinc deficiency can cause smell disorder, in some cases of olfactory epithelium disorders.
Cause of taste and smell disorders triggered by zinc deficiency continue to increase. The main causes are unbalanced diet, food additives and drug intake.

Key Words: taste disorders, smell disorder, clinical statistics, zinc treatment, animal experiments.

BACKGROUND INFORMATION ON TASTE

Taste and smell are called "chemical senses" because they are both touched off by the stimulations from chemical substances. We are aware that many of the attendants of this meeting are from fields other than medicine. So, before we explain the relations

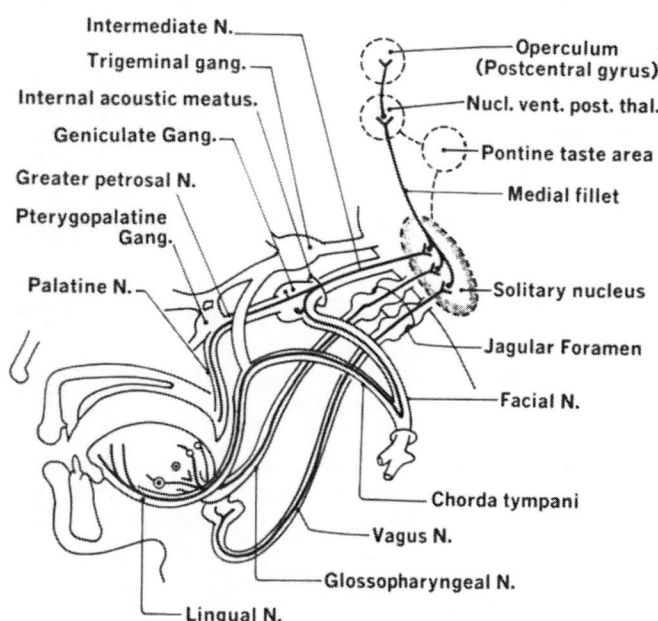

Fig. 1 Gustatory pathway

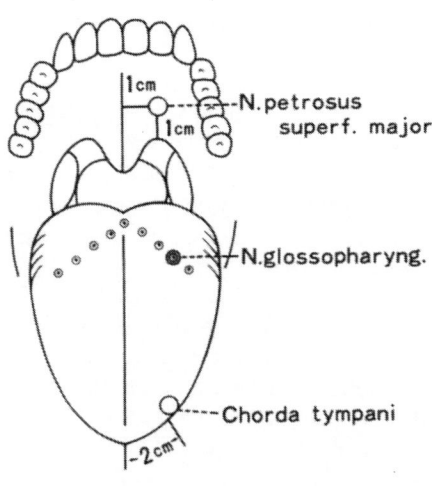

Fig. 2 Selecting the location for the measurement of each taste nerve.

Table 1   Clinical significance of gustatory function testing

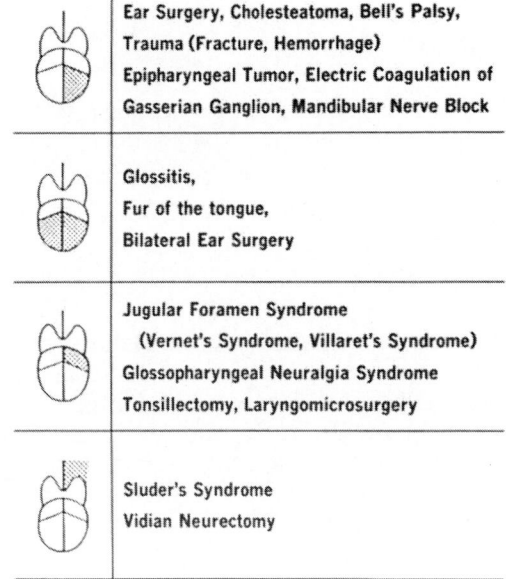

Taste Bud : Senile Degeneration, Drug-induced Zinc Deficiency, Iron Deficiency Anemia, Pellagra, Vitamin A Deficiency Hepatic & Renal Dysfunction, Diabetes **Psychogenic :** Conversion Hysteria, Depression **Congenital :** Ectodermal Syndrome, Riley-Day's Syndrome (*)	Ear Surgery, Cholesteatoma, Bell's Palsy, Trauma (Fracture, Hemorrhage) Epipharyngeal Tumor, Electric Coagulation of Gasserian Ganglion, Mandibular Nerve Block
Central Part of the Gustatory Tract : Cortical Lesion (Ipsilateral ?) **Thalamus :** Head-Holme's Syndrome **Pons :** Ipsilateral Hemorrhage **Nucl. tract. solitar. :** Tumor of the Ⅳ Ventricle **Base of Skull :** Garcin's Syndrome	Glossitis, Fur of the tongue, Bilateral Ear Surgery
	Jugular Foramen Syndrome (Vernet's Syndrome, Villaret's Syndrome) Glossopharyngeal Neuralgia Syndrome Tonsillectomy, Laryngomicrosurgery
Internal Auditory Meatus : Acoustic Tumor, Cerebello-pontine Angle Tumor, Arachnoiditis, Transverse Petrous Bone Fracture **Geniculate Ganglion :** Hunt's Syndrome	Sluder's Syndrome Vidian Neurectomy

(* : Faber-Jung's Syndrome, Henkin's Syndrome, Cronkhite-Canada Syndrome)

▨ : Lesion

between these chemical senses and trace metals, we will go into the more basic subjects of their structure and function as well as the outline of the clinical examinations for diagnosing their functional disorders.

Taste is sensed by taste buds which abound on the tongue and the soft palate. Whereas all other special senses are controlled by a pair of cranial nerves, four pairs of nerves are controlled in taste so that a person can taste by chewing, even if one or two nerves fail to function. It is therefore impossible to diagnose nerve disturbances unless the innervation area of each nerve is well understood and the location for measurement is strictly observed. The traditional "whole-mouth method" — that is, the method of examining the threshold by filling the mouth of the patient with a certain amount of taste solution, does not lead to the right diagnosis.

CLINICAL TASTE EXAMINATIONS

For the purpose of clinical taste examination, we have devised two techniques: electrogustometry and the "filter-paper disc method" as a qualitative and quantitative clinical gustometry for the taste quality (Tomita 1982, 1986).

Electrogustometry adopts decibel scale for evaluation so it excels in quantitative measurement and is useful for the diagnosis of acoustic tumor and nerve injury after operations of the middle ear or tonsils (Tomita et al. 1972).

On the other hand, the filter-paper disc method makes it possible to detect separately the threshold of four primary taste qualities in each area of gustatory innervation. In particular, taste on the soft palate which has heretofore been immeasurable can now be examined quite easily.

My taste examination for each nerve innervation area can immediately discern such group of diseases as shown in Table 1.

When these two examination methods were conducted concurrently, it was revealed that even though their correlations were comparatively good, approximately 10% of

the total cases indicated a phenomena of dissociation. In other words, the discrimination of taste qualities was poor whereas the threshold of electric taste was normal.

By analyzing this group, we found the following features: 1) Many of the cases were female. 2) The period of ailment, from the outbreak of the disorder to medical consultation, was short. 3) In many cases, the serum zinc value was comparatively low and serum copper value was high. 4) In many cases, the conditions of the disorder responded quickly to zinc treatment.

In summary, the early group of taste disorder caused by zinc deficiency can be detected by practicing the two examination methods concurrently.

It was also discovered by the filter-paper disc method that the recovery of taste bud level disorder started at the inner part of the tongue where there are more taste buds: specifically, at the area of the glossopharyngeal nerve (Tomita et al. 1986).

SPECIAL OUTPATIENT CLINIC FOR TASTE DISORDER

During the period when we were tackling with the development of clinical taste examination methods and such basic matters as the mode of oral innervation, Henkin (1969, 1970, 1971) was publishing one paper after another on the relations between taste and zinc. However, as I mentioned in my speech at the Opening Address, the Japanese are obsessed with the idea that heavy metals are harmful to one's health. The timing also coincided with the discovery in 1970 that the SMON disease (subacute-myelo-optico-neuropathy) was caused by chinoform, an intestinal antiseptics. So to start zinc treatment under such circumstances was unthinkable. It was only in 1974 that a combination of my taste examinations, measurement of serum zinc through atomic absorption spectrometry and oral administration of zinc sulfate began to be applied to the treatment of patients with taste disorder.

Meanwhile, the number of patients with main complaints of taste disorder increased gradually, making it impossible for the general outpatient clinic of otorhinolaryngology to treat all of them. As a result, a special outpatient clinic for taste disorder was opened in 1976.

The transition in the number of patients is shown in Fig. 3. The total number of patients in the second-half six years is twice as large as the number in the first-half six years.

CLINICAL STATISTICS (1981 − 1987) FOR PATIENTS WITH TASTE DIS-ORDERS

The total number of patients with taste disorder for the period of 7 years from 1981 to 1987 amounted to approximately 1,500 cases. With 570 male cases and 930 female cases, the ratio of male patients to female patients was 2:3.

The diagram (Fig. 4) is a bar graph giving the breakdown of the total number of cases by age group and by sex. The graph indicates that the number of cases as a whole reaches a vertex in the 50s age group, with the number of female and male patients peaking out in the 50s and 60s age groups, respectively.

Fig. 5 shows the incidence pattern of the disorder for each decade age of the population. The drastic curve demonstrates that the incidence pattern increased with age. The 500 cases discovered before 1980 also show a similar curve.

Fig. 6 lists up the causes of taste disturbance by distinction of sex, given in the order of the larger number of cases. However, because idiopathic disorder and zinc deficiency are both considered to result from changes in zinc metabolism, and also because the elucidation of the cause of idiopathic taste disorder is a major subject of this report, we have always placed idiopathic disorder at the top to facilitate comparison with zinc deficiency.

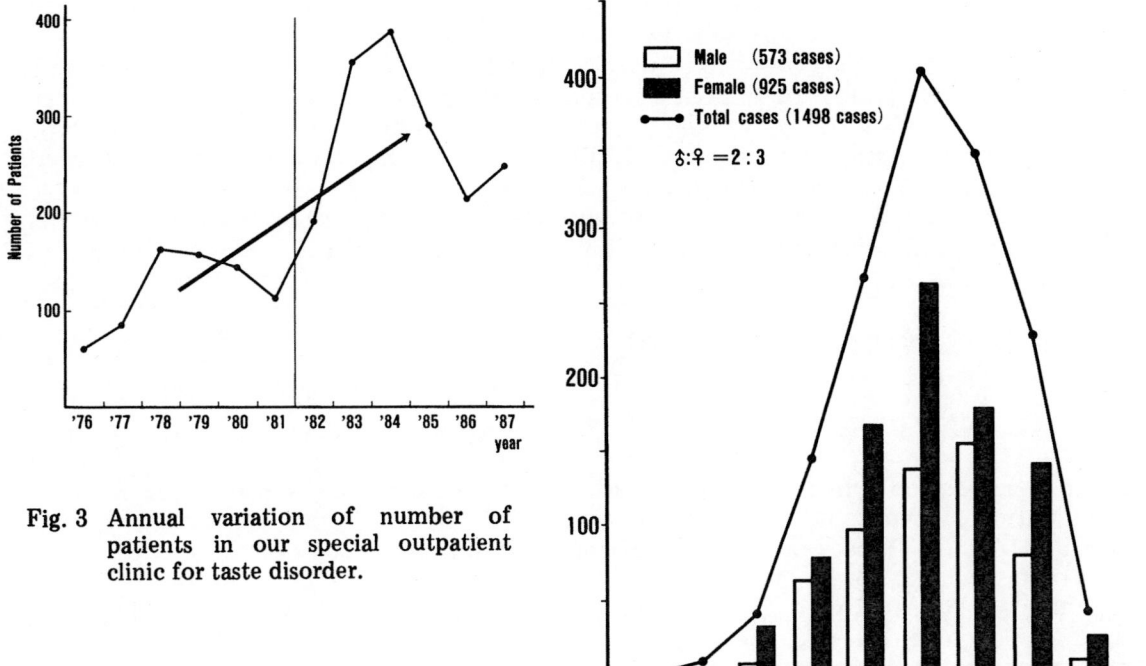

Fig. 3 Annual variation of number of patients in our special outpatient clinic for taste disorder.

Fig. 4 Distribution by age and sex of the patients with taste disorders (1981 ~ 1987).

Zinc-deficient taste disorder is a case which has no significant cause of taste disturbance other than the low serum zinc value of 59 μg/dl or below which evidently shows hypozincemia, or the boundary value of 60—69 μg/dl in serum zinc value. Cases which, for example, have hypozincemia with cirrhotic liver are not discussed under this category.

Idiopathic disorder refers to taste disturbance cases of unknown causes in which the serum zinc value maintains a normal level.

The figure indicates that there is no difference by distinction of sex for idiopathic or zinc-deficient disorder whereas male patients are more dominant for diseases of central nerve system and female patients account for a larger ratio for psychogenic disorder.

Table 2 lists up the subject matters of the examinations conducted at our special outpatient clinic for taste disorders.

What is notable here is that one can observe the fungiform papillae with a microscope of 10 magnifications. Pathognomonic findings can also be detected with changes in age. This 35 year old female with zinc-deficient taste disturbance, whose red smooth tongue is shown in the left photo, has been suffering from less sallivation and glossodynia for five years. After oral administration of 300 mg of zinc sulfate per day for six months (right photo), the condition of her tongue returned to normal as you can see, and she has also recovered her taste (Fig. 7).

However, for many patients, no specific observation by microscopy is possible except for changes with age.

Fig. 5 Pattern of the patient distribution by age — Number of the patients with taste disorders ('76—80 & '81—87) per the Japanese population ('80 & 86) —

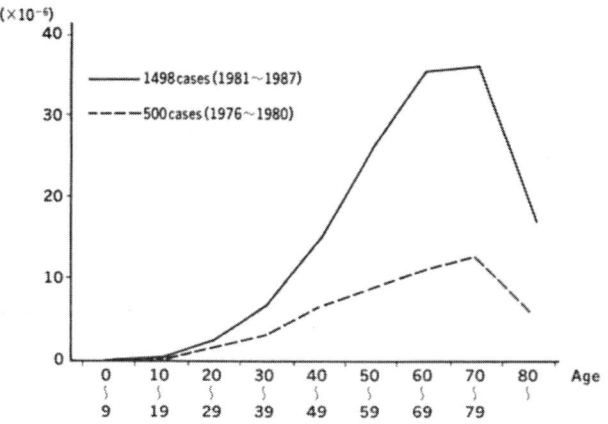

Fig. 6 Causes and distinction by sex on patients with taste disorders — Analysis of 1512 cases between 1981 ~ 1987 —

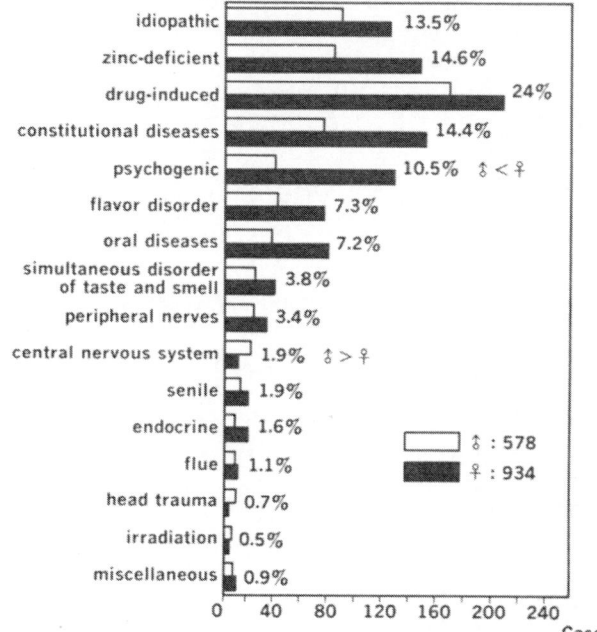

Fig. 7 The tongue and fungiform papillse (10X) of a female with zinc-deficient taste disturbance before and after zinc treatment.

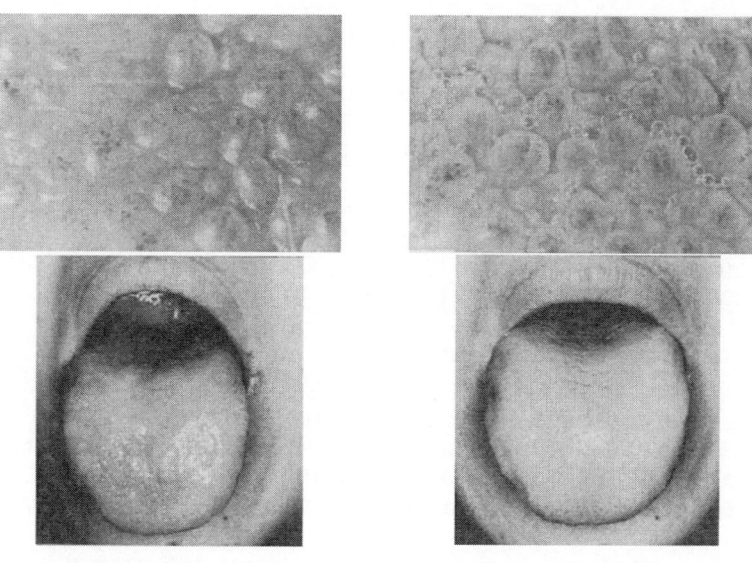

Before Treatment   **After Treatment**

EFFECT OF ORAL ADMINISTRATION OF ZINC SULFATE

We administer our patients 1 capsule containing 100 mg of zinc sulfate every day during each meal. Patients with gastroenteric complaints such as nausea are given a capsule containing 158 mg of zinc gluconate. The content of zinc is 22.5 mg per capsule, which means that the dosage amounts to 65.7 mg a day.

Zinc treatment can be applied to serious and intermediate cases of zinc-deficient taste disturbances, some of the cases of hypozincemia from drug-induced causes, liver or renal failures, and almost all cases of idiopathic taste disturbance.

785 patients, or 52% of the total, have been given zinc treatment.

Zinc treatment requires a long period of time but 85% of the cases show the effects of zinc treatment 4 months after its start. For patients who stopped coming to the clinic within 4 months after the start of treatment, we excluded them from the statistics as dropouts, except the cured cases.

Fig. 8 indicates the effectiveness of the treatment by each cause of disorder. Idiopathic disorder shows an effective ratio of 76%, which is more or less the same level as zinc deficiency. The effective ratio for the zinc treatment as a whole was 67%.

We then compared the effective ratio of zinc treatment by different period of ailment, that is, the period from the outbreak of disorder to medical consultation. Those who sought medical consultation within 1 month and 6 months showed an effective ratio exceeding 80% and 70%, respectively. On the other hand, for those who delayed consultation from 6 months to 1 year, and 10 years or more, the effective ratio dropped to 60% and 40%, respectively. This means that the longer the delay in medical consultation, the lower the effective ratio of the treatment (Table 3).

As we have just explained, human taste disorders are caused by more than 15 varying factors, which means that differential diagnosis is imperative.

Out of the factors we have mentioned, the main causes of taste bud level disorders are drug-induced, zinc deficient, constitutional and idiopathic diseases and the number of taste disorders cases caused by these four factors account for more than 70% of the total.

By its very definition, all cases of zinc deficient disease indicate hypozincemia. Forty percent of all cases of drug-induced taste disorders also indicate hypozincemia. This reveals the fact that many drugs have chelating effects of zinc. A more detailed presentation on the subject will be given by Ikeda in another chapter.

Of all the diseases of the whole body, a majority consists of diseases closely related to zinc metabolism such as renal failure, liver failure, precarious conditions after stomach operations and diabetes. Also 40% of these diseases indicate symptoms of hypozincemia.

Table 2   Routine examination of patients with taste disturbance

1) Careful questioning : circumstances of onset, appetite, bowel movements, liquor, tobacco, favorite foods, medical history, diseases affected by at present and contents of the therapy.

2) Inspection of oral cavity and tongue : xerostomia, inflammation, keratosis, atrophy, number of fungiform papillae.

3) Biomicroscopy of the fungiform papillae : form, number, capillary condition.

4) Saliva : pH, electrolytes, minerals, urea nitrogen, uric acid, creatinine.

5) Usual examination of blood and urine : anemia, syphillis, function of liver and kidney, arteriosclerosis, diabetes, albuminuria, hypothyroidism and adrenal cortical insufficiency.

6) Quantitative measurement of trace elements (Zn, Fe, Cu) in serum, hair and urine.

7) Psychoanalysis : CMI, SDS, MAS, etc.

8) Taste examination: electrogustometry, filter-paper disc method.

9) Biopsy of vallate papillae.

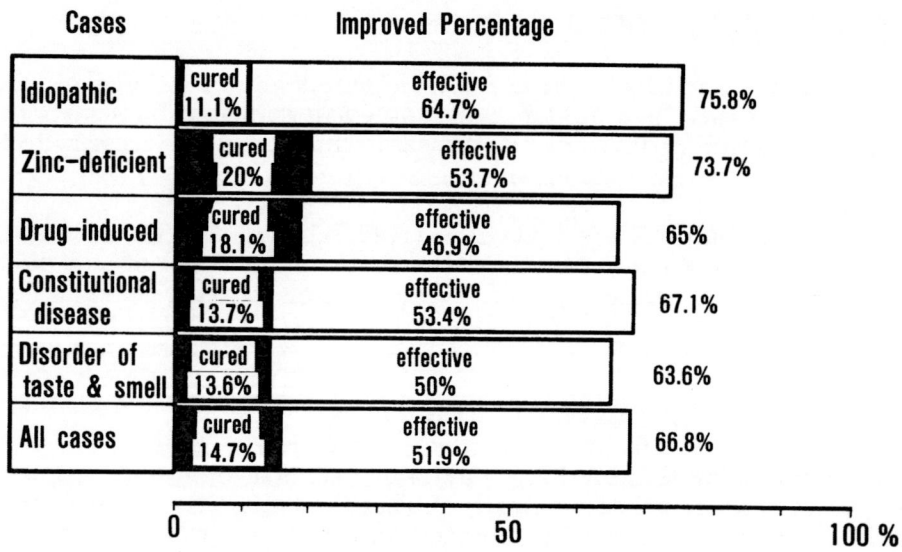

Fig. 8 Effective rate in zinc treatment by main causes of taste disorder.

Table 3 Efficacy of zinc treatment — In relation to the period from oncet to treatment —

Period	~1W	~1M	~3M	~6M	~1Y	~3Y	~5Y	~10Y	10Y~	total
cure	4	7	16	15	12	7	0	1	0	62
effective	2	32	48	52	31	35	8	8	2	218
no change	0	10	24	25	29	36	9	4	3	140
total	6	49	88	92	72	78	17	13	5	420
effectual rate	100%	79.6%	72.7%	72.8%	59.7%	53.8%	47.1%	69.2%	40%	66.7%

ANIMAL STUDY ON THE RELATIONS BETWEEN TASTE AND ZINC

The above-mentioned statistics on our taste outpatients demonstrate a close relationship between taste and zinc. Meanwhile, we have continued to study the relationship between taste and zinc, always comparing the clinical results with those of the tests conducted on rats.

CREATION OF ZINC DEFICIENT RATS

Feeding young rats immediately after weaning period with zinc deficient feeds results in suspension of growth, loss of hair and exanthemata similar to human acrodermatitis enteropathica. Since zinc deficient rats are easily infected, it is necessary to feed each of them in a separate environmental control apparatus (EBAC-S) (Kishi 1984; Hasegawa 1986).

High zinc concentration in normal rats is observed in bones, muscles, prostates, kidneys and livers. On the other hand, zinc deconcentration in the organs of zinc deficient rats is observed in the skin, prostates, testis, digestive tracts, adrenal gland and body hair. There is also significant atrophy of thymus and spleen (Hosonuma 1983).

ZINC IN TASTE BUDS (AUTORADIOGRAPHY)

The zinc deficient rat shown here was orally administered with radioactive zinc (^{65}ZuCl$_2$ 50 μCi/100g) and subsequently subjected to whole body autoradiography.

Zinc deficient rats in general show stronger density than normal rats and radioactive zinc is concentrated particularly on the tongue, digestive tracts and skin. We can safely conclude from the above that the organs that require zinc but are unable to store it perform better absorption when zinc is supplemented (Hosonuma 1983).

Fig. 9 is a microautoradiography of circumvallate papillae taste buds (intravenous injection of 926 μCi/rat of ^{65}ZuCl$_2$ was conducted through the tail vein). Silver particles are particularly concentrated on the taste buds.

Fig. 10 is a microautoradiography of filiform papillae. Dense silver particles are observed at the basal layer of the papilla and epithelia. This is a phenomenon related to the occurrence of red smooth tongue in human zinc deficiency.

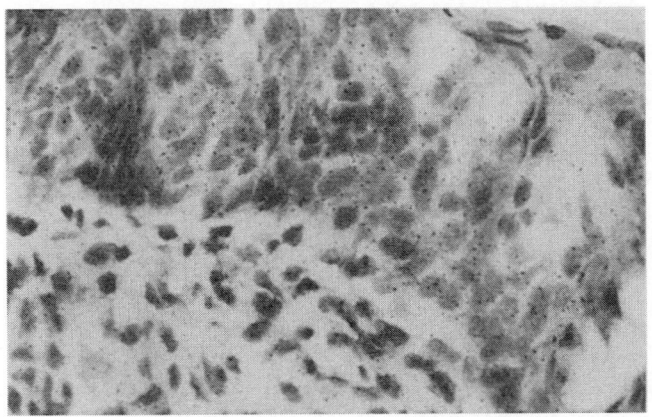

Fig. 9 Microautoradiography of vellate papillae taste buds.

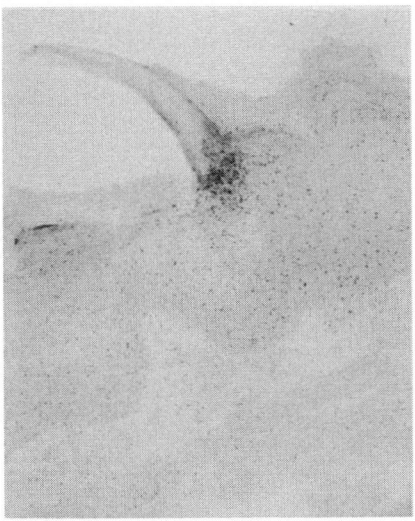

Fig. 10 Microautoradiography of filiform papilla.

ZINC ENZYME AND ZINC-ACTIVATED ENZYME IN TASTE BUDS

We then attempted to dye the zinc enzyme in taste buds.

Alkaline phosphatase (AL-P) is a zinc enzyme containing four atoms of zinc. It is secreted from taste buds and is reported to play an important role in reception of taste (Rakhawy 1963; Leuke et al. 1968).

Activity of A1-P is observed in the outer layer and middle layer of epithelial cells including the taste buds on the lateral wall of the circumvallate papillae, but is not observed in the nucleus of taste cells or basal cells.

Based on physiological experiments, Kasahara et al. (1977) drew up a hypothesis regarding the reception mechanism of taste. When tastants attach themselves to a taste receptor site, they somehow activate adenylate cyclase. Synthesis of cyclic AMP from ATP is promoted by this activated adenylate cyclase, and the increased cyclic AMP activates a kind of proteinkinase in receptor cells and releases a transmitter.

Cyclic AMP phosphodiesterase is a zinc-activated enzyme. It changes the permeability of cell membrane and plays the major role of second messenger at the junctional transmission in the synapsis.

All of the above-mentioned zinc enzyme and zinc-activated enzyme which are considered to be related to taste reception are heavily dyed by taste buds (Kishi 1984).

CAUSES OF ZINC DEFICIENT TASTE DISORDER

We have already mentioned that 15% of all patients of taste disorder indicate hypozincemia alone and no other causes.

When we try to identify these causes, the most probable one is the insufficient zinc content in our daily food. The average zinc intake per day based on the general menu of the Japanese is 9 mg (Kotake et al. 1981) and falls far behind 15 mg which is the recommended dietary allowance in the U.S.A. There are even reports that say the zinc intake per day of young Japanese women today is as little as 6 mg.

In addition to the poor intake of zinc, we must pay attention to the fact that polyphosphate, EDTA, phytic acid etc. that have metal chelating effects are now being used in large quantities in various kinds of processed foodstuffs as food additives.

So, as the next step we gave zinc deficient feed to young rats and old rats and compared them.

Prior to the feeding, we confirmed, by means of a two bottle preference test using quinine hydrochloride and distilled deionized water, that the rats to be used for the experiments were capable of discerning 10^{-6} M concentration without fail.

Both the two bottle preference test and the observation of ingestive behavior with a strong solution of quinine chloride were adopted to determine whether or not taste disorders occurred (Hasegawa, Tomita 1986).

With regard to the young rat group within the group fed with zinc deficient feed, the growth was suppressed and body weight showed no increase. As to the old rat group, decrease in body weight was observed.

Concerning serum zinc value, both young rats and old rats in the group fed with zinc deficient feed suffered from hypozincemia.

The conclusion of this experiment was that 12 out of 40 young rats, accounting for 30% of the total, were determined to have suffered from hypogeusia, whereas 12 out of 17 old rats, constituting a higher percentage of 70% of the total, were similarly determined to have hypogeusia.

TASTE DISORDER OCCURRENCE CAUSED BY FOOD ADDITIVES

We then carried out an experiment to see whether feed containing food additives caused taste disturbances in rats. The frequently used polyphosphate and phytic acid were chosen as food additives. While phytic acid in particular has metal chelating effects exceeding those of EDTA, it is commonly used in all processed foodstuffs because it is not required to be indicated as food additive in Japan on grounds that it is a natural product contained in soybeans. Also, zinc in food produces extremely insoluble complex salt in the intestinal canal through the coexistence of phytic acid and

calcium. As a result, zinc is not absorbed through the intestinal canal. It was food containing much phytic acid that caused the occurrence of male dwarfs in Egypt reported as the first case of human zinc deficiency by Prasad (1963).

The experimental groups observed are as follows:

A. Young rat group (5W. old ♂)
1) Control feed group (n=5): CE-2 (usual food for rat: Zn content. 48.4 − 56.8 ppm and Ca content 10400 − 12200 $\mu g/g$)
2) Polyphosphoric acid (PPA) added feed group (n=10): CE-2 + 10% PPA
3) Phytic acid added feed group (n=5): CE-2 + 10% phytic acid
4) Phytic acid + Ca added feed group (n=10): CE-2 + 10% phytic acid + 2% (Zn: 53.2 $\mu g/g$, Ca: 19300 − 19600 $\mu g/g$)
5) EDTA added feed group (n=5): CE-2 + 10% EDTA

B. Old rat group
1) Control feed group (n=19): CE-2. 75 − 77 W. old ♀ (n=9), 92 − 94 W. old ♀ (n=10)
2) PPA added feed group (n=12): CE-2 + 10% PPA
3) PPA + Ca added feed group (n=8): CE-2 + 10% PPA + 2% $CaCO_3$
4) Phytic acid added feed group (n=7): CE-2 + 10% phytic acid
5) Phytic acid + Ca added feed group (n=12): CE-2 + 10% phytic acid + 2% $CaCO_3$

Various physical symptoms were more prominent in the group fed with feed to which PPA was added.

Suppression of body weight increase was most significant in the group fed with feed containing PPA, for both young and old rats.

Increase of body hair zinc value was more notable in the group fed with PPA compared with the control group, for both young and old rats. Conversely, food intake increase was more prominent with the group fed with phytic acid and calcium than with the control group.

Serum zinc value of young rats differs from that of old rats. Analysis of the blood collected within 8 weeks after the start of the experiment indicated that the serum zinc value decreased with about one half of young rats in the PPA group, whereas in the group fed with phytic acid and the group given phytic acid plus calcium, the value decreased at a higher rate compared with the control group in all cases and then restored its normal value after a long period of experiment spanning 28 weeks (Shigihara et al. 1984).

In the analysis of the blood of old rats collected 9 − 13 weeks after the start of the experiment, the value decreased only with the group fed with PPA plus calcium, and other groups showed no difference compared with the control group (Abe, Tomita 1987).

Further details will be reported by Shigihara (1990), but just to summarize here, the results of measurement of zinc concentrations in duodenum, kidney, femur and body hair indicated that polyuria occurred in the group given the PPA-added feed and zinc concentration increased remarkably in the kidney and temporarily in the body hair. Renal disorder was also observed histologically.

From the above, it is possible to conclude that polyphosphate, after being absorbed from the intestine, forms a complex with zinc and precipitates in the kidney and body hair.

In the group fed with feed to which phytic acid and calcium were added, zinc concentration in the body hair was decreased. From these observations we believe that zinc deficiency is caused by the inhibition of zinc absorption.

In any case, PPA which causes renal insufficiency should not be used as food additives.

Moreover, the rats fed with feed containing 10% EDTA demonstrated refusal of food and all died within 10 days. Intoxicity of EDTA should also be re-evaluated.

Ocurrence of taste disturbance in young rats was not observed in the taste experiments. However, occurrence of taste disturbance was observed in all groups of old rats (Fig. 11).

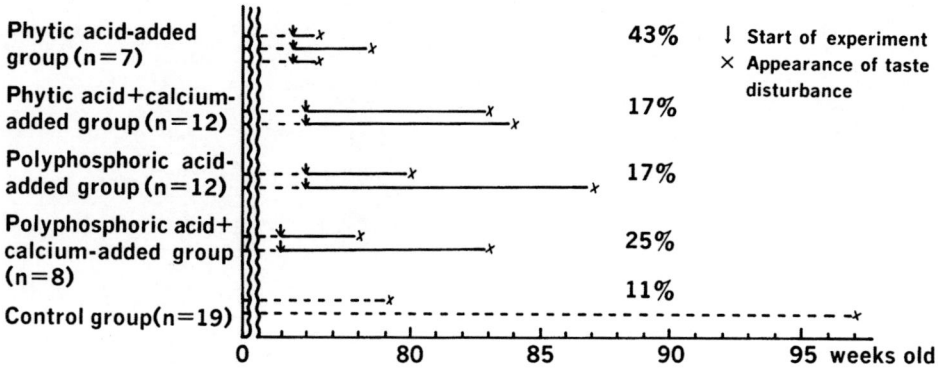

Fig. 11　Occurrence of taste disturbance caused by food additives in old rats. Determination by the two-bottle preference test. (10^{-5} M quinine hydrochloride solution and distilled water).

The group fed with the feed containing phytic acid was the most vulnerable to taste disturbance. The occurrence ratio was as high as 43% and the occurrence was also observed early, about $1-3$ weeks after the start of the experiment.

The occurrence ratios in the group fed with the feed containing both phytic acid and calcium, the group fed with the feed containing PPA, and the group fed with the feed containing both PPA and calcium were 17%, 17% and 25%, respectively. However, all of the rats that were unable to distinguish 10^{-5} mol quinine chloride were able to distinguish 10^{-4} mol solution. In other words, the degree of taste disturbance was slight.

It was also observed that, out of the 19 rats in the control groups that were aged 80 weeks or older, two rats (accounting for approximately 10%), suffered from taste disturbance. This means that senile taste disturbance can occur (Abe, Tomita 1987).

ELECTRON-MICROSCOPIC STUDY OF THE TASTE BUD OF ZINC DEFICIENT RAT WITH TASTE DIS-ORDER

As we have explained, it is positive that zinc deficient food or food additives with metal chelating effects cause taste disorders in many cases at a certain ratio as the serum zinc value decreases.

What, then, occurs to taste buds under such circumstances?

So we cut off a slice from the taste bud at a vertical section through the taste pore and observed it by electron microscope.

The photo of Fig. 12 indicates a taste bud of a normal adult rat. Microvilli, electron dense substance etc. are observed in the taste pores.

The photo of Fig. 13 is a whole image of a taste bud of a zinc deficient rat with taste disorder. The site pointed out with an arrow is a taste pore where the number of microvilli has been decreased and cut off. The dense substance that should exist among the microvilli has disappeared.

The photo of Fig. 14 is an enlarged image. The site pointed out with a bold arrow is a taste pore where no dense substance is observed. Dark granules of the type-1 cell have decreased. Furthermore, many vacuolations of various sizes can be observed in the taste bud cells. These are believed to be degeneration of the cells (Kobayashi, Tomita 1986).

Fig. 12 Longitudinal section of taste bud (control rat). Dense substance observed in the taste pore region and dark granules in cytoplasma
DS: dense substance
MV: microvilli
DG: dark granule
Arrow indicates taste pore region.

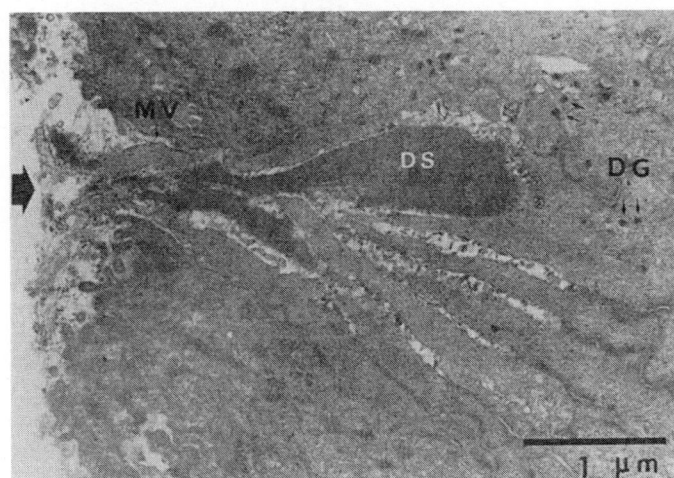

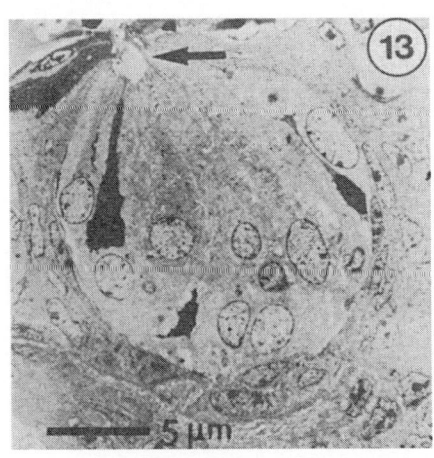

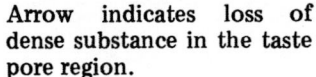

Arrow indicates loss of dense substance in the taste pore region.

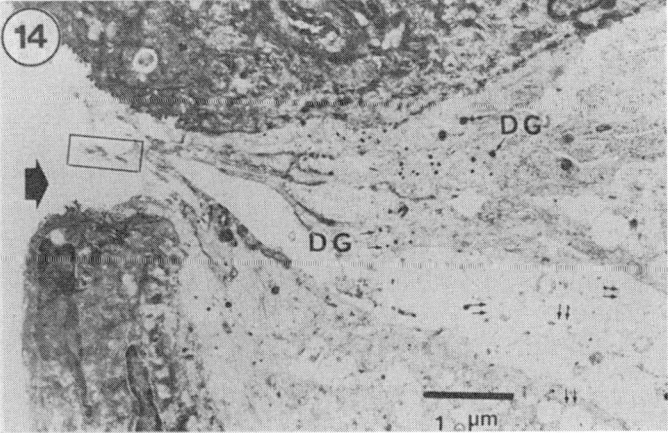

Arrow indicates loss of dense substance in the taste pore region. Double arrows indicate micro-vacuolation in cytoplasma.

Fig. 13,14 Taste bud of zinc deficient rat with taste disturbance.

We have already emphasized the importance of the human soft palate in taste (Tomita, Pascher 1964, Tomita et al. 1986). However, the changes caused by zinc deficiency of the soft palate taste buds in rats have not yet been reported.

The photo of Fig. 15 is a light microscopic image of soft palate taste buds of a young and normal rat. Each papilla has one taste bud. The object pointed out with an arrow is

The photo of Fig. 16 shows soft palate taste buds of an old zinc-deficient rat. Vacuolation of taste cells and vacant space between the taste buds and the surrounding tissues are observed.

The photo of SEM (Fig. 17) gives a taste pore of a soft palate taste bud of a young and normal rat. Accumulation-like material crammed within the taste pore is visible. It is probably the dense substance observed by transmission electron microscope.

The photo of Fig. 18 shows a taste pore of a soft palate taste bud of an old rat with taste disturbance caused by zinc deficiency. It was noted that nothing was contained in

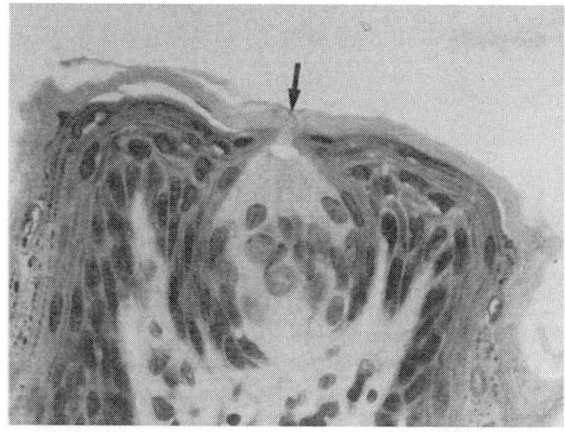

Fig. 15 Soft palate taste bud and its taste pore (arrow) of a normal young rat. (2.5x 40).

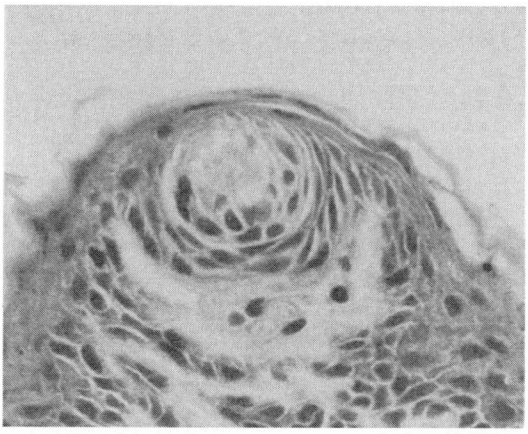

Fig. 16 Soft palate taste bud of a zinc-deficient old rat. Vacuolation of taste cells are noted. (2.4x40)

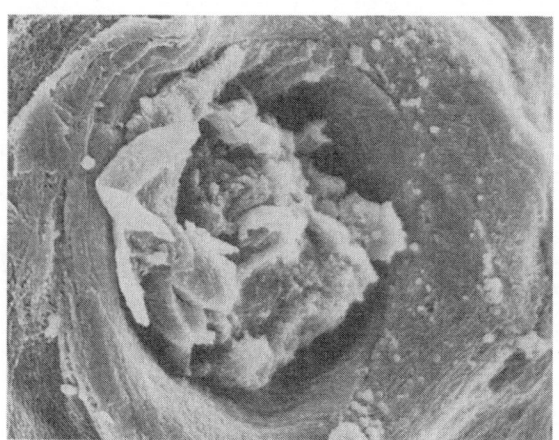

Fig. 17 Accumulation in the taste pore of papilla of the soft palate of a normal young rat.

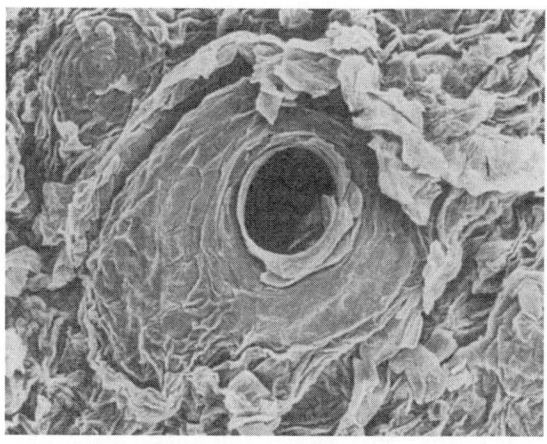

Fig. 18 Emptied taste pore of papilla of the soft palate of a zinc-deficient old rat. Microplicae is not recognized on the epithelium.

the taste pores and that the microvilli had disappeared (Naganuma et al. 1988).
From the observation outlined above, we have found that individuals with taste disturbances resulting from zinc deficiency had morphological abnormalities in their taste buds.

RENEWAL OF TASTE BUD CELLS OF RATS WITH TASTE DISTURBANCE DUE TO ZINC DEFICIENCY

Taste bud cells continue to renew themselves as long as their taste nerves are normal.

According to a report by Beidler (1965), the life span of taste bud cells of a rat is approximately 10 days.

Since zinc is involved in synthesis of DNA, nucleic acid and protein, we believe zinc deficiency extends the turnover time of the taste bud cells and that the resultant tardiness of the renewals lowers the sensitivity of taste receptors and leads to deterioration of taste acuity.

In order to test this hypothesis, we observed two groups of rats, both of which were given the same zinc deficient feeds. Only one of them was suffering from taste disorder and the other stayed normal. A third control group was added for the sake of comparison.

As we showed on the chart of Fig. 19, the turnover time of taste buds of the zinc deficient group suffering from taste disorder exceeded the turnover time of the normal group by more than 100 hours. The length of the turnover time of taste buds of the zinc deficient group not suffering from taste disorder was positioned somewhere between those of the other two groups.

When a few of the zinc deficient rats suffering from taste disturbance were fed with feed containing a suffucient quantity of zinc, they recovered their normal sense of taste. The chart of Fig. 20 demonstrates the measured length of turnover time of taste bud cells of one of these rats. The turnover time also regained its normal sphere (Ohki et al. 1988).

This experiment proved that the taste disturbance of zinc deficiency was removed through zinc treatment which normalized the tardiness of the renewals.

SHORT SUMMARY OF ANIMAL STUDIES

The following facts were revealed by measurement of taste threshold, structural variations of taste buds and measurement of renewal turnover time of the taste bud cells of rats.

1) Taste disturbance positively occurs through the zinc deficient feed or feed containing large quantities of food additives with strong metal chelating effects.
2) The occurrence ratio of taste disturbance of old rats is higher than that of young rats.
3) If rats suffering from taste disturbance caused by zinc deficiency are given feed with a sufficient quantity of zinc, the taste disturbance will disappear.
4) Taste disturbance due to zinc deficiency occurs through the abnormality of function and structure of the taste bud cells.

THE PROBLEM OF INDIVIDUAL DIFFERENCE IN TASTE DISTURBANCE OCCURRENCE

Why, then, does the same dosage of zinc deficient feed cause taste disturbance in one case but not in another?

We wish to reserve our conclusion due to the insufficient number of the cases tested. But what we did was to orally administer radioactive zinc 65 (^{65}ZnCl$_2$) to rats, collect their blood every hour, and count the radioactivities of the radioactive zinc in the plasma and blood cells.

Compared with normal rats, absorption from digestive canals differs depending on the existence or absence of taste disturbance. However, transitions from the plasma to the blood cells, with the exception of one rat with taste disturbance that died in the

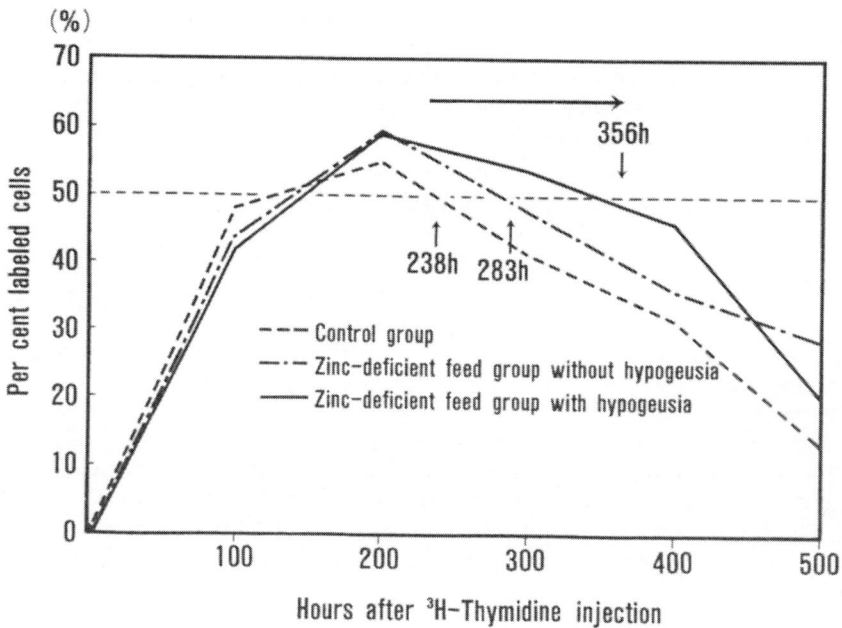

Fig. 19 Changes of turnover rates on taste bud cells in zinc-deficient rats
with or without hypogeusia.

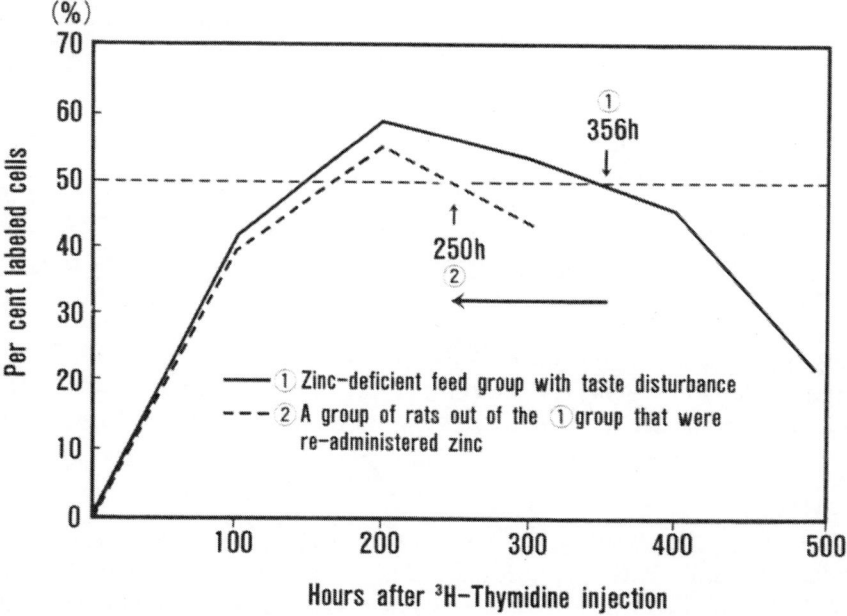

Fig. 20 Recovery of turnover rate on taste bud cells after zinc treatment
in rats with zinc-deficient taste disorders.

middle of the experiment, were good as a whole regardless of the existence or absence of taste disturbance.

From the above results, we assume that those rats to which taste disturbance occurs through zinc deficient food may be the cases possessing some inhibitory factors related to zinc absorption (Hosonuma 1983).

WHAT CAUSES IDIOPATHIC TASTE DISORDER?

As we have explained with the help of clinical statistics, idiopathic taste disorders, that is, cases displaying neither hypozincemia symptoms nor any other symptoms that cause taste disorder, account for about 14% of the total.

Observed in terms of the occurrence rate in each decade age, the peak is reached in almost the same manner as zinc-deficient taste disorder (Fig. 21).

The effective rate of zinc treatment is also the same as that of zinc deficiency.

Also in the separately conducted double-blind test administering zinc gluconate, the intake of zinc was found to be significantly effective compared with placebos, both for cases of zinc deficiency and for idiopathic taste disorder (Yoshida et al. 1989).

From these results, we can safely conclude that, although idiopathic taste disorder may include other unknown causes, it certainly includes marginal zinc deficiency to a considerable degree.

Fig. 21 Comparison between idiopathic taste disorder and zinc-deficient taste disorder reviewed in age distribution.

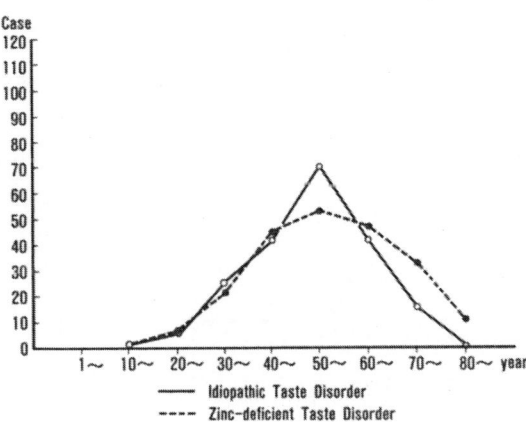

IS THERE A GOOD DIAGNOSTIC INDEX FOR SUCH A MARGINAL ZINC DEFICIENCY?

We are examining for this purpose on several laboratory tools up to the present: hair analysis, zinc in parotid saliva, red blood cell zinc, serum copper (Cu/Zn ratio), activities of zinc enzymes in serum and changes of metabolites (uric acid, free fatty acid) in blood.

1) Hair Analysis

Trace element analysis of the scalp hair is popularly used as the index for environmental pollutions and screening of insidious disease because scalp hair is easy to collect, preserve and transport as specimens and also because concentration of many trace elements is comparatively high in hair.

As the first step for obtaining the normal data of the Japanese, we adopted Inductively Coupled Plasma Emission Spectrometry (ICP) to quantity ten minerals including zinc in human hair (each weighing 0.5g — 1.0g) approximately 5cm long or shorter, col-

Table 4　Laboratory findings for main causes of taste disorder
— Changes of several diagnostic tools in blood —

	Serum Cu	Al-P	vit. A	red cell Zn	r. Zn/RBC	Cu/Zn	Fe/Zn	uric acid	free fatty acid
idiopathic	111.6±20.7 (n=119)	161.2±53.6 (n=104)	214.8±68.1 (n=78)	470.5±78.6 (n=56)	1.04±0.16 (n=40)	1.3±0.29 (n=115)	1.17±0.43 (n=115)	5.02±1.42 (n=113)	829.7±928.6 (n=49)
zinc-deficient	112.5±27.1 (n=132)	161.5±63.3 (n=123)	184.9±64.9 (n=81)	460.1±67.1 (n=75)	1.07±0.14 (n=60)	1.66±0.55 (n=129)	1.33±0.52 (n=129)	4.97±1.51 (n=130)	618.3±614.8 (n=52)
drug-induced	122.2±32.2 (n=203)	178.3±72.1 (n=186)	212.1±129.1 (n=134)	460.7±75.7 (n=100)	1.19±1.12 (n=85)	1.61±0.56 (n=193)	1.2±0.52 (n=191)	5.58±1.65 (n=200)	930.4±1435.8 (n=95)
flavor disorder	112.1±16.8 (n=36)	158.5±44.4 (n=30)	188.9±69.6 (n=18)	475.7±89.4 (n=21)	1.07±0.2 (n=19)	1.42±0.31 (n=36)	1.12±0.72 (n=36)	4.84±1.15 (n=34)	558.5±255.5 (n=14)
psychogenic	111.1±17.5 (n=69)	163.1±57.5 (n=59)	213.8±179.7 (n=47)	461.9±79.9 (n=48)	1.04±0.15 (n=39)	1.41±0.33 (n=68)	1.18±0.39 (n=64)	4.52±1.15 (n=67)	872.2±1137 (n=34)

lected from occipital scalps of about 1,500 objects free from any treatments such as permanent wave. From the data obtained, we examined the influences of sex and aging (Uchida et al. 1987).

As a result of these observations, we concluded that the normal value of zinc in the hair of a Japanese adult is between 130 μg/g and 180 μg/g.

Based on the above results, we conducted hair analysis on 182 cases of taste disturbance and also studied other trace elements with focus on zinc in hair.

We conclude that in cases of adults suffering from diseases with no symptom except for taste disturbance, measurement of hair zinc levels is not so helpful.

However, there are cases indicating abnormally high hair zinc levels of 200 ppm or more, or abnormally low levels of 100 ppm or less at the time of the first consultation regardless of the serum zinc level in 10% of the taste disordered patients, so it can be maintained that the cases are related to zinc metabolism.

It was also discovered that hair zinc levels of children aged 15 years or less had good correlations with other trace elements and could be an effective index for the circumstances of zinc intake or zinc metabolism, as was reported by Hambidge (1972).

2) Zinc Level in Parotid Saliva

Henkin (1975) emphasizes that gustin, a zinc-binding protein in parotid saliva, is associated with taste acuity. We are not in a position to comment on gustin, but we would like to mention that we looked into zinc concentration in parotid saliva and the changes in its concentration triggered by oral zinc treatment (Mikoshiba 1984).

69 patients were divided into the low serum zinc group and the normal group for the sake of comparative observation.

The correlation coefficient between zinc in parotid saliva and serum zinc at the time of the initial diagnosis was 0.23, which means there was no distinct correlation.

Although the variation of zinc in parotid saliva through zinc treatment rose in hypozincemia cases, hardly any variation was detected in the normal cases.

The results given above indicate that zinc determination in the parotid saliva and quantitative changes resulting from zinc treatment are in good agreement in the cases showing hypozincemia. However, they are not good diagnostic tools for evaluation of the prognosis for the recovery of sense of taste and seem to be much less helpful to the diagnosis of marginal zinc deficiency.

3) Changes of Several Diagnostic Tools in Blood

Blood is most commonly used as specimen for clinical examinations for evaluating the conditions of the patients' diseases.

Table 4 shows the results of comparative studies, by the causes of taste disorders, on the values obtained at the time of the initial medical consultation for the various

examination items that are considered to be closely associated with zinc metabolism. The two bottom causes on the table have little relation to zinc deficiency.

We made comparisons on serum copper, alkaline phosphatase, vitamin A, erythrocyte zinc, the copper: zinc ratio and iron: zinc ratio in the serum, uric acid and free fatty acid, none of which gave us any grounds for linking idiopathic taste disorder with marginal zinc deficiency.

Since diagnosis ex juvantibus is the only measure we have at present, oral administration of zinc should be adopted as the first choice treatment for cases of idiopathic taste disorder.

OLFACTION AND ZINC

Olfaction functions through olfactory cells in the olfactory epithelium between the nasal septum and the superior nasal concha. Olfactory cells are a variation of nerve cells.

The olfactory cell has a certain life span which is believed to be 28.6 ± 6.6 days (Moulton 1975) in the case of a mouse. At the end of the life span, the basal cell is changed into a new olfactory cell which replaces the old one.

The olfactory epithelium can be directly observed with a needle-shaped Selfox endoscope (Zusho et al. 1981). Pathologic findings of the adjacent ethmoid is also possible with X-ray tomography.

It is possible to measure the smell acuity through the standard olfactory acuity test (T & T olfactometry) to be conducted on one side at a time (Furukawa et al. 1988).

Intravenous olfaction test using Alinamin solution is applied to more serious disorders (Zusho 1978, Furukawa 1988).

Among all causes of olfactory disturbance, respiratory ones are the most prevalent. Disturbance of the olfactory epithelium after influenza, and olfactory nerve injury after traffic accidents are increasing steadily.

It is also evident that there are combined disorders of taste and smell due to zinc deficiency in Japan, although occurrences are not as frequent as was reported by Henkin (1971).

According to our clinical statistics, the group of simultaneous eruption, and probably of common causes, accounted for 3.8% of the total.

The patient is a university professor aged 62 and the symptoms are reliable. The chief complaints are hypogeusia, dysgeusia and dysosmia. About 4 months ago the sweet taste began to decline and the salty taste and sour taste changed to the taste one experiences when licking metal which spoils the taste of the meals. He would often complain to those around him of the smell of charring wood when in fact there was no such evidence. He had been diagnosed to have essential hypertension 12 years ago, coronary sclerosis, diabetes and hyperlipemia some years ago and had been taking hydrochlorothiazide etc. for the past two and a half years.

The results of the tasting test were as shown in Table 5. Although the results of electrogustometry were normal, the sweet taste was particularly indiscernible as the patient complained.

Although the olfactory threshold was not bad except for scatol, it can be maintained that isovaleric acid and recent dysosimia belonged to the same type.

The serum zinc level had decreased prominently to 35 µl/dl.

The case was diagnosed to have been caused by hypozincemia due to continual long-time dosage of medicine, particularly of hydrochlorothiazide, and was treated with oral dosage of 300 mg of zinc sulfate per day.

A week after the start of treatment, although the patient was not aware of any change, the serum zinc level increased to 58 µg/dl and recovery was noted in the tasting test as given in Table 5. Hydrochlorothiazide was removed from the dosage and the volume of spironolactone was decreased in the third week.

Table 5   Gustatory examination

Name : T.S.   Sex : Male,   Age : 62years   pH :   Date :   Examiner :

	EGM (dB)		Sweet			Salty			Sour			Bitter		
	R	L	R	L	F	R	L	F	R	L	F	R	L	F
First Exam.														
Chorda tympani N.	6/me	4/me	n.r.	n.r.		n.r.	V/sa		n.r.	IV/so		n.r.	V/bi	
Great. petrosal N.	18/me	18/me	n.r.	n.r.	V/sw	me/V	n.r.		V/so	V/so		n.r.	n.r.	
Glossopharyng. N.	8/me	8/me	n.r.	n.r.		n.r.	V/sa		n.r.	V/so		V/bi	IV/bi	
1 Week after Zinc Treatment														
Chorda tympani N.			n.r.	sa/V		III/sa	III/sa		III/so	III/so		IV/bi	III/bi	
Great. petrosal N.			n.r.	V/sw		IV/sa	IV/sa		IV/bi+so	IV/bi		V/bi	V/bi	
Glossopharyng. N.			n.r.	IV/sw		III/sa	III/sa		sw+so/III (bad taste)	sw+so/III		III/bi	III/bi	
6 Months after Zinc Treatment														
Chorda tympani N.			III/sw	III/sw		IV/sa	IV/sa		III/so	III/so		III/bi	IV/bi	
Great. petrosal N.			IV/sw	IV/sw		III/sa	III/sa		III/so	III/so		III/bi	III/bi	
Glossopharyng. N.			IV/sw	IV/sw		III/sa	III/sa		III/so	III/so		III/bi	III/bi	

sw, sweet ; sa, salty ; so, sour ; bi, bitter ; me, metallic ; n.r., no response at maximum stimulation ; F, fully spreading in the oral cavity.

The serum zinc level increased further to 96 μg/dl and both dysgeusia and dysosmia disappeared in the fourth month. The disease was totally cured after the sixth month, with the serum zinc level constantly maintaining a level of 100 μg/dl or more.

As we have explained above, in the case of concurrent occurrence of taste and olfactory disturbances, taste disturbance is more prevalent than the other. This could be attributed to the difference of the number of cells between taste cells and olfactory cells.

ZINC CONCENTRATION IN THE OLFACTORY EPITHELIUM OF ZINC-DEFICIENT RATS

We measured the zinc concentrations in the olfactory epithelium of zinc-deficient rats classified by the period of time of feeding and compared them with those of the control group.

The longer the state of zinc deficiency continued, the lower was the zinc concentration in the olfactory epithelium. Zinc concentration of the control group in the 18th week was about 4.7 times as high as that of the zinc-deficient group (Table 6).

However, as shown in Table 7, the weight of organs and zinc concentration in the rhinencephalon, the cerebrum and the pons do not differ significantly from the control group. It may also be presumed that a mechanism is working to maintain zinc concentration and organ weight of such vital organs as the rhinencephalon, the cerebrum and the pons in a zinc-deficient rat. This is because, despite the fact that body weights and serum zinc levels of the group of zinc-deficient rats are considerably lower than those of the control group, there is not much difference in the organ weight and zinc concentrations in the brain between the two groups (Mikoshiba et al. 1983).

For the results outlined above, it may be assumed that olfactory disturbance in zinc-deficiency occurs at the level of the olfactory epithelium.

Table 6 Changes of zinc concentration (μg/g) in nasal mucosa containing olfactory epithelium in process of zinc-deficient feeding

duration of feeding	16 week	17 week	18 week	20 week	21 week
Zinc-deficient Rats (n=5)	56.5	27.6 / 36.0	18.5	26.2	
Control (n=2)			86.9		97.4

Table 7 Zinc concentration in rhinencephalone, cerebrum & pons of the zinc-deficient rats

	Zinc Concentration (ng/mg)			Tissue Weight			Body Weight	Ratio to Body Weight			Serum Zinc (μg/dl)
	Rhinenceph.	Cerebrum	Pons	Rhinenceph.	Cerebrum	Pons		Rhinenceph.	Cerebrum	Pons	
Zinc-deficient Rats (n=7)	111.9 +30.0	101.8 ±10.2	78.9 ±28.6	(mg) 84.0 ±20.7	(g) 1.31 ±0.05	(g) 0.34 ±0.04	(g) 133.1 ±12.8	6.3 ×10⁻⁴	9.8 ×10⁻³	26 ×10⁻⁴	46.5 ±24.3
Control (n=2)	84.9 ±24.1	92.9 ± 4.7	76.3 ±29.1	65.0 ±21.2	1.61 ±0.06	0.36 ±0.06	486.0 ±22.6	1.3 ×10⁻⁴	3.3 ×10⁻³	7.4 ×10⁻⁴	138.0 ±24.0

TEM OF THE OLFACTORY EPITHELIUM IN ZINC-DEFICIENT RATS

Consequently, the following 6 groups were formed: 1) Young rat group in which a control diet was given (N:5), 2) Young rat group in which a zinc-deficient diet was given (N:8), 3) Old rat group in which a control diet was given (N:5), 4) Old rat group in which a zinc-deficient diet was given (N:20), 5) Young rat group in which zinc was re-administered (N:2), 6) Old rat group in which zinc was re-administered (N:3), and an experiment by transmission electron microscopy (TEM) was conducted to examine the difference due to age in the variation of olfactory epithelium caused by the dosage of zinc-deficient feed and also to see if the variation which occurred to olfactory epithelium was normalized or not as a result of repeated dosage of zinc.

In young zinc-deficient rats, the olfactory epithelium was flattened and the intercellular space was widely opened (Fig. 22). Although variations are also detected in the supporting cells, the variation of olfactory cell is stronger, the electron density in the cytoplasm is very high, and the cell organelles around the nucleus fell into degeneration, becoming unclear, with many granules of large and small sizes in dark shades. In addition, far more multivesicular bodies than normal were observed (Fig. 22). Similar degeneration was noted in the dendrites and olfactory vesicle, and morphologically there were numerous unmatured cells containing many centrioles and degenerating ones including many membrane-bound bodies.

Similar circumstances were observed in the group of old zinc-deficient rats, although in smaller degrees compared with the younger group.

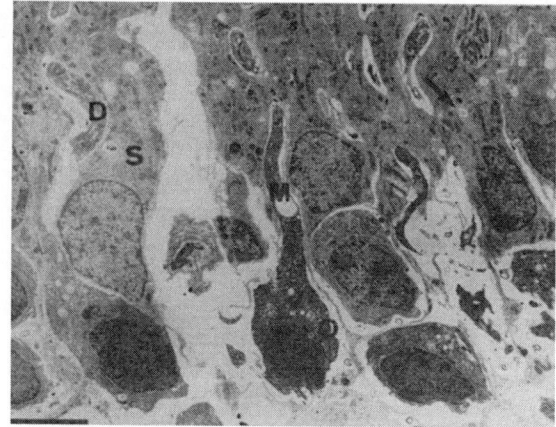

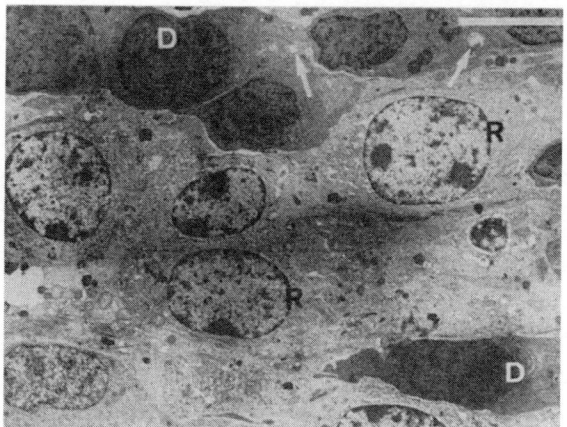

Fig. 22 Olfactory mucosa of a zinc-deficient young rat. Multivesicular bodies and other granules in the supporting cells and olfactory cells are increased. S, Supporting cell; O, olfactory receptor cell; D, degenerating dendrite; M, multivesicular body. Arrow: various vesicles in a supporting cell. Scale: 5 u.

Fig. 23 Olfactory mucosa of zinc re-administered young rats. Some groups of regenerating olactory receptor cells are present, but there are also olfactory cells which have an electron dense cytoplasm. S, Supporting cell; D, degenerating olfactory receptor cell; R, regenerating olfactory receptor cell. Large arrow: multivesicular body. Small arrow: needle-like bodies in a supporting cell. Scale: 5 u.

TEM FINDINGS IN WHICH ZINC WAS RE-ADMINISTERED

As the next step, when the young zinc-deficient rats were given zinc-sufficient diet, weight gain and improvement of such symptoms as alopecia and dermatitis were observed. The serum zinc level was also recovered. The electron microscopic image at this stage is shown in Fig. 23.

While degenerative findings still remain histologically, cell organelles are well preserved for each cell and groups of regenerated neurons from the basal cell are observed.

The group of old rats supplemented with zinc became almost the same as the control group, while nearby the basement membrane are seen many degenerating cells including needle-like structures (Shigihara, Tomia 1986).

As mentioned above, it was proved, through clinical observations on human patients and results of rat experiments, that variations occurred to olfaction through zinc deficiency.

CONCLUSION

1) Taste disorder as a chief complaint in many cases is caused by disorder of the taste bud level. However, taste examinations for each gustatory innervation is necessary for diagnosis.

2) It was proved with the results of zinc treatments and animal experiments that its cause was closely related with zinc deficiency.

3) At present, good diagnostic tools for the zinc treatment are the decrease of the serum zinc, elevation of the serum copper:zinc ratio and diagnosis ex juvantibus. There is urgent need for establishing better diagnostic aid for the marginal zinc deficiency.

4) It was also proved that olfactory disorder was caused by zinc deficiency in some part of the olfactory epithelium lesion.

5) Cases of taste and olfactory disorders are increasing. The primary cause for the above is eating habits. Use-limitation of food-additive with strong metal chelating effect and nutrition guidance including the essential trace elements are urgently called for.

REFERENCES

Abe H, Tomita H (1987) Experimental studies of taste disturbance due to food-additives. Nihon Univ J Med 29:1—10

Beidler LM, Smallman RL (1965) Renewal of cells within taste buds. J Cell Biol 27: 263—272

Furukawa M, Kamide M, Miwa T, Umeda r (1988) Significance of intravenous olfaction test using thiamine propydisulfide (Alinamin) in olfactometry. Auris·Nausus·Larynx (Tokyo) 15: 25—31

Furukawa M, Kamide M, Miwa T, Umeda R (1988) Importance of unilateral examination in olfactometry. Auris·Nasus·Larynx (Tokyo) 15: 113—116

Hambidge KM, Hambidge C, Jacobs M, Baum JK (1972) Low Levels of zinc in hair anorexia, poor growth, and hypogeusia in children. Pediatr Res 6: 868—874

Hasegawa H, Tomita H (1986) Assessment of taste disorders in rats by simultaneous study of the two-bottle preference test and abnormal ingestive behavior. Auris·Nasus·Larynx (Tokyo) 13 (Suppl 1): 33—41

Hasegawa M, Yamada K, Uchida H, Tomita H (1987) Mineral contents in the hair of taste disorder patients. Trace Metal Metabolism, Its Abnormality and Treatments XV: 109—116

Henkin RI, Bradley DF (1969) Regulation of taste acuity by thiols and metal ions. Proc Natl Acad Sci USA 62: 30—37

Henkin RI, Bradley DF (1970) Hypogeusia corrected by Ni^{++} and Zn^{++}. Life Sci 9: 701—709

Henkin RI, Schechter PJ, Mattern CFT, Hoye RC (1971) Idiopathic hypogeusia with dysgeusia, hyposmia, and dysosmia: A new syndrome. JAMA 217: 434—440

Henkin RI, Lippoldt RE, Bilstad J, Edelhock H (1975) A zinc-containing protein isolated from human parotid saliva. Proc Natl Acad Sci USA 72: 488—492

Hosonuma H (1983) Studies of the Content of zinc and the absorption of $^{65}ZnCl_2$ in each organ of normal and zinc deficient rats. J Nihon Univ Med Ass 42: 547—555

Kasahara Y, Shimotahira K, Takeda Y (1977) Effects of cyclic nucleotides on taste reception. Proc XI Jap Symp on Taste and Smell: 94—97

Kishi T (1984) Histochemical studies on the taste buds of zinc deficient rats. J Nihon Univ Med Ass 43: 15—31

Kobayashi T, Tomita H (1986) Electron microscopic observation of vallate taste buds of zinc-deficient rats with taste disturbance. Auris·Nasus·Larynx (Tokyo) 13 (Suppl 1): 25—31

Kotake Y, Ikeda S, Shibata M (1981) Studies on ingested zinc quantity of Japanese. EISOAU 34: 355—365

Leuke RW, Olman ME, Baltzer BV (1986) Zinc-deficiency in the rat: effect on serum and intestinal alkaline phosphates. J Nutrition 94: 344—350

Mikoshiba H, Hosonuma H, Kobayashi T, Ichikawa M, Tomita H (1983) A quantitative study of zinc concentrations in the olfactory epithelium of nasal cavity and the central nervous system in zinc deficient rats. Trace Metal Metabolism, Its Abnormality and Treatments XI: 1—4

Mikoshiba H (1984) Study on the movement of electrolytes and trace metals in parotid saliva of patients suffering from taste disorders. J Nihon Univ Med Ass 43: 509—518

Moulton DG (1975) Cell renewal in the olfactory Epithelium of the mouse. In: Denton DA, Coghlan JP (eds) Olfaction and Taste V. Academic Press, New York San Francisco London p111

Naganuma M, Ikeda M, Tomita H (1988) Changes in soft palate taste buds of rats due to aging and zinc deficiency — Scanning electron microscopic observation —. Auris·Nasus·Larynx (Tokyo) 15: 117—127

Ohki M, Fujita S, Tomita H, Kohbo B, Takeuchi, S (1988) Turnover of taste buds of rats with taste disorders caused by zinc deficiency. Proc XXII Jap Symp on Taste and Smell: 41—44

Rakhawy MTE (1963) Alkaline phosphatase in the epithelium of the human tongue and a possible mechanism of taste. Acta Anat 55: 323—342

Shigihara S, Hasegawa H, Kobayashi T, Takahashi Y, Yasukata J, Sekimoto K, Ichikawa M, Tomita H (1984) Time correlation on hair zinc and weight in rats supplied feeder concluding food additives (polyphosphoric acids and phytic acids). Trace Metal Metabolism, Its Abnormality and Treatments XII: 95—105

Shigihara S, Tomita H (1986) Electron microscopy of the olfactory epithelium in zinc deficient rats. Nihon Univ J Med 28: 263—280

Shigihara J (1990) Effects of food additives on zinc concentration of organs of aged and juvenile rats. J Nihon Univ Med Ass: in press

Tomita H, Pascher W (1964) Über die Geschmacksfunktion nach Ausfall der sensorischen Zungennerven. HNO 12: 163—169

Tomita H, Okuda Y, Tomiyama H, Kida A (1972) Electrogustometry in facial palsy. Arch Otolaryngol 95: 383— 390

Tomita H (1982) Methods in taste examination. In: Surján L, Bodó G (eds) Proc XIIth ORL World Congr, Budapest, Hungary, 1981, Excerpta Medica, Amsterdam Oxford Princeton, p627

Tomita H, Ikeda M, Okuda Y (1986) Basis and practice of clinical taste examinations. Auris·Nasus·Larynx (Tokyo) 13 (Suppl 1) 1—15

Uchida H, Takahashi K, Toyoda H, Takahashi J, Sugahara A, Tomita H, Ohmori H (1987) Studies on the characteristic of trace metals in the scalp hair of Japanese. Trace Metal Metabolism, Its Abnormality and Treatments XV: 103—108

Yoshida S, Endo S, Tomita H (1989) Double blind method for testing zinc gluconate on taste disorder. Proc XXIII Jap Symp on Taste and Smell: 253—256

Zusho H, Asaka H, Fukushima Y, Sanada S, Oka T, Kuroishi T, Kazama R, Okamoto M (1978) A study of olfactory threshold vs. fatigue in intravenous olfaction test. Jpn J Otol Tokyo 81: 562—568

Zusho H, Asaka H, Okamoto M (1981) Diagnosis of olfactory disturbance. Auris·Nasus·Larynx (Tokyo) 8: 19—26

Brain Aging and Trace Elements in Human: Clues into the Pathogenesis of Alzheimer's Disease

G. Chazot and E. Broussolle

C.I.O. UNESCO 9, rue du Professeur Florence, 69003 Lyon, France

ABSTRACT

The relationship between three trace elements (aluminium, zinc and selenium) and Alzheimer's disease is reviewed. Several arguments are discussed towards the role of these trace elements in the formation of the cardinal histologic lesions found in the brain, respectively, neurofibrillary tangles with aluminium, senile plaques and amyloid deposits with zinc, and lipofuschin with selenium. Aluminium may exert its primary action in the olfactory bulb. The neurotoxicity of aluminium is secondary to an alteration of calcium homeostasis, and/or of chromatin in the nucleus of neurons. Hippocampal zinc changes may result in an alteration in the activity of excitatory amino-acids and/or Cu-Zn superoxide dismutase. Finally, a decrease in brain selenium and glutathione peroxidase could lead to the deleterious effect of an increased production of free radicals. However, none of these mechanisms are mutually exclusive and the complex process of brain aging is in fact the result of multiple events.

KEY WORDS

Brain aging - Alzheimer's disease - Trace elements - Aluminium - Zinc - Selenium.

INTRODUCTION

Aging is undoubtedly one of the deciding factors in the appearance of degenerative diseases of the Central Nervous System such as Alzheimer's disease, Parkinson's disease, and even Hungtington's chorea which is nevertheless a genetic disorder. The relationship between the brain aging process and trace elements has received attention recently. However, few reliable data regarding trace elements in the human brain tissue are available, especially in the normal aged brain, and extensive studies of trace elements over the whole life span are sparse.

Several trace elements have been implicated in brain aging (Carlisle, 1974 ; Prasad, 1982 ; Solomons, 1986 a and 1986 b ; Dexter et al., 1989), but we will restrict the subject to three trace elements (aluminium, zinc and selenium) and emphasize their possible role in the pathogenesis of the most characteristic aspect of brain aging, i. e. Alzheimer's type dementia.

ALUMINIUM AND BRAIN AGING

The evidence implicating aluminium as a toxic agent in Alzheimer's disease has been reviewed recently (Crapper and Farnell, 1985 ; Birchall and Chappell, 1988).

Increased brain aluminium content in Alzheimer's disease was first reported by Crapper et al. (1976). These authors found that human brains in the age range between 50 and 65 years, completely free of neurofibrillary

degeneration (by light microscopic criteria), had an average aluminium content of 1.9 mg/kg dry weight, while the mean overall value for 585 samples from ten Alzheimer's brains was 3.8 mg/kg dry weight. The measurement by instrumental activation analysis of aluminium concentration in various human brain regions over the complete life span reveals a gradual increase with advancing age (Markesbery et al., 1984).

One of the cardinal histologic lesions in Alzheimer's brains, are the neurofibrillary tangles (NFT) which stain with thioflavin T (Mann, 1985). NFT are constituted of paired helical filaments (PHF), and result from the aggregation of microtubule associated proteins, tau (Delacourte and Defossez, 1986). Direct evidence for intranuclear aluminium in human neurons with neurofibrillary degeneration was provided by Perl and Brody (1980) in both Alzheimer's brains and brains of aged individuals considered not to be demented. However, at the ultrastructural level, the Alzheimer's type of neurofibrillary degeneration corresponds to paired helical filaments of 10 nm diameter (Wisniewski et al., 1976), whereas the aluminium-induced neurofibrillary tangle is composed of single 10 nm filaments (Terry and Pena, 1965).

Elevated brain aluminium levels, particularly in the grey matter, were reported in dialysis encephalopathy (Alfrey et al., 1976). In this peculiar syndrome, an increased blood aluminium level results in an elevated brain aluminium content, but signs of neurologic damage occur only when whole-tissue aluminium exceeds by 10-20 times the normal value (Crapper et al., 1983). Under these circumstances, aluminium accumulates only in the cytoplasm. In contrast, in Alzheimer's disease, neither CSF nor serum aluminium concentrations are increased. Furthermore, tissues other than the brain do not have elevated aluminium levels, suggesting that blood transport is not altered (Crapper et al., 1983). Therefore, two major questions should be addressed : How could aluminium reach the brain, and how could the toxicity of this metal be expressed ?

How could aluminium reach the brain ?

In the absence of obvious lesions of the blood-brain-barrier, Roberts (1986) suggested that the olfactive system could be implicated in the pathogenesis of Alzheimer's disease.

The neuroreceptors of olfaction could represent a way for aluminium to reach the brain, since they are exposed to the inhaled air, and directly connected to the olfactory glomureli. Several arguments support this hypothesis. Firstly, tracers such as horse-radish-peroxidase (HRP), when deposited inside the nasal cavity, are easily collected and transported inside the olfactory bulb and reach transsynaptically the basal nucleus of Meynert by backward axonal transport (Shipley, 1985). Secondly, a decline in smell identification ability occurs with advancing age (Doty et al., 1984), and is one of the early symptoms of Alzheimer's disease (Rezek, 1987). Thirdly, the olfactive areas in Alzheimer's dementia are always severely damaged, with histologic lesions identical to those found in the brain (Ohm and Braak, 1987 ; Talamo et al., 1989). Fourthly, the olfactive bulb is a relay for the information received from the neuroreceptors of olfaction, and is directly connected to the primary olfactive cortex, and therefore to the limbic system, which plays a major role in cognitive functions. An alteration of the olfactory bulb could therefore be responsible for a disturbance of the cholinergic system which supports the memory process. Fifthly, a decrease in the activity of the enzymes choline acetyl transferase and acetylcholine esterase has been shown in aluminium

encephalopathy (Yates et al., 1980). The use of the instrumental neutron activation analysis technique (Lai et al., 1985 ; Duckett and Galle, 1985) should permit accurate measurements of aluminium levels in the olfactory bulb.

Although a recent epidemiological study showed an increased aluminium content in drinking water in areas with higher incidence of Alzheimer's disease patients (Martyn et al., 1989), there is not yet supporting evidence of a link between a putative contamination of air by aluminium and Alzheimer's disease.

How could the toxicity of aluminium be expressed ?

Several interactions of aluminium with nuclear, cytoplasmic membrane, and body fluid have been reported.

Firstable, aluminium interferes with calcium homeostasis in the mammalian brain. Farnell et al. (1985) showed in animal a 112 % increase of cerebral calcium content 12 days after intracranial injection of a single, lethal, dose of soluble aluminium salt. At an advanced stage of the aluminium encephalopathy, animals had a average brain tissue calcium concentration that exceeded by 275 % the control values. Furthermore these authors found that the increase in whole brain tissue calcium content was associated with alterations in the eletrical properties of neurons, suggesting that part of the measured increase in total brain calcium may be associated with increased intracellular calcium ions. One mechanism by which aluminium could interfere with intracellular calcium concentration is an alteration of aluminium binding to calcium receptor proteins, resulting in a failure in calcium extrusion.

Aluminium could also exert its neurotoxic effect at the nuclear level. The transcription of a gene into mRNA by RNA polymerase occurs after a complex series of changes in the structure of DNA, which in unfolded from a compacted chromatin fiber in order to make the coding strand accessible to RNA polymerase. Divalent cations (such as magnesium, calcium, zinc and manganese) play vital roles in these reactions. Given such an essential function for divalent cations, the question arises of how other metabolic cations, particularly those derived from the environment and believed to be toxic, may interfere with these reactions.

The reactivity of aluminium and several other divalent and trivalent metallic cations toward chromatin from rat brain has been investigated recently (Walker et al., 1989). Of all the cations tested, aluminium is the most reactive to compact the chromatin fibers to the point where chromatin precipitates, and to interfere with the accessibility of exagenous structural probes (nucleases) to chromatin. These data suggest that localized aluminium increases in the nuclei of neurons could prevent the genes embodied in this altered chromatin from being expressed.

ZINC AND BRAIN AGING

Of those trace elements known to influence neurochemical functions, zinc is clearly the most well studied (Prasad et al., 1982 ; Frederickson et al., 1984 ; Sandstead, 1986).

Zinc is found in variable amounts throughout the Central Nervous System. Like most trace elements, the level of zinc is higher in grey matter compared to white matter areas, most likely reflecting the protein

distribution (Smeyers-Verbecke et al., 1974). Moreover, the region with the highest zinc content is the hippocampus (Hock et al., 1975). This unique aspect of zinc neurobiology has been thoroughly reviewed by Crawford (1983). This author showed, by using histochemical and atomic absorption spectrophotometric techniques, that in the hippocampus, zinc is specifically localized in the terminals of the excitatory mossy fiber projections. Zinc could play a role in memory processing. Indeed, in animals submitted to a brief period of zinc deprivation during the prenatal development, an impairement of long-term memory and even learning and short-term memory occurs (Halas, 1983). Piggott et al. (1974) showed a close relationship between zinc deficiency and learning abnormalities in children.

Burnet (1981) was the first to hypothesize a possible role of zinc in the pathogenesis of Alzheimer's dementia. Although not statistically significant, the CSF zinc is found decreased in Alzheimer's disease (Bourrier-Guerin et al., 1985 ; Kapaki et al., 1989). The study by Markesbery et al. (1984) of brain trace elements concentrations in various human brain areas over the complete life span, showed that zinc content remains relatively steady throughout adult life. In addition, these authors found that zinc levels in the hippocampus gradually decline from 17.8 mg/kg in the 40-59 age range to 15.2 mg/kg in the 80-99 age range. Although this difference was not statistically significant, no other brain region exhibited a similar fall in zinc levels.

These aforementioned findings raise the question of the possible link between zinc and certain histologic lesions found in Alzheimer's brain. Next to the neurofibrillary tangles (NFT), the senile plaques (SP) are the most characteristic histopathological brain lesions and the chief diagnostic features of Alzheimer's disease and senile dementia of the Alzheimer type (Mann, 1985). NFT and SP are found essentially in the hippocampus. The density of SP and NFT, which are also present in small amounts in normal aged brains, strongly correlates with the degree of dementia (Ball, 1978). The formation of senile plaques preceeds that of the neurofibrillar lesions (Constantinidis, 1988), and recent immunological investigations have shown that PHF and SP are two forms of amyloid deposit (Delacourte and Defossez, 1986).

Can a local (hippocampus) deficiency in zinc be responsible for anatomical lesions such as NFT and SP ?

One first negative hypothesis is that the decrease in zinc levels in the hippocampus is the consequence and not the cause of Alzheimer's lesions. Hence, the intracerebroventricular administration of a choline-like neurotoxic in rat produces nonspecific neuronal destruction and reduces the tissue levels of zinc in the hippocampus (Szerdahelyi, 1984)

Alternatively, endogenous zinc might serve a protective function, preventing normal levels of excitatory synaptic activity from becoming neurotoxic. Peters et al. (1987) demonstrated that zinc selectively blocks the action of N-Methyl-D-Aspartate (NMDA) on cortical neurons. A defect leading to reduced zinc release at excitatory synapses could produce gradual NMDA receptor-mediated neuronal death.

As zinc affects immunity (Prasad, 1982 ; Dardenne et al., 1982), the third hypothesis could be that local zinc deficiency in brain leads to local immune deficiency and then amyloid formation.
Finally, another possible explanation could be an alteration of the activity of the enzyme copper-zinc superoxide dismutase (Cu - Zn - SOD)

secondary to an abnormality in zinc metabolism, resulting in an excess of free radicals and then membrane lesions. This will be discussed further about selenium. In the rat brain, cytosol Cu - Zn - SOD, which represents 92 % of dismutase activity, diminishes during aging (Vanella et al., 1982). Among the first genes to be assigned to human chromosomes, was the Cu - Zn - SOD gene to chromosome 21 (Tan et al., 1973). Rapid aging is one of the features of Down's Syndrome (trisomy 21) and brain lesions are similar to that observed in Alzheimer's disease (Burger and Vogel, 1973 ; Lai and Williams, 1989). Futhermore, brain zinc content is found decreased in Down's Syndrome (Halsted et al., 1972). Nevertheless, the way how Cu - Zn - SOD might participate in the mechanisms leading to accelerated brain aging remains unclear.

SELENIUM AND BRAIN AGING

An extensive review of the neurochemical aspects of selenium has been published recently (Prohaska, 1983 and 1987).

The concentration of selenium in the human brain is depending on brain regions but averages less than 1 mg/kg dry weight (Prohaska and Ganther, 1976). Regarding the regional distribution of this trace element, selenium levels are found higher in cortical grey matter as compared to white matter areas, thus following closely the protein distribution (Prohaska and Ganther, 1976 ; Markesbery et al. 1984). Brain selenium content is low in childhood, after which it rises and maintains within a narrow concentration range throughout the adult life ; thereafter it gradually declines with age in the elderly (Markesbery et al., 1984).

The best established neurobiological function for selenium is its role in the activity of the protein glutathione peroxidase (GSH-Px). This enzyme binds to selenium in a stechiometric ratio. By using the reducing power of the tripeptide glutathione (GSH), it catalyses the breakdown of hydrogen peroxide and organic hydroperoxides. A decrease in brain selenium and GSH-Px activity leads to a high level of free radicals, molecules where an electron pair has become separated, with the two electrons having magnetic moments which are not complementary (see for review Prohaska, 1987).

In Alzheimer's disease, brain cells exhibit an age-associated increase in the content of an organelle which has been given a variety of different names, i. e., age pigment, lipopigment and lipofuscin. Lipofuscin accumulates preferentially in cortex and hippocampus. The accumulation of lipofuscin is depending on the rate of formation of free radicals. The lipofuscin is composed of polymerized products from malondialdehyde (formed by peroxidation of polyenoic acids), amine containing lipids, proteins and other components (see for review Bourre, 1988).

Free radicals are one of the major cellular cross-linkage agents. Cross linkage of both DNA and other macromolecules may be a more probable causative agent in error accumulation and they could be responsible for a cascade of errors in protein synthesis.

Alternatively free radicals could have a deleterious effect on membrane lipids. Indeed, the Central Nervous System, chief oxygen consumer in mammals, is an organ whose cellular membranes are particularly enriched in polyinsaturated fatty acids. These polyinsaturated fatty acids control the fluidity of membranes, and hence their enzymatic activities and electrophysiological properties. Peroxidation products affect the membrane assymetry which is indispensable for the correct functioning of membranes (Bourre,

1988). Moreover, since neuronal renewal is impossible, this means that the integrity of the nerve tissue depends to a very large extend on its ability to protect itself.

CONCLUSION

None of the aforementioned hypotheses are mutually exclusive. The complex process of brain aging is in fact a result of multiple events. These events give rise to the irreversible changes which take place at the cellular level throughout the life span and result in a progressive loss of adaptability, a decline in body functions, and the development of degenerative diseases. Trace elements express their neurochemical activity in many ways and most often in association with some protein that serves as an enzymatic catalyst.

However, since an accurate understanding of both processes (aging process and trace elements activity) is ultimately required, future parallel and interactive studies will determine how age-related changes can effect the central nervous system in humans.

REFERENCES

Alfrey AC, LeGendre GR, Kaehney WD (1976) The dialysis encephalopathy syndrome : Possible aluminium intoxication. N Engl J Med 294 : 184-188.

Ball MJ (1978) Neuronal loss, neurofibrillary tangles and granulovacuolar degeneration in hippocampal cortex of aging and demented patients. A quantitative study. Acta Neuropath (Berlin) 43 : 73-80.

Birchall JD, Chappell JS (1988) Aluminium, chemical physiology, and Alzheimer's disease. Lancet 1 : 1008-1010.

Bourre JM (1988) The effect of dietary lipids on the Central Nervous System in aging and disease : Importance of protection against free radicals and peroxidation. In : Bergener M, Ermini M, Stähelin HB (eds) Crossroads in aging. The 1988 Sandoz lectures in gerontology. Academic Press, London, p 141-167.

Bourrier-Guerin L, Mauras Y, Truelle JL, Allain P (1985) CSF and plasma concentrations of 13 elements in various neurological disorders. Trace Elements in Medicine 2 : 88-91.

Burger PC, Vogel FS (1973) The development of the pathologic changes of Alzheimer's disease and senile dementia in patients with Down's syndrome. Am J Pathol 73 : 457-476.

Burnet FM (1981) A possible role of zinc in the pathology of dementia Lancet 1 : 186-188.

Carlisle EM (1974) Silicon as an essential element. Fed Proc 33 : 1758-1766.

Crapper McLachlan DR, Farnell BJ (1985) Aluminium and neuronal degeneration. In : Gabay S, Harris J, Ho BT (eds) Neurology and neurobiology, vol 15, Metal ions in neurology and psychiatry. Alan R Liss, Inc, New York, p 69-87.

Crapper McLachlan DR, Farnell B, Galin H, Karlik S, Eichorn G., De Boni U (1983) Aluminium in human brain disease. In : Sarkar B (ed) Biological aspects of metals and metal-related diseases. Raven Press, New York, p 209-219.

Crapper DR, Krishnan SS, Quittkat S (1976) Aluminium neurofibrillary degeneration and Alzheimer's disease. Brain 99 : 67-80.

Crawford IL (1983) Zinc in the hippocampus. In : Dreosti IE, Smith RM (eds) Neurobiology of the trace elements, vol 1. Humana, Clifton, NJ, p 163-211.

Constantinidis J (1988) Méthodologie des corrélations anatomo-cliniques appliquées à la recherche sur la maladie d'Alzheimer. Psychol Med 20 : 1815-1820.

Dardenne M, Pléau JM, Nabarra B, Lefrancier P, Derrien M, Choay J, Bach JF (1982) Contribution of zinc and other metals to the biological activity of the serum thymic factor. Proc Natl Acad Sci USA 79 : 5370-5373.

Delacourte A, Defossez A (1986) Alzheimer's disease : Tau proteins, the promoting factors of microtubule assembly, are major components of paired helical filaments. J Neurol Sci 76 : 173-186.

Dexter DT, Wells FR, Lees AJ, Agid F, Agid Y, Jenner P, Marsden CD (1989) Increased iron content and alterations in other metal ions in brain in Parkinson's disease. J Neurochem 52 : 1830-1836.

Doty RL, Shaman P, Applebaum SL, Giberson R, Siksorski L, Rosenberg L (1984) Smell identification ability : changes with age. Science 226 : 1441-1443.

Duckett S, Galle P (1985) The application of analytical ion microscopy (secondary ion mass microanalysis) to the study of normal and pathological neural tissue. In : Gabay S, Harris J, Ho BT (eds) Neurology and neurobiology, vol 15, Metal ions in neurology and psychiatry. Alan R Liss, Inc, New York, p 345-366.

Farnell BJ, Crapper McLachlan DR, Bainbridge K, De Boni U, Wong L, Wood PL (1985) Calcium metabolism in the aluminium encephalopathy. Exp Neurol 88 : 68-83.

Frederickson CJ, Howell GA, Kasarskis EJ (1984) The neurobiology of zinc, vol 11 A and 11 B Alan R Liss, Inc, New York.

Halas ES (1983) Behavioral changes accompanying zinc deficiency in animals. In : Dreosti IE, Smith RM (eds) Neurobiology of the trace elements, vol 1, Humana, Clifton, NJ, p 213-243.

Halsted JA, Ronaghy HA, Abadi P, Haghshenass M, Amirhakemi GH, Barakat RM, Reinhold JG (1972) Zinc deficiency in man : Shiraz experiment. Am J Med 53 : 277-284.

Hock A., Demmel U, Schicka H, Kasperek K, Feinendegen LE (1975) Trace element concentration in human brain. Activation analysis of cobalt, iron, rubidium, selenium, zinc, chromium, silver, cesium, antimony and scandium. Brain 98 : 44-64.

Kapaki E, Segditsa J, Papageorgiou C (1989) Zinc, copper and magnesium concentrations in serum and CSF of patients with neurological disorders. Acta Neurol Scand 79 : 373-378.

Lai F, Williams RS (1989) A prospective study of Alzheimer disease in Down syndrome. Arch Neurol 46 : 849-853.

Lai JCK, Chan AWK, Minski MJ, Leung TKC, Lim L, Davison AN (1985) Application of instrumental neutron activation analysis to the study of trace metals in brain and metal toxicity. In : Gabay S, Harris J, Ho BT (eds) Neurology and neurobiology, vol 15, Metal ions in neurology and psychiatry. Alan R Liss, Inc, New York, p 323-343.

Mann DMA (1985) The neuropathology of Alzheimer's disease : A review with pathogenic, aetiological and therapeutic considerations. Mech Ageing Dev 31 : 213-255.

Markesbery WR, Ehmann WD, Alauddin M., Hossain TIM (1984) Brain trace element concentrations in aging. Neurobiol Aging 5 : 19-28.

Martyn CN, Osmond C, Edwardson JA, Barker DJP, Harris EC, Lacey RF (1989) Geographical relation between Alzheimer's disease and aluminium in drinking water. Lancet 1 : 59-62.

Ohm TG, Braak H (1987) Olfactory bulb changes in Alzheimer's disease. Acta Neuropath 73 : 365-369.

Perl DP, Brody AR (1980) Alzheimer's disease : X-ray-spectrometric evidence for aluminium accumulation in neurofibrillary tangle-bearing neurons. Science 208 : 297-299.

Peters S, Koh J, Choi DW (1987) Zinc selectively blocks the action of N-methyl-D-aspartate on cortical neurons. Science 236 : 589-593.

Piggott L, Caldwell D, Oberleas D (1974) Zinc deficiency, disturbed children and civil rights. Biol Psychiat 9 : 325-327.

Prasad AS (1982) Current topics in nutrition and disease, vol 6, Clinical, biochemical, and nutritional aspects of trace elements. Alan R Liss, Inc, New York.

Prasad AS, Dreosti IE, Hetzel BS (1982) Current topics in nutrition and disease, vol 7, Clinical applications of recent advances in zinc metabolism. Alan R Liss, Inc, New York.

Prohaska JR (1983) Neurochemical aspects of selenium. In : Dreosti IE, Smith RM (eds) Neurobiology of trace elements, vol 1. Humana, Clifton, NJ, p 245-268.

Prohaska JR (1987) Functions of trace elements in brain metabolism. Physiol Rev 67 : 858-901.

Prohaska JR, Ganther HE (1976) Selenium and glutathione peroxidase in developing rat brain. J Neurochem 27 : 1379-1387.

Rezek DL (1987) Olfactory deficits as a neurologic sign in dementia of the Alzheimer's type. Arch Neurol 44 : 1030-1032.

Roberts E (1986) Alzheimer's disease may begin in the nose and may be caused by aluminosilicates. Neurobiol Aging 7 : 561-567.

Sandstead HH (1986) A brief history of the influence of trace elements on brain function. Am J Clin Nutr 43 : 293-298.

Shipley MT (1985) Transport of molecules from nose to brain : Transneural anterograde and retrograde labelling in the rat olfactory system by WGA-HRP applied to the nasal epithelium. Brain Res Bull 15 : 129-142.

Smeyers-Verbeke J, Defrise-Gussenhoven E, Ebinger G, Lowenthal A, Massart DL (1974) Distribution of Cu and Zn in human brain tissue. Clin Chim Acta 51 : 309-314.

Solomons NW (1986 a) Trace elements in nutrition of the elderly, 1. Established RDAs for iron, zinc and iodine. Postgraduate Med 79 : 231-250.

Solomons NW (1986 b) Trace elements in nutrition of the elderly, 2, SADDIs for copper, manganese, selenium, chromium, molybdenum and fluoride. Postgraduate Med 79 : 251-263.

Szerdahelyi P, Kasa P, Fisher A, Hanin I (1984) Effects of the cholinotoxin, AF 64 A, on neuronal trace-metal distribution in the rat hippocampus and neocortex. Histochemistry 81 : 497-500.

Talamo BR, Rudel RA, Kosik KS, Lee VMY, Neff S, Adelman L, Kauer JS (1989) Pathological changes in olfactory neurons in patients with Alzheimer's disease. Nature 337 : 736-739.

Tan YH, Tischfield J, Ruddle FH (1973) The linkage of genes to the human interferon-induced antiviral protein and indophenol oxidase-B traits to chromosome G-21. J Exp Med 137 : 317-330.

Terry RD, Pena C (1965) Experimental production of neurofibrillary degeneration. 2. Electron microscopic, phosphatase histochemistry and electron probe analysis. J Neuropathol Exp Neurol 24 : 200-210.

Vanella A, Geremia E, D'Urso G, Tiriolo P, Di Silvestro K, Grimaldi R, Pinturo R (1982) Superoxide dismutase activities in aging rat brain. Gerontology 28 : 108-113.

Walker PR, LeBlanc J, Sikorska M (1989) Effects of aluminium and other cations on the structure of brain and liver chromatin. Biochemistry 28 : 3911-3915.

Wisniewski H, Naranny H, Terry R (1976) Neurofibrillary tangle of paired helical filaments. J Neurol Sci 27 : 173-181.

Yates CM, Simpson J, Russell D, Gordon A (1980) Cholinergic enzymes in neurofibrillary degeneration produced by aluminium. Brain Res 197 : 269-274.

Zinc in Brain Development and Function

Ivor E. Dreosti

Division of Human Nutrition Cisro, Kintore Avenue, Adelaide 5000, Australia

ABSTRACT

Zinc is critically needed during brain development and for brain function. In animals, early prenatal zinc impoverishment results mainly in cerebral teratogenesis, whereas late prenatal and postnatal deprivation principally affects mental performance. Gross teratological effects of zinc deficiency probably arise from disturbed enzyme activity during cell division and/or peroxidation of cell membranes. With respect to brain function zinc is involved in several brain enzymes and a number of neurochemical processes, particularly relating to synaptic transmission, release of neurotransmitter substances and postsynaptic neuroreceptor modulation, where it may function in an excitatory or depressant capacity.

KEY WORDS: Zinc, Brain development, Brain function

INTRODUCTION

Interest in the neurobiology of the trace elements is a phenomenon of the last two decades, which followed logically on observations by neuroscientists and by nutritionists that increasing areas of overlap existed between the two disciplines. Neurobiologists recognized the constant occurrence of several trace metals in the brain which signalled to them the need to examine their possible structural or functional significance. Nutritionists in turn noted the consistent involvement of neurological symptoms, both physical and psychological, which accompanied many trace element deficiencies and heavy metal toxicity. Probably best known historically are the early studies on copper-related enzootic ataxia in sheep and the neurological cretinism accompanying iodine deficiency in humans.

Of the dozen or so trace elements currently deemed to be essential for humans, zinc has attracted the most attention, being associated with more than 100 metalloenzymes on the one hand and supplied in limited amounts in the diet on the other. Contributing to this general interest, and central to the emerging focus on zinc in neurobiology was the finding by Hurley and Swenerton (1966) that gestational zinc deficiency in rats was highly teratogenic, especially with respect to the central nervous system. Suggestions by those workers and others that the human fetus may be similarly affected appear to be substantiated by the limited evidence currently available from studies around the world, which relate neural tube defects in infants to apparently poor maternal zinc status during pregnancy or to the inherited zinc deficiency disease acrodermatitis enteropathica (Table 1).

Further impetus for a role for zinc in neurobiology was gained following reports by Sandstead (1972) and by Halas and Sandstead (1975) of behavioural anomalies in rats following prenatal or postnatal zinc deficiency and by reference to mental dullness in zinc deficient humans studied by Prasad and coworkers in the Middle East (1961). As with

TABLE 1   Terata of the Central Nervous System Associated with Zinc
Deficiency in Humans

Terata	Condition	Reference
Anencephalus	Dietary deficiency	Cavdar et al. (1980) Stewart et al. (1981) Soltan, Jenkins (1982)
Myelomeningocoele	" "	Jameson (1976)
Spina bifida	" "	Bergman et al. (1981)
Anencephalus	Acrodermatitis enteropathica	Hambidge et al. (1975)

zinc-related neural teratology, behavioural disturbances reported in
animals appear also to have some parallel in humans in individuals who
have been exposed to suboptimal zinc status through a variety of causes,
and in patients who displayed altered zinc status in association with
several common psychological disorders (Dreosti 1984).

The last decade has seen a vast increase in the range and the complexity
of the roles attributed to zinc in the central nervous system. In broad
terms, zinc appears on the one hand to be essential for normal brain
development and on the other for proper neural function, dependant no
doubt, on its role in several important cerebral metalloenzymes and/or its
association with catecholamine and enkephalin metabolism and in the
modulation of synaptic transmission.

ZINC IN BRAIN DEVELOPMENT

The importance of zinc for normal development of the central nervous
system is especially critical at the time of closure of the neural tube,
which in rats occurs around day 10 of embryogenesis before the embryo has
rotated or the chorio-allantoic placenta has formed. The biochemical
lesion underlying defective neurulation remains obscure, although
considerable evidence supports an involvement of zinc in its various roles
during cell division and a consequent asynchrony in histogenesis and
organogenesis resulting in a distortion of normal morphogenesis.

Much attention thus far has centred on the activity of the zinc-dependant
enzyme thymidine kinase, which is widely recognized to limit the rate of
DNA synthesis through the inducible salvage pathway. A pathway which in
rapidly dividing cells assumes greater importance than de novo synthesis,
catalysed by folate-dependant thymidylate synthetase. Indeed, earlier
studies in our laboratory pointed to diminished activity of thymidine
kinase in embryonic rat brains a little after neurulation (14-day-old
fetuses) and to an accompanying diminution in incorporation of
[3]H-thymidine into fetal brain DNA, signifying a reduction in DNA
synthesis (Dreosti 1983).

Further studies in our laboratory compared the effect of zinc deficiency
in fetal rats on both the zinc-dependant salvage pathway and on the
de novo synthesis of DNA which is "zinc-independant", but can be inhibited
by methotrexate, a folate analogue which decreases the activity of
dihydrofolate reductase and in turn the activity of thymidylate
synthetase. A comparison between fetal liver and fetal brain in each case
revealed that in the liver, DNA synthesis occurred mainly de novo, and
that both pathways were affected equally by zinc depletion, whereas in the
brain the salvage pathway was relatively more important and very much more
affected by zinc deficiency. In percentage terms, overall DNA synthesis
decreased by nearly 70% and 40% respectively in zinc-deficient fetal
livers and brains. However, the decrease occurred equally in both
pathways in the liver but was essentially confined to the salvage pathway
in the brain, which suggests that this tissue is spared to some extent
from the effects of zinc deficiency on the de novo pathway, and emphasizes
the potential role of thymidine kinase in neural tube teratology (Dreosti
1984).

Subsequent studies were performed in which the outcome of neurulation was examined in 11-day-old rat fetus subjected simultaneously to maternal deficiencies of zinc and folic acid, thereby affecting both the salvage and the de novo pathway of DNA synthesis in the brain (Bremert et al. 1988). The combined effect of the treatments on neurulation was significantly greater than their individual actions as neural tube teratogens, which suggests that they may share a common biochemical lesion involving impaired synthesis of deoxythymidine monophosphate.

Coincidental studies in our laboratory investigating the teratogenic mechanism of zinc deficiency revealed that neural tube dysmorphology in fetal rats is preceded by an episode of cell death (Record et al. 1985) in those cells of the neural folds actively dividing to effect neural tube closure (Figure 1). Evidence of similar cell death was detected in the neuroepithelium of folate deficient rat embryos, with marked potentiation of necrosis in animals subjected to both zinc and folic acid deficiencies concurrently (Bremert et al. 1988). Biochemically, this zinc-folate interaction suggests an underlying lesion involving diminished DNA synthesis, which in turn may lead to necrosis in cells committed to divide, but unable to synthesize the necessary DNA to accomplish mitosis. In the human setting, the interaction is of interest, as it may in part account for the uncertainty which still surrounds the contribution of folate deficiency to the etiology of congenital neural tube defects, and the value of folic acid prophylaxis.

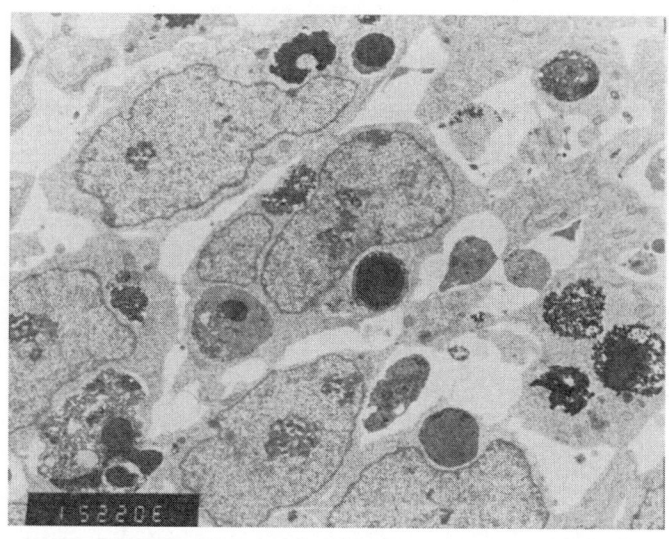

Fig. 1   Transmission electron micrograph of central nervous system of 9-day old zinc-depleted rat embryo showing dead and dying cells and large intracellular spaces (Magnification X5000).

Transmission electron microscopy of both folate and zinc-deficient embryonic neuroepitheleal cells indicated extensive distortion of sub-cellular structures and disruption of cellular membranes (Bremert et al. 1989), which raised the question of possible free radical-related peroxidative damage to membrane lipids, and focused attention on the role zinc plays as a cellular antioxidant through its participation as a prosthetic group of superoxide dismutase as well as its capacity to diminish the iron-catalysed production of highly damaging hydroxyl radicals from ubiquitous superoxide (Dreosti 1986, 1988). Brain tissue, it should be noted, is especially rich in peroxidizable polyunsaturated fatty acids, and is generally less effectively supplied with antioxidant systems than are other organs. Subsequent studies in our laboratory revealed enhanced cranial dysmorphology in fetuses subjected simultaneously to intra-uterine deficiencies of zinc and vitamin E, accompanied by severe disruption of subcellular organelles and collapse of the external head structures (Harding et al. 1987). However, unlike

deficiencies of zinc and folate which affected mainly the neuroepithelium, vitamin E deficiency principally involved mesenchymal cells, which suggests that separate, but overlapping mechanisms may operate in association with the neural tube teratogenesis accompanying these three essential micronutrients, and that while diminished DNA synthesis may be of major importance in the zinc/folate interaction, increased membrane damage may also be involved in the cellular dysfunction and necrosis associated with deficiencies of both zinc and vitamin E.

ZINC IN BRAIN FUNCTION

Apart from the critical requirement for zinc during brain development, the metal acts also in a functional capacity in the central nervous system, and a deficiency affects many aspects of behaviour in experimental animals, often in the absence of any overt indications of structural abnormalities (Dreosti 1987). Studies by Apgar (1968), Caldwell (1970). Halas (1983) and Sandstead (1985) with animals are especially relevant, and point to impaired long term memory, some reduction in short term recall and learning ability, coupled with diminished activity, less emotional control and higher susceptibility to stress. Reduced cellularity, dendritic arborization and synaptogenesis in the deficient neuropil, as well as depressed myelination and polymerization of tubulin in nerve cells undoubtedly afford some explanation for the impairment of brain function in terms of microanatomical dysmorphology, but other factors may be even more important (Dreosti 1987).

Firstly, the involvement of zinc as a prosthetic group of many enzymes results in it playing a central role in a variety of cerebral metalloenzymes (Dreosti 1987). In particular, the enzymes 2', 3'-cyclic nucleotide 3'-phosphohydrolase and alkaline phosphatase are involved with axonal myelination and the maintenance of electrical potential. Brain glutamate dehydrogenase is central to the metabolism of the putative amino acid neurotransmitter, glutamic acid. Dopamine -β-hydroxylase and phenylethanolamine-N-methyl transferase convert dopamine to norepinephrine and then to epinephrine, and may therefore affect cerebral catecholamine levels. Glutamate decarboxylase is involved in the synthesis of the inhibitory neurotransmitter -aminobutyric acid, and zinc excess injected intraventricularly in rats is epileptogenic which, it is suggested, occurs by inhibition of glutamate decarboxylase and possibly also by reducing the binding capacity of -aminobutyric acid to its receptor sites (Ebadi et al. 1984).

Much of the recent interest in zinc and brain function has centred on the hippocampus, where the metal is found to accumulate in the giant axonal boutons of the mossy fibre pathway early postnatally in rats, coincidentally with the emergence of electrophysiological function, and where it can be demonstrated histochemically by staining with silver sulphide. Recent evidence has shown that zinc is released into the extracellular space of hippocampal slices and taken up again following electrical stimulation (Howell et al. 1984), and that hippocampal electrophysiology in rats is disturbed by chronic dietary zinc deficiency (Hesse 1979). Considerable evidence indicates that zinc is involved in synaptic signalling in the hippocampus, but whether it participates directly, for example by control of the release of neurotransmitter glutamic acid, or indirectly as a neuromodulator through regulation of neuroreceptor affinity is not clear (Dreosti 1987). It is however important to recognize that while the hippocampus is undoubtedly a useful research model, zinc is present in all brain cells and the neurochemical importance of the metal almost certainly extends beyond its function in the hippocampus.

A possible relationship between zinc and the opioid peptides was raised by Stengaard-Pedersen et al. (1981), who showed that in guinea pigs the immunocytochemical staining pattern for enkephalin in the hippocampal mossy fibres precisely matched that for zinc demonstrated by the silver sulphide stain. Suggested roles for zinc in relation to enkephalin include a possible effect on enkephalin binding to opioid receptors, its involvement as a cofactor for the metalloenzyme enkephalinase and its participation in a zinc-enkephalin complex (Crawford 1983, Dreosti 1987).

Evidence for the importance of zinc in normal brain development is overwhelming and the consequences of severe deficiency during pregnancy are disastrous, with little protection afforded to the fetus against a rapid diet related fall in maternal plasma zinc levels (Dreosti 1983). In numerical terms however, the problem in humans is probably not great, as relatively few children appear to be affected in this way. Of greater significance may well be the emerging role of zinc in binding to neuroactive substances and to neuroreceptors, thereby acting to modulate synaptic activity, possibly by inhibiting excitatory synapses at physiological levels and inhibitory synapses at supranormal levels.

CONCLUSIONS

It is tempting to speculate on the part dietary zinc may play in the development and the control of several psychological disturbances to which human beings are prone. Such projections must, however, be tempered by the realization that unlike in the plasma, brain zinc is avidly retained even in animals subjected to near lethal dietary zinc impoverishment (Wallwork et al. 1983). Thus, while fetal neurogenesis may well be compromised by a severe, transient maternal zinc deficit, cerebral dysfunction is more likely to be associated with chronic zinc insufficiency. However, to what extent the behavioural effects of zinc deficiency in animals may be attributed to the accompanying inanition and calorie restriction remains to be clearly quantified, although there is little doubt that at least some of the behavioural teratology arises in this way. Nevertheless the wide and central involvement of zinc in brain development and function is indisputable. What needs to be established now is the contribution dietary zinc status may make towards attainment of maximal intellectual development in humans and its preservation into old age.

REFERENCES

Apgar J (1968) Comparison of the effects of copper, manganese and zinc deficiencies on parturition in the rat. Am J Physiol 215:428-432
Bergman KE, Makosch E, Tews KH (1980) Abnormalities of hair zinc concentration in mothers of newborn infants with spina bifida. Am J Clin Nut 33:2145-2150
Bremert JC, Dreosti IE, Tulsi RS (1988) Teratogenic interaction of folic acid and zinc deficiencies in the rat. Nutr Rep Int 39:383-390
Bremert JC, Dreosti IE, Tulsi RS (1989) A teratogenic interaction between dietary deficiencies of zinc and folic acid in rats : an electron microscope study. Nutr Res 9:105-112
Caldwell DF, Oberleas D, Clancy JJ et al (1970) Behavioural impairment in adult rats following acute zinc deficiency. Proc Soc Exp Biol Med 133:1417-1421
Cavdar AO, Arcasoy A, Baycu T et al (1980) Zinc deficiency and anencephaly in Turkey. Teratology 22:141
Crawford IL (1983) Zinc and the hippocampus. In Dreosti IE, Smith RM (eds) Neurobiology of the trace elements, Humana Press, Clifton, New Jersey, pp 169-211
Dreosti IE (1982) Zinc in prenatal development. In Prasad AS, Dreosti IE, Hetzel BS (eds) Clinical applications of recent advances in zinc metabolism. Alan R Liss, New York, pp 19-38
Dreosti IE (1983) Zinc in the central nervous system. In Dreosti IE, Smith RM (eds) Neurobiology of the trace elements, vol 1. Humana Press, Clifton, New Jersey, pp 135-162
Dreosti IE (1984) Zinc in the central nervous system : the merging interactions. In Frederickson CJ, Howell GA, Kasarskis EJ (eds) The neurobiology of zinc, part A. Alan R Liss, New York, pp 1-26
Dreosti IE (1986) Zinc-alcohol interactions in brain development. In West JR (ed) Alcohol and brain development. Oxford University Press, New York, pp 373-405
Dreosti IE (1987) Neurobiology of zinc. In Mills CF (ed) Zinc in human biology, Springer-Verlag, London, pp 235-247
Dreosti IE (1988) Antioxidants, micronutrients versus free radicals. Aust Family Physician 17:684-686

Ebadi M, Wilt S, Ramaley R et al (1984) The role of zinc and zinc binding proteins in regulation of glutamic acid deca-boxylase in the brain. In Evangelopoulos AE (ed) Chemical and biological aspects of vitamin B6 catalysis. Alan R Liss, New York, pp 307-324

Halas ES (1983) Behavioural changes accompanying zinc deficiency in animals. In Dreosti IE, Smith RM (eds) Neurobiology of the trace elements, vol 1. Humana Press, Clifton, New Jersey, pp 213-243

Halas ES, Sandstead HH (1975) Some effects of prenatal zinc deficiency on behaviour of the adult rat. Pediat Res 9:94-97

Hambidge KM, Neldner, Walravens PA (1975) Zinc, acrodermatitis enteropathica and congenital malformations. Lancet I:577-578

Harding AJ, Dreosti IE, Tulsi RS (1987) Teratogenic effect of vitamin E and zinc deficiency in the 11-day rat embryo. Nutr Rep Int 36:473-480

Hesse GW (1979) Chronic zinc deficiency alters neuronal function of hippocampal mossy fibers. Science 205:1005-1007

Howell GA, Welch MG, Frederickson CJ (1984) Stimulation-induced uptake and release of zinc in hippocampal slices. Nature 308:736-738

Hurley LS, Swenerton H (1966) Congenital malformations resulting from zinc deficiency in rats. Proc Soc Exp Biol Med 123:692-697

Jameson S (1976) Effects of zinc deficiency on human reproduction. Acta Med Scand [Suppl] 593:5-89

Kasarskis EJ (1984) Zinc metabolism in normal and zinc-deficient rat brain. Exp Neurol 85:114-127

Prasad AS, Halst JA, Nadimi M (1961) Syndrome of iron deficiency anemia, hepatosplenomegaly, hypogonadism, dwarfism and geophagia. Am J Med 31:532-539

Sandstead HH (1985) Zinc : essentiality for brain development and function. Nutr Rev 43:129-137

Record IR, Tulsi RS, Dreosti IE, Fraser FJ (1985) Cellular necrosis in zinc-deficient rat embryos. Teratology 32:397-405

Sandstead HH, Gillespie DD, Brady RN (1972) Zinc deficiency : effect on brain of the suckling rat. Pediat Res 6:119-125

Stengaard-Pedersen K, Fredens K, Larson LI (1981) Encephalin and zinc in the mossy fiber system. Brain Res 212:230-233

Stewart C, Katchen B, Collipp PJ et al (1981) Zinc and birth defects. Pediat Res 15:515

Wallwork JC, Milne DB, Sims RL et al (1983) Severe zinc deficiency : effects on the distribution of nine elements in regions of the rat brain. J Nutr 113:1895-1905

Zinc Binding Peptides from Rat Brain

EDWARD J. KASARSKIS and THOMAS C. VANAMAN

Departments of Neurology, Toxicology, and Biochemistry, Albert B. Chandler and Veterans Administration Medical Centers, University of Kentucky, Lexington, KY 40536-0084, USA

ABSTRACT

A family of acidic, zinc-binding peptides has been isolated and partially purified from the syanptosomal membrane fraction of rat brain. These peptides are composed predominantly of ASP and GLU residues and as such, represent potential "storage peptides" for transmitter aspartate and glutamate at excitatory amino acid synaptses.

KEY WORDS: Zinc, glutamate, aspartate, synapse, hippocampus

ENZYMATIC AND SYNAPTIC ZINC POOLS IN THE CENTRAL NERVOUS SYSTEM

Zinc is a ubiquitous micronutrient which has been recognized as essential for life for over 120 years.   It has been only over the last 10-20 years however that the neurobiology of zinc has been investigated with anything resembling a sustained effort (Frederickson 1984). Although zinc undoubtedly serves as a cofactor for enzymes in nervous tissue as it does elsewhere in the body (Berg 1987), a major focus of interest in zinc neurobiology has been the unusual distribution of zinc within the hippocampus and other restricted brain regions (Frederickson 1987).   The hippocampus is of great clinical importance because of its central role in memory function and when pathologically altered, in epilepsy.   Frederickson and Danscher (1988) have reviewed a large number of studies which support the concept that zinc inhabits two distinct pools within the central nervous system.   Within the first pool, zinc fulfills the role as an activator of a number of metalloenzymes.   In this capacity, zinc serves in much the same way as it does throughout the body as a cofactor for oxidoreductases, transferases, and hydrolases (Berg 1987).   In the central nervous system, zinc may also function in the process of tubulin aggregation and therefore may fulfill a structural role as well (Gaskin 1977, Larsson 1976).   It is also likely, although unproved at the present time, that zinc may regulate the transcription of neuronal and glial DNA by organizing proteins into "zinc-finger" motifs as has recently been shown in other systems (Evans 1988).

The second zinc pool is identified by its interaction with specific histochemical reagents. The well-known Timm's method and dithizone reagent both stain for a unique pool of zinc and reveal its discrete localization, most notably in the mossy fiber projection of the hippocampus (Danscher 1981, Danscher 1985).   Frederickson and Kasarskis (1987) refined a histofluorescence method for the detection of zinc based upon its chelation to various substituted 8-hydroxyquinolines).   This technique, originally suggested by the earlier work of Toropsev and Eschenko in the Soviet Union (1970), is specific for zinc.   More recently, Savage and Kasarskis (1989) have developed zinc standards to quantify the amount of zinc chelated by the TSQ reagent and have mapped the concentration of zinc in the mossy fiber projection, confirming the direct chemical analysis of zinc in microdissected subregions. Importantly, the chelation of zinc by one reagent blocks the reactivity of zinc for the other methods indicating that all three techniques are probing a single specific pool of zinc (Frederickson 1988).   These histochemical reactions have been utilized to great practical advantage in anatomical studies of hippocampal development on a light microscopic level. Examining Timm's stain material by electron microscopy reveals zinc to be associated with the syanptic vesicles, suggesting that zinc is involved in neurotransmission between the presynaptic hippocampal mossy fibers (axons of dentate granule cells) and the CA3/CA4 pyramidal cells (Frederickson 1988).

The concentration of zinc within this pool can be modified by dietary zinc restriction and is under strong homeostatic control (Kasarskis 1984, Wensink 1987).   Savage, Slevin, and Kasarskis and Frederickson et al. have exploited the TSQ histofluorescence method to investigate changes in the concentration of the hippocampal mossy fiber zinc in various experimental models of epilepsy.   Frederickson et al. (1988) have shown that mossy fiber-associated zinc decreases

with kainate-induced seizures. These findings were extended by Fukahori et al. who demonstrated a reduction of zinc in the dentate gyrus of the hippocampus in the epileptic (El) mouse (Fukahori et al. 1988). It is of interest that these dramatic changes in mossy fiber-associated zinc are not reflected in the total hippocampal zinc concentration analyzed using atomic absorption spectrometry. Thus it appears that the mossy fiber zinc pool is decreased in several animal models of seizures, supporting the concept that zinc-mediated synaptic events may play a role in the induction or perpetuation of the epileptic state.

Current thinking favors the notion that zinc in this pool is accessible to these histochemical reagents whereas the more-abundant zinc bound within metalloenzymes is not (Frederickson 1988). The weight of the evidence suggests that zinc is localized in synaptic vesicles where it might function in the storage of a neurotransmitter or neuromodulator. Zinc evidently enters a pre-synaptic pool in vivo, in tissue slice preparations, and is taken up by isolated synaptosomes in vitro (Wensink 1988). It appears to be released with membrane depolarization (Howell 1984) although whether or not this is of synaptic vesicular origin has not been proved (Wensink et al. 1988). Thus the evidence supporting the involvement of zinc presynaptically is quite good although the nature of the transmitter or neuromodulator in hippocampal mossy fibers is unknown inasmuch as the pattern of zinc staining does not map to any recognized transmitter or peptide (Frederickson 1988). A role for zinc in packaging vesicle-associated peptides has been defined in a variety of extra-cerebral secretory tissues such as salivary glands and the beta cell of the pancreas (Emdin 1980).

On the postsynaptic side, zinc inhibits the binding of many neurotransmitters to their receptors including glutamate, aspartate (Slevin 1985), GABA (Baraldi 1984), benzodiazepine (Lo 1983), muscarinic cholinergic (Hulme 1983), and opiate (Sadee 1982). Recently, zinc has also been shown to act as a non-competitive antagonist of N-methyl-D-aspartate (NMDA)-receptor-mediated EPSPs (excitatory post-syanptic potentials) in cultures of mouse hippocampal neurons (Koh 1988). Other studies by Choi's group have extended these observations and demonstrated that zinc attenuates NMDA-induced neurotoxicity of cortical neurons in culture (Choi 1988).

ZINC BINDING LIGANDS IN BRAIN

What is known regarding zinc-binding ligands from brain? As discussed above, zinc undoubtedly activates enzymes in brain and that this process probably does not differ substantially from its function in other tissues. Secondly, zinc can also serve as an allosteric regulator of calmodulin (Mills 1985) and calregulin (Khanna 1986) by occupying a non-calcium, metal-binding site of these proteins. The implications of these observations for the modulation of calcium-signalling systems by zinc in nervous tissue are largely unexplored but would likely be of considerable importance. The investigations of Baudier's group have clearly demonstrated that zinc binds to both the S-100a and S-100b proteins, induces conformational changes, and differentially affects the assembly of brain microtubules from tubulin (Baudier 1986, Deinum 1983, Baudier 1983). Recently, Oteiza et al. (1988) have observed abnormal microtubule function in brains of rats made zinc deficient by dietary restriction. The presence of metallothionein in brain has been shown by Ebadi's group (1986), Gulati et al. (1987), and Brady (1983). However the biological function of these proteins in cerebral zinc metabolism and homeostasis is not known with certainty. Other zinc-binding peptides have also been observed by Itoh et al. (1983) and Sato et al. (1984) believe that zinc may be chelated to glutathione as well. The cellular localization of these proteins/peptides has not been defined so that it is impossible to judge whether or not any of them function in synaptic events, especially in the hippocampal mossy fiber projection.

ISOLATION OF ACIDIC ZINC BINDING PEPTIDES

We began this investigation with the assumption that there were as yet unidentified zinc-binding ligands in brain to account for the unique zinc pools revealed by Timm's staining and the TSQ histofluorescence method. Moreover it appeared that metallothionein was unlikely to be this ligand because the binding of zinc to the cysteinyl moieties in metallothionein at physiological pH is strong, greater in fact than the binding of zinc at the active sites of enzymes such as carbonic anhydrase, which is not imaged by TSQ chelation (Frederickson 1988). We therefore designed a fractionation scheme to search for acidic peptides and chose the synaptosomal membrane fraction as the starting material, cognisant of the extensive literature indicating a synaptic localization for zinc. Table 1 outlines the steps which we have developed for the isolation and partial purification of these zinc-binding peptides.

Our initial attempts in isolation of novel zinc-binding peptides involved intracerebroventricular (icv) administration of tracer amounts of high-specific activity zinc-65 to rats. Identical results were obtained after spiking the partially purified peptides with zinc-65 in vitro, therefore it appears unlikely that the synthesis of these peptides was

induced by icv zinc administration, as has been demonstrated for metallothionein (Ebadi 1986). At least three closely related acidic, zinc-binding peptides were identified which were released from synaptosomal membranes by freeze/thaw/sonication. These peptides are not retained by a C2/C18 Reverse Phase Column and elute as a group. Anion exchange chromatography (Mono-Q; Pharmacia) separated these peptides into three broad fractions which were fractionated further by gel filtration chromatography. Two peptides are closely related and tentatively, the following sequences have been assigned: GLU-GLU-ASP- ? -ASP-ASP (Peptide Ia) and GLU-ASP- ? -ASP-Trp-ASP (Peptide Ib). A third zinc-binding peptide contained only Asx residues. The traditional zinc-chelating moieties, His and Cys, were conspicuously absent from all three of these peptides.

TABLE 1: ISOLATION AND PARTIAL PURIFICATION OF ACIDIC ZINC-BINDING PEPTIDES

1. Sucrose homogenate prepared with 0.5 mM Mg
2. Differential centrifugation to isolate the microsomal fraction.
3. Freeze/thaw/sonicate. Centrifuge in presence of 10 mM Mg to pellet membrane fraction.
4. Spike with zinc-65.
5. Ultrafiltration; Saved <10 kD ultrafiltrate.
6. FPLC on C2-C18 Reverse Phase Column and anion exchange column (Mono-Q).
7. Chromatography on Ultrogel AcA 202.
8. Amino acid analysis and sequencing.

Are there any precedents from the literature for such peptides? The only other peptide sequence remotely related to these peptides is a poly(L-aspartate) from the carboxyl terminus of a protein isolated from a soybean seed (Glycine max) recently reported by Odani et al. (1987). The physiological significance of this nona-aspartyl sequence is not known. Van Loon et al. (1984) have identified a 25 residue sequence in yeast ubiquinol-cytocrome c reductase which contains only glutamic or aspartic acid residues. Kinoshita and Ganther (1988) have isolated a cadmium- and zinc-binding protein from testes which has a high concentration of Glx and Asx residues although the sequence has not been reported to date. Lastly, a lead-binding protein has been isolated from rat brain by DuVal and Fowler (1989). This protein contains significant levels of GLU and ASP but weighs approximately 23 kD and contains cysteinyl residues as well. Thus it appears that our peptides may be without precedent in mammalian systems.

What is the function of these peptides in the brain? Is zinc the physiologically important cation or are other cations chelated to these peptides in situ? There are no data to provide an answer to these questions at the present time. The working hypothesis is, obviously, that these peptides represent the ligands to which zinc is bound in hippocampal mossy fibers and other brain regions stained with Timm's or TSQ. It would make good teleological sense to sequester ASP and GLU for neurotransmission by packaging them in the form of short peptides which could either function as transmitters themselves or be hydrolyzed into constituent amino acids. Should the hypothesis be supported by the results of studies in progress, then the implications would be great for enhanced understanding of hippocampal function in human health and disease. (Supported in part by the Veterans Administration Research Service and NIH grants NS25165 and NS21868.)

REFERENCES

Baraldi M, Caselgrandi E, Santi M (1984) Effect of zinc on specific binding of GABA to rat brain membranes. In: Frederickson CJ, Howell G, Kasarskis EJ (eds) The Neurobiology of Zinc Part A: Physiochemisty, Anatomy, and Techniques. Alan R Liss, New York p 59
Baudier J, Glasser N, Gerard D (1986) Ions binding to S100 proteins. J Biological Chem 261:8192-8203
Baudier J, Haglid K, Haiech A, Gerard D (1983) Zinc ion binding to human brain calcium binding proteins, calmodulin and S100B protein. Biochem Biophysical Res Comm 114:1138-1146
Berg JM (1987) Metal ions in proteins: structural and functional roles. Cold Spring Harbor Symp Quant Biol 52:579-585
Brady FO (1983) Metabolism of zinc and copper in the neonate: zinc thionein in developing rat brain, heart, lung, spleen, and thymus. Life Sci 32:2981-2987
Choi DW (1988) Glutamate neurotoxicity and diseases of the nervous system. Neuron 1:623-634
Danscher G (1981) Histochemical demonstration of heavy metals: A revised version of the sulphide silver method suitable for both light and electronmicroscopy. Histochem 71:1-16
Danscher G, Howell G, Perez-Clausell J, Hertel N (1985) The dithizone, Timm's sulphide silver and the selenium methods demonstrate a chelatable pool of zinc in CNS. Histochemistry 83:419-422

Deinum J, Baudier J, Briving C, Rosengren L, Wallin M, Gerard D, Haglid K (1983) The effect of S-100a and S-100b proteins and Zn2+ on the assembly of brain microtubule proteins in vitro. FEBS Lett 163:287-291

DuVal G, Fowler BA (1989) Preliminary purification and characterization studies of a low molecular weight, high affinity cytosolic lead-binding protein in rat brain. Biochem Biophys Res Commun 159:177-184

Ebadi M (1986) Biochemical characterization of a metallothionein-like protein in rat brain. Biol Tr Elem Res 11:101-116

Emdin SO, Dodson GG, Cutfield JM, Cutfield SM (1980) Role of zinc in insulin biosynthesis: Some possible zinc-insulin interactions in the pancreatic beta-cell. Diabetologia 19:174-182

Evans RM, Hollenberg SM (1988) Zinc fingers: Gilt by association. Cell 52:1-3

Frederickson CJ, Howell GA, Kasarskis EJ (1984) Neurobiology of Zinc. Part A: Physiochemistry, Anatomy, and Techniques. Alan R Liss, New York

Frederickson CJ, Kasarskis EJ, Ringo D, Frederickson RE (1987) A quinoline fluorescence method for visualizing and assaying the histochemically-reactive zinc (bouton zinc) in the brain. J Neurosci Meth 20:91-103

Frederickson CJ, Danscher G (1988) Hippocampal zinc, the storage granule pool: localization, physiochemistry, and possible functions. In: Morley J, Sterman MB, Walsh J (eds) Nutritional Modulation of Neural Function. Academic Press, New York, p 289

Fukahori M, Itoh M, Oomagari K, Kawasaki H (1988) Zinc content in discrete hippocampal and amygdaloid areas of the epilepsy (El) mouse and normal mice. Brain Res 455:381-384

Gaskin F, Kress Y (1977) Zinc ion-induced assembly of tubulin. J Biol Chem 252:6918-6924

Gulati S, Paliwal VK, Sharma M, Gill KD, Nath R (1987) Isolation and characterization of a metallothionein-like protein from monkey brain. Toxicol 45:53-64

Howell GA, Welch. M. G., Frederickson CJ (1984) Stimulation-induced uptake and release of zinc in hippocampal slices. Nature 308:736-738

Hulme EC, Berrie CP, Birdsall NJM, Jameson M, Stockton JM (1983) Regulation of muscarinic agonist binding by cations and guanine nucleotides. Eur J Pharmacol 94:59-72

Itoh M, Ebadi M, Swanson S (1983) The presence of zinc-binding proteins in brain. J Neurochem 41:823-829

Kasarskis EJ (1984) Zinc metabolism in normal and zinc deficient rat brain. Expt Neurol 85:114-127

Khanna NC, Tokuda M, Waisman DM (1986) Conformational changes induced by binding of divalent cations to calregulin. J Biol Chem 261:8883-8887

Kinoshita CM, Ganther HE (1988) Isolation of a novel rat testicular metalloprotein binding cadmium and zinc. Biol Tr Elem Res 17:189-206

Koh J-Y, Choi DW (1988) Zinc alters excitatory amino acid neurotoxicity on cortical neurons. J Neurosci 8:2164-2171

Larsson H, Wallin M, Edström A (1976) Induction of a sheet polymer of tubulin by Zn2+. Expt Cell Res 100:104-110

Lo MMS, Snyder SH (1983) Two distinct solubilized benzodiazepine receptors: Differential modulation by ions. J Neurosci 3:2270-2279

Mills JS, Johnson JD (1985) Metal ions as allosteric regulators of calmodulin. J Biol Chem 260:15100-15105

Odani S, Koide T, Ono T (1987) Amino acid sequence of a soybean (Glycine max) seed polypeptide having a poly(L-aspartic acid) structure. J Biol Chem 262:10502-10505

Oteiza PI, Hurley LS, Lönnerdal B, Keen CL (1988) Marginal zinc deficiency affects maternal brain microtubule assembly in rats. J Nutr 118:735-738

Sadee W, Pheiffer A, Herz A (1982) Opiate receptor: Multiple effects of metal ions. J Neurochem 39:659-667

Sato S, Frazierj. M., Goldberg AM (1984) The distribution and binding of zinc in the hippocampus. J Neurosci 4:1662-1670

Savage DD, Montano CY, Kasarskis EJ (1989) Quantitative histofluorescence of hippocampal mossy fiber zinc. Brain Res in press

Slevin JT, Kasarskis EJ (1985) Effects of zinc on markers of glutamate and aspartate neurotransmission in rat hippocampus. Brain Research 334:281-286

Toroptsev IV, Eshchenko VA (1970) Histochemical Determination of Zinc with Fluorescent 8-Arensulfonilamino-Quinolines. Tsiologia 12:1481-1484 (In Russian)

Van Loon APGM, De Groot RJ, De Haan M, Dekker A, Grivell LA (1984) DNA-sequ;ence of the nuclear gene coding for the 17-kD subunit-VI of the yeast ubiquinol-cytochrome C reductase. A protein with an extremely high content of acidic amino acids. EMBO J 3:1039-1043

Wensink J, Lenglet WJM. Vis RD, Van Den Hamer CJA (1987) The effect of dietary zinc deficiency on the mossy fiber zinc content of the rat hippocampus. A microbeam PIXE study. Histochem 87:65-69

Wensink J, Molenaar AJ, Woroniecka UD, Van Den Hamer CJA (1988) Zinc uptake into synaptosomes. J Neurochem 50:782-789

Role of Zinc in the Peripheral Nervous System[1]

BOYD L. O'DELL

Department of Biochemistry, University of Missouri, Columbia, MO 65211, USA

ABSTRACT

Zinc deficiency causes peripheral neuropathy in guinea pigs and chicks. The pathology is exhibited grossly by abnormal locomotion, stance and hypersensitivity. In both species, the motor nerve conduction velocity is decreased and can be reversed by zinc repletion. The severity of clinical signs correlate with nerve function as measured electrophysiologically in vivo and in vitro.

Key Words: Zinc, chick, guinea pig, peripheral nerve function.

Zinc is well established as an essential trace element and its metabolic role in the nervous system has received increasing research attention in recent years. Most of the reported research related to zinc has been concerned with the central nervous system. Zinc deprivation in rats during prenatal or early postnatal life leads to altered behavior (Sandstead et al 1975; Halas 1983). This may include impaired learning ability (Halas 1983; Caldwell et al 1970) as well as impaired cognitive and emotional behavior. The hippocampus is involved in the processing of memory and the integration of emotion (Sahgal 1980), and the concentration of zinc is high in the intrahippocampal mossy fiber pathway (Crawford 1983). Early research on this relationship has been reviewed (Dreosti 1984). This paper is concerned with the role of zinc in the peripheral nervous sytem (PNS).

I. PERIPHERAL NERVOUS SYSTEM

The role of zinc in the PNS has received minimal research attention, but the gross signs of deficiency suggest zinc-specific pathology in this component of the nervous system. At the neuromuscular level, it has been observed that zinc (50 μM) reduces the frequency of the miniature end-plate potentials that are induced by calcium (2 mM) in a high potassium medium (Nishimura 1987). These data suggest that zinc inhibits the entry of calcium into nerve terminals at the neuromuscular junction, thereby inhibiting transmitter release.

The gross pathology of zinc deficiency in most vertebrate species includes loss of appetite and reduced growth rate, parakeratosis and skin lesions, impaired reproduction and disease resistance, and abnormal posture and locomotor function. The latter signs directed our interest to the peripheral nervous system. Of the species studied, the growing chick and guinea pig are most prone to abnormal gait and posture (O'Dell et al 1958, 1989a; Nielsen et al 1968; Quarterman & Humphries 1983). The long bones of the zinc-deficient chick are shorter and thicker than normal and the tibiotarsal joint is greatly enlarged, but this does explain the abnormal locomotion. Besides the abnormal posture and gait, zinc-deficient guinea pigs are hypersensitive to touch and vocalize when handled even gently. In the following section, recent neurological data related to these pathological signs are presented.

----
[1] The research reported here was supported in part by NIH Grant No. NS25395 and in part by the Missouri Agricultural Experiment Station.

1. Chick (O'Dell et al 1989b)

The zinc deficient chick develops an "arthritic" gait and assumes a squat position, presumably to eliminate weight stress. This defect is readily reversed by dietary zinc repletion even though the bone and joint abnormalities persist. To investigate the physiological basis of the abnormal locomotion described above, two experiments were performed. Day-old chicks were fed low-zinc (ca 6 mg/kg) and control (50 mg/kg) diets. The control diet was fed both ad libitum or paired to the intake of deficient groups. At the end of week 3, sciatic nerve function was evaluated non-invasively in anesthetized chicks, using a contact-electrode electrodiagnostic system (Nicolet, Madison WI). This system allowed measurement of the muscle action potential (AMP) as well as the motor nerve conduction velocity (NCV). Two stimulator electrodes and a recording electrode were placed subcutaneously and a supramaximal square-wave pulse of 1 msec duration was applied once per sec. The amplitude (AMP) of the action potential was measured in mV from the baseline to the peak of depolarization; latency from the appearance of the stimulus artifact to the onset of depolarization. The distance between the proximal (P) and distal (D) stimulating electrodes was divided by the difference between proximal and distal latencies to calculate NCV.

Experiment 2 was similar to 1 but, beginning at the 4th week, two groups of deficient chicks were repleted by feeding the control diet; pair-fed controls were maintained also. At the end of the 3rd, 4th and 5th weeks, sciatic nerve function was evaluated as above. The results of experiment 1 are summarized in Figure 1.

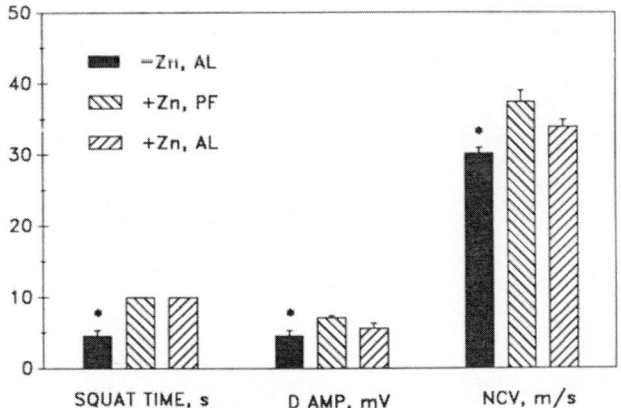

Figure 1. Comparison of squat time, amplitude (millivolts) of the action potential, determined in the distal position (DAmp), and sciatic nerve conduction velocity (NCV), measured in meters per second. Chicks were fed a low-zinc diet (-Zn; 6 mg/kg) ad libitum (AL) or a control diet (+Zn; 50 mg/kg) AL or paired (PF) to -Zn. Standard of the mean (SEM) shown by small bars; * different from AL control, p< 0.01.

Motor nerve conduction velocity was significantly lower in deficient than in pair- (p<0.001) and ad libitum-fed (p<0.07) controls. Action potential measured after distal stimulation, was lower in the zinc-deficient chicks than in controls. All nerve function parameters correlated with clinical signs, including squat time, the time before assumption of the resting position after being placed in a cage. Similar observations were made in experiment 2 (See Figure 2). The decreased nerve conduction velocity was restored to normal within 2 weeks. However, reversal did not occur within 1 week.

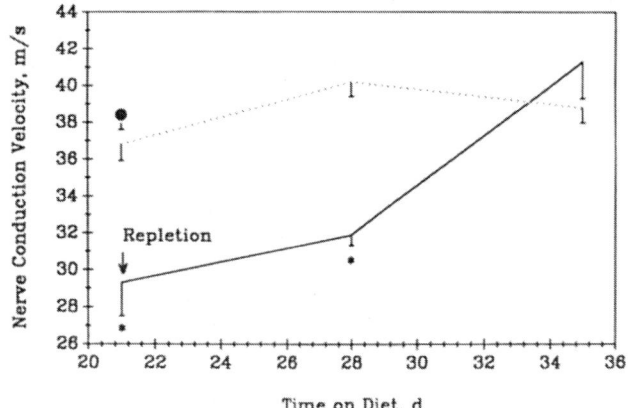

Figure 2. Effect of zinc repletion on sciatic nerve conduction velocity. NCV was measured in vivo at 3 weeks in chicks fed the low-Zn diet (————), the control diet PF (---------), and the control diet AL (Only a single point at 3 weeks). Deficient chicks were repleted by feeding the control diet for 2 weeks while pair-fed controls were maintained. SEM bars; * different from PF controls, p< 0.01.

The biochemical nature of the neuropathology is unknown, but one might speculate that it is a functional rather than structural defect. The relatively short reversal time suggests that demyelination is not involved, but this possibility has not been ruled out.

2. Guinea Pig (O'Dell et al 1989c)

Gross Pathology. Weanling guinea pigs fed a low-zinc (<1 mg/kg) diet develop an abnormal posture and stilted gait, the first signs of which appear after approximately 4 weeks (O'Dell et al 1989a). As deficiency progresses, they assume a kangaroo posture and a hopping gait, designated stage 3, based on an arbitrary scale of 0=normal and 3=severe. In the extreme, animals resist bearing weight on the hind-quarters and move about on their forelegs. Voluntary movement and handling elicit excessive vocalization, suggestive of pain. This hyperalgesia develops at the same rate as abnormal posture and locomotion. All signs are reversed by a single parenteral dose of $ZnSO_4$ (50 μmol/kg BW) within 4 days and return within 1 week if they continue to consume the low-zinc diet. This cycle of depletion and repletion has been repeated as many as 4 times (O'Dell 1989a). The neurological signs of zinc deficiency in the guinea pigs resemble those observed in the chick. Because of the guinea pig's ability to vocalize, hypersensitivity is more evident although it may be present in both species.

Electrophysiological Studies. To study the effect of zinc status on the peripheral nervous sytem, weanling guinea pigs were fed a low-zinc (<1 mg/kg) or control (100 mg Zn/kg) diet. The controls were ad libitum- or restricted-fed so as to maintain body weight near that of animals consuming the low-zinc diet ad libitum. When the low-zinc group developed stage-3 signs, nerve function was measured electrophysiologically as described above. In vivo measurement of sciatic nerve function was made in both legs of animals under anesthesia (Ketamine/Xylazine) and the results averaged. The results from a series of 16 animals per group, age 7-8 weeks, are presented in Figure 3. As in the chicks, zinc deficiency significantly decreased the motor nerve conduction velocity compared to both control groups. The action potential was lower than that of the ad libitum controls.

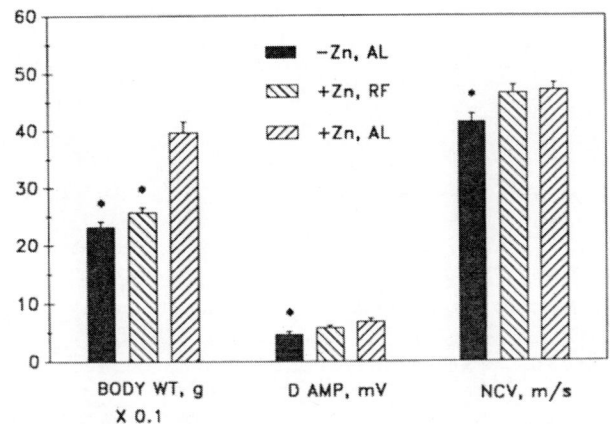

Figure 3. Body weight, action potential (DAmp), and sciatic nerve conduction velocity (NCV) in guinea pigs fed a low-Zn diet (-Zn) for 7 weeks and controls, restricted-fed (RF) or ad libitum-fed (AL) for the same period. SEM bars; * different from AL control, p< 0.01.

Sciatic nerves dissected from severely deficient guinea pigs that had been on the low-zinc diet for approximately 12 weeks, by virtue of repeated parenteral zinc injections, also showed significantly slower conduction rates. See Table 1. The dissected nerves were mounted on a series of platinum electrodes in a moist chamber maintained at 37° C. They were stimulated with variable voltage and pulse duration to obtain maximal response. The response was recorded by use of a dual beam storage oscilloscope and photographed. The photos were analyzed for conduction velocity of the A_{α}, A_{δ} and C fibers. The NCV of the A_{α} fibers was similar to that measured in vivo and showed the same slower rate in sciatic nerves from zinc deficient guinea pigs.

Table 1. Zinc status and electrophysiological function of guinea pig sciatic nerves measured in vitro.

Dietary Zn Suppl.[1]	Nerve Conduction Velocity		
	A_{α} Fibers	A_{δ} Fibers	C Fibers
mg/kg	m/s*	m/s	m/s
0, AL	28.6*	9.3	0.8
100, RF	35.3	9.5	0.8
100, AL	35.8	10.4	0.8

1. Six animals per group fed ad libitum (AL), or restricted-fed (RF) so as to maintain body weight similar to deficient group. * Different from controls, p<0.05.

Finally, a longitudinal study of nerve function was performed. Guinea pigs were fed the diets described above and measurements made beginning at week 4 when there were incipient neurological signs. Measurements were continued at weekly intervals until severe signs occurred, approximately 6 weeks. At that time a single intraperitoneaal dose of zinc was given and the adequate-zinc diet supplied. Measurements were made 3 and 10 days after repletion. The results, summarized in Figure 4, showed that nerve conduction velocities slowed as deficiency progressed, the rate becoming significantly less when stage 3 of clinical signs was reached. NCV correlated well with clinical signs, returning to normal after 2 weeks of repletion when clinical signs had disappeared.

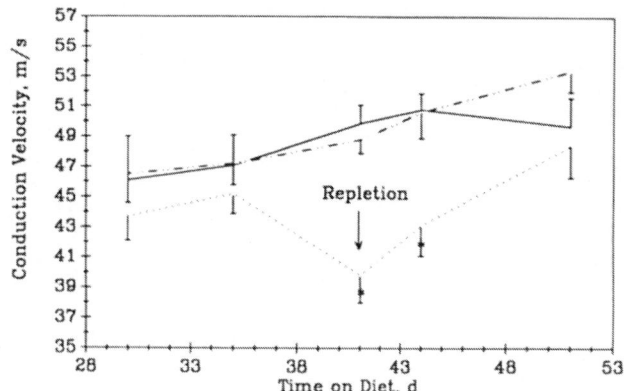

Figure 4. Sciatic nerve conduction velocity measured longitudinally in guinea pigs during the periods of depletion and repletion. For repletion animals were given 50 μmoles Zn/kg BW, ip, and changed to control diet. SEM bars; * different from both controls, p< 0.01. Low-Zn diet (......); adequate-Zn RF (————); adequate-Zn AL (--- . ---).

Although zinc is generally recognized to play a specific role in CNS function, this constitutes the first correlation of the neurological signs of zinc deficiency with impaired peripheral nerve function. The biochem- ical defect which gives rise to the decreased nerve conduction velocity is unknown, but several possiblities exist. Axonal transport depends on functional neurotubules (Jeffery & Austin 1973) and the latter is zinc dependent. Microtubule formation involves tubulin polymerization, a process which is decreased by zinc deficiency (Hesketh 1981). Decreased conduction velocity could also relate to the function of zinc at neurotransmitter receptors.

While motor nerve conduction velocity correlates with the severity of the neurological signs of zinc deficiency, a cause and effect relationship is only speculative. It is particularly pertinent to ask whether the decreas- ed pain threshold, as exhibited by excess vocalization, is caused by decreased conduction velocity. The painful neuropathy of diabetes is associated with decreased NCV (Brown et al 1976), but sensory nerve func- tion has not been studied in that syndrome or to any extent in the zinc- deficient guinea pig. In vitro measurements made here show that only the large myelinated, A$_\alpha$ fibers have different conduction velocities. More research with the sensory nerves is needed to show a relationship of conduction rate to the hyperalgesia of zinc deficiency. In any case, the zinc-deficient guinea pig could serve as a model to study peripheral neuropathy and to test therapeutic agents.

III. REFERENCES

Brown MJ, Martin JR, Asbury AK. (1976) Painful diabetic neuropathy. Arch Neurol 33:164-171
Caldwell DF, Oberleas D, Clancy JJ, Prasad AS. (1970) Behavioral impairment in adult rats following acute zinc deficiency. Proc Soc Exp Biol Med 133: 1417-1421
Crawford IL. (1983) Zinc and the hippocampus. Histology, neurochemistry, pharmacology, and putative functional relevance. In: Dreosti IE, Smith RM, eds. Neurobiology of the Trace Elements, vol 1. Humana Press, Clifton, New Jersey, pp. 163-211
Dreosti IE. (1984) Zinc in the central nervous system: The emerging inter- actions. In: Frederickson CJ, Howell Ga, Kasarkis EJ. eds. The Neuro- biology of Zinc, Part A, Alan R Liss, New York, pp. 1-26

Halas ES. (1983) Behavioral changes accompanying zinc deficiency in animals. In: Dreosti IE, Smith RM, eds. Neurobiology of the Trace Elements, vol 1. Humana Press, Clifton, New Jersey, pp. 213-243

Hesketh JE. (1981) Impaired microtubule assembly in brain from zinc-deficient pigs and rats. Int J Biochem 13: 921-926

Jeffery PL, Austin L. (1973) Axoplasmic transport. Prog Neurobiol 2: 207-255

Nielsen FH, Sunde ML, Hoekstra WG. (1968) Alleviation of the leg abnormality in zinc-deficient chicks by histamine and various anti-arthritic agents. J Nutr 94:527-533

Nishimura M. (1987) Zinc competitively inhibits calcium-dependent release of transmitter at the mouse neuromuscular junction. Pfluger's Arch 410: 623-626

O'Dell BL, Newberne PM, Savage JE. (1958) Significance of dietary zinc for the growing chick. J Nutr 65:503-524

O'Dell BL, Becker JK, Emery MP, Browning JD. (1989a) Production and reversal of the neuromuscular pathology and related signs of zinc deficiency in guinea pigs. J Nutr 119:196-201

O'Dell BL, Conley-Harrison J, Browning JD, Besch-Williford C, Hempe JM, Savage JE. (1989b) Zinc deficiency and peripheral neuropathy in chicks. Submitted

O'Dell BL, Conley-Harrison J, Besch-Williford C, O'Brien D. (1989c) Zinc deficiency in guinea pigs decreases sciatic nerve conduction velocity. Submitted

Quarterman J, Humphries WR. (1983) The production of zinc deficiency in the guinea pig. J Comp Path 93:261-270

Sahgal A. (1980) Functions of the hippocampal system. Trends Neurosci 3:116-119

Sandstead HH, Fosmire GJ, McKenzie JM, Halas ES. (1975) Zinc deficiency and brain development in the rat. Fed Proc FASEB 34:86-88

Cardiovascular Disorders

Experimental and Epidemiological Studies on the Relation of Trace Elements in the Pathogenesis and Prevention of Cardiovascular Diseases

Yukio Yamori[1], Yasuo Nara[1], Ryoichi Horie[1], Fuji Morii[2], Taichiro Nishima[3], Hiroko Nomiyama[4], and Kazuo Nomiyama[4]

[1] Department of Pathology, Shimane Medical University and WHO Collaborating Center for Research on Primary Prevention of CVD, Izumo, 693 Japan
[2] College of Analytical Chemistry, Osaka, 530 Japan
[3] Tokyo Metropolitan Research Laboratory of Public Health, Tokyo, 160 Japan
[4] Department of Hygiene, Jichi Medical School, Tochigi, 329-04 Japan

ABSTRACT

Review of previous experimental and epidemiological studies indicated some relationship of water quality with cardiovascular mortalities and the possible involvement of selenium in cardiomyopathy and of cadmium in hypertension. Present studies on rat models for hypertension, stroke and atherosclerosis revealed significant alterations in plasma and some organ contents of magnesium (Mg) among macroelements and of zinc (Zn) among microelements so far examined, suggestive of the possible involvement of these minerals in hypertension mechanisms, especially in its intracellular processes in addition to causative and preventive roles of macroelements such as sodium (Na), potassium (K) and calcium (Ca) that have been experimentally proven by our previous studies. In our world-wide epidemiological studies on cardiovascular diseases and nutrition, 24 hour urinalysis data of 5 Chinese populations showed not only correlation of Na but also inverse correlations of Ca and Mg with blood pressure. Moreover, comparative analysis on water quality in 6 populations with high and low cardiovascular mortalities has demonstrated Ca and/or Mg contents are obviously higher in those with low risks.

KEY WORDS: Hypertension, Atherosclerosis, Calcium, Magnesium, Zinc

INTRODUCTION

The involvement of some trace elements in the pathogenesis and prevention of cardiovascular diseases (CVD) has been suggested by some previous studies (Keshan Disease Research Group 1979; Klevay 1980; Kobayashi 1957; Kopp et al. 1982; Mertz 1982a,b; Salonen 1982; Schroeder 1960) demonstrated either experimentally in animals or epidemiologically in man.

In this article, studies in the related area are briefly reviewed and experimental studies in our rat models for hypertension (Okamoto and Aoki 1963; Yamori 1984a), stroke (Yamori et al. 1974; Yamori 1984b) and atherosclerosis (Yamori 1977; Yamori 1982c) are reported especially in relation to macro and trace elements. Finally, some highlights of our international cooperative epidemiological studies on nutrition and CVD, so-called WHO-CARDIAC Study (CARDIAC Study Protocol 1986; Yamori 1989c) are reported to suggest experimental results in animal models will hopefully be applied to the primary prevention of CVD in man.

I. HISTORICAL REVIEW

The possible pathogenic involvement of trace elements in CVD of humans has been suggested by these three directions of research: 1) CVD in relation to water quality, 2) selenium deficiency and 3) cadmium pollution. In each area oriental researchers have contributed a lot.

(a) Water Quality and CVD

It was Kobayashi who first reported the variabilities of water acidity in all over Japan in 1957 and suggested the close relation to the mortality of stroke. The great regional differences of stroke incidence are well known, that is, high

in the north-eastern Japan and low in the south-western Japan. Kobayashi's pioneering report provoked epidemiological studies on the regional difference in CVD mortality in relation to water hardness in the U.S. (Schroeder 1960) and Europe (Pocock et al. 1980). Some of the later studies supported the original hypothesis but others failed, because it is presently well known that other nutritional factors such as salt and fat intakes are more influential risk factors of CVD (Kannel and Sorlie 1975; Keys 1980; Sasaki 1962; Yamori 1989c) and the influence of water quality on CVD appears to be rather marginal compared with these major nutritional factors.

(b) Selenium Deficiency and Keshan Disease

The insight of Chinese scientists into the pathogenic relation of selenium (Se) deficiency to Keshan disease, an endemic cardiomyopathy was convincingly confirmed by their extensive intervention study which proved the effectiveness of Se supplementation (Keshan Disease Research Group 1979), although the influence of concomitant improvement of nutritional conditions could not be eliminated completely. Supportive evidence for the pathogenic importance of Se deficiency in heart diseases was reported from Finland (Salonen 1982), which clearly demonstrated the relative risks of coronary death (x6.9), all cardiovascular deaths (x6.2) and all myocardial infarctions (x2.3) were high in the people with lower serum Se level ($\leq 34 \mu g/l$) compared with those with the higher level ($\geq 45 \ \mu g/l$).

(c) Cadmium and Blood Pressure

As for the relation of cadmium (Cd) to hypertension, variable grades of blood pressure elevation and reduction by chronic administration of Cd have been reported by many investigators (Kopp et al. 1982; Schroeder 1965; Perry and Erlanger 1971). Kopp et al. reviewed the controversial issue and demonstrated a certain range of the concentration caused the maximum pressor response in their rat experiments (Kopp et al. 1982; Kopp 1986). Their animal experiment was consistent with the extensive epidemiological observation by Shigematsu and other Japanese investigators on the lowering of blood pressure levels and the concomitant reduction of stroke mortality in the inhabitants living in the regions with Cd pollution (Shigematsu 1981). Since Cd affects potentially various membrane sites related to Ca at rapidly exchangeable Ca binding sites, Na-Ca exchange, Ca channel, and also at Na-K ATPase, ß receptors, and cholinergic receptors (Kopp 1986), it is theoretically understandable that chronic exposure to Cd may produce elevation or lowering of blood pressure depending on the dosis administered.

Because of such complexity with interaction of macroelements, the influence of trace elements on cardiovascular system was further analyzed for their pathological implication in our animal models for hypertension, that is spontaneously hypertensive rats (SHR) (Okamoto and Aoki 1963; Yamori 1984a), and others, especially in stroke-prone SHR (SHRSP) (Yamori et al.1974; Yamori 1984b), developing stroke spontaneously in 100% nowadays.

II. EXPERIMENTAL STUDIES ON CVD

(a) Progress in Studies on Pathogenesis and Prevention of CVD

SHR and SHRSP develop moderate to severe hypertension spotnaneously without exception (Okamoto and Aoki 1963; Yamori et al. 1974; Yamori 1984a,b). Our studies on SHR up to the present have demonstrated that the initial blood pressure elevation is due to neurogenic factors, since reversible functional vasoconstriction is observed in young SHR concomitantly with the increased urinary catecholamine excretion (Yamori 1976b; Yamori 1983). However, BP rise stimulates protein synthesis of arterial walls and subsequently induce the hypertrophy or hyperplasia of vascular smooth muscle cells (VSMC) (Yamori 1983; 1976a, b; Folkow et al. 1975).

The cellular mechanism of vascular hyperplasia was further studied in cultured SMC obtained from SHR and normotensive Wistar-Kyoto rats (WKY) and cells from SHR or SHRSP always showed faster growth rate (Yamori 1983; Igawa et al. 1988). Also in these SMC from hypertensive rats Na accumulates and K reduces more quickly after the inhibition of Na, K ATPase by ouabain, indicating membrane permeability to Na and K is increased in VSMC of hypertensive rats (Yamori 1983; Yamori et al. 1982a).

Such an increase in Na permeability may affect the growth rate of the cells. After the deprivation of a growth factor, fetal calf serum from the culture medium for 18 hours, the growth factor was added to culture medium with different Na concentrations (Nabika et al. 1986). Na concentration of the culture media was proven to affect clearly the growth rate of VSMC.

Therefore, we can now briefly summarize the cellular mechanism of genetic hypertension and its interaction with environmental factors (Yamori 1983): Intracellular Na or K in VSMC tends to increase or decrease due to the permeability or transport abnormality of the cell membrane. Such intercellular electrolyte imbalance causes cytosolic ionic Ca elevation through Na-Ca exchange (Matlib et al. 1985) and functionally or structurally increase peripheral vascular resistance. Environmental factors, especially excess salt intake aggravate this process through the inhibition of Na, K ATPase by the action of ouabain or digoxin-like natriuretic factors. These VSMC are theoretically supposed to be more excitable in response to neurogenic stimuli.

(b) Possible Involvement of Macro- and Microelements in Hypertension

We actually confirmed that graded doses of norepinephrine always caused greater pressor response in the perfused mesenteric vessels from SHR than in those from normotensive WKY (Mtabaji et al. 1986). We further observed the preperfusion of Zn, 3.2 µg/ml for 15 minutes augmented the pressor response to NE only in the

mesenteric vessels from SHR (Fig. 1) and Zn did not alter the pressor response to K (Mtabaji et al. 1985). Thus, Zn appears to accelerate intracellular Ca release by NE stimulation particularly in SHR but not to stimulate K-induced entry of extracellular Ca.

Such stimulatory effect of Zn may be partly involved in the pathogenesis of genetic hypertension, since plasma Zn levels are significantly increased in 3-month-old SHR and SHRSP (Fig. 2) like patients with essential hypertension as reported by Vivoli et al. On the other hand, plasma Ca is significantly decreased in both SHR and SHRSP and other significant changes in these serum macro- and microelements commonly noted in two hypertensive strains are the reduction of Mg/Zn and Ca/Zn ratios and the elevation of Zn/Cu ratio. These common alterations in both hypertensive strains may primarily or secondarily related to hypertension.

The elevation of Zn is not limited to the serum but is also noted in the heart of 2-month-old SHRSP and in the kidney and the brain in both 2-month-old SHR and SHRSP at the incipient stage of genetic hypertension (Fig.

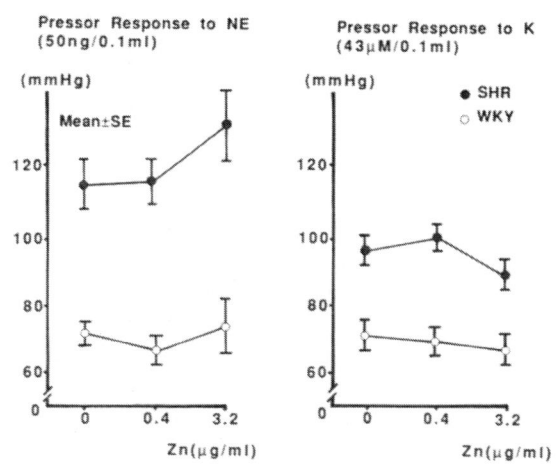

Fig. 1 Effect of Zinc (Zn) on Pressor Response to Norepinephrine (NE) and Potassium (K) in SHR and WKY

3). Therefore, such increases in Zn level may somewhat contribute to the development of genetic hypertension.

Although the implication of increased Zn content in the cardiovascular system of hypertensive strains is not known presently, the activity of alkaline phosphatase, a typical Zn containing enzyme, is significantly elevated in the mesenteric arteries and aorta of SHR after 3 months of age (Ooshima 1973) when Zn level in the heart is increased. Moreover, these alkaline phosphatase activities of blood vessels were confirmed to be related to blood pressure levels in the F_2 generation obtained from the cross-breeding between SHR and normotensive rats (Ooshima 1973). If alkaline phosphatase activity is not secondarily activated by hypertension itself, such cosegregation of the activity with blood pressure level indicates that it is genetically linked to the heredity of hypertension or involved in the genetic mechanism of hypertension.

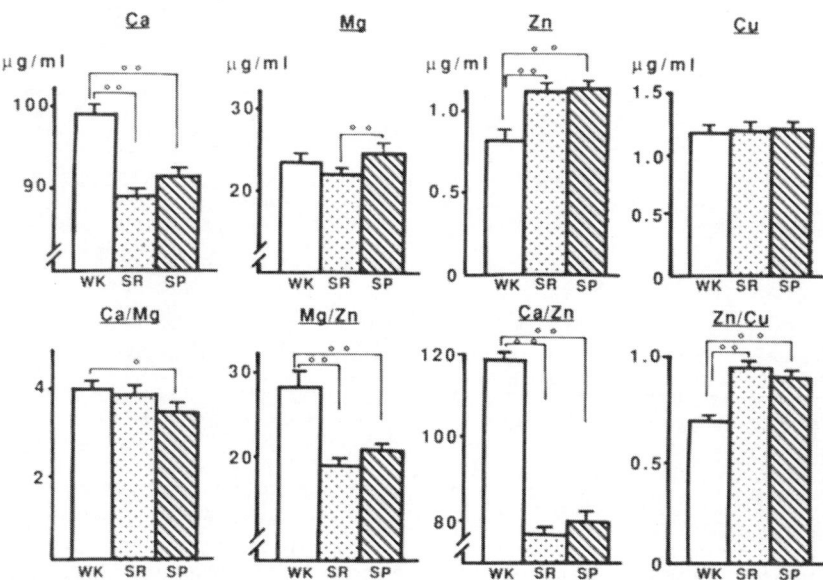

Fig. 2 Plasma Calcium (Ca), Magnesium (Mg), Zinc (Zn), Copper (Cu) and Their Ratios in 3-Month-Old SHRSP (SP), SHRSR (SR) and WKY (WK)

68

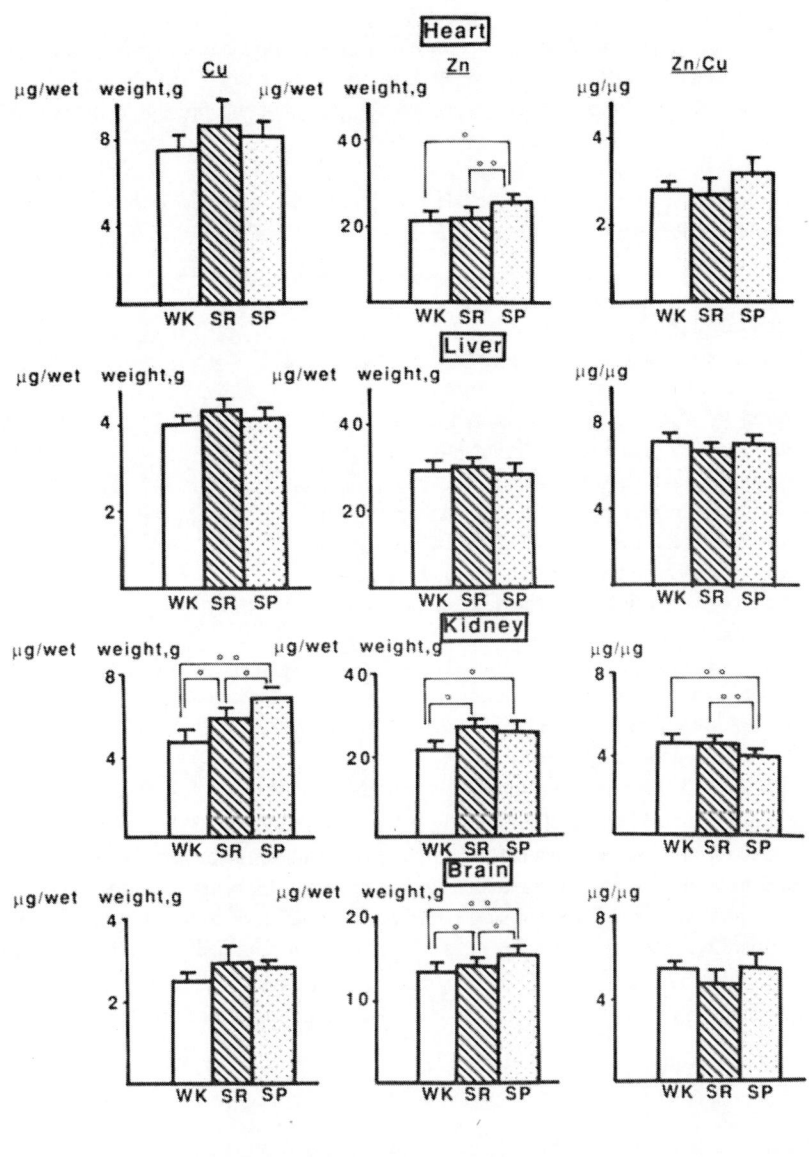

Fig. 3 Copper (Cu) and Zinc (Zn) Contents of Various Organs, 2-Month-Old SHRSP (SP), SHRSR (SR) and WKY (WK)

On the other hand, diabetes mellitus or impaired glucose tolerance is one of the risk factors of CVD other than hypertension. Diabetic tendency and hypertension are concomitantly noted in SHR (Yamori et al. 1978) as well as in essential hypertension in man (Hirata et al. 1962; Welborn et al. 1966; Dieterle et al. 1967; Wagner et al. 1971; Berglund et al. 1976).

Recently, Resnick et al. (1987, 1988) applied ^{31}P-NMR spectroscopy to measure intraerythrocyte free Mg and reported its lowering related inversely to blood pressure level, as well as close inverse relation to insulin response in glucose tolerance test. Interesting enough, Henrotte et al. (1985) reported intraerythrocyte Mg reduction in SHR and also the elevation of intracellular Zn concentration. Thus, these intracellular electrolyte imbalances, especially Mg reduction commonly noted in hypertensive patients and SHR with diabetic insulin response may be of basic importance in the pathogenesis of these diseases.

Table I Similarity between SHR and Essential Hypertension (EH) In Man

	SHR vs WKY		EH vs Normotensives	
Zn: Serum	↑	(Yamori et al. 1989)	↑	(Vivoli et al. 1989)
Erythrocyte (E)	↑	(Henrotte et al. 1985)	↑	(Frithz et al. 1979)
ALPase: Serum	↗ →	(Oshima. 1973)	↑	(Vivoli et al. 1989)
Arteries	↑		?	
Mg: E	↓	(Henrotte et al. 1985)	↓	(Resnick et al. 1988)
pH: E or VSMC	↓	(Yamori. 1988)	↓	(Resnick et al. 1987)

VSMC = Vascular Smooth Muscle Cells

Table I summarizes afore-mentioned various similarities related to Zn and Mg between SHR and essential hypertension in man, which suggest the pathogenic importance of intracellular imbalance of Zn and Mg in the pathogenic mechanism of genetic hypertension.

In order to clarify the causative role of cellular Mg reduction in the development of CVD, Mg was dietary supplemented in SHRSP given 1% NaCl water for drinking. Increased Mg content, 0.8% in total, in the diet did not so effectively attenuated the salt-accelerated hypertension in SHRSP given the control diet containing 0.2% of Mg, but interestingly enough the survival rate of SHRSP was greatly improved (Yamori 1989a,b; Yamori et al. 1989).

Ca supplementation, 1.6% in total to the control diet containing 0.7% of Ca did effectively improve the survival rate in SHRSP given 1% NaCl water for drinking and the mechanism of stroke prevention is supposedly ascribed to the acceleration of urinary Na excretion by enough Ca in the diet (Yamori 1989b, Yamori et al. 1989).

K is now well known to be effective for preventing stroke in SHRSP, since our first report in 1981 (Yamori 1981). Even a minute reduction of Na/K ratio from 0.93 to 0.61 significantly improved the survival rate in SHRSP (Yamori 1989b, Yamori et al. 1989). We previously reported Na/K ratio in the urine was positively related to the blood pressure level and the prevalence of hypertension in one Japanese community by our epidemiological survey (Yamori 1981; Yamori et al. 1982b) and suggested the importance of decreasing Na/K ratio for dietary improvement in men.

Effective dietary factors other than these minerals are fish and soy proteins which decreased stroke incidence with or without blood pressure lowering effect (Yamori1983, 1989a; Yamori et al. 1979, 1989), fish or vegetable oils (Yamori 1981; Yamori et al. 1986), and the dietary fibers, that reduced the intestinal Na absorption and thus counteracted the adverse effect of salt (Yamori et al. 1986). The effective amino acids in the protein-rich diet were further analyzed, and lysine as well as sulfur amino acids such as taurine effectively prevented stroke in SHRSP (Yamori 1983a; Yamori 1989b, Yamori et al. 1989).

Finally, a marked long term effect of Mg on the survival of SHRSP was confirmed by our experiment (Yamori 1989b). When stroke-prone SHR were given 1% NaCl for drinking at the age of 40 days, they quickly developed severe hypertension and died of stroke within a few weeks. Soy protein plus Ca with or without phosphate significantly prolonged the life span, and the maximum improvement of survival rate was noted in the group given soybean protein plus Ca and Mg.

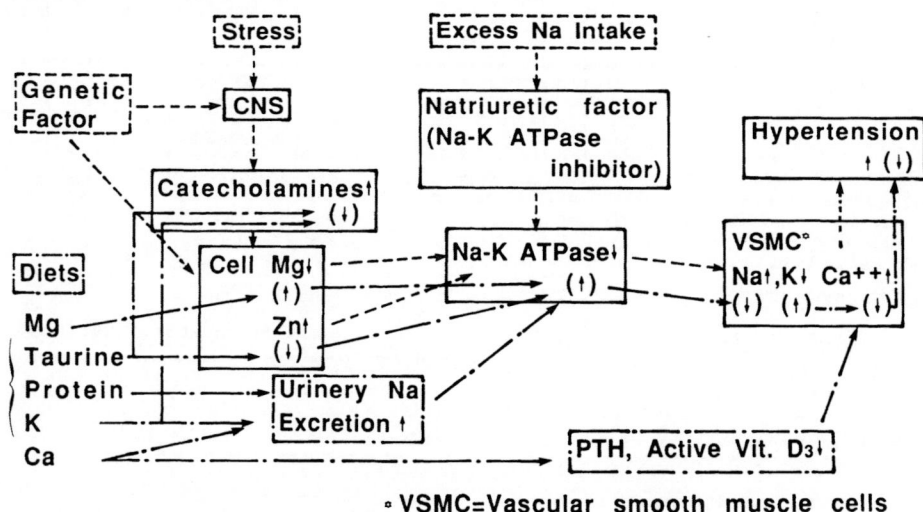

Fig. 4 Gene-Environment Interaction in Mechanisms of Genetic Hypertension

70

Here we may speculate the pathogenic and preventive mechanisms of hypertension-related CVD by emphasizing the intracellular electrolyte imbalance (Fig. 4). Cellular Mg reduction (Itokawa 1989) and increased Zn (Barbeau and Donaldson 1974), a powerful inhibitor of Na-K ATPase reduced the Na-pump activity and accelerate intracellular electrolyte imbalance to elevate blood pressure through intracellular rise of free Ca in VSMC. On the other hand, Mg supplementation (Itokawa 1989) itself as well as taurine (Barbeau and Donaldson 1974) especially by its chelation of Zn stimulate Na-K ATPase activity to improve intracellular electrolyte imbalance, thus to prevent hypertension-related CVD, together with protein, K and Ca which all increase urinary Na excretion to attenuate the adverse effect of Na.

(c) Trace Elements and Atherosclerosis

In addition to the model for stroke, we could successfully establish a rat model for atherosclerosis called ALR (Yamori 1977) from a substrain of SHRSP which quickly developed reactive hypercholesterolemia as well as fat deposition within a few weeks in mesenteric arteries (Yamori et al. 1975) and also in such small arteries as cerebrobasal arteries (Yamori et al. 1976). Similar segmental fat deposits in the latter are sometimes noted in autopsy cases of hypertensive patients.

SHRSP are different from stroke resistant SHR (SHRSR) which are resistant also to arterial fat deposition. Trace metal analyses showed both strains had higher Zn levels and lower Cu levels in the heart, and therefore, Zn to Cu ratios were significantly greater than that in normotensive strain (Fig. 5). Since Cu concentration in the liver is greater in SHRSR, Zn to Cu ratios are significantly different between SHRSP and SHR-SR; SHRSR with higher Cu level in the liver are also resistant to arterial fat deposition and reactive hypercholesterolemia when fed on a high-fat cholesterol diets. This may be consistent with the Zn to Cu hypothesis as reported by Klevay (Klevay 1980; Allen and Klevay 1978), and with Sham et al's report (1988) on an inverse relation between liver Cu and aortic atherosclerosis in human autopsy cases and turkeys.

Moreover, VSMC from SHRSP deposit more cholesterol ester than those from SHRSR even under tissue culture condition when LDL-cholesterol is added to the culture media (Yamori et al. 1984a). The possible involvement of trace elements in such cellular difference in the lipid deposition need further research.

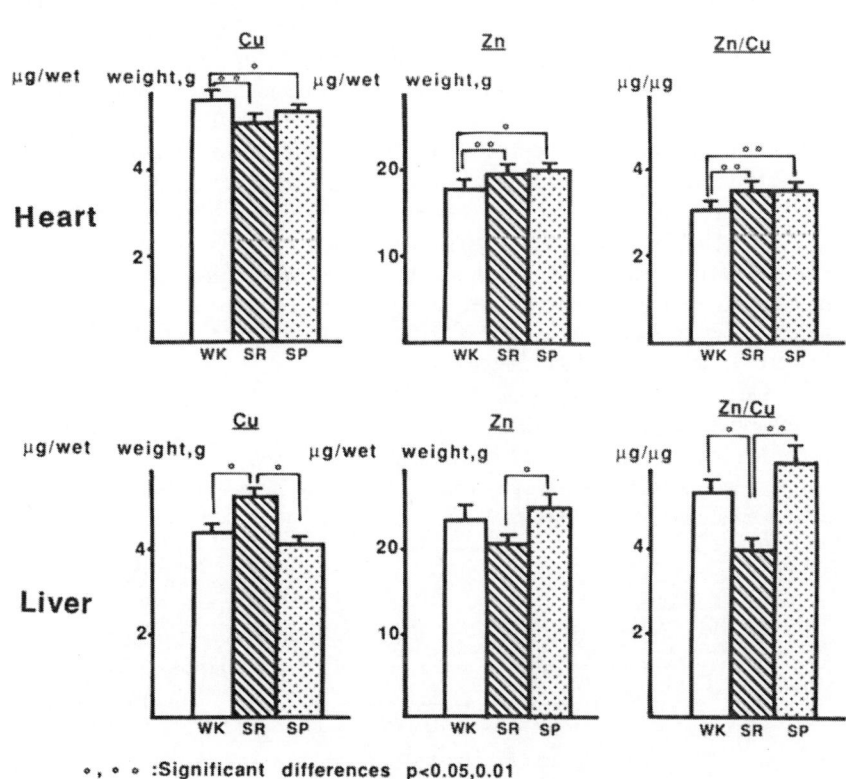

Fig. 5   Copper (Cu) and Zinc (Zn) Concentrations of the Heart and the Liver in 3-Month-Old SHRSP (SP), SHRSR (SR) and WKY (WK)

III EPIDEMIOLOGICAL STUDIES ON CVD

In order to test such relations between dietary factors and blood pressure, or hypertension-related CVD, we proposed in 1982 to WHO international cooperative study on cardiovascular diseases and alimentary comparison, so-called CARDIAC Study (CARDIAC Study Protocol 1986; Yamori 1989c), 40 centers in 20 countries so far are participating in this cooperative study to offer hopefully basic data for establishing dietary goals for nonpharmacological prevention of cardiovascular diseases.

The characteristic of CARDIAC Study is first, standardized measurement of blood pressure by an automated blood pressure measurement system (Fukuoka and Yamori 1987). This system developed in Japan has been successfully introduced in this study for the objective measurement of blood pressure among different populations.

Another characteristic is the application of biological dietary markers of 24-hour urine and blood samples. For collecting 24-hour urine, a new device, so-called 'aliquot cup' (Yamori et al. 1984b; Nara et al. 1984) was developed in Japan and successfully used all over the world.

Our cooperative study in Tanzania was successfully carried out in 3 populations, urban population of Dar es Salaam, rural one of Handeni, and typical pastoral populations of Masai in Longido. In contrast to urbanized population in Dar es Salaam, Masai people are obtaining food mainly from their cattles, that is, milk. They drink milk a lot even 3 to 10 litters of milk a day, and they have no custom to add salt to diets. The prevalence of hypertension is increased markedly with ageing process from 30's to 50's in Dar es Salaam. In contrast, the prevalence was low in Handeni and no hypertension was detected in Masai at the ages of 30-34 and 40-44. Only a few mild hypertensive were detected at the ages of 50-54. Correspondingly, Na and Na to K ratios were highest in Dar es Salaam, the second in Handeni and the third in Masai populations. It is noteworthy that hypertension is rare in Masai who are eating low Na diets containing less than 2 g of salt per day and with Na/K ratios less than 1 (Yamori 1989a,c).

Studies in 5 populations in China clearly demonstrated both systolic and diastolic blood pressures were positively related with 24-hour urinary Na excretion, Na/K ratio and inversely related to urinary taurine to creatinine ratios, good indices of the intake of fish and other sea food protein, and also inversely to urinary Ca excretion (Yamori 1988; Yamori 1989b, c). Moreover, both systolic and diastolic blood pressures were inversely related to urinary Mg excretion (Yamori 1989a) and also inversely to some extent to urinary strontium excretion (Table II). Both systolic and diastolic blood pressures showed various degrees of correlations to these macroelements and week possible correlations to trace elements in these population studies. Therefore, these macro and trace elements may possibly affect blood pressure levels in populations. Although correlation of blood pressure with Zn is low, Mg/Zn ratios are likely to be related inversely to both systolic and diastolic blood pressures (Fig. 6).

Moreover, as an optional CARDIAC Study, we analyzed water samples from some high-risk populations such as Quito in Ecuador, Altai in northern China near Russian border and Dar es Salaam in Tanzania, as well as those from low risk populations such as Hotien and Tulfan in the oasis of north-western China and Guangzhou in southern China. In these low risk communities water contents of Ca plus Mg seem to be higher (Fig. 7a,b). Although these mineral intakes from water are relatively limited compared with those from food, quality of water may affect the mineral contents of common daily food produced locally and cardiovascular mortality to some extent.

Table II Correlations between Blood Pressure and Urinary Mineral Excretions in China

	Ca	Mg	P	Fe	Sr	Zn
SBP	-0.964**	-0.794	0.551	0.738	-0.522	0.132
DBP	-0.990**	-0.893*	0.581	0.739	-0.600	0.305

*, ** Significant Differences; $p<0.05, 0.01$

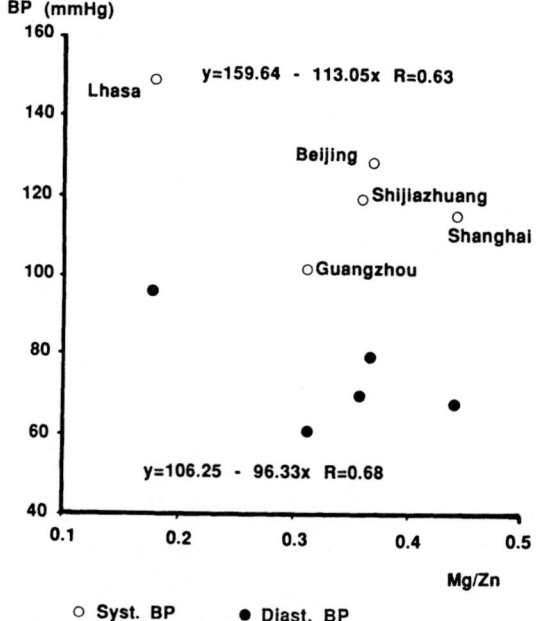

Fig. 6   Relationship of Blood Pressure and Urinary Mg/Zn Ratio in People's Republic of China

CONCLUSION

Two main fundamental disease processes of CVD, hypertension and atherosclerosis are possibly related to macroelements such as Na, K, Ca and Mg as summarized in Tables III and IV. However, trace elements such as Cd, Se, Cu and Zn together with other microelements, iodine and chromium not discussed in this review may be involved in

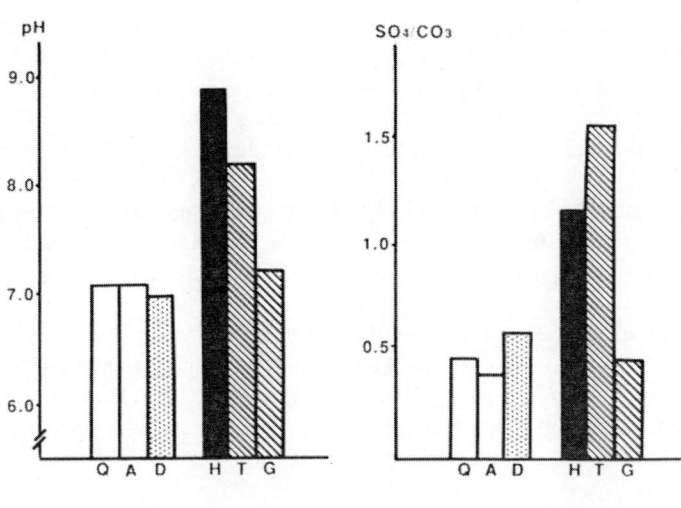

Fig. 7a
Water Analyses of Some High and
Low Risk Communities in the World:
Hydrogen Ion Concentration (pH)
and Sulfate to Carbonate (SO_4/Co_3)
Ratio

High Risk Communities
Q: Quito
A: Altai
D: Dar es Salaam

Low Risk Communities
H: Hotien
T: Tulfan
G: Guangzhou

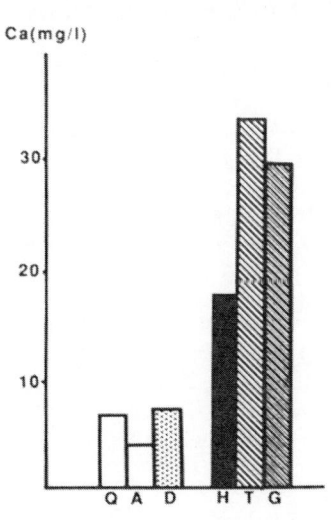
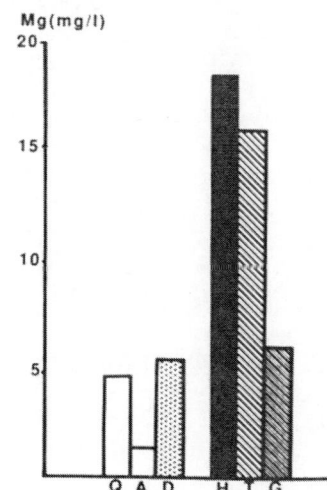

Fig. 7b
Water Analyses of Some High and
Low Risk Communities in the World:
Calcium (Ca) and Magnesium (Mg)

High Risk Communities
Q: Quito
A: Altai
D: Dar es Salaam

Low Risk Communities
H: Hotien
T: Tulfan
G: Guangzhou

Table III Minerals and Hypertension

<Macro Elements>

Na	↑	BP ↑
K	↑	BP ↓ or →
Ca	↑	BP ↓
Mg	↑	BP ↓ or →

<Trase Elements>

Cd	↑	BP ↑ or ↓
Se, Cu	↑	(BP ↑ by Cd) ↓
Zn	↑	(BP ↑ by NE) ↑

Table IV Minerals and Atherosclerosis

<Macro Elements>

Na	↑	Serum Cholesterol	↑
Ca	↑	Serum Lipids	↓
K, Mg	↑	Fat Deposits	↓

<Trace Elements>

I	↓	Serum Cholesterol	↑
Cr	↓ (↑)	Serum Cholesterol	↑ (↓) *
Cu	↓	Serum Cholesterol	↑
Zn	↑	HDL ↓, (Cholesterol by Cu ↓) ↑

*Abraham et al 1980; Newman et al 1978; Riales & Albrink 1981

these disease process so far as experimental and epidemiological studies indicated. Reduced Mg to Zn ratios noted intracellularly in man and animal models of hypertension require further study to clarify the pathogenic implications in hypertension-related CVD.

REFERENCES

Abraham AS, Sonnenblick M, Eini M, Shemesh O, Batt AP (1980) The effect of chromium on established atherosclerotic plaques in rabbits. Am J Clin Nutr 33: 2294-2298

Allen KGD, Klevay LM (1978) Cholesterolemia and cardiovascular abnormalities in rats caused by copper deficiency. Atherosclerosis 29: 81-93

Barbeau A, Donaldson J (1974) Zinc, taurine, and epilepsy. Arch Neurol 30: 52-58

Berglund G, Lasson B, Andersson O, Larsson O, Svärdsudd K, Bjorntorp P, Wilhelmsen L (1976) Body composition and glucose metabolism in hypertensive middle-aged males. Acta Med Scand 200: 163-169

CARDIAC Study protocol (1986) WHO Collaborating Center for Research on Primary Prevention of Cardiovascular Diseases, WHO, Shimane-Geneva

Dieterle P, Felm H, Ströder W, Henner J, Bottermann P, Schwarz K (1967) Asymptomatischer Diabetes Mellitus bei normalgewichtigen Hypertonikern. Dtsch Med Wochenschr 92: 2376-2381

Folkow B, Hallback M, Lundgren Y et al (1975) Importance of adaptive changes in vascular design for establishment of primary hypertension studied in man and in spontaneously hypertensive rats. Circ Res 32/33 (Suppl 1): 2

Fukuoka M, Yamori Y (1987) A proposal for indirect and objective blood pressure measurement in adults. In: Yamori Y, Lenfant C (eds) Prevention of cardiovascular diseases: an approach to active long life. Elsevier, Amsterdam, p127-137

Henrotte JG, Santarromana M, Bourdon R (1985) Concentrations en magnésium, calcium et zinc du plasma et des érythrocytes de Rats spontanément hypertendus. C R Acad Sc Paris 300: 431-436

Hirata Y, Hirano M, Ito M, Yamauchi M, Makino N, Ishimoto M, Sato T, Hososako A (1962) A diabetes detection study in Kyushu, Japan - the relation of diabetes to hypertension. Diabetes 2: 44-48

Igawa T, Yamori Y, Lewis LJ, Tarazi RC (1988) Norepinephrine-induced enlargement of nucleus in cultured myocardial cells. Heart Vessels 4: 1-5

Itokawa Y (1989) Recent aspects of magnesium and cardiovascular diseases. In: Yamori Y, Strasser T (eds) New horizons in preventing cardiovascular diseases. Elsevier, Amsterdam, p27-34

Kannel WB, Sorlie P (1975) Hypertension in Framingham. In: Paul O (ed) Epidemiology and control of hypertension. Stratton Intercontinental Medical Book Corporation, New York. pp553-592

Keshan Disease Research Group of the Chinese Academy of Medical Sciences (1979) Observations on effect of sodium selenite in prevention of Keshan Disease. Chinese Med J 92: 471-476

Keys A (1980) Seven countries: a multivariate analysis of death and coronary heart disease. Harvard University Press, Cambridge

Klevay LM (1980) Interactions of copper and zinc in cardiovascular disease. In: Levander OA, Cheng L (eds) Micronutrient interactions: vitamins, minerals, and hazardous elements. The New York Academy of Sciences, New York, p140-151

Kobayashi K (1957) Geographical relationship between the chemical nature of river water and death rate from apoplexy. Ber Ohara Inst Landwirtsch Biol Okayama Univ 11: 12-21

Kopp J (1986) Cadmium and the cardiovascular system. In: Foulkes EC (ed) Cadmium (Handbook of experimental pharmacology Vol. 80). Springer-Verlag, Berlin/Heidelberg, p.195-280

Kopp SJ, Glonek T, Perry Jr HM, Erlanger M, Perry EF (1982) Cardiovascular actions of cadmium at environmental exposure levels. Science 217: 837-839

Matlib MA, Schwartz A, Yamori Y (1985) A Na+-Ca2+ exchange process isolated sarcolemmal membranes of mesenteric arteries from WKY and SHR rats. Am J Physiol 249: C166-C172

Mertz W (1982a) Trace metals and hypertension. In: Horan MJ, Blaustein MP, Dunbar JB, Kachadorian W, Kaplan NM, Simopoulos AP (eds) NIH workshop on nutrition and hypertension. DHEW Publication, Washington DC, p271-276

Mertz W (1982b) Trace minerals and atherosclerosis. Federation Proc 41: 2807-2812

Mtabaji JP, Kihara M, Yamori Y (1985) Zinc and vascular reactivity in rat mesenteric vessels: possible altered dihomo-g-linolenic acid metabolism in spontaneously hypertensive rats. Prostagland Leukotr Med 18: 235-243

Mtabaji JP, Kihara M, Yamori Y (1986) Effects of indomethacin and prostaglandin E2 on vascular reactivity in spontaneously hypertensive rats. J Hypertension 4 (Suppl 3): S73-S75

Nabika T, Nara Y, Endo J et al (1986) Effect of Na on kinetics of cell proliferation in cultured vascular smooth muscle cells. J Hypertension 4 (Suppl 3): S303-S305

Nara Y, Kihara M, Mano M et al (1984) "Aliquot cups", simple method for epidemiological and clinical studies. In: Lovenberg W, Yamori Y (eds) Nutritional prevention of cardiovascular diseases. Academic Press, Orlando, p211-216

Newman HAI, Leighton RF, Lanese RR, Freedland NA (1978) Serum chromium and angiographically determined coronary artery disease. Clin Chem 24: 541-544

Okamoto K, Aoki K (1963) Development of a strain of spontaneously hypertensive rats. Jpn Circ J 27: 282-293

Ooshima A (1973) Enzymological studies on arteries in spontaneously hypertensive rats. Jpn Circ J 37: 497-508

Perry HMJr, Erlanger MW (1971) Metal-induced hypertension following chronic feeding of low doses of cadmium and mercury. J Lab Clin Med 83: 541-547

Pocock SJ, Shaper AG, Cook DG, Packham RF, Lacey RF, Powell P, Russell PF (1980) British regional heart study: geographic variations in cardiovascular mortality, and the role of water quality. Br Med J 280: 1243-1249

Resnick LM, Gupta RK, Sosa RE, Corbett ML, Laragh JH (1987) Intracellular pH in human and experimental hypertension. Proc Natl Acad Sci USA 84: 7663-7667

Resnick LM, Gupta RK, Gruenspan H, Laragh JH (1988) Intracellular free magnesium in hypertension: relation to peripheral insulin resistance. J Hypertension 6 (Suppl 4): S199-S201

Riales R, Albrink MJ (1981) Effect of chromium chloride supplementation on glucose tolerance and serum lipids including high-density lipoprotein of adult men. Am J Clin Nutr 43: 2670-2678

Salonen JT (1982) Association between cardiovascular death and myocardial infarction and serum Se in a matched-pair longitudinal study. The Lancet: 175-179

Sasaki N (1962) High blood pressure and the salt intake of the Japanese. Jpn Heart J 3: 313-324

Schroeder HA (1960) Relation between mortality from cardiovascular disease and treated water supplies: variations in states and 163 large municipalities in the United States. J Am Med Assoc 172: 1902-1908

Schroeder HA (1965) Cadmium as a factor in hypertension. J Chronic Dis 18: 647-656

Sham RP, Gresham GA, Howard AN, Minski MJ (1988) Studies in copper deficiency and atherosclerosis. In: Abstract Book for 8th International Symposium on Atherosclerosis, Rome Oct. 9-13, 1988

Shigematsu I, Tsuchiya K, Kitamura S, Takeuchi J, Kajikawa K, Kato T, Nomiyama K, Ishimoto F, Ogata H, Saito H, Minowa M (1981) Analysis and report of the results on the health survey on the inhabitants in environmentally contaminated areas by cadmium. In: Recent studies on health effects of cadmium in Japan. Jpn Environ Agency, Tokyo, pp319-380

Vivoli G, Bergomi M, Rovesti S (1990) Cadmium, zinc and copper in human hypertension. In: Proceedings of the Symposium on Cardiovascular Disorders and Trace Metal of the 2nd Meeting of the International Society for Trace Element in Humans.

Wagner H, Wessels F, Zierden E, Junge-Hülsing G (1971) Untersuchugen zum Verhalten von Insulin und Glucosestoffwechsel bei Hypertonie. Verh Dtsch Ges Inn Med 77: 133-136

Welborn TA, Breckenridge A, Rubinstein AH, Dollery CT, Fraser TR (1966) Serum-insulin in essential hypertension and in peripheral vascular diseases. Lancet 1: 1336-1337

Yamori Y, Nagaoka A, Okamoto K (1974) Importance of genetic factors in hypertensive cerebrovascular lesions: an evidence obtained by successive selective breeding of stroke-prone and -resistant SHR. Jpn Circ J 38: 1095-1100

Yamori Y, Hamashima Y, Horie R et al (1975) Pathogenesis of acute arterial fat deposition in spontaneously hypertensive rats. Jpn Circ J 39: 601-609

Yamori Y (1976a) Interaction of neural and nonneural factors in the pathogenesis of spontaneous hypertension. In: Julius S, Esler M (eds) The nervous system in arterial hypertension. CC Thomas, Springfield, p17-50

Yamori Y (1976b) Neural and non-neural mechanisms in spontaneous hypertension. Clin Sci Mol Med 51: 431s-434s

Yamori Y (1977) Selection of arteriolipidosis-prone rats (ALR). Jpn Heart J 18: 602-603

Yamori Y (1981) Environmental influences on the development of hypertensive vascular diseases in SHR and related models, and their relation to human disease. In: Worcel M, Bonvalet Jp, Langer SZ et al (eds) New trends in arterial hypertension (INSERM Symposium No.17). Elsevier, Amsterdam, p305-320

Yamori Y (1983) Physiopathology of the various strains of spontaneously hypertensive rats. In: Genest J, Kuchel O, Hamet P et al (eds) Hypertension, McGraw-Hill, Montreal, p556-581

Yamori Y (1984a) Development of the spontaneously hypertensive rats (SHR) and of various spontaneous rat models, and their implications. In: de Jong W (ed) Handbook of hypertension, Vol 4: experimental and genetic models of hypertension. Elsevier, Amsterdam, p224-239

Yamori Y (1984b) The stroke-prone spontaneously hypertensive rat: contribution to risk factor analysis and prevention of hypertensive diseases. In: de Jong W (ed) Handbook of hypertension, Vol 4: experimental and genetic models of hypertension. Elsevier, Amsterdam, p240-255

Yamori Y (1988) The salt balance in Asia. In: Rettig R, Ganten D, Luft F (eds) Salt and hypertension, Springer-Verlag, Heidelberg, p319-328

Yamori Y (1989a) Epidemiological studies on the role of magnesium in the pathogenesis and prevention of cardiovascular diseases. In: Itokawa Y, Durlach J (eds) Magnesium in health and disease. John Libbey & Co Ltd, p243-252

Yamori Y (1989b) Predictive and preventive pathology of cardiovascular diseases. Acta Pathol Jpn 39: 683-705

Yamori Y (on behalf of CARDIAC Study Group)(1989c) Hypertension and biochemical dietary markers in urine and blood: a progress report from the CARDIAC Study Group. In: Yamori Y, Strasser T (eds) New horizons in preventing cardiovascular diseases. Elsevier, Amsterdam, p111-126

Yamori Y, Horie R, Sato M et al (1976) Hypertension as an important factor for cerebrovascular atherogenesis in rats. Stroke 7: 120-125

Yamori Y, Ohtaka M, Nara Y, Horie R, Ooshima A, Endo T (1978) Experimental and clinical studies on the relationship between genetic hypertension and glucose metabolism. Shimane J Med Sci 2: 124-132

Yamori Y, Horie R, Ikeda K et al (1979) Prophylactic effect of dietary protein on stroke and its mechanisms. In: Yamori Y, Lovenberg W, Freis ED (eds) Prophylactic approach to hypertensive diseases. Raven Press, New York, p573-580

Yamori Y, Kanbe T, Igawa T et al (1982a) Electrolyte balance in cultured smooth muscle cells from the aorta of SHR. In: Rascher W, Clough D, Ganten D (eds) Hypertensive mechanism: the spontaneously hypertensive rats as a model to study human hypertension. FK Schattauer Verlag, Stuttgart, p281-284

Yamori Y, Kihara M, Fujikawa J et al (1982b) Dietary risk factors of stroke and hypertension in Japan. Part1: Methodological assessment of urinalysis for dietary salt and protein intakes. Part 2: Validity of urinalysis for dietary salt and protein intakes under a field condition. Part 3: Comparative study on risk factors between farming and fishing villages in Japan. Jpn Circ J 46: 933-947

Yamori Y, Kihara M, Nara Y (1982c) Myocardial-ischemic rats (MIR): coronary vascular alteration induced by a lipid-rich diet. Atherosclerosis 42: 15-20

Yamori Y, Wang H, Ikeda K et al (1983) Role of sulfur amino acids in the prevention and regression of cardiovascular diseases. In: Kuriyama K, Huxtable RJ, Iwata H (eds) Sulfur amino acids: biochemical aspects. Alan R Liss Inc, New York, p103-115

Yamori Y, Nara Y, Tagami M et al (1984a) common cellular disposition to hypertension and atherosclerosis. J Hypertension 2 (Suppl 3): S213-S215

Yamori Y, Nara, Y, Kihara M et al (1984b) Simple method for sampling consecutive 24 hour urine for epidemiological and clinical studies. Clin Exp Hypertension A6: 1161-1167

Yamori Y, Nara Y, Tsubouchi T et al (1986) Dietary prevention of stroke and its mechanisms in stroke-prone spontaneously hypertensive rats: preventive effect of dietary fiber and palmitoleic acid. J Hyeprtens 4 (Suppl 3): S449-S452

Yamori Y, Nara Y, Ikeda K et al (1989) Recent advances in experimental studies on dietary prevention of cardiovascular diseases. In: Yamori Y, Strasser T (eds) New horizons in preventing cardiovascular diseases. Elsevier, Amsterdam, 1-11

Hepatic and Gastrointestinal Disorders

Alcohol, Trace Elements, and Liver Dysfunction

J. Aaseth, J. Ringstad, H. Bell, and Y. Thomassen

Hedmark Central Hospital, N-2400 Elverum, Norway

ABSTRACT

The serum concentrations of selenium were consistently decreased in patients suffering fram alcoholic cirrhosis, usually to around 50% of control values. Serum concentrations of zinc and sulphur, as well as of prealbumin and albumin, were almost parallelly decreased in these patients. Interactions of ethanol with the metalbolism of sulphur and selenium are apparently very early events in the process leading to liver disease, as judged from the decreased liver cencentrations of glutathione and selenium. These observations are supposed to reflect an increased vulnerability of hepatocytes toward the lipoperoxidations induced by ethanol consumption.

Key words: Alcoholism, hepatotoxicity, copper, zinc, sulphur.

INTRODUCTION

It has been recognized for several years that alcohol may disturb the balance of certain essential nutrients in the body , for instance zinc (Dreosti 1984).

The pathogenetic mechanisms of alcohol-induced liver disease are not fully known. The possible involvement of some trace metals in the cellular defense mechanisms protecting against ethanol toxicity is of interest, particularly the role of trace metals as components of enzymes protecting against lipoperoxidation.

Ethanol har been found to induce free radical production in liver cells (Valenzuela et al. 1980), although it is still some controversy as to the role of lipid peroxidation as a molecular mechanism of alcoholic liver disease. The microsomal ethanol oxidizing system plays a key role for the activation of molecular oxygen. Hepatic lipoperoxidation has been demonstrated in alcohol-treated baboons by Shaw and coworkers (1981), and in rats given high doses of alcohol, by several investigators (Di Luzio 1973, Valenzuela et al.1980)

Most of the enzyme systems protecting against lipoperoxidations are metalloenzymes, like superoxide dismutases and the selenoenzyme glutathione peroxidase (GSH-PX). The activity of the latter enzyme depends on an adequate supply of sulphur amino acids necessary to synthesize the cofactor glutathione (GSH).

The whole cellular battery affording cellular protection against lipoperoxides and toxic oxygen species also depends on the nutrition as regards other micronutrients such as vitamin E, copper, zinc and manganese (see fig. 1), which will be discussed in the present paper.

SELENIUM

Fig. 2 shows selenium concentrations in serum of a group of patients with alcoholic cirrhosis and other types of liver cirrhosis. The clinical material comprised patients with alcoholic cirrhosis (n=20) admitted to Aker Hospital, Oslo, and other hospitalized patients with chronic active hepatitis (n=20) or primary biliary cirrhosis (n=19). Classification of the patients was based on histological examination of liver biopsy material. Selenium in serum was determined by electrothermal atomic absorption.

The patients with alcoholic cirrhosis had serum selenium decreased to 46% of the control values, which was a more striking reduction than that seen in other types of cirrhoses. Similar results have been reported from Finland (Valimaki et al. 1983), Italy (Casaril et al. 1989) and United States (Dworkin et al. 1985). The latter authors also reported that serum selenium was reduced to about 80% of control values in apparently well-nourished symptom-free alcoholics. Recently, a similar reduction in serum selenium was found in a

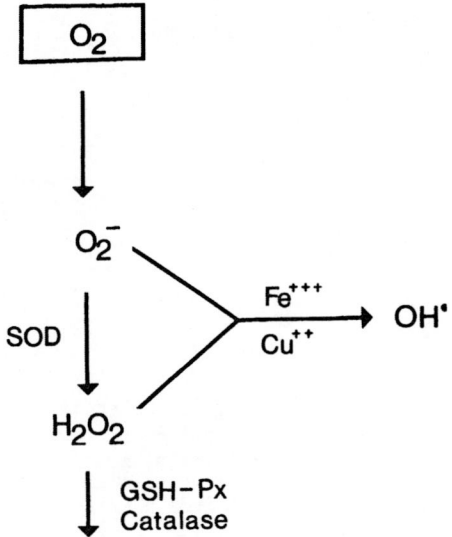

Fig. 1. The microsomal oxidation of ethanol in hepatocytes enhances the conversion of molecular oxygen (O_2) to superoxide (O_2^-). Metalloenzymes denoted SOD (see text) convert the superoxide radicale to the less reactive form H_2O_2, which is further detoxified by the metalloenzymes GSH-PX or catalase. If increased amounts of Fe or Cu occur in ionic forms, they can catalyze the generation of the highly reactive HO$^\bullet$-radicale.

Norwegian group of socially well-adapted and well-nourished alcohol abusers, as will be reported by Ringstad, in an other session here.

Thus, although it is not clear whether low serum selenium is a cause or a consequence of the liver disease, it apparently is a VERY early "event" as compared to other indicators of liver dysfunction. Here, it is also of relevance that patients subjected to intestinal bypass for obesity may develop hepatic lesions quite similar to those present in alcoholics (Peters et al. 1975). In a group of such patients we found serum selenium lowered to about 75% of the mean reference value (Aaseth et al. 1986), and a parallel lowering of the red cell glutathione concentration. Some malabsorption may also be caused by ethanol.

In their groups of alcoholics Dworkin and coworkers (1985) found a significant increase in serum bilirubin and the aminotransferase ALAT in the subgroup with the lower selenium values. An inverse correlation between serum selenium and ɣ-glutamyl transpeptidase was reported by Valimaki et al. (1983). An attractive explanation of these correlations is that selenium as a component of GSH-PX exerts protection against liver cell membrane damage, and this protection is lossed when selenium deficiency is developed. Another interesting relationship, namely between low serum selenium and hepatic fibrosis, has been indicated by Casaril at al. (1989), who reported an inverse correlation between selenium and the amino-terminal procollagen-III peptide in serum (see also Lu at al. 1986). This peptide is reported to be a good indicator of active hepatic fibrosis.

It should be noted that low serum selenium in these cases usually reflects low liver selenium values. The liver concentration of selenium in alcoholics (3.2 ± 0.4 umol/kg) as found in an autopsy material was only 65% of the control values (4.9 ± 1.0 umol/kg) (Aaseth et al. 1982).

Selenium in serum and in liver occurs essentially in proteinbound forms. By using gel filtration (Aaseth et al. 1986) we found that at least half of the serum selenium was recovered in the 50,000 - 100,000 molecular weight range, with an apparent enrichment localized around 80,000. This latter enrichment disappeared from the chromatograms upon development of liver cirrhosis. Although selenium is reported to be bound to nonalbumin proteins, we found the lowering of selenium to be almost parallel to that of albumin, and strongly positively correlated to prealbumin (r=0.86). This has been reproduced by others (Dworkin et al. 1985).

Our clinical observations precipitated animal studies, using experimental groups of rats given ethanol in the diet, making up around 30% of the energy intake, for 6 - 7 weeks, and control groups given isocaloric diets supplied with lipids or carbohydrates. The diets contained similar amounts of selenium. Liver concentrations of selenium in the ethanol-fed rats were reduced to 53 - 57% of control values, which were accompanied by an almost parallel decrease in the measured protein synthesis rate, to around 70% of the control values. These parameters were positively correlated (r=0.54) (Aaseth at al. 1986).

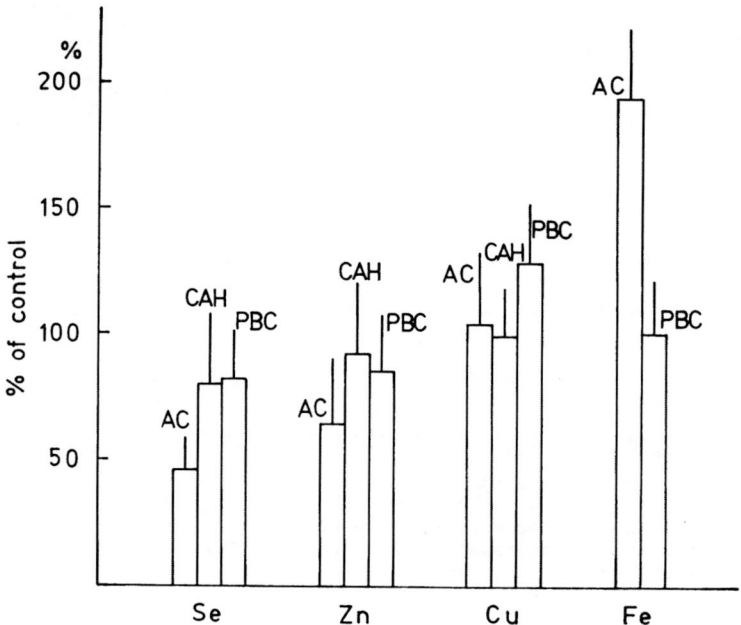

Fig. 2. Serum levels of trace elements, given as percent of average control values, in groups of patients with alcoholic cirrhosis (AC), chronic active hepatitis (CAH), and primary biliary cirrhosis (PBC) (see text).

Some discussion of important sulphur containing compounds may be helpful when trying to interpret our results.

SULPHUR

Sulphur amino acids are essential nutrients that are incorporated into numerous proteins and into glutathione (GSH). The latter tripeptide which is the major intracellular thiol is critical for the activity of GSH-PX.

Strubelt and collegues (1987) found that experimental depletion of GSH from rat liver cells, led to increased ethanol toxicity as judged from the release of the aminotransferase ALAT. Their observations favor the hypothesis that ethanol exerts its hepatotoxic action, at least partly, via activation of molecular oxygen. In accordance with this the ethanol-induced lipoperoxidation which has been observed in the liver of baboons and humans was accompanied by lowering of hepatic glutathione (Shaw et al. 1981 & 1983). In chronic alcohol abusers Jewell et al.(1986) found hepatic GSH lowered <u>before</u> cirrhosis developed. The depletion of GSH induced by ethanol intake is reflected by increased escape into bile of oxidized GSH, formed through the action of GSH-PX (Sies et al. 1979) in addition to the formation of acetaldehyde-GSH conjugates. The depletion of sulphur amino acids following long-term abuse of ethanol has been referred to as the "methionine-wasting effect of alcohol".

Compounds like GSH, cysteine or methionine, and also D-pencillamine, that can restaure or increase the hepatic GSH concentrations, can decrease the acute toxicity of ethanol, at least in mice or rat experiments (Mc Donald et al. 1977, Torielli et al. 1978), again illustrating the critical protective role of the GSH-PX- system.

The "methionine-wasting" effect of ethanol is expected to be reflected by decreased levels of sulphur-containing serum proteins. Recently, we could demonstrate that total serum sulphur in a group of alcoholics with cirrhosis (n=20) was significantly lower than in a control group, the ICP emission spectroscopy technique was used for the sulphur determination. A positive correlation was found between serum sulphur and selenium (r=0.44) in the material (n=68) of patients with alcoholic cirrhosis, fatty lever and other chronic liver diseases.

Ethanol seems to inhibit the incorporation in proteins of both sulphur amino acids and selenium amino acids, but the mechanism of the inhibition is not fully known.

The interactions of alcohol with the sulphur and the selenium metabolism are apparently EARLY events in the development of hepatic dysfunction. Further studies of these interactions may provide a key to deeper insight into hepatotoxicity of ethanol.

ZINC

Increased urinary Zn excretion, combined with poor nutrition, accompanies excessive alcohol intake, and explains the reported decrease in tissue zinc concentrations (Weismann 1985). In a group of patients with alcoholic cirrhosis, we found mean hepatic Zn concentration reduced to 42% of the control values (Aaseth et al. 1982), and serum Zn values were also low in such patients (fig. 2). A positive correlation between albumin and serum zinc has been reported (Weismann 1985). Increased sensitivity of zinc deficient rats to toxic amounts of ethanol (Dreosti 1984) as well as to carbon tetrachloride, indicates that Zn is involved in the cellular protection, probably through a direct stabilization of

biological membranes as suggested by Chvapil (1973). Treatment of rats with ethanol induces increased hepatic levels of the manganese-containing superoxide dismutase (SOD), whereas the Cu/Zn-SOD was not changed significantly (Dreosti 1984). Whether Zn supplementation will improve the results in the clinical treatment of alcoholic cirrhosis remains to be answered, as only very small patient groups have been studied (Reding et al. 1984).

IRON

Excess of iron may lead to accumulation in the liver, as in hemochromatosis or hemosiderosis. Reference values of liver iron are about 800 ug/g dry weight. Around two-fold higher levels are often seen in alcoholic cirrhosis. Ethanol may reduce the transport of iron and of transferrin from liver cells to plasma (Aisen 1985). Pronounced iron accumulation characterizes primary hemochromatosis (Chapman et al. 1982). The mechanism of the liver cell damage in hemochromatosis may involve iron-catalyzed formation of reactive oxygen species.

COPPER

Free copper ions, like iron, can catalyze the formation of oxygen radicals (Aaseth at al.1987), which may contribute to the cirrhosis seen in Wilson's disease and advanced biliary cirrhosis. Alcoholics, however, have almost always normal liver copper concentrations. It is however, of interest to note here that cirrhoses associated with copper accumulation very seldom lead to hepatic malignancy, whereas hepatocellular carcinoma develops in 10 - 20 % of patients with alcoholic cirrhosis. It has been hypothesized that copper protects against hepatic malignancy, whereas selenium depletion may possess a carcinogenic potential.

CONCLUSION

In conclusion, alcohol abuse leads to oxygen radical formation and lipoperoxidation, which may contribute to the hepatotoxicity. This may explain the protective effect of vitamin E and glutathione precursors in animal experiments. The reactive oxygen species may be toxic mediators capable of depressing or distorting the protein syntesis, leading to low serum levels of selenoproteins and sulphur-containing proteins, whereas the collagen formation is increased. It is still unclear if the low serum selenium seen in alcohol abusers accelerates the progression of cirrhosis, and also if these selenium values make a potential for malignant transformation of the liver cells.

REFERENCES

Aaseth J, Alexander J, Thomassen Y, Blomhoff JP, Skrede S (1982) Serum selenium in liver
diseases. Clin Biochem 15: 281-283
Aaseth J, Smith-Kielland A, Thomassen Y (1986) Selenium, alcohol and liver diseases. Ann
Clin Res 18:43-47
Aaseth J, Bochev PG, Ribarov SR (1987) The interactions of copper (Cu++) with the
erythrocyte membrane and 2,3-dimercaptosulphonate in vitro - a source of activated
oxygen species. Pharmacol Toxicol 61: 250-253
Aisen P (1985) Transferrin and the alcoholic liver. Hepatology 5: 902-903
Casaril M, Stanzial AM, Gabrielli GB, Capra F, Zenari L, Galassini S, Moschini G, Liu NQ,
Corrocher R (1989) Serum selenium in liver cirrhosis: correlation with markers of
fibrosis. Clin Chim Acta 182: 221-228
Chapman RW, Morgan MY, Bell R, Sherlock S (1983) Hepatic iron uptake in alcoholic liver
disease. Gastroenterology 84: 143-147
Chvapil M (1973) New aspects in the biological role of zinc: a stabilizer of
macromolecules on biological membranes. Life Sci 13: 1041-1049
Di Luzio NR (1973) Antioxidants, lipid peroxidation and chemical- induced liver injury.
Fed Proc 1973: 287-290
Dreosti EE(1984) Interactions between trace elements and alcohol in rats. Ciba Found Symp
105: 103-123
Dworkin B, Rosenthal WS, Jankovski RH, Gordon GG, Haldea D. Low blood selenium levels in
alcoholics with and without advanced liver disease. Dig Dis Sci 30: 838-844
Jewell SA, DiMonte D, Gentile A, Guglielmi A, Altomare E, Olbano O (1986). Decreased
hepatic glutathione in chronic alcoholic patients 3: 1-6
Lu W, Bantok I, Desai S, Lloyd B, Hinks L, Tanner AR (1986) Aminoterminal procollagen III
peptide elevation in alcoholics who are selenium and vitamin E deficient. Clin shim
Acta 154: 165-170
Mc Donald CM, Dow J, Moore MR (1977) A possible protective vole for sulphhydryl compounds
in acute alcoholic liver injury. Biochem Pharmacol 26: 1529-1531
Peters RL, Gay L, Reynolds TB (1975) Postjejunoileal bypass hepatic disease. Its
similarity to alcoholic hepatic disease. Am J Clin Pathol 63: 318-331
Reding P, Duchateau J, Bataille C (1984) Oral Zinc supplementation improves hepatic
encephalopathy Lancet 2: 493-494
Shaw S, Jayatilleke E, Ross WA, Gordon ER, Lieber CS (1981) Ethanol-induced lipid
peroxidation: potentiation by longterm alcohol feeding and attenuation by
methionine. J Lab Clin Med 98: 417-424
Shaw S, Rubin KP, Lieber CS (1983) Depressed hepatic glutathione and increased diene
conjugation in alcoholic liver disease, evidence of lipid peroxidation.
Dig Dis Sci 28: 585-589.
Sies H, Koch OR, Martino E, Boveris A (1979) Incrased glutathione disulphide release in
chronically ethanol-treated rats. FEBS Lett 103: 187-290
Strubelt O, Younes M, Rentz R (1987) Enhancement by glutathione depletion of
ethanol-induced acute hepatic toxity in vitro and in vivo. Toxicology 45: 213-223
Torielli MV, Gabriel L, Dianzani MU (1978) Ethanol-induced hepatotoxicity, experimental
observations on the role of lipid peroxidation. J Pathol 126: 11-25
Valenzuela A, Fernandez N, Ugarte G, Videla LA (1980) Effect of acute ethanol ingestion on
lipoperoxidation and on the activity of the enzymes related to peroxide metabolism in
rat liver. FEBS Lett 111: 11-13
Valimaki MJ, Harju KJ, Ylikahri RM (1983) Decreased serum selenium in alcoholics - a
consequence of liver dysfunction. Clin Chim Acta 130: 291-296
Weismann K (1985) Zinc metabolism in alcoholic cirrhosis. In: Bostrøm H, Ljungstedt N.
Trace Elements in Health and Disease, Almquist & Wiksell, Stockholm, pp 221-241

Regional Variations of Eight Trace Elements in Normal Human Liver

MOHAMED MOLOKHIA[1] and ANAT MOLOKHIA[2]

[1] Department of Dermatology, Royal Liverpool Hospital, Liverpool, UK
[2] Medical Department, Mersey Regional Health Authority, Liverpool, UK

KEY WORDS: Human liver, Trace elements, Neutron activation analysis

INTRODUCTION

The liver plays a major role in trace element metabolism and is closer to a single cell type than many other organs in the body. It has also a high storage capacity for essential elements like iron and copper which may cause disease in overload states. It is conceivable that a similar effect may result from accumulation of other elements in excessive amounts.

Most of the trace element studies in liver disease have been concerned with levels in serum or urine which do not necessarily reflect tissue levels. The use of needle biopsy of liver has been invaluable in the identification of disease processes and has provided more accurate assessment of overload states in that organ. The usual method uses the percutaneous route through the right intercostal spaces and the specimen is taken from the right lobe.

Earlier studies by Lievens and Co-Workers (1977) have shown that the coefficient of variation for non-essential trace elements within the liver was much higher than that for essential ones. Others, like Perry et al. (1977) have indicated that the choice of a sampling site within the liver was not important.

The purpose of this study was to assess the normal levels of some trace elements, both essential and non-essential, and to test the hypothesis that a single small liver sample can be used to give a representative picture of the trace element status of the whole organ.

MATERIALS AND METHODS

Autopsy liver samples were obtained from 30 adults (15 males and 15 females) dying from causes other than liver disease in the North Western Region of England. The majority died suddenly as a result of coronary heart disease or road traffic accident (90%). Their ages ranged from 19 to 86 with a mean of 64 years. Liver samples were taken from 4 anatomical sites; namely the upper part of right lobe, lower part of right lobe, left lobe and quadrate lobe. Stainless steel instruments were used to cut pieces weighing approximately 2.0 g from each of the above sites and an effort was made to exclude visible inhomogeneities like blood vessels or ligaments as much as possible. Samples were then dried at 70° C until they reached a steady weight. Dried liver samples weighing between 50 and 100 mg were individually wrapped in aluminium foil of analytical grade. Standard reference materials such as orchard leaves (NSB) and bovine liver as well as standards of single elements were irradiated along with liver samples for 30 hours in the Universities Research Reactor at Risley in a thermal neutron flux of 2.10 neutrons/cm^2/s. Following a cooling period to allow the decay of short lived activities, samples and standards were transferred to polythene tubes and their gamma spectra were counted by a Ge(Li) detector connected to a 4096 multichannel analyser system. Computer programmes were used to calculate the concentration of cobalt, caesium, iron, rubidium, antimony, scandium, selenium and zinc in 120 liver samples.

RESULTS

The concentrations of the 8 trace elements studied are summarised in Table 1.
There were no significant differences between the mean concentrations of these
elements in the two sexes and data were therefore grouped together. Apart from
Sc and Sb all other elements were detectable in all the samples analysed.

There were no significant differences between the concentrations of elements
in upper and in lower parts of the right lobe. However, the left lobe was
found to contain significantly higher levels of Fe (P<0.001) and Co (P<0.02)
and significantly lower levels of Cs (P<0.001).

Positive correlations were found between the concentrations of some elements
and those which reached significance are shown in Table 2. Negative
correlation was found only between Cs and Sb. Wide variations were noticed
between trace element concentrations in different livers and these were most
noticeable for Sb and Fe and least for Rb and Se as shown in Table 3. Although
individual livers showed also variability of trace element content in
different regions, this was much smaller than that between livers.

Table 4 shows that the concentrations of Fe and Rb were negatively correlated
with age.

DISCUSSION

The concentrations of the 8 trace elements in liver found in this study are
comparable to those obtained by other workers (Lievens et al. 1977, Milman et
al. 1986, and Subramanian et al. 1985) studying different populations. Our
results indicate that although the variability in trace element concentrations
in the 4 sites within a liver is generally much less than variability between
different livers, it can sometimes be significant; such as for Fe, Co, Rb and
Cs. The left lobe was found to have the highest concentrations of Fe, Zn, Se,
and Rb.

Needle liver biopsy is normally taken from the right lobe which is about 6
times larger than the left lobe. This makes any differences between the 2
lobes less significant in assessing the metal content of the liver as a whole.

TABLE 1

TRACE ELEMENT CONCENTRATIONS IN DIFFERENT ANATOMICAL LIVER SITES

Element	No. of Samples	Concentrations in $\mu g.g^{-1}$ dry weight (Mean $\pm$ S.D.)			
		Rt. lobe Upper	Rt. lobe lower	Lt. lobe	Quadrate lobe
Iron	30	846 $\pm$ 671	880 $\pm$ 596	1011 $\pm$ 794	935 $\pm$ 766
Zinc	30	242 $\pm$ 117	236 $\pm$ 93	249 $\pm$ 104	236 $\pm$ 85
Selenium	30	1.52 $\pm$ 0.42	1.54 $\pm$ 0.40	1.62 $\pm$ 0.41	1.58 $\pm$ 0.32
Cobalt	30	0.17 $\pm$ 0.09	0.20 $\pm$ 0.09	0.21 $\pm$ 0.10	0.22 $\pm$ 0.08
Rubidium	30	21.45 $\pm$ 5.94	19.78 $\pm$ 6.15	22.32 $\pm$ 7.46	21.08 $\pm$ 6.55
Caesium	30	0.09 $\pm$ 0.06	0.07 $\pm$ 0.05	0.06 $\pm$ 0.03	0.05 $\pm$ 0.02
Scandium	16	0.006 $\pm$ 0.003	0.008 $\pm$ 0.005	0.008 $\pm$ 0.005	0.007 $\pm$ 0.006
Antimony	14	0.11 $\pm$ 0.25	0.07 $\pm$ 0.16	0.11 $\pm$ 0.23	0.12 $\pm$ 0.20

TABLE 2

SIGNIFICANT CORRELATIONS BETWEEN TRACE ELEMENT CONCENTRATIONS IN LIVER (P<0.05)

Metal pair	No. of Samples	Correlation Co-efficient
Se : Zn	30	0.421
Fe : Rb	30	0.417
Co : Se	30	0.401
Rb : Se	30	0.400
Cs : Se	22	0.467
Cs : Rb	30	0.403
Co : Zn	30	0.401
Cs : Sc	16	0.670
Sc : Zn	16	0.631
Sc : Se	16	0.495
Fe : Zn	30	0.323
Cs : Sb	14	-0.705

TABLE 3

QUOTIENT OF HIGHEST/LOWEST VALUES OF TRACE ELEMENT CONCENTRATIONS IN INDIVIDUAL LIVERS AND IN ALL LIVERS

Element	Individual Liver	All livers
Iron	3.5	34
Zinc	2.2	7
Selenium	1.5	3
Cobalt	2.6	11
Rubidium	1.4	4
Caesium	3.0	11
Scandium	2.7	19
Antimony	3.2	104

TABLE 4

CORRELATION BETWEEN AGE AND TRACE ELEMENT CONCENTRATIONS IN LIVER SAMPLES

Element	Correlation Co-efficient	P
Iron	-0.353	0.028
Rubidium	-0.390	0.016

The most variable trace element concentrations between livers were those of Sb and Fe. While this is not surprising for Sb, a non essential element thought to be an environmental contaminant, the results of Fe variability were unexpected. These wide variations in Fe concentrations may however reflect the potential for liver as a depot for this metal. Selenium and zinc on the other hand show a much narrower range of concentrations and this would indicate that the liver does not hold any significant stores of these two essential elements.

SUMMARY

* Instrumental neutron activation analysis was used to measure the concentrations of Co, Cs, Fe, Rb, Sb, Sc, Se, and Zn in 4 anatomical sites of normal liver.

* There was a wide variation in concentrations especially with regard to Fe and Sb. Anatomical sites showed much less variability although this could be as high as 3 fold.

* Significantly positive correlations were found between Zn and Se and between Fe and Rb, while negative correlation was found between Sb and Cs.

* There was a significant decrease in Fe and Rb concentrations with age.

ACKNOWLEDGEMENT

Preliminary results of this study was part of a Ph.D. thesis, submitted to Salford University, by one of the authors (A. Molokhia). The authors are grateful to the generous assistance of Dr. G. R. Gilmore and all the staff at URR, Risley, U.K.

REFERENCES

1- Lievens P, Versieck J, Cornelis R, and Hoste J. (1977)
   The distribution of trace elements in normal human liver determined by semi-automated radiochemical neutron activation analysis. J. Radioanal. Chem. 37:483.

2- Milman N, Laursen J, Podenphant, and Asnaes S. (1986)
   Trace elements in normal and cirrhotic human live tissue. I- Iron, copper, zinc, selenium, manganese, titanium and lead measured by X ray fluorescence spectometry. Liver. 6:111-77.

3- Perry H M, Perry E F, and Hixon BB. (1977)
   Trace metal concentrations in normal human liver: methods to cope with marked variability. Sci. Total Environ. 9:125.

4- Subramanian K S, Meranger J C, and Burnett R T (1985)
   Kidney and liver levels of some major, minor and trace elements in two Ontario communities. Sci. Total Environ. 42:223

The Role of Copper and Zinc as Pathogenic Factors in Liver Disease

A. Sawa and K. Okita

First Department of Internal Medicine, School of Medicine, Yamaguchi University, 1144 Kogushi, Ube, 755 Japan

ABSTRACT

In order to clarify the roles of copper and zinc in a progress of chronic liver diseases, levels of copper and zinc in both sera and liver tissues were measured. As a result, serum concentration of zinc decreased significantly, in accordance with an agravation of liver disease, while serum and hepatic levels of copper increased. The ratio of serum copper to zinc elevated significantly in parallel with a progress of liver disease. Patients with the ratio exceeding 2.0 generally suffered from cirrhosis or chronic active hepatitis. This ratio co-incided with changes of liver function test which reflected liver fibrosis and residual liver function.

Key words; copper to zinc ratio, chronic liver disease, liver fibrosis

INTRODUCTION

Recently, various studies concerning the role of trace elements in liver disease were sccumulated. Trace elements are well known as an important components of enzymes and other proteins, and the liver is a center organ of metabolism of such proteins. Moreover, it has been shown that the liver is the main site of storage and excretion of these trace elements. From these point of view, it seems extremely important to investigate the relationship between the liver disease and trace elements.

MATERIALS AND METHODS

The subjects were 71 patients with chronic liver disease whose diagnosis were confirmed histologically, but in some cirrhotic patients, their diagnosis were introduced only by bio-camical examinations and imaging studies. There were 20 patients of cirrhsis(LC), 25 of chronic active hepatitis(CAH), and 26 of chronic inactive hepatitis(CIH). In these patients, serum Zn and Cu concentrations were measured in the blood samples collected early in the morning while fasting. As a normal controls(NC), levels of Zn and Cu were also determined in 20 healthy adults having no history of liver disease or blood transfusion. In 58 subjects out of 71 patients(14 LC, 16 CAH, 19 CIH and 9 NC), the relationship between serum Cu and Zn concentration and various liver function tests were studied (Table 3).

Liver tissue was obtained by needle biopsy from 19 patients with chronic viral liver disease. There were 5 patients of LC, 8 of CAH, and 6 of CIH. As a normal controls, liver specimen was obtained by liver biopsy in the surgical operation from 4 patients with no histological finding. The levels Zn and Cu in the liver were determined using these samples.

The serum concentration of Zn was measured by atomic absorption spectroscopy by adding 2.25 ml of 0.1N hydrochloric acid solution containing 6% n-butanol to 0.25ml of serum. To measure the serum Cu level, 1.0ml of 0.8N HCI and 1.0ml of 10% TCA added to 0.5ml of serum. After centrifugation for 5 minutes at 3,000 rpm, the Cu concentration of supernatant was examined.

We also measured the Cu and Zn content of liver tissue. The tissue samples were cleaned from blood, and dried at 60℃ for 24 hours in a test tube, and the dry weight of the sample was measured. After adding 1.0ml of nitric acid, the sample was heated at 50℃ for 1 hour, and it was again ignited and vaporized at 85℃ . Sample was then dissolved in 5.0ml of pure water. Thereafter, the Cu and Zn levels in this solution was measured by atomic absorption spectroscopy.

RESULTS

1) Zn levels in both serum and liver tissue in normal subjects and patients with chronic liver disease (Table 1, Table 2): Serum level of Zn fell significantly with a progression to cirrhosis. On the other hand, patients with LC tended to have a lower hepatic Zn content, as compared with CIH and CAH, but the statistical difference was not significant.

2) Relationship between serum Zn concentration and serum vareous liver function tests in normal subjects and patients with chronic liver disease (Table 3) : Serum Zn concentration was significantly correlated with the levels of serum albumin, prealbumin, cholesterol and choline esterase. On the other hand, serum concentration of Zn had a significant negative correlations with the serum levels of globulin, γ-globulin, P Ⅲ P, laminin and ICG R15.

3) Cu levels in both serum and liver tissue in normal subjects and patients with chronic liver disease (Table 1, Table 2) : Both serum and hepatic Cu levels in cirrhosis were significantly higher as compared with normal controls.

4) Relationship between serum concentration of Cu and serum liver function tests in normal subjects and patients with chronic liver disease (Table 3) : Between serum levels of Cu and albumin, a negative correlation was noted, and there was a positive correlation with ceruloplasmin, globulin and γ-globulin. But, no significant correlation was detevted with the markers of cholestasis.

5) Cupper to zinc(Cu/Zn) ratio in both serum and liver tissue in normal subjects and patients with chronic liver disease (Fig.1) : Cu/Zn ratio in serum rose significantly with a progress of liver disease. Most of cases with the ratio over 2.0 were patients with LC or CAH. Cu/Zn ratio in liver increased significantly according to an aggravation of liver disease. In LC group, the ratio was over 0.40.

6) Relationship between serum Cu/Zn and serum liver function tests in normal subjects and patients with chronic liver disease (Table 3, Table 4) : Serum Cu/Zn ratio presented a negative correlation with serum albumin and cholesterol levels. On the other hand, it had positive correlations with serum globulin, γ-globulin, P Ⅲ P and laminin levels. Incidentally, the ratio showed a positive correlation with ICG R15. In 15 patients out of 17 cases with the ratio exceeding 2.0 (88.2%), ICG R15 was 15% or more. In contrast, in 12 out of 14 cases with less than 15% of ICG R15 (85.7%), the ratio was 2.0 or less.

Table 1. Serum concentration of Zn & Cu in patient with chronic liver disease

Diagnosis	Serum concentration (μg/dl)	
	Zn (Mean $\pm$ SD)	Cu (Mean $\pm$ SD)
NC (n=20)	96.4$\pm$17.6	96.5$\pm$11.0
CIH (n=26)	78.4$\pm$19.5	102.2$\pm$24.5
CAH (n=25)	61.1$\pm$16.6	103.4$\pm$27.1
LC (n=20)	54.1$\pm$13.3	120.0$\pm$33.0

* p<0.05, **p< 0.01, ***p< 0.001

Table 2. Liver content of Zn Cu in patient with chronic liver & disease

Diagnosis	Liver content (μg/g.d.w.)	
	Zn (Mean $\pm$ SD)	Cu (Mean $\pm$ SD)
NC (n=4)	211.6 $\pm$ 42.6	21.9 $\pm$ 4.6
CIH (n=6)	269.8 $\pm$109.3	47.4 $\pm$31.1
CAH (n=8)	187.3 $\pm$ 66.7	60.3 $\pm$30.1
LC (n=5)	166.8 $\pm$ 22.2	80.1$\pm$25.3

** p< 0.01, d.w.=dry weight

Table 3. Correlations between Zn, Cu or Cu/Zn in the serum & various parameters

	No.	Zn	Cu	Cu/Zn		No.	Zn	Cu	Cu/Zn
Albumin	58	0.364**	-0.268*	-0.565**	T.bili.	58	-0.316*	NS	NS
Prealbumin	48	0.554**	NS	NS	γ-GTP	58	NS	NS	NS
Ch-E	58	0.465**	NS	NS	ALP	58	-0.399**	NS	0.333*
Chole.	58	0.316*	NS	-0.395**	TBA	58	-0.467**	NS	NS
Cp.	50		0.716**		P Ⅲ P	55	-0.575**	NS	0.477**
Globulin	58	-0.328*	0.393**	0.306*	Laminin	56	-0.433**	NS	0.433**
γ-Globulin	48	-0.483**	0.386**	0.471**	PH	50	NS	NS	NS
GOT	58	NS	NS	NS	ICG R15	35	-0.496**	NS	0.392*
GPT	58	NS	NS	NS					

*p<0.05, **p< 0.01,Abbreviations:NS=not significant;Ch-E=choline esterase;Cp=ceruloplasmin; Chole=eholesterol;T.bili=total bilirubin;TBA=toral bile acids;PⅢP=procollagen Ⅲ peptide; PH=prolyl hydroxylase.

Table 4. Relationship between serum
Cu/Zn & ICG R15 in chronic liver disease

ICG R15	serum Cu/Zn		Total no. of patients
	< 2.0	≧2.0	
< 15%	12	2	14
≧15%	6	15	21
Total no.	18	17	35

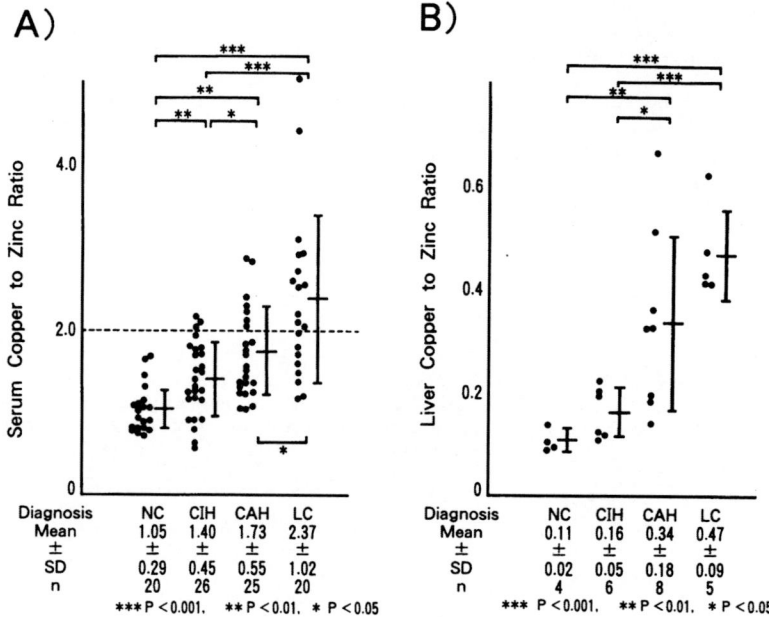

Fig.1 The Cu/Zn in both serum & liver
tissue in patients with chronic liver disease.

DISCUSSION

The serum concentration of Zn decreased significantly, when liver disease advanced from CH to LC. And, a significant positive correlation was found between serum Zn level and serum level of albumin, choline esterase and cholesterol. Zn is present in the blood as a conbined substance with macromlecular ligands, such as albumin and α_2-macroglobulin, and with micromolecular ligands, such as amino acids and polypeptides. Nearly 70% of serum Zn binds with albumin, majority of the remainder is bonded with α_2-macrogloblin, and only a very small portion of Zn bonded with micromolecular ligands[3]. In cirrhotic patients, binding substance of Zn changes from macromolecular ligands to micromolecular ligands, because of decrease of serum albumin level. Increase of Zn excretion into urine is ragarded as a major factor in the lowering of serum Zn level[1,4,6]. Our results are consistent with this explanation. Our data suggested that the lowering of the serum Zn level in patients with liver disease might be partly caused by their poor nutritional state, due to decreased oral intake of food and impaired absorption of nut rients from the intestines. On the other hand Schölemrich, et al (1983) mentioned that the Zn deficiency in patients with liver disease is mainly due to increased portosystemic shunting and reduced hepatointestinal extraction[5]. In this sutdy, negative correlation between ICG R15 and serum Zn level suggested the presence of portosystemic shunts related to the decrease of Zn levels.

In patients with LC,Cu levels in both sera and liver tissue were significantly higher than those in normal subjects. The main pathway of Cu excretion is into bilialy tract. In patients with sevier and long-term obstructive jaundice or primary biliary cirrhosis, serum and hepatic Cu level is increased. Correlations have been noted between the serum Cu level and serum markers of cholestasis among cirrhotic patients[6]. But, in our study, there was no significant relation between serum Cu level and these markers. This means no patient with sevier jaundice in our cases, or suggests that the higher level of Cu concentration might be due to not only cholestasis, but also some other conditions.

Zn and Cu are thought to be closely related with liver fibrosis. As an example, lysly oxidase which responds to closslinking of covalent bonds in collagen peptides prossesses Cu as a coenzyme. But, Zn is known to suppress these procedures[2,4]. Moreover, Zn is necessary to keep the activity of collagenase and its maintenance[6]. As serum Zn level showed significant negative correlations with serum globulin, γ-globulin, P Ⅲ P and laminin, it seemed that Zn might be related with supression for liver fibrosis. On the other hand, as serum Cu level was positively correlated with globulin and γ-globulin, copper also appeared to be related with liver fibrosis.

Changes of Cu and Zn levels in both liver and serum with an aggravation of liver disease were in reverse. Cu and Zn may be competitive in their absorption and also bonding with metallothionein[6]. Additionally, Cu and Zn showed an acomplicated relationship with the liver fibrosis. Then, the serum Cu/Zn ratio significantly increased with a progress from CH to LC. In 90% of NC group, the ratio was below 1.5. But, majority of cases with the ratio over 2.0 was occupied by the patients with LC or CAH. Furthermore, all of cases exceeding 3.0 were patients with LC. Serum Cu/Zn ratio presented a negative correlation with the items associated with the synthesizing capacity of the liver. While, it showed a positive correlations with markers of liver fibrosis. In a correlation with ICG R15, this ratio closely related. As a conclusion, Cu/Zn ratio appears to be a useful marker for estimating the stage of the progress of chronic liver disease.

REFERENCES

1. Boyett JD, Sullivan JF (1970) Distribution of protein-boud zinc in normal and cirrhotic serum. Metabolism 19:148-157
2. Burch RE, Hahn HKJ, Sullivan JF (1978) Zinc deficiency and cirrhotic process. In : Powell LW (ed) Metals and the liver, vol.1. Marcel Decker, New York and Basel, 342-355
3. Kameda Y (1985) Zinc and disease of the liver. Kan Tan Sui(Japan) 10:915-912
4. Nagai K, (1988) Studies of metabolic abnormality of zinc in chronic liver disease. Japanese Journal of Gastroenterology 85:2618-2623
5. Schölmerich J, Becher MS, Köttgen E, Rauch N, Häussinger D, Löhle E, Gerok.W, Vuilleumier JP, (1983) The influence of portosystemic shunting on zinc and vitamine A metabolism in liver cirrhosis. Hepato-gastroenterol 30:143-147
6. Wakiyama K, (1987) A pathophysiological significance of trace metals in rats with experimental liver cirrhosis. Japanese Jour. of Gastroenterology 84:27-35

Zinc and Gastrointestinal Diseases

CRAIG J. MCCLAIN and LUIS MARSANO

Division of Digestive Diseases and Nutrition, Room MN 654, University of Kentucky Medical Center, Lexington, KY 40536-0084, USA

ABSTRACT

Zinc is an essential trace element whose metabolism is frequently disturbed in gastrointestinal diseases. Zinc deficiency may manifest in different ways, e.g. crusting skin lesions, anorexia, hypogonadism, or altered immune function. Mechanisms for this altered zinc metabolism include stress-related redistribution, poor dietary intake, impaired absorption and increased excretion through the urinary and GI tract. Thus, it is important that health care providers be aware that altered zinc metabolism is a frequent and correctable problem in many gastrointestinal diseases. Furthermore, this zinc deficiency may manifest itself in multiple different ways and challenge the diagnostic acumen of the health care provider.

KEY WORDS

Zinc, gastroenterology, intestine, zinc deficiency

INTRODUCTION

At the first meeting of the International Society for Trace Element Research in Humans, we reviewed zinc and the gastrointestinal system. In order not to duplicate that manuscript, we will attempt to provide an overview of new concepts in zinc and gastrointestinal diseases. We will review: 1) GI disease processes discussed at the initial meeting, 2) manifestations of zinc deficiency, 3) two processes added to the list, and 4) mechanisms of zinc deficiency.

Zinc is an essential trace element required for protein synthesis and the function of over 200 zinc metalloenzymes. The RDA for zinc is 15 mg for males and 12 mg for females. Approximately 1/3 of ingested zinc is absorbed. The amount of ingested zinc that is absorbed varies with factors including competing influences such as phytic acid, zinc status of the subject, and underlying intestinal diseases which may impair zinc absorption or increase intestinal zinc loss. Zinc is absorbed mainly from the small intestine, with colonic absorption being negligible. In studies performed in stable patients receiving TPN, it appears that approximately 3 mg of zinc daily is required intravenously. Zinc is excreted mainly in the feces, with urinary zinc excretion normally being negligible (< 600 ug/24 hr). Zinc is normally bound in the plasma; approximately 70% loosely to albumin, 25% more tightly bound to alpha-2 macroglobulin, and the remaining bound to peptides and amino acids.

GI DISEASES

There are a host of gastrointestinal diseases or processes having altered zinc metabolism. Below is the list of diseases that was presented at the meeting in 1986. New diseases or processes that have been added are highlighted in italics at the bottom.

Gastrointestinal Diseases or Processes with Altered Zinc Metabolism

---

Intestinal processes
        Acrodermatitis enteropathica
        Inflammatory Bowel (Crohn's, Ulcerative Colitis)
        Short-bowel syndrome
        Jejunoileal bypass
        Sprue
        Nonspecific diarrhea
    Pancreatic disease
        Alcoholic pancreatic insufficiency
        Cystic fibrosis
        Shwachman's syndrome
Acute alcoholic pancreatitis
    Liver disease
        Alcoholic hepatitis or cirrhosis
        Viral hepatitis and Cryptogenic cirrhosis
        Primary biliary cirrhosis
    Total parenteral nutrition
    SEPSIS, TRAUMA, MULTISYSTEM ORGAN FAILURE,
    EATING DISORDERS (ANOREXIA NERVOSA, BULIMIA NERVOSA),
    AGING

MANIFESTATION OF ZINC DEFICIENCY

Much of our knowledge concerning the metabolic role of zinc in man is derived from manifestations of zinc deficiency either in zinc deficient animals or in patients with acrodermatitis enteropathica (a hereditary disease of impaired zinc absorption) or in patients with acquired zinc deficiency due to an underlying disease process. Ten major functional consequences of zinc deficiency are presented below, and clinical examples have been presented in previous reviews (McClain 1988; McClain 1985).

Functional Consequences of Zinc Deficiency

---

1. Skin lesions
2. Anorexia (with possible alterations in taste and smell acuity)
3. Growth retardation
4. Depressed wound healing
5. Hypogonadism
6. Altered immune function
7. Impaired night vision, altered Vitamin A metabolism
8. Diarrhea
9. Depressed mental function
10. Teratogenesis

It is clear that clinical and biochemical zinc deficiency usually occurs when some type of stress is placed on the organism. Otherwise, man appears to adapt to differing amounts of zinc intake, often even in the presence of competing factors such as phytic acid. An example of this is a study done by Dr. Essatara, in collaboration with our group, of healthy, free living Moroccan subjects. The mean Moroccan zinc intake was 5-8 mg a day (thus only approximately 1/3-1/2 of the RDA), and their diet had a very high phytic acid content (approximately 2 gm). However, there was only limited biochemical or physical evidence of zinc deficiency in these subjects. Thus, it is interesting that even though these patients had markedly decreased zinc intake and had reasons to absorb zinc poorly, they seemed to be in amazingly good zinc status. This supports the concept that people often do well as regards zinc status until some "stress state" tips them over into zinc deficiency. Many gastrointestinal diseases fall into the category of "stress states". We will review two new diseases or processes associated with zinc deficiency: eating disorders and the elderly, and we will cover multiple trauma in the discussion of mechanisms for zinc deficiency. We will then review potential mechanisms for the development of zinc deficiency.

Zinc and Eating Disorders, Aging

The first new disease processes are eating disorders. We initially evaluated 86 inpatients and outpatients undergoing treatment in our Eating Disorder Unit using fasting serum levels and 24 hr urine zinc excretion (Humphries, in press). No patient was receiving zinc therapy at the time of the study or received zinc supplementation for the month preceding this study. A total of 22 female and 2 male anorexia nervosa patients were studied, with a mean age of 17. A total of 61 female and 1 male bulimia nervosa patients were studied with a mean age of 20 years. Normal serum zinc concentrations in our laboratory are 70-120 ug/dl and 24 hr urine zinc excretion 200-700 ug/24 hr. Patients were defined as possibly zinc deficient if they had a serum zinc concentration of < 70 ug/dl and/or a 24 hr urine zinc excretion < 200 ug/24 hr. Of the 24 patients with anorexia nervosa, 6 had serum zinc concentrations below 70 ug/dl and 13 of the 62 bulimics had depressed serum zinc concentrations. Nine of the 24 patients with anorexia nervosa had urine zinc excretions below 200 ug/24 hr and 14 of the 62 bulimics had decreased 24 hr urine zinc excretions. Overall, 13 of 24 anorectic patients (54%) and 25 of 62 bulimic patients (40%) met biochemical criteria for zinc deficiency. This initial study prompted an ongoing prospective randomized study of zinc supplementation versus placebo in patients with eating disorders. Thusfar, we have entered 33 patients in this study, 15 with anorexia and 18 with bulimia. Patients were all hospitalized in an Eating Disorder Unit for one month and were randomized to either oral zinc supplementation or placebo. As with the initial study, several patients had biochemical evidence of zinc deficiency on admission. However, the most important data demonstrate that anorexia nervosa patients in the placebo group all had a 24 hr urine zinc excretion under 250 ug/24 hr on discharge. Thus, during aggressive refeeding with a regular hospital diet and enteral nutritional support products, the stress of refeeding and anabolism precipitated biochemical zinc deficiency. This zinc deficiency theoretically could provide a biochemical mechanism for continued anorexia and altered eating habits and initiate a vicious cycle. In conclusion, it is clear that many patients with eating disorders have altered zinc metabolism that should be monitored and treated in appropriate situations.

The vast majority of studies performed in the elderly demonstrate depressed zinc intake, both in those living in the community and those in continuing care nursing home situations (Field 1987; Bunker 1984). Zinc intake correlates directly with protein intake, and in the elderly protein intake may be low due to inadequacy of meal size and food quantity. Studies of zinc absorption have shown depressed uptake in the elderly. However, one study also showed a compensatory decrease in endogenous losses of zinc (Turnland 1986). The elderly frequently use diuretics which may enhance renal loss of zinc. A variety of the previously discussed manifestations of zinc deficiency may relate to the elderly, such as impaired wound healing after surgery and impaired immune function. Possibly the most interesting problem relating to zinc deficiency in the elderly is that of macular degeneration. We previously noted that zinc is important for normal night vision (McClain 1985). Our group has shown that severe zinc deficiency can cause anatomic defects in the retina (McClain 1985). A recent investigation by Newsome and coworkers showed that zinc supplementation was helpful in preventing macular degeneration (Newsome 1988). Macular degeneration is the major cause of vision loss in the elderly in this country and thus of great clinical consequence to the elderly. This was a prospective, randomized double-masked, placebo-controlled study of the effects of zinc supplementation in 151 patients with drusen or macular degeneration (Newsome 1988). Although some patients in the zinc treated group lost vision, this group had significantly less vision loss than the control group at 12 and 24 months. Men had less vision loss than women for unexplained reasons. Although patients were given zinc sulfate 100 mg bid, there was no significant increase in the serum zinc level over the study period, an observation of some concern. Also of concern was the relatively severe vision loss in the control group which enhanced the positive findings of the zinc therapy group. This important study requires confirmation by other laboratories.

MECHANISMS FOR ZINC DEFICIENCY

Mechanisms for zinc deficiency can be divided into 2 main categories: 1) redistribution and 2) true deficiency due to traditional mechanisms such as poor dietary intake, increased excretion, etc. Redistribution is an area of increasing recognition whose importance needs to be further clarified. Clearly, the hypozincemia and altered zinc metabolism in a variety of gastrointestinal states is due to inflammation with a redistribution of zinc. This would occur in GI diseases such as inflammatory bowel disease or a variety of types of hepatitis including alcoholic hepatitis (McClain 1986a). This redistribution of zinc is mediated by the cytokines such as interleukin-1 (IL-1), tumor necrosis factor (TNF) and probably most importantly interleukin-6 (IL-6) (and probably to a lesser extent the hormone ACTH). Cytokines injected into experimental animals or man cause a marked decrease in the serum zinc concentration and an internal redistribution of zinc, with zinc going into certain tissues such as the liver. A variety of diseases or processes seen by gastroenterologists show increased levels of cytokines which mediate the inflammatory response and hypozincemia. For example, we have recently reported increased monocyte TNF production in patients with alcoholic hepatitis, both in the spontaneous state and after stimulation with endotoxin, a potent inducer of TNF production (McClain 1989). We also have reported markedly elevated IL-6 levels in severe trauma patients that improved as the patient clinically improved. The other form of zinc deficiency is the more classically accepted form, that of zinc deficiency caused by decreased dietary intake, poor absorption, etc.

There are probably multiple mechanisms for the zinc deficiency and altered zinc metabolism in patients with gastrointestinal disease processes. First, many of these patients have depressed intake. This is typical of patients with alcoholic liver disease (McClain 1986a). They classically have anorexia, their diet is not well balanced, and alcohol calories are empty calories as far as zinc is concerned. This is opposed to patients with regional enteritis who often eat voraciously and have a very adequate intake of zinc, but still have hypozincemia. Another patient population with obvious inadequate intake is the eating disorder patient population. Indeed, virtually all of the patients we evaluated had dietary intake of zinc below the RDA of 15 mg. The elderly also have poor dietary zinc intake in almost every study that has been done (both in freeliving, healthy subjects and those that are institutionalized).

Zinc absorption is impaired in a variety of GI disease states. An example is the impaired zinc tolerance in patients with alcoholic liver disease as shown by our group and in nonalcoholic liver disease as shown by other investigators (McClain 1986a). Impaired zinc absorption in Crohn's disease using either the zinc tolerance test or radioisotope measurements has been documented (McClain 1980). Impaired zinc absorption in multiple other gastrointestinal diseases such as pancreatic insufficiency, eating disorders or even in the elderly has been reported. This impaired absorption can be due to intestinal mucosal problems as in Crohn's disease or to inhibitors of absorption such as dietary phytate.

Increased urinary zinc excretion is a frequently unrecognized cause of zinc deficiency. In many patients in the ICU, there is markedly increased urinary zinc excretion. This is classically seen in patients with sepsis, trauma and multisystem organ failure. It was initially thought that much of the increased urinary zinc excretion was due to TPN formulas. However, the stress state itself is usually the major cause of this zinc excretion. We have shown in both head injured patients and thermal injury patients that there is a marked increase in urinary zinc excretion that occurs over a long period of time (McClain 1986b). We evaluated possible mechanisms for this increased urinary zinc excretion. Fell and coworkers suggested that after surgery or trauma, there is increased muscle breakdown which leads to the urinary zinc excretion (Fell 1973). We evaluated a noninvasive marker of muscle catabolism, urinary 3-methylhistidine, and showed that it did not correlate with urinary zinc excretion. Similarly, two amino acids that are known to bind zinc tightly (cystine and histidine) also did not correlate with urinary zinc excretion. It is well documented that during stress states there is renal tubular dysfunction. Therefore, we evaluated a noninvasive marker of

renal tubular function, urinary amylase, and showed that it did relate temporally to the increase in urinary zinc excretion. Thus, we think that there is probably a renal tubular defect that plays a role in the urinary zinc excretion. Furthermore, altered binding of zinc in the serum would allow zinc to be delivered to the renal tubule. As we discussed previously, zinc is loosely bound (about 70%) to albumin, about 25% to alpha-2 macroglobulin quite tightly, and a small amount to peptides in amino acids. Giroux, et al. have shown that albumin in plasma from patients with a variety of disease states (such as decompensated liver disease) has less affinity for zinc than normal plasma albumin (Giroux 1977). Thus, probably in most inflammatory states there is an alteration in zinc binding, with more zinc bound to low molecular weight peptides or amino acids which can now be filtered by the glomerulus and then lost through impaired renal tubular reabsorption.

Increased gastrointestinal losses of zinc also may precipitate zinc deficiency or cause negative zinc balance, as reviewed by other investigators at this conference such as Dr. Jeejeeboy and Dr. Solomons. Thus, patients with inflammatory bowel disease, short bowel syndrome, or severe diarrhea may have large losses of fecal zinc.

In summary, multiple mechanisms account for the altered zinc metabolism observed in many GI processes including redistribution of zinc due to the effect of cytokines, poor dietary intake of zinc, malabsorption of zinc, and increased losses through GI or urinary routes. Clinicians must be aware of the metabolism of zinc in normal and pathologic states and the manifestations of zinc deficiency in order to prevent or correct the easily treatable problem of zinc deficiency.

REFERENCES

Bunker VW, Hinks LJ, Lawson MS, Clayton BE (1984) Assessment of zinc and copper status of healthy elderly people using metabolic balance studies and measurement of leukocyte concentrations. Am J Clin Nutr 40:1096-1102

Fell GS, Cuthbertson DP, Morrison C, Fleck A, Queen K, Bessent RG, Husain SL (1973) Urinary zinc levels as an indication of muscle catabolism. Lancet 1:280-282

Field HP, Whitley AJ, Srinivasan TR, Walker BE, Kelleher J (1987) Plasma and leukocyte zinc concentrations and their response to zinc supplementation in an elderly population. Internat J Vit Nutr Res 57:311-317

Giroux E, Schechter PJ, Schoun J, Sjoerdsma A (1977) Reduced binding of added zinc in serum of patients with decompensated hepatic cirrhosis. Eur J Clin Invest 7:71-73

Humphries L, Vivian B, Stuart M, McClain CJ (In press) Zinc deficiency and eating disorders. J Clin Psych

McClain CJ, Adams L, Shedlofsky S (1988) Zinc and the gastrointestinal system. Essential and Toxic Trace Elements in Human Health and Disease, pp 55-73

McClain CJ, Antonow DR, Cohen DA, Shedlofsky SI (1986a) Zinc metabolism in alcoholic liver disease. Alcoholism:Clin Exp Res 10:582-589

McClain CJ, Cohen DA (1989) Increased tumor necrosis factor production by monocytes in alcoholic hepatitis. Hepatology 9:349-351

McClain CJ, Kasarskis EJ, Allen JJ (1985) Functional consequences of zinc deficiency. Prog in Food Nutr Sci 9:185-226

McClain CJ, Souter C, Zieve L (1980) Zinc deficiency: A complication of Crohn's disease. Gastroenterology 78:272-279

McClain CJ, Twyman DL, Ott LG, Rapp RP, Tibbs PA, Norton JA, Kasarskis EJ, Dempsey RJ, Young B (1986b) Serum and urine zinc response in head-injured patients. J Neurosurg 64:224-230

Newsome DA, Swartz M, Leone NC, Elston RC, Miller E (1988) Oral zinc in macular degeneration. Arch Ophthalmol 106:192-197

Turnlund JR, Durkin N, Costa F, Margen S (1986) Soluble isotope studies of zinc absorption and retention in young and elderly men. J Nutr 116:1239-1247

Pathophysiological Significance of an Oral Zinc Tolerance Test in Patients with Chronic Hepatic Diseases

YASUYUKI ARAKAWA, KAZUTOMO SUZUKI, KEIKO SUZUKI, NAOHIDE TANAKA, YUTAKA MATSUO, and SHIGEO TAKEUCHI

Third Department of Internal Medicine and Department of Chemistry, Nihon University School of Medicine, 30-1 Oyaguchi, Itabashi-ku, Tokyo, 173 Japan

Key Words: Zinc tolerance test, Chronic hepatitis, Serum Zn/Cu ratio

Introduction

The significances of the biochemical and nutritional roles of metals, particularly trace metals, is widely recognized because they are found in vivo as constituent components of many metalloproteins and metalloenzymes. In addition, a close correlation between trace metal metabolism and the pathology of hepatic diseases has been recognized based on the fact that such metalloproteins and metalloenzymes are synthesized and react mainly in the liver and trace metals such as iron, copper, zinc, magnesium and manganese act as cofactors against hepatic fibrosis in chronic hepatic diseases, particularly the biosynthesis of collagen.
In the present study, the authors determined the concentrations of serum metals in patients with various chronic hepatic diseases, and the pathophysiological significance of these concentrations was evaluated.

Subjects and Methods

1.Subjects

The subjects comprised 138 patients with chronic hepatic diseases diagnosed at the Third Department of Internal Medicine, Nihon University School of Medicine. As shown in Table 1, they consisted of 14 patients with chronic inactive hepatitis, 50 with chronic active hepatitis, 33 with compensatory liver cirrhosis, 26 with decompensatory liver cirrhosis and 15 with primary biliary cirrhosis (PBC). As healthy controls, 220 adults who exhibited no abnormalities on general biochemical tests of their blood were used.

Table 1. Characteristics of the subjects studied

Diagnosis	No. of cases	Sex		Age (yrs)	
		Male	Female	Mean±S.D.	Range
Healthy controls	220	124	96	45±99	24−69
Chronic inactive hepatitis	14	8	6	47±18	20−79
Chronic active hepatitis	50	39	11	48±11	26−72
Liver cirrhosis	59	42	17	56±88	32−75
compensatory	33	23	10	57±88	32−75
decompensatory	26	19	7	55±88	32−66
Primary biliary cirrhosis	15	4	11	53±88	37−70
asymptomatic	5	0	5	48±66	37−52
symptomatic	10	4	6	55±88	44−70

2.Oral administration of zinc (zinc tolerance test)

Oral administration of zinc was performed in 6 healthy adults, 11 patients with chronic hepatitis and 17 patients with liver cirrhosis to evaluate the ability of the digestive organs in patients with chronic hepatic diseases to absorb zinc. That is 300 mg of zinc sulfate powder ($ZnSO_4$ $7H_2O$)--equivalent to 68 mg of zinc--was dissolved in 200 ml of physiological saline solution, and the subjects received oral administration of the solution in a fasting condition during the early morning. Blood samples were collected from the cubital vein before the administration , and the concentrations of serum zinc were determined after isolation of the serum, according to the usual method.

3.Methods for determining the concentrations of serum metals

Blood samples from healthy controls and patients with hepatic diseases were collected from the cubital vein into metal-free test tubes and the serum was isolated by the usual method. Serum specimens were kept at -20°C in a freezer until assay. When determining the metals, 3 ml of nitric acid was added to 2 ml of serum and the mixture was allowed to undergo partial hydrolysis at room temperature. Another 3 ml of nitric acid was then added, and the mixture was heated at 80°C for 3 h. Subsequently, 1 ml of hydrogen peroxide was added and the mixture was heated at 90°C for 1 h. This procedure was repeated twice, and the concentrations of metals in the specimens were finally determined with an inductively coupled plasma (ICP) emission spectrometry analyzer (JY 48P System, Seiko). All values were expressed in ppm.

Results

1.Concentrations of serum metals in chronic hepatic diseases

As shown in Table 2, statistically significant differences were observed in the mean (+SD) concentrations of serum metals between patients with various diseases and healthy controls. The concentration of Ca was low (P<0.001) in patients with liver cirrhosis, whereas it revealed no significant variation in chronic hepatitis or PBC. The concentration of Mg was low in liver cirrhosis and PBC (P<0.001, P<0.01), but it showed no significant variation in chronic hepatitis. The concentration of P was low as the disease condition developed from chronic hepatitis to liver cirrhosis (P<0.01, P<0.001), whereas it was high in PBC (P<0.001). The concentration of Cu was high only in liver cirrhosis gradually as chronic active hepatitis developed to liver cirrhosis (P<0.001), but it showed no significant changes in chronic inactive hepatitis or PBC.

Table 2. Concentrations of serum metals in patients with chronic liver diseases

Metal	Concentration of serum metal (mean±S.D., ppm)				
	Healthy controls	Chronic inactive hepatitis	Chronic active hepatitis	Liver cirrhosis	Primary biliary cirrhosis
Ca	92.2±4.0	92.9±8.3	91.4±6.8	86.5±8.4***	91.7±3.1
Mg	19.9±1.4	19.7±1.2	19.7±2.2	18.1±2.2***	19.0±1.4**
P	124.6±15.8	114.4±15.3*	118.2±20.4*	111.7±21.5***	186.5±47.0***
Fe	1.41±0.47	1.31±0.63	1.70±0.69**	1.50±0.83	1.26±0.48
Cu	0.98±0.2	1.03±0.18	1.05±0.24	1.18±0.27***	1.46±0.36***
Zn	0.88±0.1	0.85±0.13	0.79±0.2***	0.60±0.18***	0.80±0.18

*P<0.05, **P<0.01, ***P<0.001 as compared with healthy controls.

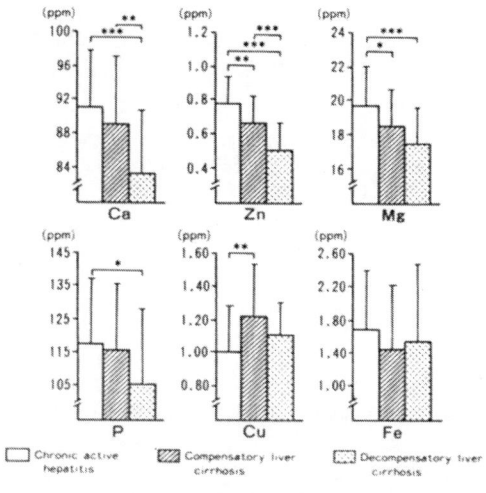

Fig.1.Comparison of serum metal concentrations in patients with chronic
active hepatitis and liver
cirrhosis

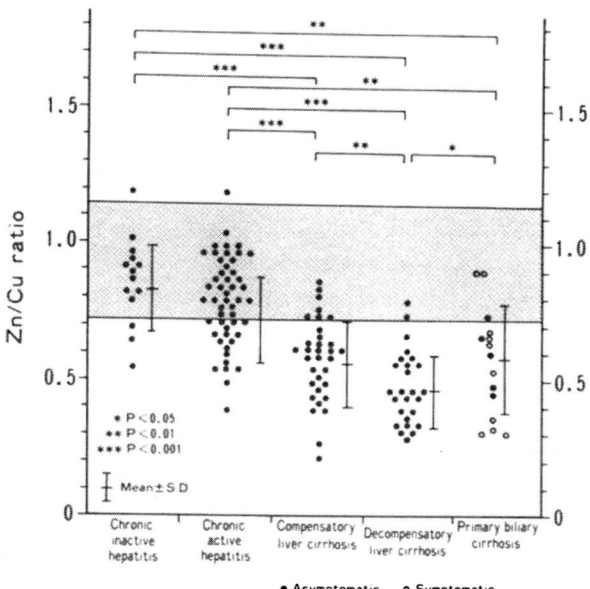

Fig.2.Comparison of serum Zn/Cu ratio
in patients with primary biliary
cirrhosis and other chronic
liver diseases

Figure 1 presents the results of a comparison of the serum metal concentrations in chronic active hepatitis, compensatory liver cirrhosis and
decompensatory liver cirrhosis. The concentrations of serum Ca and Zn showed
marked decreases (P<0.01, P<0.001) as compensatory liver cirrhosis changed to
decompensatory liver cirrhosis. However, there were no significant differences
in the serum concentrations of Mg, P, Cu and Fe between the compensatory and
decompensatory stages of liver cirrhosis.
Figure 2 shows the rations of the concentration of serum Zn to that of serum
Cu (Zn/Cu ratio) in chronic hepatic diseases. The ratio decreased significantly as the morbidity of these chronic hepatic diseases advanced, particularly
in the case of liver cirrhosis and PBC. The Zn/Cu ratio in the healthy
controls ranged from 0.73 to 1.1, whereas none of the patients with chronic
inactive hepatitis revealed a level of less than 0.5. Such low levels were
observed in 2 (4%) of 50 patients with chronic active hepatitis, 10 (30%) of
33 patients with compensatory liver cirrhoses, 16 (62%) of 26 patients with
decompensatory liver cirrhosis, and 6 (40%) of 15 patients with PBC. Hence,
when the Zn/Cu ratio is less than 0.5 in chronic hepatic diseases, liver
cirrhosis is highly likely.
In the healthy controls the ratio of the serum concentration of Ca to P and
that of P to Mg ranged from 0.60 to 0.78 and from 5.5 to 0.7, respectively,
but there were no significant differences in these ratios between chronic
hepatic and liver cirrhosis. The Ca/P and P/Mg ratios in PBC were lower and
higher (P<0.001), respectively, than those in other chronic hepatic diseases.

2. Expression of imbalance in the concentrations of serum metals on metallographs

In order to take full advantage of the ICP emission spectrochemical analysis,
i.e., simultaneous determination of multiple elements, the concentrations of
metals were standardized and expressed as metallographs on a radar chart. In
this procedure, the metallograph for the healthy controls was displayed as a
regular hexagonal pattern. The metallograph for chronic active hepatic
revealed slight decreases in P and Zn and an increase in Fe, as illustrated in
Figure 3, but there was no marked distortion of the hexagonal pattern. For
liver cirrhosis, however, the metallograph revealed a hexagonal pattern which
was greatly distorted by marked decreases in Ca, Mg, P and Zn, as shown in
Figure 4. Such distortion became increasingly extreme as compensatory liver
cirrhosis changed to decompensatory liver cirrhosis. In PBC, the concentrations formed an almost rhombic metallograph due to marked increases in P and
Cu, as illustrated in Figure 5.

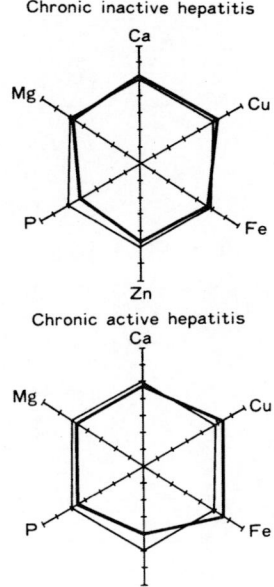

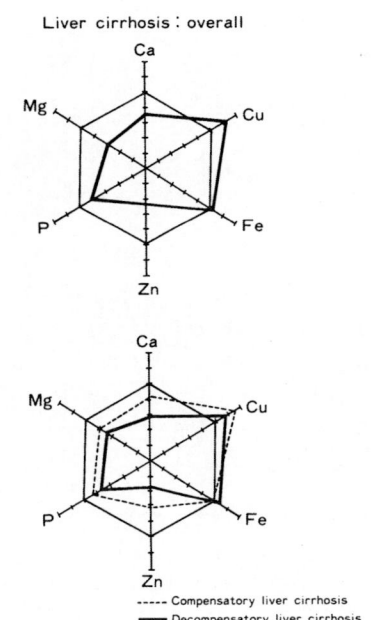

Fig.3.Serum metallogram in patients with chronic hepatitis

Fig.4.Serum metallogram in patients with liver cirrhosis

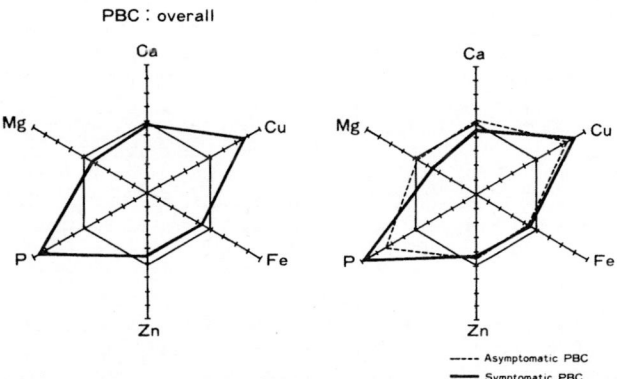

Fig.5.Serum metallogram in patients with primary biliary cirrhosis

3. Correlation between concentrations of serum metals and metal-binding proteins

The concentrations of serum metalloproteins, including prealbumin, albumin, transferrin,α_2-macroglobulin, and retinol-binding protein (RBP), were significantly decreased in liver cirrhosis, particularly that of the decompensatory type. In this connection, patients with chronic hepatic diseases (chronic hepatitis and liver cirrhosis) were examined for possible correlations between these major metal-binding proteins and the concentrations were observed between Ca ($P<0.05$) and Zn ($P<0.01$) and prealbumin, albumin and transferrin, between Mg and albumin ($P<0.05$), between P and retinol-binding protein ($P<0.01$), and between Cu and ceruloplasmin ($P<0.01$).

4. Correlations between the serum concentration of zinc, and the Fischer's BCAA/AAA molar ratio and serum III type procollagen N-terminal peptide (PC-III-NP)

The serum concentration of zinc was positively correlated with branched chain amino acids (BCAA) such as valine, leucine and isoleucine (plasma free amino

acids), but not with aromatic amino acids (AAA) such as tyrosine and phenylalanine. There was thus a positive correlation between the zinc concentration
and Fischer's ratio (BCAA/AAA molar ratio).
On the other hand, a negative correlation was observed between the zinc
concentration and PC-III-NP concentration.

5. Oral zinc tolerance test in patients with chronic hepatic diseases

Figure 6 presents results for the oral administration of zinc in patients
without hepatic diseases who underwent resection of 2/3 of the stomach and
those who underwent duodenal resection. In the former patient, the serum zinc
concentration at one hour after administration was high, suggesting that zinc
absorption might be facilitated. In contrast, the absorption of zinc was
extremely poor in the latter patient. It is clear therefore that the small
intestine, particularly the duodenum, is important for zinc absorption.
Figure 7 shows results for the oral zinc tolerance test in healthy controls
and patients with chronic hepatic diseases. The peak serum concentration at 2
h. after administration and the patients with liver cirrhosis were lower than
those in the healthy controls and patients with chronic hepatis (P<0.05,
P<0.01).
Figure 8 compares the areas under the serum zinc curve (AUC) between hepatic
diseases following oral zinc administration. The AUC at 1 h after administration (AUC1) in patients with liver cirrhosis was significantly (P<0.01) less,
by about 13%, than that in healthy controls. The AUC at 2 h after administration (AUC2) in patients with liver cirrhosis was significantly less (P<0.01),
by about 33%, than that in healthy controls. In a similar manner, the AUC at 3
h after administration (AUC3) in patients with liver cirrhosis was significantly less (P<0.01), by about 28%, than that in healthy controls. In
addition, the AUC1, AUC2 and AUC3 in patients with liver cirrhosis were
significantly (P<0.01) less than the corresponding areas in patients with
chronic hepatitis. The above results suggest that the ability of patients with
liver cirrhosis to absorb zinc is less than that of healthy controls and
patients with chronic hepatitis.

Discussion

Metals have frequently been discussed in terms of environmental pollution from
the stand point of toxicology. However, it has recently become clear that
metals exist _in_ _vivo_ at various concentrations and are important for the

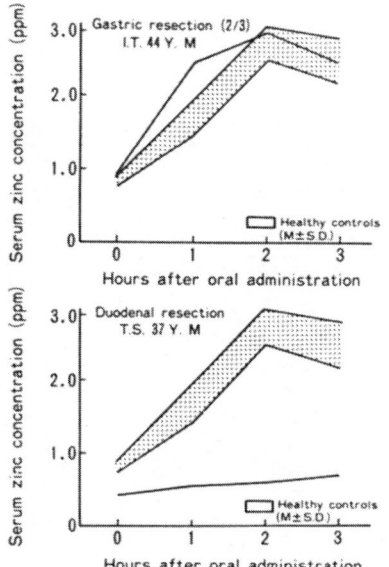

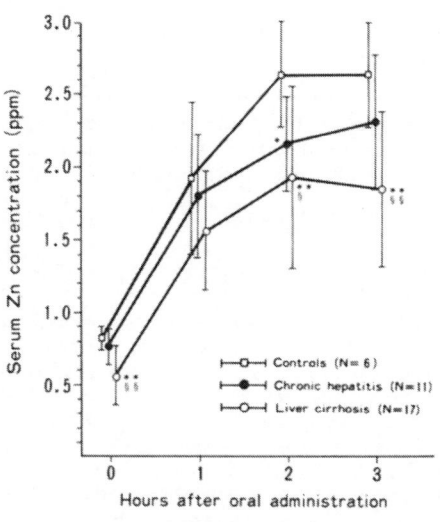

Fig.6.Zinc tolerance test of an oral
    dose of 300 mg zinc sulfate
    in patients with gastric and
    duodenal resections

Fig.7.Zinc tolerance test of an oral
    dose of 300 mg zinc sulfate in
    healthy controls and patients

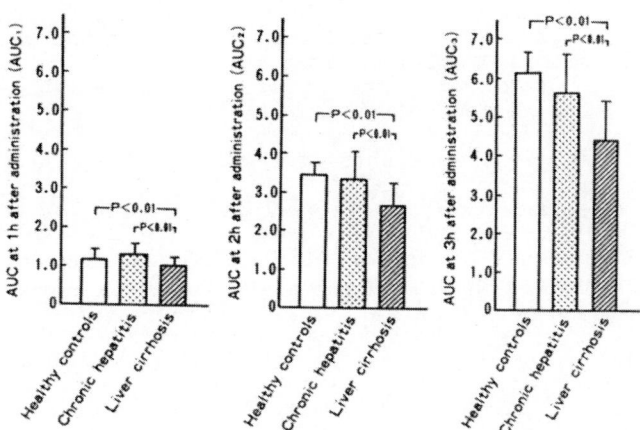

Fig.8. Areas under the serum zinc curve (AUC) in healthy controls and chronic liver paitents with a zinc tolerance test of an oral dose of 300 mg zinc sulfate

maintenance of viral functions. As a result, biochemical and nutritional studies on metals have increased. In the study of trace metals in particular, the initial emphasis on the existence and quantity of "metals" has shifted to an interest in the chemical structure or function and metabolism of metals. Thus, enzymological studies on the metabolism of compounds containing trace metals are expected to advance rapidly.
Hemochromatosis and Wilson's disease have long been known to induce excesses of iron and copper in the liver, the major organ of protein metabolism. However, metabolic abnormalities induced by excesses or deficiencies of other trace metals have also been increasingly recognized.
The authors determined the serum concentrations of various metals (Ca, Mg, P, Zn, Fe and Cu) in 220 healthy controls and 138 patients with hepatic diseases including chronic hepatitis, liver cirrhosis and primary biliary liver cirrhosis, by the ICP emission spectrometry analysis to clarify any abnormalities of serum metals in chronic hepatic diseases, and also carried out oral zinc tolerance tests in some patients. As the disease condition advanced from chronic hepatitis to liver cirrhosis, Ca, Mg, P and Zn were found to decrease, while Cu increased. The P and Cu concentrations were characteristically increased in primary biliary liver cirrhosis. The oral zinc tolerance tests revealed a decreased Zn absorption in the digestive tract of patients with chronic hepatic diseases, particularly liver cirrhosis. It appeared that the decrease might be caused partly by digestive tract lesions related to portal hypertension.
The main methods used previously for the determination of metals have been activation analysis and atomic absorption spectrophotometry. ICP emission spectrochemical analysis, which is highly sensitive and accurate, has recently been applied to measurements in vital specimens, enabling the simultaneous determination of several metals. To understand the effects of _in vivo_ metals on the living body, it is necessary to evaluate the interactions of two or more coexisting metals. This underlines the importance of the ICP emission analysis system in clinical medical practice.
In general, when the need for essential metals is larger, the mechanism for homeostasis is more developed in terms of providing a stable internal environment. It is important to take into account abnormalities of _in vivo_ metallic metabolism in patients with chronic hepatic diseases, especially those of the absorption, transport, storage, and biosynthesis of metallo-proteins and metalloenzymes. Furthermore, attention should be paid to insufficient intake of metals in patients with liver cirrhosis, which is, in general, frequently associated with a decrease in the reserve force of the liver and malabsorption of the digestive tract, since dietary intake tends to decrease with decrease in the energy requirement. It is also necessary to bear in mind that homeostasis of the _in vivo_ metallic metabolism is occasionally affected by the chelating effect of the therapeutic drugs used to treat such patients..
The serum concentrations of Ca, P and Mg are decreased in certain hepatic diseases, particularly liver cirrhosis, and some reports have shown that the decrease is approximately proportional to the severity of biliary engorgement

and hepatoparenchymatous disorder. Ca, Mg and P do not exist independent of each other in vivo; they coexist as bone minerals and are metabolized via a close interaction as extracellular and intracellular ions or by renal tubules. Ca-regulating hormones are also involved, but again, they do not act on individual metals alone. It is important to consider their physiological interactions.

Ca, which displays a marked difference in its extracellular and intracellular concentrations (10,000:1), play an important role in cellular division, proliferation, contraction, motility, agglutination, secretion, metabolism and excitability. As observed in experiments on the hepatopathy induced by carbon tetrachloride and galactosamine, an increase in the quantity of Ca in the cells and a decrease in the gradient of the extracellular and intracellular concentrations of Ca induce cellular hypofunction, eventually leading to cellular necrosis. To maintain the in vivo distribution of Ca, it is necessary to have active regulation based on energy and to control the Ca level with Ca-regulating hormones, i.e., parathyroid hormone, calcitonin and activated vitamin D (V_D). The decease in serum Ca concentration in liver cirrhosis, particularly in decompensatory liver cirrhosis, is strongly affected by the concentration of albumin and other factors including malabsorption of Ca and V_D by the intestinal tract, which is accompanied by a decrease in the pooled bile acids, a decrease in the concentration of 25 hydroxyvitamin D_3 (25 OHD_3), etc.

In addition, phosphate participates in energy storage, exchange, transport and information conduction, and exerts a body fluid buffer function, as demonstrated by the sugar phosphorylated materials in saccharometabolism, phospholipids as a membrane component, intracellular nucleic acid compounds and phosphoprotein enzymes, etc. Hepatic hypofunction occurs in patients with hypophosphatemia, and some reports have indicated that patients with liver cirrhosis also showed hypophosphatemia with aggravation of hepatic function. Parathyroid hormone is important as a humoral factor controlling the serum P concentration, but it is possible that in liver cirrhosis, accelerated excretion of P from the kidney resulting from an increase in blood concentration due to V_D deficiency, malabsorption of by the intestinal tract and inactivation of estrogen in the liver, may further aggravate hypophosphatemia.

Mg is an important intracellular cation like Ca. It is distributed mainly in the cytoplasm, where it plays an important role in the catalysis of enzyme reactions. Since Mg is essential for enzyme reactions involving phosphate conduction and ATP, it is necessary for membrane transport, amino acid activation, nucleic acid synthesis, protein synthesis, oxidative phosphorylation, muscle contraction, and maintenance of the morphology of erythrocytes and platelets. The actions of individual organelles in hepatocytes are dependent on Mg, and this metal is important for maintaining the structure and function of microsomes and mitochondria. Mg deficiency may thus induce hypofunction of hepatocytes, as in P deficiency. Mg deficiency in rats is accompanied by metabolic abnormalities of protein metabolism occur and appear as marked decreases in liver enzymes such as ornithine transcarbamylase and alginosuccinas. Clinically, decreased Mg levels cause muscle spasm and tremor, tetany, perspiration, tachycardia, pyrexia, and mental symptoms such as anxiety and excitement. These phenomena are considered to occur when the inhibitory effect of Mg on the central and peripheral nerves is decreased through a decrease in Mg. In this connection, the excitability of nerves and muscles, which can be expressed by the formula $Na \times K^+/Ca^{2+} \times Mg^{2+}$, is increased by decreases in the bivalent ions, Ca^{2+} and Mg^{2+}, and increases in the monovalent ions, Na^+ and K^+. Conversely, increases in the bivalent ions and decreases in the monovalent ions reduce the excitability. It is known that "twist"--localized muscle cramp--occurs at a high incidence in patients with chronic hepatic diseases, suggesting that decreases of Ca and Mg may be involved, in addition to abnormalities in electrolytes such Na and K.

Since zinc deficiency was first reported in humans by Prasad et al. in 1961, zinc has been increasingly shown to possess various physiological properties necessary for vital maintenance, and its role as an essential trace metal in vivo has been extensively investigated. Seventy or more zinc-binding metalloenzymes are known, and it is considered that decreases in their activity may cause serious metabolic disorders in the metabolism of lipids, including protein and carbohydrate, as well as in nucleic acid biosynthesis. Many studies have demonstrated that hypozincemia and hyperzincuria are present in chronic hepatitis and liver cirrhosis, since Vallee et al. reported that the serum zinc level in severe alcoholic liver cirrhosis (66 ± 19 ug/dl) was markedly less than that (120 ± 19 ug/dl) in healthy controls. In the present study also, the serum zinc concentration in chronic hepatic diseases, particularly liver cirrhosis, was distinctly less than that in healthy adults, and

this tendency became increasingly marked as the condition progressed. Hypozincemia is thought to involve multiple factors such as a decreased dietary zinc intake, decreased zinc absorption by the small intestine, decreased zinc-binding protein, decreased zinc storage in the liver, and increased urinary excretion of zinc. Serum zinc consists mainly of the type binding to macro-molecular ligands such as albumin, α_2-macroglobulin, etc. and the type binding to micro-molecular ligands including amino acids, polypeptides, etc. It is composed of 3 fractions: about 2% consists of amino acid (histidine, cysteine)-binding zinc called diffusive zinc, about 66% is albumin-binding zinc which is in equilibrium with the diffusive zinc, and about 32% is α_2-macroglobulin-binding zinc. The letter fraction is not in equilibrium with the other two and is formed and degraded in the liver alone. The number of zinc bindings is as large as 10^{10}, and it is not believed to be directly related to zinc nutrition. As demonstrated by the present results, significant correlations are observed between the blood level of zinc in chronic hepatic diseases and the concentrations of prealbumin, albumin and transferrin. However, there was no correlation between the level of zinc and the α_2-macroglobulin concentration. It is considered therefore that the amino acid-binding zinc is increased by decreases in the amounts of prealbumin, albumin and transferrin, resulting in an increased urinary excretion of zinc in liver cirrhosis. Such dysequilibrium of zinc-binding ligands may cause hypozincemia in liver cirrhosis. The zinc levels in the serum and liver tissue are decreased in patients with liver cirrhosis. Wakiyama et al. analyzed the zinc-binding materials in the cytoplasm of liver cells by Sephadex G-75 gel chromatography in carbon tetrachloride-induced liver cirrhosis in rats, and found that the high-molecular void volume fraction containing zinc and the Zn-SOD fraction showed markedly low levels. The former fraction is thought to correspond to the zinc thioneine fraction. Zinc thioneine, which regulates the zinc pool in liver tissue, usually undergoes mobilization and biosynthesis when it receives physical or chemical stimulation. Such biosynthesis may be decreased in liver cirrhosis. Although the significance of zinc in the development of chronic hepatitis and liver cirrhosis is not always clear, zinc deficiency is considered to be closely related to a decreased rate of protein synthesis in the liver, cell membrane instability and cell injury due to increased lipid peroxidation of the cell membrane, and hepatic fibroplasia. In the present study, the serum zinc level was confirmed to show a negative correlation with the serum concentration of serum III type procollagen N-terminal peptide, which is an index of hepatic fibrosis. In the liver, zinc inhibits lysyl oxidase, a copper enzyme, cross-linking of the covalent bond of collagen polypeptide and hepatic fibrosis.

Zinc is absorbed by the small intestine, mainly by the duodenum and jejunum, and is affected by various absorption accelerators and inhibitors. The absorption is considered to involve a homeostatic process, since it varies markedly with the vital demand for zinc. It is also thought to involve a negative feedback system; i.e., zinc absorption by the intestinal tract increases with *in vivo* zinc deficiency and decreases when the *in vivo* level of zinc is sufficient. Although the details of the mechanism of absorption via transcellular transport have not yet been elucidated, the uptake of zinc by the intestinal tract appears to consist of a fast phase and a subsequent slow phase, as demonstrated by our simple oral zinc tolerance test. The absorption is thought to be carried out by different mechanisms in the two phases; i.e., the former phase may involve passive absorption by diffusion based on the intra- and extra-intestinal tract zinc concentrations, while the latter may involve active absorption against the gradient of the concentrations. Maintenance of active transport seems to involve one or more of the following stages via the epithelial cells: 1) intracellular transport via the chorion from the intestinal cavity, 2) intracellular movement, and 3) transport from the basal membrane to the blood. It is currently considered to involve low-molecular zinc-binding factors secreted from the pancreas and metallothionein in the intestinal epithelium. In the present study, we undertook oral zinc tolerance tests in healthy controls and patients with chronic hepatic diseases, and found that the patients with liver cirrhosis showed less absorption of zinc via the intestinal tract than did the healthy controls and patients with chronic hepatitis. In the patients with liver cirrhosis, the zinc absorption level reached a maximum at 2 h after administration, and was decreased at 3 h after administration. The decrease in serum zinc level during the period between 2 and 3 h after administration is thought to have been caused by poor absorption of zinc by the intestinal tract, but other findings, including an increase in the urinary excretion of absorbed zinc and an increase in the uptake of zinc by each organ, are also considered to be reflected in the zinc concentration curves. The small intestinal membrane in

patients with liver cirrhosis is generally irregular and distinctly lobular in form, i.e., there is a decrease in height, increase in width, and decrease in area of the villi on the small intestinal membrane, lymphostasis in the submucosal layer, and dilatation of capillary vessels. Such atrophic lesions of the small intestinal membrane are considered to reflect abnormalities in the hepato-intestinal circulatory dynamics due to portal hypertension secondary to liver cirrhosis. These morphological abnormalities of the small intestinal membrane induce a decrease in the area of absorption of the villi and lead to poor absorption of various nutrients.

Sullivan et al. carried out zinc sulfate tolerance tests by orally administering 220 mg of zinc sulfate ($ZnSO_4$ $7H_2O$; containing 50 mg of zinc) to 16 patients with alcoholic liver cirrhosis. They reported that, whereas in 16 healthy controls, the preload level of serum zinc, 99 ± 9 ug/dl, increased to 280 ± 45 ug/dl at 2 h after the load but decreased to 145 ± 39 ug/dl at 6 h after the load, in the patients with alcoholic liver cirrhosis the preload level, which was only 54 ± 15 ug/dl, increased slightly to 98 ± 31 ug/dl at 2 h after the load, and the level was restored to the preload level at 6 h after the load. Prohit proposed a dose of 5 mg for oral administration because 50 mg of zinc is more than the required daily amount of 15 mg and is not a physiological dose. According to his report, the serum zinc level increased 53% at one hour after oral administration of 22 mg of zinc sulfate (containing 5 mg of zinc), which was sufficient for clinical evaluation. Miyata et al., however, reported that the serum zinc level increased only slightly following oral administration of 5 mg of zinc, from which it was impossible to evaluate the ability of the intestinal tract to absorb zinc. Zinc sulfate (100-600 mg) has been administered orally to patients with taste disorders associated with zinc deficiency, and no serious side effects have been reported, although some patients did show nausea, vomiting and abdominal discomfort after the administration. The authors performed zinc tolerance tests by oral administration of 300 mg of zinc sulfate and occasionally recognized symptoms similar to those described above accompanying an increased blood concentration of zinc in healthy controls and some patients who had the ability to absorb zinc. The authors therefore attempted further oral zinc loading tests with 100 mg of zinc sulfate. In comparison with the test using 300 mg, the 100 mg tolerance test yielded a lower incidence of side effects and more highly significant differences in the serum zinc concentration curve and AUC between healthy controls, patients with chronic hepatitis and patients with liver cirrhosis. It was considered sufficient, therefore, to administer 100 mg of zinc sulfate to evaluate the ability of the intestinal tract to absorb zinc in zinc tolerance tests.

The amount of copper present in vivo is large among the in vivo heavy metals, following iron and zinc. The liver, brain, heart and kidney are known to contain copper in especially large amounts. The majority of in vivo copper is bound to protein, mostly to enzymes with an oxidative function. These copper proteins and enzymes are involved in a wide range of in vivo catabolic reactions including (1) oxygen transport, (2) electron transfer, (3) oxidation-reduction, and (4) oxygen addition. The amount of in vivo copper in the human adult is about 100 mg, and the liver which contains approximately 1/10 of the amount, is a major organ for the dynamics and metabolism of copper, acting as a site for the biosynthesis of various copper proteins, in addition to storing and excreting copper. Of the serum copper, 95% is present as ceruloplasmin-binding copper, while the remaining 5% is albumin- and amino acid-binding copper. Although the ceruloplasmin-binding copper is relatively stable, the albumin-binding copper is liable to become free. This is regarded as the "available" type of copper. Thus, changes in serum copper frequently occur in close parallel with those of ceruloplasmin-binding copper, as shown by the results of our present study.

Copper is mainly adsorbed by the duodenum, binds to serum albumin and amino acids to reach the lung, and is rapidly incorporated by liver cells. Copper is metabolized in a free from within the cells and released at all times as a result of synthesis with ceruloplasmin in the microsomes. It is also incorporated by various copper proteins and copper enzymes. The copper distribution within the hepatocytes is 20% in the nucleus, 20% in the mitochondria and lysosomes, 10% in the microsomes, and the remainder in the from of superoxide dismutase (SOD) and copper-metallothionein in the soluble fraction. Extra-corporeal excretion of copper is carried out mostly via the bile, but an extremely small amount of copper is also excreted in the urine, intestinal fluid and sweat. The copper metabolism is therefore strongly influenced by disturbed excretion of copper in the bile-its main excretory route--, e.g., in cholestasis.

Precipitation of excessive copper in the liver tissue exerts a cytotoxic effect. The present results showed that the serum copper concentration was significantly increased in liver cirrhosis and primary biliary liver cirrhosis. It is known that the concentration of copper in the liver tissue is increased in Wilson's disease and primary biliary liver cirrhosis. Some reports have demonstrated that an increase is observed in other hepatic diseases and that there is a correlation between the copper concentration in the liver tissue and hepatic diseases. The causes of such abnormalities in the blood and liver tissue concentrations of copper in hepatic diseases mainly include 1) increased absorption of copper by the dibestive tract, 2) bile excretion disorders due to biliary acclusion or passage disorders, 3) disturbance of the copper excretion mechanism from the hepatocyte to the cholangiole, and 4) metabolic disturbance of copper in the hepatocytes. The maintenance and blood transfer of the copper in hepatocytes are considered to be regulated by the copper supply to copper enxymes, blood excretion of copper incorporated by ceruloplasmin, and the accelerated binding to excessive copper caused by induced synthesis of intracellilar metallotionein. In usual chronic hepatic diseases, the blood concentration of copper is closely correlated with the ceruloplasmin concentration. The synthesis of ceruloplasmin is thereby increased as a sacondary compensatory mechanism, copper release from the liver is enhanced, and the amount of urinary copper excreted is increased in disease conditions where copper is accumulated in the liver tissue due to copper excretion disorder. Furthermore, the ratio of serum copper to serum ceruloplasmin in primary biliary liver cirrhosis associated with a markedly increased blood concentration of copper is more than that in liver cirrhosis, indicating that increased amounts of copper compared to the ceruloplasmin-binding type may be present in the blood in the disease. This suggests release of copper within the liver tissue exceeding the synthetic activity of ceruloplasmin, probably reflecting enhancement of the compensatory route for the excretion of copper accumulated in the liver. For these reasons, usual liver cirrhosis is possibly different from primary biliary liver cirrhosis is possibly different from primary liver cirrhosis in terms of the abnormality in copper metabolism.

Conclusions

The authors determined the serum concentration of metals using an ICP emission spectrochemical analysis system in patients with chronic hepatic diseases. The following conclusions were drawn.
1) With advance of the disease condition from chronic hepatitis to liver cirrhosis, the serum levels of Ca, Mg, P and Zn decreased, while that of Cu increased. The serum Ca and Zn levels in decompensatory liver cirrhosis were significantly less than those in compensatory liver cirrhosis. The increases in serum P and Cu levels were characteristic in primary biliary liver cirrhosis. These abnormalities in the metabolism of serum metals are assumed to be closely related to the disease conditions of hepatic hypofunction, hepatic fibroplasia, and cholestasis.
2) The finding that abnormalities of metallic metabolism advanced with the morbidity of chronic hepatic diseases indicates that it is useful for the differential diagnosis of such diseases to standardize and express each metal concentration determined with the ICP system as a metallographs on a radar chart, since it is possible to assess the correlations between the concent-ration of many metals at the same level.
3) The serum Zn concentration was found to be negatively correlated with serum III type porcollagen N-terminal peptide and positively correlated with the BCAA/AAA molar ratio.
4) Dysequilibrium of metal-binding ligands, which is based on hyposythesis metalloprotein, is an important cause of metabolic adnormalities of the serum metals in chronic hepatic diseases.
5) Zinc tolerance tests in patients with chronic diseases revealed significant decreases in the peak level at 2 h and the level at 3 h after administration in patients with liver cirrhosis in comparison with those in healthy adult controls, confirming decreased zinc absorption. This suggests that digestive tract lesions may be related to portal hypertension secondary to liver cirrhosis.

Zinc Absorption in Normal Subjects and in Gastrointestinal Diseases

G.C. STURNIOLO, M.C. MONTINO, L. ROSSETTO, R. D'INCÀ, A. MARTIN, A. D'ODORICO, and R. NACCARATO

Division of Gastroenterology, University of Padua, Via Giustiniani 2 35128, Padua, Italy

ABSTRACT

We studied zinc absorption by means of the zinc tolerance test in healthy volunteers and in patients with liver cirrhosis, inflammatory bowel diseases and chronic pancreatitis. In normal subjects age, sex and gastric secretion have a significant effect on absorption. Zinc absorption is significantly reduced in all gastrointestinal and liver disorders examined. The administration of ligands, such as picolinic acid, citric acid and prostaglandin inhibitors, did not modify zinc absorption in healthy subjects and in patients with cirrhosis and inflammatory bowel diseases. On the contrary, citric acid significantly enhanced absorption in chronic pancreatitis.

KEY WORDS

Zinc, absorption, human, ligands, gastrointestinal disease.

INTRODUCTION

Zinc deficiency has been demonstrated in various liver and gastrointestinal diseases. Moreover this alteration of zinc metabolism can be marginal and detected only by biochemical tests, but in some occasions, such as during the acute, severe or convalescent stages of the disease, the deficiency can become clinically evident. The zinc deficiency can possibly derive from increased elimination, low intake, increased requirements or impaired absorption. Only a few studies have investigated zinc absorption in normal subjects and even fewer have carried out studies in pathological conditions. Several different methods have been employed, including radioactive probes (^{69}Zn, ^{65}Zn), cold isotopes (^{70}Zn), zinc balance, oral zinc tolerance test (ZnTT) and intestinal segmental perfusion. Some of these tecniques are rather invasive, expensive and hazardous.

In the studies reported here, we assessed zinc absorption using the oral ZnTT, because of the advantages of reproducibility, safety and cost.

Our study aimed to determine the influence of age and sex on zinc absorption and to investigate some of the physiological mechanisms of absorption such as gastric secretion, zinc supplementation, effect of ligands and different doses of zinc. Furthermore zinc absorption and the role of ligands have been evaluated in pathological conditions involving the liver and the gastrointestinal tract, We studied patients affected by Chronic Liver Diseases (CLD)of different etiology, Inflammatory Bowel Diseases (IBD) and Chronic Pancreatitis (CP).

METHODS

Zinc absorption was evaluated by the oral ZnTT: zinc sulphate was administered orally at the dose of 55 mg (12.5 mg of Zn^{++}) when evaluating the influence of sex and ligands and in all the studies involving patients. Fifty mg of elemental zinc were employed in the elderly and in the evaluation of the influence of gastric acidity. Different doses (6, 12.5, 25 and 50 mg of Zn^{++}) were

tested in the same subjects to evaluate zinc homeostasis. Blood samples were obtained at 0' time and every hour over a 6 hour period. Plasma zinc levels were assessed by AAS. The areas under the plasma concentrations curves (AUC; μmol.min/dl) were calculated. The effect of ligands was studied by administering picolinic acid (PA) 148 mg, citric acid (CA) 94 mg, together with a capsule of $ZnSO_4$, and prostaglandin inhibitor indomethacin (I) 50 mg tid po for 2 days and 100 mg rectally 1h before the test. All the ligands were tested in both healthy subjects and patients in random order at least one week apart from each other. The influence of gastric acidity was tested administering H_2-blockers, cimetidine (200 mg tid for 3 days and 200 mg 1 h before the test) and ranitidine (150 mg bid for 3 days and 300 mg 1 h before the test). Gastric pH was measured by titration in gastric juice samples collected hourly over the same time period of the ZnTT.

RESULTS

The physiological studies on healthy subjects showed that zinc absorption is significantly reduced in the elderly (78-84 years; n 15) compared to adults (22-45 years; n 10),(AUC 1016$\pm$552 vs 5387$\pm$2345; p<0.005). Females (n 5) absorbed significantly more zinc than males (n 5) (AUC 7239$\pm$1471 vs 3170$\pm$1942; p<0.005). The inhibition of gastric secretion by cimetidine or ranitidine as confirmed by the increased pH significantly reduced zinc absorption (AUC: basal 5374$\pm$2256, cimetidine 3495$\pm$487, ranitidine 2480$\pm$663; Anova two way p<0.05) (Fig. 1).

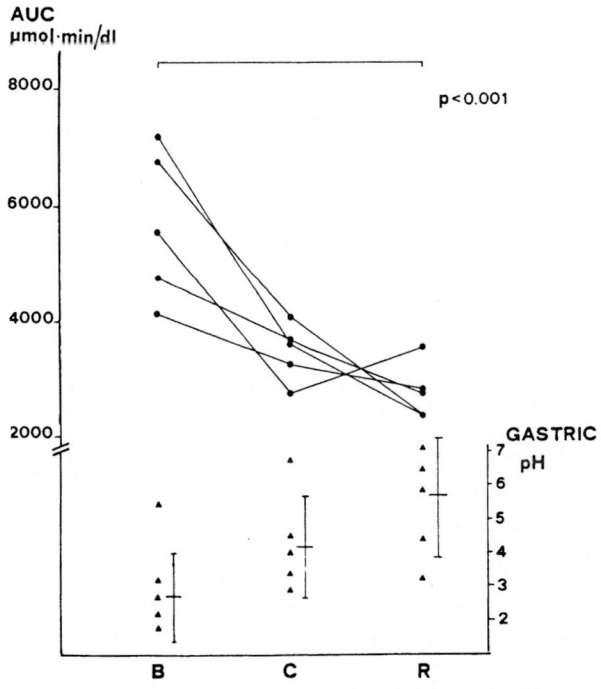

Fig.1 Effect of cimetidine (C) and ranitidine (R) on ZnTT and gastric pH. (B= basal values).

Ligands administration, PA, CA and I had no effect on absorption in physiological conditions.
A 3 week supplementation with 220 mg $ZnSO_4$ tid, did not modify zinc absorption.
As shown in Fig. 2 the administration of increasing doses of zinc did not result in proportional increments of absorption. The rate of zinc absorption increased proportionally up to a dose of 2 mmol and with a dose of 8 mmol, while 4 mmol did no achieve the same linear increase.

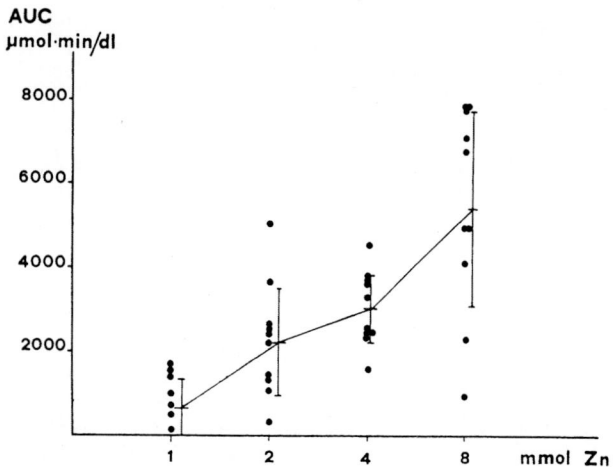

Fig.2 Zinc absorption after increasing doses of zinc.

The oral ZnTT was employed to assess zinc absorption in CLD, ulcerative colitis (UC), Crohn's disease (CD) and CP. In all these conditions absorption was
significantly reduced in comparison to normal subjects (AUC: normal subjects,
n 10: 2552±1327; CLD, n 24: 1033±180, p<0.001; UC, n 14: 1350± 1085, p<0.001;
CD, n 17: 1379±876, p<0.001 and CP, n 9: 965±692, p<0.001).
The role of ligands (CA and PA) tested in all the patients effectively improved zinc absorption only in CP. The basal AUC (965±692) significantly increased CA administration (1540±777; p<0.05). PA improved zinc absorption, but the
difference did not reach statistical significance (1314±532).
The influence of alcoholic etiology and the severity of cirrhosis did not represent a worsening factor for the reduced zinc absorption, but patients with
albumin levels less than 3 g/l had a significantly reduced zinc absorption,
when compared with patients with higher albumin levels (AUC 810±217 vs
1033±180; p<0.005).

DISCUSSION

Our study demonstrates the causal importance of gastric secretion for an adequate zinc absorption. Gastric acid suppression was proportional to the dose
and potency of the drug administered (Fig. 1) and this effect was followed by
a proportional decrease in zinc absorption. The reduction in zinc absorption
observed with both drugs probably indicates a direct effect on gastric acidity
rather than the formation of a complex between drug and zinc. A lower pH could
facilitate the absorption through the ionization of the zinc molecule. Alterna
tively, there could be changes in the structure of the zinc transporting substances occurring in the intestinal lumen or of the link between these ligands
and the metal.
A reduction of zinc absorption was seen in the healthy elderly. Previous studies hypothesised a role for marginal zinc deficiency in explaining the impaired cellular immunity and the frequent recurrence of infections seen in the
elderly. Our findings indicate in the impaired zinc absorption one of the poss
ible explanatory factors for zinc deficiency.
Animal studies hypothesised a homeostatic mechanism of zinc absorption
(Jackson, 1981), however, in our experimental conditions, a 3 week supplementation with zinc did not reduce significantly zinc absorption.
According to the hypothesis of a regulation by the metallothionein (Cousins,
1989) an increased synthesis of this protein can account for the intracellular
entrapment of the zinc in excess. Unfortunately, our test does not explore
zinc uptake or the intracellular concentration of the metal thus preventing
a demonstration for this hypothesis.

The pattern of zinc absorption after doubling the dose of zinc administered,
seems to suggest both a passive and a carrier-mediated mechanism, confirming
the results on segmental perfusion of human small intestine (Lee, 1989).
The studies performed in gastrointestinal pathological conditions demonstrate
a significant reduction of zinc absorption. Thus a contributory role for this
alteration is suggested in order to explain the deficiencies seen in these
chronic diseases. This simple test could be employed as a method for detecting
a potential deficiency of zinc and on the basis of this assessment, a supplementation can be started before the deficiency becomes clinically evident.
Zinc absorption in CLD is significantly more impaired in patients with low levels of albumin. This can be explained either by a more severe disease and /or
by a reduced efficiency in zinc transport mainly bound to albumin from the mucosal to the vascular compartments. The latter mechanism seems more likely sin
ce the patients with the decompensated cirrhosis did not have a significantly
lower zinc absorption than compensated patients. Alcohol is a well known factor which can affect zinc metabolism. Our alcoholic patients showed a reduction in absorption in comparison with cirrhotic patients of viral etiology,
but the difference was not statistically significant. Earlier experimental stu
dies demonstrated the existence of low molecular weight ligands which play a
role in zinc uptake in the intestinal mucosa.It has been suggested that PA,
CA and prostaglandins might act as ligands, therefore we studied the effect
of these substances both in normal subjects and in patients with gastrointestinal disorders. Our results indicate that the ligands supplementation improves zinc absorption in pancreatic patients. CA reverted the absorption to
normal, while only a trend was seen with PA. The same ligands administered to
patients with CLD or IBD did not influnce zinc absorption. No role for any of
the ligands was seen in healthy volunteers. A possible explanation for these
results could be that these subjects have an adequate intestinal concentration
or pancreatic secretion of ligands.

CONCLUSIONS

These studies contribute to a better comprehension of our knowledge of the
mechanisms and factors affecting zinc absorption. A previous unknown role of
gastric acidity, sex and age was demonstrated. Furthermore, it was shown that
altered intestinal absorption of zinc may be the key mechanism of zinc deficiency in several gastrointestinal tract disorders. Finally, ligands such as
citric acid may be deficient: in such cases, ligands might be employed in the
treatment of these conditions of zinc deficiency.

REFERENCES

Cousins RJ. (1989) Theoretical and practical aspects of zinc uptake and absorp
    tion. Adv Exp Med Biol 249:3-12
Jackson MJ, Jones DA, Edwards RHT. (1981) Zinc absorption inthe rat. Br J Nutr
    46:15-27
Lee HH, Prasad AS, Brewer GJ, Owyang C. (1989) Zinc absorption in human small
    intestine. Am J Physiol 256:G87-G91

The Relationship Between Plasma Iron Tolerance Curves and Fractional Intestinal Absorption of Iron

CH. HANSEN, E. WERNER, P. ROTH, and J.P. KALTWASSER

GES für Strahlen- u. Umweltforschung, Paul-Ehrlich-Str.20, Frankfurt/M D-6000, F.R. Germany

ABSTRACT

It was the aim of this study to investigate the relationship between plasma iron tolerance curves (PITC) after high and low iron load and the fractional intestinal absorption of iron. A total of 39 subjects were given iron doses between 1 and 100 mg of iron tagged with stable Fe-54 and/or radioactive Fe-59. PITC were determined either from plasma iron increase, Fe-59 activity or changes of Fe-54 abundance in blood plasma. Intestinal iron absorption was derived from Fe-59 retention on day 14 after application by means of whole body counting. Significant correlations were obtained for values of intestinal iron absorption and maximum increase of plasma iron and maximum tracer content in blood plasma.

Key words: iron absorption, radioiron, stable iron isotopes, whole body counting, plasma iron tolerance curves

INTRODUCTION

Iron is an essential trace element with manifold physiological functions. Although its abundance in the human environment is rather high, iron deficiency is a common finding not only in developing but also in industrialized countries. Iron deficiency is a problem of balance: on the one hand the fraction of body iron lost per day is very low unless there is increased menstrual bleeding in females or blood losses from other sources but on the other hand the uptake of iron from food is always incomplete. The body iron status is regulated by the adaptation of the fractional intestinal absorption of available iron from food. To maintain the balance it is therefore necessary not only to have a sufficient iron supply with food, but also the food iron is required in a form available for intestinal absorption. The availability of iron is strongly dependent on a number of factors such as form of iron i.e. vegetable or animal origin, food composition, manipulations during food preparation etc. (Bothwell 1979). It is therefore still necessary to study the properties of intestinal food iron absorption.

It has been clearly shown that balance measurements of intestinal iron absorption are expensive, inconvenient, and only of very limited accuracy (Bothwell 1979). Therefore, most investigations of intestinal iron absorption have been performed applying marker substances. Up to now almost exclusively radioisotopes of iron have been used as markers. Fe-59 and Fe-55 are the isotopes most often used in such investigations. But in recent years more and more restrictions have been imposed on the in vivo administration of radioactive substances to humans especially to subjects at risk for iron deficiency such as children, adolescents, and menstruating or pregnant women. Therefore, alternative methods are currently under investigation. In this context the use of stable iron isotopes as tracers for studies of human iron metabolism is investigated in several research centers (Janghorbani 1980, King 1978, Lehmann 1988, Turnlund 1982). There are four stable iron isotopes with very different natural abundances. Fe-56 with 91.8% is the most abundant isotope. Thus, Fe-54 with an abundance of 5.8%, Fe-57 with an abundance of 2.2%, and Fe-58 with an abundance of 0.3% may be used in a highly enriched form as marker substances.

This study was aimed to compare results of intestinal iron absorption as measured by whole body counting with those obtained from plasma iron tolerance curves using either serum iron increase or tracer concentrations in plasma after administration of radioactive or stable iron isotopes.

METHODS

Investigations were carried out in a total of 39 subjects from 18 to 72 years. Twentythree patients with chronic iron deficiency anaemia and 16 male subjects who were phlebotomized weekly until development of a frank iron deficiency anaemia with haemoglobin concentrations between 90 and 110 g/l. The patients were given test doses containing between 1 and 100 mg of iron in an aqueous solution either as ferrous sulphate with ascorbic acid or as ferric poly-maltose complex labelled with radioactive Fe-59. In 10 investigations the iron content was between 1 and 3 mg, in 40 studies 50 mg of iron and in the other 5 investigations iron content of the test dose was 100 mg. In 6 subjects part of the stable iron was replaced by the same amount of highly enriched stable isotope Fe-54. Blood samples were drawn at 15, 30, 60, 120, 180, and 240 minutes after the test dose to follow plasma iron tolerance curves. Serum iron concentration was measured colorimetrically (ICSH 1978), Fe-59 activity was determined in blood plasma in a well type sodium iodide detector with multichannel analyzer, and Fe-54 concentration was obtained by means of fast atom bombardment mass spectrometry. From these curves the maximum increase of serum iron concentration, the maximum of radioiron concentration in blood plasma, and the maximum of Fe-54 concentration in plasma were obtained (Werner 1983). Whole body counting was performed before and after administration of the test dose and after 14 days. The fractional absorption was calculated from the ratio of Fe-59 activities retained in the body on day 14 after administration of the test dose and the orally administered activity (Werner 1983). A total of 55 investigations were carried out in the 39 subjects.

RESULTS

On figure 1 the relation of the maximum increase of serum iron concentration and the absorbed amount of iron, derived from whole body counting, is given. There is a close correlation between these two parameters. However, a broad range of the 95% confidence level for the determination of absorbed amount of iron from the maximum increase of serum iron is observed. For example, for an individually measured serum iron increase of 200 ug/dl the amount of iron absorbed lies between 11 and 22 mg. Moreover, the estimation of the amount of iron taken up in the intestine is only possible for therapeutic iron doses. For small iron doses, even if the percentage absorption is high, the relative uncertainty is very large. One has also to consider, that this relation is more or less only useful for well bioavailable non-haem iron. For the assessment of food iron absorption this method seems not very practicable since usually no significant increase of serum iron concentration is expected because of the low iron content of a meal of less than 5 mg and its limited iron uptake. Figure 2 shows the relation between intestinal iron absorption measured by whole body counting and the maximum increase of radioiron in plasma, given as the fraction of dose in total plasma. There is a similar close correlation as in figure 1. However, since here is correlated not the amount of iron absorbed with the maximum increase but the percentage intestinal absorption, also for low iron doses a measurable increase of the tracer in plasma could be observed. Applying the radioisotope Fe-59 as tracer this method can be used to estimate iron absorption from very low iron doses in the

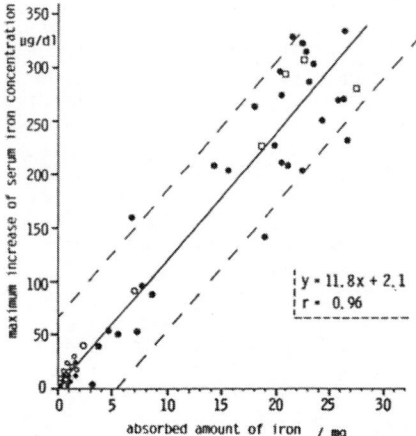

Fig. 1: Correlation between absorbed amount of iron measured by whole body counting and maximum increase of serum iron concentration

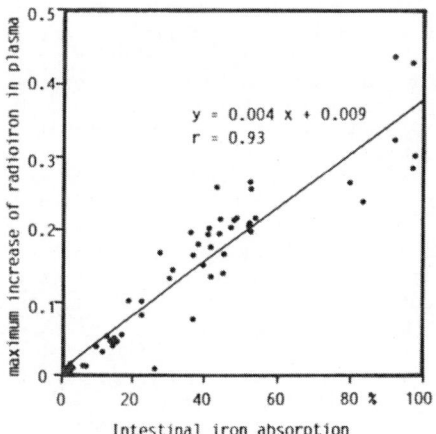

Fig. 2: Correlation between intestinal iron absorption and maximum increase of radioiron in blood plasma

test meal. Because the specific activity of the radiopharmaceutical is in the order of 300 kBq/μg Fe and for an absorption test an activity of less than 50 kBq is required, the amount of the marker substance is less than 1 μg. The assessment of food iron absorption applying this method seems easily be possible by just adding the tracer iron to the meal since it has been shown that there is no difference in iron uptake from intrinsically and extrinsically labelled foodstuffs (Bothwell 1979). On figure 3 the same correlation but for the 6 subjects who received the stable isotope Fe-54 is given. By contrast to Fe-59 the amount added to the test meal is higher than 1 μg. It depends very much on the particular type of study but in general may be less than 1 mg of Fe-54. Again for a given maximum increase of tracer in blood plasma there is a considerable uncertainty of the actual fractional absorption. A direct comparison of the figures of the maximum fractions of the tracers in blood plasma for the radioactive Fe-59 and the stable Fe-54 is shown in figure 4. In 4 of these 6 subjects there is a slight underestimation of absorption as derived from the Fe-54 tracer. This may be due to a contamination of the samples during work up for the mass spectrometric measurement with natural iron from reagents or containers.

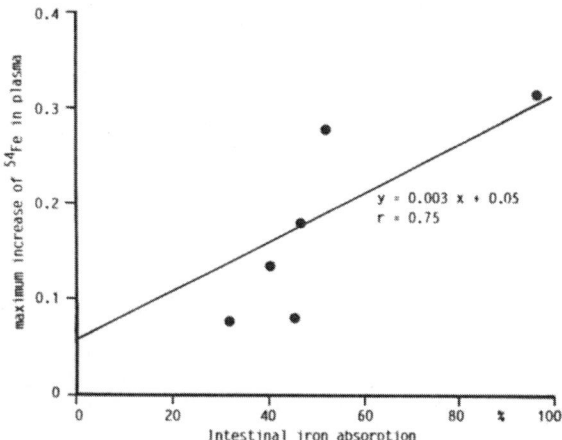

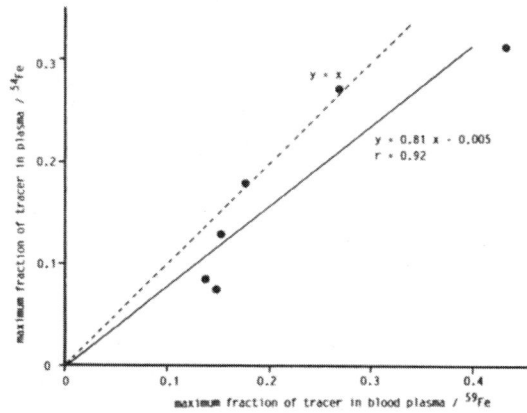

Fig. 3: Correlation between intestinal iron absorption and maximum increase of Fe-54 in blood plasma

Fig. 4: Intraindividual comparison between maximum increase of radioiron (Fe-59) and maximum increase of stable iron (Fe-54) in blood plasma

CONCLUSIONS

The data obtained in these studies show that the iron absorption from foodstuffs or whole meals may be determined by the measurement of plasma iron tolerance curves applying stable iron isotopes as tracers. The inaccuracy of the figures obtained for individuals may be rather high, but the mean value of groups correlates well with the results derived from whole body counting after radioactive labelling. The application of the stable isotopes provides safe methods to the subject and to the investigator, however, considerable high expenses for material, for sample work up procedures, and for the measurement of the isotopes. Up to now the development of suitable procedures for the use of stable isotopes in such studies is not finally completed.

REFERENCES

Bothwell TH, Charlton RW, Cook JD, Finch CA (1979) Iron Metabolism in Man, Blackwell Scientific Publications, Oxford, London, Edingburgh, Melbourne
International Committee for Standardization in Haematology (1978): Recommendations for measurement of serum iron in human blood. Br J Haematol 38:291-294
Janghorbani M, Young VR (1980) Use of stable isotopes of minerals in human diets using the method of fecal monitoring. Amer J Clin Nutr 33: 2021-2030
King JC, Raynolds WL, Margen S (1978) Absorption of stable isotopes of iron copper, and zinc during oral contraceptive use. Amer J Clin Nutr 31:1198-1203
Turnlund JR, Michel MC, Keyes WR King JC, Margen S (1982) Use of enriched stable isotopes to determine zinc and iron absorption in elderly men. Amer J Clin Nutr 33:1033-1040

Lehmann WD, Fischer R, Heinrich HC (1988) Iron absorption in man calculated from erythrocyte incorporation of the stable isotope iron-54 determined by fast atom bombardment mass spectrometry. Anal Biochem 172:151

Werner E, Roth P, Hansen Ch, Kaltwasser JP (1983) Comparative evaluation of intestinal iron absorption by four different methods in man. In: Urushizaki I, Aisen P, Listowsky I, Drysdale JW, Structure and Function of iron storage and transport proteins, Elsevier Science Publishers, Amsterdam, New York, Oxford, p403-408

Regulators of Iron Absorption in the Small Intestine

Marcel E. Conrad, Jay N. Umbreit, Raymond D.A. Peterson, and
Elizabeth G. Moore

University of South Alabama, Mobile, AL 36688, USA

ABSTRACT

The absorption of iron can not be explained solely by transferrin and iron. Therefore, a search was undertaken to discover other iron binding proteins in homogenates of rat duodenum. Using heat, ammonium sulfate precipitation, serial chromatography and immunologic methods, we have identified four iron binding substances which are both biologically and immunologically distinct from ferritin and transferrin. Two of these iron binding complexes are water soluble with molecular sizes of about 520,000 and 60,000 daltons respectively. One of the water insoluble proteins is soluble in triton X-100 and is probably a membrane protein with a molecular size of about 60,000 daltons. Insoluble mucin is believed to be the fourth newly identified iron binding complex.

KEY WORDS

Iron binding proteins, iron absorption, duodenum, regulation, mucin.

The regulation of iron absorption in the proximal small intestine is not adequately explained solely by transferrin and ferritin. Ferritin appears to be a storage protein which protects the cell from the oxidative damage of iron. Transferrin probably plays an important role in the transfer of iron from the mucosal cell into the body. It is unlikely that it is the sole regulator of iron uptake by absorptive cells in the intestinal mucosa because (1) humans with atransferrinemia develop iron overloading rather than iron deficiency (2) the gene for transferrin is not expressed in intestinal mucosal cells (3) neither transferrin nor transferrin receptors can be identified in the apical portions of intestinal mucosal cells where iron binding substances have been identified (4) there are a number of physiologic states in which iron absorption is altered without changes in the quantity of mucosal transferrin and (5) the absorption of several divalent metal cations which do not bind transferrin _in vivo_ vary inversely with the state of iron repletion in humans and animals. Therefore, we searched for other iron binding proteins in the duodenal mucosa of rats using classical biochemical methodology.

Using heat, ammonium sulfate precipitation, serial chromatography and immunologic techniques, we have identified four iron binding substances from small intestinal homogenates which are both biochemically and immunologically distinct from ferritin and transferrin. The steps providing separations are outlined and described below. While our standard isolations are obtained from homogenized duodenal mucosa of rats following intraluminal administration of radioiron-59 and radioiron-55; similar labeling was accomplished by adding the radio-isotope to the crude homogenate in _vitro_ (Fig. 1).

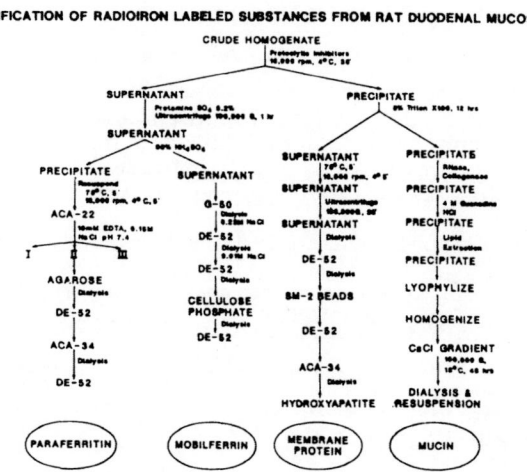

Fig. 1

PREPARATIVE METHODS AND RESULTS

Wistar, pathogen free rats (175-200g) were obtained and made iron deficient by a 4 ml bleeding and placing them on an iron deficient diet for two weeks. Iron deficient rats were used to minimize apoferritin production within the mucosa and possibly to stimulate production of substances which might be increased by iron deficiency. Under pentobarbital anesthesia, a celiotomy was performed and the duodenum was isolated with umbilical tape. Then 10 μCi of ^{59}Fe and ^{55}Fe in 1MM of $FeCl_2$ was injected into the lumen of the isolated duodenum. After 10 minutes incubation, rats were perfused to death by rapid injection of 0.15M NaCl into the left ventricle after transection of the hepatic vein (viviperfusion markedly diminishes contamination of mucosal preparations with plasma transferrin and hemoglobin). Isolated segments were excised, opened with an iris scissors and washed thoroughly in two changes of cold 0.15M NaCl, 10mM Hepes, pH 7.4. Mucosa was scraped from the mucosa into 1.5 M Hepes, pH 7.4 containing proteolytic inhibitors. Mucosal preparations were homogenized in a Virtis homogenizer followed by a Porter-Elvehjem homogenizer with a teflon pestle. Mucosal homogenates were centrifuged at 16,000 X G for 20 minutes. The pellet was washed in 1.5 M Hepes buffer and recentrifuged three times to diminish contamination with water soluble substances. Approximately 60 percent of radioiron was precipitated in the pellet with the remainder being found in water soluble fractions. Since ferritin and transferrin are water soluble, we investigated the insoluble fraction. Approximately 60 percent of radioactivity in the insoluble fraction is associated with mucin. The remainder (40 percent) is soluble in triton-X100 and appears to contain a heat stable iron binding substance with a molecular size of about 60,000 daltons. Water soluble fractions of mucosa were treated with 0.2 percent protamine sulfate to diminish contamination of the preparations with nucleic acid fragments. This markedly reduced the radioiron in Peak I and considerably sharpened peaks II, III and IV in early experiments (Fig 2,3).

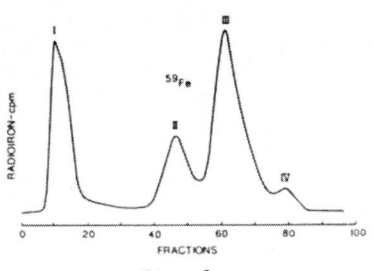

ACA22 COLUMN - SOLUBLE ^{59}Fe FRACTIONS FROM DUODENAL MUCOSA IRON DEFICIENT RATS

Fig. 2

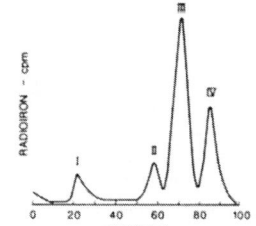

ACA22 COLUMN - SOLUBLE ^{59}Fe FRACTIONS FROM DUODENAL MUCOSA IRON DEFICIENT RATS

Fig. 3

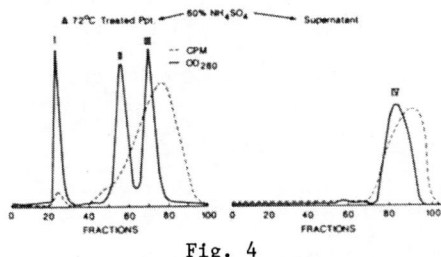

ACA22 COLUMNS OF SOLUBLE ^{59}Fe FRACTIONS FROM DUODENAL MUCOSA

Fig. 4

The use of 60% NH_4SO_4 prior to separations on Ultrogel ACA22 columns showed that Peak IV remained soluble, whereas, peaks I, II & III were precipitated by 60 percent NH_4SO_4 (Fig. 4)

This provided a simple method of separating Peak IV from other fractions. Using the outlined isolation schema; yields of all iron binding fractions have been satisfactory and reproducible. More specific observations related to each of these iron binding isolates are provided below.

Water Soluble Iron Binding Substances

ACA22 Ultrogel permits separation of iron binding substances into four peaks. These become more distinct if preparations are treated with protamine sulfate, heat and 60 percent ammonium sulfate. Peak I occurs at the void volume of the column and contains cellular debris and nucleic acids with a molecular size in excess of one million daltons. The radioiron trapped in this peak and its protein concentration is markedly diminished by treatment of specimens with protamine.

Peak II contains complexes between about 100,000 and 1 million daltons and was shown on autoradiographs to contain two distinct bands of radioiron on native PAGE gels (Fig. 5). Similar preparations of liver from rats which had received both intravenous radioiron and iron-59 labeled hemoglobin showed only a single band of radioactivity which was presumed to be ferritin. This is similar to observations by Pearson in a posthumous publication (J. Nutr.,1969).

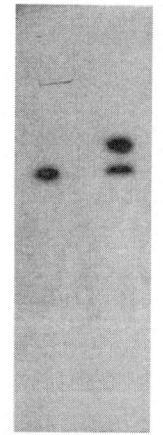

Fig. 5

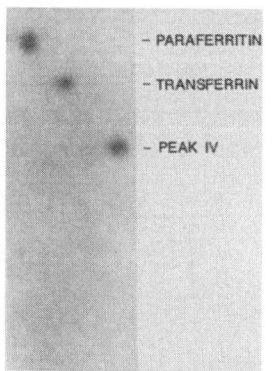

Fig. 6

These findings led us to investigate intestinal preparations for a large molecular weight protein within this peak which was distinct from ferritin. Iron deficient animals were used in most of our experiments to minimize the ferritin concentrations in this fraction. Peak III has a molecular size of about 75,000 daltons. Electrophoresis of this fraction showed a single laddered band characteristic of transferrin. Peak IV contains an iron binding complex which we subsequently showed had a molecular size smaller than albumin and larger than ovalbumin (60,000 daltons) [Autoradiograph, Fig. 6].

1. Peak II Protein (Paraferritin)

This peak has an affinity for nucleic acid which was separated using EDTA and an ACA22 Ultrogel column. Serial chromatographic steps using agarose gel, DE52 and hydroxyapatite permitted isolation of an iron binding complex which was not reactive with rat liver antiferritin by an ELISA method. Molecular sizing of the peak II protein on ACA22 columns with comparison of proteins of known molecular size indicated it had a molecular size of about 520,000 daltons and was slightly larger than ferritin (Fig.7).

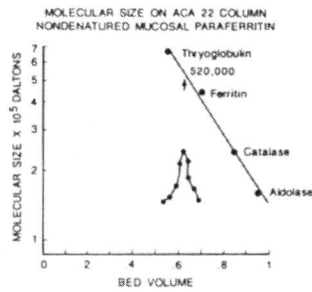

Fig. 7

On native gels, multiple bands were observed (labelled P) which were different than those observed for ferritin (F) or transferrin (T) [Fig.8]. On SDS PAGE gels the new putative protein from peak II had several major bands. That these bands were subunits of a single molecule was shown by the addition of dimethyl-subermidate (DS) to the specimen before SDS & heating (Fig.9).

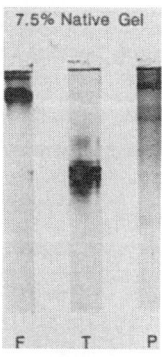

Fig. 8

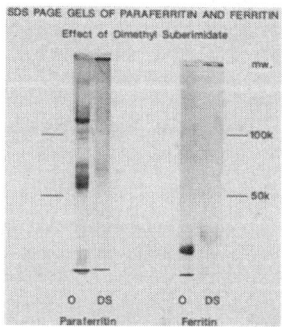

Fig. 9

This prevents subunit formation but does not permit aggregation between different molecules. In these studies paraferritin showed only a dense band near the point of application similar to observations with ferritin which breaks down into 18,000 and 22,000 dalton subunits.

Polyclonal antibodies to paraferritin have been raised in rabbits which were multiply immunized with the high molecular weight band excised from acrylamide gels. Antibody has been identified which is reactive with rat intestinal mucosa but not liver, lung, spleen or brain. In sections of both rat and human duodenum, this antibody reacts with the apical portions of intestinal epithelial cells but not other anatomic portions of the gut.

2. Peak IV

Peak IV was easily isolated
from other soluble fractions
containing radioiron because it
remains soluble after treatment
with 60% NH_4SO_4. It has been
purified further following
chromatographic separation on
G50 Sephadex, DE52 and hydroxyapatite. On SDS Page gels and
on an ACA44 Ultrogel column
with comparison to proteins of
known molecular size; this
complex has a molecular size of
about 60,000 daltons and
migrates midway between albumin
(66,000) and ovalbumin (46,000)
standards (Fig.10,11).

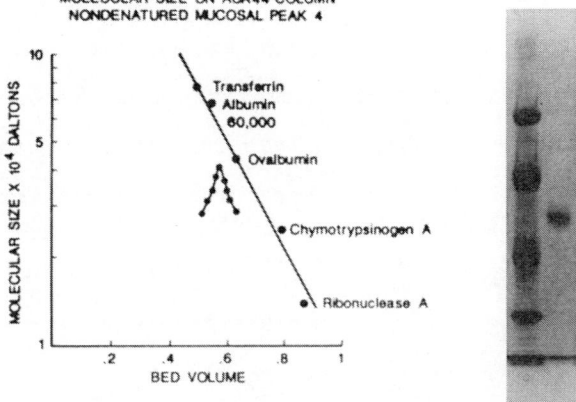

Fig. 10   Fig. 11

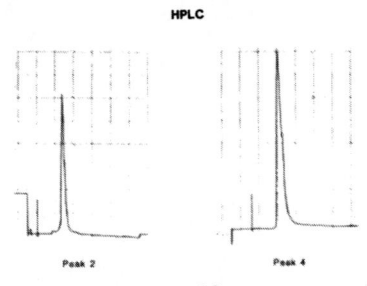

HPLC

Fig. 12

Autoradiographs of native gels have shown
migrations dissimilar to transferrin and
ferritin (Fig.6). Preparations containing this
iron binding complex were immunologically
distinguished from ferritin and transferrin
using an ELISA assay. Examination of Peak II
and Peak IV protein isolates showed single
peaks on high performance liquid chromatography
(HPLC) which on computer analysis were more
than 95 percent pure (Fig. 12). Polyclonal
antibodies were raised in rabbits to the 60,000
dalton protein which react immunologically
using an immunogold method with the apical

cytoplasm of duodenum. Significantly less reactivity was observed in the
distal small intestine with no reactivity in tissues from other body organs.

Water Insoluble Iron Binding Substances

Since about half of the radioactivity in the mucosal homogenates was found in
water insoluble precipitates, we investigated these fractions. Most of the
radioiron in water insoluble fractions was isolated in mucin despite extensive
washings of the duodenal specimens prior to preparation (60 percent). In
addition to mucin, radioiron was solubilized in triton-X100 (40%) and this was
purified by heat and chromatographic steps which included Biorad SM 2 beads,
DE52 and hydroxyapatite. Based upon solubility and staining characteristics,
we postulate that the triton-X100 soluble iron binding complex is a membrane
associated protein.

1. Membrane Heat Stable Iron Binding Complex

In mucosal homogenates from rats, we identified heat
stable triton soluble iron binding complexes with
molecular sizes of about 60,000 daltons (H_2) and
150,000 daltons (H_1) on SDS PAGE gels respectively
(Fig. 13). This constitutes about 40 percent of the
radioiron in the insoluble fractions. This fraction
stains with Schiff's reagent suggesting it contains
carbohydrate. These complexes are soluble in a number
of detergents other than triton-X100 but not in
chloroform, butanol or methanol.

Polyclonal antibodies raised in rabbits to the
60,000 dalton and 150,000 dalton membrane proteins
were crossreactive with each other but not with
ferritin or transferrin. These antibodies reacted
with the apical surface of duodenal rat mucosal cells
but not with tissues from other organs.

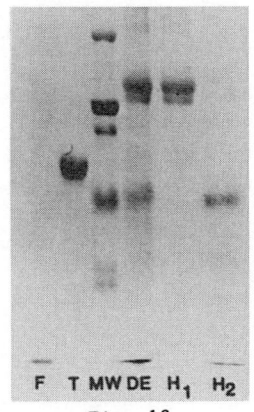

Fig. 13

2. Mucin

Approximately 60 percent of the radioiron in water insoluble fractions can not be solubilized with detergents, guanadine hydrochloride and lipid solubilizing agents. This material has a high fucose concentration and a density in cesium chloride equilibrium centrifugation similar to that reported for mucins (Fig. 14). In vitro binding of ferric iron to mucin occurs only at an acid pH when iron is soluble. Once the iron binds to mucin it remains soluble at neutral or alkaline pH. This was demonstrated by placing mixtures of radioiron and mucin which were combined either acid or alkaline pH and subsequently neutralized on a sucrose cushion and ultrocentrifuging the specimens. Mucin bound radioiron remained above the cushion; whereas unbound iron was centrifuged through the cushion as precipitated ferric iron (Fig. 15). Commercially purchased mucins bound iron with only slightly less efficiency than duodenal isolates. Mucin was shown to bind other divalent metal cations and this was inhibited in a competitive manner by iron. This observation may explain, in part, the competition which exists between metals for absorption in the small intestine.

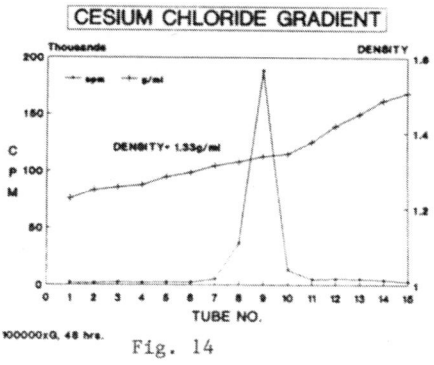

Fig. 14

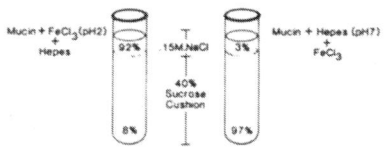

EFFECT OF pH UPON Fe BINDING TO MUCIN

Percentage of ^{59}Fe bound to mucin following ultracentrifugation 100,000XG-1 hr.

Fig. 15

COMMENT

Although there is a need for considerable additional information to determine the role of each of our isolates in the regulation of iron absorption, we can provide a hypothetical model which will need revisions with the accumulation of additional data.

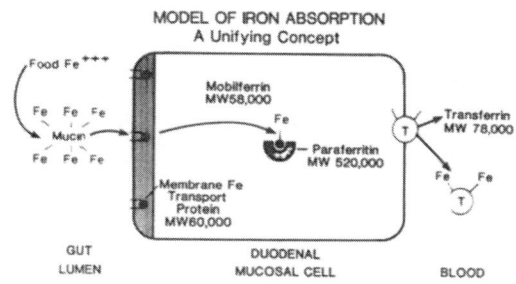

Fig. 16

Ferric and ferrous iron are delivered into the duodenum in a soluble form because of the acidity of gastric hydrochloric acid. Binding to mucin prevents precipitation of iron with the increases in pH which occur in the duodenum. This keeps both ferric and ferrous iron in an available form at a near neutral pH so that it can transit the mucosal barrier. Achlorhydric subjects would slowly become iron deficient because inorganic iron could not bind mucin. The membrane iron binding complex accepts iron from mucin and provides it a mechanism for entering the mucosal absorptive cell. The water soluble 60,000 dalton and 520,000 dalton proteins regulate iron traffic within the cytosol of the intestinal mucosal cell and deliver it to transferrin at the corporeal surface of the absorptive cell. Ferritin serves as a protective storage protein; little is synthesized in cells of iron deficient animals, whereas considerable amounts of ferritin are produced in cells of iron replete animals. Regulation of mucosal uptake of iron from the gut lumen could be affected by either synthesis of additional iron binding apoprotein or by saturation of available iron binding sites on these proteins or both. These proteins explain changes in the absorption of other nonferrous metals in various states of iron repletion because cobalt, zinc, and lead bind competitively with iron to several of them. While the experiments in this publication describe findings in the rat; observations have been possible in duodenal specimens from patients undergoing Whipple procedures for pancreatic carcinoma which have shown that human duodenum contains homologies for each of our rat derived proteins.

Dialysis and Renal Disorders

Aluminum Intoxication

Donald J. Sherrard

Division of Nephrology, Seattle Va Medical Center, 111A 1660 South Columbian Way, Seattle, WA 98108, USA

ABSTRACT:

Aluminum toxicity is a serious disorder of medical progress which occurs mostly in patients with renal failure. Aluminum enters the body either internal or parenterally. Principle toxicities are bone, brain and blood (anemia). Prevention is important - - - by limiting or avoiding aluminum. Treatment with the chelator, deferozamine is effective but hazardous.

Key words: Aluminum, Encephalopathy, osteopathy, anemia, uremia

INTRODUCTION

Aluminum toxicity is associated with several distinct clinical disorders. As aluminum accumulates in various organs it interferes with function and leads to disease. Recent information on aluminum metabolism has improved our understanding of its toxicity.

Although aluminum is primarily toxic in patients with renal disease, it is associated with disease in patients with normal renal function. Up to now most such patients have accumulated excess aluminum as a result of aluminum loading which exceeded the renal excretory ability. Patients on total parenteral nutrition with aluminum contaminated solutions (Ott, 1983) or receiving large amounts of albumin (Milliner, 1985) which is known to be heavily contaminated with aluminum are at risk. Two patients with severe liver disease have also been described who accumulated aluminum apparently due to both large oral doses and loss of hepatic excretion of aluminum (Williams, 1986). Whether renal excretion was also impaired is not clear. All of these patients with normal renal function only developed bone disease.

In contrast, patients with renal failure who accumulate aluminum develop disease in several systems other than the bones, including the hematopoietic marrow (Swartz, 1987) the central nervous system (Alfrey, 1986B) and possibly the heart (Coburn, 1988). Usually disease in these other systems is accompanied by bone disease (Sherrard, 1985). It may be that aluminum does not achieve high levels in other tissues until the bone is highly saturated.

ALUMINUM METABOLISM

Our understanding of aluminum metabolism has been markedly increased by the appearance of disorders related to aluminum accumulation. Aluminum enters the body either through an intravenous route or the gastro-intestinal tract. Normal diets contain about 25 mg and of this about 0.1% is absorbed and largely excreted by the kidney (Alfrey, 1986A). If oral intake is markedly increased absorption appears to be proportionately increased - - - this is presumed to be the case because of the prompt concomitant increase in urinary excretion. Even in normal subjects, however, aluminum may accumulate in the body if high oral doses are sustained for many years (Recker, 1977).

A modest amount of evidence suggests that there is increased bowel permeability in certain situations which can lead to a toxic accumulation of aluminum. Infants, for instance, accrue aluminum and develop toxicity much more quickly than adults (Andreoli, 1984). Along the dame line diabetic patients on dialysis, whose only aluminum exposure was from oral antacids, developed bone aluminum toxicity much more rapidly than other dialysis patients (Andress, 1987). A possible explanation for this could relate to the bowel dysfunction seen in diabetics. Finally, we have observed excessive uptake of aluminum in three subjects with Crohn's disease. One of these subjects had only modest exposure to aluminum containing antacids for a relatively brief period and normal renal function. The other two had renal insufficiency, but much greater aluminum accumulation than other patients with similar severity and duration of renal disease who did not have bowel disfunction.

A further mechanism for enhancing gut absorption of aluminum involves citrate. Various citrate salts and citric acid have been shown to enhance gut absorption of aluminum (Slanina, 1986). Two major interactions have been identified. Shohl's solution contains citrate and citric acid and is used to treat renal tubular acidosis. If aluminum containing antacids are also given, toxic levels of aluminum may be quickly achieved. Recently calcium citrate has been recommended as an alternative to aluminums for binding phosphate in patients with renal failure. Should aluminum be added as adjunctive therapy the result could be fatal, as nearly occurred in a patient we were recently asked to see.

The other route by which aluminum enters the body is the bloodstream. As noted above a variety of solutions have been found to be contaminated with aluminum. Prolonged exposure to these will lead to serious bone disease. The solution which has been best described in this regard is the dialysis fluid used to perfuse artificial kidneys. Today, it is recognized that this fluid needs to be essentially aluminum free (< 10 ug/L) to be safe for dialysis. Because of technical problems in measuring aluminum it was many years after the recognition of dialysis dementia that its cause could definitely be linked to aluminum (Alfrey, 1986b). This, almost uniformly fatal, encephalopathy devastated a number of dialysis programs before it was identified and prevention and treatment clarified.

Aluminum gets into the dialysate water from the municipal water supply. Either as a result of industrial contamination or from natural sources municipal water may have levels much too high for safe dialysis. In addition, water utilities often add alum to reservoirs to precipitate out organic material. Unfortunately the levels resulting from all these sources may be quite variable. Therefore, several determinations of aluminum levels must be lade before one can be confident of what the real situation is.

Once aluminum enters the body it is bound to transferrin in the blood and transported to various tissues. Only about 5% of plasma aluminum is unbound (Alfrey, 1986a). Therefore, once absorbed, it is not easily excreted even by normal kidneys.

In addition to transporting iron and other metals including aluminum, transferrin also is involved in the cellular uptake of iron (Alfrey, 1986a). It may well be that cellular uptake of transferrin bound aluminum occurs by a similar mechanism. For iron this involves the binding of transferrin to its receptor at the cell surface (virtually all cells have such receptors). The transferrin-receptor complex is then internalized, the iron released and the unsaturated transferrin returned to the extra-cellular fluid. Whether aluminum enters cells by this or some other mechanism is unknown; it is clear that aluminum does enter cells.

ALUMINUM TOXICITY

Bone toxicity may result from at least 3 different effects of aluminum (Goodman, 1989). First, aluminum enters osteoblasts and inhibits both osteoblast function and proliferation. Second, it binds to the surface of bone and, in vitro, has been shown to inhibit growth of apatite crystals. Third, it inhibits release of parathyroid hormone from the parathyroid glands, thus reducing levels of this potent osteoblast trophic factor.

The fact that anemia results from aluminum toxicity is well established (Swartz, 1985). The mechanism is not clear. One proposal is that aluminum blocks the intra-marrow transport of iron by an effect on marrow macrophages, the site where it is commonly seen.

The central nervous system toxicity is also well documented. The mechanism is perhaps even less well understood, through a number of enzyme systems and messenger systems which exist within cells are disturbed (Goodman, 1989). It is clear that exposure to aluminum produces a specific neurologic syndrome which is cured by chelation therapy to remove aluminum.

TREATMENT OF ALUMINUM TOXICITY

Early reports of the management of aluminum toxicity were discouraging. The neurologic syndrome, particularly, was usually rapidly fatal. With the recognition that the iron chelator, deferoxamine (DFO), also bound aluminum effective treatment became available (Swartz, 1985).

Initial favorable reports of the response of bone disease and encephalopathy to DFO led to its widespread use. Anemia usually responded to withdrawal of aluminum exposure and chelation therapy was not widely used in its management. After initial enthusiasm for DFO treatment recognition of two serious toxicities has made physicians more circumspect in its use. The first of these toxicities, infections with siderophelic organisms, particularly Rhizopus, was usually fatal (Windus, 1987). The second, precipitation or worsening of encephalopathy, was often fatal, but at least caused serious morbidity (Sherrard, 1988).

While this toxicity tempered the enthusiasm for DFO use it did not prevent the need for it. The current, practical approach is to eliminate aluminum exposure from patients who are asymptomatic or whose only symptoms are due to anemia. For those with symptoms of bone disease (bone pain or fracture) or encephalopathy DFO is still indicated. Currently doses in the range of 5-10 mg/kg given once weekly are recommended. In conjunction with the elimination of aluminum exposure, such doses appear effective and safe.

PREVENTION OF ALUMINUM TOXICITY

Aluminum salts are used to bind phosphate in the gut. Limiting dietary phosphate intake will reduce the need for "phosphate binders". Calcium salts (e.g. calcium carbonate) are somewhat less effective than aluminum salts, but can be used as an alternative. With newer more efficient dialyzers phosphate clearance is improved; thus levels can be better controlled now by dialysis.

The other sources of aluminum must also be minimized. Water used to prepare dialysate must be screened and treated as needed to keep levels of aluminum below 10 ug/L. Parenteral fluids which are given for any prolonged period must also be screened and their aluminum content kept to a minimum.

CONCLUSION

Aluminum toxicity is a serious disease of medical progress. In this discussion I have outlined the metabolism of aluminum and the mechanisms of toxicity as well as its prevention and treatment. Further study of this curious disorder may well be of value in our understanding of a number of metabolic disorders of bone and brain.

REFERENCES

Alfrey AC (1986) Aluminum metabolism. Kidney Int 295:8-11.

Alfrey AC (1986b) Dialysis encephalopathy. Kidney Int 295:53-57.

Andreoli SP, Bergstein JM, Sherrard DJ (1984) Aluminum intoxication from aluminum-containing phosphate binders in children with azotemia not undergoing dialysis. N Engl J Med 310:1079-1084.

Andress DL, Kopp JB, Maloney NA, Coburn JW, Sherrard DJ (1987) The early deposition of aluminum in bone in diabetic patients on hemodialysis. N Engl J Med 316:292-296.

Coburn JW, Norris KC, Sherrard DJ, Bia M, Llach F, Alfrey AC, Slatopolsky E (1988) Toxic effects of aluminum in end stage renal disease. Am J Kidney Dis 12:171-184

Goodman WG (1988) The spectrum of aluminum intoxication in man and potential mechanisms of tissue injury. Proceedings 8th International Conference on Calcium Regulating Hormones (in press).

Kachny WD, Hegg P, Alfrey AC (1977) Gastrointestinal absorption of aluminum from aluminum-containing antacids. V Engl J Med 296:1389-1390.

Millner DS, Shinaberger JH, Shuman P, Coburn JW (1985) Inadvertent aluminum administration during plasma exchange due to aluminum contamination of albumin-replacement solutions. N Engl J Med 312:165-167.

Ott SM, Maloney NA, Klein GL, Alfrey AC, Ament ME, Coburn JW, Sherrard DJ (1983) Aluminum is associated with low bone formation in patients receiving chronic parenteral nutrition. Am Intern Med 98:910-914.

Recker RR, Blotcky AJ, Leffler JA, Rack EP (1977) Evidence for aluminum absorption from the gastrointestinal tract and bone deposition by aluminum carbonate ingestion with normal renal function. J Lab Clin Med 90:810-817.

Sherrard DJ, Ott SM, Andress DL (1986) Pseudohyperparathyroidism. Am J Med 79:127-130.

Sherrard DJ, Walker JV, Boykin JL (1988) Precipitation of dialysis dementia by deferoxamine treatment of aluminum-related bone disease. Am J Kidney Dis 12:126-130.

Slanina P, French W, Ekstrom L, Loof L, Slorach S, Cedergren A (1986) Dietary citric acid enhances absorption of aluminum in antacids. clin Chem 32:539-541.

Swartz RD (1985) Deferoxamine and aluminum removal. Am J Kidney Dis 6:358-364.

Swartz R, Dombrouski J, Burnatowska, Hledin M, Mayor G (1987) Microcytic anemia in dialysis patients; Reversible marker of aluminum toxicity. Am J Kidney Dis 9:217-223.

Williams JW, Vera SR, Peters TG, Luther RW, Bhattacharya S, Spears H, Graham A, Pitcock JA, Crawford AJ (1986) Biliary excretion of aluminum in aluminum osteodystrophy with renal disease. Am Intern Med 104:782-785.

Windus DW, Stokes TJ, Julian BA, Fenves AZ (1987) Fatal rhizopus infections in hemodialysis patients receiving deferoxamine. Am Intern Med 107:678-680.

Search for the Unknown Trace Element Abnormalities in Uremia

YUSUKE TSUKAMOTO[1], SHIGERU IWANAMI[1], OSAMU ISHIDA[2], and FUMIAKI MARUMO[2]

[1] Department of Medicine, Kitasato University School of Medicine, Kitasato Biochemical Laboratory, SMI Bristoles, 1-15-1 Kitasato, Sagamihara, Kanagawa, 228 Japan
[2] Second Department of Medicine, Tokyo Medical and Dental University, 1-5-45 Yushima, Bunkyo-ku, Tokyo, 113 Japan

ABSTRACT

Deterioration of renal function produces various abnormalities. of trace element metabolism in the patients. The therapy of uremia such as an aluminum containing phosphate binder and hemodialysis also causes these abnormalities in the patients with chronic renal failure. These abnormalities could be an important causality of uremia.

KEY WORDS: Hemodialysis, Chronic renal failure, Vanadium, Fluoride, Hair

INTRODUCTION

Study on trace element abnormalities is extremely important in the patients with renal failure for following reasons. First of all, kidney is a main excretory organ for many elements such as Al, F and V. Naturally, a failure in renal function could cause an accumulation of these elements. Secondly, hemodialysis therapy could affect serum concentration of every elements which are present as a free form at any proportion in a blood. Since dialysate are prepared with water, any contamination of water by metals would causes its accumulation. Aluminum is the typical example for such an incidence. On the other hand, some essential elements might be removed extensively by dialysis and causes a deficiency in these elements because dialysate contains only 6 major electrolytes. Zinc is the example for such a deficiency (figure 1). These facts obligate us to study the distribution of every essential trace elements as many as possible in the patients with renal failure.

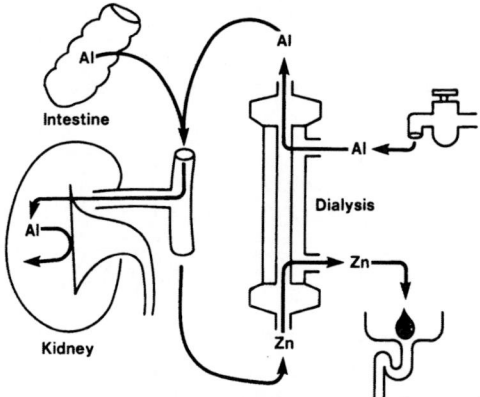

Fig. 1. Etilogy of abnormal trace elements metabolism in uremia.

ANALYTICAL METHODS

We used non-destructive neutron activation analysis for the measurement of elements in biological tissues(1). For liquid material, such as blood, urine and water, we used flameless atomic absorption(2)(3). We also used an ion selective electrode for fluoride measurements in serum, urine and bone.

RESULTS

1. Determination of trace elements in hair and fingernails of patients with CRF.

Difficulty in the diagnosis of deficiency or excess of trace elements is that serum level of these elements does not always reflect the distribution in tissues. For this reason, analysis on nail and hair has been widely employed for the diagnosis on many diseases caused by metal intoxication or deficiency such as arsenics and lead. Table 1 shows the concentrations of 9 trace elements in the hair of patients with chronic renal failure. Nondialyzed patients were at conservative therapy stage and their serum creatinine levels varied between 5.7 and 12.1 mg/dl. Hemodialyzed patients had been maintained on chronic hemodialysis therapy from 3 to 68 months, with a mean duration of 21 months. Al and Ca increased in the hair of both nondialyzed and hemodialyzed patients. V increased only in the hair of hemodialyzed patients. Decreased Zn content was only found in nondialyzed patients.

Table 1. Concentration of trace elements in the hair of patients with CRF (1).

	Mg µg/g	Br µg/g	V pg/g	Al µg/g	Ca µg/g	Zn µg/g	Mn pg/g	Cu µg/g	As pg/g
Control	132±11 (13)	5.33±1.20 (13)	433±15 (14)	61.6±3.1 (14)	351±62.4 (11)	155±29.5 (9)	417±70 (8)	32.8±2.6 (8)	294±199 (8)
Nondialyzed	180±28 (9)	11.2±4.1 (8)	740±223 (10)	157±52*** (9)	642±132* (10)	32.5±9.9** (8)	697±161 (8)	40.2±11.0 (7)	336±264 (8)
Hemodialyzed	159±19 (13)	7.85±1.86 (11)	661±57*** (14)	78.6±7.1* (12)	1020±192** (13)	173±30.3 (10)	946±282 (9)	57.8±17.2 (9)	563±393 (9)
Hemofiltered	204±49 (8)	8.93±4.20 (7)	449±109 (8)	64.6±15.7 (9)	735±284 (9)	—	—	—	—

*p<0.05, **p<0.01, ***p<0.001, Figures in parentheses are numbers of samples.

Table 2 shows the concentrations of 6 elements in the fingernails of the patients. As and Cu were extremely high in the nails of nondialyzed patients. Mn was also high in the nails of nondialyzed patients. However, finger nails of hemodialyzed patients did not show any significant difference compared to normals.

Table 3 summarizes the results of trace element measurements in plasma, hairs and fingernails. Serum Zn levels decreased in both patients group, but decreased level was found only in the hair of nondialyzed patients. Cu levels increased in plasma and nails of both groups. In these results, we have to pay special attention on increased arsenic levels in the nails of

Table 2. Concentration of trace elements in the fingernails of patients with CRF (1).

	Al µg/g	Ca µg/g	Zn µg/g	As pg/g	Mn pg/g	Cu µg/g
Control	13.8±4.3 (6)	458±114 (6)	186±80.4 (10)	333±176 (7)	348±135 (9)	19.0±6.3 (8)
Nondialyzed	—	—	192±120 (8)	1150±675** (8)	722±302* (6)	120±76* (6)
Hemodialyzed	25.0±15.5 (6)	603±468 (6)	171±82.2 (11)	647±422 (8)	709±590 (8)	39.2±31.4 (10)
Hemofiltered	21.8±8.9 (6)	453±175 (6)	—	—	—	—

*$p<0.05$, **$p<0.01$, Figures in parentheses are numbers of samples.

Table 3. Summary of trace element abnormalities in plasma, hair and fingernails

		Zn	Mn	Cu	Al	Ca	Mg	Br
PLASMA	Non-HD	↓↓	→	↑↑	↑↑	→	→	→
	HD	↓↓	→	↑↑	↑↑	↓	↑↑	↓↓
HAIR	Non-HD	↓↓	→	→	↑	↑↑	↑	→
	HD	→	↑	→	↑↑	↑↑	↑	→
NAILS	Non-HD	→	↑↑	↑↑				
	HD	→	↑	↑	↑	→		

nondialyzed patients and also increased vanadium levels in the hair and serum of hemodialyzed patients. I will show more details on vanadium accumulation later.

2. Abnormal accumulation of fluoride in the patients with CRF.

Recently, we focus our research on two particular trace elements. Those are F and V. Serum F concentration was higher in both nondialyzed (43.6±22.8µg/l, n=19) and hemodialyzed(28.8±14.4µg/l, n=20) patients with CRF than in normal controls (7.9±2.6µg/l, n=50). Renal fluoride clearance positively correlated with creatinine clearance (r=0.934, n=10). F contents in biopsied iliac bone were also very high (464±106µg/g, n=11) in hemodialyzed

patients compared to normals (44±35µg/g, n=9) These results indicate that decreased renal clearance of F causes the retention of this element in the patients with renal failure. Although dialysis by the low F containing water could reduce serum F levels, intake of F must exceed the amount of removal and caused accumulation in the bone. We do not know whether F accumulation in the bone is beneficial or toxic to the patients at present time.

3. Vanadium accumulation in hemodialyzed patients with CRF.

Serum V levels were extremely high in hemodialyzed patients of our dialysis center located in the suburb of Tokyo, Kanagawa-ken. Hemodialyzed patients had an average of 23.9±11.3ng/ml (n=43) of serum V, which was almost 100 times higher than those of normal persons (0.12±0.05ng/ml, n=42). There was the positive correlation between duration of hemodialysis and skin content of V (r=0.9115, n=8).

We measured V concentration in tap water from 26 cities in Japan and 9 cities in the U.S.. Unfortunately, tap water from our prefecture, Kanagawa-ken contained extremely high amount of V (22.6ng/ml) and this value was the highest. The cities in which tap water contained vanadium more than 1 ng/ml were Baltimore, Houston, Miami and LA in the U.S. , and Muroran-shi, Kitami-shi, Kumamoto-shi, Kagoshima-shi and Miyazaki-shi in Japan.

SUMMARY

1. It may be possible to diagnose the accumulation of V and Al, Zn deficiency and metastatic calcification by the measurement of these elements in the hair of the patients.
2. The contents of As, Mn and Cu increased in the fingernails of nondialyzed patients.
3. F concentration increased in the serum and bone of the patients with CRF.
4. Contamination of tap water with V causes its accumulation in hemodialyzed patients.
5. In order to prevent V accumulation, reverse osmosis is the absolute requirement for hemodialysis even in the area in which tap water contains low V.

REFERENCES

1. Marumo F, Tsukamoto Y, Iwanami S, Kishimoto T, Yamagami S (1984) Trace element concentrations in hair, fingernails and plasma of patients with chronic renal failure on hemodialysis and hemofiltration. Nephron 38:267-272

2. Tsukamoto Y, Iwanami S, Marumo F (1980) Disturbances of trace element concentrations in plasma of patients with chronic renal failure. Nephron 26:174-179

3. Ishida O, Kihira K, Tsukamoto Y, Marumo F (1989) Improved determination of vanadium in biological fluids by electrothermal atomic absorption spectrometry. Clin Chem 35: 127-130

Neoplasma

Selenium and Cancer

RAYMOND J. SHAMBERGER

Ciba-Corning Diagnostics, 132 Artino St., Oberlin, OH 44074, USA

ABSTRACT

The vast majority of about 60 carcinogenesis experiments using dietary selenium have shown a significant inhibitory effect in 10 of 11 organ systems. Selenium inhibited tumor development in the epithelial tissues of colon, mammary gland, liver, skin, stomach and esophagus. Tumors in these experiments were induced by carcinogens, ultraviolet light and mouse mammary tumor virus. The inhibition of cancer by selenium may be a supranutritional effect, but more research is needed at nutritional levels. Selenium seems to modulate initiation and promotion and interacts with other trace elements. Selenium affects transformation and sodium selenite may react through a variety of mechanisms: sodium selenite may affect chromosome breakage; DNA and RNA polymerases; phases of the cell cycle; amino acid synthesis; carcinogen metabolism; and also affects enzymes such as protein kinase C, ornithine carboxylase and DNA methylase. Sodium selenite also enhances production of IgG and IgM antibodies.

KEY WORDS: Cancer, Selenium

INTRODUCTION

Summaries of about 60 experiments found that in the vast majority of the experiments, there was an inhibitory effect by selenium supplementation in 10 of 11 organ systems. (Medina 1988) Eight experiments showed no effect and four experiments showed enhancement. These experiments are summarized and updated in Table 1. Selenium inhibited tumor development in epithelial tissues of colon, mammary gland, liver, skin, stomach and esophagus. (Shamberger 1989) Tumors were induced by several types of carcinogens, ultraviolet light, and mouse mammary tumor virus. Because of the chemopreventive effect by selenium of the colon, mammary and stomach cancer, selenium has interest because of its possible value against human cancer. Colon, mammary and stomach cancer are major causes of cancer death in humans. It is also interesting that selenium prevented cancer induced by carcinogens, virus and ultraviolet irradiation. Therefore, selenium may inhibit cancer through the same molecular mechanism. The key steps of viral, carcinogen and irradiation induced cancer that is inhibited by selenium may be the same for all three types of carcinogenic stimulus. These observations may be ultimately important in identifying the key steps of cancer inhibition.

SUPRANUTRITIONAL OR CHEMOPREVENTIVE EFFECT?

Most of the experiments in Table 1 used sodium selenite at dietary levels of 0.5 and 1.0 ppm. These levels should be considered supranutritional because the rodent requirement for selenium is about 0.1 ppm. Selenium in the organic forms are also effective inhibitors of tumorigeneses but are usually utilized at greater dietary levels than the inorganic forms. Selenomethionine and p-methoxybenzeneselenol inhibit mammary and stomach tumorigeneses (Medina 1988).

TABLE 1. EFFECT OF SELENIUM SUPPLEMENTATION ON TUMORIGENESIS
(MEDINA 1988)

Organ	Species	Carcinogen	Effect on tumorigenesis		
			decrease	increase	no effect
Liver	rat	DAB,AAF, DEN, AFB1	8	2	0
Colon	rat	DMH,MAM, AOM, BOP	10	0	0
Colon	mouse	DMH	1	0	0
Mammary gland	rat	DMBA MNU, Ad-9	13	1	0
Mammary gland	mouse	DMBA MMTV	11	0	3
Skin	mouse	DMBA,UV	5	1	0
Stomach	mouse	BP	1	0	1
Stomach	rat	MNNG	1	0	0
Esophagus	rat	MBN	2	0	0
Oral Cavity	hamster	DMBA	2	0	0
Trachea	hamster	MNU	1	0	2
Pancreas	hamster	BOP	1	0	1
Kidney	rat	AOM	1	0	0
Lung	mice	urethan	1	0	1

DAB=3-methly-4-diaminoazobenzene;AAF=2-acetylaminofluorene;
DEN=diethylnitrosamine; AFB1=aflatoxin B1; DMH=1,2-dimethylhydrazine;
MAM=methylazoxymethanol acetate; AOM=azoxymethane;
BOB=bis(2-oxopropyl)nitrosamine; DMBA=7,12-dimethylbenzanthracene;
MNU=methylnitrosourea; Ad-9=adenovirus type 9; MMTV=mouse mammary tumor virus;
BP=benzo(a)pyrene; MMNG=N-methyl-N'-nitro-N-nitrosoguanidine;
MBN=methylbenzylnitrosamine.

There have been several studies investigating the effects of selenium
deficient diets on carcinogenesis. In general, a selenium deficiency does
not affect carcinogenesis in the colon or liver of the rat (Combs, 1985).
Selenium deficiency enhances carcinogenesis in mouse skin and rat mammary
glands. Even though the effect seems to be supranutritional, there should
be more research to study the effects of marginal selenium diets on
carcinogenesis in animals.

DOES SELENIUM MODULATE INITIATION OR PROMOTION?

Ip, et al (1981a) have also studied the effect of varying the timing of feeding sodium selenite to animals dosed with DMBA. Experimental groups were
supplemented with 5 ppm for various periods of time: -2 to 24 weeks;
-2 to +2; +2 to +24; +2 to +12; +12 to +24; and -2 to +12. Selenium
inhibited tumorigenesis to various degrees at all time periods, but selenium
fed throughout the experiment exhibited a maximal inhibition. In general,
selenium seems to affect initiation more so than the promotion phase. It is
difficult to separate the biochemical effects because the initiation and
promotion effects might overlap. The initiator and promotor are given only
two or three weeks apart. Derepression and repression effects of the
initiator and promotor might also overlap.

INTERACTIONS WITH OTHER TRACE ELEMENTS

Schrauzer and Ishmael (1974) have found that selenite and arsenite lowered the incidence of mouse mammary tumor virus induced mammary cancer, but the growth of transplanted mammary tumors was greatly enhanced by arsenite. Ip and Ganther (1988) found that arsenite reduced the chemopreventative effect of selenium in the DMBA-induced rat mammary cancer system, but together with trimethylselenonium metabolite also had a chemopreventative effect. Arsenite may be blocking the methyltransferase step, but when fed with 40 ppm of trimethylselenonium instead of 3 ppm selenite, may enhance demethylation and make a metabolic return to a more active methylated form such as dimethylselenide. Ip (1988) has also found that by feeding a high level of methionine that greater amounts of selenium can be fed to rats. In this way higher amounts of selenium could be fed to animals without toxicity, thereby enhancing the chemopreventative effect of selenium. Selenium may also interact with other elements such as cadmium and zinc. Cadmium itself seems to be a carcinogen. Its effects are known to be modulated by selenium.

EFFECTS OF SELENIUM ON TRANSFORMATION

Sodium selenite has inhibited the transformation of mammary cells in organ culture of the whole mammary glands from BABL/c female mice (Chatterjee 1982). Selenium may act by preventing progression of the transformed cells to potentially neoplastic lesions in the glands "in vitro". This model may become an important tool to study the mechanism of cancer.

MECHANISMS OF SELENIUM CHEMOPREVENTION

DNA Effect

Wortzman, et al. (1980) has found that dietary selenium reduces the appearance of DNA fragments induced by aminoacetylfluorene (AAF) as determined by sucrose gradient centrifugation. These breaks were repaired 24 hours after injection of AAF, but under the same conditions, AAF failed to damage the DNA in the livers of selenium-supplemented rats. Lawson and Birt (1981) have found that 5 ppm selenium helped reduce single strand breaks induced by N-Nitroso(2-oxopropyl)amine. Human chromosome breakage induced by 7,12-dimethylbenzanthracene has been reduced in the cultures with incubated selenium (Thompson 1980). Peroxidations may increase damage to DNA. Shamberger (1972) and Ip (1981b) have shown there is an association between an increase of peroxidation and skin and mammary tissue carcinogenesis especially when dietary fat is increased. Lipid peroxidation may be the primary mechanism by which dietary fat increases cancer development.

Effect on DNA and RNA Polymerases

Frenkel et al (1987) have found that DNA and RNA polymerases were inhibited by 1 mm selenotrisulfide. Both enzymes are necessary for cellular growth and both enzymes are activated by zinc. Selenium may be inactivating the zinc on the enzymes.

Effect on the Cell Cycle

Medina (1986) has examined the effect of selenium on DNA synthesis in a mammary cell line grown in vitro. Low doses of selenium stimulated cell growth and high doses inhibited cell growth. When cell growth was decreased, there was also a decreased cell number, reduced uptake of 3H-thymidine into DNA, decreased DNA labeling index and decreased DNA synthesis. A delay in the G2 and S-phase of the cell cycle was observed by flow cytofluorometry of selenium treated mammary cells. The biochemical basis for this effect remains to be determined.

Amino Acid Synthesis

Vernie et al (1983) have observed the reaction products of selenite and thiols have an inhibitory effect on protein synthesis in intact P815 and LI210 cells. Some of the inhibition may be attributed to toxicity, but it is also possible that selenium at nutritional levels might inhibit protein synthesis and cellular growth.

Effect on Carcinogen Metabolism

High-performance liquid chromatography analyses have shown that selenium supplemented animals dosed with AAF excreted less N-hydroxy-AAF but more 5-hydroxy-AAF. Dietary selenium may be protecting against AAF induced hepatogenesis by at least inpart to its ability to inhibit the in vivo production of N-hydroxyl-AAF, which is the proximate carcinogenic metabolite of AAF (Besbris 1982). Milner et al (1985) have found that sodium selenite inhibits the binding of DMBA to DNA in tertiary cultures of fetal mouse cells. Sodium selenite seems to be selective in inhibiting the anti-dihydro-diol-epoxide product formation at certain times in the presence of an induction of a DMBA-activating enzyme system. Once induction has occured, sodium selenite seems to be no longer capable of inhibiting the DMBA-DNA binding.

Effect on Protein Kinase C

Selenium dioxide, selenious acid, and selenic acid have been found to inhibit protein kinase C from acute myelocytic leukemia cells (Su 1986). Protein kinase C has been shown to play an important role in tumor promotion, cell regulation and in membrane signal transduction. The amounts of selenium used to inhibit protein kinase C is relatively large and may not have physiological significance.

Effect on Ornithine Decarboxylase

Sodium selenite, GSH and vitamin E all decreased phorbol ester induced ornithine decarboxylase activity and increased glutathione peroxidase activity in mouse epidermis in vivo (Perchellet 1987). The ornithine decarboxylase step is a key step in tumor promotion and reducing the enzyme level seems to also reduce the carcinogenic effect.

DNA Methylase

Sodium selenite is a good inducer of hemoglobin production in Friend erythroleukemic cells (Cox 1986). DNA is hypomethylated and sodium selenite at a concentration of 20 um will produce hemoglobin in 70-80% of the cells. DNA methylase is very sensitive to selenite inhibition. Some data indicates that alterations in the normal methylation process plays an important role in the cancer process. These studies support the hypothesis that the regulatory process of gene expression may be dependent on the methylation of DNA. More research using other experimental systems is needed to establish the relationship between sodium selenite and hypomethylation of DNA.

Effect on Immunoglobulin Antibodies

Dietary supplementation and injections of selenium have been shown to enhance production of IgG and IgM anti-sheep red blood cell (SRBC) antibodies in immunized mice (Spallholz 1973). Selenium supplementation has been observed to cause the enhanced expression of spontaneous killer cell cytotoxicity in spleen cells and of specific cytotoxic T-lymphocyte cytotoxicity in peritoneal exudate cells (Petrie 1989).

CONCLUSION

Selenium has a marked inhibitory affect on the cancer process in many tumor systems. More work is needed to establish whether or not this is a supranutritional or a nutritional effect. The number of possible mechanisms makes it difficult to predict the best mechanism. However, study of the mechanism is important because insight can be gained into the cancer process itself.

REFERENCES

Besbris H J, Wortzman M S, Cohen A M (1982) Effects of dietary selenium on the metabolism and excretion of 2-acetylaminofluorene in the rat. J Toxicol Environ Health 9: 63-76

Chatterjee M, Banerjee M R (1982) Selenium mediated dose-inhibition of 7,12-dimethylbenzanthracene-induced transformation of mammary cells in organ culture. Cancer Lett. 17: 187-195

Combs Jr. G F, Clark L C (1985) Can dietary selenium modify cancer risk? Nutr Rev 43: 325-331

Cox R, Goora H A (1986) A study of the mechanism of selenite-induced hypomethylated DNA and differentiation of Friend erythroleukemic cells. Carcinogenesis 7: 2015-2018

Frenkel G D, Walcott A, Middleton C (1987) Inhibition of RNA and DNA polymerases by the product of the reaction of selenite with sulphydryl compounds. Molec Pharmacol 31: 112-116

Ip C (1981a) Modification of mammary carcinogenesis and tissue peroxidation by selenium deficiency and dietary fat. Nutr Cancer 2: 136-142

Ip C (1981b) Prophylaxis of mammary neoplasia by selenium supplementation in the initiation and promotion phases of chemical carcinogenesis. Cancer Res 41: 4386-4390

Ip C Ganther H (1988) Efficiacy of trimethylselenonium versus selenite in cancer chemoprevention and its modulation by arsenite. Carcinogenesis 9: 1481-1484

Ip C (1988) Differential effect of dietary methionine on the biopotency of selenomethionine and selenite in cancer chemoprevention. J Nat Cancer Inst 80: 258-262

Lawson T, Birt D (1981) BOP induced damage of pancreas DNA and its repair in hamsters pretreated with selenium. Proc Am Assoc Cancer Res 22: 93

Medina D (1986) Selenium and murine mammary tumorigenesis. In Volume II, Diet, Nutrition and Cancer: A critical Evaluation. CRC Press, Boca Raton, Fla 23-42

Medina D, Morrison D G (1988) Current ideas on selenium as a chemopreventative agent. Pathol Immunopathol Res 7: 187-199

Milner J A, Pigott M A, Dipple A (1985) Selective effects of sodium selenite on 7,12-dimethylbenzanthracene DNA binding in fetal mouse cell cultures. Cancer Res 45: 6347-6354

Perchellet J P, Abney N L, Thomas R M, Guislair Y L, Perchellet E M (1987) Effects of combined treatments with selenium, glutathione and vitamin E on glutathione peroxidase activity, ornithine decarobxylase induction, and complete and multistage carcinogenesis in mouse skin. Cancer Res 477-485

Petrie H T, Klassen L W, Klassen P S, O'Dell J R, Kay, H D (1989) Selenium and the immune response: 2. Enhancement of murine cytotoxic T-lymphocyte and natural killer cell cytotoxicity in vivo. J Leuk Biol 45: 215-220

Schrauzer G N, Ishmael D (1974) Effects of selenium and of arsenic on the genesis of spontaneious mammary tumors in inbred C3H mice. Ann Clin Lab Sci 4: 411-417

Shamberger R J (1972) Increase of peroxidation in carcinogenesis J Nat Cancer Inst 48: 1491-1497

Shamberger R J (1989) Selenium and vitamin E in cancer prevention. Proceeding of the Second International Meeting for Vitamin and Nutritional Oncology. In Press

Spallholz J E, Martin J L, Gerlach M L, Heinzerling R H (1973) Immunological responses of mice fed diets supplemented with selenite selenium. Proc Soc Exp Biol Med 143: 685-689

Su Huai-De, Shoji M, Mazzei G J, Vogler W R, Kuo J F (1986) Effects of selenium compounds on phospholipid/Ca2+dependent leukemic cells. Cancer Res 46: 3684-3687

Thompson H J, Becci P J (1980) Selenium inhibition of N-methyl-N-nitrosourea-induced mammary carcinogenesis in the rat. J Nat Cancer Inst 65: 1299-1301

Vernie L N, Vries M D, Karreman L, Topp R J, Bont W S (1983) Inhibition of amino acid incorporation in a cell-free system and inhibition of protein synthesis in cultured cells by reaction products of selenite and thiols. Biochem et Biophys Acta 739: 1-7

Wortzman M S, Besbris H J, Cohen A M (1980) Effect of dietary selenium on the interaction between 2-acetylaminofluorene and rat liver DNA in vivo. Cancer Res 40: 2670-2676

Kinetics of Trace Elements in Cancer Patients

KAZUO SAITO

Department of Hygiene and Preventive Medicine, Hokkaido University School of Medicine, Nishi 7, Kita 15, Kita-ku, Sapporo, 060 Japan

ABSTRACT

Blood copper, zinc, manganese, selenium, magnesium, calcium, superoxide dismutase (SOD), catalase (CAT) and glutathion peroxidase (GSH-px), albumin, alpha 2-macrogrobulin and ceruloplasmin were determined in patients with stomach cancer. Significantly high blood copper, manganese and magnesium, low blood zinc, selenium and calcium concentrations, high CAT and low GSH-px activities in erythricytes were observed for stomach cancer patients, but SOD activity in erythrocytes did not show any significant difference as compared with normal subjects. The higher ceruloplasmin levels and low albumin and alpha 2-macrogrobulin were characteristic observations in the advanced stage of stomach cancer.

KEY WORDS: Trace elements, Stomach cancer, Metalloenzymes

INTRODUCTION

Trace elements may play an important role as either inhibitory or causative agents of cancer, but the role is as yet unknown. It is suggested that these trace elements regulates activities of the enzymes of superoxide dismutase (SOD), catalase (CAT) and glutathion peroxidase (GSH-px). However, the metabolism and the reduction of superoxide anions are not clearly explained in cancer. Also, it is not clear how do these trace elements connect with metal binding proteins of serum albumin, alpha 2 macrogrobulin and ceruloplasmin. Therefore, it is very important to know the role of these three factors in the pathogenesis of cancer through the changes in trace elements, the related metalloenzymes and the metal binding proteins.

The purpose of this study is to make clear the relationships among changes in Cu, Zn, Mn, Se, Mg and Ca concentrations in blood and tissues, the activities of SOD, CAT and GSH-px, and the changes in metal binding proteins of albumin, alpha 2 macrogrobulin, and ceruloplasmin in serum with special reference to stomach cancer.

MATERIALS AND METHODS

Eighty-one stomach cancer patients (52 males and 29 females) between the ages of 30 and 83 (average 58.6 years old) and 128 normal subjects (78 males and 50 females) between the ages of 18 and 89 (average 45.4 years old) were engaged in the measurement of Cu, Zn, Mn, Se, Mg and Ca concentrations in blood, in normal and abnormal gastric tissues obtained from stomach cancer patients by surgical operation, and in the determination of SOD, CAT, and GSH-px activities, ceruloplasmin, albumin, and alpha 2-macroglobulin concentrations in blood.

Diagnosis of patients with stomach cancer was confirmed by histologic examination of a post-operative specimens. These patients were classified into four stages of malignancy, I to IV, according to the rules laid down for surgical and pathologic treatment of stomach cancer by Japanese Research for Gastric Cancer (1985) and grouped into three categories of metastases.

Blood Cu, Zn, Mn, Mg and Ca, and tissue Cu, Zn and Mn levels were determined by atomic absorption spectrophotometry. Blood and tissue Se concentrations were determined essentially by the method of Watkinson (1966). SOD activity in erythrocytes was assayed by the method of Beauchamp and Fridovich (1971). Human erythrocyte CAT activity was assayed by Sinha's method (1972). GSH-px activity in erythrocytes was determined by a modified form of Paglia's method (1967). Plasma ceruloplasmin and alpha 2-macroglobulin were measured by the single radial immunodifusion method of Mancini et al (1965), and plasma albumin was measured by the method of cellulose acetate electrophoresis.

RESULTS

1. Blood Cu, Zn, Mn, Se, Mg, and Ca levels

Blood Cu concentration of the patients with stomach cancer in stage I and II, the nonmetastatic group, and groups with metastasis to only the lymph nodes showed a significantly lower level than in normal subjects. In contrast, these values in stage IV tended to increase. The group with metastasis to another organ, e.g., liver, spleen, peritoneum, showed a remarkable increase in comparison with normal subjects. The blood Cu level in stage IV was significantly higher than in stage I and II. The group with marked metastases to other organs showed a significantly higher level than those of the non-metastatic group and the group with metastasis to only the lymph nodes. Blood Zn level of the stomach cancer patients, in all groups at all stages, in non-metastatic or metastatic groups, showed a significant decrease compared with normal subjects. However, no significant difference within the stages of malignancy, or between metastatic and non-metastatic groups was observed.

Blood Mn concentration of the stomach cancer patients with metastases to other organs showed a significantly higher level than that of normal subjects. The same tendency was recognized in the groups of other stages in stomach cancer. Blood Se concentration of the patients with stomach cancer was significantly lower than that of normal subjects. The same results were observed at stage I, at stage II and among the non-metastatic group. Blood Mg concentrations of the patients at stage I and II were significantly higher than those of the other stages and normal subjects. These Cu, Zn, Mn, Se and Mg concentrations of the stomach cancer patients were almost the same results in whole blood and in blood plasma. Plasma Ca concentration of the patients was significantly lower than that of normal subjects, but Ca concentration in the whole blood was significantly higher than that of normal subjects.

As to relationships among blood Cu, Zn, Mn and Se in patients with stomach cancer, there were negative correlation between Cu and Zn, and positive correlation between Cu and Mn concentrations in whole blood. On the contrary no correlation among Cu, Zn, Mn and Se was recognized in normal subjects respectively. In correlations between Se concentration and Cu, Zn or Mn concentration in whole blood, correlation coefficients between Se and Cu or Mn were negative, but that between Se and Zn was positive (Saito, et al 1981, 1984, 1984).

That is, characteristic changes in blood trace elements of the patients with stomach cancer were low Cu, Zn and Se, high Mg levels in stage I. The blood trace element levels during stage II were low Cu, Zn and Se levels, and that during stage III was low Zn, and high Ca. However, high Cu and Ca levels, and low Zn and Se levels, during stage IV in the advanced type of patients with stomach cancer (Fig. 1).

2. Tissue Cu, Zn, Mn and Se concentrations

There was no statistically significant difference in mean gastric tissue Cu concentration, but the Zn content showed a decrease in comparison between normal and abnormal tissues obtained from the stomach cancer patients. Manganese concentrations in abnormal tissues with stomach cancer of the stage III and IV of malignancy were statistically significant low level in comparison with normal tissues of the same patients (Fig. 3). Se concentration in gastric tissue of stomach cancer showed significantly higher level than those of their normal tissues (Fig. 2, 3).

	Total	Stage I	Stage II	Stage III	Stage IV	Metastasis −	Metastasis +	Metastasis ++
Cu	↓	⬇	⬇	−	⇡	⬇	⬇	↑
Zn	⬇	⬇	⬇	⬇	⬇	⬇	⬇	⬇
Mn	⇡	−	−	−	⇡	−	−	⬆
Se	↓	−	−	−	⇣	⇣	−	⇣
Mg	−	↑	↑	−	−	−	−	−
Ca	⬆	−	−	⬆	⬆	−	⬆	⬆
SOD	−	−	−	−	−	−	−	−
CAT	↑	−	↑	↑	−	−	−	−
GSH-Px	⬇	⬇		⬇	⬇	⬇	⬇	⬇

⬆⬇: $p<0.01$ ↑↓: $p<0.05$ ⇡⇣: $p<0.10$

Fig. 1 Changes in Cu, Zn, Se, Mg and Ca concentrations in whole blood, and erythrocyte SOD, CAT and GSH-px activities of the patients with stomach cancer. Arrow marks show significant change as compared with normal subjects.

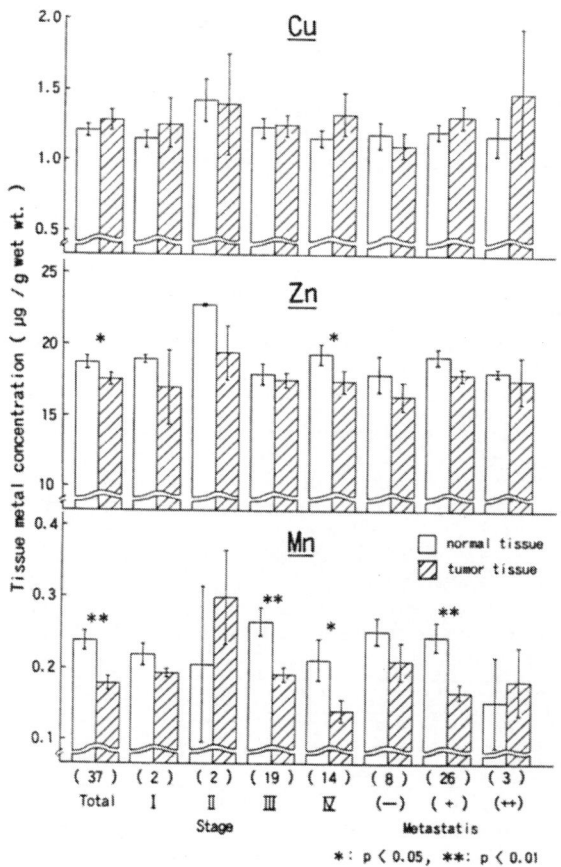

Fig. 2 Comparison between Cu, Zn and Mn concentrations in normal and stomach cancer tissues of the patients with stomach cancer. Values show mean and standard error..pa

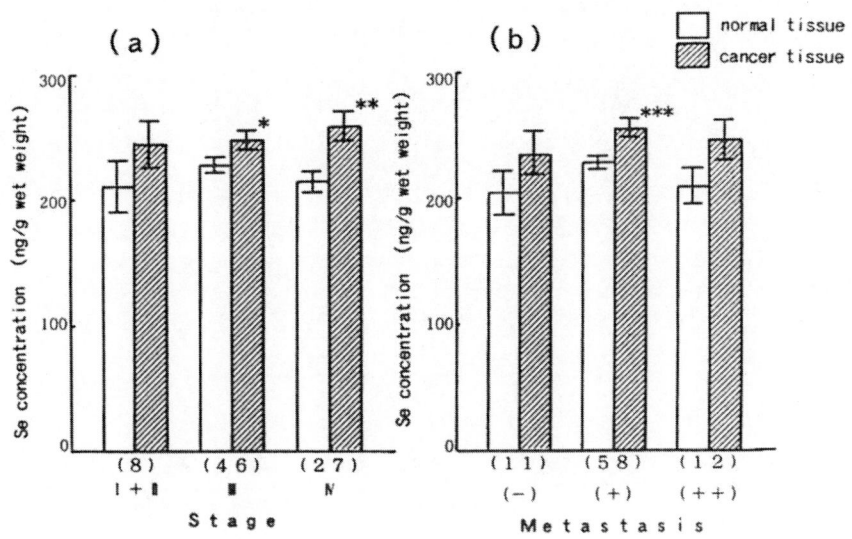

(* : p<0.05, ** :p<0.01,*** :p<0.001)

Fig. 3 Comparison between Se concentrations in normal and stomach tumor tissues of the patients with stomach cancer. Values show mean and standard error.

3. SOD, CAT and GSH-px activities

SOD activity in erythrocytes of the stomach cancer patients did not show any significant difference within the stages of malignancy and within the groups of metastasis, and as compared with that of normal subjects. CAT activity in erythrocytes of these patients showed no difference within the stages and the groups, but it showed a significantly higher activity than that in normal subjects. GSH-px activity in erythrocytes of these patients showed a significantly lower activity than that that of normal subjects (Fig. 1)(Saito, et al 1985, 1985, 1987).

4. Relationship between blood trace elements and SOD, CAT or GSH-px

There were significantly negative correlations between Cu level and SOD or CAT activities in blood of the stomach cancer patients. Decrease in Zn concentration in blood of the stomach cancer patients significantly correlated to decrease in SOD and CAT activities, and to increase in GSH-px activity. Decrease in Se concentration of the stomach cancer patients significantly correlated to decrease in GSH-px activity and to increase in SOD activity. However, there were no significantly correlations between blood Cu level and GSH-px activity, and between Blood Se level and CAT activity in patients with stomach cancer (Saito et al 1985).

5. Ceruloplasmin, albumin and alpha 2-macroglobulin in plasma

Plasma ceruloplasmin concentration showed significant decreases in stage I and nonmetastatic group in compared with normal subjects (Fig. 4). Plasma albumin concentrations were significantly lower in the stage of IV and metastatic groups than that of normal subjects (Fig. 5). Alpha 2-macroglobulin concentration showed a marked decrease in stomach cancer patients (Fig. 6).

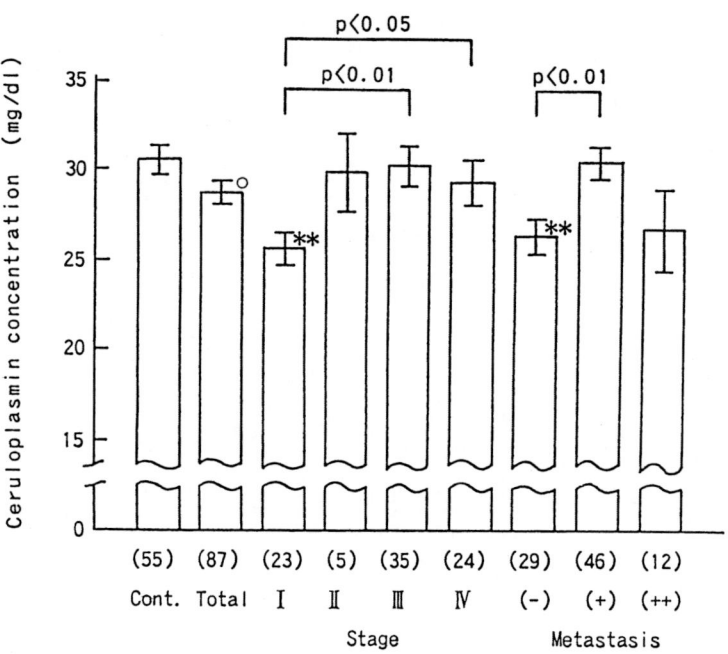

Fig. 4 Ceruloplasmin concentrations of the patients with stomach cancer and normal subjects. Number of subjects is shown in parentheses. An open circle, one and two asterisks mean significant difference from the control at 10 %, 5 % and 10 % level, respectively.

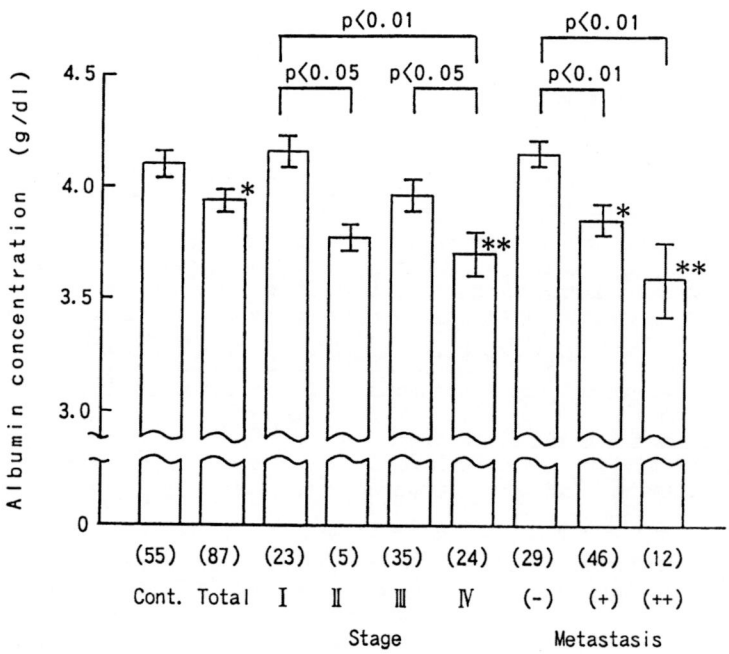

Fig. 5 Albumin concentrations of the patients with stomach cancer and the control.

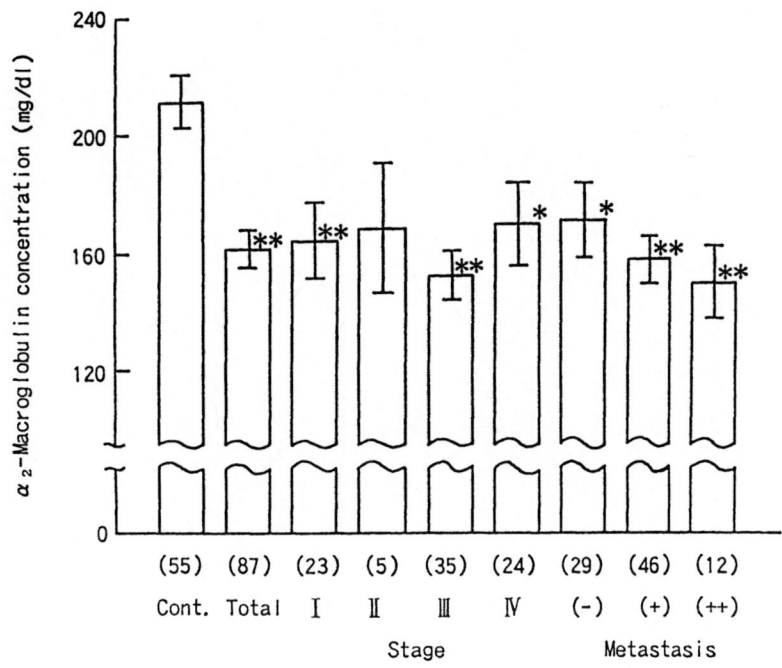

Fig. 6 Alpha 2-macroglobulin concentrations of the patients with stomach cancer and the control.

DISCUSSION

It was revealed that blood Cu, Zn and Se concentrations decreased and Mg concentration increased in early types of stage I and II, and that a decrease in GSH-px activity in stage I, an increase in CAT activity in stage II and both in stage III were recognized in the patients with stomach cancer. These results show that increases in blood glutathion (GSH) levels and/or decreases in hydrogen peroxide levels are provoked in the early stage I or II and III of stomach cancer. In contrast, the advanced type of stage IV shows high blood Cu, Mn and Ca levels, low blood Zn and Se levels, and a decrease in GSH-px activity. It is well known that the transport of Cu and Zn may depend upon plasma proteins of ceruloplasmin, albumin and alpha 2-macroglobulin. These protein concentrations of stomach cancer patients were almost in a proportion to each stage of blood Cu and Zn concentrations. In the relationship between changes in trace element concentrations and metabolism of superoxide anions in stomach cancer, an increase in GSH level caused by lowering of GSH-px activity is provoked at the early stage, and an increase in superoxide anions and hydrogen peroxide consecutively add to this increase in GSH at the II and III stages. High blood Cu, Mn and Ca, and low blood Zn and Se levels are characteristic signs in the advanced type of stomach cancer. These signs of high Cu and low Zn, and low Se are connected to the decrease in SOD, CAT and GSH-px activities with significant correlation coefficients, respectively. Changes in these enzymes show that there are increases in GSH, superoxide anions and hydrogen peroxide levels in stomach cancer.

These characteristics may suggest that there are independent enzyme reactions to the changes of blood Cu, Zn, Mn, Se, Mg and Ca at the early stage between metabolisms of GSH due to GSH-px and of both superoxide anions due to SOD and hydrogen peroxide due to CAT. Correlations between low Zn levels and an increase in GSH-px activity, and between low Se levels and an increase in SOD activity may show a compensatory local defense reaction involving increase in gastric cancer tissue Se concentration, which means initial response. And then the intermediate and terminal reactions in the body with the findings of

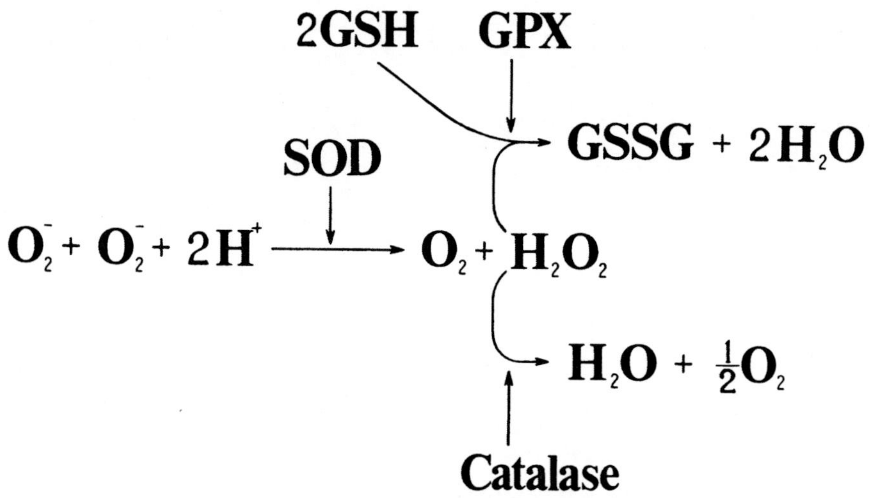

$$2\text{GSH} \quad \text{GPX}$$

$$\text{SOD} \quad \rightarrow \text{GSSG} + 2\,\text{H}_2\text{O}$$

$$\text{O}_2^- + \text{O}_2^- + 2\text{H}^+ \longrightarrow \text{O}_2 + \text{H}_2\text{O}_2$$

$$\rightarrow \text{H}_2\text{O} + \tfrac{1}{2}\text{O}_2$$

$$\text{Catalase}$$

Fig. 7 Destiny of superoxide anion

high Cu, low Zn and Se, high Mn, and low GSH-px will follow to the initial local defense reaction. Finally, increases in hydrogen peroxide and superoxide anions will inhibit SOD and CAT activities. However, it is not clarified how is the dynamics of trace elements and the related enzymes taken place in abnormal gastric tissues with cancer (Fig. 7).

CONCLUSION

It was confirmed that low or high blood Cu, Zn, Mn, Se, Mg and Ca levels by stages of malignancy, and low Mn level in abnormal gastric tissues were found in patients with stomach cancer, and that the metabolic flows of superoxide anions and glutathion are stagnant in stomach cancer due to decrease in activities of the related enzymes and metal binding proteins of ceruloplasmin, albumin and alpha 2-macroglobulin.

ACKNOWLEDGMENTS

The author is grateful to the collaborators of Department of Hygiene and Preventive Medicine, Hokkaido University School of Medicine and the Graduate School of Environmental Science, Hokkaido University.

REFERENCES

Beauchamp C, Fridovich I (1971) Superoxide dismutase: improved assays and on assay applicable to acrylamide gels. Analytical Biochemistry 44:276-287
Japanese Research Society for Gastric Cancer (1985) The general rules for the gastric cancer study, 11th edn. Kanehara, Tokyo
Mancini G, Carbonera AO, Heremans JF (1965) Immunochemical quantitation of antigens by single radial immunodiffusion. Immunochemistry 2: 235-254
Paglia DE, Valetine WN (1967) Studies on the quantitative and qualitative characterization of erythrocyte glutathione peroxidase. Journal of Laboratory and Clinical Medicine 70:158-169
Saito K, Fujimoto S, Sasaki T, Saito T, Kurasaki M, Kaji H (1981) Relationship between gastric cancer and blood trace metal levels. In: Hemphill DD (ed) Trace substances in Environmental Health XV. University of Missouri, Coloumbia, p 35-44

Saito K, Saito T, Hosokawa T,Fujimoto S, Sasaki T (1984) Changes of blood copper, zinc, and manganese in stomach cancer. Trace elements in medicine 1: 24-28

Saito K, Saito T, Hosokawa T, Ito K (1984) Blood selenium level and the interaction of copper, zinc and manganese in stomach cancer. Trace elements in medicine 1: 148-152

Saito K, Saito T, Ito K, Fujimoto S, Sasaki T (1985) Interaction of copper, zinc, manganese, and selenium to superoxide dismutase, catalase, and glutathione peroxidase in stomach cancer. Trace Elements in Man and Animals-TEMA 5, p 805-807

Saito K, Saito T, Kurasaki M, Ito K, Sasaki T (1985) Interaction of zinc, copper and selenium with superoxide dismutase, catalase and glutathion peroxidase in stomach cancer. Nutrition Research Suppl I: 714-724

Saito K, Saito T, Kurita T, Kobayashi M, Ito T (1987) Blood and tissue trace elements and destiny of superoxide anion in patients with stomach cancer. Environmental Health 20: 177-180

Sinha AK (1972) Colorimetric assay of catalase. Analytical Biochemistry 47:389-394

Watkinson, JH (1966) Fluorometric determination of selenium in biological material with 2,3-diaminonaphthalene. Analytical Chemistry 38:92-97

Trace Elements in Cancer Diagnosis

SHUICHI KIMURA[1], KUNIHISA IWAI[1], KAZUKO SAITOH[1], MIEKO KAWAMURA[1],
TATSUO IDO[2], REN IWATA[2], KIICHI ISHIWATA[2], and MOTONOBU KAMEYAMA[3]

[1] Laboratory of Nutrition, Faculty of Agriculture, Tohoku University, 1-1 Tsutsumidori,
Amamiya-machi, Sendai, 981 Japan
[2] Division of Radiopharmaceutical Chemistry, Cyclotron and Radioisotope Center, Tohoku
University, Aoba, Aramaki, Sendai, 980 Japan
[3] Division of Neurosurgery, Institute of Brain Disease, Tohoku University School of Medicine,
2-1 Seiryo-cho, Sendai, 980 Japan

ABSTRACTS

The positron emitting metal-chelates were synthesized, and they were
evaluated as a tumor imaging agents in tumor-bearing animals. [^{48}V]vanadyl-
pheophorbide a (^{48}V-Pheo) was synthesized by insertion of ^{48}V into
pheophorbide (Pheo), a decomposition of product from chlorophyll.
Furthermore, as a new derivative, ^{48}V was inserted into Chlorine e$_6$Na (Chl)
which was similar to Pheo but it has hydrophilic properties. The time for
attaining the maximum uptake of [^{48}V]vanadyl-chlorin e6Na [^{48}V-Chl] in tumor
tissue after injection was 12 hrs, while that of ^{48}V-Pheo was 24 hrs. These
results suggested that ^{48}V-Chl may be useful as a positron emitting agent
for diagnosis and phototheraphy for tumors. On the other hand, it was shown
that [^{45}Ti] titanium ascorbate (^{45}Ti-AsA) is also useful diagnostic agent
for brain tumor. Both ^{48}V- and ^{45}Ti- chelate compounds may be applied to
positron tomography as tumor imaging agent.

KEY WORDS

[^{45}Ti]titanium-ascorbate (^{45}Ti-AsA), [^{48}V]vanadyl-pheophorbide a (^{48}V-Pheo),
[^{48}V]vanadyl-chlorin e$_6$Na (^{48}V-Chl), tumor imaging agent, positron emitter

Some radioactive trace elements produced by a cyclotron, such as titanium-45
[^{45}Ti] and vanadium-48 [^{48}V], are positron emitters and have a relatively
short life span. Besides, titanium and vanadium are classified as
bioelements which are distinct from gallium, indium and thalium. They also
would be expected to hold some role in vivo although there is not so much as
the evidence of these two elements being an essential element for man or
animals. Our interest has been focussed on these characteristics, and a
positron emmision tomographic agent for imaging distribution-behavior of
trace elements in the animal organs or tissues has been developing. This
study was taken to evaluate metal-chelate compounds for tumor imaging.

Titanium, the ninth most abundant element in the earth's crust, is widely
distributed. However, the role in vivo remains to be elucidated. We have
been carry on the studies for the nutritional modification of metabolism and
speciation of this trace element. As a part of this work, titanium-45
chloride (^{45}TiCl$_4$) and titanium-45 ascorbate (^{45}Ti-AsA) were prepared. ^{45}Ti
is a positron emitter having a half life of 3.09 h. In tissue distribution
studies, the ^{45}Ti-AsA uptake was increased with time in the liver, while the
^{45}TiCl$_4$ was gradually cleared from one (Fig. 1). The chromatographic results
presented here for plasma taken from ^{45}Ti-AsA treated rats show clearly that
the radionuclide is bound to the the transferrin (Fig. 2). In the course of
this study, an interesting autoradiographic image of a rat brain bearing a
glioma (EA-285 S/C) was obtained by administration of ^{45}Ti-AsA (Fig. 3). The
target to nontarget ratio was 26.7. It seems that ^{45}Ti-AsA may be useful
diagnostic agent for brain tumors.

On the other hand, we have been working the photodynamic substances, which is Pheophorbide a (Pheo) and its derivatives. Pheo, a porphyrin produced from chlorophyll by elimination of phytyl and magnesium (Fig. 4). Lipson and Baldes (1960) prepared a hematoporphyrin derivative (HPD) from hematoporphyrin. They revealed that HPD was retained by malignant tumors. It was possible to diagnose malignant tissues by detecting the fluorescence of HPD (Lipson et al. 1961; Sandersons et al. 1972). Since Dougherty et al.(1975, 1979) reported the efficiency of photodynamic theraphy (PDT) in malignant tumors using HPD, this therapeutic modality has been applied in case of cancer in various organs (Hayata et al. 1984). Pheo has a photodynamic action and an affinity for malignant tumor tissues (Kimura et al. 1982; Kimura 1987). It has also a stronger tumor cell-killing effect than hematoporphyrine(Iwai et al. 1986). However, Pheo is unstable for cancer diagnosis using its fluorescence from a tumor, because the fluorescent lifetime of Pheo is one third of HPD (Sato et al. 1986).

In our interest to the applicability of Pheo and usefulness of positron emitting nuclide for cancer diagnosis, ^{48}V-labeled Pheo was synthesized. ^{48}V (a positron emitter having a half life of 15.97 days) was inserted into the porphyrin ring, as a tumor imaging agent for positron-emission tomography, etc. The tissue distribution studies with [^{48}V]vanadyl-pheophorbide (^{48}V-Pheo) in mice bearing FM3A tumor showed that at 2 hrs high concentration

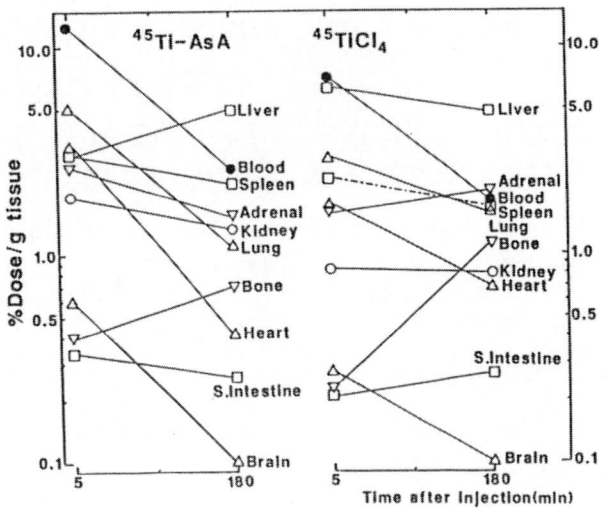

Fig.1 Tissue distribution of ^{45}Ti-AsA and ^{45}TiCl$_4$ in Wistar rats.

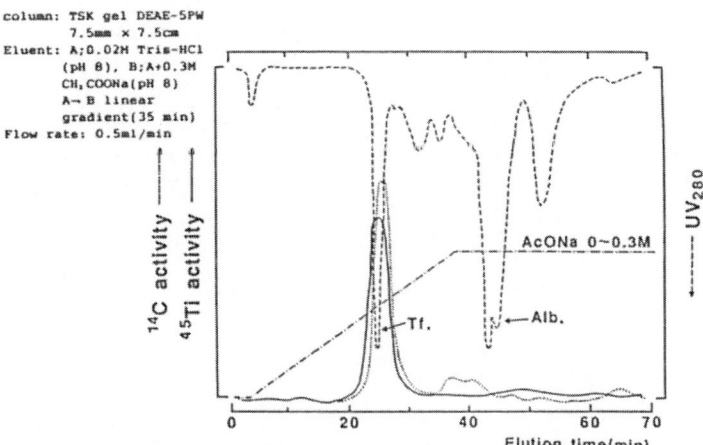

Fig.2 Elution patterns of Wistar rat plasma after 10 minutes incubation with ^{45}Ti-[^{14}C]-AsA obtained from TSK DEAE-5PW ion-exchange column.

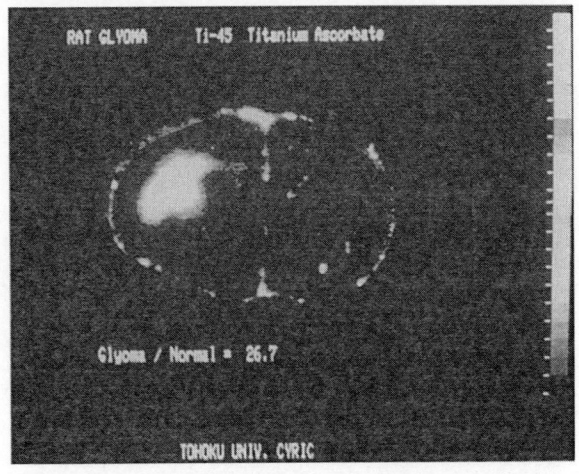

Fig.3 Autoradiogram of a rat brain tumor.

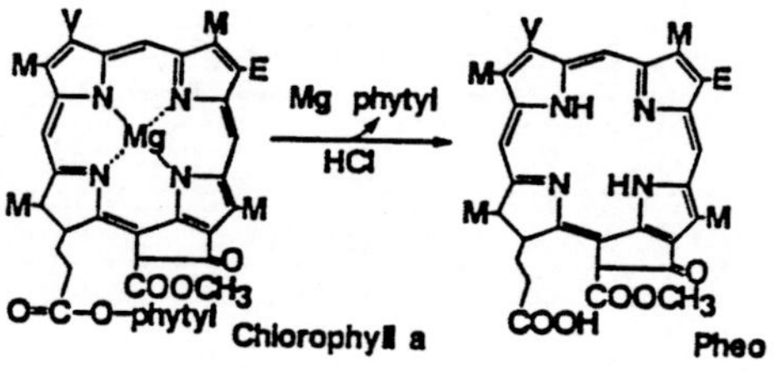

was found in the blood, liver and kidney (Table 1). Contrary to normal organs, the levels of ^{48}V in FM3A tumor increased from 2 hrs to 24 hrs. In the case of inorganic ^{48}V, the uptake in tumor is low at 2 hrs after injection and the level decreased gradually with time (Table 2). The tumor-to-organ ratios of inorganic ^{48}V and ^{48}V-Pheo at 24 hrs after injection are presented (Table 3). The ratios of inorganic ^{48}V were less than one, except for tumor-to-muscle and tumor-to-skin ratios. On the other hand, with the exception of the tumor-to-liver and tumor-to-kidney ratios, the ratios of ^{48}V-Pheo were greater than one. These ratios of ^{48}V-Pheo increased with the time and reached maximum at 24 hrs. All ratios of ^{48}V-Pheo were significantly higher than those of inorganic ^{48}V. Beside ^{48}V-Pheo and inorganic ^{48}V, [^{48}V]vanadyl sulfate (^{48}V-SO$_4$), [^{48}V]vanadyl ascorbate (^{48}V-AsA), and [^{48}V]vanadyl-tetraphenylporphyrine (^{48}V-TPP) were also prepared to compare with ^{48}V-Pheo(Table 4). Again, ^{48}V-Pheo imaged clearly the tumor tissue. The ability of ^{48}V-Pheo to localize into FM3A tumor was compared with other types of tumor tissues, such as hepatoma (MH 134) and sarcoma (S 180) (Table 5). The concentrations in these tumor tissues increased with time and were maximum at 24 hrs after administration. No difference was observed among ^{48}V-Pheo levels in the three tumor species at each period. The nonspecific ability of ^{48}V-Pheo to localize in different tumor tissues was showed.

Fig.4 Structures of pheophorbide
M, -CH$_3$; E, -CH$_2$CH$_3$; V, -CH=CH$_2$

Table 1 Tissue distribution of ^{48}V-Pheo in C3H mice bearing FM3A tumor.

Tissue	% Dose/g tissue				
	2 h	8 h	16 h	24 h	48 h
Blood	11.98 ± 0.74	9.42 ± 0.09	6.34 ± 0.42	4.54 ± 0.31	2.11 ± 0.13ᵃ
Liver	11.05 ± 0.66	8.25 ± 0.77	6.84 ± 0.47	5.94 ± 0.43	4.10 ± 0.23
Kidney	9.10 ± 0.62	7.51 ± 0.63	5.79 ± 0.39	4.82 ± 0.35	3.78 ± 0.28
Spleen	4.01 ± 0.26	3.59 ± 0.23	2.78 ± 0.22ᵇ	2.10 ± 0.20ᶜ	1.54 ± 0.11ᶜ
Lung	3.35 ± 0.24	2.85 ± 0.24ᵇ	1.62 ± 0.15ᶜ	1.07 ± 0.12ᶜ	0.72 ± 0.07ᵇ
Intestine	3.59 ± 0.21	3.05 ± 0.28ᵃ	2.16 ± 0.17ᶜ	1.83 ± 0.25ᶜ	1.22 ± 0.12ᶜ
Muscle	1.33 ± 0.12ᶜ	0.86 ± 0.12ᶜ	0.67 ± 0.07ᶜ	0.56 ± 0.03ᶜ	0.43 ± 0.05ᶜ
Skin	1.87 ± 0.12ᶜ	1.48 ± 0.11ᶜ	1.08 ± 0.11ᶜ	0.88 ± 0.06ᶜ	0.56 ± 0.06ᶜ
Tumor¹	3.62 ± 0.11	4.22 ± 0.31	4.47 ± 0.33	4.98 ± 0.36	3.11 ± 0.28

¹FM3A tumor was implanted subcutaneously into the back of C3H mice. Values are means ± SE of 5 mice.
Significant difference from tumor, ᵃ$P < 0.05$; ᵇ$P < 0.01$; ᶜ$P < 0.001$.

Table 2 Tissue distribution of inorganic ^{48}V in C3H mice bearing FM3A tumor.

Tissue	% Dose/g tissue				
	2 h	8 h	16 h	24 h	48 h
Blood	9.55 ± 0.23	7.72 ± 0.61	5.83 ± 0.07	4.03 ± 0.08	1.92 ± 0.11
Liver	6.42 ± 0.35	5.57 ± 0.25	3.98 ± 0.17	3.08 ± 0.10	2.02 ± 0.06
Kidney	8.14 ± 0.32	6.47 ± 0.33	4.84 ± 0.14	4.37 ± 0.11	3.06 ± 0.25
Spleen	2.42 ± 0.21	2.22 ± 0.18	1.83 ± 0.11	1.30 ± 0.08	1.01 ± 0.06
Lung	6.72 ± 0.26	6.18 ± 0.31	5.54 ± 0.18	4.19 ± 0.13	3.16 ± 0.12
Intestine	2.25 ± 0.22	2.06 ± 0.21	1.76 ± 0.20	1.49 ± 0.13	0.97 ± 0.15
Muscle	1.12 ± 0.08	1.04 ± 0.13	0.92 ± 0.08ᵃ	0.76 ± 0.04	0.62 ± 0.04ᵃ
Skin	1.36 ± 0.11	1.25 ± 0.13	1.02 ± 0.06	0.94 ± 0.09	0.73 ± 0.05
Tumor¹	1.51 ± 0.16	1.36 ± 0.14	1.19 ± 0.12	0.97 ± 0.09	0.77 ± 0.04

¹FM3A tumor was implanted subcutaneously into the back of C3H mice. Values are means ± SE of 5 mice.
ᵃSignificant difference from tumor, $P < 0.05$.

Table 3 Tumor-to-organ ratios of inorganic ^{48}V and ^{48}V-Pheo in C3H mice bearing FM3A tumor at 24h after injection

Tumor/organ	Inorganic ^{48}V	^{48}V-Pheo
Tumor/blood	0.24 ± 0.03	1.10 ± 0.10[a]
Tumor/liver	0.31 ± 0.03	0.84 ± 0.08[b]
Tumor/kidney	0.22 ± 0.02	0.88 ± 0.09[a]
Tumor/spleen	0.75 ± 0.06	2.37 ± 0.14[b]
Tumor/lung	0.23 ± 0.02	4.65 ± 0.32[b]
Tumor/intestine	0.65 ± 0.05	2.72 ± 0.19[b]
Tumor/muscle	1.28 ± 0.11	8.89 ± 0.50[b]
Tumor/skin	1.03 ± 0.08	5.66 ± 0.44[b]

Values are means ± SE of 5 mice.
Significant difference between inorganic ^{48}V and ^{48}V-Pheo.
[a]$P < 0.01$; [b]$P < 0.001$.

Table 4 Tumor-to-muscle ratios of ^{48}V-compounds

^{48}V-compound	Ratio of tumor/muscle		
	Time after injection		(hr)
	2	12	24
$^{48}V(IV)O$-Pheo	1.922 ± 0.149	4.504 ± 0.409	5.307 ± 0.622
$^{48}V(IV)O$-TPP	2.792 ± 0.251	3.703 ± 0.472	3.314 ± 0.325
$^{48}V(IV)O$-SO$_4$	1.292 ± 0.123	3.703 ± 0.472	3.314 ± 0.325
$^{48}V(IV)O$-AsA	2.359 ± 0.366	2.355 ± 0.143	2.160 ± 0.225

Values are means ± SD of four mice.
Each compound was injected intravenously into mice bearing FM3A tumor.

^{48}V-Pheo can be used for tumor diagnosis, except for cancer of the liver and kidney. ^{48}V-Pheo seems to be more suitable as a tumor-imaging agent instead of Pheo fluorescence and can be used in conjunction with photodynamic therapy using Pheo. However, because of its hydrophobic property

Table 5 Uptake of ^{48}V-Pheo into FM3A, MH134 and S180 tumors

Tumor	% Dose/g tumor tissue				
	2 h	8 h	16 h	24 h	48 h
FM3A[1]	3.62 ± 0.11[a]	4.22 ± 0.31	4.47 ± 0.33	4.98 ± 0.36	3.11 ± 0.28
MH 134[1]	3.23 ± 0.20	4.11 ± 0.22	4.58 ± 0.49	5.01 ± 0.48	3.34 ± 0.31
S 180[2]	2.58 ± 0.19[a]	3.21 ± 0.21	4.10 ± 0.24	4.92 ± 0.31	2.83 ± 0.19

[1]FM3A or MH 134 tumor was implanted subcutaneously into the back of C3H mice.
[2]S 180 tumor was implanted subcutaneously into the back of ddY mice. Values are means ± SE of 5 mice.
[a]Significant difference $P < 0.01$.

and of its high accumulation in liver Pheo is not necessarily the desirable tumor-imaging agent. Therefore, we decided to examine Chlorin e$_6$Na (Chl), a new derivative of chlorophyll, for use in a tumor-imaging agent. This material is structurally similar to Pheo, but it has hydrophilic properties (Fig. 5).

Chl had an affinity to tumor and a photodynamic action [9]. We synthesized ^{48}V-Chl, in which ^{48}V was inserted into the porphyrin ring of Chl. ^{48}V-Chl was rapidly excreted from normal tissues (Fig. 6). The time for attaining the maximum uptake of ^{48}V-Chl in tumor tissue after injectiion was 12 hrs, while that of ^{48} V-Pheo was 24 hrs. All tumor-to-organ ratios of ^{48}V-Chl at 24 hrs after injection were higher than those of ^{48}V-Pheo and inorganic ^{48}V (Table 6). Especially the tumor-to-blood and tumor-to-liver ratios of ^{48}V-Chl were over twice larger than those of ^{48}V-Pheo. In the whole-body

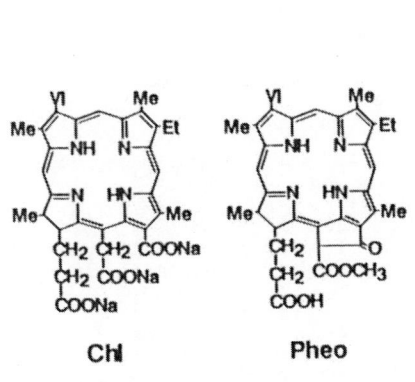

Fig.5 Structures of Chl and Pheo.
Me,-CH$_3$;   Et,-CH$_2$CH$_3$;
Vi,-CH=CH$_2$

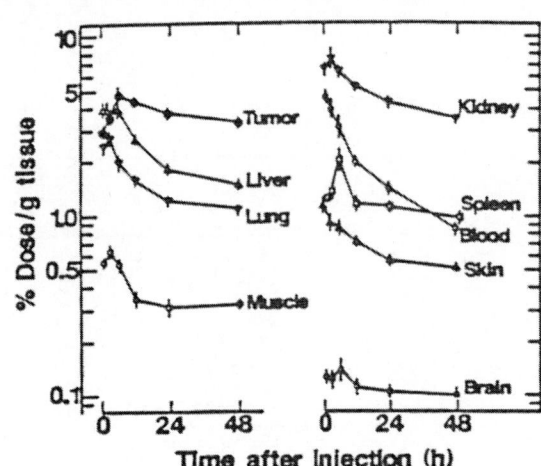

Fig.6 Biodistribution of ^{48}V-Chl in C3H mice bearing FM3A tumor.

autoradiograms, 48V-Chl showed the tumor image most distinctly in 48V-compounds (Fig. 7). These results suggest that 48V-Chl is more suitable as a positive tumor imaging agent than 48V-Pheo. We concluded that 48V-Chl provided more good basis for its clinical application than 48V-Pheo as a tumor imaging agent except in kidney and bone tumors.

Table 6 Tumor-to-organ ratios of inorganic ^{48}V, ^{48}V-Pheo and ^{48}V-Chl in C3H mice bearing FM3A tumor at 24h after injection

Tumor/organ	Inorganic ^{48}V	^{48}V-Pheo	^{48}V-Chl
Tumor/Blood	0.24±0.03***	1.10±0.10***	2.65±0.18
Tumor/Brain	12.21±0.97***	39.27±2.38*	46.75±2.11
Tumor/Liver	0.31±0.03***	0.84±0.08***	2.10±0.19
Tumor/Kidney	0.22±0.02***	0.88±0.09	0.89±0.07
Tumor/Spleen	0.75±0.06***	2.37±0.14*	3.31±0.26
Tumor/Lung	0.23±0.02***	4.65±0.32	3.12±0.17
Tumor/Muscle	1.28±0.11***	8.89±0.50**	12.06±0.48
Tumor/Skin	1.03±0.08***	5.66±0.44	6.68±0.37

Values are means ± SE of five mice. Significant difference from ^{48}V-Chl, *p<0.05; **p<0.01; ***p<0.001.

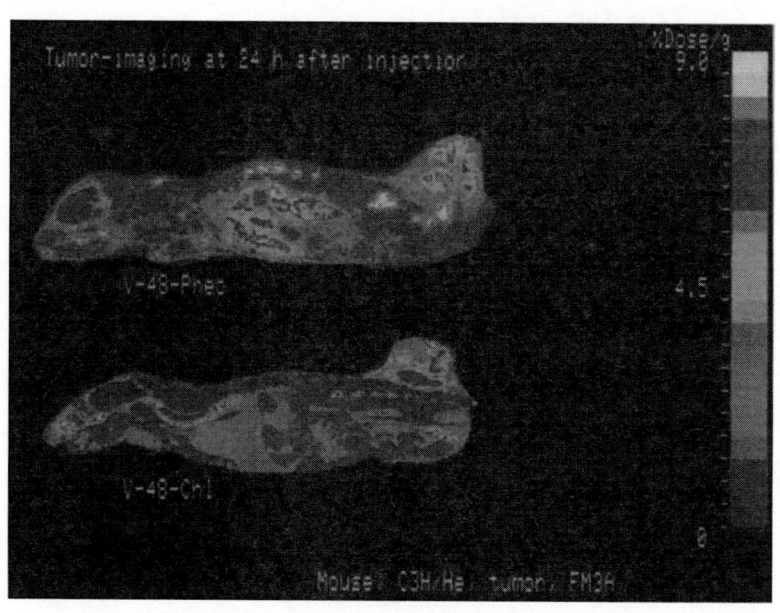

Fig.7 Autoradiograms of FM3A mice 24h after injection of ^{48}V-Pheo and ^{48}V-Chl.

ACKNOWLEDGEMENTS

The authors wish to thank the cyclotron operation group for carrying out the irradiations. Special thanks are due to Mr. H. RAI (Tama Sekagaku Co., Ltd.)for supply of Pheo and Dr. S. SAITOH (Dep. Exptl. Oncol., Res. Inst. Tubercul. Cancer, Tohoku Univ.) for the supply of tumor cells.

REFERENCES

Dougherty T J, Grindey G B, Fiel R, Weishaupt K R, Boyle D G (1975) Photoradiation therapy: II. Cure of animal tumors with hematoporphyrin and right. J Natl Cancer Inst 55:115-121

Dougherty T J, Lawrence G, Kaufman J H, Boyle D G, Weishaupt K R, Goldfarb A (1979) Photoradiation in the treatment of recurrent breast carcinoma. J Natl Cancer Inst 62:231-237

Hayata Y, Kato H, Konaka C, Ono J, Amemiya R, Kinoshita K, Sakai H, Yamada R (1984) Hematoporphyrin derivative and photoradiation therapy in early stage lung cancer. Laser Surg Med 4:39-47

Iwai K, Horigome M, Kimura S (1986) Studies on photodynamic effects of porphyrins and phycobilins. Photomed Photobiol 8:25-26

Kimura S (1987) Photobiology of pheophorbide. Photomed Photobiol 9:35-43

Kimura S, Isobe A, Sai T, Takahashi Y (1982) The role of lipid peroxidation on the development of pheophorbide a. In: Lipid peroxids in Biology and Medicine, Academic Press New York, p 243-253

Lipson R L, Baldes E J (1960) The photodynamic properties of a particular hematoporphyrin derivatives. Arch Dermatol 82:508-516

Lipson R L, Baldes E J, Olsen A M (1961) The use of a derivative of hematoporphyrin in tumor detection. J Natl Cancer Inst 26:1-11

Sanderson D R, Fontana R S, Lipson R L, Baldes E K (1972) Hematoporphyrin as a diagnostic tool a preliminary report of new techniques. Cancer 30:1368-1372

Sato S, Kikuchi S, Inaba F, Taguchi Y, Kasai M, Kimura S (1986) Time-resolved fluorescence spectroscopy and lifetime measurement of organic dyes in cancer cells by N_2 laser excitation J Jpn Soc Laser Med 6:139-142

Trace Elements in Cancer Therapy

Nobumasa Imura, Masahiko Satoh, and Akira Naganuma

School of Pharmaceutical Sciences, Kitasato University, 5-9-1 Shirokane, Minato-ku, Tokyo, 108 Japan

ABSTRACT

We have recently found that selenium, an essential trace element could efficiently depress the adverse side effects of cis-diamminedichloroplatinum (cis-DDP) in mice without compromising its antitumor activity. More recently we have developed a novel method to protect the animals against toxicity of cis-DDP by preadministering heavy metals which induce metallothionein (MT) in its target tissues. Preadministration of MT inducer has been proved to be effective not only for reducing toxicity of several anticancer drugs but also for preventing side effects of γ-ray irradiation. Among the MT inducers tested bismuth appears to be most useful for depressing the side effects of various anticancer drugs and irradiation without affecting their antitumor activity.

KEY WORDS

Cancer chemotherapy, radiotherapy, anticancer drugs, selenium, metallothionein

INTRODUCTION

A large number of anticancer drugs have been developed for chemotherapy, but their severe side effects prevent their effective use in clinical treatment. Previously we demonstrated that selenium depressed acute toxicity of inorganic mercury through the interation which formed a high molecular weight complex containing equimolar amount of mercury and selenium with proteins. During the course of this study we found that selenium might interact with platinum in animal body. This fact prompted us to examine a potency of selenium to modify the severe toxicity of cis-DDP.

On the other hand, it is well known that the toxicity of various heavy metals was reduced by MT, a cysteine rich metal binding protein of low molecular weight. Since we could confirm that platinum administered as cis-DDP to mice was detected in the renal cytosol fraction as a form bound to MT, we attempted to use this metal binding protein as another tool for depressing side effects of cis-DDP.

The results so far obtained are of great interest, promising that trace elements can provide unique and useful techniques for specifically reducing the toxic side effects of anticancer drugs and γ-irradiation in cancer therapy.

RESULTS AND DISCUSSION

I. Selenium Coadministration Depressed Cis-DDP Toxicity

Being encouraged by the results of a preliminary experiment that co-administration of sodium selenite with lethal dose of cis-DDP dramatically reduced the lethal toxicity of this anticancer agent, precise experiments were carried out to establish the optimal administration schedule for the reduction of cis-DDP toxicity.

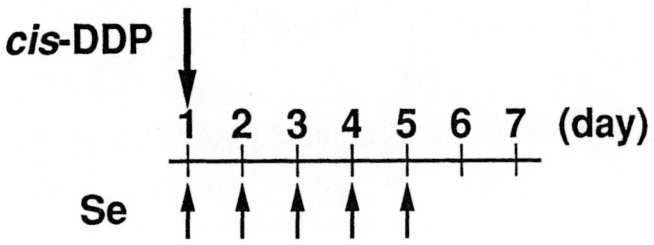

Fig.1 Administration Schedule of cis-DDP and Selenite.

Finally we found that the simultaneous administration of selenite with cis-DDP at the molar ratio of 1 to 3.5 on the first day and subsequent daily injection of the same amount of selenite for further 4 days completely depressed the lethal toxicity of cis-DDP given to mice.

Selenite coadministration according to this schedule (Fig. 1) markedly reduced the renal toxicity indicated by blood urea nitrogen value, intestinal toxicity indicated by incidence of diarrhea and lethal toxicity caused by cis-DDP in mice. Although the data are not shown, we confirmed that the coadministration of selenite did not affect the antitumor activity of cis-DDP at all using tumor bearing mice inoculated with Ehrlich tumor cells, colon 38 adenocarcinoma cells or P-388 leukemic cells (Satoh et al. 1989a).

Since in the actual clinical treatment, cis-DDP is administered repeatedly to the patients, we further examined the effect of selenite co-administration on the toxicity caused by the repeated injection of cis-DDP. The administration schedule shown in Fig. 1 with some combination doses of cis-DDP and selenite was repeated seven times over the total period of seven weeks. It was confirmed that the selenium coadministration efficiently depressed the renal, hepatic and bone marrow toxicities caused in mice by repeatd administration of cis-DDP.

These data were already presented at the 1st meeting of this Society held in Palm Springs in December, 1986, and a part of them have already been published (Naganuma et al. 1983, and Satoh et al. 1989).

In order to approach to the mechanism of the protective action of selenite against cis-DDP toxicity, we examined the tissue distributions of both platinum and selenium when the mice were treated with the combination of ^{195m}Pt-labeled cis-DDP and ^{75}Se-labeled selenite. Compared with the case of separate administration of cis-DDP or selenite alone, we could not see any change in the platinum tissue distribution on the simultaneous administration of cis-DDP and sodium selenite.

Table 1. Characteristics of Metallothionein (MT)

Low Molecular Weight
High Metal Content
Characteristics Amino Acid Composition 　1. High Cysteine Content 　2. No Aromatic Amino Acids Heat Stability
Metal Thiolate Clusters
Inducibility
Reactivity 　1. Binding to Heavy Metals 　2. Scavenging Free Radicals 　3. Highly Reactive to Alkylating Agents

On the other hand, as for the distribution of selenium, significant increase in selenium level was observed in the liver, kidney and lung 24hr after the combined administration of cis-DDP and selenite. This increase in selenium level in these tissues observed on the concurrent injection, may play a significant role in reducing the side effects of cis-DDP. This profile of alterlations in tissue distributions of Pt and Se on their concurrent administration is different from that observed in the case of interation of selenium with inorganic mercury or cadmium. In the cases of interation with these heavy metals, tissue distribution of the heavy metal was markedly altered by the simultaneous administration of selenite.

Thus, we have so far been unable to clarify the detailed mechanism of the protective action of selenite against cis-DDP toxicty. Characterization of the platinum and selenium compounds in the tissues of animals given both cis-DDP and selenite is now going on in our laboratory.

II. Introduction of Metallothionein Prevents the Adverse Effects of Anticancer Drugs and γ-irradiation

1. Cis-DDP
Our next attempt to specifically reduce the cis-DDP toxicity was to induce MT in its target tissues. MT is cytoplasmic protein of low molecular weight and well known for its characteristic properties listed in Table 1. Since it has many cysteine residues which account for approximately 30% of the total amino acids in its molecule, MT strongly binds to heavy metals and has recently been reported to have high reactivity to free radicals and to alkylating agents.

We found by preliminary experiments that administration of some MT inducing metals prior to cis-DDP injection effectively depressed the lethal toxicity of this anticancer agent (Naganuma et al.1985). Among the metals effective for reducing the lethal effect of cis-DDP bismuth appeared to be preferable, because bismuth induces MT in the kidney which is a major target of dose-limiting toxicity of cis-DDP. Moreover, some bismuth compounds have been used as an antidiarrheal drug, such as bismuth subnitrate and bismuth subgallate.

After many trials to find the optimal adminitsration schedule, we finally established a protocol for reduction of cis-DDP toxicity by bismuth subnitrate (BSN) as shown in Fig. 2. Daily oral preadministration of 50mg/kg of BSN according to this schedule completely depressed the lethal effect of cis-DDP observed during 20 days after cis-DDP injection. The BSN preadministration was further proved to be effective for reducing the renal, intestinal and bone marrow toxicities indicated by the blood urea nitrogen (BUN) value, the incidence of diarrhea and the number of total lenkocytes, respectively. We then examined the effect of BSN-pretreatment on the antitumor activity of cis-DDP. 2×10^6 cells of Ehrich tumor cells were inoculated subcutaneously to the mice one day prior to the first administration of 50mg/kg of BSN. Subsequent treatment of these mice with varying amounts of cis-DDP according to the administration protocol markedly

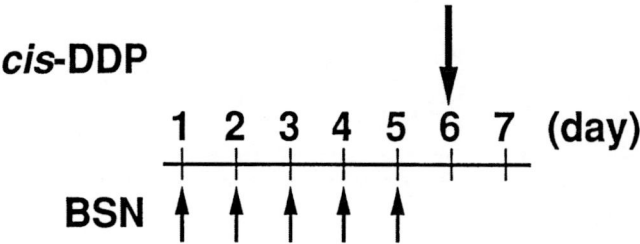

Fig. 2 Administration Schedule of cis-DDP and BSN.

depressed the tumor growth regardless of the pretreatment with BSN. Until day 12 all the mice died when they were treated with the higher doses of cis-DDP but they could survive by the BSN pretreatment. BUN values increased by cis-DDP dose-dependently, but markedly reduced by the BSN pretreatment. Namely, preinduction of MT by BSN efficiently depressed the toxicity of cis-DDP without affecting its antitumor activity as in the case of selenite co-administration (Naganuma et al. 1987).

2. Other anticancer drugs

As described previously, MT contains 20 thiols of cysteine residues in its molecule which consists of 61 amino acids. This may account for its ability to scavenge free radicals as recently reported and for its high reactivity to the alkylating agents. On the other hand, there are many anticancer drugs acting as free radical inducers or as alkylating agents. Thus, the effects of MT induction on the toxicity of various anticancer drugs of different action mechanisms were examined. Interestingly, as we expected, MT induction by bismuth or zinc compounds prevented the lethal effect of some anticancer drugs. Surprisingly, even the lethal toxicity of 5-fluorouracil, an antimetabolite, was efficiently depressed besides those of the free radical inducers and the alkylating agents (Table 2).

In the case of adriamycin, for example, the administration schedule was the same as the case of cis-DDP. Adriamycin is an anthracycline antibiotic and also extensively used as an anticancer agent and has often been used in combination with cis-DDP in cancer chemotherapy. Adriamycin is considered to possibly function by generating superoxide radicals, but this mechanism also seems to be responsible for its high toxicity especially toward the heart and bone marrow. After confirming the efficacy of bismuth preadministration in reducing the lethal effect of adriamycin without compromising its antitumor activity (Naganuma et al. 1988), we examine the effect of

Table 2. Effect of Pretreatment with Bismuth Subnitrate (BSN) or $ZnCl_2$ on Lethal Toxicity of Anticancer Drugs in Mice

Drug	Dose	Survival rate on day 20 (%) Pretreatment		
		None	BSN	$ZnCl_2$
Control	–	100	100	100
cis-DDP	35 µmol/kg	0	100	63
Adriamycin	35 µmol/kg	0	63	57
Mitomycin C	60 µmol/kg	0	14	29
Peplomycin	65 µmol/kg	0	43	71
Cyclophosphamide	2.5 mmol/kg	0	0	86
Vinblastine	20 µmol/kg	0	0	29
5-Fluorouracil	4.0 mmol/kg	0	86	43
Bleomycin	600 mg/kg	43	100	–

Mice were pretreated with BSN (50mg/kg;p.o.) once a day for 5 days or $ZnCl_2$ (400 µmol;s.c.) once a day for 2 days.

Anticancer drugs were injected i.p. 24 hr after the last administration of BSN or $ZnCl_2$.

this method on the dose-limiting cardiotoxicity of adriamycin. The extent of lipid peroxidation was determined in the heart of mice administered adriamycin with or without preadministration of bismuth as an indicator of cardiotoxicity by quantitating malondialdehyde as thiobarbiturate-reacting substance and conjugated diene. The values of both indicators increased by adriamycin were reduced by the bismuth dose-dependently and showed clear inverse correlations with the level of cardiac MT induced by bismuth. Thus, the preinduced MT might prevent the lipid peroxidation, probably by scavenging oxygen free radicals generated from adriamycin (Naganuma et al. 1988, Satoh et al.1988).

Next, the protective effects of three different metals, Bi, Zn and Cu, as MT inducers on the coardiotoxicity of adriamycin were compared using malondialdehyde and conjugateddienes as indicators of the toxicity.

Bismuth, subcutaneously administered as bismuth nitrate, and zinc markedly depressed the lipid peroxidation, but copper couldn't reduce the level of indicators. Then the effects of preadministration of these three metals on the antitumor activity of adriamycin were studied using mice subcutaneously inoculated with colon 38 adenocarcinoma cells. Bismuth preadministration did not affect the antitumor activity of adriamycin indicated by the reduction of tumor weight. However, zinc and copper diminished the antitumor activity of the drug.

The characteristics of these metal compounds observed in their effects on the antitumor activity and toxicity of adriamycin can be explained by the ability of individual metal to induce MT in the target tissues of adriamycin toxicity and in the tumor tissues. Namely, Bi significantly induced MT in some normal tissues, including kidneys and heart, but not in the tumor tissues. Contrary, Zn induced MT in the tumor tissues as well as in the heart and the other organs. Cu induced MT in the tumor tissues, but not in the heart. Thus the specific depressing effect of Bi-pretreatment on the adverse effect of adriamycin may be attributable to the fact that Bi does not induce MT in the tumor tissues, but it does induce MT in the normal tissues, especially those which are target of the toxicity of the anticancer drug.

Further, it is of great interest that not only the survival rate but also incidence and grade of pulmonary fibrosis, which is a dose-limiting toxicity of peplomycin, were significantly improved in mice administered this free radical inducing anticancer drug subsequently to the pretreatment with BSN. In addition, the bone marrow toxicity of 5-fluorouracil, an antimetabolite, indicated by the decrease in total leukocytes in mice was also improved by the pretreatment with BSN in a dose-dependent fashion. However, in this particular case, we cannot explain the mechanism of the reduction of 5-FU toxicity by bismuth preadministration.

3. Radiation Therapy

Considering the fact that MT may act as a free radical scavenger, we can expect its protective effect against the lesions caused by irradiation, because the damage by irradiation is known to be caused through intracellular radical formation. Actually, a few papers so far published suggested a possible protective ability of MT against irradiation (Bakka et al.1982, Matsubara et al. 1987). Thus, the effect of bismuth pretreatment on the lesions caused by γ-irradiation was examined using mice.

The mice were irradiated with ^{60}Co 24hrs after the last administration of bismuth as in the cases of cis-DDP and adriamycin. The survival rate of mice determined at 30 days after the irradiation (6 or 9 gy/mouse) was markedly improved by the preadministration of BSN in a dose-dependent fashion (Fig.3)(Satoh et al. 1989b). Moreover, the decrease in number of total leukocytes in irradiated mice, counted 10 days after the irradiation as an indicator of the lesion in bone marrow, was significantly improved by the preadministration of relatively high dose of BSN. At the same time, the extent of lipid peroxidation induced in the bone marrow cells of irradiated mice was efficiently reduced by BSN pretreatment. At the time point of the γ-irradiation, a significant increase in MT level was observed in the bone marrow of the mice treated with bismuth. Thus, bismuth preadministration induced MT in mouse bone marrow. This induced MT might

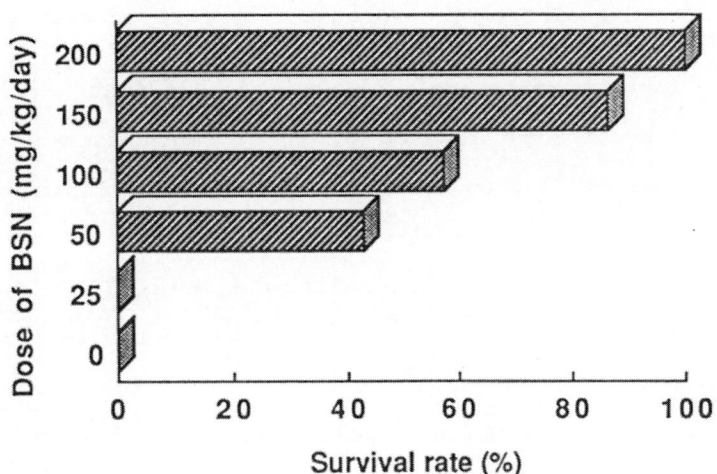

Fig. 3   Effect of Pretreatment with Bismuth Subnitrate (BSN) on Lethal
Effect of γ-ray (⁶⁰Co) Irradiation in Mice.   Survival rate was determined
30 days after γ-ray (⁶⁰Co) irradiation.   Mice were pretreated p.o. with BSN
once a day for 5 days.   γ-Ray (⁶⁰Co) was irradiated 24 hr after the last
administration of BSN.

scavenge the free radicals formed by irradiation, resulting in the
protection of bone marrow cells from peroxidation.   In the radiotherapy
using solid tumor bearing mice we could confirm that bismuth treatment did
not diminish the antitumor effect of the irradiation at all.
From the experimental results described above we may conclude that the
appropriate use of bismuth can provide a new and highly efficient technique
not only for cancer chemotherapy but also for cancer radiotherapy.

As a major bismuth compound BSN has been used orally to induce MT.   We also
confirmed that subcutaneous injection of bismuth nitrate gave exactly the
same results as those obtained with BSN.   Quite recently, we have sometimes
suffered from extremely low intestinal absorption rate of bismuth from
bismuth subnitrate in animal experiment using mice.   The reason of this
trouble has not yet been clarified.   Although, as long as we use bismuth
nitrate subcutaneously, we can always get constant induction of MT in
appropriate tissues, the absorption problem of bismuth subnitrate has to
be solved as soon as possible.

4. Clinical Applications
Several groups of physicians and surgeons were interested in our method to
specifically depress the toxic effects of anticancer agents and some
clinical trials using protocols similar to ours have been carried out.
For example, Takahashi et al. (1989) recently reported a result obtained by
applying our protocol for treatment of patient having lung cancer with
cis-DDP combined with 5-fluorouracil and vindesine.   They examined the
effect of BSN preadministration on the renal function of the patients
indicated by levels of ß-2-microglobulin(ß-MG) and N-acetyl-ß- -D-glucosamini-
dase(NAG) excreted into the urine collected during 15 days after injection
of anticancer drugs.   It was obvious that the values of both ß-MG and NAG
obtained from the patients pretreated with bismuth were significantly lower
than those without bismuth pretreatment.   They concluded that in the
clinical treatment bismuth preadministration was effective for reduction of
renal toxicity of cis-DDP.

5. Possible Role of MT as a Factor Manipulating Efficacy of Cancer Chemo-
and Radiotherapy
Considering the facts mentioned above together with the recent findings of
Imura et al. (1988) and Kelly et al. (1988) that cultured cells containing
unusually high level of MT were resistant against various anticancer drugs,
MT may act as a multi-drug-resistance factor in cancer chemotherapy when MT

is induced in tumors. Further, we have recently found that the administration of MT in both normal and tumor tissues in mice. Therefore, we can speculate as follows:
1. If we can reduce MT level in tumors during cancer therapy, the antitumor effects of the drugs and irradiation would be amplified.
2. If we can keep a substantial level of MT specifically in normal tissues during cancer therapy, the patients may be protected from carcingenesis caused by secondary effects of anticancer agents or irradiation. The experiment to examine this possibility is now going on in our labortory.

REFERENCES

Bakka A, Johnsen AS, Endresen L, Rugstad HE (1982) Radioresistance in cells with high content of metallothionein. Experientia 38:381-383
Imura N, Naganuma A, Satoh M (1989) Metallothionein as a resistance factor for antitumor drugs. Jpn J Cancer Chemother 16:599-604
Kelley SL, Basu A, Teicher BA, Hacker MP, Hamer DH, Lazo JS (1988) Overexpression of metallothionein confers resistance to anticancer drugs. Science 241:1813-1815
Matsubara J, Tajima Y, Karasawa M (1987) Promotion of raeioresistance by metallothionein induction prior to irradiation. Environ Res 43:66-74
Naganuma A, Satoh M, Yokoyama M, Imura N (1983) Selenium efficiently depressed toxic side effect of cis-diamminedichloroplatinum. Res Commun Chem Pathol Pharmacol 42:127-134
Naganuma A, Satoh M, Imura N (1984) Effect of selenite on renal toxicity and antitumor activity of cis-diamminedichloroplatinum in mice inoculated with Ehrich ascites tumor cell. J Pharm Dyn 7:217-220
Naganuma A, Satoh M, Koyama Y, Imura N (1985) Protective effect of metallothionein inducing metals on lethal toxicity of cis-diamminedichloroplatinum in mice. Toxicol Lett 24:203-207
Naganuma A, Satoh M, Imura N (1987) Prevention of lethal and renal toxicity of cis-diamminedichloroplatinum by induction of metallothionein synthesis without compromising its antitumor activity in mice. Cancer Res 47:983-987
Naganuma A, Satoh M, Imura N (1988) Specific reduction of toxic side effects of adriamycin by induction of metallothionein in mice. Jpn J Cancer Res 79:406-411
Satoh M, Naganuma A, Imura N (1988) Metallothionein induction prevents toxic side-effects of cisplatin and adriamycin used in combination. Cancer Chemother Pharmacol 21:176-178
Satoh M, Naganuma A, Imura N (1989a) Optimum shcedule of selenium administration to reduce lethal and renal toxicities of cis-diamminedichloroplatinum in mice. J Pharmacobio-Dyn 12:246-253
Satoh M, Miura N, Naganuma A, Matsuzaki N, Kawamura E, Imura N (1989b) Prevention of adverse effects of -ray irradiation after metallothionein induction by bismuth subnitrate in mice. Eur J Cancer Clin Oncol, in press
Takahashi A, Takagi M, Hishida H, Sakamoto Y, Takagi N, Amano H, Ogura Y (1989) The pharmacokinetics of cisplatin and its influence on renal function according to different infusion methods (Report II) - alleviation of renal impairment by bismuth subnitrate combined with Ginseng and Tan-Keui Ten -

Immunity

Modulation of the Immune Response by Trace Elements

Motoyasu Ohsawa, Fuminori Otsuka, and Kazuko Takahashi

Faculty of Pharmaceutical Sciences, Teikyo University, 1901-1 Suarashi, Sagamiko, Kanagawa, 199-01 Japan

ABSTRACT

The immune response comprizing multi-step processes is greatly susceptible to an alteration in nutritional and environmental factors including trace elements. Recently research concerns on the immune response and trace elements have been directed to clarify (a) their marginal levels in nutritional deficiency or excess environmental exposure, (b) immunological implications in pathogenesis of diseases caused in their deficiency or excess, and (c) their specific or critical significance in immune function. Current findings are briefly summarized from these aspects. Moreover Cd-induced autoantibody formation associated with its immunostimulative effects is described as an example for modulation of the immune response by direct interaction of a trace element with immune cells. Its mechanism is further discussed as follows: Cd as well as Hg may act on B lymphocytes, as a polyclonal activator and via involvement of T lymphocytes, consequently to induce production of autoantibodies such as anti-nuclear and anti-DNA antibodies.

Key words: nutritional deficiency, environmental exposure, cadmium, autoantibodies, polyclonal activation

1. INTRODUCTION

Immune response is known to be highly susceptible to an alteration in nutritional and environmental factors including trace elements (Good et al., 1980; Koller, 1980; Dean et al., 1982; Chandra, 1983; Gershwin et al., 1985). There are some suppositions which may explain the high susceptibility of the immune response to nutritional deficiency or toxic insult as following. 1) The immune response is comprized of multi-step processes including rapid proliferation and differentiation of many types of cells. Cells in proliferation require sufficient nutrients and are more susceptible to chemical insult. 2) The immune response is regulated by highly organized network of cellular and humoral components. Any imbalance of this regulatory network resulting from depletion or excess of a component may be expressed as hypo- or hyperfunction in immunity. 3) Lymphocytes, immunocompetent cells, have a low ability to detoxify toxic chemicals. Moreover adverse modulation of the immune response, immunosuppression or unusual immunoenhancement can be a potential risk to develop diseases such as infectious disease, cancer, allergy and autoimmune disease. Research aspects and current findings on the immune response and trace elements are briefly overviewed here in consideration of such features of the immune response. Recently increased concerns are focused on a direct and specific interaction of a trace element with immune cells. Therefore, our study on induction of autoantibodies by Cd associated with its immunostimulative effects is described as an example for modulation of the immune response by direct interaction of a trace element with immune cells.

2. A BRIEF OVERVIEW ON THE IMMUNE RESPONSE AND TRACE ELEMENTS

(1) Research Aspects on the Immune Response and Trace Elements

Recent researches on the immune response and trace elements have been directed mainly to the following three aspects. The first is nutritional and toxicological aspect to estimate the marginal level of trace elements for nutritional intake or environmental exposure, due to the high susceptibility of the immune response to an alteration in nutritional or environmental factors. It is known that Zn is essential for lymphocyte proliferation induced by a mitogen (Williams and Loeb, 1973), and that thymic weight loss is a sensitive response in marginal Zn deficiency (Dardenne et al., 1984; Gershwin et al., 1985). Regarding toxic heavy metals, moreover, it may be one of explanations for the susceptibility of the immune response that lymphocytes have low and/or slow inducibility of metallothionein (Durnam and Palmiter, 1981; Koizumi et al., 1987; Sone et al., 1988) which binds heavy metals such as Hg and Cd to detoxify them.

We developed a sensitive method for measuring metallothioneins by employing SDS-polyacrylamide gel electrophoresis of carboxymethylated metallothioneins followed by the silver stain (Otsuka et al., 1988). By use of this method contents of metallothioneins in cytosol fractions were compared among various types of cell cultures including mouse spleen cells, FM3A, V79 and HeLa S3 cells. When the same amount of protein samples from Cd-treated cells were applied for the gel electrophoresis, metallothionein could not be detected only in spleen cell cultures. When spleen cell cultures were treated simultaneously with Cd and Con A, a T-lymphocyte mitogen, metallothionein was detected at last by use of spleen cell samples containing ten-times more protein than others. Time course study on metallothionein induction in cytosols after addition of Cd and Con A to spleen cell cultures showed that it took much time, more than 15 hours, for initiation of induction of detectable metallothioneins. Thus the low inducibility of metallothionein in lymphocytes may be responsible for the susceptibility of the immune response to heavy metals. As mentioned above, unusual immunological changes can be a useful indicator to know the marginal level of trace elements in deficiency or excess.

The second is etiological aspect on immunological implication in pathogenesis of diseases caused in deficiency or excess of trace elements. Adverse modulation of the immune response is a potential risk to result in profound health effects as mentioned already. Critical implication of immunological modulation in pathogenesis has been exemplified in the repeated infection in Zn deficiency (Barnes and Moyanahan, 1973; Chandra, 1980) and immune complex-type glomerulonephritis induced by Hg (Druet et al., 1982). The third is immunological aspect to explain a specific or critical significance of trace elements in the immune function. Although definite roles of most trace elements in the immune response are still not clear, there are good examples like Zn as an essential component in thymulin required for thymus development and T lymphocyte function (Dardenne et al., 1982) and Fe in production of hydroxy radicals (Repine et al., 1981) and hypochloric ions (Thomas, 1979) contributed to bacterial killing of neutrophils. These research concerns have accumulated clinical and experimental findings on immunologic manifestations and related health effects in relation to trace elements.

(2) Trace Elements and Their Immunologic Manifestations

Trace elements known or presumed to produce immunologic manifestations and related health effects are summarized in TABLE 1. Immunologic and health findings in deficiency, or supplement or excess of trace elements are selected from documents on animal experiemnts or clinical examinations. Immunologic manifestations are classified in cellular and humoral immunity (specific immunity), and phagocyte function and natural killer cell activity (non-specific immunity). Health effects except allergy triggered by metals as a hapten are summarized on host resistance to infection and malignancy, and expression of autoimmune syndrome. In deficiency of nutritional trace elements immune functions are suppressed in general. Consequently host resistancy against infectious or carcinogenic agents seems to decrease, as observed

TABLE 1. TRACE ELEMENTS KNOWN OR PRESUMED TO PRODUCE IMMUNOLOGIC MANIFESTATIONS IN EXPERIMENTAL ANIMALS OR MAN[a]

(A) DEFICIENCY

Element	Cellular immunity	Humoral immunity	Phagocyte function	NK[b] activity	Infection	Tumorigenesis[c] T	S/V	C	Autoimmune syndrome
Fe	↓	↓	↓	↓	↑				
Zn	↓	↓	↓	↓	↑		↑∿↓	↑	↓*
Cu	↓	↓	↓	-∿↓	↑	↑			
Se	↓	↓	↓	↓	↑		↑**		
Mn		↓							
I			↓						

(B) SUPPLEMENT/EXCESS

Element	Cellular immunity	Humoral immunity	Phagocyte function	NK activity	Infection	Tumorigenesis T	S/V	C	Autoimmune syndrome
Fe					-				
Zn		↓	↓		-	↑/↓		↑/↓	
Cu							-	↓	
Se	↑/↓	↑/↓	-/↑∿-	↑∿-		↓	-/↓	-/↓	↓*
Mn		↓			↑				
Co		↑			↑				
Ni		↑					-	↑∿-	
As		↓			↑	↓	-∿↓	↓	
Cd	↑/-/↓	-/↓	↑∿-		↑/↓		↑	-	↑
Pb	↓	↓	↑	-	↑		↑		
Hg		-∿↓			↑				↑
Au									↑
Organic Pb		↓							
Organic Hg		↓	-		↑∿-		-	↑	↑
Organic Sn	↓	-∿↓	-		↑				

↑: increase, ↓: decrease, -: no effect
a. Hypersensitivity is excluded.
b. NK: natural killer cell
c. T: transplanted tumor, S/V: spontaneous or virus-induced tumor, C: chemical induced tumor
* Data on autoimmune mice with a genetic predisposition
** Data from epidemiological studies

(Reproduced from Ohsawa, 1989)

in an increased incidence of infection and tumorigenesis of many types of tumors. To speculate a critical role of each trace element in the immune response from these findings, however, we must be cautious to exclude indirect effect due to protein-energy malnutrition resulted presumably from severe deficiency of a trace element, and to clarify that a single supplementation of a trace element can recover from the immune dysfunction.

In supplement or excess of trace elements a variety of effects are reported dependent upon the doses, duration of intake or exposure, genetic suscepti-bility and so on. The findings shown here are selected mainly from experi-ments on rodents orally exposed to subchronic or chronic doses of trace ele-ments for a long term. Supplementation of Se seems to enhance host resis-tancy against tumorigenic factors, in association with enhancement of immune functions. However, it must be cautious to supply Se, because the dose range of Se for the immunostimuatory effects may be narrow, as observed that a little higher dose of Se conversely suppressed functions of both cellular and humoral immunity. Many researches report that exposure to non-essential heavy metals such as Cd, Pb and Hg is suppressive for cellular or humoral immune response. Correspondent with their immunosuppressive effects, animals exposed to such metals are more susceptible to infection and tumorigenesis. However, recent researches show that these metals also have immunostimulatory effects to develop autoimmune response or autoimmune glomerulonephritis. So far Hg, Cd and Au are known to produce autoimmune reaction in mice and rats (Druet et al., 1982; Gleichmann et al., 1989). Especially Hg and Cd can produce such reaction by oral exposure to them for a long term.

Among a variety of immunological manifestations produced by trace elements we have interested in autoimmune response induced by heavy metals, because it can be abnormality in quality of immune response, and may closely relate to development of diseases.

3. CADMIUM AND AUTOIMMUNE RESPONSE

(1) Induction of Circulating Anti-nuclear Antibodies in Mice Exposed to Cd

In animals exposed to Cd at subchronic or chronic dose abnormal immunostimu-lation has been reported, which included glomerular amyloidosis in rabbits (Castano, 1971), immune-complex nephritis (Joshi et al., 1981) and induction of autoantibodies against two components of glomerular basement membrane, laminin and type-IV procollagen in rats (Bernard et al., 1984). Lauwerys and his co-workers (1984) found that the prolonged oral administration of Cd resulted mainly in the development of the glomerular type proteinuria. They suggested that immunological reaction might be involved in the pathogene-sis of the glomerular type or glomerular and tubular mixed-type proteinuria found in Cd-workers. Recently we found that oral exposure of mice to Cd could induce circulating anti-nuclear antibodies (ANA) which contributed to pathogenesis of the lupus-type nephritis (Ohsawa et al., 1988a).

When male ICR mice were fed with the drinking water containing 3, 30 or 300 ppm Cd as $CdCl_2$ for 10 weeks, they showed neither clinical symptoms nor toxic manifestations in diet intake, body weight gain, organ weights and hematologi-cal analysis. However, a tendency toward an increase in number of spleen cells (Ohsawa et al., 1983) and serum IgG level was observed in Cd-exposed mice. Moreover, proliferative response of their spleen cells to mitogens or allogeneic cells was not suppressed, but even enhanced (Ohsawa, 1987). Subsequently immunological alterations were further examined as shown in TABLE 2 (Ohsawa et al., 1988a). The delayed-type hypersensitivity reaction as an indicator of cell-mediated immunity in Cd-exposed mice was comparable to controls. However, direct plaque forming cell (PFC) response to sheep red blood cells (SRBC), as an indicator of humoral immunity, was affected by oral exposure to Cd. PFC response of spleen cells after the priming with the specific antigen SRBC was reduced in mice exposed to Cd at higher concent-rations. On the contrary, the PFC response under no priming with SRBC (un-primed PFC) was enhanced in Cd-exposed mice. So far we have no clear evidence to explain these different effects of Cd exposure in antibody forming res-ponse. However, as immunostimulatory effects were observed in Cd-exposed mice without the priming of specific antigen, it is suggested that Cd exposure may primarily stimulate the immune response for antibody formation in mice.

TABLE 2. ALTERATIONS IN IMMUNOLOGICAL FUNCTION IN ICR MICE EXPOSED TO CADMIUM IN THE DRINKING WATER FOR 10 WEEKS

	Concentration of Cd in the drinking water (ppm)			
	0	3	30	300
DTH reaction to SRBC (10^{-2}mm)	184 (12)	162 (15)	192 (12)	170 (12)
PFC response to SRBC After priming with SRBC (PFC/10^6 spleen cells)	1010 (43)	1011 (81)	850 (96)	768 (67)¶
(PFC x 10^{-5}/spleen)	1.99 (0.25)	2.31 (0.27)	1.54 (0.24)	1.61 (0.19)
Without priming (PFC/10^6 spleen cells)	40 (12)	104 (19)*	102 (32)	145 (44)*
(PFC x 10^{-3}/spleen)	1.30 (0.28)	2.43 (0.29)§	1.90 (0.29)	1.90 (0.20)

Results represent the mean (standard error). *$P<0.05$; §$P<0.01$; ¶$P<0.001$.
DTH: delayed-type hypersensitivity estimated by the footpad swelling test.
PFC: direct plaque-forming cells.
SRBC: sheep red blood cells.

TABLE 3. INDUCTION OF ANTI-NUCLEAR ANTIBODIES IN ICR AND BALB/c MICE EXPOSED TO CADMIUM IN THE DRINKING WATER

	Concentration of Cd in the drinking water (ppm)			
	0	3	30	300
ICR	1/12 (8%)	5/10 (50%)*	8/9 (89%)**	9/10 (90%)**
BALB/c	1/10 (10%)	1/9 (11%)	2/10 (20%)	9/10 (90%)**

Results indicate number of positive mice/number tested.
*$P<0.05$; **$P<0.001$.

It was further supported by our finding on induction of circulating autoantibodies, ANA, in Cd-exposed mice. As shown in TABLE 3, mouse sera with 40 and more of ANA titer are expressed as positive for ANA induction. Enhanced induction of ANA was observed in all Cd-exposed groups of ICR mice. The

ANA induction by Cd was observed also in BALB/c mice, however, which seemed to be less sensitive to the induction than ICR mice. Thus these findings in ICR mice indicate that Cd exposure can have immunostimulatory effects including induction of autoantibodies in primary, and that the effects are observed as sensitive toxic responses.

ANA were also induced in ICR mice received subcutaneous injections of subtoxic dose of Cd (5 times a week), whithin 2 weeks after the first injection of Cd (Ohsawa et al., 1988b). Moreover, no genetic differences in the susceptibility to the ANA induction by the Cd administrations were observed qualitatively among tested mouse strains (BALB/c, C3H/He, C57BL/6 and ICR) with different H-2 haplotypes. When ICR mice were injected with equimolar doses (8.9 moles/kg) of some heavy metal compounds in a similar way, the ANA induction was positive for Cd and Hg, not significant but seemingly positive for Pb, but negative for Zn and Fe(II).

(2) Evidence for Induction of Autoantibodies as Direct Effect of Cadmium on Immune Cells

To answer the question whether the autoantibodies were induced by direct effect of Cd on immune cells, we further investigated inducibility of auto-antibodies in cultured mouse spleen cells treated with Cd in vitro. Studies showed that Cd around 1 µM increased ANA and anti-DNA (native and denatured DNA) IgG antibody levels in the culture media and unprimed PFC IgM response, not accompanied with significant cell toxicity. These results may indicate that Cd directly acts on immune cells and activates polyclones of B lympho-cytes including autoreactive clones to produce autoantibodies. Similar results were obtained in Pb-teated spleen cells, but not in Zn- or Hg-treated cells. Except Hg which was highly toxic to mouse spleen cells, these results were well consistent with those of in vivo experiments described above. Hg is known to cause polyclonal B lymphocyte activation in rat spleen cell cultures (Hirsch et al., 1984), therefore, these heavy metals may produce induction of autoantibodies in a similar mechanism.

(3) A Cellular Mechanism for the Polyclonal Effect of Cd

A cellular mechanism for the polyclonal effect of Cd was investigated by use of the unprimed PFC response as the indicator of polyclonal effect on B lymphocytes. First we examined T lymphocyte requirement for the polyclonal effect of Cd by employing spleen cells from athymic nude mice. Enhancement by Cd of unprimed PFC response observed in spleen cell cultures from normal BALB/c mice was not reproduced in the cultures from BALB/c-nu/nu mice which were deficient in T lymphocytes. In contrast, the reconstituted cultures with nude spleen cells and non-adherent spleen cells to Ig-coated beads (T rich cells) from normal BALB mice restored Cd-induced unprimed PFC response. Thus the polyclonal effect of Cd on B lymphocytes may require the presence of T lymphocyte population, as well as that of Hg (Hirsch et al., 1984). Moreover, the mixed culture of untreated BALB spleen cells with Cd-treated syngeneic spleen cells after X-ray irradiation could produce the induction of unprimed PFC response. This means that the polyclonal effect of Cd can be partially reproduced by Cd-modified spleen cells. Consequently it appear less possible that Cd directly activates B lymphocytes. In summary a cellular mechanism for the polyclonal effect of Cd can be proposed as follows: 1) Cd directly acts on immune cells to induce polyclonal activation of B lympho-cytes, thereby producing autoantibodies including ANA and anti-DNA antibodies. 2) The polyclonal effect of Cd may require the presence of T lymphocytes and can be induced directly or indirectly by Cd-modified spleen cells.

In the future study on immunological implications of environmental trace elements, it is expected that autoimmune responses induced by heavy metals such as Cd and Hg can be a good model for the mechanism study of autoimmune diseases, and that induction of autoantibodies such as ANA and anti-DNA anti-bodies may be useful for predicting autoimmune manifestations in animals and human exposed to heavy metals, especially those with a genetic predis-position to autoimmune disease.

REFERENCES

Barnes PM, Moynahan EJ (1973) Zinc deficiency in acrodermatitis enteropathica: multiple dietary intolerance treated with synthetic diet. Proc Royal Soc Med 66:327-329
Bernard A, Lauwerys R, Gengoux P, Mahieu P, Foidart JM, Druet P, Weening JJ (1984) Anti-laminin antibodies in Sprague-Dawley and Brown Norway rats chronically exposed to cadmium. Toxicology 31:307-313
Castano P (1971) Chronic intoxication by cadmium experimentally induced in rabbits. A study of kidney structure. Path Microbiol 37:280-301
Chandra RK (1980) Acrodermatitis enteropathica: zinc levels and cell-mediated immunity. Pediatrics 66:789-791
Chandra RK (1983) Trace element and immune response. Immunol Today 4:322-325
Dardenne M, Pleau J-M, Nabarra B, Lefrancier P, Derrien M, Choay J, Bach J-F (1982) Contribution of zinc and other metals to the biological activity of the serum thymic factor. Proc Natl Acad Sci 79:5370-5373

Dardenne M, Savino W, Wade S, Kaiserlian D, Lemonnier D, Bach J-F (1984) In vivo and in vitro studies of thymulin in marginally zinc-deficient mice. Eur J Immunol 14:454-458

Dean JH, Luster MI, Boorman G (1982) Immunotoxicology. In: Sirois P, Rola-Plezczynski M (eds) Immunopharmacology, Elsevier Biomed Press, Amsterdam, p 349-397

Druet P, Bernard A, Hirsch F, Weening JJ, Gengoux P, Mahieu P, Birkeland S (1982) Immunologically mediated glomerulonephritis induced by heavy metals. Arch Toxicol 50:187-194

Durnam DM, Palmiter RD (1981) Transcriptional regulation of the mouse metallo-thionein gene by heavy metals. J Biol Chem 256:5712-5716

Gershwin M, Beach RS, Hurley LS (1985) Nutrition and Immunity. Academic Press, Inc, Orland, p 190-227

Good RA, West A, Fernandes G (1980) Nutritional modulation of immune response. Fed Proc 39:3098-3104

Gleichmann E, Kimber I, Purchase IFH (1989) Immunotoxicology: suppressive and stimulatory effects of drugs and environmental chemicals on the immune system. Arch Toxicol 63:257-273

Hirsch F, Conderc J, Sapin C, Fournie G, Druet P (1984) Polyclonal effect of $HgCl_2$ in the rat, its possible role in an experimental autoimmune disease. Eur J Immunol 12:620-625

Joshi BG, Dwivedi C, Powell A, Holscher M (1981) Immune complex nephritis in rats induced by long-term oral exposure to cadmium. J Comp Path 91: 11-15

Koizumi S, Sone T, Kimura M, Otsuka F, Ohsawa M (1987) Metallothioneins of monocytes and lymphocytes. In: Kagi JHR, Kojima Y (eds) Metallothionein II. Birkhäuser Verlag, Basel, p 519-523

Koller LD (1980) Immunotoxicology of heavy metals. Int J Immunopharmac 2:269-279

Lauwerys RR, Bernard A, Roels HA, Buchet J-P, Vian C (1984) Caracterization of cadmium proteinuria in man and rat. Environ Health Perspect 54:147-152

Ohsawa M (1987) Immunotoxicity of cadmium. In: Hashimoto K, Minami M (eds) Seminars of Toxicity Mechanisms-1. Centers for Acad Publications Tokyo, Tokyo, p 57-64

Ohsawa M (1989) Immunological disturbance and heavy metals. Eiseikagaku (Japan J Toxicol Environ Health) in press

Ohsawa M, Sato K, Takahashi K, Ochi T (1983) Modified distribution of lympho-cyte subpopulation in blood and spleen from mice exposed to cadmium. Toxicol Lett 19:29-35

Ohsawa M, Takahashi K, Otsuka F (1988a) Induction of anti-nuclear antibodies in mice orally exposed to cadmium at low concentrations. Clin Exp Immunol 73:98-102

Ohsawa M, Takahashi K, Otsuka F (1988b) Induction of antinuclear antibodies by in vivo and in vitro treatments with cadmium. Int J Immunopharmac 10 (Suppl):145

Otsuka F, Koizumi S, Kimura M, Ohsawa M (1988) Silver staining for carboxy-methylated metallothioneins in polyacrylamide gels. Anal Biochem 168:184-192

Repine JE, Fox RB, Berger EM (1981) Hydrogen peroxide kills staphylococcus aureus by reacting with Staphylococcal iron to form hydroxyl radical. J Biol Chem 256:7094-7096

Sone T, Koizumi S, Kimura M (1988) Cadmium-induced synthesis of metallothio-neins in human lymphocytes and monocytes. Chem-Biol Interactions 66:61-70

Thomas EL (1979) Myeloperoxidase, hydrogen peroxide, chloride antimicrobial system: nitrogen-chlorine derivatives of bacterial components in bacteri-cidal action against Echerichia coli. Infect Immun 23:522-531

Williams RO, Loeb LA (1973) Zinc requirement for DNA replication in stimulated human lymphocytes. J Cell Biol 58:594-601

Trace Element Regulation of Immunity and Infection

Ranjit Kumar Chandra

Departments of Pediatrics and Medicine, Memorial University of Newfoundland,
Janeway Child Health Centre, St. John's, Newfoundland, A1A 1R8, Canada

ABSTRACT

Epidemiological studies over the past decades have documented the important
relationship between malnutrition and infection. Recent data have shown
that impaired immunocompetence may be a critical factor in increasing the
susceptibility of malnourished subjects to infection. This applies also to
deficiencies of several trace elements, including zinc, copper, selenium,
and iron.

KEY WORDS

Cell-mediated immunity, phagocytes, antibodies, trace elements.

INTRODUCTION

The clinical observation that malnutrition increases the severity of infection
has been confirmed in careful epidemiological studies. It has been shown,
for instance, that diarrhoeal illness occurs 40% more frequently in under-
nourished infants than among healthy well nourished controls living in the
same community with similar sanitation, housing and socioeconomic status.
The duration of each episode of diarrhoea is at least twice as long in the
wasted group; this compounds the problem of malnutrition. Among other
factors, impaired immunocompetence may be a critical etiopathogenetic factor.
Since the first published account of this (Chandra, 1972), a number of
investigators have confirmed that protein-energy malnutrition results in
impaired cell-mediated immunity, phagocyte dysfunction, reduced levels of
complement components C3 and Factor B, reduced mucosal sIgA antibody
response, and finally decreased affinity of antibody.

It was logical to extend these observations to deficiencies of single
nutrients. It was shown that vitamin A deficiency reduced cell-mediated
immunity and increased the number of bacteria attaching to respiratory
epithelial cells. Vitamin B6 decreased thymic hormone activity. Folic
acid deficiency reduced lymphocyte response to mitogens. Vitamin E
deficiency in rodents impaired T cell-dependent antibody formation.

DEFICIENCIES OF ZINC, SELENIUM AND COPPER

Clinical deficiencies of these elements are rare in isolation. In children
with the inherited syndrome of acrodermatitis enteropathica, serum zinc
is low and there are profound changes in cell-mediated immunity. Delayed
cutenaous hypersensitivity reactions are depressed and chemotactic
migration of leukocytes is slow. In addition, there are interesting
changes in T cell-dependent antibody response. This if illustrated in
Fig. 1. If cocultures of T and B lymphocytes are set up and stimulated
with poke weed mitogen, the production of IgG is decreased in zinc
deficient subjects. However, if you mix T cells of zinc-replete controls
with B cells of the deficient subjects, there is significant improvement
in response, whereas the reverse combination is not as satisfactory.
Zinc deficiency also reduces thymulin activity (Fig. 2).

Copper deficiency in rodents reduces reticuloendothelial cell function.
In addition, these animals are very susceptibile to Salmonella and
Listeria infection. The morbidity and mortality due to cardiac and hepatic
necrosis is enhanced.

IgG Plaques/10⁶ Cells

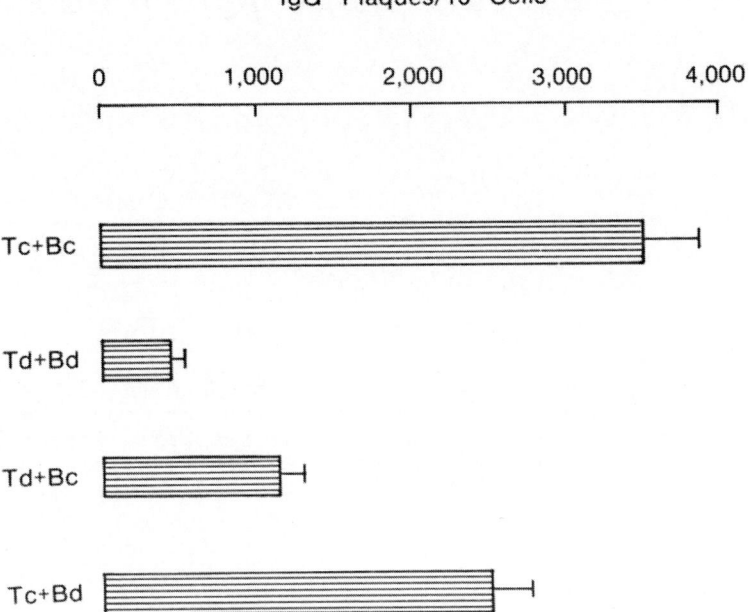

Fig. 1. Coculture experiments using T and B cells from zinc-deficient or
control subjects, employing the reverse hemolytic plaque technique.

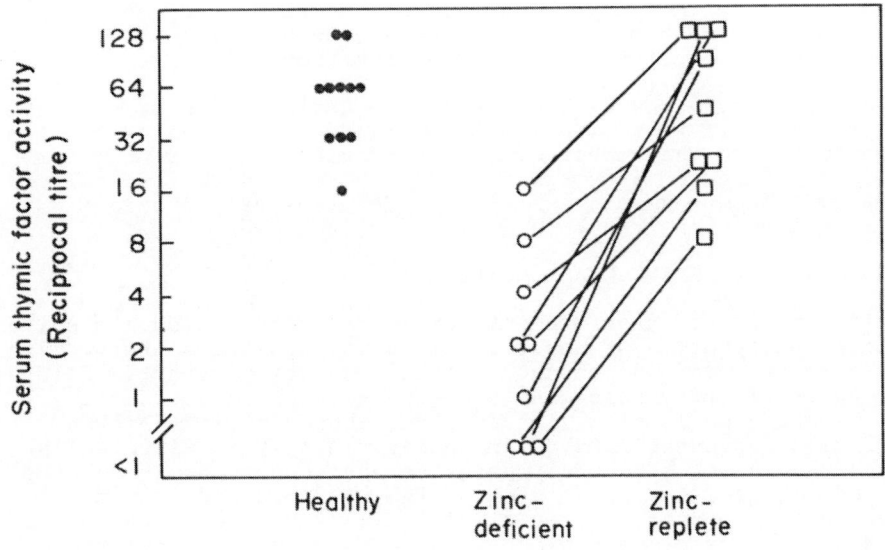

Fig. 2. Serum thymulin activity measured by bioassay in zinc-deficient
children, before and after correction of the deficieny, and in
healthy controls.

Selenium deficiency has been linked to Keshan cardiomyopathy. Our studies in rodents suggest that an additional precipitating factor, eg Coxsackie virus is an essential pathogenetic factor.

IRON DEFICIENCY

Iron is an interesting micronutrient in that it is essential for the growth of bacteria with the sole exception of Lactobacilli and at the same time, it is a cofactor for many critical enzymes that are present in neutrophils and lymphocytes. Unfortunately, many of the published studies on iron status and risk of infection had very poor design and the results cannot be accepted without reservation. Clinical data suggest that parenteral therapy with iron may increase the chance of making a latent bacterial infection symptomatic, whereas oral iron therapy is safe. In fact, the correction or prevention of iron deficiency is associated with reduced incidence of diarrhoeal and respiratory diseases. It may be noted that in states of chronic iron load, such as sickle cell disease or hemachromatosis, most deaths are not related to iron excess producing infection. In the former, the infections are most frequent in the first decade of life when iron stores are not excessive, and in the latter, the most common causes of death are liver failure, heart problems, and diabetes, whereas infection accounts for less than 15% of deaths.

CONCLUDING REMARKS

The critical role of many trace elements for optimum immunity has been proven. It is likely that trace element deficiencies contribute to the immunologic decline seen in old age (Chandra, 1989). At the same time, it would be wise to keep in mind the dictum that if a moderate amount of a nutrient is beneficial for the immune system, a larger amount is not necessarily better. A more detailed discussion of this topic appears elsewhere (Bendich & Chandra, 1990).

REFERENCES

Chandra RK (1972) Immunocompetence in undernutrition. J Pediatr 81:1192-1200.
Chandra RK (1989) Nutritional regulation of immunity and risk of infection in old age. Immunology 67:141-147.
Chandra RK & Bendich A (1990) Micronutrient effects on immune functions. NY Academy of Sciences, New York (in preparation)

Zinc and Thymulin

MIREILLE DARDENNE[1], ANANDA PRASAD[2], and JEAN-FRANÇOIS BACH[1]

[1] INSERM U25 and CNRS UA122, Paris, France
[2] Wayne State University School of Medicine, Detroit, MI 48201, USA

ABSTRACT

Thymulin is a thymic hormone produced by thymic epithelial cells (TEC) known to induce intra- and extra thymic T cell differentiation. It is a nonapeptide whose biological activity and antigenicity depend upon the presence of zinc, which also induces conformational changes in the molecule as demonstrated by RMN studies. The presence of zinc and metallothionein have been demonstrated with TEC which produce the peptide, suggesting that the molecule is secreted in its active zinc-containing form.

In order to define the zinc/thymulin relationship, we studied various models of mild zinc deficiency in humans. First, we showed that serum thymulin activity was decreased as a result of zinc deficiency, and was corrected by in vivo and in vitro zinc supplementation, suggesting that this parameter could be a sensitive indicator of zinc deficiency in humans. Secondly, these observations, when considered together with the parallel variations seen in T-cell subpopulations and lymphokine production, could provide a possible explanation of the role of zinc on T cell functions.

Key words = Zinc - Thymulin - Thymic function - Zinc deficiency.

INTRODUCTION

It is now widely accepted that the trace element zinc exerts a powerful and apparently specific influence on the thymus, on T lymphocytes and on cellular immunity, resulting in a strong immune modulatory activity on cell-mediated responses, in both human and animal systems (Bach, 1981 ; Good, 1981 ; Prasad, 1985).

It is apparent that Zn^{2+} deprivation induces a marked thymic atrophy the effects of whcih closely resemble those of thymectomy : either natal thymectomy or young adult thymectomy when performed in adulthood (Nash et al, 1979).

The mechanisms underlying the early thymic atrophy observed in zinc-deprived mice are of considerable interest. One hypothesis is that the rapidly dividing lymphocytes present in the thymus are particularly sensitive to the abnormal function of zinc dependent enzymes necessary for lymphocyte proliferation. Alternatively, several investigators have evoked the possibility that the epithelial function of the thymus may be specifically altered by such nutritional deficiency of zinc (Chandra et al. 1980 ; Iwata et al. 1979). They reported that a Zn-deficit induces a significant lowering of the serum level of the thymic hormone, thymulin (previously called serum thymic factor (FTS), both in animal models such as mice and rats and in patients with common variable immunodeficiency (Cunningham-Rundles et al., 1981) chronic renal failure (Travaglini et al. 1989) or sickle cell anaemia (Prasad et al. 1988).

Moreover, Zn administration restored normal hormone levels. Data from our laboratory favored the view that thymulin was directly dependent on zinc, a finding which which could explain many of the observations reported above (Dardenne et al,1982). In the present review we shall focus our attention on the presence of zinc within the thymus and its relationship with the biological activity and secretion of thymulin, a chemically defined, zinc dependent thymic hormone, which by itself is able to induce T-cell markers and functions our immature lymphocytes.

PRESENCE OF ZINC WITHIN THE THYMUS

Direct evidence for the presence of zinc within the thymus has come from our previous observations using electron microprobe analysis on ultrathin sections of mouse thymus. Using this technique we detected zinc inside cytoplasmic vacuoles of thymic epithelial cells (TEC), and further experiments demonstrated that this metal can be accumulated within these subcellular structures in thymuses from mice previously injected with zinc chloride (Nabarra et al. 1984). These vacuoles were also shown to contain thymulin.

These data clearly indicate that zinc could be stored within TEC by some specific mechanism. One hypothetical candidate for a role in the intra-thymic storage of zinc could be metallothionein. This low molecular weight protein has a high cysteine content and is known to bind, with high avidity, classe II B transitional metals, namely zinc, cadmium and copper (Kai et al. 1980). Moreover, it has been previously shown that it is highly expressed in cells that can store zinc (Danielan et al. 1982). These findings led us to look for metallothionenin within thymus, and we demonstrated by immunocytochemical means the presence of this protein in thymic epithelial cells both in vivo and in vitro (Savino et al. 1984). By in vitro experiments we showed that zinc-loaded metallothionein released zinc which was eventually trapped by thymulin. In this respect our findings support the notion that metallothionein can function as a zinc acceptor/donor.

PRESENCE OF ZINC IN BIOLOGICALLY ACTIVE THYMULIN

General aspects of thymulin

Thymulin is a thymic hormone isolated from serum, which has been characterized by its capacity to induce T cell markers on T cell precursors (Bach, 1983).

Thymulin is absent in the serum of nude or thymectomized (Tx) mice and reappears after thymus grafting. Chemical analysis showed that it is a peptide of molecular weight 847.

Based on aminoacid analysis and sequence studies on the intact peptide and on the peptide treated with proteolytic enzymes by Edman's method, the aminoacid sequence proposed for thymulin is the following :

< Glu-Ala-Lys-Ser-Gln-Gly-Gly-Ser-Asn-OH.

There is no apparent species specificity, as the aminoacid analysis of calf and human thymulin is identical to that of porcine thymulin.

On the basis of this sequence, a peptide has been synthesized according to two methods, using solid-phase synthesis (Merrifield's technique) and classical solution methods by Bricas et al (1972) and P. Lefrancier et al (1984). The synthetic material showed full biological activity and chromatographically displayed characteristics identical to those of natural thymulin in several chromatography systems.

Data obtained in our laboratory demonstrated that thymulin binds to high affinity receptors (Pléau et al. 1980), induces several T cell markers and promotes T-cell functions including allogeneic cytotoxicity, suppressor function and interleukin-2 production (Bach et al. , 1980).

Direct demonstration of zinc in the molecule

The fortuitous preparation of inactive or unstable lots of thymulin suggested that the peptide could exist in two forms : one biologically active and the other inactive (Dardenne et al.1982). Data have recently been collected showing that the active form contains a metal, probably zinc, whereas the inactive form lacks metal. We have demonstrated that thymulin used in its synthetic or natural form loses its biological activity after treatment with a metal ion chelating agent, Chelex 100. This activity is restored by the addition of zinc salts and to a lesser extent by several other metals, notably aluminium and gallium. The specificity of the zinc effect was assessed in a rosette assay, which demonstrated the absence of biological activity of zinc used alone. The interaction between zinc and thymulin was directly shown by gel chromatography of a mixture of Chelex 100-treated [3H]-FTS and $^{65}Zn^{2+}$ on Bio-Gel P-2. The [3H]-FTS and bound $^{65}Zn^{2+}$ were precisely coeluted with the peak of thymulin biological activity. The presence of zinc in active lots of synthetic thymulin has been confirmed by atomic absorption spectrometry.

Gel filtration studies in ligand equilibrium showed that the nonapeptide could strongly bind one Zn^{2+} ion at pH 7.5 with a Kd of about $5 \pm 2 \times 10^{-7}$M. This binding site is relatively specific for Zn^{2+} although Al^{3+}, Cu^{2+}, and Mn^{2+} are also good competitors for Zn^{2+} thymulin binding. Furthermore, there is a good correlation between the metal binding affinity to thymulin and the capacity of the metal to reactivate the Chelex-treated peptide (Gastinel et al. 1984).

Conformation states of thymulin studied by nuclear magnetic resonance (NMR)

Studies were performed on a zinc free nonapeptide, zinc-coupled thymulin as well as aluminium or copper complexed thymulin, in aqueous medium and in dimethyl sulfoxide (DMSO) solution by means of one or two dimensional ^{1}H, ^{13}C-NMR and circular dichronism. These studies strongly suggested that the free nonapeptide in aqueous solution is flexible and in rapid equilibrium between multiple conformations. In DMSO-d_6, the peptide adopts a partially folded structure stabilized by three intramolecular hydrogen bounds (Laussac et al. 1986).

Complexation with Zn^{2+} induces important NMR changes : analysis of the spectra suggests that Zn^{2+} has a specific binding site involving the C-terminal carboxylate and the hydroxyl groups of the two serine residues in position 4 and 8 , in a tetrahedral structure. The changes caused by Zn2+ addition are quite different from those observed when zinc is replaced by other metals, suggesting that the coordination of each ion induces a different and unique conformation (Cung et al. 1988 ; Laussac et al 1987, 1988).

Monoclonal antibody defined zinc-dependent epitope in the molecule of thymulin

In the past few years, we produced monoclonal antibodies (MAb) directed either against synthetic or natural (intracellular) thymulin (). Both kinds of antithymulin MAb were screened for their ability : a) to inhibit the activity of synthetic or seric thymulin in the rosette assay routinely used to measure thymulin biological activity, and b) to bind specifically to TEC in an immunofluorescence assay.

More recently, we observed that these MAb recognized the thymulin molecule only if it was coupled with zinc (Dardenne et al. 1985). Thus, for example, when synthetic or seric thymulin was subjected to a chelation procedure , it was no longer detectable by the antibodies in both the rosette and the immunofluorescence assays.

However, when zinc was added to the chelated peptide, the anti-thymulin MAb were again able to bind to the hormone.

These data suggest that the presence of zinc in the thymulin molecule determined a spatial configuration which in addition to being necessary to its biological activity, yields a new epitope that can be specifically recognized by monoclonal antibodies.

ZINC DEFICIENCIES AS BIOLOGICAL MODELS TO DEMONSTRATE ZINC DEPENDENCY OF THYMULIN

Experimental zinc deprivation

The zinc/thymulin relationship was further investigated in our laboratory using two models of in vivo zinc deprivation. First, we studied active thymulin levels in sera from mice subjected to a long term marginally zinc-deficient diet. In spite of the absence of thymic atrophy, we observed a significant decrease in the serum levels of thymulin as early as two months after beginning the diet. However, these levels could be consistently restored after in vitro addition of $ZnCl_2$ (Dardenne et al, 1984). These findings strongly suggest that the non-active zinc-deprived peptide is secreted in these experimental conditions. Interestingly, analysis of thymuses from Zn-deprived mice showed that, in spite of the absence of major changes in thymic epithelial cells, there was a progressive increase in the number of thymulin-containing cells, suggesting an increase in the production of the hormone. These observations indicate a compensatory phenomenon, similar to a classical feedback response. We have recently shown in normal mice that an experimental decrease of circulating thymulin, obtained by injections of antithymulin monoclonal antibodies, induces an increase in the number of thymulin-containing cells (Savino et al. 1983). In the case of marginal Zn-deficiency, this feedback effect could be stimulated by the decline of the hormone in its active form, suggesting that the biological regulatory system only recognizes the peptide containing zinc. This hypothesis is in agreement with the fact that all the thymulin functions investigated to date require the presence of zinc in the molecule.

More recently, we made similar observations in human volunteers submitted to zinc-deficient diets. Their thymulin levels, which were in a normal range at the beginning of the diet, dropped very rapidly during zinc deprivation, in close parallel with zinc plasma levels and were restored to normal after in vivo zinc repletion. In addition, the low thymulin levels observed during zinc deprivation could also be restored to normal values by in vitro addition of zinc chloride (Prasad et al, 1988). These results represent a further argument in favor of natural coexistence of the active (zinc containing) and inactive (zinc-deprived) forms of thymulin in the serum, the latter being predominant in zinc-deficient conditions.

Zinc deficiency in pathological situations

We recently studied thymulin levels in various pathological situations with zinc deficiency : the first study was performed in young children with nephrotic syndrome where we observed low thymulin levels, compared to age-matched subjects : these low levels of thymulin activity were restored to normal after remission of the disease or after in vitro addition of zinc salts to the serum under study (Bensman et al., 1984). In addition, we performed similar studies in subjects with sickle cell anemia. Studies have shown that several parameters of cellular immune function may be altered in such patients and related to a zinc deficiency. In these patients, thymulin levels were found to be significantly lower than in age matched healthy subjects. however, as previously observed in human volunteers in whom restricted zinc intake induced a mild specific deficiency of zinc, paralleled with low thymulin levels, we observed that zinc supplementation or in vitro zinc activation restored normal thymulin values (Prasad et al. 1988). Similar results have been recently reported in patients with chronic renal failure and zinc deficiency in which zinc therapy significantly increased zinc-bound thymulin (Travaglini et al, 1989). These observations that zinc is required to render thymulin biologically active suggest that functional thymulin deficiency could contribute significantly to the immune deficiency induced by lack of zinc intake.

CONCLUSIONS

The bulk of data reported in the last few years and summarized above strongly indicate that, at least part of the multifaceted intra-thymic functions of the trace element zinc, may occur via the activation of the thymic hormone thymulin.

As judged by our findings, the zinc/thymulin coupling probably takes place before the hormone releease by thymic epithelial cells. In this context, the mature (zinc-containing) form of the hormone would be able, immediately after being secreted, to influence surrounding thymocytes through its zinc -dependent active site.

REFERENCES

Bach JF (1981) The multi-faceted zinc dependency of the immune system Immunol Today 2 : 225-228.

Bach JF (1983) Thymic hormones. Clin Immunol Allergy 3 : 133-155.

Bensman A, Dardenne M, Morgant C, Vasmant D, Bach JF (1983) Decreased biological activity of serum thymic hormone (thymulin) in children with nephrotic syndrome. Int J Pediatric Nephrol : 201-204.

Bricas E, Martinez J, Blanot D, Auger G, Dardenne M, Pléau JM (1977) In Goodman M, Meinenhofer J (Eds) : Peptides. Proceedings of the 5th American Peptide Symposium. J. Wiley and Sons, New York p. 564-569.

Chandra RK, Heresi G, Au B (1980) Serum thymic factor activity in deficiencies of calories, zinc, vitamin A and pyrodoxine. Clin Exp Immunol 42 : 332-337.

Cunningham-Rundles C, Cunningham-Rundles S, Iwata T, Incefy G, Garofalo JA, Menendez-Botet C, Lewis V, Twomey JJ, Good RA (1981). Zinc deficiency, depressed thymic hormones and T lymphocyte dysfunction in patients with hypogammaglobulinemia. Clin Immunol Immunopathol 21 : 387-391.

Cung MT, Marraud M, Lefrancier P, Dardenne M, Bach JF, Laussac JP (1988) NMR study of a lymphocyte differentiating thymic factor. An investigation of the Zn (II)- nonapeptide complexes (thymulin). J Biol Chem 263 : 5574-5580.

Danielan KJ, Ohi S, Hayang PC (1982) Immunochemical detection of metallothionein in specific epithelial cells or rat organs. Proc Natl Acad Sci USA 79 : 2301-2305.

Dardenne M, Pléau JM, Nabarra B, Lefrancier P, Derrien M, Choay J, Bach JF (1982) Contribution of zinc and other metals in the biological activity of the serum thymic factor. Proc Natl Acad Sci USA 79 : 5370-5379.

Dardenne M, Savino W, Wade S, Kaiserlian D, Lemonnier D, Bach JF (1984). In vivo and in vitro studies of thymulin in marginally zinc-deficient mice. Eur J Immunol 47 : 454-458.

Dardenne M, Savino W, Berrih S, Bach JF (1985) Evidence for a zinc dependent antibody binding site on the molecule of thymulin, a thymic hormone. Proc Natl Acad Sci 82 : 7035 7041.

Gastinel LN, Dardenne M, Pléau JM, Bach JF (1984) Studies on the zinc binding to the serum thymic factor. Biochim Biophys Acta 797 : 147-155.

182

Iwata T, Incefy GS, Menendez-Botet TG, Phi K, Good RA (1983) Circulating thymic hormone in zinc deficiency. Cell Immunol 47 : 100-105.

Kaga JH, Kohima Y, Kissling MM, Leich K (1980). Metallothionein : an exceptional metalthiolate protein. Ciba Foundation Sym 72 : 223-226.

Laussac JP, Cung MT, Pasdeloup M, Haran H, Marraud M, Lefrancier P, Dardenne M, Bach JF (1986) NMR studies of thymulin, a lymphocyte differentiating peptide 1. Conformational states of free peptide in solution. J Biol Chem 261 : 7784-7790.

Laussac JP, Pasdeloup M, Lefrancier P, Dardenne M, Bach JF. (1987) Study of the complexation of copper (II) with a peptidic hormone from thymus naturally occurring as a zinc (II) complex (thymulin). New J Chem 11 : 67-74.

Laussac JP, Lefrancier P, Dardenne M, Bach JF, Marraud M, Cung MT (1988) Structural and conformational study of the aluminum-thymulin complex using 1-D and 2-D NMR techniques. Am Chem Soc 27 : 4094-4099.

Lefrancier P, Derrien M, Amiot JL, Choay J. (1984) Large scale synthesis of serum thymic factor (FTS) or thymulin. In Almquist and Wigzell (Eds) Peptides. Stockholm Sweden p.251-258.

Nabarra B, Halpern S, Kaiserlian D, Dardenne M (1984) Localization of zinc in thymic reticulum of mice by electron probe microanalysis. Cell Tissue Res 238 : 209-212.

Nash L, Iwata T, Fernandes G, Good PA, Incefy G (1979) Effect of zinc deficiency on autologous rosette forming cells. Cell. Immunol. 48 : 238-241.

Pléau JM, Fuentes V, Morgat JL, Bach JF (1980) Specific receptor for the serum thymic factor (FTS) in lymphoblastoid cultured cell lines. Proc. Natl. Acad. Sci. USA 77 : 2861-2866.

Prasad AS, Meftah S, Abdallah J, Kaplan J, Brewer GJ, Bach JF, Dardenne M (1988) Serum thymulin in human zinc deficiency. J Clin Invest 82 : 1202-1210.

Prasad AS (1985) Clinical endocrinological and biological effects of zinc deficiency. Clinics in Endocrin and Metab 14 : 567-578.

Savino W, Dardenne M, Bach JF (1983) Thymic hormone containing cells. III. Evidence for a feed-back regulation of the secretion of the serum thymic factor (FTS) by thymic epithelial cells. Clin Exp Immunol 52 : 7-11.

Savino W, Huang PC, Corrigan A, Berrih S, Dardenne M (1984) Thymic hormone containing cells . V. Immunohistological detection of metallothionein within the cells bearing thymulin (a zinc containing hormone) in human and mouse thymuses. J Histochem Cytochem 32 : 942-946.

Travaglini P, Moriondo P, Togni E, Venegoni P, Bochicchio D, Conti A, Ambroso G, Ponticelli C, Mocchegiani E, Fabris N, Faglia G (1989) Effect of oral zinc administration on prolactin and thymulin circulating levels in patients with chronic renal faillture. J Clin Endocrin Metab 68 : 186-192.

Stress

Effects of Exercise, Physical Trauma, and High Sugar Intake on Chromium, Copper, and Zinc Metabolism

RICHARD A. ANDERSON

USDA, Human Nutrition Research Center, Beltsville, MD 20705, USA

ABSTRACT

The stresses of strenuous exercise, physical trauma and high sugar intake have all been shown to alter chromium (Cr), zinc (Zn) and copper (Cu) metabolism. Strenuous exercise (10 K run or treadmill exercise to exhaustion) led to elevated urinary Cr and Zn losses. Serum Cr was elevated immediately following exercise, serum zinc decreased two hours post exercise and serum Cu was not altered significantly by acute strenuous exercise. Basal urinary Cr and Cu losses of trained runners were greater than those of sedentary control subjects. Physical trauma (primarily motor vehicle accidents) led to several-fold increases in urinary Cr concentration and marked changes in serum Zn. Hypozincemic values (<30 ug/dl) remained below normal four days following trauma. The stress of a high sugar diet also increased Cr losses. Urinary Cr losses were associated with the insulinogenic properties of the sugars consumed. In summary, physical and dietary stresses alter trace metal metabolism and changes in trace metal metabolism are often independent of changes in macroelements.

KEY WORDS: Stress, Chronium, Copper, Zinc

Stress, in various forms, is associated in a causal manner with most diseases and leads to unexplained morbidity and mortality of individuals of all ages. The exact role stress plays in the onset or duration of most diseases is not known.

One factor that needs to be considered in the alleviation of and in combating stress is the role of proper trace metal nutrition. Suboptimal trace metal status is associated with many similar diseases and disease states as those associated with stress (Tables 1, 2 and 3). Additional suggestive evidence linking stress, trace metal status and certain disease states is that stress exacerbates trace metal deficiency and trace metal intake appears to be marginal.

SIGNS OF CHROMIUM, COPPER AND ZINC DEFICIENCY

The clinical signs of Cr, Cu and Zn deficiency are shown in Tables 1, 2 and 3, respectively. Upon closer examination, it becomes evident that deficiency of these trace elements either individually or collectively leads to some of the most prevalent causes of morbidity and mortality in most industrialized societies namely maturity-onset diabetes and coronary artery diseases. Impaired nerve function, decreased sperm count and impaired wound healing are also common occurrences associated with trace metal deficiency. However, the clinical signs of Cr, Cu and Zn deficiency are not unique to these trace elements and there are other factors that may contribute to many of the clinical symptoms listed in Tables 1-3.

Table 1: Clinical Signs of Chromium Deficiency

Impaired Glucose Tolerance	Increased Incidence of Aortic Plaques
Elevated Circulating Insulin	Elevated Cholesterol and Triglycerides
Glycosuria	Nerve and Brain Disorders
Fasting Hyperglycemia	Decreased Fertility and Sperm Count

Table 2: Clinical Signs of Copper Deficiency

Hypercholesterolemia   Impaired Glucose Tolerance
Abnormal Cardiac Function   Anemia?
Hyperglycemia   Decreased RBC Superoxide Dismutase Activity

Table 3: Clinical Signs of Zinc Deficiency

Growth Retardation   Impaired Taste and Smell
Delayed Sexual Maturation in Males Impaired Wound Healing
Low Sperm Count   Immune Deficiencies
Skin Rashes   Nightblindness

DIETARY INTAKE OF CHROMIUM, COPPER AND ZINC

Additional evidence that Cr, Cu and Zn may play a role in lesions associated with impaired carbohydrate and lipid metabolism and related abnormalities is the overwhelming evidence that dietary intake of these trace elements is consistently and substantially below the recommended or suggested intakes.

Dietary chromium intake is approximately 50 percent for females and 60 percent for males of the minimum suggested adult daily intake of 50 uq (Anderson and Kozlovsky 1985). Reliable studies reporting Cr intake from several countries indicate that the Cr intake in the U.S. is similar to that in England, Finland, New Zealand and other developed countries.

Recent studies also suggest that dietary copper intake is often in the region of 1 mg/day compared to the suggested safe and adequate daily intake for adults of 2-3 mg (Davis and Mertz 1987). While older studies reported higher Cu intakes, these studies appeared to be flawed due to analytical problems.

Zinc intake also appears to be below the RDA of 15 mg for adults. Mean daily zinc intake from self-selected diets in the United States, Japan, West Germany, England, Scotland, Canada and other industrialized countries ranges from 8 to 14 mg/day (Hambidge et al. 1986).

Therefore, dietary intake of Cr, Cu and Zn is suboptimal based upon the suggested or recommended intake. However, it may also be possible that the suggested or recommended intakes are too high. However, this does not appear to be the case since studies involving supplementation of these trace elements, especially for Cr and Zn, have reported beneficial effects on subjects consuming normal self-selected diets. In addition, if dietary intake is suboptimal and these trace elements are required in reactions activated in response to stress, stress may lead to marginal signs of deficiency or a depletion of stores that may not be able to be replenished simply from diets low in trace minerals.

STRESS EFFECTS ON TRACE ELEMENT METABOLISM.

A. High Sugar Diet

High sugar diets are often considered "normal diets" and are usually not assumed to be a source of stress. The average American consumes about 50 percent of carbohydrates in the form of simple sugars. This amount represents more than 30 teaspoons a day or 115 pounds per year as contrasted to a 4 pound per person intake 100 years ago (McArdle et al. 1986).

To ascertain the effects of a diet high in simple sugars, subjects were given a nutritionally well-balanced diet either high in simple or complex carbohydrates. Thirty-seven subjects, 19 men and 18 premenopausal women consumed a diet formulated by nutritionists to contain optimal levels of protein, fat, carbohydrates and other nutrients for 12 weeks followed by a diet high in simple sugars for six weeks. Diets high in complex or simple sugars were similar with respect to protein, fat, total carbohydrate and other nutrients except the high complex carbohydrate diet was comprised of 35% of the total calories as complex and 15% as simple sugars. The high sugar diet was similar except 35% of the total calories were derived from simple sugars and 15% from complex carbohydrates (Hallfrisch et al. 1987; Kozlovsky et al. 1986).

Chromium

Consumption of the high simple sugar diet increased urinary Cr losses from 10 to 300% for 27 of 37 subjects compared to the control diet high in complex carbohydrates (Kozlovsky et al. 1986). Chromium contents of the two diets were similar (16 ug/1000 calories). Simple sugars or combinations of sugars that increase circulating insulin levels have recently been shown to also increase Cr losses in proportion to the sum of the circulating insulin (Anderson et al., 1990). Following an overnight fast, 11 male and 9 female adult subjects were given one of the following five carbohydrate drink combinations (per/kg body wt) on each of five mornings at intervals of at least two weeks: 1) 1.0 g glucose 2) 0.9 g uncooked cornstarch 3) 1.0 g glucose followed 20 min later by 1.75 g fructose 4) 0.9 g uncooked cornstarch followed 20 min later by 1.75 g fructose and 5) deionized water followed 20 min later by 1.75 g fructose per kg body wt. Glucose plus fructose was the most insulinogenic followed by glucose alone, starch plus fructose, starch alone and water plus fructose. Urinary chromium losses of control subjects with good glucose tolerance and insulin response followed the same pattern. However, subjects with slightly impaired glucose tolerance (elevated insulin response) did not display elevated urinary Cr losses for the most insulinogenic carbohydrates (glucose alone and glucose plus fructose). Subjects with the highest concentrations of circulating insulin displayed decreased ability to mobilize Cr in response to the added stresses of the most insulinogenic carbohydrates. In both studies male and female subjects responded similarly.

Copper

While consuming the control complex carbohydrate diet, women were in positive carbohydrate balance, 0.13 ± 0.05 mg/day, but were in negative balance, -0.20 ± 0.06 mg/day, while consuming the high simple sugar diet. Copper intake of 1.65 mg/day was identical in both diets. The negative copper balance during the high simple sugar portion of the study was due primarily to increased fecal loss rather than differences in urinary losses. Men also were in positive balance while consuming the control diet and in negative balance during the high simple sugar diet period but differences were not significant (Hallfrisch et al., 1987). In studies involving experimental animals, copper metabolism and requirements have been shown to be adversely affected by simple sugars, primarily fructose (Fields et al. 1986). Copper deficiency anemia is also more severe in rats fed sucrose or fructose rather than starch (Johnson and Gratzek 1986).

Zinc

Women were in positive zinc balance during both the control and high sugar portions of the study. Men were also in positive balance during both portions of the study but men retained two times more zinc during the high sugar portion of the diet compared to the control diet, 2.00 ± 0.68 and 0.86 ± 0.47 mg/day, respectively. Differences between the two diet periods were not significantly different but the zinc balance during the high sugar period was significantly greater than zero balance.

B. Strenuous Exercise

Running and other forms of aerobic exercise are becoming seemingly every increasingly popular activities. Aside from the obvious benefits of helping to control weight, exercise is linked to clinical changes that may lead to improved health. For example, epidemiologic data support the postulate that high density lipoprotein (HDL) cholesterol, which is elevated by long term exercise, protects against coronary artery disease. However, long term exercise or training should be differentiated from acute exercise especially for those who seldom exercise. Acute exercise may be extremely stressful for poorly trained individuals.

Table 4: Effect of a 10K Run on Urinary Excretion of Chromium and Zinc

Element	Run	Rest	Ratio
Chromium	0.37 ± 0.08^b	0.20 ± 0.04^a	2.09 ± 0.4
Zinc	711 ± 497^b	489 ± 343^a	1.57 ± 0.1

Values in same row with different superscripts are significantly different at $p < 0.05$.

Table 5: Basal Urinary Chromium, Copper and Zinc Excretion of Trained and Untrained Runners

Element (ug/day)	Trained	Untrained
Chromium	0.09 ± 0.01^a	0.21 ± 0.03^b
Zinc	828 ± 85	754 ± 71
Copper	9.5 ± 0.6^a	12.2 ± 0.09^b

[a,b] Values in same row with different superscripts are significantly different at $p < 0.05$.
[+] Runners were divided into trained and untrained groups based upon VO_2max. Runners classified in the VO_2 max category or above average or above were considered trained and those in the average or below were designated as untrained (Anderson et al., 1988b).
Values for basal urinary losses were for samples collected during a seven day period that subjects consumed a constant diet.

Table 6: Effects of a 10K Run on Serum Chromium, Copper and Zinc.

	Chromium (ng/ml)	Copper (ug/dl)	Zinc (ug/dl)
Before	0.12 ± 0.02^a	93 ± 5	81 ± 4^b
Immediately after	0.17 ± 0.03^b	94 ± 4	85 ± 4^b
2 Hours postrun	0.19 ± 0.03^b	94 ± 4	75 ± 4^a

[a,b] Values in same column with different superscripts are significantly different at $p < 0.05$.
Subjects ran a 10K run at their maximal capacity. Subjects ran to the blood drawing area after the run and samples were collected immediately.

Chromium

We presented evidence previously that urinary chromium concentration increased from 1.5 to 8.6-fold within two hours of completion of a 10K run. Values were similar when corrected for creatinine excretion. Total urinary chromium losses were more than double on the run day compared to a nonrun day

(Table 4). In a follow up study involving controlled dietary intake and controlled bouts of exercise (Anderson et al., 1988a) mean chromium losses of trained subjects was less than half of that of sedentary subjects, 0.09 $\pm$ 0.01 and 0.21 $\pm$ 0.03 ug/day, respectively (Table 5). The lowered Cr losses of trained subjects may be suggestive of a depletion of Cr due to frequent increased Cr losses associated with frequent exercise bouts or adaptive mechanisms to conserve Cr. For example, there may be a redistribution of Cr to the tissues. That postulate is supported by the work of Vallerand et al. (1984) who reported that exercise trained rats retained significantly higher Cr concentrations in the heart and kidneys compared with respective tissues of sedentary controls. However, adaptive mechanisms associated with training to conserve Cr stores would not be operational for individuals who exercise strenuously but infrequently. Exercised rats fed a low chromium diet for six weeks show increased signs of Cr deficiency compared to sedentary control animals but adapt to the low Cr diet after 18 weeks (Campbell et al. 1989).

Copper

Reports on the effects of acute exercise on serum copper levels are varied. Some reports cite increases (Haralambie 1981; Ohla et al. 1982), decreases (Rusin et al. 1980) or no change (Anderson et al. 1984) in blood copper level due to acute exercise. Effects of training on blood Cu levels are also confusing but most studies reported increased or no change in blood Cu levels associated with training (Ohla et al. 1982; Haralambie 1981). However, Dressendorfer and Sockolov (1980) reported that there was no relationship between training mileage and serum Cu. Training due to swimming also has been reported to lead to decreased Cu levels (Dowdy and Burt 1980).

Serum Cu was not altered immediately following a 10 K run and remained unchanged two hours after the run (Table 6). Basal urinary copper excretion was greater for trained vs untrained subjects. However, biliary excretion is the primary route of excretion of absorbed Cu and usually much less than 10% of the absorbed Cu is excreted in the urine. Therefore, the physiological significance of small changes in urinary Cu excretion of trained vs untrained runners is questionable.

Zinc

Strenuous exercise whether acute or chronic usually leads to decreases in serum zinc. Hypozincemic values (<65 ug/dl) have been reported in 23% of the male runners with an average weekly training distance of only 22 miles (Dressendorfer and Sockolov 1980). A similar incidence of hypozincemia has also been reported in male runners who trained one and half to two hours four times per week and more than 40% incidence in similarly trained female runners (Haralambie 1981).

Following acute exercise, changes in serum zinc are time dependent and do not follow the same time course observed for chromium (Table 6). Serum zinc, immediately following, strenuous running was essentially identical to that observed prior to running. However, two hours following exercise serum zinc concentrations dropped. Leucocyte endogenous mediator, a hormone-like substance released by phagocytes in response to stress such as infection or exercise leads to a redistribution of zinc from the blood to the liver (Pekarek et al. 1972). This may be responsible for the decreased zinc levels two hours post exercise.

Accompanying the hypozincemia associated with strenuous exercise, there is an increase in urinary zinc losses (Anderson et al. 1984). Urinary losses were more than 1.5-fold higher on an exercise day compared to a rest day (Table 4). Urinary zinc losses are very dependent upon the intensity and duration of exercise. In some of our studies involving running at a more leisurely pace, we do not observe significant increases in urinary zinc losses. We also did not observe hypozincemia or increased urinary zinc losses following repeated bouts of acute exercise for 30 seconds at 90% of VO_2 max followed by 30 seconds of rest to exhaustion in our study involving trained and untrained subjects (Anderson et al., 1988b).

C. Physical Trauma

Of the three forms of stress we are discussing namely, high sugar intake, strenuous exercise and physical trauma, physical trauma is the most severe and leads to the largest changes in chromium and zinc parameters. The physical trauma subjects in our study are victims whose injuries are complex enough to necessitate their admittance to a local shock trauma unit. Patients are selected for our studies based on the following criteria: presence of injuries such that the probable stay in the Neurotrauma Intensive Care Unit would enable sample collection for a week, absence of renal disease or injury and a Foley catheter and an indwelling arterial line to facilitate sample collection of urine and blood (Borel et al. 1984).

Chromium

Mean urinary Cr concentration of trauma patients was 10.3 $\pm$ 2.5 ng/ml, (range 2.3-22 ng/ml) in the first four period following admission to the shock trauma unit (normal, 0.2 $\pm$ 0.01) and remained 10-fold above normal even 72 hours post admission. Some of the elevated Cr losses is due not only to stress associated with trauma but to intravenous solutions, that are high in Cr, given initially in the treatment of the multiple injuries. For example, whole blood is low in Cr containing approximately 0.2 ng/ml while plasma protein fraction given during treatment contains more than 30 ng of Cr per ml. These elevated chromium concentrations above values for whole blood appear to be due to contamination during collection and processing. Chromium intake from intravenous solutions administered during the initial 24 hours of treatment of the five patients in one of our studies was 37.1 $\pm$ 12.4 ug/day compared to 0.12 $\pm$ 0.012 ug in the ensuing 24 hour period. Mean Cr intake from foods is usually in the range of 25 to 35 ug/day and 2% or less of this is absorbed and enters the blood (Anderson and Kozlovsky 1985) while essentially all of the Cr from the i.v. solutions would enter the blood supply. Chromium supplementation of trauma patients did not improve glucose and related parameters but this may have been due to the elevated Cr intake due to Cr contamination of the parenteral fluids or the lack of the ability of the trauma patient to convert inorganic chromium to a useable form (Anderson et al. 1988a).

Zinc

Mean serum zinc concentration of eight male patients 24 hours post admission was 41 $\pm$ 4 ug/dl compared to 87 $\pm$ 8 ug/dl for control subjects. Two trauma patients had serum zinc values less than 30 which is usually associated with severe zinc deficiency in nontraumatized subjects.

Urinary zinc was 645 $\pm$ 127 ug/day in the one to four day period following admission and increased to almost 1500 ug between four and seven days and increased further in the next five to ten days to 1778 $\pm$ 189 ug/day (Moser et al., 1985). Mean urinary zinc losses for 12 male control subjects was 590 $\pm$ 140 ug/day. Mean urinary creatinine values were also still elevated ten days following injury. Muscle breakdown and resynthesis may lead to elevated zinc values. Similarly, tissue breakdown due to fasting leads to elevated zinc values (Spencer et al. 1982).

SUMMARY

In summary, the stresses associated with high sugar intake, acute exercise and physical trauma all lead to increased burden on chromium, copper and zinc stores. The severity of the stress seems to be roughly correlated with the magnitude of the change in the serum and urinary losses. Since dietary intake of Cr, Cu and Zn appears to be suboptimal, the added burden associated with stress, both chronic and acute, may exacerbate already marginal status of these trace elements. This is a very important area of research and additional studies are required to define the role and consequences of stress on trace metal metabolism and the effects of optimal trace metal status on combating stress.

REFERENCES

Anderson RA, Borel JS, Polansky MM, Bryden NA, Majerus TJ, Moser PB (1988a) Chromium intake and excretion of patients receiving total parenteral nutrition: effects on supplemental chromium. J Trace Element Exp Med 1:9-18.

Anderson RA, Bryden NA, Polansky MM, Deuster PA (1988b) Exercise effects on chromium excretion of trained and untrained men consuming a constant diet. J Appl Physiol 64:249-252.

Anderson RA, Bryden NA, Polansky MM and Reiser S (1990) Urinary chromium excretion and insulinogenic properties of carbohydrates. Am J Clin Nutr (in press).

Anderson RA and Kozlovsky AS (1985) Chromium intake, absorption and excretion of subjects consuming self-selected diets. Am J Clin Nutr 41:1177-1183.

Anderson RA, Polansky MM, Bryden NA (1984) Strenuous running: Acute effects on chromium, copper, zinc and selected clinical variables in urine and serum of male runners. Biol Trace Element Res 6:327-336.

Borel JS, Majerus TC, Polansky MM, Moser PB, Anderson RA (1984) Chromium intake and urinary chromium excretion of trauma patients. Biol Trace Element Res 6:317-326.

Campbell WW, Polansky MM, Bryden NA, Soares JH Jr, Anderson RA (1989) Exercise training and dietary chromium effects on glycogen, glucagon synthase, phosphorylase and total protein in rats. J Nutr 119:653-660.

Davis GK, Mertz W (1987) Copper. In: Mertz W (ed) Trace Elements in Human and Animal Nutrition, 5th ed. Academic Press, Inc., vol. 1, pp. 301-364.

Dowdy RP, Burt J (1980) Effect of intensive, long-term training on copper and iron nutriture in man. Fed Proc 39:786 (Abstract).

Dressendorfer RH, Sockolov R (1980) Hypozincemia in runners. Phys Sports Med 8:97-100.

Fields M, Holbrook J, Scholfield D, Rose A, Smith JC, Reiser S (1986) Development of copper deficiency in rats fed fructose or starch: weekly measurements of copper indices in blood. Proc Soc Exp Biol Med 181:120-124.

Hallfrisch J, Powell A, Carafelli C, Reiser S, Prather ES (1987) Mineral balances of men and women consuming high fiber diets with complex or simple carbohydrates. J Nutr 117:48-55.

Hambidge M, Casey CE, Krebs NF (1986) Zinc. In: Mertz W (ed) Trace Elements in Human and Animal Nutrition, 5th ed. Academic Press, Inc., vol. 1, pp. 1-138.

Haralambie G (1981) Serum zinc in athletes in training. Int J Sports Med 2:135-138.

Johnson MA, Gratzek JM (1986) Influence of sucrose and starch on the development of anemia in copper- and iron-deficient rats. J Nutr 116:2443-2452.

Kozlovsky AS, Moser PB, Reiser S, Anderson RA (1986) Effects of diets high in simple sugars on urinary chromium losses. Metabolism 35:515-518.

McArdle WD, Katch FI, Katch VI (1986) Exercise Physiology, Energy, Nutrition and Human Performance 2nd ed., Lea and Febiger, Philadelphia, PA, p. 8.

Moser PB, Borel JB, Majerus T, Anderson RA (1985) Serum zinc and urinary zinc excretion of trauma patients. Nutr Res 5:253-261.

Ohla AE, Klissouras V, Sullivan JD, Shoryna SC (1982) Effect of exercise on concentration of elements in serum. J Sports Med 22:414-425.

Pekarek RS, Wannemacker RW, Beisel WR (1972) The effect of leucocytic endogenous mediator (LEM) in the tissue distribution of zinc and iron. Proc Soc Exp Biol Med 140:685-688.

Rusin VY, Nasaolodin VV, Varobev VA (1980) Iron, copper, manganese and zinc metabolism in athletes under high physical pressure. Vapr Pitan 4:15-19.

Spencer H, Kramer L, Perakis E, Norris C, Osis D (1982) Plasma levels of zinc during starvation. Fed Proc 41:347 (Abstract).

Vallerand AL, Cuerrier JP, Shapcott D, Vallerand RJ, Gardiner PF (1984) Influence of exercise training on tissue chromium concentrations in the rat. Am J Clin Nutr 39:402-409.

Zinc: The Role and It's Implication in Clinical Surgery

Akira Okada, Yoji Takagi, Riichiro Nezu, and Shoko Lee

Department of Pediatric Surgery, Osaka University Medical School, 1-1-50 Fukushima, Fukushima-ku, Osaka, 553 Japan

ABSTRACT

Among clinical medicine, surgery or surgical science may be defined as a specialty aimed at controlling disease processes by means of surgical operation. Much of the efforts have been directed not only to the question of how to perform surgical treatment effectively, but also to minimize the surgical stress involved and to expedite recovery from the outcome of the attack. As considering these points, it appears reasonable to say that no trace elements have been more closely related to surgery than zinc.
This communication discuss its significance, implications or role of zinc in the areas such as : (1) wound healing, (2) total parenteral nutrition, (3) specific pathological conditions and (4) surgical stress.

KEY WORDS: Zinc, Wound healing, Total parenteral nutrition, Stress, Zinc tolerance test

(1)WOUND HEALING AND ZINC

It was Pories,[1] who took a leading part in studies on zinc in wound healing. In 1953, Pories and Strain[2] incidentally found wound healing much more rapid in rats fed a phenylalanine- containing diet than otherwise. A close analysis revealed that it was not phenylalanine but zinc as a contaminant of the diet that accelerated wound healing. Thenafter, he also found more rapid wound healing by zinc supplementation in postoperative patients with pilonidal sinus lesions.
Since then, the role of zinc in wound healing was extensively investigated until 1970's. Although most investigators affirmed the usefulness of zinc, there were some who reported in the negative.(Table 1) In this connection, Pories[3] summarized that zinc might not always be responsible for retarded wound healing, and is useful in promoting wound healing only in the presence of zinc deficiency.

Table 1   **ZINC AND WOUND HEALING**

positive
Strain WH and Pories WJ	(1953)	trauma and burns(rat)
Pories WJ and Henzel JH	(1967)	pilonidal sinus
Husain SL and Bessant RG	(1069)	leg ulcer
Greaves MW and Skillen AW	(1970)	leg ulcer
Serjeant GR	(1970)	Sickle cell anemia
Larson DL	(1970)	burns
Pullen FW	(1971)	tracheal granuloma
Hallböök T and Lanner E	(1972)	leg ulcer (double blind study)

negative
Brewer RD	(1966)	decubitus
Myers MB and Cherry G	(1970)	leg ulcer
Clayton RJ	(1972)	leg ulcer (double blind study)

(2)TOTAL PARENTERAL NUTRITION AND ZINC

The development of total parenteral nutrition (TPN) has induced great changes in clinical medicine during these 20 years.[4] Severely emaciated patients who otherwise would not survive because of prolonged starvation can be cured by

giving hypertonic nutrient solutions from central vein. But, in early 1970's, the formulas employed in most instituions show that there is no concern whatever about trace elements.[5]

In around 1973 when TPN became reasonably safe procedure and there was an increasing number of patients on TPN for a month or longer, some of the patients developed progressive skin lesions. Those lesions were localized initially in the face and perianal area, then got worsened and extending to the extremities. Those signs were usually accompanied by enteritis-like symptoms such as abdominal pain and diarrhea mimicking acrodermatitis enteropathica. We experienced 6 those cases in which the nature, distribution pattern, and course of skin lesions were very similar each other. In the first 4 patients, the rapid improvement of these lesions occurred in response to oral feeding suggested the possibility of some nutritional defect. And in the latter 2 cases, that was identified as zinc deficiency, since those lesions responded to parenteral zinc supplementation quickly, reflected by the significant changes of plasma zinc level. In the histological findings of skin biopsies, a parakeratotic lesion characterized by impaired keratinization with the nuclei retained in the stratum corneum was observed, a skin defect is also present at places. The basal cell layer is edematous with cells aranged in a random fashion.

It was in 1975 that we reported (acute) zinc deficiency associated with TPN.[6][7] And in about the same periods, similar cases were reported by Kay and his associates in New Zealand. Several other reports supporting this have been published thenafter. Although there was certain evidences of zinc being effective in the treatment of skin lesions associated with TPN, many questions remained to be answered. (1)What abnormality in systemic zinc level is really reflected ? (2)Is there any difference between diseases as to the incidince of zinc deficiency ? etc.

These questions induced us to undertake the following clinical investigations.[9] A TPN solution not containing zinc was administrated to 47 patients. Of these patients 11 developed clinical manifetations of zinc deficiency. The figure 1 shows the concentrations of zinc in plasma, erythrocytes and urine in those patients. At the onset of the zinc deficiency, the average plasma zinc level is significantly lower than in the controls, the level before TPN and the level at the time of symptomatic relief after zinc administration. The erythrocyte zinc level is slightly lower than in the controls, the level before TPN and the level after zinc administration, but statistically not significant. The urinary zinc level at the onset of zinc deficiency seems lower than that before TPN and controls, but because of great variance, statistically not siginificant. So that, it may be concluded that among 3 parameters of zinc, the plasma level correlated most with the clinical symptoms.

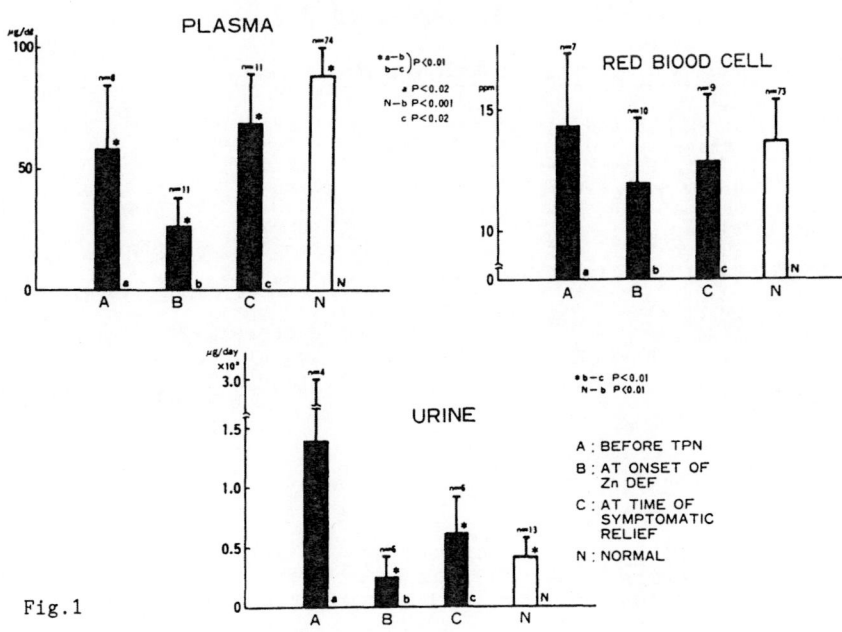

ZINC LEVELS AT ONSET OF
ZINC DEFICIENCY

Fig.1

A time course of plasma and urinary zinc levels during TPN in a typical case is shown in Figure 2. A marked increase in urinary excretion of zinc occurs with the start of TPN. Shortly, this is followed by a rapid decrease, when the eruptions develop. The plasma zinc level is also reduced concurrently. The urinary excretion of zinc is increased again as the skin lesions regress with zinc supplementation.

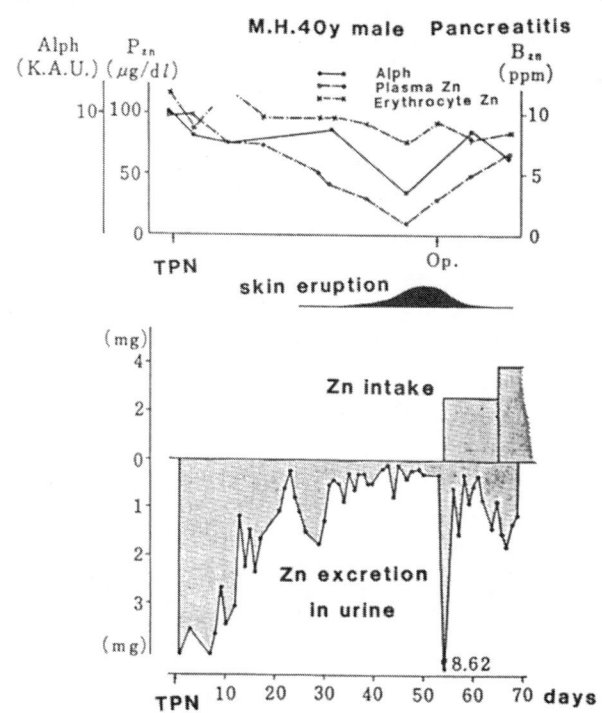

Fig.2 **Clinical course of a case exhibiting zinc deficiency**

The reason why urinary excretion of zinc was increased during TPN is explained by the work of Freeman.[10] According to him, when glucose is added to an amino acid mixture, sugar-amino acid complexes will form, which in turn chelate zinc, thereby resulting in mobilization and increased urinary excretion. Another point of note is contamination of IV infusates. We have been wondering why there are such different incidences of TPN-associated zinc deficiency among countries, since, in Japan, certain number of patients developing zinc deficiency have been experienced. Our examination has shown that TPN solution, notably amino acid solutions, commercially available in Japan contained only few microgram-per-dl amounts of zinc.[11] The concentrations of zinc in these solutions were approximately 1/40 to 1/100 of those in corresponding foreign preparations.(Table 2)

Now, another question rises: "Why, do skin eruptions occur with TPN-associated zinc deficiency ? Does TPN cause any effects upon zinc metabolism of skin ? " To answer these questions, an experiment using an animal model of parenteral nutrition was performed.[12] The experiment was designed to determine the effect of zinc-free TPN solution in malnourished rats fed a protein-free diet for 8 weeks. Sprague Dowley rats were divided into 2 groups, i.e. normally fed and protein malnourished groups. And, they received central venous catheterization. Postoperatively, each group animals were further divided into two groups by TPN solution with or without supplementary zinc. All animals were sacrificed on the 7th postoperative day.

Table 2   Zinc concentration in amino acids solutions

FreAmine 8.5 per cent	400 ug/dl	U.S.A.[4]
Casein Hydrolysate 8 per cent	210	
Aminofusion 850	117	Australia[5]
Aminofusion 1000	61	
Aminofusion Forte	85.6	
Ispol 12 per cent	8	Japan
Moriamin SN 10 per cent	2	
Ami-U (Essential)	2	

The figure 3 summarizes the results of the plasma and organ zinc levels at the time of sacrifice in these animals. It was only in protein-malnourished animals maintained on zinc-free TPN solution that there was a marked fall in plasma and skin zinc levels. In a way, hepatic and blood cells zinc levels were not decreased in contrast to the reduced levels of plasma and skin.
The data indicate that TPN induced re-distribution of zinc in several tissues and a decrease in skin zinc level may contribute to the occurrence of skin lesions.

Zinc Dymanics in Body Tissue During TPN

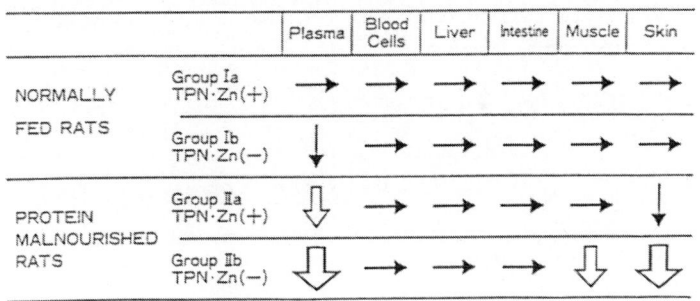

Fig.3

(3)SPECIFIC PATHOLOGIC CONDITIONS AND ZINC

In a previous series of TPN patients, the time of onset of zinc deficiency varied considerably from case to case. In addition, a considerable proportion of patients was free from the deficiency during that period of time. In order to examine for a correlation between the development of the deficiency and the specific pathologic conditions, a comparative study was made for a period of 4 weeks just following the start of zinc-free TPN in 32 randomly selected patients. Five (15.6%) of these patients developed zinc deficiency in 4 weeks, and all had a benign disease of the gastrointestinal tract, such as ulcerative colitis or Crohn's disease, as the underlying pathology. Of the remaining 27 patients, who were free from the deficiency, 22 had gastrointestinal disease, benign in 6 and malignant in 16. The minimal zinc level in each parameter was assessed in relation to the symptoms. As a result, the deficiency occurred in none of the patients with a plasma zinc level of 50 μg/dl or higher, in 5 out of 10 with less than 50 μg/dl, and in all 3 with a level below 30 μg/dl of zinc in plasma.(Figure 4) No such definite tendency was observed for erythrocytes and urinary zinc level. The figure 5 shows the plasma zinc levels by disease in this series of patients. Significantly, it was in patients with benign and malignant diseases of the GI tract and not in those with other diseases that plasma zinc level was reduced progressively during the course of TPN until the signs of zinc deficiency became manifest.

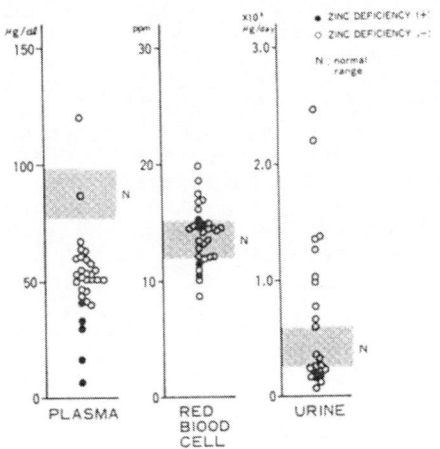

Fig.4. The lowest plasma, red blood cell, urinary zinc levels during the initial 4 wk of TPN.

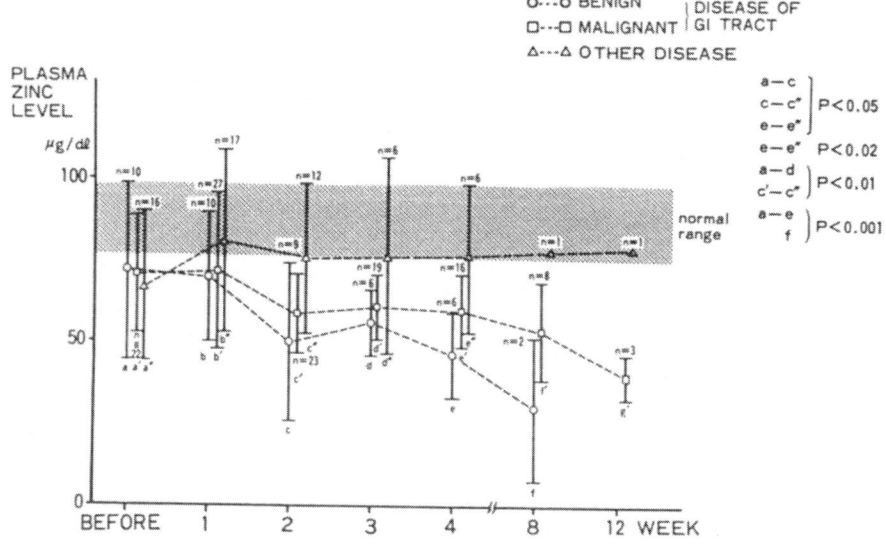

Fig.5

It is possible that the disease of the GI tract causes pre-existing deficiency of zinc before TPN because of long-lasting malabsorption. So, in the next study, patients with Crohn's disease, were examined. Zinc-free TPN was started in those patients for a certain period. As a result, 5 out of 6 patients (83%) developed clinical signs of zinc deficiency in 14th-65th day of TPN. Plasma zinc levels showed less than 50 μg/dl in all cases at the onset of zinc deficiency.(Table 3)

An oral zinc tolerance test was performed in the hope that it might be of help in detecting zinc deficiency somewhat earlier.[13] As a result, the plasma zinc level following an oral dose of 50mg of zinc was much lower than normal in patients with Crohn's disease.(Figure 6) The area under the plasma zinc concentration-time curve (AUC) was much smaller in patients with Crohn's disease of small bowel than in normal controls or patients with the disease of colon, suggesting that zinc deficiency might occur secondary to a lesion of small intestine as the principal site of digestion and absorption.

CASES DEVELOPING ZINC DEFICIENCY DURING TPN
IN CROHN'S DISEASE

Table 3

case		location	onset (TPN) (days)	plasma zinc levels*			therapy (ZnSO4 iv)
				pre TPN	onset	post Rx	
48	M	small intestine & colon	14	38	7	35	80 μmol/day $\times$ 6
26	M	small intestine & colon	30	—	23	30	20 μmol/day $\times$ 18, 40 μmol/day $\times$ 20
25	M	ileum & ascending colon	28	59	33	72	120 μmol/day $\times$ 6
22	F	terminal ileum	22	—	24	96	120 μmol/day $\times$ 7
59	F	transverse colon	65	—	—	124	80 μmol/day $\times$ 7

* μg/dl (nor 88 $\pm$ 11)

ZINC TOLERANCE TEST
IN CROHN'S DISEASE
(Zn 50mg po)

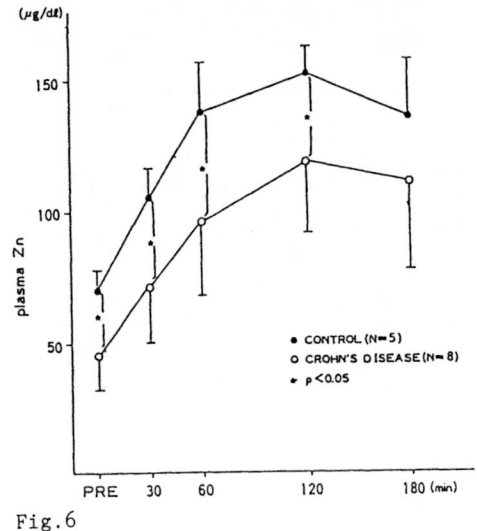

Fig.6

(4)SURGICAL STRESS AND ZINC

From the above discussion, it is apparent that in a case of TPN, zinc deficiency can be easily induced by cooperation of various factors. There was also an evidence of TPN resulting in increased hepatic zinc level, which suggests the occurrence of redistribution of body zinc among various organs. It also became clear that surgical stress as well as TPN causes a decrease in plasma zinc level reflected by an increased urinary zinc level.[14)15)] Why does surgical stress produce such a change in zinc status ? Recently, attention has been paid to the early response of plasma zinc to surgical stress. Surgical stress triggers the release of stress hormones, such as growth hormone, glucagon, cortisol and catecholamines, which in turn exert diverse effects on the metabolism of nutrients, presumably including zinc. The release of leukocyte endogenous mediator (LEM) or endotoxin also takes place -- and indeed earlier than the release of stress hormones. Stimulation by these and other mediators results in a redistribution of body zinc, as indicated by increased hepatic zinc sequestration and decreased plasma and skin zinc levels.[16)] It may be that the liver functions as the site of synthesis of proteins that serve the purpose of homeostatic regulation by protectiong against a surgical stress. Metallothionein is among the proteins that take part in such a reaction, which appears to occur very early. With respect to the possibility of endotoxin acting as a mediator of increased body zinc loss, it was experimentally demonstrated that no response was evoked by treatment with endotoxin in rats made tolerant by its repeated daily administration.[17)] It appears that distinction must be made between LEM and endotoxin; the mode of action appears to be direct for the former and indirect for the latter. Interleukin-1 (IL-1), formerly called LEM, has provided a subject of intensive study as a mediator synthesized in response to stress. Klasing, in his experimental study, determined plasma and liver zinc and liver metallothionein levels in rats treated with IL-1.[18)] IL-1 injection resulted in reduced plasma zinc levels with increased liver zinc and metallothionein levels. This response was early in onset. Should the response serve homeostatic purposes, then zinc might be of great biologic significance. IL-1 is reported to be produced by macrophages and monocytes in response to an attack on the organism. IL-1 appears to act in 2 different modes: direct and indirect. Treatment with IL-1 resulted in decreased skin and plasma zinc levels with increased liver zinc levels, whereas the immune response, particularly natural killer activity was steadily depressed.[18)] All these changes reversed following zinc supplementation. Incidentally, monocyte cytotoxicity responded in the opposite direction. The implications of zinc in the immune response will provide a subject of interest in the future.

CONCLUSION

It was often by chance that surgeons became interested in the study of trace elements. Surgeons have the advantages of having more stressed or malnourished patients in their area as an invaluable source of information about nutrition research, specially in the area of trace elements resarch. However, to elucidate the biologic significance of trace elements is a task enough to require great cooperation and advice of those who are engaged in the study of basic medicine or other bio-technological field.

REFERENCES

1)Pories WJ, Strain WH (1966) Zinc and wound healing. In: Prasad AS (ed) Zinc Metabolism. Charles C Thomas, Springfield, p378-391
2)Strain WH, Dutton Am, Heyer HB, Pories WJ (1954) Acceleration of Burns and Wound Healing with Metionine Zinc. University of Rochester Reports, Rochester
3)Pories WJ, Strain WH (1974) Zinc sulfate therapy in surgical patients. In: Pories Wj, Strain WH, Hsu JM, Woosley RL (eds) Clinical Applications of Zinc Metabolism. Charles C Thomas, Springfield, p139-157
4)Dudrick SJ (1967) Long term total parenteral nutrition with growth in puppies and positive nitrogen balance in patients. Surgical Forum 18:356-357
5)Dudirck SJ, MacFadyen BV Jr, VanBuren CT (1972) Parenteral hyperalimentation. Ann Surg 176:259-264
6)Okada A, Takagi Y, Itakura T (1975) Zinc deficiency during intravenous hyperalimenation. Igaku no Ayumi (Jpn ed) 92:436-442
7)Okada A, Takagi Y, Itakura T (1976) Skin lesions during intravenous hyperalimentation. Zinc deficiency. Surgery 80:629-635
8)Kay RG, Tasman-Jones C, Pybus J (1976) A sundrome of acute zinc deficiency during total parenteral alimentation in man. Ann Surg 183: 331-340
9)Takagi Y, Okada A, Itakura T (1986) Clinical studies on zinc metabolism during total parenteral nutrition as related to zinc deficiency. JPEN 10:195-202
10)Freeman JB, Stegink LD, Meyer PD (1975) Excessive urinary zinc losses during parenteral alimentation. J Surg Res 18:463-469
11)Okada A, Takagi Y (1976) An importance of zinc in total parenteral nutrition solution. JPN J Surg 189:6-8
12)Nezu R, Takagi Y, Okada A (1988) Effect of total parenteral nutrition on distribution of zinc in murine tissues. Trace Nutrients Research 4:115-120
13)Nezu R, Takagi Y, Okada A (1989) unpublished data
14)Fell GS, Fleck A, Cuthbertson DP (1973) Urinary zinc level as an indication of muscle catabolism. Lancet 1:280-282
15)Hallbook T, Hedelin H (1977) Zinc metabolism and surgical trauma. Br J Surg 64:271-273
16)vanRij AM, Hall MT, Bray JT, Pories WJ (1981) Zinc as an integral component of the metabolic response to trauma. SGO 153:677-682
17)Pekarek RS, Beisel WR (1971) Charanterization of the endogenous mediators of serum zinc and iron depletion during infections and other stress. Proc Soc Exp Biol Med 138:728-732
18)Allen JI, Perri RT, MacClain CJ (1983) Alterations in human natural killer activity and monocyte cytotoxicity induced by a zinc deficient patient J Lab Clin Med 102:577-589

Total Parenteral Nutrition

Trace Elements in Total Parenteral Nutrition

KHURSHEED N. JEEJEEBHOY

Division of Gastroenterology, Toronto General Hospital, EN9-223, 200 Elizabeth Street, Toronto, Ontario, M5G 2C4, Canada

KEY WORDS: Total parenteral nutrition, Requirement of essential trace elements, Requirement of vitamins

Trace Elements

Cotzias (1) defined an essential trace element as one which has the following characteristics:

(1) Present in healthy tissues of all living things.

(2) Constant tissue concentration from one animal to the next.

(3) Withdrawal leads to a reproducible functional and/or structural abnormality.

(4) Addition of the element prevents the abnormality.

(5) The abnormality is associated with a specific biochemical change.

(6) The biochemical change is prevented and/or cured along with the observed clinical abnormality.

Animal studies show that 15 elements are essential for health. They are iron, zinc, copper, chromium, selenium, iodine, cobalt, manganese, nickel, molybdenum, fluorine, tin, silicon, vanadium and arsenic. However, using the strict criteria suggested by Cotzias, only the first seven are necessary for health in man. Of these, cobalt is only essential as part of the corrin ring in vitamin B_{12}. The possibility that nutritional edema may be a sign of vanadium deficiency has been raised by Golden and Golden (2, and a case of molybdenum deficiency has been described (3)

General observations

Trace elements are absorbed as inorganic substances and as organic compounds. In natural foods the latter often predominate. Since the absorption of the two forms may differ, results of studies with inorganic test substances cannot be equated with the availability of the same elements from food. For example, heme iron is absorbed very easily (4) and the availability of this form of iron from red meat cannot be judged from studies with elemental iron. Similarly, Cr as an inorganic salt is poorly absorbed (5) but Cr in yeast is well absorbed (6).

Absorbed trace elements circulate as protein-bound complexes which are not always in free equilibrium with tissue stores. For example, the exchangeable plasma copper is present in very small amounts bound to

albumin (7). In contrast, the major form of circulating copper, ceruloplasmin, is not freely exchangeable (8). For this and other reasons circulating levels may not reflect the status of the element available for nutritional needs.

Tissue stores of a trace element may not be available to meet the needs during deficient supply because of two factors. First, they may be bound to enzyme proteins from which exchange does not occur. Second, during anabolism there is net inflow of trace elements into cells, and cellular stores cannot be mobilized to meet needs of other tissues. For example, in hypercatabolic states, even though zinc is being lost and the patient is in negative zinc balance, deficiency does not occur and plasma zinc is normal because of net outflow (9).

When nutritional support is given and protein synthesis occurs, there is a positive zinc balance but plasma levels fall and deficiency will result unless exogenous zinc is given (2). Because the action of trace elements depends upon other metabolic and nutritional factors such as age, anabolic or catabolic state and availability of agonists and antagonists, clinical deficiency cannot be predicted by a simple demonstration of a level of the element. For example, with selenium the interaction with vitamin E may alter the appearance of a deficiency syndrome. Children with selenium-responsive Keshan's disease do not have lower levels of plasma selenium than do children with phenylketonuria receiving artificial diets and not showing clinical deficiency (10).

These findings make it imperative to look for subclinical functional changes as means for defining needs. For example, despite the lack of clinical deficiency, Wolman et al. (9) have shown that a negative zinc balance reduces nitrogen retention and carbohydrate tolerance, thus justifying the need for maintaining balance. Finally, the route of excretion of most trace elements, excepting chromium, is mainly through the gastrointestinal tract. This raises the possibility that abnormal gastrointestinal losses may raise requirements in patients with disease of the G.I. tract; on the other hand, renal disease does not reduce the needs for these elements.

Iron

Distribution. Iron is an essential constituent of porphyrin-based compounds bound to protein, such as hemoglobin and myoglobin. In addition, smaller amounts of tissue iron associated with enzymes and mitochondria have important metabolic functions. Iron is also found in storage or transport forms bound to protein as ferritin and transferrin. In storage another protein-free form, hemosiderin, also occurs. The proportion of ferritin to hemosiderin in the liver depends upon the iron concentration. At lower concentrations ferritin predominates and at higher concentrations hemosiderin predominates (11).

Control of body iron stores. There is little physiological excretion of iron. About 0.2-0.5 mg is excreted in feces per day (12), 0.2 mg in urine (13) and variable amounts in sweat. However, the latter may be high in the tropics where sweating is profuse (14). Thus in normal man losses amount to 0.6-1.0 mg/day. However menstrual losses increase output in women by 0.5-0.8 mg/day and even more. Hence control of iron stores is by the control of absorption from the gastrointestinal tract. In man, as has been proven in animal studies (15,16), absorption and transfer probably occur mostly from the duodenum. In this connection it is of note that in two stable patients receiving home parenteral

nutrition, who have only duodenum remaining, iron stores have been maintained for several years without intravenous supplementation.

The following factors influence iron absorption:

(1) Haeme iron is absorbed directly, and its absorption is not sensitive to gastric pH and intestinal factors.

(2) Normal gastric secretion is necessary to release protein-bound iron in food.

(3) Ascorbic and organic acids enhance absorption.

(4) Trace elements in large amounts may reduce iron absorption.

(5) Copper deficiency reduces iron absorption.

(6) Iron deficiency, hypoxia and anemia from various causes increase iron absorption.

Abnormal iron losses. Blood loss from the gastrointestinal tract, inflamed surfaces, and bile or small bowel drainage enhance iron losses. In addition, venesection for laboratory tests in hospital adds to this load. Thus in seriously sick patients the need for iron is increased.

Assessment of iron status. Iron status is best determined by chemical assay of the iron content of bone marrow (17). A less quantitative method is to evaluate the stainable hemosiderin in the marrow. Ferritin levels in plasma reflect iron stores in general but are unreliably high in patients with inflammation and liver disease (18). If there is no inflammation then 1 microgram/litre of ferritin corresponds to 140 microgram/kg body weight of stored iron (4). In addition, the circulating ferritin levels are raised after iron-dextran injections.

Transferrin saturation determines the flow of iron to tissues, and when it falls below 16% then iron supply is suboptimal (19). Unfortunately this fall can occur as a result both of deficiency and inflammation (4). Transferrin levels and serum iron are unreliably low when infection and malnutrition reduce protein synthesis and alter the distribution of iron. Iron released from erythrocytes is taken up by macrophages and then transferred to transferrin. With inflammation, macrophages do not release iron, thus reducing circulating serum iron levels and saturation. In malnutrition, transferrin synthesis falls and thus circulating iron levels are reduced.

Iron requirements. In normal males the requirements are about 1 mg of absorbed iron per day. In menstruating females this increases to 2 mg/day. In TPN patients the needs may well be enhanced because of abnormal losses. However, no carefully controlled study of iron needs has been done. Brennan and his associates (20) found that up to 25 mg of iron given per week to patients on TPN significantly reduced the need for blood transfusion. The problem with patients on a parenteral nutrition regimen is the combination of factors altering the availability of iron to the tissues. Deficiency in tissues can result both from a reduction of iron stores and reduced macrophage transfer despite normal stores. In addition, reduced transferrin levels reduce the availability of total circulating iron. Only the first can be treated by giving more iron.

Recommendations for intravenous iron. A dilute solution of iron-dextran can be used to provide 1 mg of iron per day to meet physiologic losses (21). This should be increased to 2 mg/day in females. Additional needs due to abnormal losses should be met by infusing a calculated amount of iron-dextran. In home TPN (HTPN) patients we give 1-2 mg/day and monitor the bone marrow at intervals of 1-2 years.

Zinc

Zinc is a widely distributed element in the human body and is now identified as a part of about 120 enzymes (22). Among them are carbonic anhydrase, carboxypeptidase, alkaline phosphatase, oxido-reductases, transferases, ligases, hydrolases, lyases and isomerases. However, the syndrome of zinc deficiency cannot be identified with the deficiency of any one enzyme. On the other hand, zinc deficiency has a pronounced effect on nucleic acid metabolism, thus influencing protein and amino acid metabolism.

Zinc is an integral constituent of DNA polymerase, reverse transcriptase, RNA polymerase, tRNA synthetase and the protein chain elongation factor (22). Thus zinc deficiency can alter protein synthesis at a number of different points, and it is not surprising that growth is arrested in the absence of zinc (23). Furthermore, zinc deficiency is teratogenic, as determined by animal studies and observations in untreated patients with acrodermatitis enteropathica (24). This finding suggests that zinc deficiency may affect gene expression. In experimental studies in unicellular organisms, zinc deficiency changes the nature of RNA polymerase and the base composition of mRNA. The translated peptides contain a preponderance of arginine-rich peptides which can bind to anions such as phosphate groups in nucleic acids and alter their action. Such an alteration could affect the synthesis of histones, proteins which are known to reduce the activity of DNA as a template (22).

These experimental findings are interesting in view of the clinical observation that a number of functions dependent on protein synthesis are suppressed by zinc deficiency. These include:

(1) Growth (22).
(2) Cellular immunity (25,26).
(3) Fertility (34).
(4) Hair growth.
(5) Wound healing (27).
(6) Plasma protein levels.

Thus it is obvious that zinc deficiency leads to profound disturbances of protein synthesis.

Distribution. Zinc is widely distributed in all soft tissues (28), blood cells, bone and teeth (29). However, zinc at these sites is firmly bound to protein, and during deficiency and refeeding (with the exception of blood, milk, hair and liver) the concentrations of zinc do not change significantly (30). Endogenous stores of zinc are mobilized in the fasting state but do not meet metabolic needs during anabolism, because the net movement of zinc is into tissues and there is little free zinc circulating.

Control of body zinc. Zinc absorption has a significant effect on body zinc. Zinc is absorbed by a process which involves binding to an acceptor, followed by uptake into the enterocyte (31). The process is saturable and the efficiency of absorption decreases at high zinc intakes. In the enterocyte, zinc is bound to a metallothionein, the proportion bound depending on the metallothionein content of the enterocyte. Since the bound zinc is not transferred to plasma, this binding inhibits absorption. Subsequently the bound zinc is lost into the bowel lumen when the enterocyte is shed. When body zinc is high there is a stimulation of metallothionein synthesis, inhibiting absorption. In addition, absorption is influenced by the following factors:

(1)   Binding to a ligand secreted by the pancreas enhances absorption.

(2)   Luminal amino acids bind zinc and prevent its precipitation by substances such as phosphates and phytates.

(3)   Pregnancy, corticosteroids, endotoxin and leucocyte endogenous mediator (LEM) all enhance absorption.

(4)   Phytates, phosphates, iron, copper, lead and calcium inhibit absorption.

Excretion.   This occurs mainly in the feces and a smaller amount in the urine (13). The fecal losses rise with increased intake, as they consist mainly of unabsorbed zinc. In contrast, urinary losses are not influenced by intake.   In the tropics, significant losses may occur in sweat but such losses diminish with deficiency (32).

Abnormal losses.   Wolman et al. (9) showed that diarrhea, and stomal and fistula losses were the major source of enhanced abnormal losses of zinc from endogenous sources in patients kept NPO. Increased losses also occurred in urine in hypercatabolic individuals. Amino acid infusions also increase urinary zinc losses.   In the kidney, zinc infusions enhance distal resorption of zinc, while amino acid infusion increases proximal secretion (18,33).

Metabolism.   Absorbed zinc is bound to albumin (34) and an alpha-2 macroglobulin in circulation. From the circulation it is taken up by the liver and other tissues. Infection increases uptake of zinc by the liver (35).   This process is stimulated by leucocyte endogenous mediator (LEM) (35).   Enhanced uptake of zinc into the liver reduces plasma concentrations, and so circulating zinc concentrations may be reduced by factors other than deficiency (36-38).

Assessment of zinc status. While circulating zinc levels fall in the deficient state, other causes of low circulating zinc levels make this measurement unreliable.   Hair zinc levels are low when there is low-grade chronic deficiency, but in acute deficiency hair does not grow and with profound deficiency hair loss occurs and the remaining hair has normal zinc concentrations (39).   It has recently been claimed that leucocyte zinc levels are a reliable indicator of zinc deficiency, but this is not an easy measurement to perform, and its validity has not been tested under different clinical circumstances.   Currently the best way to assess zinc status and requirements is by multiple clinical parameters. The presence of abnormal G.I. losses, hypercatabolism or amino acid infusions indicates the need for zinc supplementation.   The clinical syndrome of acrodermatitis enteropathica confirms the need for zinc.

Requirements. Adults should ingest 15 mg of zinc per day. Based on a mean absorption of 7%, this is about 1 mg of absorbed zinc per day. In patients receiving parenteral nutrition which included intravenous amino acids, Wolman et al. (9) found that about 2.5 mg/day was required for balance in patients without diarrhea. Increased catabolism and gastrointestinal losses increased requirements by 12 mg/L of small intestinal fluid and 17 mg/L of stool, measured in the NPO state.

Requirements in infants. In addition to meeting losses, infants need zinc for growth. This is especially true of preterm infants to whom 2/3 of their zinc is transferred during the last 10-12 weeks of gestation. About 0.5-0.75 mg of zinc is taken up per day during the last 3 weeks of gestation and the first 3 weeks of postnatal life in babies weighing 1.5 kg, making the requirements about 300-500 micrograms/kg/day (40). James et al. (41) found that infants required 300 micrograms/kg/day to maintain balance. In older children, 50 micrograms/kg/day maintained normal serum levels (42), and 100 micrograms/kg/day is a safe intake for growth. In addition, supplementation is required for abnormal gastrointestinal losses, but these have not been determined experimentally.

Copper

Copper is widely distributed in human tissues and is a part of enzymes such as cytochrome-C oxidase, superoxide dismutase, dopamine beta-hydroxylase, monoamine oxidase and lysyloxidase (43). In addition, 90% of the plasma copper is in the form of ceruloplasmin (44). The major effects of copper deficiency are a result of ceruloplasmin and lysyloxidase deficiency (45,46), although abnormalities in catecholamine metabolism have also been described as a result of its association with dopamine beta-hydroxylase (47). Ceruloplasmin is an iron oxidase. Iron released by red-cell breakdown is taken up by macrophages. It is then released from macrophages and bound to transferrin for transport to iron-requiring tissues. Similarly, stored iron in the liver is in equilibrium with transferrin. Ceruloplasmin oxidizes ferrous iron and aids in transfer of iron from stores to transferrin (48). It is believed that iron in cells is reduced by riboflavin to the ferrous form, to cross the cell membrane; after crossing, it is reoxidized to the ferric form so that it can bind to transferrin (49). Therefore copper deficiency results in conditioned iron deficiency.

Mature collagen and elastin are characterized by the presence of cross-links formed from a precursor peptide called alpha-aminoadipic acid delta-semialdehyde, or allysine (50). This substance is formed by the oxidative deamination of peptidyl lysine residues (51).

```
      l                                              l
      C=O      H H                                   C=O      H H
      l        l l                    lysyl-         l        l l
      HC-(CH2)2-C-C-NH2+O2+H2O-----------> HC-(CH2)2-C-C=O
      l        l l                    oxidase        l        l
      HN       R H                                   HN       R
      l                                              l

                                   + NH3 + H2O2
```

This process depends on the copper-containing enzyme lysyloxidase (46).

In addition, to the effects on iron and collagen, another result of copper deficiency is the phenomenon of leucopenia in infants (52).

Distribution. Copper is mainly concentrated in the liver and the brain, with smaller amounts in the heart, kidneys, spleen and the skeleton (53). Out of a total of 23 mg of copper in the human body, 16 mg is found in the liver and brain. In the blood, copper circulates as ceruloplasmin and is bound to albumin (54). The latter form is the true transport copper and exchanges with tissue.

Control of body copper. The control of body copper is also likely to be through altered absorption and biliary secretion (55,56). Copper absorption is enhanced by deficiency and is depressed by phytates, ascorbic acid and cadmium in the diet. Zinc inhibits copper absorption by promoting intestinal metallothionein synthesis, which binds copper and prevents its transfer to the circulation (57). The dietary intake has been estimated to be 2-5 mg/day, of which about 32% is absorbed (0.6-1.6 mg/day) (44). However, in normal man, balance studies indicate that about 1.2 mg/day is sufficient (58,59). The major route of excretion is the bile, in which about 0.5-1.3 mg is excreted per day. The copper excreted in bile is believed to be completely excreted and there is no enterohepatic circulation of copper (44). Urinary losses are only about 10-60 micrograms/day (44). The intravenous infusion of copper does not increase urinary excretion (60). Loss of copper in sweat is variable and is estimated to be 0.34 ± 0.24 mg/day (61).

Abnormalities of copper excretion. Diarrhea increases copper losses but the increase is not related to volume of diarrhea. Abnormalities of liver function reduce copper losses (60). Adrenocortical deficiency reduces urinary copper excretion and vice versa (62). In injured patients urinary copper excretion rises to a mean of 256 micrograms/day (63).

Metabolism. Absorbed copper is bound to albumin and amino acids such as histidine, threonine and glutamine in the circulation, and from there it is taken up by the liver and bone marrow, amongst other sites. In the liver, copper is incorporated into ceruloplasmin and released into the circulation or excreted into the bile. In the marrow it is incorporated into erythrocuprein and released as red cells. Ceruloplasmin also donates copper for incorporation into enzymes such as superoxide dismutase, lysyl oxidase and cytochrome oxidase (64).

Assessment of copper status. Plasma copper is reduced in deficiency but is also affected by a variety of factors altering the level of ceruloplasmin. These include deficient production due to protein-calorie malnutrition and increased loss in nephrosis (65). Infections and inflammatory conditions, leukemia, and Hodgkin's disease increase the levels of serum copper (66). Oral contraceptive agents likewise increase plasma copper levels to 300 ± 7 micrograms/dl, from a mean normal of 118 ± 2 micrograms/dl (67). For reasons already noted, hair copper is also not a reliable index of deficiency (68).

Requirements. The normal diet supplies 2-4 mg/day, and on this intake deficiency has never been observed in adults. In contrast, infants gain a major part of body copper during the last 10 weeks of gestation

(40) and it is estimated that a premature or neonatal infant will retain 100-130 microgram/kg/day over a 3-week period. Malnourished infants require 40-135 micrograms/kg/day.

In adult patients on a parenteral nutrition regimen, Shike et al. (60) found that 0.3 mg/day was sufficient in patients without diarrhea; the requirements rose to 0.5 mg/day in those with diarrhea. In contrast, the requirements fell to 0.1 mg/day in patients with abnormal liver function. These figures compare well with the findings of Jacobson and Western (69), who obtained a positive balance with all patients but one when they infused 0.24-0.29 mg of copper per day. In critically ill patients Phillips and Garny (70) have recommended 0.5 mg/day.

In infants, the requirements based on balance are 50 micrograms/kg/day (41). However, the range from infant to infant is very wide, varying from 10 to 50 micrograms/kg/day. Hence caution should be exercised to avoid overload, and the lower figure of 20 micrograms/kg/day has been suggested. In older children 20 microgram/kg/day is sufficient to meet their needs (42).

Chromium

Chromium deficiency in animals causes a syndrome of glucose intolerance similar to that of clinical diabetes. The abnormalities are correctable by giving chromium (71). This element is important in promoting insulin action in peripheral tissues (72). In vitro chromium enhances insulin stimulation of glucose oxidation and lipogenesis in adipose tissue (72). In the muscle it also increases insulin-induced glycogenesis (73). Insulin-stimulated amino acid transport is also positively influenced by chromium (73).

These observations in animals are supported by the finding that the intravenous administration of chromium increased glucose clearance in a human patient with chromium deficiency (74). Chromium administration to this patient also increased the fall in circulating leucine levels in response to a glucose load (74). Since the insulin response to the glucose load was normal, the above observation suggests that chromium enhances insulin-stimulated tissue uptake of leucine. The respiratory quotient (R.Q.) was low and the plasma free fatty acid levels high prior to the administration of chromium, showing that fat mobilization and oxidation were continuing despite normal insulin levels. Administration of chromium reduced free fatty acid levels, increased the R.Q., and promoted glucose oxidation for energy. Thus chromium is one of the factors which influences insulin sensitivity.

Distribution. Chromium is distributed throughout the human body and its concentration declines with age (75). If chromium deficiency is the cause of increased glucose intolerance with age, then chromium substitution should improve glucose tolerance. Offenbacher and Pi-Sunyer (76) found that older persons receiving a yeast containing chromium in a bound form had improved glucose tolerance, as compared with controls given a yeast (Torula) poor in chromium. Hence the decline in body chromium may contribute to glucose intolerance in the elderly.

Control of body chromium. Chromium is probably absorbed in organic compounds since less than 3% of a dose of inorganic chromium is absorbed (5). In contrast, 10-25% of an oral dose of yeast chromium is absorbed (6). Absorption is inhibited by zinc and phytate. However,

since absorption studies were done with inorganic chromium it is not clear how these findings relate to chromium in food. In natural foods chromium exists in a dinicotinato-glutathionine complex (77) called glucose tolerance factor (GTF) (78). This organic complex appears to be the form in which chromium becomes available from food. As indicated earlier, it is absorbed easily and animal studies have confirmed its ability to enhance insulin action on glucose metabolism (72,78,79).

Excretion, occurring chiefly in the urine (80), is considerably increased in insulin-dependent diabetics (81), but not in adult-onset ones (82), and by glucose loading in insulin-dependent diabetics (79) and normals (83). Quite small amounts are lost in stool (80). In one patient, receiving total parenteral nutrition, losses amounted to about 20 micrograms per day (74).

Assessment of chromium status. This area is as yet not clearly defined. Plasma chromium levels are reduced in deficiency and in acute infection (84). Elevated plasma levels have been found after insulin injection in the rat (85) and after glucose loads (86) although Anderson (87) found no such change. Urinary chromium excretion in response to a glucose challenge has been proposed as an indirect index of deficiency. However at present it appears that in individual cases the only convincing way of assessing chromium deficiency is to demonstrate prolonged glucose clearance that responds to chromium supplementation.

Requirements. In the adult, oral chromium requirements have not been determined. Deficiency in patients receiving total parenteral nutrition (74,88) may be due to continuous glucose loading, resulting in a higher urinary excretion, which in turn increases requirements. It is estimated that in one such patient the needs may have been increased to 10-20 micrograms/day (74). In infants, balance studies by James et al. (41) indicated a requirement of 0.14-0.2 micrograms/kg/day. However, balance studies in more patients with a spectrum of clinical conditions need to be done in order to obtain further information.

Selenium and vitamin E

These two substances are interrelated in their action, and the deficiency of one can be partially corrected by giving the other. To understand the need for these two micronutrients it is necessary to examine the alternative means by which oxygen is reduced in biological systems. Normally the enzyme cytochrome oxidase accepts electrons from cytochrome-c at one end exposed to the cytosol, and discharges them by reacting with 4H+ to form water (89). The alternative path involves the monovalent addition of electrons to form superoxide (90). The superoxide, if left unaltered, dissociates to H_2O_2 and oxygen. The H_2O_2 can also react by the Haber-Weiss reaction with superoxide to form hyroxyl ions. Thus, in the absence of appropriate controls, a number of reactive peroxide and hydroxyl radicals can be formed and damage the cell. In well-nourished cells, superoxide dismutase (SOD) catalyses the conversion of superoxide to H_2O_2, and the peroxide so formed is reduced by glutathione peroxidase (GSHpx) to water.

$$O_2 \ \text{--------------------} \xrightarrow{e^-} \ O_2^- \ \text{----------------} \xrightarrow{e^-} \ O_2-$$

$$O_2^- + H_2O_2 \ \text{--------------------} \longrightarrow \ O_2 + OH^- + \ \cdot OH$$

$$O_2^- + H^+ \ \text{--------------} \xrightarrow{SOD} \ H_2O_2 \ \text{--------------} \xrightarrow{GSHpx} \ H_2O$$

Glutathione peroxidase is an enzyme made up of four subunits, each containing selenocysteine as an integral part of the molecule (91). In association with superoxide dismutase it controls the the levels of superoxide and peroxide in the cell. This in turn affects lipid peroxidation of polyunsaturated fatty acids in cell membranes. Vitamin E is the second line of defense and controls the formation of hydroperoxides in the fatty acid residues of phospholipids, a process which depends on both the antioxidant role of the vitamin and its structural relation with the membrane phospholipids (92). Finally, any lipid hydroperoxides may be reduced by GSHpx to hydroxyacids.

Abnormal losses   In patients living in a low-selenium area (New Zealand), the concentration of selenium may be as high as 130 micrograms/litre in wounds and pus and 100-380 micrograms/litre in fistula fluids during oral supplementation with selenite-Se (98,99). Similar data are not available for other parts of the world.

Requirements.   The dietary intake of selenium varies from a low of 18-26 micrograms/day in New Zealand (100) to a high of 150-220 micrograms/day in Canada (101). Human metabolic studies suggest a minimum intake of 20 micrograms/day (97), and in studies from China 30 micrograms was considered a minimal intake (102). In North American volunteers, the requirements were estimated to be 54 micrograms/day.

Patients receiving parenteral nutrition may develop selenium deficiency with muscle pains (98), and cardiomyopathy has also been described in three patients receiving long-term TPN (103,104,105). However, the need for selenium will be conditioned by other factors such as vitamin E status, heavy-metal intake, abnormal losses and the presence of oxidants and other anti-oxidants. These aspects need resolution.

Selenium-Vitamin E Interaction

In animals it has been shown that the myonecrosis induced by a diet deficient in selenium and vitamin E can be corrected by giving either to the animal (106). They protect equally well against several experimentally induced cardiomyopathies in animals. In patients receiving TPN the amount of vitamin E required to prevent functional changes appears to be five times as great as the recommended dietary allowances (107). This could be due to the fact that selenium was not added to the TPN solutions used.

Assessment of vitamin E status. Vitamin E status can be assessed by the lower levels of this vitamin in circulation and functionally by observing the red cell hemolysis induced by peroxide (108), and the platelet aggregation response to the addition of ADP (109). Experimentally, vitamin E deficiency increases lipid peroxidation, increasing ethane and pentane in the breath (110). Monitoring breath ethane and pentane may be a means of assessing lipid peroxidation. Recently we have proposed such a breath pentane method for use in humans. We showed a good correlation between the breath pentane output and the blood alpha-tocopherol level (111). Measurement of erythrocyte malondialdehyde in vitro has also been described as a functional measure of vitamin E status (112).

Peroxidation cleaves the polyunsaturated fatty acid molecule at the first double bond from the methyl end. The size of the alkane thus released depends on the omega-number of the fatty acid involved.

peroxy fatty acid

$$CH_3-(CH_2)_3-CH_2-\underset{\underset{\cdot}{\overset{|}{O}}}{\overset{|}{C}}H-CH=CH-R \longrightarrow CH_3-(CH_2)_3-\overset{|}{\underset{|}{C}}H_2 + \overset{\overset{\cdot}{H}}{\underset{O}{C}}-CH=CH-R$$

$$+ .H$$

$$CH_3-(CH_2)_3-CH_3 \qquad \text{lipid peroxide}$$

pentane

Requirements for vitamin E. The RDA of vitamin E is 10 I.U. of alpha-tocopherol per day. However, in TPN patients Thurlow and Grant (107) found that correction of resistance of peroxide-induced hemolysis and normalization of platelet aggregation with ADP required about 50 I.U. per day. This requirement has not been studied with different amounts of selenium in the infusion. The same amounts were required to maintain normal levels in patients receiving TPN at home (113). This area needs further study.

Manganese

The role of this trace element is not well defined in the human. It is important for the action of glycosyltransferases. In this role its deficiency leads to abnormality of cartilage growth in young animals. In addition it appears to be necessary for the action of vitamin K in adding the carbohydrate component of prothrombin to the preprothrombin protein. In this regard Doisy (114) described a patient in whom vitamin K could not correct prothrombin levels until the patient was given manganese. The mitochondria are very rich in manganese and it is an essential component of mitochondrial superoxide dismutase (115).

Distribution. The human body contains 12-20 mg of manganese which is mainly distributed in the mitochondria (116).

Control of body manganese. Only 3-4% of an oral dose of manganese is absorbed (117). Manganese is very efficiently excreted by the intestine. This occurs mainly through the bile with lesser amounts through

the pancreatic juice and the intestinal wall. There is negligible excretion in the urine. The absorption of manganese is enhanced with low intakes and depressed when intake is high.

Metabolism. Manganese in circulation is bound to a beta-globulin, transferrin. It is taken up rapidly by the mitochondria and at a slower rate by the nuclei. The manganese in cells is in dynamic equilibrium with that in the circulation. Intake in excess of needs increases excretion into the gastrointestinal tract (118).

Requirements. The dietary intake varies but is between 2-3 mg per day (119). The amount retained is about 50-400 micrograms/day (120). Patients receiving home TPN which contained 2.0 mg/day had elevated levels of whole-blood manganese (113). However, requirements have not been determined for human TPN.

Molybdenum

Molybdenum is an essential component of xanthine oxidase (121), sulfite oxidase (122) and aldehyde oxidase (123). Xanthine oxidase catalyzes the conversion of oxypurines to uric acid. In its absence the levels of oxypurines will rise and those of uric acid fall. Sulfite oxidase similarly influences the conversion of sulfite to sulfate. The lack of sulfite oxidase is responsible for neurologic abnormalities (124), and it is of interest that Abumrad et al. (3) described a TPN patient who developed coma when infused with amino acid solutions containing sulfite. The coma was reversed by supplementing the TPN solutions with 300 micrograms of molybdenum per day. The concurrent findings of hyperoxipurinemia, hypouricemia, and low sulfate excretion, also corrected by giving molybdenum, are supportive evidence for molybdenum deficiency as a cause of this syndrome.

Distribution. Human liver and kidney have the highest concentrations of molybdenum (53).

Control of body molybdenum. This element in the molybdate hexavalent form is easily absorbed from salts and from herbage (125). Excretion is mainly in the urine but the urinary excretion rises as sulfate intake or endogenous sulfate production increases. Interestingly, urinary copper excretion rises in man on a high molybdenum diet. Hence copper requirements will be influenced by molybdenum intake.

Abnormal losses. Patients with Crohn's disease have been found to excrete 600 micrograms of molybdenum in stools per day.

Requirements. These are not known but preliminary balance studies indicate that individuals may be in equilibrium with as little as 48-96 microgram/day. However, if patients have abnormal losses then requirements may become much larger. Abumrad et al. (3) gave 300 micrograms/day to restore the metabolism of the patient referred to earlier with Crohn's disease and short bowel. More studies are required.

All other trace elements including fluorine, tin, arsenic, silicon, vanadium, cadmium, lead and mercury are not of dietary importance in the human, and some, like lead, aluminium, cadmium and mercury, may be toxic. Clearly much more needs to be done to assess the requirements for substances like selenium, molybdenum and chromium in TPN patients, especially those with GI losses and those with acute illness.

Vitamins

Vitamins are essential nutrients which are active in minute quantities. While it is obvious that these substances have to be given in any regimen of TPN to avoid deficiency, with the exception of vitamin D, the optimum dose and frequency of administration have not been studied in detail in patients receiving TPN. The currently available studies have been simple observations of plasma or blood levels during a given regimen.

Vitamin A

This vitamin is essential for the integrity of epithelial surfaces and synthesis of retinal pigments, and is also of importance in protecting against infection. It is fat-soluble and is stored in the liver. In patients on long-term TPN, 2500 IU maintains normal circulating levels for periods of several years. When all vitamin A was withdrawn for six months these patients did not show a fall in plasma levels, indicating that 2500 IU per day had maintained adequate stores. Furthermore, despite a series episode of sepsis in three of these patients, there was no fall in plasma vitamin A levels and no difficulty in recovering from infection (126). Hence 2500 IU is recommended as the daily dose of this vitamin by the parenteral route.

Vitamin C

Vitamin C is a strong reducing agent and is required for redox reactions, collagen synthesis and normal immune functions. In deficiency the states the syndrome of scurvey occurs with perifollicular hemorrhages in the skin, gingivitis and infections. In children subperiosteal hemorrhages may occur. About 300-500 mg per day will maintain normal to high blood levels of this vitamin in patients receiving TPN.

Vitamin D

In long-term TPN, 250 IU maintains normal plasma levels of 5-hydroxycholecalciferol, a metabolite of this vitamin. However, observations by Shike et al. (127) indicate that patients receiving long-term TPN may develop a syndrome of hypercalciuria, intermittent hypercalcemia, and osteomalacia with bone pains and fracture of the axial skeleton. In the short-term TPN patient, the hypercalcemia is associated with pancreatitis. All these changes are reversed and clinical healing of fractures occurs by simply withdrawing vitamin D. This improvement occurs despite the fact that an active metabolite of this vitamin, 1,25 dihydroxycholecalciferol, is low in these patients. Hence, intravenous vitamin D is undesirable during TPN for reasons not entirely clear at this time. Furthermore, in patients observed over one year, the withdrawal of vitamin D did not result in subnormal plasma levels of the 25 monohydroxy derivative. On the basis of these findings, intravenous vitamin D is not recommended; instead, long-term TPN patients are encouraged to expose themselves to sunlight to maintain normal D levels from natural sources.

Vitamin E

This vitamin has been discussed earlier in connection with selenium (q.v.).

Vitamin K

This vitamin is required for the synthesis of four factors necessary for coagulation. While there are no data about the precise needs for this vitamin, administering 10 mg per week of Synkavite[R], a water-soluble analogue of this vitamin, maintains normal coagulation parameters over several months in patients receiving long-term TPN. A recent study has shown that antibiotics, particularly cephalosporins, can impair vitamin K function in patients receiving total parenteral nutrition. Supplementation (up to 10 mg K_1/day) is required in this situation (128).

Thiamine

Thiamine is an integral part of the cocarboxylase enzyme complex which is necessary for the metabolism of alpha-keto acids such as pyruvate. Cells such as neurons which depend exclusively on carbohydrate as an energy substrate need this vitamin and are especially vulnerable to thiamine deficiency. Thiamine requirements are about 1 mg/1000 kcals of energy intake. Based on observations of the thiamine-dependent enzyme activity in such patients, Kishi et al. (129) found that 5 mg/day meets the requirement in patients receiving short-term TPN, 16 mg of thiamine per day resulted in blood levels which were three times higher than the mean normal value for this vitamin. Hence, 5 mg/day should be sufficient for most patients receiving TPN. It is of note that severe metabolic acidosis has been described in 3 patients receiving parenteral nutrition with no vitamin supplementation (130). All responded promptly to the administration of intravenous thiamine. This seems appropriate since without thiamine, glucose cannot enter the Krebs cycle in order to be completely oxidized for energy production, and therefore accumulates as lactic acid.

Riboflavin

This vitamin is a component of coenzymes flavin mononucleotide and flavine adenine dinucleotide, which are necessary for hydrogen transfer in redox systems. Riboflavin deficiency causes photophobia, glossitis, cheilosis and pruritus of the skin with inflammation especially around the anogenital area. In patients receiving TPN 5 mg/day is recommended as a safe amount in order to avoid deficiency.

Niacin

Niacin is a component of nicotine adenine dinucleotide (NAD) and its phosphate (NADP), which are necessary for several dehydrogenation reactions required for carbohydrate and protein metabolism and for cell respiration. In deficiency states the syndrome of pellagra develops, characterized by dark-red erythema on the exposed areas of the body, with pigmentation and cracking of the skin, glossitis, stomatitis and diarrhea. In addition, there are neurologic symptoms of delirium and confusion. On the basis of observed circulating levels of this vitamin in long-term TPN patients, about 50 mg/day is a safe intake which will avoid deficiency.

Pantothenate

Pantothenic acid is a component of coenzyme A. The deficiency of this vitamin is poorly characterized in man but appears to result in

fatigue, mental disturbances, paresthesias and epigastric discomfort. Studies of blood levels of this vitamin in patients on long-term TPN suggest that 15 mg/day is a suitable intake.

Pyridoxine

This vitamin is a coenzyme in many reactions concerned with amino acid metabolism, including decarboxylation, transamination and dehydroxylation. In the adult human the symptoms of deficiency are not not very specific and consist of dermatitis, intertrigo, seborrhea, irritability, somnolence and neuropathy. An intake of 5 mg per day is sufficient to maintain normal blood levels of this vitamin in long-term TPN patients.

Folic acid and vitamin B$_{12}$

These vitamins are important for the synthesis of nucleic acids. Deficiency results in megaloblastic anemia and glossitis. Severe vitamin B$_{12}$ deficiency will result in neuromyelopathy. The recommended intake parenterally is 600 micrograms of folic acid and 12 micrograms of vitamin B$_{12}$ per day.

Biotin

Adults receiving long-term TPN have been shown to have quite subnormal serum levels of biotin without signs and symptoms of deficiency (113). It is possible that special circumstances contributed to even lower levels in the 7 documented cases of clinical deficiency recently listed (131). The AMA has recommended administration of 60 ug/day for TPN patients (132).

References

1. Cotzias GC: Trace Substances Environmental Health - Proceedings of the University of Missouri Annual Conference. 1967; p 5

2. Golden MHN, Golden BE: Trace elements. Br Med Bull 1981; 37:31-36

3. Abumrad NN, Schneider AJ, Steele D, et al: Amino acid intolerance during prolonged total parenteral nutrition reversed by molybdate therapy. Am J Clin Nutr 1981; 34:2551-2559

4. Finch CA, Huebers H: Perspectives in iron metabolism. N Engl J Med 1982; 306:1520-1526

5. Donaldson RM, Barreras RF: Intestinal absorption of trace quantities of chromium. J Lab Clin Med 1966; 68:484-493

6. World Health Organization, WHO Tech Rep Ser 1973; #532

7.   Bush JA, Mahoney JP, Gubler CJ, et al: Studies on copper metabolism: transfer of radio-copper between erythrocytes and plasma. J Lab Clin Med 1956; 47:898-906

8.   Sternleib I, Morell AG, Tucker WD, et al: The incorporation of copper into ceruloplasmin in vivo: studies with copper-64 and copper-67. J Clin Invest 1961; 40:1834-1840

9.   Wolman SL, Anderson GH, Marliss EB, et al: Zinc in total parenteral nutrition: requirements and metabolic effects. Gastroenterology 1979; 76:458-467

10. Diplock AT: Metabolic and functional defects in selenium deficiency. Phil Trans R Soc Lond B 1981; 294:105-117

11. Shoden A, Gabrio BW, Finch CA: The relationship between ferritin and hemosiderin in rabbits and man. J Biol Chem 1953; 204:823-830

12. Dubach R, Moore CV, Callender ST: Studies in iron transportation and metabolism: excretion of iron as measured by isotope technique. J Lab Clin Med 1955; 45:599-615

13. Robinson MF, McKenzie JM, Thomson CD, et al: Metabolic balance of zinc, copper, cadmium, iron, molybdenum and selenium in young New Zealand women. Br J Nutr 1973; 30:195-205

14. Foy H, Kondi A: Anaemias of the tropics; relation to iron intake, absorption and losses during growth, pregnancy and lactation. J Trop Med Hyg 1957; 60:105-118

15. Wheby MSA, Jones LRG, Crosby WH: Studies on iron absorption. Intestinal regulatory mechanics. J Clin Invest 1964; 43:1433-1442

16. Osterloh K, Forth W: Determination of transferrin-like immunoreactivity in the mucosal homogenate of the duodenum, jejunum and ileum of normal and iron-deficient rats. Blut 1981; 43:227-235

17. Morgan EH, Walters MNI: Iron storage in human disease. Fractionation of hepatic and splenic iron into ferritin and hemosiderin with histochemical correlations. J Clin Path 1963; 16:101-107

18. Lipschitz DA, Cook JD, Finch CA: A clinical evaluation of serum ferritin as an index of iron stores. N Engl J Med 1974; 290:1213-1216

19. Bainton DF, Finch CA: The diagnosis of iron deficiency anemia. Am J Med 1964; 37:62-70

20. Peters ML, Maher M, Brennan MF: Minimal IV iron requirements in TPN. (Abst.) JPEN 1980; 4(6): 601

21. Kong KW, Tsallas G: Dilute iron dextran formulation for addition to parenteral nutrient solutions. Am J Hosp Pharm 1980; 37:206-210

22. Vallee BL, Falchuk KH: Zinc and gene expression. Phil Trans R Soc Lond B 1981; 294:185-197

23.   Prasad AS: Zinc in Human Nutrition. Boca Raton, Fl, CRC Press 1979; pp 1-80

24.   Hambidge KM, Neldner KH, Walravens PA: Zinc, acrodermatitis enteropathica and congenital malformations. Lancet 1975; i:577-578

25.   Golden MHN, Golden BE, Harland PSEG, et al: Zinc and immunocompetence in protein-energy malnutrition.   Lancet 1978; i:1226-1227

26.   Fernandes G, Nair M, Onoe K, et al: Impairment of cell-mediated immunity functions by dietary zinc deficiency in mice. Proc Natl Acad Sci USA 1979; 76:457-461

27.   Golden MHN, Golden BE, Jackson AA: Skin breakdown in kwashiorkor responds to zinc. Lancet 1980; i:1256

28.   Tipton IH, Cook MJ: Trace elements in human tissue. II. Adult subjects from the United States. Health Phys 1963; 9:103-145

29.   Underwood EJ: Zinc. Ch. 8 in Trace Elements in Human and Animal Nutrition, 4th ed. New York, Academic Press 1977; pp 196-242

30.   Kirchgessner M, Roth H, Weigand E: Trace elements in human health and disease.   In: Zinc and Copper, Vol. 1. Prasad AS (Ed) New York, Academic Press, 1976; pp 189-225

31.   Davies NT: Studies on the absorption of zinc by rat intestine. Br J Nutr 1980; 43:189-203

32.   Prasad AS, Schulert AR, Sandstead HH, et al: Zinc, iron and nitrogen content of sweat in normal and deficient subjects. J Lab Clin Med 1963; 62:84-89

33.   Abu-Hamdan DK, Migdal SD, Whitehouse R, et al: Disparate urinary zinc (ZN) handling in response to ZN infusion and amino acids. (Abst.) Kidney Int 1979; 16:818

34.   Smith KT, Cousins RJ: Quantitative aspects of zinc absorption by isolated, vascularly perfused rat intestine.   J Nutr 1980; 110:316-323

35.   Biesel WR, Pekarek RS, Wannemacher RW Jr: The impact of infectious disease on trace-element metabolism of the host. In: Trace Element Metabolism in Animals, Vol. 2.   Hoekstra WG, et al (eds). Baltimore, University Park Press, 1974; pp 217-240

36.   Talbot TR, Ross JF: The zinc content of plasma erythrocytes of patients with pernicious anemia, sickle cell anemia, polycythemia vera, leukemia and neoplastic disease. Lab Invest 1960; 9:174-184

37.   Vallee GL, Wacker WEC, Bartholomay AF, et al: Zinc metabolism in hepatic dysfunction. Ann Int Med 1959; 50:1077-1091

38.   Vikbladh I: Studies on zinc in blood. Scand J Clin Lab Invest 1950; . 2:143-148

39. Hambidge KM: Zinc deficiency in man: its origin and effects. Phil Trans R Soc Lond B 1981; 294:129-144

40. Widdowson EM, Dauncey J, Shaw JCL: Trace elements in foetal and early postnatal development. Proc Nutr Soc 1974; 33:275-284

41. James BE, MacMahon RA: Balance studies of 9 elements during complete intravenous feeding of small premature infants. Aust Ped J 1976;. 12(3):154-162

42. Ricour C, Duhamel J-F, Gross J, et al: Estimates of trace element requirements of children receiving total parenteral nutrition. Arch Fr Pediat 1977; 34:92-100

43. Mason KE: A conspectus of research on copper metabolism and requirements in man. J Nutr 1979; 109:1979-2066

44. Cartwright GE, Wintrobe MM: Copper metabolism in normal subjects. Am J Clin Nutr 1964; 14:224-232

45. Evans JL, Abraham PA: Anemia, iron storage and ceruloplasmin in copper nutrition in the growing rat. J Nutr 1973; 103:196-201

46. O'Dell BL: Roles for iron and copper in connective tissue biosynthesis. Phil Trans R Soc Lond B 1981; 294:91-104

47. Fell BF: Pathological consequences of copper deficiency and cobalt deficiency. Phil Trans R Soc Lond B 1981; 294:153-169

48. Osaki S, Johnson DA, Freiden E: The possible significance of the ferrous oxidase activity of ceruloplasmin in normal human serum. J Biol Chem 1966; 241:2746-2751

49. Golden MHN: Trace elements in human nutrition. Hum Nutr Clin Nutr 1982; 36C:185-202

50. Pinnell SR, Martin GR: The cross-linking of collagen and elastin. Proc Natl Acad Sci USA 1968; 61:708-714

51. Siegal RC, Pinnell SR, Martin GR: Cross-linking of collagen and elastin. Properties of lysyl oxidase. Biochemistry 1970; 9:4486-4492

52. Cordano A, Baertl JM, Graham GG: Copper deficiency in infancy. Pediatrics 1964; 34:324-336

53. Hamilton EI, Minski MJ, Cleary JJ: The concentration and distribution of some stable elements in healthy human tissue from the United Kingdom. Sci Total Environ 1972-73; 1:341-374

54. Gubler CJ, Lahey ME, Cartwright GE, et al: IX. Studies on copper metabolism: transportation of copper in blood. J Clin Invest 1953; 32:405-415

55. Bremner I: Absorption, transport and storage of copper. In: Biological Roles of Copper. Ciba Foundation Symposium #79. Excerpta Medica Amsterdam 1980; pp 23-48

56.   Owen CA: Absorption and excretion of ^{64}Cu-labelled copper by the rat. Am J Physiol 1964; 207:1203-1206

57.   Hall AC, Young BW, Bremner I: Intestinal metallothionein and the mutual antagonism between copper and zinc. J Inorg Biochem 1979; 11:57-66

58.   Guidelines for essential trace element preparations for parenteral use. A statement by an expert panel. JAMA 1979; 242:2051-2054

59.   Sandstead HH: Copper bioavailability and requirements. Am J Clin Nutr 1982; 35:809-814

60.   Shike M, Roulet M, Kurian R, et al: Copper metabolism and requirements in total parenteral nutrition. Gastroenterology 1981; 81:290-297

61.   Jacob RA, Sandstead HH, Munoz JM, et al: Whole body surface loss of trace metals in normal males. Am J Clin Nutr 1981; 34:1379-1383

62.   Henkin RI: On the role of adrenocorticosteroids in the control of zinc and copper metabolism.   In: Trace Element Metabolism in Animals, Vol. 2.   Hoekstra WG, et al (ed). Baltimore, University Park Press, 1974; pp 647-651

63.   Askari A, Long CL, Murray RL, et al: Zinc and copper balance in severely injured patients. (Abst.) Fed Proc 1979; 38:707

64.   Bremner I, Mills CF: Absorption, transport and tissue storage of essential trace elements. Phil Trans R Soc Lond B 1981; 294:75-89

65.   Kovalsky VV: The geochemical ecology of organisms under conditions of varying contents of trace elements in the environment. In: Trace Element Metabolism in Animals.   Mills, CF, Vol. 1. Edinburgh, Livingstone, 1970; pp 385-397

66.   Wintrobe MM, Cartwright GE, Gubler CJ: Studies on the function and metabolism of copper. J Nutr 1953; 50:395-419

67.   Halsted JA, Hackley BM, Smith JC: Plasma zinc and copper in pregnancy and after oral contraceptives. Lancet 1968; ii:278-279

68.   Hambidge KM: Increase in hair copper concentration with increasing distance from the scalp. Am J Clin Nutr 1973; 26:1212-1215

69.   Jacobson S, Western PO: Balance study of twenty trace elements during total parenteral nutrition in man.   Brit J Nutr 1977; 37:107-126

70.   Phillips GD, Garnys VP: Parenteral administration of trace elements to critically ill patients. Anaesth Intensive Care 1981; 9:221-225

71.   Schwarz K, Mertz W: Chromium (III) and the glucose tolerance factor. Arch Biochem Biophys 1959; 85:292-295

72. Mertz W, Roginski EE, Schwarz K: Effect of trivalent chromium complexes on glucose uptake by epididymal fat tissue of rats. J Biol Chem 1961; 236:318-322

73. Roginski EE, Mertz W: Effects of chromium 3+ supplementation on glucose and amino acid metabolism in rats fed a low protein diet. J Nutr 1969; 97:525-530

74. Jeejeebhoy KN, Chu RC, Marliss EB, et al: Chromium deficiency, glucose intolerance and neuropathy reversed by chromium supplementation in a patient receiving long-term total parenteral nutrition. Am J Clin Nutr 1977; 30:531-538

75. Schroeder HA, Balassa JJ, Tipton IH: Abnormal trace metals in man. J Chron Dis 1962; 15:941-964

76. Offenbacher EG, Pi-Sunyer FX: Beneficial effects of chromium-rich yeast on glucose tolerance and blood lipids in elderly subjects. Diabetes 1980; 29:919-925

77. Toepfer WW, Mertz W, Polansky MM, et al: Synthetic organic chromium complexes and glucose tolerance. J Agri Food Chem 1977; 25:162-166

78. Anderson RA, Mertz W: Glucose tolerance factor: an essential dietary agent. Trends in Biochem Sci 1977; 2:277-279

79. Schroeder HA: The role of chromium in mammalian nutrition. Am J Clin Nutr 1968; 21:230-244

80. Hopkins LL Jr: Distribution in the rat of physiological amounts of injected Cr^{51} (III) with time. Am J Physiol 1965; 209:731-735

81. Hambidge KM: Chromium nutrition in the mother and the growing child. Ch. 9 in Newer Trace Elements in Nutrition. Mertz E, Cornatzer WE (eds). New York, Dekker, 1971; pp 169-194

82. Doisy RJ, Streeten DHP, Souma ML, et al: Metabolism of ^{51}chromium in human subjects, normal, elderly and diabetic subjects. Ch. 8 in Newer Trace Elements in Nutrition, Mertz E, Cornatzer WE (eds). New York, Marcel Dekker, 1971; pp 155-168

83. Anderson RA, Polansky MM, Bryden NA, et al: Urinary chromium excretion of human subjects: effects of chromium supplementation and glucose loading. Am J Clin Nutr 1982; 36:1184-1193

84. Pekarek RS, Hauer EC, Bayfield EJ, et al: Relationship between serum chromium concentrations and glucose utilization in normal and infected subjects. Diabetes 1975; 24:350-353

85. Mertz W, Roginski EE, Reba RC: Biological activity and fate of trace quantities of intravenous chromium (III) in the rat. Am J Physiol 1965; 209:489-494

86. Glinsmann WH, Feldman FJ, Mertz W: Plasma chromium after glucose administration. Science 1966; 152:1243-1245

87. Anderson RA, Bryden NA, Polansky MM: Serum chromium of human subjects: effects of chromium supplementation and glucose. Am J Clin Nutr 1985; 41:571-577

88. Freund H, Atamian S, Fischer JE: Chromium deficiency during total parenteral nutrition. JAMA 1979; 241:496-498

89. Chance B, Sies B, Boveris A: Hydroperoxide metabolism in mammalian organs. Physiol Rev 1979; 59:527-605

90. Hill HAO: The superoxide ion and the toxicity of molecular oxygen. In: New Trends in Bio-inorganic Chemistry. Williams RJP, Da Silva JRRF (eds). London, Academic Press, 1978; pp 173-208

91. Rotruck JT, Pope AL, Ganther HE, et al: Selenium: biochemical role as a component of glutathione peroxidase. Science 1973; 179:588-590

92. Diplock AT, Lucy JA: The biochemical modes of action of vitamin E and selenium: a hypothesis. FEBS Lett 1973; 29:205-210

93. Dickson RC, Tomlinson RN: Selenium in blood and human tissues. Clin Chim Acta 1967; 16:311-321

94. Underwood EJ: Selenium. Ch. 12 in Trace Elements in Human and Animal Nutrition. 4th ed. New York, Academic Press, 1977; pp 302-346

95. Thompson CD, Stewart RDH: The metabolism of ^{75}Se-Selenite in young women. Br J Nutr 1974; 32:47-57

96. Levander OA, Sutherland B, Morris VC, et al: Selenium balance in young men during selenium depletion and repletion. Am J Clin Nutr 1981; 34:2662-2669

97. Stewart RDH, Griffiths NM, Thomson CD, et al: Quantitative selenium metabolism in normal New Zealand women. Brit J Nutr 1978; 40:45-54

98. van Rij AM, Thomson CD, McKenzie JM, et al: Selenium deficiency in total parenteral nutrition. Am J Clin Nutr 1979; 32:2076-2085

99. van Rij AM, McKenzie JM, Thomson CD, et al: Selenium supplementation in total parenteral nutrition. JPEN 1981; 5:120-124

100. Robinson MF, McKenzie JM, Thomson CD, et al: Metabolic balance of zinc, copper, cadmium, iron, molybdenum and selenium in young New Zealand women. Brit J Nutr 1973; 30:195-205

101. Thompson JN, Erdody P, Smith DC: Selenium content of food consumed by Canadians. J Nutr 1975; 105:274-277

102. Chen X, Yang G, Chen J et al: Studies on the relations of selenium and Keshan disease. Biol Trace Elem Res 1980; 2:91-107

103. Johnson RA, Baker SS, Fallon JT, et al: An occidental case of cardiomyopathy and selenium deficiency. N Engl J Med 1981; 304:1210-1212

104. Fleming CR, Lie JT, McCall JT, et al: Selenium deficiency and fatal cardiomyopathy in a patient on home parenteral nutrition. Gastroenterology 1982; 83:689-693

105. Quercia RA, Korn S, O'Neill D, et al: Selenium deficiency and fatal cardiomyopathy in a patient receiving long-term home parenteral nutrition. Clin Pharm 1984; 3:531-535

106. Van Vleet JF: An evaluation of protection offered by various dietary supplements against experimentally induced selenium-vitamin E deficiency in ducklings. Am J Vet Res 1977; 38:1231-1236

107. Thurlow PM, Grant JP: Vitamin E, essential fatty acids and platelet function during total parenteral nutrition. (Abst.) JPEN 1981; 4:586

108. Losowsky MS, Leonard PF: Evidence of vitamin E deficiency in patients with malabsorption of alcoholism and the effects of therapy. Gut 1967; 8:539-543

109. Ali M, Gudbranson CG, McDonald JWD: Inhibition of human platelet cyclooxygenase of alpha-tocopherol. Prostaglandins Med 1980; 4:79-85

110. Hafeman DG, Hoekstra WG: Lipid peroxidation in vivo during vitamin E and selenium deficiency in the rat as monitored by ethane evolution. J Nutr 1977; 107:666-672

111. Lemoyne M, Van Gossum A, Kurian R, et al: Breath pentane analysis as an index of lipid peroxidation: a functional test of vitamin E status. AM J Clin Nutr 1987; 46:267-272

112. Cynamon HA, Isenberg JN, Nguyen CH: Erythrocyte malondialdehyde release in vitro: a functional measure of vitamin E status. Clin Chim Acta 1985; 151:169-176

113. Jeejeebhoy KN, Langer B, Tsallas G, et al: Total parenteral nutrition at home: studies in patients surviving 4 months to 5 years. Gastroenterology 1976; 71:943-953

114. Doisy EA Jr: Micronutrient controls on biosynthesis of clotting proteins and cholesterol. In: Trace Substances in Environmental Health, Hemphill, DD, ed. Vol. VI. Columbia, MO. University of Missouri Press, 1973; pp 193-199

115. Weisiger RA, Fridovich I: Superoxide dismutase (organelle specificity). J Biol Chem 1973; 248:3582-3592

116. Cotzias GC: Manganese in health and disease. Physiol Rev 1958; 38:503-532

117. Greenberg DW, Copp DH, Cuthbertson EM: Studies in mineral metabolism with the aid of artificial radioactive isotopes. J Biol Chem 1943; 147:749-757

118. Bertinchamps AJ, Miller ST, Cotzias GC: Interdependence of routes excreting manganese. Am J Physiol 1966; 211:217-224

119. Wenlock RW, Buss DH, Dixon EJ: Trace nutrients. 2. Manganese in British foods. Br J Nutr 1979; 41:253-261

120. McLeod BE, Robinson MF: Metabolic balance of manganese in young women. Br J Nutr 1972; 27:221-227

121. de Renzo EC, Kaleita E, Heytler P, et al: Identification of the xanthine oxidase factor as molybdenum. Arch Bioch Biophy 1953; 45:247-253

122. Cohen HJ, Fridovich I, Rajagopalan KV: Hepatic sulfate oxidase. A functional role for molybdenum. J Biol Chem 1971; 246:374-382

123. Mahler HR, Mackler B, Green DE, et al: Studies on metalloflavoproteins. III. Aldehyde oxidase: a molybdoflavoprotein. J Biol Chem 1954; 210:465-480

124. Cohen HJ, Drew RT, Johnson J, et al: Molecular basis of the biological function of molybdenum: the relationship between sulfite oxidase and the acute toxicity of bisulfite and SO_2. Proc Natl Acad Sci USA, 1973; 70:3655-3659

125. Underwood EJ: Molybdenum: Ch. 4 in Trace Elements in Human and Animal Nutrition, New York, Academic Press, 1977; pp 109-131

126. Shike M, Harrison JE, Sturtridge WC, et al: Metabolic bone disease in patients receiving long-term total parenteral nutrition. Ann Int Med 1980; 92:343-350

127. Shike M, Sturtridge WC, Tam CS, et al: A possible role of vitamin D in the genesis of parenteral-nutrition-induced metabolic bone disease. Ann Int Med 1981; 95:560-568

128. Hands LJ, Royle GT, Kettlewell MCW: Vitamin K requirements in patients receiving total parenteral nutrition. Br J Surg 1985; 72:665-667

129. Kishi H, Nishii S, Ono T: Thiamin and pyridoxine requirements during intravenous hyperalimentation. Am J Clin Nutr 1979; 32:332-338

130. Velez RJ, Myers B, Guber MS: Severe acute metabolic acidosis (acute beriberi): an avoidable complication of total parenteral nutrition. JPEN 1985; 9:216-219

131. Matsusue S, Kashihara S, Takeda H, et al: Biotin deficiency during total parenteral nutrition: its clinical manifestation and plasma nonesterified fatty acid level. JPEN 1985; 9:760-763

132. Guidelines for multivitamin preparations for parenteral use. American Medical Association, Department of Foods and Nutrition, Chicago, Ill, 1975.

Effect of Zinc Redistribution on Wound Healing during Total Parenteral Nutrition in Rats

Riichiro Nezu[1], Yoji Takagi[2], Akira Okada[2], and Yasunaru Kawashima[1]

Department of Surgery I[1], and Department of Pediatric Surgery[2], Osaka University Medical School, 1-1-50 Fukushima, Fukushima-ku, Osaka, 553 Japan

ABSTRACT

Wound healing in anastomosed intestine, sutured muscle layer and skin layer was examined as related to the redistribution of zinc during total parenteral nutrition(TPN). Gr.I(normally fed rats) and Gr.II(protein-malnourished rats) received TPN with zinc(Gr.Ia,IIa) or without zinc(Gr.Ib,IIb) for 6 days post-operatively. As a result, there was no significant difference between Gr.Ia and Ib in bursting pressure of intestinal anastomosis, tensile strength(TS) of muscle wound and skin wound, collagen-hydroxyproline (HYP) in each wound and zinc concentration in each tissue, whereas, the reduction of TS and HYP in skin wound was marked in Gr.IIb compared to Gr.IIa accompanied with the decrease of skin zinc concentration. In conclusion, redistribution of zinc was manifested in protein-malnourished rats during TPN and it might affect wound healing.

Key Words: wound healing, zinc redistribution, total parenteral nutrition

INTRODUCTION

Zinc deficiency during total parenteral nutrition (TPN) was first described in 1975 (Okada, Kay), and is, at present, recognized as an independent entity though much remains unclear in its pathogenesis. It was usually observed during repletion, i.e., anabolic state during TPN especially in severely depleted patients, supposed to be induced by an increased demand for zinc, since zinc is known to be intimately involved in the process of mitosis and protein synthesis (O'Dell 1974). Our previous experiment showed that redistribution of zinc in several tissues was induced in protein-malnourished rats receiving zinc-free TPN and a marked decrease in skin zinc concentration may contribute to the skin lesion observed as a common onset symptom in patients with zinc deficiency during TPN (Nezu 1988). In order to determine whether the redistribution of zinc in tissues during TPN may affect wound healing, present study was undertaken to evaluate wound healing in anastomosed intestine, sutured muscle layer and skin layer with zinc concentration in their respective tissues.

MATERIALS AND METHODS

Thirty male Sprague-Dawley rats weighing about 200g were used and divided into two groups. Group I (normally fed rats; N=15) was fed with regular diet containing 60.5ppm of zinc ad libitum, and Group II (protein-malnourished rats; N=15) was fed with protein-free diet containing 37ppm of zinc ad libitum, which was specially made with cornstarch to replace casein of the regular diet (Oriental Yeast Co.,Ltd., Tokyo, Japan) used in Group I. After 4 weeks of feeding, 5 rats served as preoperative control in each group, remaining rats underwent laparotomy and were further divided into two subgroups according to TPN formula as shown in Table 1: Group Ia and IIa received TPN with 2.5micromol/kg/day of zinc, whereas Group Ib and IIb received zinc-free TPN.

Table 1. Composition of TPN solution

Dextrose	208.3g	Vitamin B_1	8.3mg	Zn	10.0μmol
Amino acid[1]	40g	B_2	8.3mg	Mn	3.3μmol
Na^+	92mEq	B_6	5mg	Cu	0.83μmol
K^+	25mEq	B_{12}	50ug	I	0.17μmol
Mg^+	5mEq	Nicotinic acid	33.3mg	Fe	5.8μmol
Ca^+	13.3mEq	Folic acid	1.7mg		
Cl^-	62mEq	Biotin	3mg		
SO_4^{2-}	5mEq	Vitamin C	167mg		
Phosphate	13.3mmol	Pantothenic acid	20mg		
Acetate	42mEq	Vitamin A	4167IU		
Gluconate	13.3mEq	D	333IU		
		E	25mg		
		K	3.3mg		

Each content was provided per 1,000ml (993.3kcal).
1. Proteamin 12 (Tanabe Ph. Co., Tokyo, Japan): Non-protein Cal/N=130.2
2. Supplemented in Group Ia and IIa

Under ether anesthesia, a 6cm midline abdominal skin incision was made and then 3cm of lower half abdominal wall incision was made through muscle, fascia and peritoneum. Terminal ileum was divided 7-8cm from the ileocecal junction and then reanastomosed with interrupted 6-0 Nylon sutures. Then, abdominal wall and skin were closed in each layer with continuous 4-0 silk sutures. Each rat underwent catheterization of the superior vena cava with a silicone rubber catheter, tunneled subcutaneously to the dorsum and connected to a swivel apparatus through a catheter shield, and received a constant infusion of 250kcal/kg/day in a metabolic cage following TPN system of Steiger (1972). On the 6th postoperative day, each rat was sacrificed with exsanguination. The center of the abdominal wall wound and upper half of the skin wound were excised which measured 1.5cm wide by 4-5cm long. The ends of the strips of abdominal wall and skin were then grasped in needle forceps on either side and tensile strength was evaluated with the method of Crawford (1963). The segment of terminal ileum, about 6cm in length, containing the anastomosis was resected and bursting pressure was measured by air inflation leak test. After these measurements, the biochemical active zone (BAZ) of these wounds (Adamsons 1966), i.e., the strips of abdominal wall, skin and the segment of ileum, within 5mm from wound margin were analyzed for their zinc concentration and collagen-hydroxyproline (HYP) content. Plasma and red blood cells were directly diluted with redistilled water, and tissues were dissolved in nitric acid and diluted with redistilled water to desired concentrations for zinc analysis (Parker 1967). Zinc determination was made by flameless atomic absorption spectrophotometry using Zeeman Model Hitachi 170-70, Tokyo, Japan. Collagen-hydroxyproline content was determined by the method of Kivirikko (1967).
Data were expressed as means±SD. Multiple comparison were performed after analysis of variance with the Student-Newman-Keules test. Differences were considered to be significant when p<0.05.

RESULTS

Zinc concentrations in plasma, red blood cells, liver, pancreas, spleen, intestine, muscle, skin, and wounds (BAZ) of intestine, muscle and skin are shown in Table 2. In Group I, zinc concentration in pancreas was significantly increased in Group Ia and plasma zinc concentration was significantly decreased in Group Ib compared with preoperative control values, whereas no significant changes was observed in other tissues. In Group II, zinc concentration in plasma, pancreas, spleen, muscle and skin was significantly decreased in Group IIb, with the lowest in skin (-50%). Zinc concentration in red blood cells, liver and intestine showed no significant changes irrespective of zinc supplementation in both Group I and II.

Table 2. Zinc concentrations in tissues

	Preoperative control (N=5)	Group Ia TPN:Zn(+) (N=5)	Group Ib TPN:Zn(-) (N=5)	
Plasma	107.2±15.3*	94.0±16.0#	66.8±22.8*#	µg/dl
Blood Cells	9.1± 1.9	8.9± 1.1	8.0± 1.4	PPM
Liver	28.2± 4.2	26.0± 4.0	24.8± 2.5	µg/g
Pancreas	20.4± 2.8*	31.4± 6.5*#	16.9± 2.6#	µg/g
Spleen	19.4± 3.2	15.7± 1.5	17.0± 2.6	µg/g
Intestine	20.2±10.6	21.3± 7.0	15.2± 1.7	µg/g
-BAZ		47.9±25.0#	20.9± 6.6#	µg/g
Muscle	11.0± 1.3	11.5± 2.0	12.5± 2.2	µg/g
-BAZ		10.6± 2.7	13.0± 2.5	µg/g
Skin	12.7± 4.6	10.4± 2.0	12.1± 3.1	µg/g
-BAZ		10.2± 3.8	14.4± 3.2	µg/g
	Preoperative control (N=5)	Group IIa TPN:Zn(+) (N=5)	Group IIb TPN:Zn(-) (N=5)	
Plasma	97.4±19.7*	47.0±13.8*	29.1± 9.9*	µg/dl
Blood Cells	10.0± 0.7	8.8± 0.7	9.7± 1.9	PPM
Liver	26.4± 7.6	22.6± 5.0#	20.2± 4.5#	µg/g
Pancreas	19.1± 6.1*	19.4± 2.3#	13.1± 1.0*#	µg/g
Spleen	20.5± 4.0*	20.0± 1.9#	13.6± 1.1*#	µg/g
Intestine	15.0± 2.5	14.8± 2.8	12.5± 2.1	µg/g
-BAZ		13.9± 1.6	12.0± 1.1	µg/g
Muscle	14.3± 2.9*	14.1± 2.4	10.6± 1.6*	µg/g
-BAZ		15.5± 2.5#	9.6± 2.0#	µg/g
Skin	14.1± 2.0*	12.1± 3.0#	7.1± 1.9*#	µg/g
-BAZ		14.1± 5.4#	8.0± 1.7#	µg/g

*: $p < 0.05$ compared with preoperative control
#: $p < 0.05$ between Group Ia and Ib, or IIa and IIb

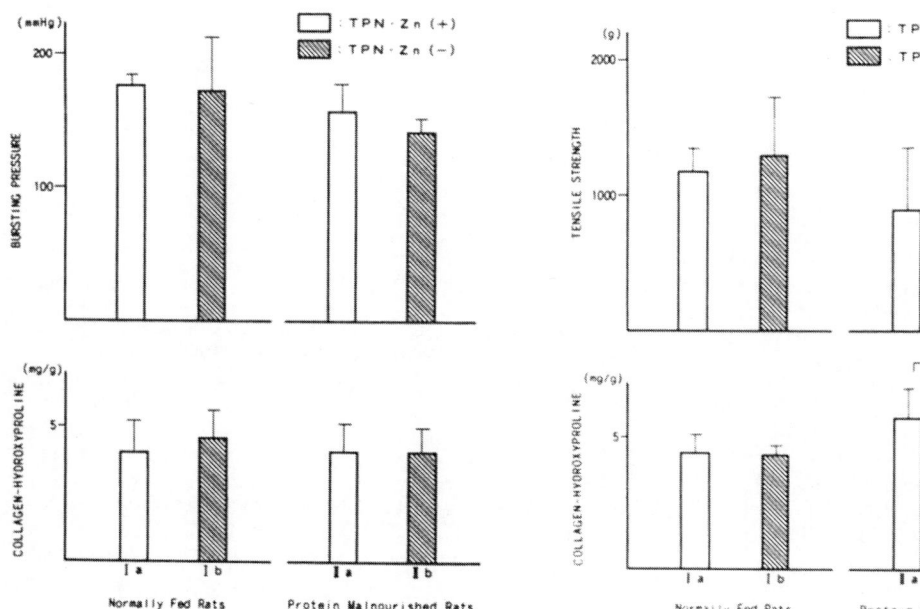

Fig.1 Bursting pressure and collagen-hydroxyproline content in intestinal wound

Fig.2 Tensile strength and collagen-hydroxyproline content in muscle wound

Bursting pressure and HYP in ileal anastomosis in each group are shown in Fig.1. There was no significant difference in the four groups. Tensile strength and HYP in abdominal muscle wound are shown in Fig.2. There was no significant difference in tensile strength in the four groups, whereas, HYP in Group IIa was significantly higher than the value in Group IIb. Tensile strength and HYP in skin wound are shown in Fig.3. In Group I, HYP in Group Ib was significantly lower than the value in Group Ia, whereas, there was no significant difference in tensile strength between these two groups. In Group II, both tensile strength and HYP in Group IIb were significantly lower than the value in Group IIa, and there was significant correlation between zinc concentration and tensile strength (r=0.728) or HYP (r=0.564) in skin wound.

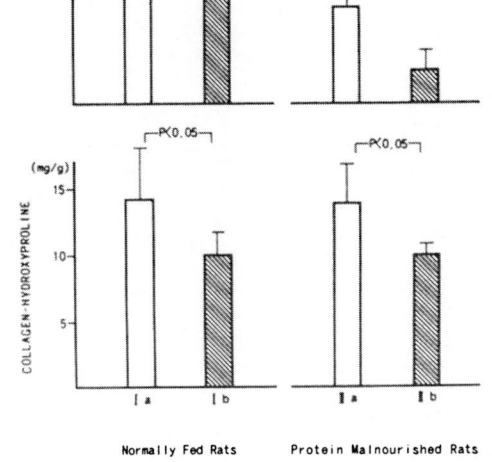

Fig.3 Tensile strength and collagen-hydroxyproline content in skin wound

SUMMARY

In this study, redistribution of zinc in tissues during TPN and its effect on wound healing was investigated, since zinc is required for protein synthesis and cell growth and therefore for wound healing.
In conclusion, redistribution of zinc in tissues was manifested in protein-malnourished rats during zinc-free TPN and the decreased zinc concentration may contribute to the impaired wound healing especially in skin.

REFERENCE

Adamsons RJ, Musco F, Enquist IF (1966) The chemical dimensions of a healing incision. Surg Gynecol Obstet 123:515-521
Crawford DT, Bains JW, Ketcham AS (1965) A standard model for tensiometric studies. J Surg Res 6:265-269
Kay RG, Tasman-Jones C (1975) Acute zinc deficiency in man during intravenous alimentation. Aus NZ J Surg 45:325-330
Kivirikko KI, Laitinen O, Prockop DJ (1967) Modification of a specific assay for hydroxyproline in urine. Anal Biochem 19:249-255
Nezu R, Takagi Y, Okada A, Inoue Y, Miyata M, Kawashima Y (1988) Effect of total parenteral nutrition on distribution of zinc in murine tissues. Trace Nutrients Research 4:115-120
O'Dell BL (1974) Role of zinc in protein synthesis. In: Pories WJ (ed) Clinical application of zinc metabolism. Charles C Thomas, Springfield, p5-8
Okada A, Takagi Y, Itakura T, Satani M, Manabe T, Iida Y, Nose O, Iwasaki M, Kasahara N (1975) Zinc deficiency during intravenous hyperalimentation. Igaku-no-Ayumi(Advan.Med.) 92:436-442
Parker MM, Humoller FL, Mahler DJ (1967) Determination of copper and zinc in biological materials. Clin Chem 13:40-48
Steiger E, Vars HM, Dudrick SJ (1972) A technique for long-term intravenous feeding in unrestrained rats. Arch Surg 104:330-332

Clinical Study on Zinc Deficiency Receiving Total Parenteral Nutrition

Masaharu Nishi, Hiroo Takehara, Yoshitaka Kita, and Nobuhiko Komi

First Department of Surgery, School of Medicine, The University of Tokushima,
3-18-15 Kuramoto-cho, Tokushima, 770 Japan

ABSTRACT

During the past decade, 5 subjects with inflammatory bowel disease (IBD) and 3 with recurrent gastrointestinal cancer (RGC) developed symptomatic zinc deficiency during long term total parenteral nutrition (TPN). Characteristic skin lesions appeared earlier in IBD than in RGC (27.3 ± 13.6 vs 57.3 ± 4.9 days ; $p < 0.05$). The plasma zinc levels at the onset were lower in RGC cases than those with IBD (12.0 ± 4.2 vs $23.8 \pm 6.8 \mu$ g/dl). Doses Above 40 μ mol/day of parenteral zinc were necessary to improve plasma zinc levels.

Key Words: zinc deficiency, recurrent gastrointestinal cancer, inflammatory bowel disease, TPN.

INTRODUCTION

Long term TPN is being used with increasing frequency for the nutritional management of patients with gastrointestinal diseases. Despite the obvious benefits of TPN, many problems associated with this therapy have been identified, including metabolic derangements and trace metal deficiencies (1,4). This may be in part due to the fact that, complete nutritional requirements for the long term TPN has not yet been well established (3). Trace metal deficiencies and particularly zinc deficiency in patients receiving TPN,has been reported by many authors (6,8). The causes of this deficiency are multiple and many factors may take part for its development.

PATIENTS AND METHODS

During a ten years period (1979-1988), among 546 patients who underwent TPN in our department,eight patients developed zinc deficiency. There were five cases of IBD, including one case of intestinal tuberculosis, three cases of Chron's disease and one case of ulcerative colitis. The other three cases were of RGC, one from the colon and two from the stomach. Although, some other patients suffering from other illnesses,were also treated with long term TPN,there were no more cases who developed symptomatic zinc deficiency. In these eight patients the clinical signs and symptoms, the day of the onset, plasma zinc levels and zinc supplementation dose were ascertained. Further studies were performed in some other patients with gastric cancer at different stages, whose plasma zinc level was measured,and in surgical patients whose plasma zinc and urinary zinc was examined.

RESULTS

Among those cases who had developed zinc deficiency,only one case had received zinc at a dose of 15μ mol/day, before the skin lesions had appeared, but all

232

other cases did not received any zinc supply before the onset. The patient who
received oral zinc supplementation, showed the longest recovery time, in spite
of receiving higher doses than cases with a parenteral zinc supply "Fig.1".

| Case | Zn Supplementation (μmol/day) | | Time of Recovery |
	Before Onset	After Onset	
1	—	60	
2	—	20 (110~170)	
3	—	20~40	
4	—	20~40	
5	—	20~40	
6	15	60	
7	—	20~100	
8	—	40	

(Oral Administration)

0   20   40   60 Days

"Fig. 1" Zinc supplementation doses and individual
patient response time to therapy

In cases of IBD, zinc deficiency symptoms became apparent at 27.3 ± 12.6 days
of TPN while, in RGC cases they were documented at 57.3 ± 4.9 days "Fig.2".
When the characteristic skin lesions had appeared,plasma zinc levels showed to
be between 6 to 30μ g/dl. Mean zinc levels were at 23.8 ± 6.8μ g/dl in IBD
cases and 12.0± 4.2μ g/dl in RGC cases "Fig.3".

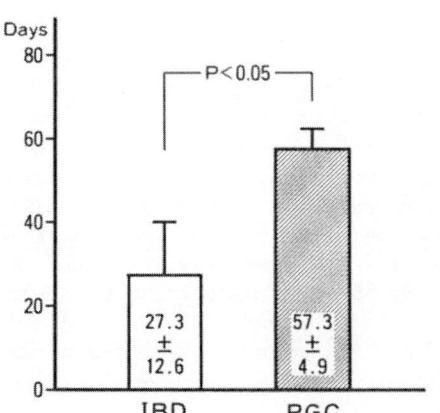

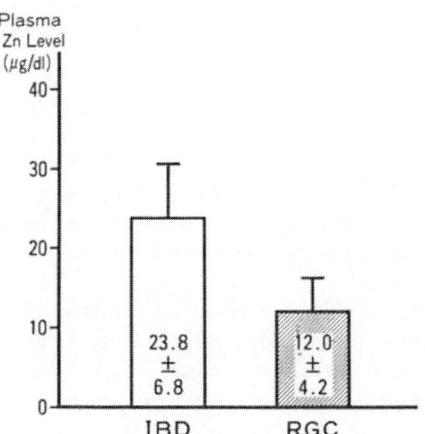

"Fig. 2" Appearance time of the
skin lesions during
TPN

"Fig. 3" Plasma zinc levels at
the time of skin
lesions appearance

Zinc supplementation at different doses were correlated with the changes of the plasma zinc levels . This correlation showed that doses of 20, 40, 60 and 100 μ mol/day produced values of -0.8, + 3.1, + 5.9 and + 10.5μ g/ dl/ day respectively "Fig.4".

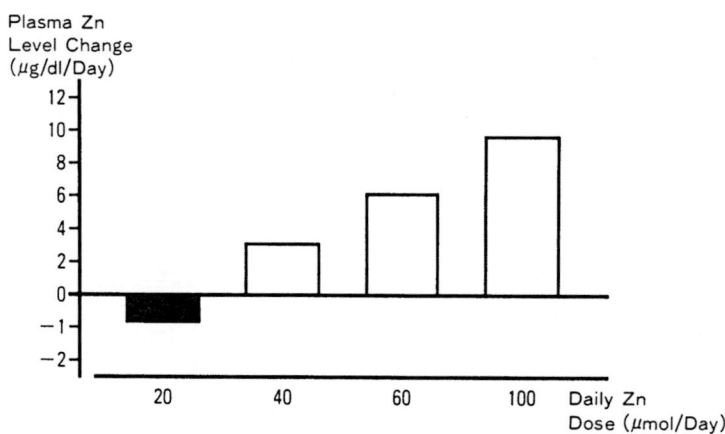

"Fig. 4" Relation between zinc dose and daily changes of plasma zinc level in zinc deficient patients

As for gastric cancer patients,plasma zinc levels at stage IV were significantly lower than at stage I and II, and those with recurrent cancer had significantly lower values than cases at any other stage "Fig.5".
The zinc urinary excretion in surgical patients increased about two to three fold the first postoperative day and returned to basal values after ten days. Parenteral zinc supply at doses of 20 or 30 μ mol/day in gastric cancer patients postoperatively, were enough to achieved positive zinc balance, but this was more positive with a dose of 30 μ mol/day.

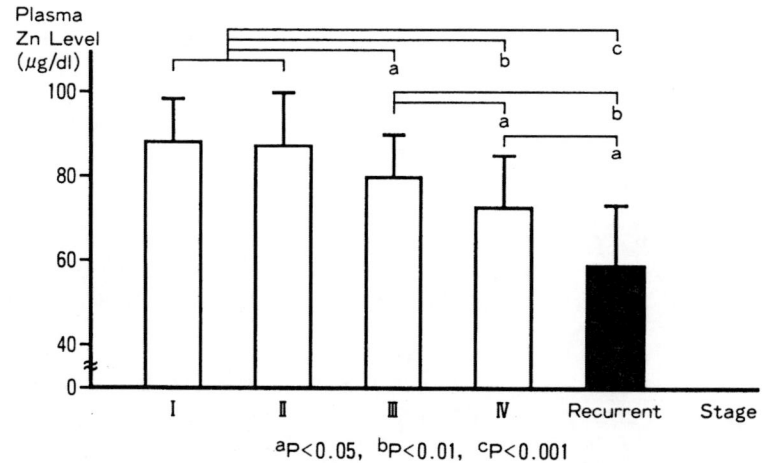

^{a}P<0.05, ^{b}P<0.01, ^{c}P<0.001

"Fig. 5" Plasma zinc levels in gastric cancer patients

DISCUSSION

Symptomatic zinc deficiency had appeared earlier during TPN in IBD cases than in RGC cases, suggesting that IBD, which are usually associated with diarrhea or intestinal fistulas, may produce large amounts of zinc losses in a brief period of time (2),which may be responsible for the earlier appearence of zinc deficiency in those cases. The plasma zinc levels, at the time the characteristics skin lesions had appeared, were lower in RGC than IBD cases. Furthermore, plasma zinc levels of recurrent gastric cancer patients were significantly lower than any other stage cases. It suggests that, RGC led to a chronic depletion of zinc (7) which allows some metabolic adjustments to decrease the requirements of zinc in those functions in which the presence of this element is necessary.
The oral zinc supplementation led to a longer period of time for recovery than parenteral zinc supplementation.It is clear that intestinal absorption of zinc was impaired (2,5), leading to a lack of response to oral zinc administration.

Our correlation of the changes in plasma zinc levels showed that, parenteral doses of 20 μ mol/day may produce values in the negative range. Therefore, we think that doses above 40 μ mol/day of a parenteral zinc supplement may be necessary to improve plasma zinc levels of patients with zinc deficiency. This data also suggest that patients with IBD and RGC are more prone to develop zinc deficiency during TPN because,they have already marginal zinc deficiency. We suggest that TPN regimes should contain adecuate amounts of trace elements, particularly for patients with IBD or RGC.

REFERENCES

1.Hankins DA, Riella MC, Scribner BH, Babb AL (1976) Whole blood trace element concentrations during total parenteral nutrition. Surgery 79 : 674-6 77
2.Hendricks KM, Walker WA (1988) Zinc deficiency in inflammatory disease. Nutr Rev 46 : 401-408
3.Hull RL (1974) Use of trace elements in intravenous hyperalimentation solutions. Am J Hosp Pharm 31 : 759-761
4.James BE, MacMahon RA (1970) Trace elements in intravenous fluids. Med J Aust 19 : 1161-1163
5.Jeejeeboy KN (1984) Micronutrients - state of the art. In : Ogoshi S,Okada A (ed) Parenteral and enteral hyperalimentation. Exerpta Medica, Asterdam New york Oxford, p 93
6.Kay RG, Tasman-Jones C(1975) Acute zinc deficiency in man during intravenous alimentation. Aus NZ J Surg 45 : 325-330
7.Nishi M, Hiramatsu Y, Hatano T, Hioki K, Yamamoto M (1988) Effect of nutritional support as an adjunct to the treatment of esophageal cancer.In : Siewert JR, Holscher AH (ed) Disease of the esophagus. Springer, Berlin Heidelberg New york, p 2877
8.Okada A, Takagi Y, Itakura T, Satani M, Manabe H, Iida Y, Tanigaki T,Iwasaki M, Kasahara N (1976) Skin lesions during intravenous hyperalimentation : zinc deficiency. Surgery 80 : 629-653

Molecular and Biochemical Aspects of Selenium Deficiency

HARVEY J. COHEN and NELLY AVISSAR

Department of Pediatrics, University of Rochester Medical Center, 601 Elmwood, Rochester, NY 14642, USA

ABSTRACT

Selenium deficiency results in a decrease in both cellular and plasma glutathione peroxidase activities. This decrease was found to occur in a number of individuals receiving intravenous hyperalimentation. With selenium replacement, there was a rapid increase in plasma glutathione peroxidase activity and a much slower increase in red blood cell glutathione peroxidase activity. Using polyclonal antibodies to cellular glutathione peroxidase, it was found that all of the glutathione peroxidase activity in red blood cells was due to the selenium-dependent glutathione peroxidase protein and that with selenium deficiency there was an absence of both activity and protein. With selenium replacement, there was a one to one correspondence in the appearance of activity and protein. Plasma glutathione peroxidase was not immunoprecipitated by antibodies against the cellular enzyme. Antibodies were prepared against purified plasma glutathione peroxidase and these did not cross-react with the cellular enzyme. In the absence of selenium, both the plasma glutathione peroxidase activity and protein were diminished. Using these mutually non-cross reactive antibodies, it was determined that the liver cell line, Hep G2, like other cells, synthesized the cellular enzyme. In addition, however, this cell synthesized and secreted the plasma enzyme. The activity of cellular glutathione peroxidase in this liver cell line could be enhanced three-fold by the addition of exogenous selenium.

KEY WORDS

Selenium, Glutathione Peroxidase, Synthesis Secretion, Antibodies

INTRODUCTION

Selenium in the form of selenocysteine has been found to be a component of [1] and responsible for the catalytic activity of [2] glutathione peroxidase. Selenium deficiency results in a decrease in glutathione peroxidase activity in both cells [3] and plasma [4]. We have studied the effects of selenium deficiency on the development of a deficiency in glutathione peroxidase in cells and in plasma, and the relationship between activity and protein content in deficient states. We have also examined the sites of synthesis and processing of these proteins utilizing non-cross reacting antibodies. The results are presented below.

RESULTS

We followed the levels of plasma and red blood cell glutathione peroxidase activities in approximately 200 patients receiving parenteral nutrition. We have documented the development of a deficient state in approximately 10 of these patients. Although there are many factors that might account for the development of glutathione peroxidase deficiency in patients receiving defined formula feeding that is deficient in selenium, those individuals that were found to be deficient in this enzyme activity, were more likely to have increased gastrointestinal loss in addition to poor or no oral intake of food. The deficiency in glutathione peroxidase activity was found to develop between 6 months and 2 years on parenteral nutrition [5]. The time course of the development of the deficient state was similar for both plasma and glutathione peroxidase activities.

Using polyclonal antibodies directed against the cellular glutathione peroxidase, it was found that there was direct relationship between cellular glutathione peroxidase activity and protein content during the course of the development of the deficient state. Antibodies against cellular glutathione peroxidase do not cross react with plasma glutathione peroxidase [6]. Utilizing antibodies made specifically against plasma

glutathione peroxidase,[7] it was found, that there was a decrease for plasma glutathione peroxidase protein content in the selenium deficient state.

With repletion of selenium to the intravenous fluids (400 ug selenium/day) there was a rapid increase in plasma glutathione peroxidase activity with a zero order rate constant of approximately 10% per day. For red cells there was a much slower rate of increase in enzyme activity with a 2-week lag occuring prior to any increase in red cell enzyme activity. A zero order rate constant for the repletion of red cell glutathione peroxidase activity was found to be approximately .75% per day, similar, to the normal red blood cell production rate of 0.8% per day.[8]

In order to determine the sites of synthesis and secretion of various glutathione peroxidase activities, a primary culture of endothelial cells, a human myeloid leukemic cell line and a human hepatic tumor cell line were incubated for 3 days with [75] Se selenious acid. The cells and the media were then separated, and after lysis of the cells, the gluathione peroxidase from both the cells and the media were immunopurified with antibodies against cellular and plasma glutathione peroxidase proteins. It was present that only the cellular form of glutathione peroxidase protein was found in the cells of the endothelial cells and the human leukemic cell line. There was no cellular form of glutathione peroxidase found in the medium and there was no evidence for the presence of the plasma form of the enzyme either in the cells or in the medium. Utilizing the Hep G2 cell line, it was found that the cellular form of the enzyme was present in the cells but not in the medium. The plasma form of glutathione peroxidase was initially only found in the medium. Using longer exposure times for autoradiography of gels, a small amount of plasma glutathione peroxidase could be found in cells. In the presence of monensin less plasma glutathione peroxidase protein was found in the medium. At the same time, a greater amount of plasma protein was found in the cells. The plasma form of glutathione peroxidase appears to have intracellular disulfide bonds and migrates more rapidly in nonreducing SDS gels than in reducing SGS gels.[9] Since the reduced proteins migrate similarly, it was important to determine that the protein from the cells, immunopurified by the antiplasma antibody, was indeed the plasma protein. Therefore, nonreducing gels were run. In the absence of dithiothreitol, the protein present in the cells, and immunopurified by anti-plasma glutathione peroxidase, migrated more rapidly. This confirms the conclusion that the protein that was precipitated from the cells by antibodies against the plasma enzyme was indeed plasma glutathione peroxidase. SDS gel electrophoresis of all the selenium containing proteins present in both cells and medium a variety of cells incubated with [75] Se selenious acid demonstrated, was that approximately 10-15% of the radioactivity in the protein are due to glutathione peroxidase.

DISCUSSION

Selenium has been recognized to be an essential element for approximately 30 years.[10] It is only recently that the molecular and cellular basis for this deficiency has been recognized. These studies indicate that in the absence of selenium, there is a concomitant decrease in both glutathione peroxidase activity and protein.[11] The gene for mammialian glutathione peroxidase has recently been cloned and the codon for selenocysteine has been found to be UGA.[12] It has previously been shown that the carbon source for selenocysteine, in glutathione peroxidase, was serine.[13] These data, together with the molecular data, indicate that selenium is important for the translation of the glutathione peroxidase m-RNA. There are conflicting data in the literature concerning the effect of selenium on glutathione peroxidase, m-RNA. In rat liver there have been reports that there was a direct relationship between glutathione peroxidase m-RNA and activity, [14] and that there was no change in glutathione peroxidase m-RNA in the selenium deficient and glutathione peroxidase deficient state.[15] In the human leukemic cell line, HL-60, it has been found that there is a 25-fold change in the amount of protein in cells grown in the absence of selenium, with an approximately 3-fold difference in the m-RNA glutathione peroxidase.[16] Nuclear run-on studies in this system showed no change in the rate of synthesis in the glutathione peroxidase m-RNA. Thus, at this time, it is not clear whether the control of glutathione peroxidase protein synthesis by selenium is at the level of protein synthesis alone, or also at the level of m-RNA, either synthesis or stability. Any of these mechanisms are consistent with the results described. It was somewhat surprising to find that there were two immunologically distinct forms of glutathione peroxidase present in cells and serum respectively. We have previously shown that the plasma form of glutathione peroxidase is a glycoprotein, whereas the cellular form is not.[17] This led us to examine various cells for both the synthesis and secretion of cellular and plasma forms of glutathione peroxidase. Since many serum proteins are made by the liver, it was not surprising to find that a liver cell line, both made and secreted the plasma form of glutathione

peroxidase. Monensin, an inhibitor of processing of proteins made for secretion, was able to increase the intracellular concentration of plasma glutathione peroxidase and decrease its secretion. We are currently attempting to determine the amino acid sequence for plasma glutathione peroxidase to detect structural differences between the cellular and plasma forms of the enzyme. We are also attempting to obtain better information on the processing of this selenoprotein.

REFERENCES

1) Forstrom, J.W., Zakowski, J.J., and Tappel, A.L. Biochemistry 17, 2639-2644 (1978).

2) Zakowski, J.J., Forstrom, J.W., Condell, R.A., and Tappel, A.L. Biocehmi Biophys Res Commun 84, 248-255 (1979).

3) Lane, H.W., Dudrick, S. and Warren, D.C. Proc Soc Exp Biol Med 167, 383-390 (1981).

4) van Rij, A.M., Thompson, C.D., Mackenzie, J.M., et. al. Am J Clin Nutr 32, 2076-2085 (1979).

5) Cohen, H.J., Brown, M.R., Hamilton, E., et. al. Am J Clin Nutr 49, 132-139 (1989).

6) Takahashi, K. and Cohen, H.J. Blood 68, 640-645 (1986).

7) Avissar, N., Whitin, J.C., Allen, P.Z., Palmer, I.S. and Cohen, H.J. Blood 73, 318-323 (1989).

8) Ashby, W. Blood 3, 486-500 (1948).

9) Takahashi, K., Avissar, N., Whitin, J. and Cohen, H.J. Arch Biochem Biophys 256, 677-686 (1987).

10) Schwarz, K. and Foliz, C.M. J Am Chem Soc 79, 3292-3293 (1957).

11) Takahashi, K., Newburger, P.E. and Cohen H.J. J Clin Invest 77, 1402-1404 (1986).

12) Mullenbach, G.T., Tabrizi, A., Irvine, B.D., Bell, G.I., and Hallewell, R.A. Nucleic Acid Res 15 5484 (1987).

13) Sunde, R.A. and Evenson, J.K. J Biol Chem 262, 933-937 (1987).

14) Saedi, M.S., Smith, C.G., Frampton, J., Chambers, I., Harrison, P.R. and Sunde, R.A. Biochem Biophys Res Commun 153, 855-861 (1988).

15) Reddy, A.P., Hsu, B.L., Reddy, P.S., et. al. Nucleic Acid Res 16, 5557-5568 (1988).

16. Newburger, P.E., Chada, S., Casey, L. and Le Beau, M.M. Ped Res 25, 143a (1989).

Platelet Glutathione Peroxidase Activity in Long-term Total Parenteral Nutrition with/without Selenium Supplementation

Kinya Sando[1], Riichiro Nezu[2], Yoji Takagi[2], and Akira Okada[2]

[1] Department of Pediatric Surgery, Kure National Hospital, 3-1 Aoyama-cho, Kure, 737 Japan
[2] Department of Pediatric Surgery, Osaka University Medical School, 1-1-50 Fukushima, Fukushima-ku, Osaka, 553 Japan

ABSTRACT

Platelet glutathione peroxidase (GSH-Px) activity in six patients receiving prolonged total parenteral nutrition (TPN) with chronic gastrointestinal diseases was studied to investigate its usefullness for an indicator of Se status. During Se depletion and repletion, blood Se indices were meaured. Platelet GSH-Px activity significantly decreased and increased in one week. Plasma Se indices significantly changed in three weeks. Erythrocyte Se indices showed no remarkable changes during study period. Platelet GSH-Px activity may be served as one of the most sensitive Se indices and useful for assessing the short-term Se status in TPN patients.

KEY WORDS

Total parenteral nutrition, Selenium, Selenium status, Glutathione peroxidase, Platelet.

INTRODUCTION

Although the importance of Se in human nutrition has been recognized (Rotruck 1973), little is known about its metabolism during TPN and there is no consensus for the use of Se (Levander 1986). It is now required to investigate the form, amount and route of Se supplementation and the means to monitor the responce. The purpose of this study is to determine the effect of Se supplementation on the Se status, especially platelet GSH-Px acitivity in long-term home TPN patients.

MATERIALS AND METHODS

Patients
Six long-term home TPN patients with chronic gastrointestinal diseases were studied. They took no oral intake or little. The profiles of these patients are shown in TABLE 1.

TABLE 1.   Profiles of patients

CASE	(AGE	SEX)	DIAGNOSIS	DURATION OF TPN
1.	38y	F	Short bowel syndrome	156 months
2.	27y	F	Short bowel syndrome	36
3.	7y	F	Intractable diarrhea in infant	87
4.	4y	M	Intractable diarrhea in infant	37
5.	43y	F	Nonspecific multiple ulcer of small intestine	81
6.	36y	F	Chronic idiopathic intestinal pseudo-obstruction syndrome	51

Experimental Design
Before discontinuation of Se supplementation (adult : 200, children : 100μg/day
as selenite) , Se level and GSH-Px activity in plasma and erythrocyte had been
maintained within normal range during 127 weeks. Se was discontinued for 12
weeks, and then Se was resupplementated for the following 12 weeks. TPN
solution for this study contained 3.3% amino acid solutiuon, 21% glucose,
macro elements, trace element mixture (adult: Zn 60μmol, Mn 20μmol, Cu 5μmol,
I 1μmol, Fe 2mg/day, child : half volume) and multiple vitamin infusion. Se
concentration is lower than 10ppb in Se-free TPN solution. (Fig.1)

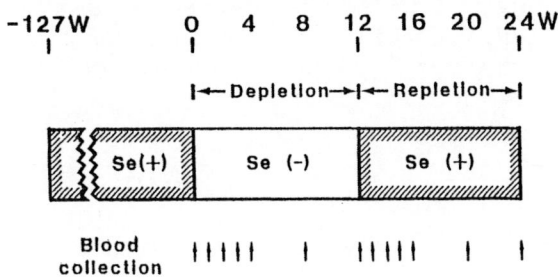

Fig.1 Experimental Design

Laboratory Analysis
GSH-Px activity in platelet, Se concentration and GSH-Px activity of plasma
and erythrocyte were measured for indices of Se status. The preparation of the
specimen for analysis of platelet GSH-Px was modified from that of Levander
(1983). Se concentration and GSH-Px activity in plasma and erythrocyte were
determined by fluorometric (Watkinson 1979) method and the method of Paglia
and Valentein (1967) respectively. The protein concentration of the platelet
sample was determined by the method of Lowry (1951).

TABLE 2. Laboratory Analysis

BLOOD SELENIUM INDICES

1) Plasma selenium concentration
2) Plasma GSH-Px activity
3) Erythrocyte selenium concentration
4) Erythrocyte GSH-Px activity
5) Platelet GSH-Px activity

METHODS OF ANALYSIS

Platelet preparation	: The modified method of Levander
GSH-Px activity	: The method of Paglia & Valenteine
Selenium level	: Fluorometric method
Protein concentration	: The method of Lowry

RESULTS

Plasma Se concentration was gradually decreased after discontinuation of Se
supplementation, significantly changed from 136± 28 (mean± SD) μg/L to 75± 14
in tree weeks, and then appeared to plateau. Three weeks of resupplemantaion
with Se resulted in significant increase from 61± 22 to 125± 33. (Fig.2)
Plasma GSH-Px was also decreased from 236± 50 U/L to 140± 36, and increased
128± 32 to 220± 64 in 3 weeks. (Fig.3)

In constrast to the response of plasma, erythrocyte Se indices showed no
remarkable changes during study period. (Fig.4,5)

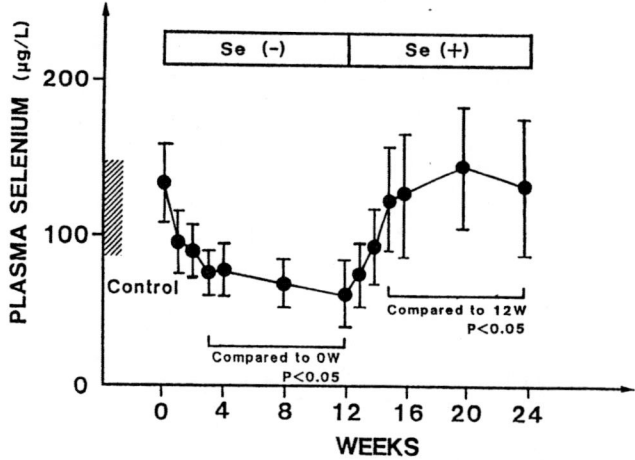

Fig.2 The change in plasma Se concentration
during Se depletion and repletion.

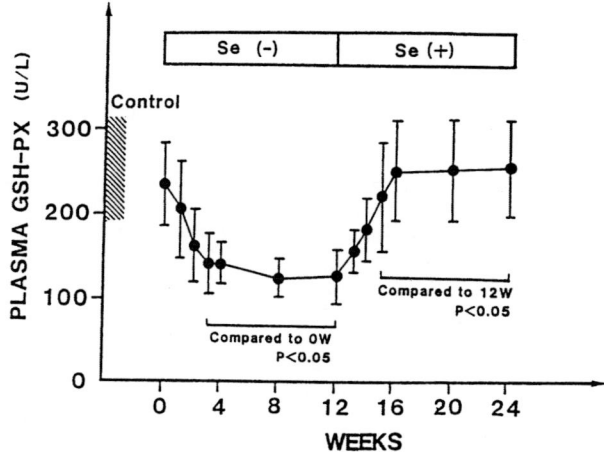

Fig.3 The change in plasma GSH-Px activity
during Se depletion and repletion.

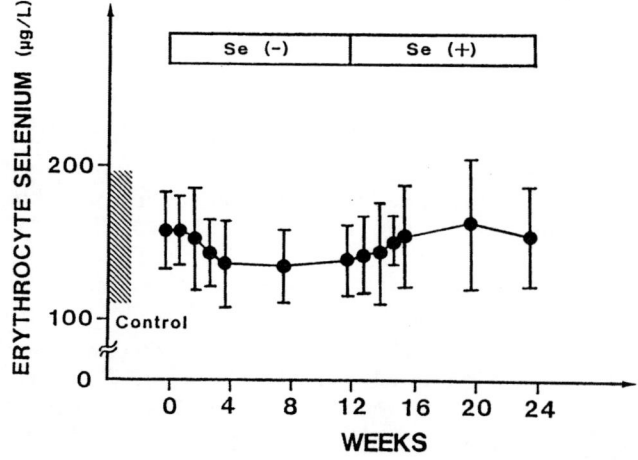

Fig.4 The change in erythrocyte Se concentration
during Se depletion and repletion.

A significant decrease of platelet GSH-Px activity from 64 ± 7 U/gProtein to
39 ± 5 was rapidly observed in only 1 week. Once Se resupplementation began,
platelet GSH-Px activity also significantly increased from 44 ± 9 to
65 ± 10 in 1 week. (Fig.6)

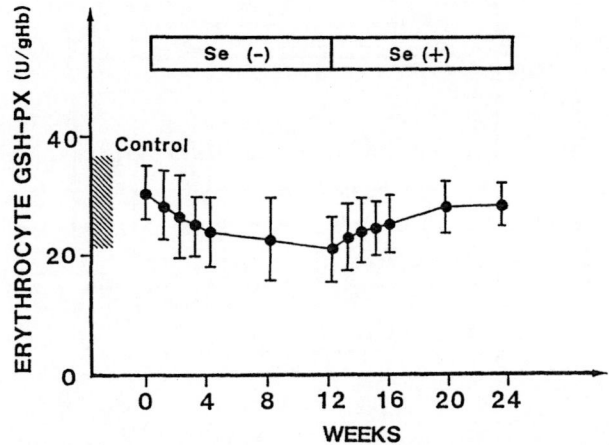

Fig.5 The change in erythrocyte GSH-Px activity
during Se depletion and repletion.

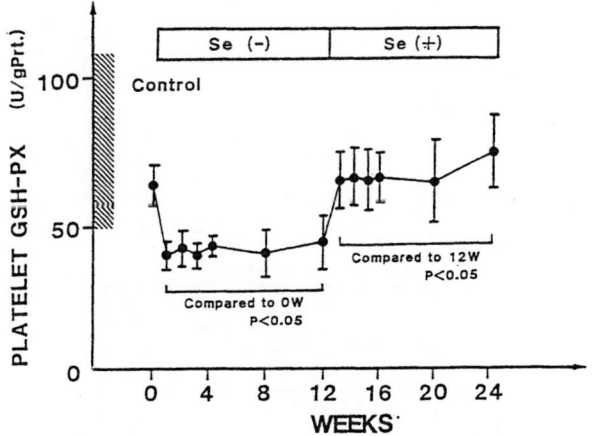

Fig.6 The change in platelet GSH-Px activity
during Se depletion and repletion.

There were no clinical signs of Se deficiency and toxicity in these TPN
patients.

SUMMARY

During Se depletion and repletion, platelet GSH-Px activity decreased and
increase rapidly in one week. Plasma Se indices significantly changed in three
weeks. Erythrocyte Se indices showed no remarkable changes during study
period. Platelet GSH-Px activity may be served as one of the most sensitive
indices of Se status and useful for assessing the short-term Se status in TPN
patients.

REFERENCES

Levander OA,Burk RF (1986) Report on the 1986 A.S.P.E.N. research workshop on
    selenium in clinical nutrition. J Parenter Enteral Nutr 10:545-549
Levander OA,DeLoach DP,Morris VC et al(1983) Platelet glutathione peroxidase
    activity as an index of selenium in rats. J Nutr 113:55-63
Lowry OH,Rosebrough NJ,Farr AL et al (1951) Protein measurement with the Folin
    phenol reagent. J Biol Chem 193:265-275
Paglia DE and Valentine WN (1967) Studies on the quantitative characterization
    of erythrocyte glutathione peroxidase. J Lab ClinMed 70:158-69
Rotruck JT,Pope AL,Ganther HE et al (1973) Selenium:Biochemical role as a
    component of glutathione peroxidase. Science 179:588-590
Watkinson JH (1979) Semi-automated fluorimetric determination of nanogram
    quantities of selenium in biological material. Anal Chim Acta 105:319-325

Cardiomyopathy and Selenium Deficiency in a Caucasian Adolescent

GILLIAN LOCKITCH, PENELOPE DISON, GLENN TAYLOR, LAURENCE WONG, GEORGE DAVIDSON, and DAVID RIDDELL

Departments of Pathology and Pediatrics, University of British Columbia and B.C. Children's Hospital, Vancouver, British Columbia, Canada

ABSTRACT

A 17 year old girl with chronic idiopathic pseudo-obstruction and short gut, had been on home parenteral nutrition for 17 months. Following diagnosis of severe selenium deficiency, she received intravenous supplemental selenium for the last seven months. She developed septic shock followed by cardiac arrest. At autopsy, the heart was enlarged, soft and flabby, and the histological findings were suggestive of cardiomyopathy due to prolonged selenium deficiency. Sepsis may precipitate cardiomyopathy in marginal selenium deficiency in a susceptible subject.

KEY WORDS: Selenium, parenteral nutrition, cardiomyopathy

INTRODUCTION

The importance of severe selenium deficiency as a cause of dilated cardiomyopathy and sudden death in humans, was first recognised in China. Keshan Disease, an endemic cardiomyopathy mainly seen in young children, occurred in an area characterized by extreme selenium deficiency. The introduction of sodium selenite supplementation has largely eradicated this disease (Chen 1980).

Symptomatic selenium deficiency has been associated with unsupplemented parenteral nutrition. Van Rij (1979) showed that parenteral nutrition unsupplemented with selenium could result in muscle pain associated with severe selenium deficiency. Johnson (1981) described a case of non-endemic cardiomyopathy in an adult male patient who developed profound selenium deficiency after a prolonged period of parenteral nutrition, unsupplemented with selenium. The cardiac histology was reported to be similar to that described for Keshan Disease.

Following that report, there have since been 6 further case reports of non-endemic cardiomyopathy associated with selenium deficiency (Collip 1981; Fleming 1982; Stanley 1982; Quercia 1984; Sriram 1986; Volk 1986). Six of the 7 patients were on parenteral nutrition. Three cases were fatal, 4 responded to selenium supplementation. We report a case of a 17 year old girl who died of cardiac arrest following septic shock, after 17 months on parenteral nutrition, the latter 7 months with added selenium.

CASE REPORT

The girl was a healthy teenager, when at age 14 she developed a febrile illness with lymphadenopathy and lethargy diagnosed as mononucleosis. Six months later she was referred for investigation of persistent fatigue, lymphadenopathy and rash. She had high IgG and a raised antibody titre to Epstein-Barr virus. Lymphocyte studies were abnormal indicating persistent EB viral infection.

Lethargy and lymphadenopathy persisted and she developed abdominal pain, vomiting and poor appetite. Extensive investigation revealed no abnormalities other than persistently raised titres to EBV, with abnormal lymphocyte subsets. She continued to lose weight and a barium swallow revealed duodenal obstruction. A laparotomy was performed to relieve congenital bands causing malrotation, and she required short term parenteral nutrition.

Over the next 6 months severe abdominal pain, vomiting and weight loss were major problems. She had multiple episodes of parenteral nutrition and nasogastric elemental diet but was unable to maintain her weight on enteral diet. Twenty months after initial presentation, an ileostomy with resection of caecum and terminal ileum was performed for a diagnosis of chronic idiopathic pseudo-obstruction. Home parenteral nutrition was commenced at this stage.

Problems with severe abdominal pain were not alleviated by treatment with various analgesics including subcutaneous Morphine, the most effective. Coeliac nerve block was performed. A further bowel resection was done to alleviate symptoms, with a gastrostomy to facilitate gastric drainage and relieve vomiting. Weakness and lethargy remained a major problem but no neuromuscular abnormality could be identified despite electromyography and CAT scan.

After 10 months of continual TPN, during which time she had no oral intake, she was found to have severe selenium deficiency. The plasma selenium level, determined by recovery studies, was below the detection limit of the assay ie. < 0.07 µmol/L. Analysis of frozen samples previously collected for zinc analysis revealed that her plasma selenium had been at this level for at least two months. Plasma glutathione peroxidase activity was very low. Vitamin E at 30 µmol/L was above the reference range of 13 - 24 µmol/L. Selenium supplementation was commenced and plasma selenium gradually increased on an intake of up to 4 ug/kg/day (200 µg/day). A parallel increase in glutathione peroxidase was seen. She remained fairly stable over the next 6 months.

During an admission for severe nausea and vomiting, she developed acute septic shock secondary to Enterobacter cloacae and had a cardiac arrest. Following the arrest she developed respiratory distress, acute renal failure and required dopamine support for a poorly functioning myocardium. At this stage her selenium level fell from 1.13 to 0.47 µmol/L and did not increase despite increasing her supplement from 4 ug/kg/day to 9 ug/kg/day. She died 14 days after the initial arrest.

At post mortem examination the heart was enlarged, soft and flabby. Microscopically it showed, in addition to acute myocardial infarction of a papillary muscle, myocardial oedema, myocardiocyte loss and replacement fibrosis, all involving the interventricular septum. Hypertrophic change was noted in both ventricles. The coronary arteries were patent and normal.

The cardiac pathology in this patient was similar to that previously ascribed to iatrogenic selenium deficiency. Typically the heart is enlarged, "flabby" and shows subepicardial and midmural serpiginous foci of myocytolysis or replacement fibrosis, predominantly affecting the left ventricle and interventricular septum (Johnson 1981; Fleming 1982; Quercia 1984). Although this girl suffered the sequelae of an hypoxic/ischemic episode, including acute myocardial infarction of a papillary muscle, the distribution of the myocardial scars differed from that of "watershed" lesions, which are predominantly subendocardial and occur in the papillary muscles, free wall of the left ventricle and interventricular septum (McGovern 1980).

DISCUSSION

This patient was severely selenium deficient for several months before diagnosis as determined by retrospective analysis of stored serum samples. Despite intravenous supplementation with 200 µg/day selenium, she never attained serum levels within the reference range for her age. There are no generally accepted recommendations for selenium supplementation in parenteral nutrition patients. Levander (1984) has suggested that adults on parenteral nutrition in stable state would probably require 50 to 60 µg/day selenium while children would require 1.4 ug/kg/day and infants up to 3 µg/kg/day. We provided 4 µg/kg/day as selenious acid. Although a steady increase in plasma selenium was seen in this patient, even 200 µg/day was not sufficient to bring selenium levels into the normal range for age, probably due to loss of selenium in gastrostomy fluids.

Cardiomyopathy associated with selenium deficiency in parenteral nutrition has been described in only 6 patients excluding our present case despite numerous reports of selenium deficiency. It is puzzling why there may be no cardiac or skeletal muscle symptoms in some patients despite prolonged periods of selenium deficiency. Clinical signs of selenium deficiency probably are precipitated by additional factors such as concurrent vitamin E deficiency or sepsis. In two cases, multiple episodes of sepsis, with candida or bacteroides species preceded the symptomatic cardiac failure (Johnson 1981; Stanley 1982). In our case, multiple episodes of sepsis occurred and the final cardiac arrest appeared to be precipitated by septic shock.

Our patient, despite supplementation with selenium, remained in a state of marginal selenium deficiency. We speculate that the susceptibility to infection resulting from her "short gut" syndrome and indwelling central venous line, led to multiple episodes of sepsis which provided ongoing stress to the cardiac muscle and the final impetus for cardiac arrest.

REFERENCES

Chen X,Yang G,Chen J et al (1980) Studies on the relations of selenium and Keshan Disease. Biol Tr Elem Res 2:91-107

Collipp PJ,Chen SY (1981) Cardiomyopathy and Selenium Deficiency in a Two-Year-Old Girl. N Eng J Med 304:1304-1305

Fleming CR,Lie JT, McCall JT et al (1982) Selenium Deficiency and Fatal Cardiomyopathy in a Patient on Home Parenteral Nutrition. Gastroenterology 83:689-693

Johnson RA,Baker SS,Fallon JT et al (1981) An Occidental Case of Cardiomyopathy and Selenium Deficiency. New Eng J Med 304 :1210-1212

Levander OA,Morris VC (1984) Dietary selenium levels needed to maintain balance in North American adults consuming self-selected diets. Am J Clin Nutr 39:809-815

McGovern VJ (1980) Hypovolemic shock with particular reference to the myocardial and pulmonary lesions. Pathology 12:63-72

Quercia RA,Korn S, O'Neill D et al (1984) Selenium Deficiency and Fatal Cardiomyopathy in a Patient Receiving Long-term Home Parenteral Nutrition. Clin Pharmacol 3:531-535

Sriram K,Peterson JK, O'Gara J, Hammond JM (1986) Clinical improvement of congestive heart failure after selenium supplementation in total parenteral nutrition. Acta Pharmacol Toxicol 59 (suppl VII):361-364

Stanley JC,Alexander JF,Nesbitt GA (1982) Selenium deficiency during total parenteral nutrition: a case report. Ulster Med J 51:130-2

Van Rij AM,Thomson CD, McKenzie JM,Robinson MF (1979) Selenium deficiency in total parenteral nutrition. Am J Clin Nutr 32:2076-85

Volk DM,Cutliff SA. (1986) Selenium Deficiency and Cardiomyopathy In a Patient With Cystic Fibrosis. J Kentucky Med Assoc 84:222-224

The full case report has been submitted for publication in American Journal of Clinical Nutrition.

Drugs and Trace Elements

Nephrotoxicities of Gold and Cisplatinum

Kidney Center University Hospital of OCCUP and Enviro Health, Yahatanishi-ku,
Kitakyushu, 807 Japan

ABSTRACT

(1) Nephrotoxicity of gold: in this case of rheumatoid arthritis, gold deposition was determined as the electron dense filamentous structures in lysosomes of mainly proximal tubules using x-ray energy dispersive analysis. This filamentous structures was a characteristic feature of gold deposition.
25 mg of gold sodium thiomalate was injected into rats in order to clarify the nephrotoxicity of gold. Severe necrosis of pars recta of proximal tubules was induced. Electron dense particles were also observed in the proximal tubules.

(2) Nephrotoxicity of cisplatinum: cisplatinum has well known to be a principal chemotherapeutic agent in the treatment of solid cancer. Chief limit to its greater efficacy, however, is its nephrotoxicity. In this study, the comparison of nephrotoxicity between cisplatinum and transplatinum was investigated.
Cisplatinum induced massive necrosis of pars recta of proximal tubules, whereas transplatinum did not show any morphological change. Urinary enzyme activities (ALP, LAP, γ-GTP, NAG, AAP) markedly increased after injection of cisplatinum. Therefore, the daily measurement of urinary enzyme could suggest potential values which indicate not only the degree but also the site of the damaged tubules.

KEY WORDS: gold, nephrotoxicity, cisplatinum, transplatinum, urinary enzymes

1. NEPHROTOXICITY OF GOLD

Gold therapy for rheumatoid arthritis is well known to result in nephropathy. Nephrotic syndrome and/or proteinuria are well recognized complications of gold therapy for rheumatoid arthritis. About 3% of patients who were treated with this therapy become nephrotic syndrome (POLLAK 1962). The typical symptoms are edema, heavy proteinuria and hypoalbuminemia which develops during the therapy and will regress over months after the drug is discontinued. The renal pathological findings are usually similar to membranous nephropathy (SILVERBERG 1970, KATZ 1973, TORNROTH 1974).
It should be noted that membranous nephropathy does also occur in rheumatoid arthritis unrelated to gold therapy. Pathologically, it has been suggested that the renal lesion seen with gold therapy is due to immune complex deposition (SAMUELS 1977, IESATO 1982). But, gold has not been demonstrated to exist in subepithelial deposits or glomerular basement membrane.

(A) A Clinical Case Report

The patient is a 62 year old female suffering from rheumatoid arthritis since 1974. Intramuscular injection of gold sodium thiomalate was started in 1977, and was discontinued in 1980, since proteinuria had appeared. In June, 1980, this patient was admitted to the hospital and the next month, open renal biopsy was performed. The amount of urinary protein was 0.8 g/day and the

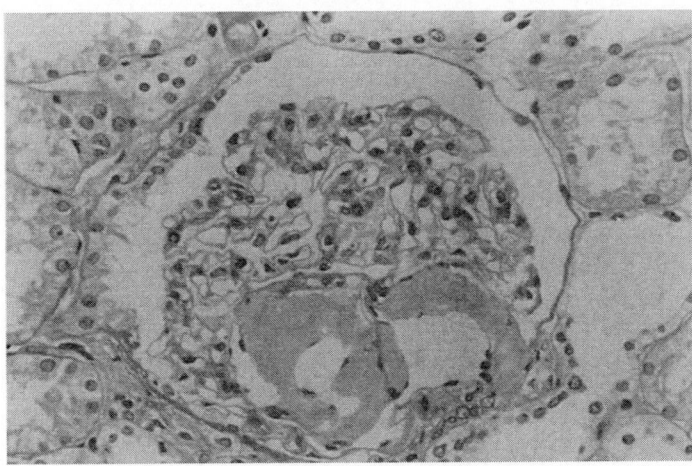

Fig.1: Light microscopic finding of kidney
tissue obtained byrenal biopsy

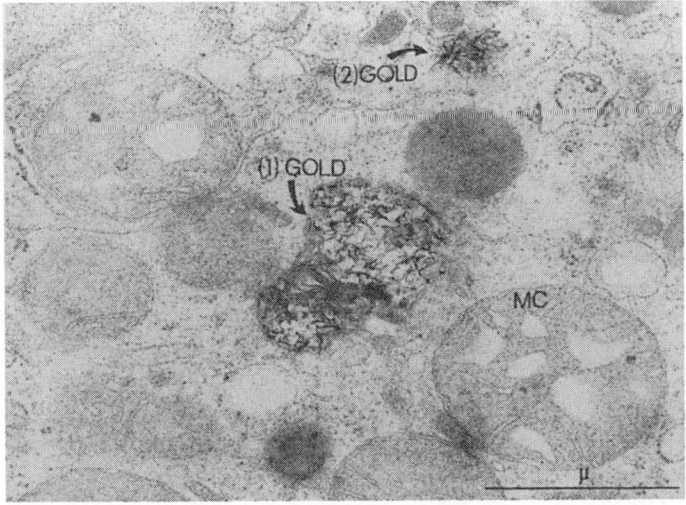

Fig.2: Electron dense filamentous structures of
lysosomes in the proximal tubular cells

renal function was not decreased. Open renal biopsy was performed 2 months after the discontinuation of gold therapy. Light microscopy revealed the deposition of homogeneous eosinophilic material in 70% of glomeruli (Fig.1). This material was confirmed to be amyloid by Congo red staining and electron microscopy. Glomerular basement membrane thickening was not observed. Electron microscopy revealed characteristic electron dense filamentous structures mainly in the proximal tubular cells (Fig.2). Characteristic electron dense filamentous structures were also observed in lysosomes of proximal tubules. X-ray energy dispersive analysis (XEDA) confirmed that these electron dense particles were gold.

(B) Animal Experiment Studies

25 mg of gold sodium thiomalate (which contained 12.1 mg of gold itself) was injected into Sprague-Dawley rats intraperitoneally. The rats were divided into five groups. Group C are the control rats injected with saline. Group 1 was sacrificed 6 hours after gold salt was injected; Group 2 after 24 hours; Group 3 after 72 hours; and Group 4 after 144 hours. Gold contents in organs were studied using a flameless atomic absorption method. Among organs, only in kidneys were remarkable changes found. Swelling and ischemic changes of kidneys were observed. Blood urea nitrogen and serum creatinine levels of Group 1 had already increased, and reached maximum values in Group 3. Serum gold concentration was maximum in Group 1 and decreased rapidly in the other groups. In light microscopy, massive necrosis of the proximal tubules, especially in pars recta was observed (Fig.3). Electron microscopy revealed many electron dense particles in lysosomes in the proximal tubules (Fig.4). Many electron dense precipitates aggregated together to form various dense particles in large inclusion bodies. No fusion of foot process, subepithelial deposits thickening of basement membrane was observed. Gold nephropathy has been diagnosed clinically by the presence of a nephrotic syndrome and the histological finding membranous nephropathy after administration of gold in patients with rheumatoid arthritis (1). Using XEDA, gold deposition in tissues could be confirmed. Gold itself has not been proved to induce a membranous change in patients with rheumatoid arthritis directly. In order to make a diagnosis of gold nephropathy, the finding of gold deposition in the kidney tissue must be demonstrated. There was no demonstration of gold deposition in the sub-

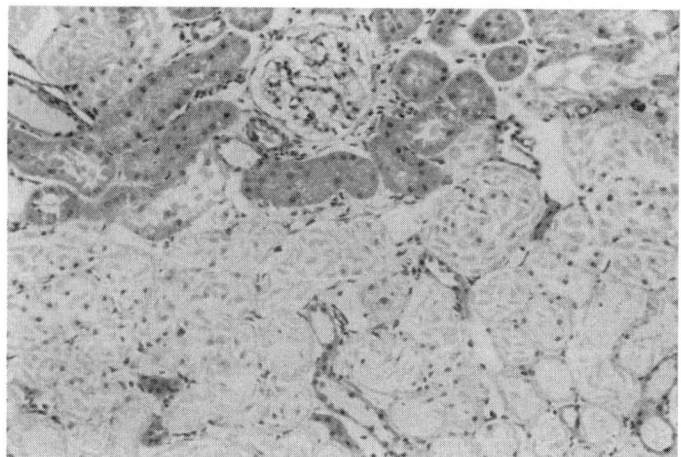

Fig.3: Necrosis of pars recta of proximal
tubules in rat was observed by 25 mg
of gold injection intraperitoneally

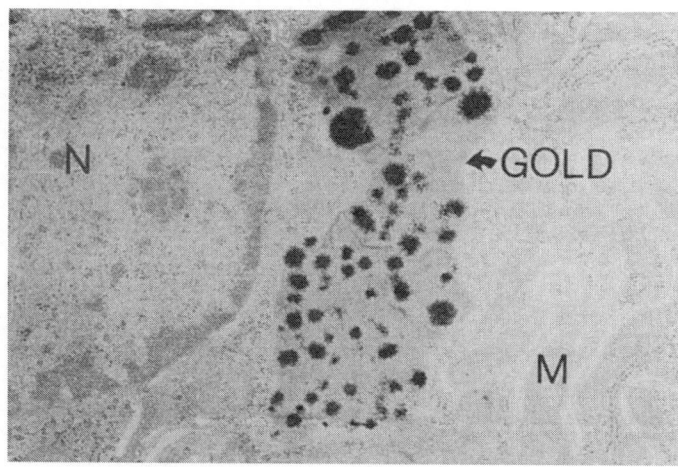

Fig.4: Electron dense particles determined to
be gold using X ray energy dispersive
analysis were observed in lysosomes of
proximal tubular cells of the rat

epithelial dense deposits of membranous nephropathy cases. It might be suggested that gold deposition in human biopsied kidney tissue is characterized by filamentous structures. Gold was absorbed quite rapidly from the peritoneum, and the serum gold concentration reached a maximum 6 hours later. In addition, about 74% of the gold contents of four organs (kidney, liver, lung, spleen) in Group 1 was accumulated in the kidneys. These data suggest that the kidney is the major target organ of gold. Electron microscopy revealed electron dense particles in the lysosomes of the proximal tubules; these particles were confirmed to be gold by XEDA. In order to reveal the first place of gold deposition in the proximal tubules, the rats were sacrificed 20 minutes after injection of gold salts. Although these data are not presented in this paper, many small electron dense particles identified as gold by XEDA were observed in the mitochondria. As gold particles were contained in the lysosomes, mitochondria would be the first site of gold deposition, as Stuve (1970) suggested. The mitochondria is thought to possess the capacity to accumulate ions (NORTON 1968). As more and more mitochondria are disrupted, the cells energy supply is disturbed and structural disintegration begins to occur. There was massive necrosis of mitochondria in the proximal tubular cells. It is still unknown whether gold neph ropathy can induce membranous nephropathy. The term "gold nephropathy" should be used only in cases with both demonstration of gold deposition in the kidney and renal dysfunction caused by gold deposition itself.

2. NEPHROTOXICITY OF CISPLATINUM

Cisplatinum has been well known to be a principal chemotherapeutic agent in the treatment of otherwise resistant solid tumor, and it is currently one of the most widely used agents in the chemotherapy of cancer. The chief limit to its greater efficacy however, is its nephrotoxicity (GOLDSTEIN 1981, MEIJER 1982). There is convincing evidence that the primary target of cisplatinum in cancer cells is DNA in the cells (HARDER 1970, ROBERTS 1979). On the other hand, transplatinum is the transisomer of cisplatinum and does not possess any antineoplastic activity.

(A) The Comparison Of Nephrotoxicity Between Cisplatinum And Transplatinum

Sprague-Dawley rats were divided into 3 groups. Group 1 (Cis), Group 2 (Trans) and Group 3 (Control) rats which received 5 mg/kg cisplatinum, 5 mg/kg of transplatinum and physislogic saline respectively. 24-hour urine collections at 4 ℃ were performed during the experiment. The volume of urine of Cis increased much more than that of Trans. This increase of urine volume of Cis was observed on the 5th day. In biochemical analysis, increase of blood urea nitrogen and serum creatinine levels were noticed also on the 5th day after injection. In contrast, no change was observed in Trans. The concentrations of platinum in both serum and urine were measured using flameless atomic absorption method. A much higher amount of platinum was excreted into urine and serum concentration of platinum was much lower in Cis than those in Trans (Fig.5). The distributions of platinum in organs were examined. The platinum concentration in the kidney was much higher in Trans than that in Cis.

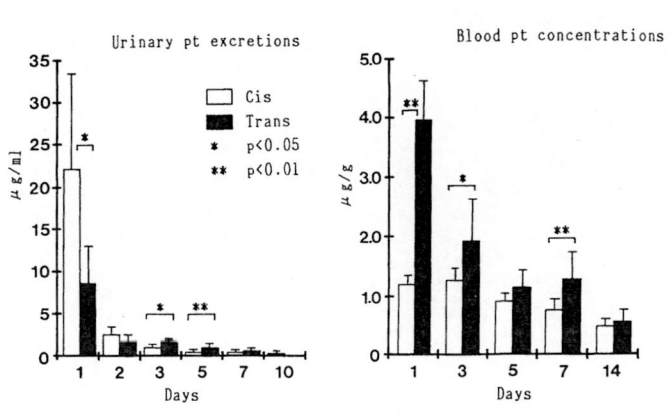

Fig.5: The changes of platinum concentration in the serum and urine after injection of 5 mg/kg of cisplatinum and transplatinum into rats

Histological examination revealed necrosis of pars recta of proximal tubules like gold nephrotoxicity.

In summary, cisplatinum exhibited the nephrotoxicity characterized by azotemia, polyuria, tubular reabsorption defect and histologically, the tubular necrosis of pars recta of proximal tubules. In contrast, the transplatinum, which does not possess an antineoplastic action, did not show the nephrotoxicity in a dose of 5 mg/day, although it has the property to accumulate into kidney more than cisplatinum. It is speculated, therefore, that the nephrotoxic action of cisplatinum may be related to the antineoplastic action.

(B) The Charge Of Urinary Enzyme Activities After Injection Of Cisplatinum In Rats.

Urinary enzymes, such as alkaline phosphate (ALP), leucine aminopeptidase (LAP), alanine aminopeptidase (AAP), γ-glutamyl transpeptidase (γ-GTP), and N-acetyl-β-D-glucosaminidase (NAG) were measured by PNPP, Tuppy, Mondorf, Orlowski, and MCP-NAG methods respectively, in order to determine the tubular damage after cisplatinum injection into rats, since it is known that NAG locate in lysosomes of tubular epithelial cells, and the other enzymes locate in brush border of proximal tubules. The changes of urinary enzyme activities were examined before and after injection of cisplatinum. ALP, LAP, AAP, and γ-GTP activities in urine markedly increased on the 3rd to 5th day after injection (Fig.6). On the other hand, the increase pattern of NAG activity was quite different from those of the other enzymes. Although NAG activity increased gradually, the amount of increase was not large (Fig.7).

(C) Enzyme-histological Studies In Proximal Tubules

Enzyme histological studies were performed in order to examine the changes of enzyme locations in proximal tubular cells before and after injection of cisplatinum. ALP activity was observed widely over proximal tubules,

particularly in pars recta in the normal rat kidney. On the other hand, NAG activity was observed to be limited to proximal tubules of the cortex. After cisplatinum injection, ALP, γ-GTP and LAP activities in pars recta markedly

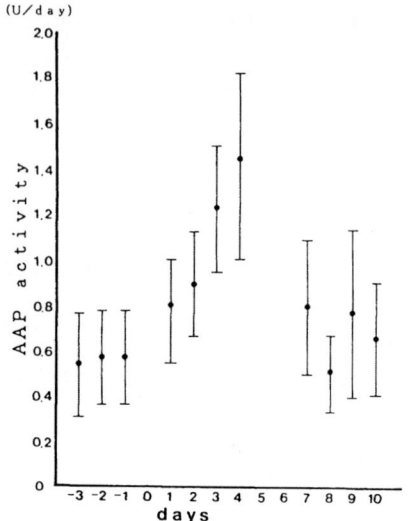

Fig.6: Effects of cisplatinum on urinary AAP activity in the rats

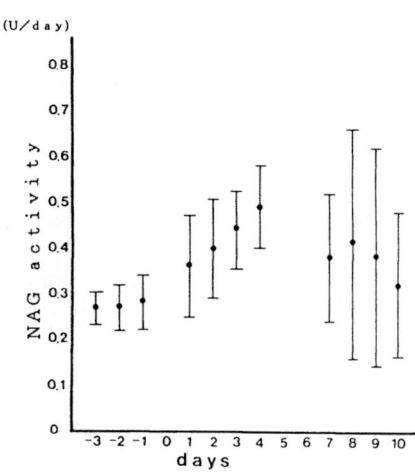

Fig.7: Effects of cisplatinum on urinary NAG activity in the rats

decreased. In contrast, NAG activity in pars recta did not change at all, since the NAG activity was limited to proximal tubules of the cortex, and it had not been observed in pars recta before the injection. This is the reason why urinary NAG showed different pattern from the other enzymes.

(D) The Relationship Between Urinary Enzyme Activities And Histological Changes Of The Pars Recta

The relationship between urinary enzyme activities and histological changes of pars recta was examined. In order to examine the histological changes quantitatively, IBAS (interactive image analysis system) produced by Zeiss was used. The number of tubular nuclei and percent area of detached tubular cells in pars recta of proximal tubules in the outer stripe of the outer medulla were studied. There was a significant correlation between the number of tubular nuclei and the percent area of detached tubular cells of the pars recta. This result indicates that the reduction of the number of tubular nuclei and percent increase of detached tubular cells measured by IBAS were able to be considered as the quantitative parameters of the tubular damage caused by cisplatinum injection. There was a significant correlation between the number of tubular nuclei and the mean of AAP activity. Correlation between the percent area of detached tubular cells and the mean of AAP activity was also observed. These relationships were also found among LAP, ALP, NAG and γ-GTP (Table 1).

	N		A	
	Correlation coefficient	P value	Correlation coefficient	P value
ALP	-0.25	N.S.	0.48	<0.05
ALP/Cr	-0.29	N.S.	0.57	<0.01
LAP	-0.82	<0.001	0.79	<0.001
LAP/Cr	-0.84	<0.001	0.82	<0.001
AAP	-0.79	<0.001	0.77	<0.01
AAP/Cr	-0.65	<0.01	0.75	<0.001
γ-GTP	-0.81	<0.001	0.82	<0.001
γ-GTP/Cr	-0.20	N.S.	0.09	N.S.
NAG	-0.72	<0.001	0.78	<0.001
NAG/C	-0.65	<0.01	0.75	<0.001

N : The number of tubular nuclei in P_3 segment

A : The percent area of detached tubular cells in P_3 segment

Table 1: Correlation between the activity of urinary enzymes and the degree of the tubular damage measured by IBAS (interactive image analysis system)

(E) Summary

Enzymes in tubules (ALP, LAP, AAP, γ-GTP, NAG) are easily excreted into urine by proximal tubular damage. The urinary enzyme activities seem to reflect the degree of the proximal tubular damage. It seems also possible to speculate on the site of damaged tubules by measuring each enzyme respectively, because each enzyme has its own distribution pattern along the proximal tubules. The daily measurement of urinary enzymes, therefore, could suggest potential values which indicate not only the degree but also the site of the damaged tubules.

REFERENCES

GOLDSTEIN MH, SAFIRSTEIN R (1981):
FUNCTIONAL CHARACTERISTICS OF CHRONIC TUBULO INTERSTITIAL (TI) AFTER REPEATED DOSES OF CIS PLATINUM (PT).
ABSTRACTS; 8TH INTERNATIONAL CONGRESS OF NEPHROLOGY P229

HARDER HC, ROSENBERG B (1970):
INHIBITORY EFFECTS OF ANTI-TUMOR PLATINUM COMPOUNDS ON DNA,RNA AND PROTEIN SYNTHESES IN MAMMALIAN CELLS IN VITRO.
INT J CANCER 6:207-16

IESATO K, MORI Y, UEDA S, WAKASHIN Y, WAKASHIN M, MATSUI N, INOUE S, OKUDA K (1982):
RENAL TUBULAR DYSFUNCTION AS A COMPLICATION OF GOLD THERAPY IN PATIENTS WITH RHEUMATOID ARTHRITIS.
CLIN. NEPHROL. 17(1):46-52

KATZ A, AND LITTLE AH (1973):
GOLD NEPHROPATHY
ARCH. PATHOL. 96:133-136

MEIJER S, MULDER NH, SLEIJFER DT, DE JOHNG PE, SLUITER WJ, SCHRAFFORDT KOOPS H, VAN DER HEM GK (1982):
NEOHROTOXICITY OF CIS-DIAMMINEDICHLOIDE PLATINUM (CDDP) DURING REMISSION-INDUCATION AND MAINTENANCE CHEMOTHERAPY OF TESTICULAR CARCINOMA.
CANCER CHEMOTHER PHARMACOL. 8:27-30

NORTON WL, LEWIS DC, AND ZIFF M (1968):
ELECTRON-DENSE DEPOSITE FOLLOWING INJECTION OF GOLD SODIUM THIOMALATE AND THIOMALIC ACID.
ARTHRITIS AND RHEUMATISM 11:436-443

POLLAK VE, PIRANI CL, STECK IE, KARK RM (1962):
THE KIDNEY IN REHEUMATOID ARTHRITIS : STUDIES BY RENAL BIOPSY.
ARTHRITIS AND RHEUMATISM 5:1-9

ROBERTS JJ, THOMSON AJ (1979):
THE MECHANISM OF ACTION OF ANTITUMOR PLATINUM COMPOUNDS.
PROG. NUCLEIC ACID RES. MOL. BIOL. 22:71-133

SAMUELS B, LEE JC, ENGLEMAN EP, HOPPER J Jr (1977):
MEMBRANOUS NEPHROPATHY IN PATIENTS WITH RHEUMATOID ARTHRITIS : RELATIONSHIP TO GOLD THERAPY.
MEDICINE 57:319-327

SILVERBERG DS, KIDD EG, SHNITKA TK (1970), et al.:
GOLD NEPHROPATHY, A CLINICAL AND PATHOLOGICAL STUDIES.
ARTHRITIS AND RHEUMATISM 13:812-25

STUVE J, AND GALLE P (1970):
ROLL OF MITOCHONDRIA IN THE HANDLING OF GOLD BY THE KIDNEY. A STUDY BY ELECTRON MICROSCOPY AND ELECTRON PROBE MICROANALYSIS.
J. CELL BIOL 44:667-676

TORNROTH T, SKRIFVARS B (1974):
GOLD NEPHROPATHY PROTOTYPE OF MEMBRANOUS GLOMERULONEPHRITIS.
AMER. J. PATH. 75:573-90

Drug-Induced Taste Disorder: Due to Zinc Chelation?

MINORU IKEDA, HIROSHI TOMITA, and KUNIHIKO SEKIMOTO

Department of Otolaryngology, Nihon University School of Medicine, 30-1 Oyaguchi, Itabashi-ku, Tokyo, 173 Japan

ABSTRACT

Our clinical investigation elucidated that taste disturbance due to drugs was one of the most common causes of taste disorder. It was found that 24.0% cases out of 1512 patients with taste disorder were suspected to be induced by administered drugs. Taste disorder triggered by some drugs has been suspected to be considerably affected by the zinc chelation activity of the drugs. Accordingly, the ability of zinc chelation of several agents which had been considered as the cause of a majority of drug-induced taste disorder was studied experimentally. The results of our studies positively support the conception that the zinc chelation activity of the drugs is an important causative factor of this type of taste disorder.

KEY WORDS: Zinc, Taste Disorder, Drug, Zinc Chelation

I. INTRODUCTION

A number of drugs have been reported to induce taste disorder.[1] However, in most of the cases, it has not been elucidated how these drugs act on the taste acuity and induce taste dysfunction.

On the other hand, clinical and experimental studies on the relationship between zinc and taste function have been reported, and it is widely accepted that zinc deficiency causes taste disturbance.[2,3]

Henkin[2] reported that penicillamine can cause taste disturbance. The patients who was examined in his report was administered penicillamine because of his cystinuria, and showed taste disturbance. The penicillamine treatment on the patient was discontinued and was replaced by zinc administration, and improvement of his taste acuity was recognized. Thereafter, the penicillamine administration was resumed and zinc administration was stopped for the patient, and this was followed by recurrence of taste disturbance. Henkin supposed that the chelation of zinc by the SH group of penicillamine was the cause of this taste disturbance. His document has thrown light on the relationship between zinc and drug-induced taste disorder. The chelate compound of penicillamine and zinc is supposed as illustrated in Fig 1.

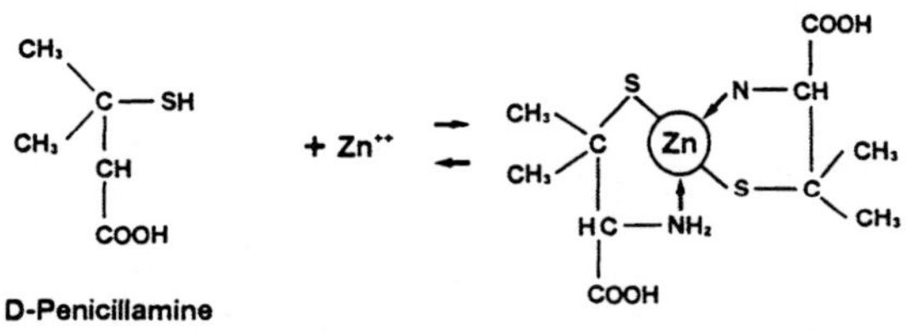

Fig 1. Penicillamine-zinc chelate compound

Thiocarbamide, which is an agent for hyperthyroidism has been reported to induce taste disorder. In this case, as well, zinc administration was efficacious. Extrication of zinc from the tissue by the SH group of thiocarbamide was assumed to be the cause of this taste disorder.[4]

Table 1 shows the drugs suspected of being the cause of taste disorder. They are mainly composed of the drugs which were documented in the literatures, and some of them were observed in our clinic. The drugs marked with asterisks indicate that they have been suspected of having the ability of zinc chelation, either bibliographically or by our experimental studies. The details of our experiment are described. below.

Table 1. Drugs suspected of being the cause of taste disorder

HYPOTENSIVE AGENTS
  THIAZIDE DIURETICS*
  FULOSEMIDE*
  SPIRONOLACTONE
  ETHACRYNIC ACID
  METHYLDOPA*
  HYDRALAZINE HCL
  CAPTOPRIL*
VASOCONTRICTORS
  ETILEFRINE HCL
VASODILATORS
  OXIFEDRINE HCL
  DIPYRIDAMOLE
  NIFEDIPIN
  DILTIAZEM
CARDIOTONICS
  DIGOXIN

ANTIARRHYTHMIC AGENTS
  PROPRANOLOL
AGENTS USED FOR ARTERIOSCLEROSIS
  CLOFIBRATE
  PANTETHIN
AUTONIMIC AGENTS
  GAMMA-ORYZANOL
  ANTI-CHOLINERGIC AGENTS
ANTI-PERKINSONISM AGENTS
  LEVODOPA*
  TRIHEXYPHENIDYL HCL
MYORELAXANS
  PHENPROBAMATE
  CHLORMEZANONE
ANTIEPILEPTICS
  PHENITOIN
  CARBAMAZEPINE

ANTI-DIABETES AGENTS
  SULFONYL UREAS
  BIGUANIDE
  GLYBUZOLE
ANTI-THYROIDISM AGENTS
  THIAMAZOLE
  THIOURACIL
HORMONS
  GLUCO-CORTICOID
VITAMINS
  VITAMIN D
  VB2 BUTYRIC ACID ESTER
ANTIPYRETIC-ANALGESICS AGENT
  MEFENAMIC ACID
  ASPIRIN
  INDOMETHACIN
  PHENYLBUTAZONE

ANTITUBERCULOUS DRUGS
  ETHAMBUTOL*
  ISONIAZID*
  PAS*
  ETHIONAMIDE
ANTIMOLDS AGENTS
  AMPHOTERICIN B
  GRICEOFULUVIN
ANTIPROTOZOAL AGENTS
  METRONIDAZOLE
OTHER AGENTS
  D-PENICILLAMINE*

ANTIEMETICS
  METOCLOPRAMIDE*
  FENIPENTOL
ANTI-PEPTIC ULCER AGENTS
  L-GLUTAMINE
  GEFARNATE
  SUCRALFATE
AGENTS USED FOR LIVER DYSFUNCTION
  GLUTATHIONE*
  GLYCYRRHIZIN
  MERCAPTOPROPIONILGLYCIN
  GLUCRONIC ACID
  PROTOPORPHYRIN SODIUM
ANTI-NEOPLASTIC AGENTS
  FLUOROURACIL*
  METHOTREXATE
  VINCRISTINE
  ADRIAMYCIN

ANTIBIOTICS PREPARATIONS
  AMPICILLIN
  SULBENICILLIN
  STREPTOMYCIN
  CLINDAMYCIN
  CEPHEMS
  LINCOMYCIN*
  TETRACYCLINE*
  CARINDACILLIN
AGENTS USED FOR GOUT
  COLCHICINE
  ALLOPURINOL
ANTIHISTAMICS
  CHLORPHENIRAMINE MALEATE
  PROMETHAZINE HCL
IMMUNOSUPRESSIVE AGENTS
  AZATHIOPRINE

MAJOR TRANQUILIZER
  TRIFLUOPERAZINE
MINOR TRANQUILIZER
  DIAZEPAM
  CHLORDIAZEPOXIDE
  OXAZOLAM
  MEDAZEPAM
ANTI-DEPRESSIVE AGENTS
  NORTRIPTYLINE HCL
  IMIPRAMINE
  AMITRIPTYLINE
  MELITRACEN
  TRIMIPRAMINE MALEATE
ANTISPASMODICS
  METHYLSCOPOLAMMONIUM METHYLSULFATE
ANTIVIRAL AGENTS
  IDOXURIDINE (IDU)

II. CLINICAL OBSERVATIONS OF DRUG-INDUCED TASTE DISORDERS

According to the list of drugs shown in Table 1, 1512 patients with taste disorder who visited our clinic from 1981 to 1987 were investigated. It was found that the drug-induced taste disorder accounted for 363 cases, 24.0%, and was the most common cause of taste disorder. It was especially remarkable in case of those over 60 years old.

Apart from this investigation, we analyzed 196 cases with taste disorder.[5] Fifty-seven, 29.1%, of those were suspected to be due to the drugs which the patients had been administered for their original diseases. The names and incidences of the drugs are shown in Table 2.

Thiazide diuretics and methyldopa which are anti-hypertensive agents, clofibrate for arterio-sclerosis, 1-glutamine as anti-peptic ulcer agents, metocloplamide as anti-emetics, agents for liver dysfunction such as glutatione, and various kinds of psycho-tropic agents appeared with high incidence.

III. EFFECT OF ZINC ADMINISTRATION ON THE CASES OF DRUG-INDUCED TASTE DISORDER

In due consideration of the documents in the literatures, taste disorder triggered by drugs was suspected to be considerably affected by the zinc chelation activity of the drugs. Accordingly, for the purpose of investigating the influence of drugs, we examined the changes in the serum zinc level of the cases suspected of drug-induced taste disorder.

Table 2. Drugs causing taste disturbance of 57 cases

Agents	No. of cases	Agents	No. of cases
Hypotensive agents		Anti-diabetes agents	
Thiazide diuretics	19	Sulfonyl ureas	2
Furosemide	3		
Hydralazide	4	Antipyretic-analgesic-anti-inflammatory agents	
Methyldopa	9	Mefenamic acid	3
β-Recepter blocking agents	7	Diclofenic sodium	2
		Ibuprofen	2
Vasodilators			
Dipyridamole	2	Psychotropic agents	
Nifedipin	1	Haloperidol	3
Diltiazem	1	Diazepam	2
		Medazepam	1
Agents used for arteriosclerosis		Clotiazepam	1
Clofibrate	3		
Pantethin	1	Antibiotic agents	
		Cefems	2
Anti-peptic ulcer agents			
L-Glutamine	3	Antituberculous agents	
Sucralfato	2	Ethanbutol	1
Antiemetics		Anabolic steroid hormones	
Metocloplamide	4	Predonisolone	2
Agents used for liver dysfunction		Other agents	
Mercaptopropionylglycine	4	D-Penicillamine	2
Glutathione	2		
Glycyrrhizin	2		

Note : Include overlaped cases.

The subjects for this investigation were a total of 24 cases with the chief complaint of taste disturbance. Fourteen cases were not using any drugs and were hence categorized as the non-drug induced (NDI) group. The remaining 10 cases were being administrated the drugs which can be the cause of taste dysfunction, and were categorized as the drug-induced (DI) group. Since the absorption of zinc might depend on the age, these 2 groups were age-matched to be from 50 to 59 years old. Moreover, since the purpose of this investigation was to see the change in the serum zinc level following zinc administration, the patients were also matched to have a low serum zinc level. The mean values of the serum zinc were 61 µg/dl (NDI group) and 62 µg/dl (DI group).

Administration of zinc was carried out at 300mg per a day, and the changes in the serum zinc value were followed for 5 months. The mean values in the DI group were lower than in the NDI group. (Table 3) The data obtained at the second month showed a statistically significant difference ($p < 0.05$). The means of the highest value of serum zinc level were 117.6 ± 25.1µg/dl in the NDI group and 102.0 ± 10.7µg/dl in the DI group, and there was a statistically significant difference ($p < 0.1$). The period it took to reach the highest serum zinc level was longer in the DI group (4.8 ± 3.1 months) than in the NDI group (3.1 ± 2.1 months).

Accordingly, it might be possible to say that the administrated drugs are exerting a considerably harmful influence on the zinc metabolism of the patients and are inducing taste disorder.

Table 3. Changes of serum zinc values following zinc administration

	Before Therapy	1 Month	2 Months	3 Months	4 Months	5 Months
Non-Drug-Induced Group	61.1 ± 5.7 (n=14)	95.5 ± 26.3 (n=13)	96.2 ± 22.9* (n=12)	99.0 ± 27.9 (n=11)	92.2 ± 21.0 (n=10)	91.4 ± 23.4* (n=11)
Drug-Induced Group	62.4 ± 7.2 (n=10)	79.9 ± 13.1 (n=9)	77.4 ± 10.4* (n=9)	82.6 ± 11.3 (n=9)	93.1 ± 12.8 (n=7)	72.8 ± 8.1* (n=6)

(* $P < 0.05$) (* $P < 0.1$)

IV. EXPERIMENTAL STUDIES ON ZINC CHELATION OF DRUGS

The authors experimentally studied the ability of zinc chelation of several agents which had been considered as the cause of a majority of drug-induced taste disorder assumed by clinical investigations. The methods employed were pH titration and polarography. In some cases, absorption curves were also examined.

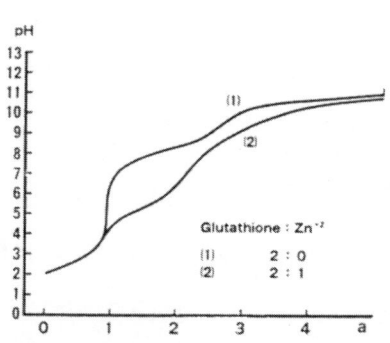

Fig 2. pH titration curves of glutathion and zinc nitrate

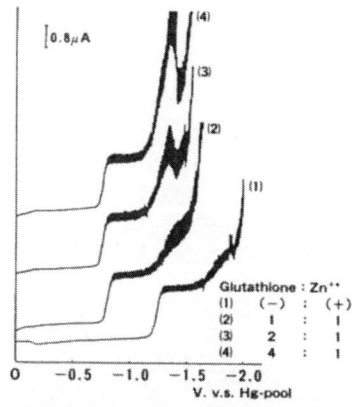

(supporting electrolyte : 0.15N KNO₃, pH 5.3)

Fig 3. DC-polarograms of glutathion-zinc

1). Glutathione

Glutathione is well known as an agent for liver dysfunction. It is composed of peptide linkage of glutamic acid, cystaine, and glicin. All of the components have been known to be metal chelating agents. Li et al[6] have reported the ability of zinc chelation of this drug. pH titration curves showed an obvious decline in pH as shown in Fig 2. In the polarograms, half-wave voltages shifted to the positive direction of almost 0.5 volt. (Fig 3) These findings were considered to be due to the formation of glutathione and zinc chelate compound.

2). Alpha-Mercapto-Propionyl Glycine

This is also a well-known agent for liver dysfunction. Funae et al[7] have reported its ability of zinc chelation. pH titration curves showed decline of pH within the range of pH 3 to pH 9. (Fig 4) The stability constants, pK1 and pK2 were 5.77 and 5.05, respectively. The polarograms showed a shift of the half-wave voltages to the positive direction of almost 0.5 volt. (Fig 5) These findings indicated the formation of zinc chelate compound.

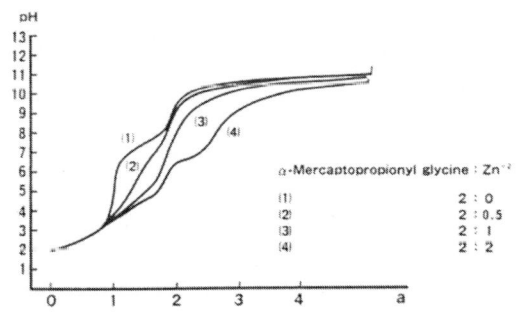

Fig 4. pH titration curves of alpha-mercapto-
        propionyl glycine and zinc nitrate

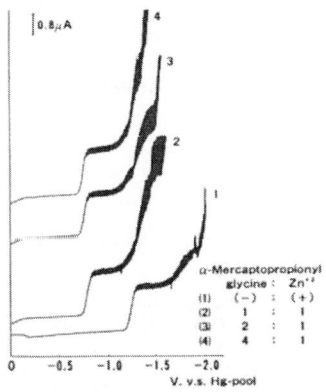

Fig 5. DC-polarograms of alpha-mercapto-
        propionyl glycin-zinc

3). Furosemide

Furosemide is a popular diuretic agent. Wester[8] has reported that the administration of furosemide results an increased urinary excretion of zinc. In the pH titration curve, a decline in pH was observed after the addition of zinc. (Fig 6) This was considered to be due to the formation of furosemide and zinc chelate compound. When the ratio of furosemide and zinc was 2 to 1, the decline in pH was the greatest. This result might be attributable to the 2 to 1 chelate formation. The structure of the chelate compound was considered to be as illustrated in Fig 7. The stability constants were calculated as 2.23 for pK1 and 2.01 for pK2.

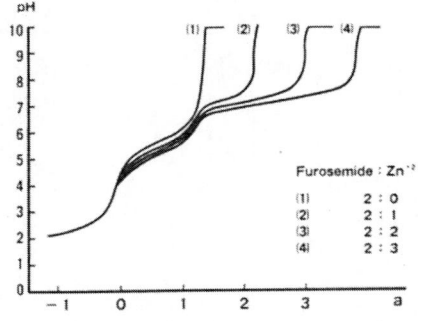

Fig 6. pH titration curves of furosemide
        and zinc nitrate

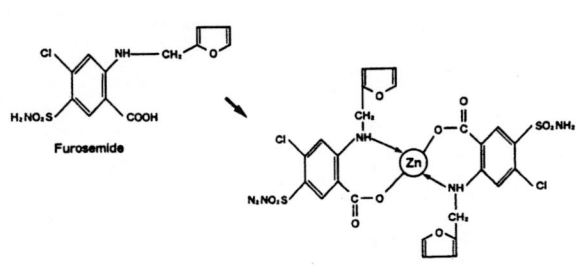

Fig 7. Frosemide-zinc chelate compound

4). Methyldopa

This drug is also a diuretic agent, and its absorption curve is shown in Fig 8.
The absorption curve of methyldopa alone, which is the curve A, has a maximum
at 280nm, while that of methyldopa with added zinc, which is the curve B, has
its maximum absorption point shifted to 297nm. This charge is considered to be
due to the formation of methyldopa-zinc chelate compound.[5]

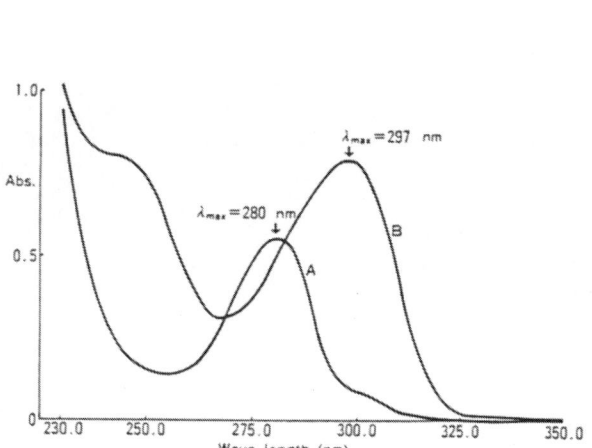

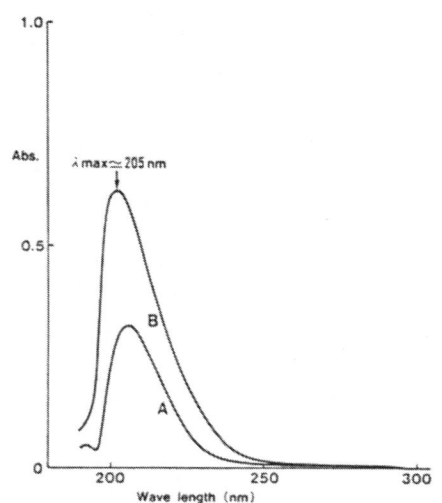

Absorption curves of methyldopa and methyldopa–Zn^{2+} chelate compound (pH = 5.5,
μ = 0.1, t = 20°C; 10-mm cell).
Curve A: Absorption curve of methyldopa (1×10^{-4} mol/1).
          The maximum absorption point (λ_{max}) is at 280 nm.
Curve B: Absorption curve of methyldopa (1×10^{-4} mol/1) plus Zn (ClO_4)$_2$ ($1/2$
          $\times 10^{-4}$ mol/1) . The maximum absorption point (λ_{max}) is shifted to 297 nm.
          This trace is considered to represent the absorption curve of methyldopa
          Zn^{2+} chelate compound.

Fig 8.

Absorption curves of captopril and captoril–Zn^{2+} chelate compound (pH =
7.3, μ = 0.1, t = 20°C; 10-mm cell).
Control :   Water + buffer.
Curve A :   Absorption curve of captopril (1.6×10^{-4} mol/1)
Curve B :   Absorption curve of captopril (1.6×10^{-4} mol/1)
           plus Zn (ClO_4)$_2$ (0.8×10^{-4} mol/1).
The maximum absorption points of curves A and B were both at about 205
nm. However, curve B showed an increased absorption. This was considered
to be due to the formation of captopril-Zn^{2+} chelate compound.

Fig 9.

5). Captopril

Captopril is also a diuretic. Its absorption curve is as shown in Fig 9. Curve
A is captopril alone, and Curve B is captopril with added zinc. Both curves
had a maximum absorption point at about 205nm, although the curve of captopril
with added zinc revealed an increased absorption. This was considered to be
due to the formation of captopril-zinc chelate compound.[5]

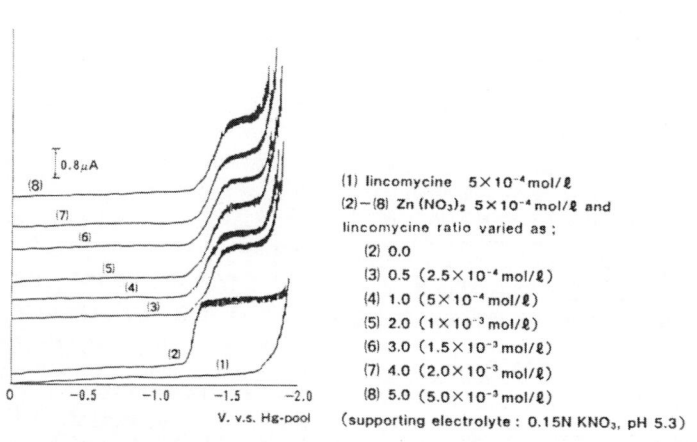

(1) lincomycine 5×10^{-4} mol/ℓ
(2)−(8) Zn $(NO_3)_2$ 5×10^{-4} mol/ℓ and
lincomycine ratio varied as :
    (2) 0.0
    (3) 0.5 (2.5×10^{-4} mol/ℓ)
    (4) 1.0 (5×10^{-4} mol/ℓ)
    (5) 2.0 (1×10^{-3} mol/ℓ)
    (6) 3.0 (1.5×10^{-3} mol/ℓ)
    (7) 4.0 (2.0×10^{-3} mol/ℓ)
    (8) 5.0 (5.0×10^{-3} mol/ℓ)
(supporting electrolyte : 0.15N KNO_3, pH 5.3)

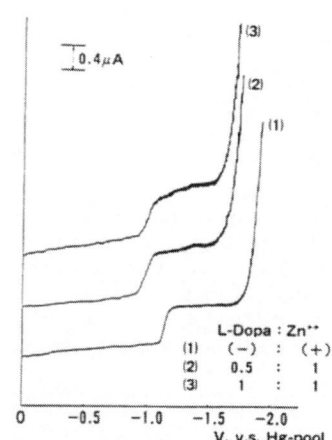

L-Dopa : Zn^{++}
(1)   (−) : (+)
(2)   0.5 : 1
(3)   1   : 1

(supporting electrolyte : 0.15N KNO_3, pH 8.1)

Fig 10. DC-polarograms of lincomycin-zinc

Fig 11. DC-polarograms of levodopa-zinc

6). Metochlopramide

This drud is an anti-emetic agent and is frequently used in combination with other drugs to the patients with taste disorder. Metochlopramide is a derivative of salicylic acid which is known to have the ability of zinc chelation. In pH titration, decline of pH could not be clarified. The polarograms, however, showed a shift of the voltage to the negative direction, and the ability of zinc chelation of this drug was suspected.

7). Lincomycin

Many of the antibiotics have been known to have the ability of combining with ionic metals. Accordingly, it could be suspected that the taste dysfunction caused by antibiotics is due to the chelation of zinc. In case of lincomycin, pH titration method did not indicate the formation of chelate compound. However, it was supposed that lincomycin had the ability of zinc chelation by the polarography. (Fig 10)

8). Levodopa

Levodopa is an agent for perkinsonism, and has been reported to induce taste disorder. However, its ability of zinc chelation has not yet been clarified. In our experiment, the half-wave voltage of the polarography shifted 0.2 volt to the positive direction (Fig 11), and chelate formation with zinc was suspected.

9). Methimazole

Methimazole is an agent for hyperthyroidism, and its SH group has been supposed to chelate zinc and to induce taste disorder.[4] pH titration did not clarify the formation of chelate compound, however, the half-wave voltage of polarography shifted 0.3 volt to the positive direction (Fig 12), and the formation of zinc chelate was suspected.

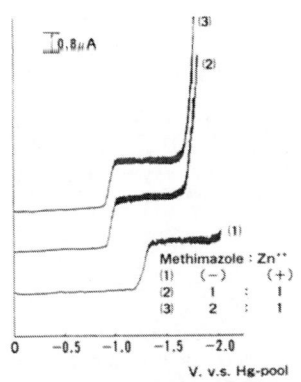

Fig 12. DC-polarograms of methimazole-zinc

V. CONCLUSION

We pointed out that taste disturbance due to drugs was one of the most common causes of taste disorder, and that it was especially remarkable in the case of the elders. It was indicated that the drugs which could induce taste disorder were abundant and widely varied. As the cause of this type of taste disorder, the existence of zinc chelation by the drugs have been highly suspected, bibliographically. The results of our clinical and experimental studies which have been shown above positively support this conception.

REFERENCES

1).Rollin H;Drug-related gustatory disorders,Ann Otol 87,37-42,1978.
2).Henkin RI,et al:Hypogeusia corrected by N^{2+} and Zn^{2+}.Life Sci 9,701-9,1970.
3).Hasegawa H,et al:Assessment of taste disorders in rats by simultanious study of the two-bottle preference test and abnormal ingestive behavior. Auris-Nasus-Larynx(Tokyo)13(Suppl 1),33-41,1986.
4).Erikssen J,et al:Side-effect of thiocarbamides.Lancet Jan 25,231-32,1975.
5).Sekimoto K,et al:Zinc chelation capacity of hypotensive agents causing taste disturbance. Nihon Univ J Med 28,233-30,1986.
6).Li NC, et al:Stability of zinc complexes with glutathione and oxidised glutathione.J Amer Chem Soc 76,225-30,1954.
7).Funae Y,et al:Metal complexes of 2-mercaptopropionulglycine.Chem Pharm Bull 19,1618,1971.
8).Wester PO:Zinc during diuretic treatment.Lancet Merch 8,578,1975.

Antiarthritic, Antiulcer, and Analgesic Activities of Copper Complexes

JOHN R.J. SORENSON

College of Pharmacy, Slot 522, University of Arkansas for Medical Sciences, 4301 West Markham, Little Rock, AR 72205, USA

ABSTRACT

Copper is presented as an essential metalloelement required by all cells for normal function based upon its presence in all tissues and its requirement for copper-dependent enzyme activity. Copper complexes of non-steroidal anti-arthritic drugs are shown to be more effective than their parent drugs as antiinflammatory and analgesic agents, consistent with the successful use of copper complexes to relieve human arthritic disease and pain. These same copper complexes are also shown to be potent antiulcer agents. All of these activities support the hypothesis that the pharmacological use of copper complexes represent a physiological approach to treatment of arthritic diseases, ulcers, and pain.

KEY WORDS

antiarthritic, antiulcer, analgesic, copper, complexes

INTRODUCTION

Copper is an essential metalloelement and as such it is required by all human cells for normal metabolism (Underwood 1977). Amounts found in tissues (Tipton, Cook 1963; Iyengar, Kollmer, Bowen 1978) are listed in Table 1 to point out direct variation with metabolic need.

Table 1. Mean concentration[a] of copper in tissues and fluids

Adrenal	210	Milk	
Aorta	97	colostrum	0.35-0.50 μg/ml
Bile	547	mature	0.20-0.50 μg/ml
Blood (total)	1.01 μg/ml	Muscle	85
erythrocytes	0.98 μg/ml	Nails	23 μg/g
plasma	1.12 μg/ml	Omentum	190
serum	1.1 9 μg/ml	Ovary	130
Bone	5 μg/g	Pancreas	150
Brain	370	Pancreatic Fluid	105
Breast	6 μg/g	Placenta	4 μg/g
Cerebrospinal Fluid	0.22 μg/g	Prostate	110
Diaphragm	150	Saliva	0.08 μg/ml
Esophagus	140	Skin	120
Gall Bladder	750	Spleen	93
Hair	19 μg/g	Stomach	230
Heart	350	Sweat	0.55 μg/ml
Intestine		Testes	95
Duodenum	300	Thymus	4 μg/g
jejunum	250	Thyroid	100
ileum	280	Tongue	4.6 μg/g
cecum	220	Tooth	
sigmoid colon	230	dentine	2 μg/g
rectum	180	enamel	10 μg/g
Kidney	270	Trachea	65
Larynx	59	Urinary Bladder	120
Liver	680	Urine	0.04 μg/ml
Lung	130	Uterus	110
Lymph Node	60		

a μg/g of tissue ash or as shown

Brain and heart contain more copper than all other tissues except liver which is a major copper storage organ. Gall bladder and bile also contain large amounts of copper which has been attributed to their putative roles in excretion. However, the gall bladder may also serve as a storage tissue and bile may contain a mobile storage form of copper complexes suitable for intestinal reabsorption. The large kidney copper content, when compared with the very small urine copper content, suggests a conservatory role for the kidney. Remaining tissues have lesser amounts of copper but it is just as important for normal metabolism in these tissues as it is in all others.

Ionic copper has a particularily high affinity for organic molecules (ligands) capable of bonding with it. A consequence of this is that all measurable copper in biological systems exists as complexes or chelates composed of copper bonded to organic components of these systems. Calculated amounts of ionic copper suggested to be present in biological systems [10^{-18}M in plasma (May, Linder, Williams 1976)] are too small to be measured using the most sensitive instrumentation available. As a result, measurable tissue copper reflects content of copper complexes and these complexes account for absorption, distribution, and biologically active forms including recognized and recently suggested copper-dependent enzymes listed in Table 2 (Anonymous 1985; Boyadzhyan 1985; Fridovich 1986; Frieden 1986; Hamilton 1981; McCord, Fridovich 1969; Marklund 1982a, 1982b; O'Dell 1976; Sorenson 1982, 1984).

Table 2. Copper-dependent mammalian enzymes and their chemical function

Enzyme	Function
Cytochrome c Oxidase	Reduction of Oxygen: $O_2 \xrightarrow{H^+,e^-} HO\cdot_2 \xrightarrow{H^+,e^-} H_2O_2 \xrightarrow{H^+,e^-} H_2O + HO\cdot \xrightarrow{H+,e^-} H_2O$
Superoxide Dismutases	Disproportionation of superoxide in prevention of its accumulation: $2O_2^{-}\cdot + 2H^+ \longrightarrow O_2 + H_2O_2$
Tyrosinase	Hydroxylation of tyrosine in melanin synthesis: DOPA
Dopamine-β-Hydroxylases Extremely Acidic Copper-Containing Proteins	Hydroxylation of dopamine in catecholamine and synthesis: NOREPINEPHRINE
Lysyl Oxidases	Oxidation of terminal amino group of lysyl amino acids in procollagen and proelastin to an aldehyde group: Peptidyl$(CH_2)_3CH_2-NH_2 \longrightarrow$ Peptidyl-$(CH_2)_3-\overset{\text{O}}{\underset{}{C}}H$
Amine Oxidases	Oxidation of primary amines to aldehydes in catecholamine and other primary amine metabolism: $R-CH_2-NH_2 \longrightarrow R-\overset{\text{O}}{\underset{}{C}}H$
Ceruloplasmin	Mobilization and utilization of stored iron: Ferroxidase, Fe(II)$\longrightarrow$Fe(III); Copper transport; SOD-like activity; serum amine oxidase activity; and angiogenic activity.
Factors V and VII	Blood clotting
Peptidyl α-Amidating Monooxygenases	Synthesis of neuroendocrine peptides (hypothalamic thyrotropin releasing hormone, a-melanocyte stimulating hormone from anterior pituitary, oxytocin and vasopressin from the posterior pituitary, gastrin from stomach, and choleocysto-kinin from the small intestine): Peptidyl-N-CH$_2$$\overset{\text{O}}{\underset{}{C}}$OH $\xrightarrow{\text{Vit C}}$ peptidyl-NH$_2$ + H$\overset{\text{O}}{\underset{}{C}}$$\overset{\text{O}}{\underset{}{C}}$OH H

Ingested copper complexes representing the recommended 3 mg (47 μmol) daily intake follow a pathway leading ultimately to incorporation into copper-dependent metalloenzymes or metalloproteins (Sorenson 1989). It is important to remember that gastric digestion is enzyme catalyzed and not acid catalyzed and, since gastric pH is likely to range from 6 when empty to 3 with the ingestion of a meal (Davenport 1982), some of the originally ingested complex may be absorbed intact. Copper complexes in the duodenal chyme, pH 7.0, would also be expected to be absorbed intact. Pharmacologic doses of copper complexes have an antisecretory effect and thus prevent lowering of the normal empty stomach pH, 6.0. Since excessive copper storage has never been reported in any normal population and copper complex absorption must vary widely in these populations, efficient homeostatic mechanisms must regulate retention and excretion of varying amounts of absorbed copper complexes in all normal, non-Wilson's disease, individuals.

Since copper is needed for normal metabolism and prevention of disease great care should be taken to assure that dietary intake provides required amounts of copper. Unfortunately, many or nearly all modern diets studied do not supply required amounts of this and other essential metalloelements which may lead to decreased enzyme activity and manifestations of acute arthritic diseases, ulcers, and sporatic pain in the short term or chronic arthritic diseases, ulcers, and pain in the long term (Sorenson, 1989).

ANTIINFLAMMATORY ACTIVITIES

Two to four times normal concentrations of blood copper complexes are associated with various chronic inflammatory disease states (Sorenson 1982, 1989) and it is likely that this association also exists for acute and chronic gastrointestinal ulcer diseases which are, in part, inflammatory diseases of gastric or intestinal tissues. The elevation usually found in inflammatory diseases is generally associated with the active disease phase. With disease remission plasma copper complex concentration returns to normal. The fact that small molecular weight copper complexes have antiinflammatory activity in Man (Sorenson, Hangarter 1977) and animal models of inflammation is offered as evidence that the associated increase in blood copper-containing components is a physiologic response which has a role in mediating remission when remission occurs.

To date, over 140 copper complexes have been studied as antiinflammatory agents. Results of these studies, which have been reviewed (Sorenson 1989), confirm as well as extend original observations that copper complexes of inactive ligands and active antiinflammatory drugs are more active than their parent ligands (drugs) and inorganic copper (Sorenson 1989) or mixtures of these ligands and inorganic copper (Sorenson 1976; Korolkiewicz, Hac, Gagalo, Gorczyca, Lodzinska 1989). The following is a very brief presentation of a small amount of data from the original report suggesting that copper complexes are active metabolites of antiarthritic drugs (Sorenson 1976).

Copper(II)$_2$(acetate)$_4$(H$_2$O)$_2$ was found to be active in an initial test (carrageenan paw edema) for antiinflammatory activity but inactive in the two follow-up antiinflammatory screens (cotton wad granuloma and adjuvant arthritis) as shown in Table 3.

Table 3. Antiinflammatory activities of some copper complexes

Compound[a]	Carrageenan Paw Edema	Cotton Wad Granuloma	Adjuvant Arthritis	% Copper
Cu(II)$_2$(anthranilate)$_4$	A at 0.01	A at 0.035	A at 0.002	18.9
Cu(II)$_2$(3,5-DIPS)$_4$	A at 0.01	A at 0.005	A at 0.001	12.5
Aspirin	A at 0.36	A at 1.11[b]	A at 0.033	
Cu(II)$_2$(aspirinate)$_4$	A at 0.01	A at 0.01	A at 0.001	15.0
D-Penicillamine	I at 1.34	I at 0.67	I at 0.201	
Cu(I)(D-pen)	A at 0.03	A at 0.04	NT	26.7
Cu(II)$_2$(D-pen disulfide)$_2$	A at 0.01	A at 0.03	A at 0.040[c]	15.4

[a]Compounds given subcutaneously or [b]intragastrically and doses expressed as m mol/kg of body weight; A-lowest active dose tested; I-inactive; NT-not tested; [c]only dose tested.

Copper chloride had no activity in any of these models of inflammation at the very high initial screening doses of 1.18, 0.59, and 0.18 m mol/kg respectively. Ligands such as anthranilic acid and 3,5-diisopropylsalicylic acid (3,5-DIPS) which were anticipated to be inactive were found to be inactive. However, their copper complexes were found to be potent antiinflammatory agents in all three models of inflammation as shown in Table 3. These observations supported the notion that complexed copper is a more active antiinflammatory form of copper and led to the suggestion that copper complexes of antiarthritic drugs might be more active than their parent antiarthritic drugs.

Representative data from the original report comparing aspirin, a standard in arthritis therapy, and D-penicillamine, an arthritic disease modifying agent, with their copper complexes are also presented in Table 3. These data show that $Cu(II)_2(aspirinate)_4$ is at least 30 times as effective as aspirin and copper complexes of penicillamine are many times more effective than penicillamine. These results along with data provided by many others, see review by Sorenson 1989, support the hypothesis that active metabolites of antiarthritic drugs are their copper complexes. Since amounts of copper in these complexes do not appear to correlate with activity, as shown in Table 3, it is suggested that pharmacologic activity may be better correlated with physicochemical properties of copper complexes.

Acute toxicity studies were done early in the course of this work (Sorenson 1976; Sorenson 1978; Sorenson, Rolniak, Chang 1984). These studies demonstrated that antiinflammatory copper complexes were less toxic than inorganic forms of copper as well as their parent antiarthritic drugs. The projected human dose of $Cu(II)_2(aspirinate)_4$, for example, is 3 to 6 μmol/kg daily with a decrease to some maintenance dose with remission. This regimen seems to be safe enough to correct what may be, in part, a copper deficiency or inadequacy disease since the recommended safe daily intake of copper is 47 μmol/day for healthy individuals.

Two general chemical mechanisms are immediately apparent possibilities in accounting for antiinflammatory activities of copper complexes. Small molecular weight copper complexes may serve as transport forms of copper that allow activation of copper-dependent enzymes, such as peptidyl lysyl oxidase and copper-dependent superoxide dismutases (Cu-ZnSOD), or they may have chemical reactivities such as superoxide disproportionation that facilitate correction of the chemical problem that led to the disease state.

Antiinflammatory activities of copper complexes partially explain earlier observations by Fenz, Forestier, Kuzell, and Hangarter that copper complexes are effective in treating arthritic and other human degenerative diseases (Sorenson, Hangarter 1977). Also, copper complexes were found to have potent antiulcer activity and they are not ulcerogenic.

ANTIULCER ACTIVITIES

To date, over 75 copper compounds have been shown to have potent antiulcer activity in six different models of gastric ulcer (Sorenson 1989). Representative data from the original report (Sorenson 1976) are presented in Table 4.

Table 4. Oral antiulcer activity of non-steroidal antiinflammatory agent copper complexes in the Shay rat

Compound	Antiulcer Activity ED_{50} (μmol/kg)	% Copper
$Cu(II)_2(acetate)_4(H_2O)_2$	53	32
$Cu(II)_2(aspirinate)_4$	13	15
$Cu(II)_2(salicylate)_4$	6	15
$Cu(II)(butazolidine)_2$	7	9
$Cu(II)_2(D\text{-}pen\ disulfide)_2\cdot$	6	16
$Cu(II)_2(3,5\text{-}DIPS)_4$	5	13
$Cu(II)_2(niflumate)_4$	4	10
$Cu(II)_2(indomethacin)_4$	3	8

Copper$(II)_2(acetate)_4$ had only very weak antiulcer activity. However, copper complexes of non-steroidal antiarthritic agents were found to be potent antiulcer agents. There are no more potent antiulcer agents in this model of

ulcer. It is also clear that antiulcer activity is not directly related to copper content. It is most likely that antiulcer activity of copper complexes is also related to their physiochemical properties. It is most interesting to note that the amount of copper in these ED_{50} doses represents one-fourth to one-fifteenth the recommended daily intake of copper.

A subsequent comparison of antiulcer activities of three different penicillamine complexes and amino acid complexes in the Shay ulcer model revealed that the water soluable mixed valence penicillamine complex, $Na_5Cu(I)_8Cu(II)_6$ (Penicillamine)$_{12}$ Cl, and $Cu(II)(glycinate)_2$ were the most effective of all amino acid complexes studied (Sorenson, Ramakrishna, Rolniak 1982; Kishore, Rolniak, Ramakrishna, Sorenson 1982; West 1982a).

The principal physiological mechanism of action used to account for these observations is antihistaminic-antisecretory activity (Sorenson 1976; Marletta, Rizzarelli, Mangiameli, Alberghina, Brogna, Sammartano, Blasi 1977; Alberghina, Brogna, Mangiameli, Marletta, Rizzarelli, Sammartano 1982). This mechanistic action has been confirmed and extended by West and his colleagues with reports showing that $Cu(II)_2(aspirinate)_4$ and $Cu(II)_2(salicylate)_4$ were more effective in the aspirin exacerbated Shay ulcer model than H_1(Mepyramine) or H_2(Metiamide) histamine blockers (Hayden, Thomas, West 1978). It was also shown that $Cu(II)_2(salicylate)_4$ was more effective than these antihistaminic compounds and cimetidine in the Cold-Stress ulcer model (Hayden, Thomas, West 1978; West 1982b) and $Cu(II)_2(salicylate)_4$ was more effective than cimetidine in preventing aspirin and indomethacin induced ulcers (Table 5) (West 1982b).

Table 5. Comparison of copper(II)$_2$(salicylate)$_4$ and cimetidine, given orally, in protecting against gastric lesions produced by oral ulcerogenic doses of aspirin (2.22 mmol/kg), indomethacin (56 μmol/kg), or Cold-Stress in rats (West 1982)

Ulcerogen	Antiulcer Agent	Dose (μmol/kg)	% inhibition (M$\pm$SE)
Aspirin	Cimetidine	148	19 $\pm$ 5
	Cimetidine	297	16 $\pm$ 4
	$Cu(II)_2(Salicylate)_4$	11	21 $\pm$ 6
	$Cu(II)_2(Salicylate)_4$	22	61 $\pm$ 10[a]
Indomethacin	Cimetidine	148	10 $\pm$ 3
	Cimetidine	297	24 $\pm$ 6
	$Cu(II)_2(Salicylate)_4$	11	37 $\pm$ 7[a]
	$Cu(II)_2(Salicylate)_4$	22	50 $\pm$ 5[a]
Cold-Stress	Cimetidine	148	38 $\pm$ 8[a]
	Cimetidine	297	61 $\pm$ 9[a]
	$Cu(II)_2(Salicylate)_4$	11	58 $\pm$ 8[a]
	$Cu(II)_2(Salicylate)_4$	22	79 $\pm$ 7[a]

[a]Significant at $p<0.05$ (when compared with non-treated rats).

These observations are consistent with observations that copper markedly increases (50-fold) specific cimetidine bonding to brain membrane H_2 receptors (Kendall, Ferkany, Enna 1980; Kawai, Nomura, Segawa 1984) and that copper decreases compound 48/80 and concanavalin-A induced releases of histamine from peritoneal mast cells (Jande, Sharma 1986). Further explanation of the antisecretory effect comes from the work of Boyle, Freeman, Goudie, Magan, Thompson (1976) showing that the antiulcer and antisecretory activity of non-steroidal antiinflammatory agent copper complexes modulate syntheses of prostaglandins E_2 and $F_{2\alpha}$, which is consistent with other reports of copper complex modulation of prostaglandin systheses (Lee, Lands 1972; Maddox 1973; Vargaftig, Tranier, Chignard 1975; Swift, Karmazyn, Horrobin, Monku, Karmali, Morgan, Ally 1978; Cunnane, Zinner, Horrobin, Monku, Morgan, Karmali, Ally, Karmazyn 1979).

Finally, Townsend demonstrated that $Cu(II)(tryptophan)_2$ and $Cu(II)_2(aspirinate)_4$ increased the rate of healing of surgically placed glandular gastric ulcers using histochemical methods (Townsend, Sorenson 1981). These complexes prevented wound regression, increased the rate of reepithelization, allowed replacement of connective tissue components such that these tissues could not be distinguished from connective tissues of normal non-operated rats, and prevented spleen, pancreas, and liver adhesions to the stomach, a constant feature associated with non-treated surgically placed glandular gastric ulcers.

Superoxide dismutase-mimetic activity reported by many research groups, and reviewed (Sorenson 1989), for antiulcer copper complexes also offers an accounting for the reduction in number of ulcers and reduced severity of remaining ulcers in the Shay rat ulcer model, and the apparent absence of wound regression in treated surgically placed gastric wounds. Superoxide dismutase-mimetic activity of $Cu(II)_2(3,5-DIPS)_4$ was also used to account for its abolition of desoxycholate-induced colonic epithelial proliferation and to suggest an oxyradical etiology for inflammatory bowel disease (Craven, Pfanstiel, DeRubertis 1986). Based upon these observations it seems reasonable to suggest that a reduction in intestinal and gastric tissue copper-dependent superoxide dismutase (Cu-ZnSOD) and the accumulation of superoxide and other oxyradicals have an etiologic or pathogenic role in gastric and intestinal ulceration. Lipophilic copper complexes may be effective in crossing lipid cell membranes and either inducing or facilitating de novo synthesis of Cu-ZnSOD or disproportionating superoxide as a result of their own chemical reactivity.

Maintenance and repair of duodenal and gastric collagen and elastin connective tissue components also seems to be an important copper-dependent enzyme function in preventing or repairing duodenal or gastric ulcers. Induction or facilitation of de novo synthesis of peptidyl lysyl oxidase by copper complexes (Harris, DiSilvestro, Balthrop 1982) merits consideration in accounting for the observed rapid and normal replacement of connective tissue components in the surgically placed gastric ulcer model (Townsend, Sorenson 1981).

Recently recognized roles for copper-dependent peptidyl α-amidating enzymes in intestinal and gastric tissues for synthesis of choleocystokinin and gastrin, respectively, may also be important in understanding normal gastrointestinal physiology as well as accounting for antiulcer activity of copper complexes (Anonymous 1985).

ANALGESIC ACTIVITIES

The question as to whether or not copper complexes have analgesic activity has been answered in the affirmative. It has recently been reported that copper complexes of non-steroidal antiarthritic agents are more effective analgesics than their parent drugs (Okuyama, Hashimoto, Aihara, Willingham, Sorenson 1987). As shown in Table 6, copper complexes of salicylic acid, 3,5-DIPS, aspirin, niflumic acid, and indomethacin were more effective analgesics than their parent compounds and copper chloride or copper acetate.

Table 6. Analgesic effects of ligands and complexes on acetic acid-induced writhing pain in mice and adjuvant-induced arthritis pain in rats

Compound	Writhing Pain ED_{50} (mmol/kg)[a] [95% C.L.]	Adjuvant Arthritic Pain ED_{50} (mmol/kg)[a] [95% C.L.]
Salicylic acid	>2.17	1.83 [0.49-6.81]
$Cu(II)_2(salicylate)_4$	1.53 [0.61-3.80]	0.25 [0.13-0.48]
3,5-DIPS	0.4 [0.29-0.67]	>0.90
$Cu(II)_2(3,5-DIPS)_4$	0.22 [0.10-0.45]	>0.20
Aspirin	0.48 [0.22-1.04]	1.41 [0.50-3.96]
$Cu(II)_2(aspirinate)_4$	0.14 [0.09-0.25]	0.09 [0.04-0.21]
Niflumic acid	0.21 [0.11-0.40]	0.48 [0.19-1.20]
$Cu(II)_2(niflumate)_4$	0.10 [0.05-0.20]	>0.16
Indomethacin	0.01 [0.01-0.03]	0.02 [0.01-0.03]
$Cu(II)_2(indomethacin)_4$	0.002[0.001-0.003]	0.002[0.001-0.003]
$Cu(II)(chloride)_2$	-----	>2.24
$Cu(II)_2(acetate)_4$	-----	>0.83
Morphine Hydrochloride[b]	0.002[0.001-0.003]	0.002[0.001-0.005]

[a]administered orally in 5% propylene glycol and 1.4% polyvinyl alcohol in water or [b]subcutaneously in saline.

In addition, the copper complex of indomethacin was found to be as effective as morphine in both pain models. Amounts of copper in these ED_{50} doses range down to one-twenty-fifth the recommended safe daily intake of copper. These observations also support the hypothesis that copper complexes are active analgesic forms, formed in vivo, of the non-steroidal antiarthritic drugs and the hypothesis that these complexes activate opioid receptors.

CONCLUSION

Potent antiarthritic and antiulcer activities without ulcerogenic activity (Williams, Walz, Foye 1976; Milanino, Concari, Conforti, Marrella, Franco, Moretti, Velo, Rainsford, Bresson 1988), coupled with potent analgesic activity, distinguish copper complexes as a unique class of antiarthritic agents since all other antiarthritic drugs do not have antiulcer activity but are ulcerogenic. All of the above mechanistic considerations offer support for the possibility that pharmacologic uses of copper complexes represent a physiologic approach to treatment of inflammatary diseases, ulcers, and pain. A search for more bio-effective copper complexes would seem to be a valid approach to developing more effective antiarthritic drugs.

ACKNOWLEDGEMENTS

I am indebted to the International Copper Research Association, Max and Victoria Dreyfus Foundation, Denver Roller Corporation, and the Elsa U. Pardee Foundation for financial support.

REFERENCES

Alberghina M, Brogna A, Mangiameli A, Marletta F, Rizzarelli E, Sammartano S (1982) Copper(II) complexes of amino acids: Gastric acid anti-secretory activity in rats. Il Farmaco 12:805-814

Anonymous (1985) Newly found roles for copper. Nutr Revs 43:117-119

Boyadzhyan AS (1985) Purification of dopamine-β-monooxygenase and the extremely acidic copper-containing protein from the adrenal medulla: The extremely acidic copper-containing protein as an electron donor for dopamine-β-monooxygenase. Biochem 50:75-81

Boyle E, Freeman PC, Goudie AC, Magan FR, Thomson M (1976) Role of copper in preventing gastrointestinal damage by acidic antiinflammatory drugs. J Pharm Pharmacol 28:865-868

Craven PA, Pfanstiel J, DeRubertis FR (1986) Role of reactive oxygen in bile salt stimulation of colonic epithelial proliferation. J Clin Invest 77:850-859

Cunnane SC, Zinner H, Horrobin DF, Monku MS, Morgan RO, Karmali RA, Ally AI Karmazyn M (1979) Copper inhibits pressor responses to noradrenaline but not potassium: Interactions with prostaglandins E_1, E_2, and I_2 and penicillamine. Can J Physiol Pharmacol 57:35-40.

Davenport HC (1982) Physiology of the digestive tract, 5th ed. Year Book. Medical Publishers, Chicago, pp 132-133

Fridovich I (1986) Biological effects of the superoxide radical. Arch Biochem Biophys 247:1-11

Frieden E (1986) Perspectives on Copper Biochemistry. Clin. Physiol Biochem 4:11-19

Hamilton GA (1981) Oxidases with monocopper reactive sites. In: Spiro TG (ed) Copper Proteins. Wiley, New York, pp 193-218

Harris ED, DiSilvestro RA, Balthrop JE (1982) Lysyl oxidase, a molecular target of copper. In: Sorenson JRJ (ed) Inflammatory diseases and copper. Humana Press, Clifton, NJ, pp 183-198

Hayden LJ, Thomas C, West GB (1978) Inhibitors of gastric lesions in the rat. J Pharm Pharmacol 30:244-246

Iyengar GV, Kollmer WR, Bowen HJM (1978) The elemental composition of tissues and body fluids. Springer, New York

Jande MB, Sharma SC (1986) The effect of cupric sulfate on compound 48/80 and concanavalin-A induced release of histamine from rat peritoneal mast cells. Brit J Pharmacol 89:570P

Kawai M, Nomura Y, Segawa T (1984) Elevation of [^{3}H]cimetidine binding by $CuCl_2$ in brain membranes of rats. Neurochem. Internat 6:563-568

Kendall DA, Ferkany JW, Enna SJ (1980b) Properties of ^{3}H-cimetidine binding in rat brain membrane fractions. Life Sci 26:1293-1302

Kishore V, Rolniak TM, Ramakrishma K, Sorenson JRJ (1982) The antiulcer activities of copper complexes. In: Sorenson JRJ (ed) Inflammatory diseases and copper. Humana Press, Clifton, NJ, pp 363-373

Korolkiewicz Z, Hac E, Gagalo I, Gorczyca P, Lodzinska A (1989) The pharmacologic activity of complexes and mixtures with copper and salicylates or aminopyrine following oral dosing in rats. Agents and Actions 26:355-359

Lee RE, Lands WEM (1972) Cofactors in biosynthesis of prostaglandins F_1-alpha and F_2-alpha. Biochim Biophys Acta 260:203-211

McCord JM, Fridovich I (1969) Superoxide dismutase. An enzymatic function for erythrocuprein (hemocuprein). J Biol Chem 244:6049-6055

Maddox IS (1973) Role of copper in prostaglandin synthesis. Biochim Biophys Acta 306:74-81

Marklund S (1982a) Human copper-containing superoxide dismutase of high molecular weight. Proc Natl Acad Sci 79:7634-7638

Marklund S, Holme E, Hellner L (1982b) Superoxide dismutase in extracellular fluids. Clin Chem Acta 126:41-51

Marletta F, Rizzarelli E, Mangiameli A, Alberghina M, Brogna A, Sammartano S, Blasi A (1977) Atti XXI Congress Naz Soc Ital Gastroenterol, Bologa, Rendic Gastroenterol 9 (Suppl 1) 35

May PM, Linder PW, Williams DR (1976) Ambivalent effect of protein binding on computed distributions of metal ions complexed by ligands in blood plasma. Experientia 32:1492-1493

Milanino R, Concari E, Conforti A, Marrella M, Franco L, Moretti U, Velo G, Rainsford KD, Bressan M (1988) Synthesis and antiinflammatory effects of some bis(2-benzimidazolyl) thioethers and their copper(II) chelates, orally administered to rats. Eur J Med Chem 23:217-224

O'Dell BL (1976) Biochemistry and physiology of copper in vertebrates. In: Prasad AS, Oberleas, D (eds) Trace elements in human health and disease, Vol. I, Zinc and copper. Academic Press, New York, pp 391-413

Okuyama S, Hashimoto S, Aihara H, Willingham WM, Sorenson JRJ (1987) Copper complexes of non-steroidal antiinflammatory agents: Analgesic activity and possible opioid receptor activation. Agents and Actions 21:130-144

Sorenson JRJ (1976) Copper complexes as possible active forms of the antiarthritic agents. J Med Chem 19:135-148

Sorenson JRJ (1978) Copper complexes, a unique class of antiarthritic drugs. Prog in Med Chem 15:211-260

Sorenson JRJ (1982) The antiinflammatory activities of copper complexes. In: Siegel H (ed) Metal ions in biological systems Volume 14, Marcel Dekker, New York, pp 77-124

Sorenson JRJ (1989) Copper complexes offer a physiological approach to treatment of chronic diseases. In: Ellis GP, West GB (eds) Progress in Medicinal Chemistry, Volume 26, Elsevier Science Publishers, Amsterdam, pp 437-568

Sorenson, JRJ, Hangarter W (1977) Treatment of rheumatoid and degenerative diseases with copper complexes. Inflammation 2:217-238

Sorenson JRJ, Ramakrishma K, Rolniak TM (1982) Actiulcer activity of D-penicillamine copper complexes. Agents and Actions 12:408-411

Sorenson JRJ, Oberley LW, Crouch RK, Kensler TW (1984) Pharmacologic activities of SOD-like copper complexes. In: Bors W, Saran M, Tait D (eds) Oxygen radicals in chemistry and biology, Walter de Gruyter and Co., Berlin, pp 821-830

Sorenson JRJ, Rolniak TM, Chang LW (1984) Preliminary chronic toxicity study of copper aspirinate. Inorg Chim Acta 91:L31-L34

Swift A, Karmazyn M, Horrobin DF, Monku M, Karmali RA, Morgan RO, Ally AI (1978) Low prostaglandin concentrations cause cardiac rhythm disturbances: Effect reversed by low levels of copper or chloroquine. Prostaglandins 15:651-657

Tipton IH, Cook MJ (1963) Trace Elements in Human Tissue. II: Adult subjects from the United States. Health Phys 1:103-145

Townsend SF, Sorenson JRJ (1981) Effect of copper aspirinate on regeneration of gastric mucosa following surgical lesion. In: Rainsford KD, Brune K, Whitehouse MW (eds) Trace elements in the pathogenesis and treatment of inflammatory conditions, Agents and Actions Supplement 8, Birkhauser Verlag, Basel, pp 389-398

Underwood EJ (1977) Trace elements in human and animal nutrition, 3rd edn. Academic Press, New York

Vargaftig BB, Tranier Y, Chignard M (1975) Blockade by metal complexing agents and by catalase of effects of archidonic acid on platelets-relevance to study of antiinflammatory mechanisms. Eur J Pharmacol 33:19-29

West GB (1982a) The copper problem and amino acids. In Sorenson JRJ (ed) Inflammatory diseases and copper. Humana Press, Clifton, NJ, pp 319-327

West GB (1982b) Testing for drugs inhibiting the formation of gastric ulcers. J Pharmacol Methods 8:33-37

Williams DA, Walz DT, Foye WO (1976) Synthesis and Biological Evaluation of tetrakis-μ-acetylsalicylatodicopper(II). J Pharm Sci 65:126-128

Genetic Disorders

Genetic Disorders of Copper Transport: Menkes' Disease, Occipital Horn Syndrome, and Wilson's Disease

DAVID M. DANKS

Murdoch Institute, Royal Children's Hospital, Melbourne 3052, Australia

KEY WORDS: Menkes' disease, Occipital horn syndrome, Wilson's disease, Copper

Unfortunately, these three genetic diseases have yet to contribute to our understanding of copper transport the revelations that we geneticists expect from "experiments of nature". Nonetheless, I still have faith that the resolution of the basic defects in these conditions will eventually play a critical role in giving us a complete understanding of this process. Therefore we battle on.

The involvement of copper as the cause of liver and brain damage in Wilson's disease (WD) was first proposed in the 1930's and has been well established for over 40 years. In 1973, soon after we found a disturbance of copper transport in Menkes' disease (MD), I was rash enough to suggest that we would soon identify the basic defect. This was because we had found expression of the defect in cultured fibroblastic cells and this seemed to give us a great advantage over those studying WD in which the secret seemed to be locked away in the liver. When David Hunt showed that the brindled mice had a similar disturbance in copper transport, we seemed ideally placed. What more could one ask than an in vitro system and an animal model. Even the X-chromosomal location of both genetic defects offered an advantage. My only excuse is that I forgot the grave drawback of working with an isotope with a 12 hour half-life. Apart from this, we have just not proved clever enough. Fifteen years later we know little more about Menkes' disease than we knew in 1974.

Progress with WD has also been slow and, if I were a betting man, I would still put my money on MD to be the first of the two to be fully understood.

The two have been joined by a third genetic defect of copper transport - the occipital horn syndrome (OHS) previously called X-linked cutis laxa or Ehlers-Danlos syndrome type IX . However, it may prove to be merely a variant of MD.

It seems very likely that the transport of copper between organs and within cells involves many more than three active processes, so there should be many more than three genetic diseases due to defects in copper transport.

I will give my views of the current understanding of the three diseases, comment on their treatment and then mention some other relevant disorders of copper transport in man and animals. Only new or contentious findings will be referenced. For a general review, see Danks (1989).

MENKES' DISEASE AND RELATED CONDITIONS.

Classical Menkes' Disease

The disease described in 1962 by John Menkes has now been recognized in most populations with a frequency of approximately 1 in 70,000 to 1 in 100,000 births. It is a progressive disease of early infantile (even neonatal) onset, generally lethal within three years, although a few patients have lived on with profound brain damage for as long as 14 years (Gerdes et al, 1988). These long survivors of the classical form of the disease should not be confused with two boys with a mild form of the disease (see below) or with three boys whose disease has been partly controlled by treatment with copper histidinate (Nadal and Baerlocher, 1988; Sherwood et al (in press).

Premature birth due to early rupture of membranes is the first indication of the connective tissue abnormality which is due to lysyl oxidase deficiency and responsible for the lax skin, loose joints, arterial elongation and rupture, bladder diverticulae, hernias, emphysema detected at autopsy and probably for the osteoporosis, metaphyseal flaring, fractures and Wormian bones. Lysyl oxidase deficiency is demonstrable in skin samples and in fibroblasts in culture in which it is not corrected by addition of copper to the medium.

The neurological degeneration is the other major feature of the disease and it is not clear how much of this is attributable to the three candidate culprit enzymes - cytochrome oxidase, dopamine-B-hydroxylase and superoxide dismutase. I would guess that this listing is in descending order of importance. The reason for the very pronounced damage to Purkinje cells is not known, but is seen in other diseases affecting cellular energy availability. Functionally less important symptoms like pili torti and depigmentation can be attributed to the role of copper in disulphide bonding in hair keratin, and in tyrosinase.

The defective transport of copper is already established in utero and some of the features of the disease are present at birth. Copper transport across the placenta is impaired in the mutant mice, but less impaired than intestinal absorption so that the access of the whole body to copper diminishes after birth.

The copper is distributed abnormally between the organs - concentrated in the kidney and intestinal mucosa, grossly reduced in liver and reduced to a lesser extent in brain and other tissues. After injection of copper the levels in most tissues, other than the brain, rise above normal. Levels in the liver do rise to normal eventually and can be raised to supranormal levels if enough is administered. Most of these statements are based on studies in the mutant mice, but some come from experience in humans. In organs with high copper levels and in cultured cells which accumulate excessive levels of copper, the metal is found to be bound to metallothionein (MT) after homogenisation and is probably so bound in life.

The simplest interpretation of these findings, and the one I still favour, is that most organs other than the liver, express a defect which diverts copper from copper enzymes to MT and that copper which would normally be stored in the liver ends up in the various other organs according to their access to the metal - greatest in intestinal mucosa and in renal tubules (as a consequence of filtration and reabsorption, probably bound to amino acids). This explanation seems to bring together the findings in the whole mouse (and human as judged by less detailed evidence) and the observations in cultured fibroblastic cells, lymphocytes and mouse kidney epithelial cells.

Several of the possible mechanisms which might explain such a defect can be dismissed. The defect is not in the MT structural gene (chromosome 16 in man and chromosome 8 in mouse, not X-chromosomes) and we have found no disturbance in control of MT mRNA production has been found in liver, kidney or cultured cells. Initial uptake of copper into lymphocytes is normal and less reliable methods haved failed to show any defect in egress from the pool of available cellular copper. (Herd et al, 1987) A defect in intracellular transport which leaves copper in a form which displaces zinc from MT and induces MT production seems most probable.

Some believe that the low copper level in the liver is primary and that a defect in the mechanism of storage in this organ allows copper to redistribute to other organs where the excess is bound by MT. If this is true, then the defect must prevent hepatic storage without affecting availability to copper enzymes which are normal in activity despite the very low copper levels in brindled and blotchy mouse liver (Phillips et al, 1986) Caeruloplasmin levels are very low, but injected copper quickly restores these to normal in patients and mutant mice. Copper levels in liver can be restored to normal (Danks, 1988) and I have seen levels as high at 600ug/g dry weight achieved in a baby with the disease by overzealous administration of copper parenterally.

Nonetheless, one must take note of a recent observation that at 55kD copper binding protein is lacking in brindled mouse liver (Palida et al, 1988) and that the same research group of Murray Ettinger has reported reduced copper accumulation in primary hepatocyte preparations from brindled mice. (Darwish et al, 1983). Still one is left having to explain the excessive accumulation of copper in three classes of cells in culture. If expression of the defect in liver is important, it cannot be the only organ primarily affected.

Mild Menkes' Disease

The two boys who have been described as having a mild form of MD presented much later, and in a quite different manner. Our patient presented at two years with gross ataxia, pili torti and mild skin and joint changes. The ataxia made it difficult to assess intellectual development. He is now 10 years old and of borderline normal intellect. (Danks, 1988) Major arteries were elongated and mildly dilated. Serum copper was only slightly reduced, but liver and intestinal mucosal levels were low and high, respectively, as expected, and results in cultured fibroblasts could not be distinguished from those in the classical form.

The second case from USA presented with intestinal and urinary symptoms, has severe connective tissue changes involving skin, joints, herniae, bladder diverticulae (Westman et al, 1988). He has abnormal arteries. His intellectual abilities are also borderline normal, but he has very little ataxia. One might even ask whether this boy has a more severe variant of the OHS.

Both these boys are isolated cases in their families, but placental copper levels were high in a subsequent female pregnancy in our family as is found in heterozygous females with the classical MD mutation.

Occipital Horn Syndrome (OHS)

I have no personal clinical experience of this condition and find most of the publications frustrating for their failure to evaluate all the features of copper deficiency which need to be demonstrated or excluded. The patients have inguinal herniae, bladder diverticulae and urinary infection, lax skin and joints, occipital exostoses beside the foramen magnum (giving the syndrome its name), an unusual hammer-like expansion of the lateral ends of the clavicles and a wavy outline to the cortex of many long bones. (Sartoris et al, 1984) Arterial elongation, tortuosity and stenoses has been described in one patient (Sartoris et al, 1984). Intellect is apparently low normal, and hair is apparently abnormal (I. Kaitila, personal communication), but published details are lacking. Pigmentation has not been mentioned. Inheritance is X-chromosomal.

Serum copper has been slightly reduced in most cases, but neither liver nor intestinal copper levels has been reported. Lysyl oxidase is greatly reduced in skin and cultured fibroblasts, but other copper enzymes have not been assayed. Copper therapy does not seem to have been used.

Relationship Between Classical MD, Mild MD And OHS

If one accepts the high placental copper in the sister of our boy with mild MD as sufficient evidence, all three conditions are X-linked, and may therefore be allelic. The linkage data on classical MD is inconclusive and compatible with location on the proximal p or q arm of the X-chromosome. The occurrence of classical MD in a girl with an X-autosome translocation with breakpoint at Xq13 strongly suggests this as the locus of this gene (Kapur et al, 1987). This would put the gene close to that for phosphoglycerate kinase (PGK), as is true of the mottled locus in the mouse. No linkage data has been published on OHS (Kuivaniemi et al, 1985).

We have been unable to find any difference in the phenotype in cultured fibroblasts between classical MD, mild MD (both cases) and OHS (Finnish and Californian cases). The Finnish group has reported similar results in comparing classical MD and OHS. The same is true with the mouse mutants, brindled and blotchy, in our hands. However, the copper changes in the tissues do differ quantitatively, being more severe (both the reduction in liver and the increase in kidney) in the brindled male which has severe and fatal neurological symptoms, but no connective tissue changes (even when kept alive long beyond the usual 14 day age of death by a copper injection on day 7). By contrast, the blotchy male has no neurological symptoms, but does develop emphysema, and arterial disease.

It has been usual to speak of the "mottled series of allelic mutants", but there is no proof that these mutants are allelic, as opposed to occurring at closely linked loci. Six mutants have been known for some years - mottled, dappled, tortoise-shell, brindled, viable - brindled and blotchy - listed in descending order of severity. Recently, confusion has been introduced by the description from Japan of a macular mutant which sounds very like brindled. The authors have largely ignored the importance of determining whether it is allelic to the other mutants or maps at another locus altogether - this point would be easily determined because PGK is close to Mo in the mouse.

While allelism of all 7 mouse mutants and of the 3 human mutants and homology of the loci concerned in the two species is most likely, there are some strange differences which may point to two loci perhaps controlling two sequential steps in intracellular copper transport - one delivering copper to lysyl oxidase only, the other important for all copper enzymes.

Treatment

Copper histidinate seems the most logical form to administer by injection as it is probably the form actually taken up by the liver and it seems to have given the best results. Solubility at an acceptable pH is the main problem, but can be achieved with appropriate buffers although the solution is stable for only four weeks.

Results of treatment of classical MD after the usual presentation with severe neurological symptoms have never been satisfactory. Some physical benefit may occur, but no control or reversal of neurological damage is seen. It is better not to treat these babies with copper.

Treatment started soon after birth has been much more successful, producing children with low normal intellectual ability and some mild skin and joint laxity. Two groups have used penicillamine along with copper histidinate - I cannot understand the logic behind this regime. Of course, the opportunity to treat in this way arises only in families with a previous affected child and a strong opposition to termination of pregnancy. In my view, the results are not good enough to put copper injections forward as a serious alternative to termination of pregnancy, except in this circumstance.

Brindled mice recover well if injected with copper on day 7, but not when treated on day 10. This may indicate a phase of critical copper requirement for brain development between these stages or may merely be telling us that the mice have irreversible general effects by day 10 (usual survival, untreated, 14 days). We have treated one boy with copper histidinate from day 2 after delivery induced at 35 weeks. The boy (a half-brother of a baby who died of classical MD) is doing well, but has some ataxia and mild delay in development at the age of 2½ years. Now that we have some idea of the dose needed at this stage of development, we might try intrauterine copper injection at 30 weeks if the opportunity presents again. We have used serum copper to tell if we are giving enough copper and liver copper to check that we are not giving too much. Sherwood et al (in press) measured urinary copper excretion. Given that Menkes' disease causes cupruria, it is hard to know what level of excretion to use as a target. The ideal would be to monitor several different copper enzymes, as well as copper levels.

We have treated our mild case with copper histidinate with similar monitoring. It is not possible to know whether it has been effective or necessary (Danks, 1988). The other case has started treatment recently following demonstration of arterial abnormalities. None of the OHS cases seems to have been treated.

Future Directions - Pursuit Of The Basic Defect

One would have hoped that the group with access to cells from the girl with the X-autosome translocation would have cloned the breakpoint during the last 3 years or have shared the cells with others. (Kapur et al, 1987) Neither has happened. Establishment of a close linkage to the MD locus might allow chromosome walking , or examination for deletions by pulse field electrophoresis. Few genes are known in the Xq13 region. PGK is the most obvious starting point, but has not been informative in families available to us.

Several groups continue to search for proteins which are altered in, or missing from, cells or tissues in MD.

Related Conditions

We have studied copper uptake and release in cultured fibroblasts from a number of patients with unusual and unclassified connective tissue disorders, but all have proved normal. In several patients we have been misled by the coexistence of a moderate degree of hypocaeruloplasminaemia, but it has proved to be inherited from a normal parent and was presumably present by coincidence.

WILSON'S DISEASE AND RELATED CONDITIONS

Wilson's Disease

It has long been apparent that the two most fundamental disturbances in copper transport are a decrease in biliary excretion and a reduction in the rate of incorporation into caeruloplasmin. Studies of predicted copper compartments in rat and human liver have not provided an explanation of this combination and difficulty of access to human liver samples has prevented a search for a mutant protein in liver or for a protein missing from the liver. For a time it seemed that a phenotype which could be exploited for research was present in cultured fibroblastic cells, but none of the three groups who described this phenotype has been able to use it in this way. The existence of any phenotypic effect in these cells remains surprising to me because everything else suggests a liver specific defect with "overflow" to other organs.

The most interesting recent finding is the description of proteolytic fragments of caeruloplasmin in bile and the lack of these fragments in the WD (Iyengar et al, 1988). This implies that uptake of aged caeruloplasmin by the liver, partial proteolysis and excretion of the resulting fragments in the bile may be an important part of copper homeostasis. Evidence of uptake of desialated glycoprotein by liver has been available for some years so this mechanism is not totally unexpected. This process could allow a primary defect in caeruloplasmin synthesis to interfere with both incorporation of copper into caeruloplasmin and biliary excretion. The primary defect cannot be in the caeruloplasmin gene itself because it is on chromosome 3 in humans and WD maps to chromosome 13. Also, there are some patients with normal levels of caeruloplasmin.

An alternative could be a defect in the mechanism of uptake and excretion of caeruloplasmin by the liver with secondary suppression of synthesis. This hypothesis might fit better with the very variable levels of caeruloplasmin found in WD. It would predict a prolonged life for caeruloplasmin in the circulation in WD which is logically necessary to explain the combination of low copper incorporation with near normal plasma levels of caeruloplasmin in some patients.

Of course, the other major advance of recent years has been the mapping of the WD gene to 13q14 close to esterase D and the locus affected in retinoblastoma. Several groups are trying to walk to the gene, but this is not easy to do.

The final new hope of recent years has been the recognition of the toxic milk mouse (Rauch, 1986) which seems more likely to be homologous to WD than the liver disease of the Bedlington terrier. Studies in these mice and in several other systems are pointing to either a copper transport role for superoxide dismutase, or the existence of another hepatic protein of similar molecular weight which is important in transport. The disturbance of secretion of copper in the milk of these mice highlights a deficiency of information about the copper content of milk in WD.

Treatment

Penicillamine has suffered some setbacks as the drug of choice because of more widespread recognition of acute worsening of neurological symptoms in the first days of treatment (Starosa-Rubinstein et al, 1987) and doubt about the degree of decoppering of the liver achieved. The exacerbation of neurological symptoms has been recognised by some for many years, but others had strenuously denied it until recently. Very gradual introduction over several weeks may avoid this effect, but also delays benefit even longer than is already the case with this very slowly acting drug. This phenomenon has not been described yet with Trien (triethylamine diamine), but it has not been used extensively as the initial drug. Our experience in one case suggests a possible role for tetrathiomolybdate (TTM). This was a young woman with severe dystonia due to copper retention secondary to cholestatic liver disease, who deteriorated abruptly after one day on penicillamine and recovered well on TTM (see below).

The suggestion that penicillamine may not fully decopper the liver was raised because many WD patients who have ceased treatment after many years have relapsed into liver failure surprisingly quickly (in less than two years). A small series of liver biopsy assays after some years on treatment seemed to support this view (Scheinberg et al, 1987), but more data is needed. Our studies of hepatocytes in culture showed neither removal of copper by penicillamine nor prevention of uptake by these cells, but some induction of MT which may reduce the toxicity of the metal. (McArdle et al, 1989). These studies were performed on normal mouse hepatocytes and need to be repeated on toxic milk hepatocytes.

Zinc sulphate is gaining credibility in the treatment of WD, even for initial therapy. It is probable that zinc would induce hepatic MT more powerfully than penicillamine and may act at this level, as well as by blocking intestinal absorption.

Tetrathiomolybdate (TTM) is an extremely potent copper binder and certainly does reduce hepatic copper levels - from greater than 4500ug/g dry weight to 176ug/g over 18 months in our patient with cholestatic liver disease. Walshe (1986) has found it effective in five patients, but has warned of bone marrow depression. By our experience, his dosage was excessive and may have induced copper deficiency. TTM is capable of causing copper deficiency in a few days in normal animals and must be used with great care and close monitoring. It should be regarded as a promising experimental agent. We have used it in acute hepatic failure in WD with some definite benefit, but have yet to achieve survival in this desperate phase of the disease.

Liver transplantation is being used more and more in WD and can be life saving in acute liver failure, or end stage chronic liver damage. It seems unfortunate to approach such a treatable disease with such a crude method. On the other hand, the degree of recovery of neurological damage after transplantation is impressive.

Related Conditions

Copper accumulation secondary to cholestatic liver disease has been known for many years, especially in primary biliary cirrhosis, in which the benefits of penicillamine therapy have been dubious. Overflow of copper to other organs has been limited to Kayser-Fleischer rings in a few cases in this condition. We have now seen severe copper accumulation in, and damage to, basal ganglia in two young people with cholestatic liver disease. A girl who had cholestasis and lymphoedema from birth died at $5\frac{1}{2}$ years of chronic liver disease and destruction of basal ganglion. The copper accumulation was demonstrated in life, but parents refused treatment. Basal ganglion copper content was 256ug/g dry weight at autopsy. The disease resembled Aagenaes syndrome except that progressive liver disease is rare in this Norwegian condition and hepatic copper accumulation has not been found (O. Aagenaes, personal communication).

Recently we have seen severe progressive dystonia in a young woman who had suffered cholestatic liver disease from three months to 15 years, losing itch and jaundice after puberty. We had documented copper accumulation in liver (> 2000ug/g dry weight) and were treating her with penicillamine. She move to an adult hospital and later penicillamine was stopped. Neurological symptoms began at 16 years. We saw her again at 19 years, confined to a wheelchair. After acute deterioration on penicillamine she has done well on TTM and is now walking again. Liver copper was 4500 ug/g after several months on TTM and is now 176ug/g.

Indian childhood cirrhosis is now attributed to high copper intake and it seems that this high intake must start during the physiological phase of inefficient biliary copper excretion in early infancy if this

consequence is to occur. The condition has been described in Australian and German families. An additional genetic defect in hepatic copper excretion in Indian children has not been absolutely excluded, but is unlikely.

Toxic milk mice have been mentioned already, as have Bedlington terriers. Little new have been reported about the latter in recent years.

REFERENCES

Danks DM (1988) The mild form of Menkes disease: progress report on original case. Am J Med Genet 30: 859-864

Danks DM (1989) Disorders of copper transport. In: Scriver CR, Beaudet AL, Sly WS, Vallee D (eds) The metabolic basis of inherited disease, 6th edn. McGraw Hill, New York

Darwish HM, Hoke JE, Ettinger MJ. (1983) Kinetics of Cu(II) transport and accumulation by hepatocytes from copper-deficient mice and the brindled mouse model of Menkes disease. J of Biol Chem. 258, pp 13621-13626.

Gerdes AM, Tonnesen T, Pergament E, Sander C, Baerlocher KE, Wartha R, Guttler F, Horn N. (1988) Variability in clinical expression of Menkes syndrome. Eur J Pediatr; 148: 132-5

Herd SM, Camakaris J, Christofferson R, Wookey P, Danks DM. (1987) Uptake and efflux of copper-64 in Menkes' disease and normal continuous lymphoid cell lines. Biochem J 247: 341-347.

Iyengar V, Brewer GJ, Dick RD, Owyang C. (1988) Studies of cholecystokinin-stimulated biliary secretions reveal a high molecular weight copper-binding substance in normal subjects that is absent in patients with Wilson's disease. J Lab Clin Med, 111: 267-74.

Kapur S, Higgins JV, Delp K, Rogers B. (1987) Menkes syndrome in a girl with X-autosome translocation. Am J Med Genet 26: 503-510.

Kuivaniemi H, Peltonen L, Kivirikko KI. (1985) Type IX Ehlers-Danlos syndrome and Menkes syndrome: the decrease in lysyl oxidase activity is associated with a corresponding deficiency in the enzyme protein. Am J Hum Genet 37: 798-808

McArdle H, Gross SM, Creaser I, Sargeson AM, Danks DM. (1989) Am J Physiol 256: G667-G672

Nadal D, Baerlocher K. (1988) Menkes' disease: long-term treatment with copper and D-penicillamine. Eur J Pediatr; 147: 621-5.

Palida F, Waldrop G, Plonergan, Ettinger M. (1988) Cytosolic Cu-binding components and the brindled mouse defect. In: Hurley LS, Keen CL, Lonnerdal B, Rucker RB. (eds) Trace elements in man and animals 6. Plenum Press, New York p. 297-298.

Phillips M, Camakaris J, Danks DM. (1986) Comparisons of copper deficiency states in the murine mutants blotchy and brindled. Changes in copper-dependent enzyme activity in 13 day old mice. Biochem J 238: 23-27.

Rauch H, Dupuy D, Stockert RJ, Sternlieb I. (1986) Hepatic copper and superoxide dismutase activity in toxic milk mutant mice. In: Rotilio G, (ed) Superoxide and superoxide dismutase in chemistry, biology and medicine.   Elsevier.

Sartoris DJ, Luzzutti L, Weaver DD, Macfarlane JD, Hollister DW, Parker BR (1984) Type IX Ehlers-Danlos syndrome: a new variant with pathogenic radiographic features. Radiology 152:665-670

Scheinberg IH, Sternleib I, Schilsky M, Stockert RJ, (1987) Penicillamine may detoxify copper in Wilson's disease. Lancet 1: 95.

Sherwood G, Sarkar B, Sasskortsak A. Copper histidinate therapy in Menkes' disease: prevention of progressive neurodegeneration. J Inher & Metab Dis (in press)

Starosta-Rubinstein S, Young AB, Kluin K. (1987) Clinical assessment of 31 patients with Wilson's disease. Arch Neurol 44, 365-370.

Walshe JM (1986) Tetrathiomolybdate (MoS$_4$) as an anti-copper agent in man. In: Scheniberg IH, Walsh JM (eds) Orphan diseases/orphan drugs. Manchester Uni. Press.

Westman JA, Richardson DC, Rennert OM, Morrow G. (1988) Atypical Menkes steely hair disease. Am J Med Genet 30: 853-8.

Prevention Possibility for Brain Dysfunction in Menkes' Disease by Maternal Administration of Trace Elements

Harumi Tanaka, Tooru Kasama, and Masataka Arima

Division of Mental Retardation and Birth Defect Research, National Institute of Neuroscience, NCNP, 4-1-1 Ogawahigashi Kodaira, Tokyo, 187 Japan

ABSTRACT

Effects of prenatal supplementations of trace elements on fetal and neonatal developments in the mouse model of Menkes' disease are investigated. The maternal administration of zinc, vitamin E and copper resulted in good effects on fetal and neonatal deaths and cerebral copper levels of offspring, especially those of hemizygous males. The development of Menkes' offspring is influenced by the interaction of abnormal copper homeostasis and oxygen radical disturbances.

KEY WORDS: Menkes' kinky hair disease, brindled mouse, copper, antioxidants, prenatal-treatment

The basic metabolic defect in Menkes' disease (Menkes et al. 1962) is considered to be an aberrant copper distribution in several organs. The therapeutic administration of copper to patients, however, has so far failed to influence the course of the disease. On the other hand, the brindled mouse (Hunt 1974) is an allele in x-linked mottled mutant mice. Hemizygous males (by/y), as animal model of this disease, that die at approximately 2 weeks of age without treatment, survived with copper supplementation. One reason for the discrepant effect of copper treatment between humans and mice may be the time of treatment. The purpose of this study is to examine the effectiveness of prenatal treatment of heterozygous mothers as to the viability of hemizygous males, without affecting normal offspring, by using the brindled mutant. Two examinations are conducted ; A and B.

A. Maternal Abnormal Movements of Brindled Mutant Mouse Heterozygotes as Related to the Development of the Offspring

MATERIALS AND METHODS - A

The female heterozygous and normal mice were mated with normal males (Fig. 1). Pregnant mice received four kinds of drinking fluid , ad libitum ; 1) tap water, 2) 20 ppm zinc, and 3) 0.004% α-tocopherol acetate were given throughout pregnancy and lactational period, and 4) 6 ppm copper was given from gestational day 13 to delivery. After spontaneous delivery, the offspring were examined as to litter size and genotypes at 5 days after birth.

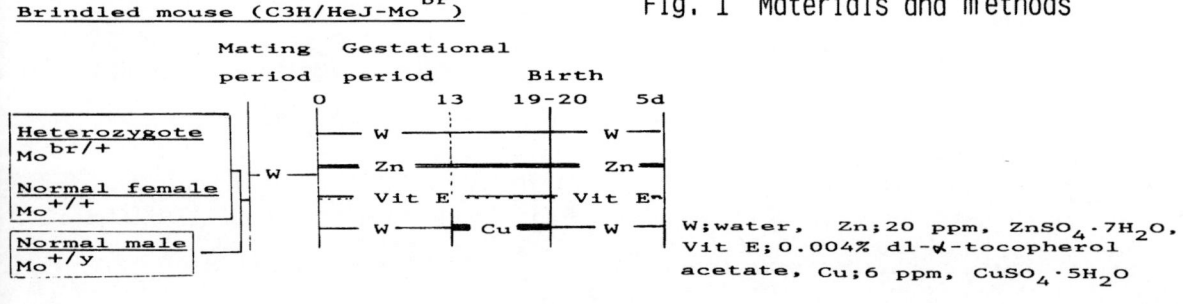

Fig. 1 Materials and methods

RESULTS - A

Live offspring at 5 days after birth of heterozygous females are compared as to four drinking fluids (Table 1). Dam with live offspring was decreased in water as compared to in the other three kinds of treatments. Furthermore, as to the ratios of the four genotypes in the dams with live offspring, the number of hemizygous males was greatly decreased in the W group. Therefore, the viability as to hemizygous males was better in heterozygous pregnant females treated with zinc, vitamin E and copper than in those with water.

Table 1   Effects of maternal treatments on live offspring
at 5 days after birth of heterozygotes

Treatment	Percent dams with live offspring	Mean live offspring	Percent ratios of genotypes			
			br/y	br/+	+/y	+/+
W	38%	4.3	8	32	20	40
Zn	86%	5.3	29	17	30	24
Vit E	100%	4.3	27	13	18	42
Cu	92%	4.7	29	32	23	17

In the heterozygotes without offspring, we observed abnormal movements, after 3 months of age, at or after delivery. These abnormal movements comprised rotatory movements in the same direction or circling behavior of high speed accompanied by slight tremor and sometimes an ataxic gait.

Lipid peroxide formation in cerebral homogenate of heterozygous females at zero time and after 30 min, at 37° C was significantly higher in hetero-zygous cerebrum with abnormal movements than in that with normal movement (Fig. 2). This finding shows that abnormal movements relate to an increase in cerebral lipid peroxidation in the heterozygotes.

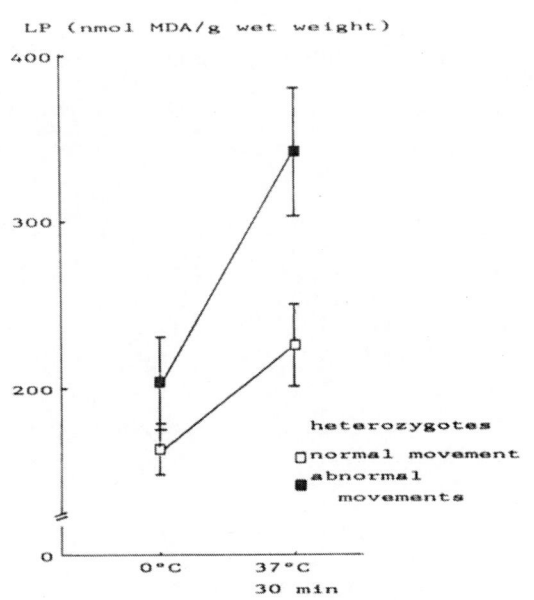

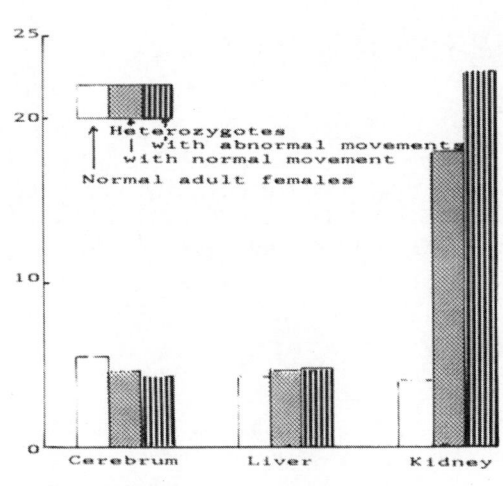

Fig. 2   Abnormal movements in heterozygous females and cerebral
lipid peroxide formation (LP)

Fig. 3   Mean copper concentrations in
cerebrum, liver and kidney
(μg/g wet weight)

The copper concentrations in cerebrum, liver and kidney are compared in normal adult females, and heterozygotes with or without abnormal movements (Fig. 3). There were no significant differences in the cerebral and hepatic copper concentrations among the three groups. However, in kidney the mean copper concentration in heterozygote was greatly higher than that in normal female. Furthermore, when heterozygotes with abnormal movements were compared to those with normal movement, a significant increase in the copper concentration in kidney was also observed in heterozygotes with abnormal movements. This finding shows that abnormal movements relate to an increase in renal copper concentration in heterozygote.

Activities of copper-containing enzymes ; Cu,Zn-SOD and cytochrome oxidase in cerebrum, liver and kidney did not differ in heterozygotes between normal movement and abnormal movements (data not shown).

B. Effects of Oral Copper Administration to Pregnant Heterozygous Brindled Mice on Fetal and Neonatal Developments

MATERIALS AND METHODS - B

Two experiments were attempted (Fig. 4). In the experiment I, the effects of copper administration to pregnant normal mice on neonatal mice were examined to know the safe levels of copper for mothers and neonates as described previously (Kasama and Tanaka 1988). In the experiment II, the effects of copper administration to pregnant heterozygous mice on their fetuses and neonates were investigated. In the both experiments, concentrations of copper were 6 ppm from 13 days gestation to delivery and 5 ppm during lactational period.

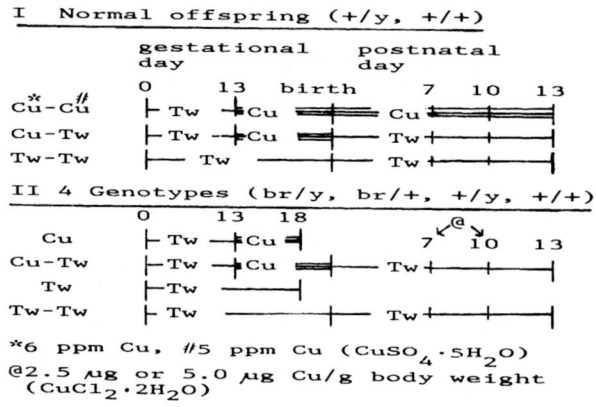

Fig. 4 Materials and methods (Tw:tap water)

RESULTS - B

I. Normal Offspring

The changes in the mean copper concentrations of normal neonatal mice differed with the tissues ; cerebrum, liver and kidney. At 1 day of age, maternal administration of copper increased the copper concentration in the cerebrum (data not shown).

Neonatal cerebrum of normal mouse still contained a high copper concentration at, at least 13 days after birth, by maternal copper during pregnancy, independent of what the mothers drank during the lactational period (Fig. 5).

As shown by the body weights (Fig. 6), normal neonatal mice of the Cu-Cu group grew less well than those of the other two groups.

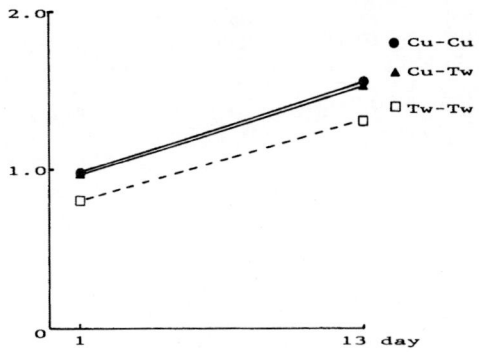
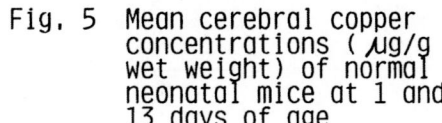

Fig. 5 Mean cerebral copper concentrations (μg/g wet weight) of normal neonatal mice at 1 and 13 days of age

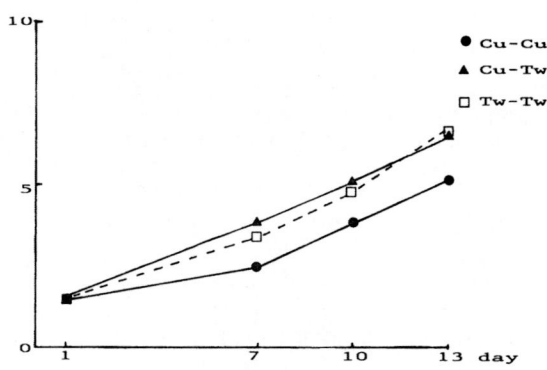

Fig. 6 Mean body weights (g) of normal male mice at 1, 7, 10 and 13 days of age

II. Four Genotypes

The effect of maternal copper administration on mean live fetus on gestational day 18 was clear (Table 2). The heterozygous mothers given water had small number of live fetuses. Therefore, the copper administration resulted in an increased live fetus from heterozygous mother.

Table 2 Effect of maternal copper administration on litter size of fetuses on gestational day 18

Dam	Treat-ment	No of dams	Mean litter size	Mean live fetus
br/+	Cu	6	6.7	5.2(78%)
	Tw	5	5.2	2.4(46%)
+/+	Cu	3	6.3	5.7(90%)
	Tw	4	6.3	5.8(92%)

However, on gestational day 18, there was no difference in body weight among different genotypic fetuses (data not shown). Therefore, administration of copper to the heterozygous mothers was not considered to affect the fetal growth.

Figure 7 shows copper concentrations of hemizygous, heterozygous and normal fetuses from heterozygous or normal mothers given copper or water. The cerebral copper concentrations of hemizygous and heterozygous fetuses were the same levels and significantly lower than that of normal fetuses, regardless of the maternal drinking fluid. The copper administration increased the cerebral copper concentrations of all three kinds of fetuses. However, the cerebral copper of hemizygous and heterozygous fetuses did not reach normal levels. The hepatic copper concentrations of hemizygous and heterozygous fetuses were also the same and significantly lower than that of normal fetuses. An increase in the hepatic copper concentration with copper administration was observed only in normal fetuses. On the other hand, in the kidney, the copper concentration of hemizygous fetuses was significantly higher than those of heterozygous and normal fetuses. The copper administration had no effect on fetal renal copper concentrations.

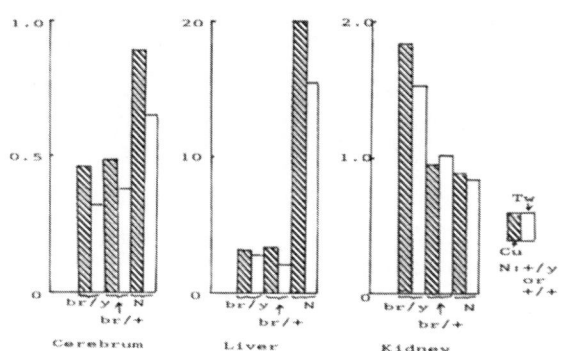

Fig. 7 Effect of maternal copper administration on mean
copper concentrations (µg/g wet weight) of fetuses
on gestational day 18

Furthermore, on gestational day 18, placental copper concentration was
inversely proportional to the fetal cerebral copper concentration, and
directly proportional to the fetal renal copper concentration (data not shown).

The effects of maternal copper administration on neonatal hemizygotes were
examined. At 8 days of age, copper-containing enzyme activity was low,
especially in cerebrum ; Cu,Zn-SOD and cytochrome oxidase were 74% and 8% of
normal values, respectively. Furthermore, an intra-peritoneal injection of
5.0 µg copper per g body weight at 7 days of age resulted in slightly
increased Cu,Zn-SOD and also cytochrome oxidase activities in the cerebrum at
10 days of age. Therefore, after birth, treatment with a greater dose of
copper than 5.0 µg per g body weight is required for the survival of hemizygotes.

DISCUSSION AND CONCLUSION

When we attempted to treat Menkes' kinky hair disease, we need precise
information on heterozygous mothers. However, even adult heterozygous
brindled mice have not been investigated so much. We found, unexpectedly,
abnormal movements in heterozygous mothers related to a decrease in live
offspring. Maternal abnormal movements resulted in non-viable offspring,
which showed some relation to maternal drinking fluids. That is, zinc or
vitamin E or copper gave better effects on viability of offspring, especially
hemizygous males than water only. It is now accepted that vitamin E and zinc
protect the damage due to the radical process, but copper ions generate
free-radicals and act as a promotor of hydroxyl radical formation. Therefore,
these findings suggest that the development of hemizygous male mice may be
influenced by both copper and oxygen radical metabolisms.

There have been few reports of the administration of copper to fetuses of
brindled mice. We report here, that copper administration improved the copper
deficiency in the cerebrum of affected fetus. Our results suggest that oral
copper administration to pregnant heterozygotes could improve an abnormal
copper distribution in hemizygous and heterozygous fetuses without affecting
fetal growth. However, in addition to prenatal treatment, postnatal copper
injection is needed for hemizygous male to survive.

REFERENCES

Hunt D M (1974) Primary defect in copper transport underlies mottled mutants
  in the mouse. Nature 249:852-854
Kasama T, Tanaka H (1988) Effects of copper administration on fetal and
  neonatal mice. J Nutr Sci Vitaminol 34:595-605
Menkes J H, Alter M, Steigleder G K, Weakley D R, Sung J H (1962) A
  sex-linked recessive disorder with retardation of growth, peculiar hair,
and focal cerebral and cerebellar degeneration. Pediatrics 29:764-779

Human and Canine Inherited Copper Toxicosis: Copper Balance Regulation and Molecular Genetics of Its Impairment

GEORGE J. BREWER, VILMA YUZBASIYAN-GURKAN, and DOH-YEEL LEE

Department of Human Genetics, University of Michigan Medical School, 4708 Medical Science Building II, Box 0618, Ann Arbor, MI 48109-0618, USA

ABSTRACT

In this paper we briefly summarize the evidence supporting the view that zinc acetate is an excellent, fully efficacious, non-toxic therapy for the maintenance treatment of Wilson's disease. The recommended dose is 50 mg three times per day with each dose separated from food and beverages, except water, by at least one hour. Zinc therapy is also excellent for the treatment of presymptomatic Wilson's disease patients and for the pregnant patient. We also summarize our hypothesis and supporting data that Wilson's disease results from a failure to secrete a copper-rich, protease-resistant fragment of ceruloplasmin into the bile. We have updated our research in the area of molecular genetics of both the human gene and the canine copper toxicosis gene.

KEY WORDS: Wilson's disease, Copper toxicity, Zinc therapy, Canine copper toxicosis, Ceruloplasmin

Wilson's disease is an autosomal recessive disorder with the gene now known to be located on the long arm of chromosome 13 (Bowcock 1987; Yuzbasiyan-Gurkan 1988). Copper accumulates due to a failure of biliary excretion, and leads to liver and brain disease. A number of treatments are available (Brewer 1987). The oldest is penicillamine, but it has a number of toxicities. There is much less experience with trien, another chelating drug. We have spent time over the last 10 years perfecting the use of zinc acetate as a treatment (Brewer 1983, Hill 1987, 1986; Brewer 1987, 1987; Yuzbasiyan-Gurkan 1989; Brewer 1989; Lee 1989). Zinc works in a completely different manner than the chelating drugs; it acts by inducing intestinal cell metallothionein, which blocks the intestinal uptake of copper (Richards 1976; Hall 1979). We will begin by briefly reviewing our progress with zinc therapy of Wilson's disease.

Copper balance is a key tool in our work. In each balance, stool and urinary copper are assayed and subtracted from carefully assayed dietary copper over a 10 day period during a 14 day hospital admission. Data from 61 balance studies are reviewed here. The expected observed positive balance in untreated Wilson's disease is the sum of the unmeasured losses of about 0.35 mg from the skin surface (Jacob 1981), and the true positive balance in Wilson's disease based on copper accumulation over a period of a few decades, estimated to be about 0.25 mg, or a total of about 0.6 mg/day positive balance. We have found that the actual data in six untreated patients are very close to this expectation.

Zinc even in doses as small as 25 mg/day has a significant effect on copper balance. At 75 mg/day the effect becomes such as to produce consistent control of copper balance. There is an interesting difference between 25 mg 3 x /day and 75 mg 1 x /day, with the latter being much less effective. Our recommended dose is 50 mg 3 x /day, or 150 mg total, providing a safety margin. In extensive studies we have shown consistent control of copper balance with this dose (Hill 1987).

Another way of evaluating the efficacy of zinc therapy is to evaluate blood uptake of radioactivity after an orally administered dose of [64]copper (Hill 1986; Van den Hamer 1984). Before zinc treatment, an average uptake of about 6% is seen. Zinc therapy is very effective in reducing this value to the target range, of below 1.2% (Hill 1986). A few patients exhibit compliance problems, but with counseling all are subsequently brought under control.

Another way of assessing zinc efficacy is to follow the 24 hour urine copper (Brewer 1987). In the absence of chelation therapy, the 24 hour urine copper is a good reflector of body copper status. Here we discuss data on two kinds of patients, one in which zinc is the initial therapy, usually in presymptomatic patients, and one in which zinc has been introduced after decoppering by another therapy, usually penicillamine. In the untreated patients the initial urinary copper is higher. As time goes on, urinary copper is reduced, consistent with the decoppering effect of zinc. In all patients zinc therapy easily controls urinary copper, with the exception of compliance problems which are easily identified and controlled by counseling. In addition to following urine copper, we follow plasma copper to monitor zinc therapy. This too is brought under excellent control by zinc (Brewer 1987). Hepatic copper levels determined from repeat biopsies in patients who had been decoppered at the time of initiation of zinc is stable during zinc maintenance therapy (Brewer 1987).

A large number of clinical variables are also monitored, and no patient has ever deteriorated clinically while on maintenance zinc therapy in our program. By making sure that the copper variables are controlled, we essentially protect the patient against clinical worsening.

The cumulative number of patients who are on zinc maintenance therapy since the beginning of our studies is over 60. The longest treated is about 8 years. 27 patients are at about the 4.5 year followup at this point. We have found essentially no toxicity of zinc therapy.

In addition to maintenance therapy, we think zinc is excellent for the treatment of the presymptomatic patient. We have treated 14 such patients so far. All of the copper related variables are controlled in this type of patient by zinc therapy, with one interesting difference (Brewer 1989).

Hepatic copper values determined on repeat biopsies on 7 patients increased in 4 patients. Even though the copper levels increase in some patients, there is no evidence of increasing liver damage. To the contrary, it appears that the liver disease, if such is evident, comes under good control during this period. We have modeled this situation in rats loaded with copper by diet (Lee 1989). When we introduce zinc after sufficient copper loading to produce liver damage, as measured by elevated SGPT levels, zinc therapy then controls and reduces the damage, compared to the control animals. When we examine the copper binding proteins in the livers of these rats, we believe we find the explanation. It is due to binding of copper by metallothionein. Using sephadex column chromatography of liver material from control rats, almost all of the copper is in high and low molecular weight fractions, and very little is in the metallothionein fraction. In contrast, the zinc treated animals have a very much higher peak of copper in metallothionein and very much less in other fractions. Thus, we believe that zinc therapy, by inducing high levels of hepatic metallothionein, allows the liver to sequester copper from other parts of the body.

The mechanism of zinc action as shown in animal studies, involves the induction of intestinal metallothionein which blocks the uptake of copper. Copper sequestered by intestinal metallothionein is eventually sloughed as a complex when the intestinal cell turns over and is itself sloughed into the stool. In collaboration with Dr. Robert Cousins, we have carried out [64]copper uptake studies concomitant with intestinal biopsies to measure metallothionein during early, intermediate and late phases of zinc therapy. Our preliminary data show that as zinc therapy causes blockade of oral [64]copper uptake, there is an increase in intestinal metallothionein. Thus the data in the human are consistent with the metallothionein hypothesis.

Finally, zinc therapy appears to be the almost ideal therapy for the pregnant Wilson's patient. Unlike the chelators, zinc is not known to be teratogenic. Four healthy babies have been produced while their mothers were on zinc therapy.

While we believe zinc is the drug of choice for the treatment of presymptomatic patients and patients in the maintenance phase, we do not think it is ideal for the initial treatment of acutely ill patients because of its slow action. At least for neurologically affected patients we do not believe penicillamine is a good choice either, because of the syndrome of neurologic worsening. In a retrospective survey we've found that about half of patients treated with penicillamine, get worse neurologically (Brewer 1987). Of that number, another half, or 1/4 of the original sample, never recover to prepenicillamine baseline.

For initial therapy of such patients we have been using ammonium tetrathiomolybdate (TM). TM acts by complexing copper with food proteins and preventing its absorption. In contrast to zinc, it acts immediately and it acts on the entire GI tract. TM which is absorbed complexes with copper and albumin in the blood, rendering the copper unavailable for cellular uptake. We have now treated 5 acute neurological patients, with TM, and have had successful results in avoiding initial worsening.

Some of our work on Wilson's disease has involved studies of the defect in the bile which results in a failure to excrete copper (Iyengar 1988). In our protocol we intubate the patient and aspirate duodenal secretions, while injecting cholecystokinin to cause a bile dump. We then carry out G-75 sephadex chromatography on the samples. There is a high molecular weight copper binding fraction in the void volume in the samples from normal subjects, missing in Wilson's disease. This is a consistent observation. The amount of copper in this fraction calculates to be about 0.25 mg/day, as expected if it is the regulatory copper. When we supplement normal subjects with copper to about 6 times the previous intake, we see a marked increase in this peak. If we use bile salts in the column to prevent salting out of proteins, we see that this peak moves substantially, so that its probable true molecular weight is somewhere between 15-25,000. Biliary samples from normal, but not from Wilson's disease patients, show a line of partial immunological identity with ceruloplasmin, using antibodies to ceruloplasmin. We believe the material we are seeing in the bile is the 19,000 molecular weight fraction of ceruloplasmin which is very copper rich and resistant to proteolytic digestion. We believe that this fragment serves as a packaging mechanism for excretion of biliary regulatory copper in normal subjects, and is missing in Wilson's disease.

Another area of our research activity has involved the molecular genetics of Wilson's disease. The Wilson's disease gene is located in the same area of chromosome 13 as esterase D and retinoblastoma (Bowcock 1987; Yuzbasiyan-Gurkan 1988). Because of the interest in retinoblastoma, a number of other probes, now mapped to this general area, have also been identified. Because of the proximity of the retinoblastoma locus, possibly as close as 2 centimorgans, the use of probes from this locus using RFLP's is becoming very useful for genetic counseling and assistance in diagnosis.

In the last few years we have extended our studies to include canine copper toxicosis (CT). The gene for copper toxicosis is very prominent in Bedlington terriers. Canine CT is quite similar to Wilson's disease in many regards, including copper accumulation, liver disease, and failure of biliary copper excretion. There are also a few differences. It is a reasonable candidate as a model for Wilson's disease, but of course could also be some type of phenocopy. We have postulated that if in fact the genes are homologous, the very common occurrence of linkage homology, which refers to the holding together of blocks of genes over long periods of evolution, will have held these genes together (Brewer 1989). In this case we will find linkage of the CT gene to the markers which are linked to the Wilson's gene in the human. Our preliminary data are encouraging that the genes are linked.

A number of good things will happen if we confirm linkage of canine CT to the same markers that are linked in Wilson's disease. Most important, it allows the virtual elimination of a disease gene in a domestic species such as this (Brewer 1989). This technology is really quite powerful in veterinary medicine, as opposed to the situation in the human where reproductive behavior is rarely affected. Because of the control of which animals we use for breeding, breeds can be freed of a harmful disease gene very rapidly once markers such as those under discussion are established. Thus this study, if

positive, will set several precedents. These include the use of this "short cut" approach of using linkage knowledge and probes from another species to eliminate a veterinary disease. It is also a way of validating an animal model.

In addition to molecular genetic work in the dog we are also establishing zinc as a therapy for canine CT and evaluating the dietary copper requirement for the dog. At present there appears to be too much copper added to commercial dog food, generally speaking, at least in the U.S.

Our work for the future in the human involves the final establishment of zinc as an effective therapy for Wilson's disease, including licensing of the drug in the United States as a first line therapy. We are also interested in cloning the Wilson's disease gene and are undertaking a number of approaches towards that end. The objective here of course, is to express the gene and find out what type of product it makes and how that product is defective in Wilson's disease.

In summary, we believe zinc acetate is the treatment of choice for the maintenance therapy of Wilson's disease, for the complete treatment of the presymptomatic patient, and for the pregnant patient. We think there is no ideal solution at present for the pregnant of the acutely ill neurologically affected patient. We are studying the use of ammonium tetrathiomolybdate for this purpose, and are quite encouraged. We hypothesize that copper balance is normally regulated by excretion of a copper rich, protease resistant, ceruloplasmin fragment in the bile. Probes linked to the Wilson's disease gene are proving quite useful for genetic counseling and assistance in diagnosis. Our preliminary data are encouraging that the canine copper toxicosis gene is linked to the same gene to which Wilson's disease is linked, indicating genetic homology between the canine CT and human Wilson's disease genes.

REFERENCES

Bowcock AM, Farrer LA, Cavalli-Sforza LL, Herbert JM, Kidd KK, Frydman M, Bonné-Tamir B (1987) Mapping the Wilson disease locus to a cluster of linked polymorphic markers on chromosome 13. Am J Hum Genet 41:27-35

Brewer GH, Hill GM, Dick RD, Nostrant TT, Sams JS, Wells JJ, Prasad AS (1987) Prevention of reaccumulation of hepatic copper. J Lab Clin Med 109:526-531

Brewer GJ, Hill GM, Prasad AS, Cossack ZT, Rabbani P (1983) Oral zinc therapy for Wilson's disease. Annals of Internal Medicine 99:314-320

Brewer GJ, Hill GM, Prasad AS, Dick RD (1987) Treatment of Wilson's disease with zinc: IV. Efficacy monitoring using urine and plasma copper. Proc Soc Exper Biol Med 7:446

Brewer GJ, Terry CA, Aisen AM, Hill GM (1987) Worsening of neurologic syndrome in patients with Wilson's disease with initial penicillamine therapy. Arch Neurol 44:490-493

Brewer GJ, Yuzbasiyan-Gurkan V (1989) Fighting disease with molecular genetics. Am Kennel Club Gazette 106:66-75

Brewer GJ, Yuzbasiyan-Gurkan V, Lee D-Y (1989) The treatment of Wilson's disease with zinc VI. Initial treatment studies. J Lab Clin Med (in press)

Brewer GJ, Yuzbasiyan-Gurkan V, Young AB (1987) Treatment of Wilson's disease. Sem Neuro 7:209-220

Hall AC, Young BW, Bremner I (1979) Intestinal metallothionein and the mutual antagonism between copper and zinc in the rat. J Inorg Biochem 11:57-66

Hill GM, Brewer GJ, June JE, Prasad AS, Dick RD (1986) Treatment of Wilson's disease with zinc: II. Validation of oral [64]copper uptake with copper balance. Am J Med Sci 12:344-349

Hill GM, Brewer GJ, Prasad AS, Hydrick CR, Hartmann DE (1987) Treatment of Wilson's disease with zinc: I. Oral zinc therapy regimens. Hepatology 7:522

Iyengar V, Brewer GJ, Dick RD, Owyang C (1988) Studies of cholecystokinin-stimulated biliary secretions reveal a high molecular weight copper-binding substance in normal subjects that is absent in patients with Wilson's disease. J Lab Clin Med 111:267-274

Jacob RA, Sandstead HH, Munroz JM, Klevay IM, Milne DB (1981) Whole body surface loss of trace metals in normal males. Am J Clin Nutr 34:1379-1383

Lee D-Y, Brewer GJ, Wang Y (1989) The treatment of Wilson's disease with zinc VII. Protection of the liver from copper toxicity by zinc induced metallothionein in a rat model. J Lab Clin Med (in press)

Richards MP, Cousins RJ (1976) Metallothionein and its relationship to the metabolism of dietary zinc in rats. J Nutr 106:1591-1599

Van den Hamer CJA, Hoogenraad TU (1984) ^{64}Cu loading tests for monitoring zinc therapy in Wilson's disease. Trace Elements Med 1:84-87

Yuzbasiyan-Gurkan V, Brewer GJ, Abrams GD, Main B, Giacherio D (1989) Treatment of Wilson's disease with zinc: V. Changes in serum levels of lipase, amylase and alkaline phosphatase in Wilson's disease patients. J Lab Clin Med (in press)

Yuzbasiyan-Gurkan V, Brewer GJ, Boerwinkle E, Venta PJ (1988) Linkage of the Wilson Disease gene to chromosome 13 in North-American Pedigrees. Am J Hum Genet 42:825-829

Gene

Finger-Loop Domains and Trace Metals

F. WILLIAM SUNDERMAN JR.

University of Connecticut Medical School, Farmington, CT 06032, USA

1. ABSTRACT

Certain DNA-binding proteins that regulate gene expression contain 'Zn-fingers',– single or multiple copies of short polypeptide sequences, approximately 30 residues long, containing four cys or his residues at defined spacing, so that Zn^{2+} is complexed in tetrahedral coordination with thiol-sulfur or imidazole-nitrogen atoms. The Zn^{2+} serves as a 'strut' that stabilizes folding of the domain into a 'finger-loop', which is capable of site-specific binding to double-stranded DNA. Since finger-loops provide a molecular means to control genetic transcription, finger-loop domains have been highly conserved during evolution. This paper reviews (a) the several classes of finger-loops that have been identified, (b) the essentiality of Zn^{2+} for the physiological functions of finger-loops, (c) the evidence that foreign metal ions (*e.g.*, Cd^{2+}, Co^{2+}, Ni^{2+}) can substitute for Zn^{2+} in finger-loop domains, and (d) the speculation that such substitutions might represent a molecular basis for genotoxicity and carcinogenicity of certain metal ions. Finally, this paper suggests that, since Zn finger-loops are present in DNA-binding regions of the human glucocorticoid-, mineralocorticoid-, estrogen-, progesterone-, androgen-, thyroid-, vitamin D-, and retinoic acid-receptor proteins, as well as the recently discovered 'testis determining factor', depletion of Zn^{2+} from such domains could be involved in the endocrine manifestations of zinc deficiency.

Key words: DNA-binding proteins, gene regulation, metal toxicity, transcription factors, Zn-fingers

2. INTRODUCTION

The structure, biochemical significance, and biological functions of finger-loop domains in proteins that regulate gene expression have been discussed in earlier reviews (Berg 1986, Klug and Rhodes 1987, Evans and Hollenberg 1988, Payre and Vincent 1988, Sunderman and Barber 1988). The best-studied finger-loop protein is Transcription Factor IIIA ('TFIIIA'), a Zn-metalloprotein that binds to the internal control region of the gene for 5S RNA in the African frog, *Xenopus laevis*, stimulating the synthesis of 5S RNA. TFIIIA also binds to the product, 5S RNA, forming a 7S ribonucleoprotein particle that is abundant in homogenates of *Xenopus* oocytes (Brown 1984). Soon after the amino acid sequence of TFIIIA was reported by Ginsberg *et al.* (1984), two groups of investigators (Brown *et al.* 1985, Miller *et al.* 1985) noted that the sequence contains tandem repeats of segments of approximately 30 residues, including pairs of cysteine and histidine molecules arranged in a regular pattern; they speculated that tetrahedral coordination of Zn^{2+} ions to thiol and imidazole ligands of the cys or his residues might stabilize the repetitive segments in finger-loop configurations that could bind to specific loci on double-stranded DNA. This speculation was soon confirmed experimentally. Two basic models have been proposed for the interaction of the finger-loops of TFIIIA with DNA (Fairall *et al.* 1986, Berg 1988, Gibson *et al.* 1988). In one model, TFIIIA is located on one side of the double-helix, with fingers projecting into the major groove; in the other model, the TFIIIA wraps around the double-stranded DNA, lying in the major groove with an α-helical region on one side of each finger interacting with the hydrogen bonds, and the linker region between the fingers contributing to a turn, so that the protein conformation resembles, in certain respects, the well-established helix-turn-helix class of DNA-binding structures. Recent evidence derived by NMR spectroscopy supports the second model (Párraga *et al.* 1988, Lee *et al.* 1989), but the three-dimensional relationship of finger-loop domains to the DNA helix has not been definitely established. The structure of the DNA-binding site of TFIIIA appears to be intermediate between A-form and B-form DNA (Fairall *et al.* 1989).

TABLE 1. ILLUSTRATIVE PROTEINS WITH PUTATIVE FINGER-LOOP DOMAINS

Class	Genus, species	Protein	Finger-loop pattern							Repeats	Reference
I	Xenopus laevis	Transcription Factor IIIA	C	-X2,4,5-	C	-X12-	H	-X3,4-	H	9	Miller et al. '85
		Xfin (Krüppel prot homolog)	C	-X2-	C	-X12-	H	-X3-	H	35	Ruiz i Altaba.. '87
	Drosophila	Serendipity gene product	C	-X2-	C	-X12-	H	-X3,4-	H	13	Vincent et al. '86
	melanogaster	Krüppel gene product	C	-X2-	C	-X12-	H	-X3-	H	4	Schuh et al. '86
		Hunchback gene product	C	-X2-	C	-X12-	H	-X3,5-	H	6	Tautz et al. '87
		Snail gene product	C	-X2-	C	-X12-	H	-X3-	H	4	Boulay et al. '87
	Trypanosoma	TRS-1 (rev transcriptase)	C	-X2-	C	-X12-	H	-X4-	H	5	Pays/Murphy '87
	Rheovirus T3	ps3 (outer capsid protein)	C	-X2-	C	-X12-	H	-X3-	H	1	Schiff et al. '88
	Saccharomyces	pRPO21 (RNA polymerase)	C	-X2-	C	-X9-	H	-X2-	H	1	Allison et al. '85
	cerevisiae	ADR1 (ADH gene activator)	C	-X2-	C	-X12-	H	-X3,4-	H	2	Hartshorne... '86
		SWI5 (HO gene activator)	C	-X4-	C	-X12-	H	-X3-	H	2	Nagai et al. '88
	Mouse	mKr-1 (Krüppel pr homolog)	C	-X2-	C	-X12-	H	-X3-	H	7	Chowdhury... '87
		mKr-2 (Krüppel pr homolog)	C	-X2-	C	-X12-	H	-X3-	H	9	Chowdhury... '88
		Krox-20 (SRE of 3T3 cells)	C	-X2,4-	C	-X12-	H	-X3-	H	3	Chavrier et al. '88
		Egr-1 (EGF-induced TF)	C	-X2,4-	C	-X12-	H	-X3-	H	3	Seyfert et al. '89
	Rat	NGFI-A (NGF-induced TF)	C	-X2,4-	C	-X12-	H	-X3-	H	3	Changelian... '89
	Human	Sp1 (transcription factor)	C	-X2,4-	C	-X12-	H	-X3-	H	3	Kadonaga... '88
		EGR2 (EGF-induced TF)	C	-X2,4-	C	-X12-	H	-X3-	H	3	Joseph et al. '88
		HF-10 (Krüppel pr homolog)	C	-X2-	C	-X12-	H	-X3	H	10	Pannuti et al. '88
		pZFY (testis determ factor)	C	-X2-	C	-X12-	H	-X3,4-	H	13	Mardon/Page '89
		pZFX (homolog of pZFY)	C	-X2-	C	-X12-	H	-X3,4-	H	13	Schneider-... '89
		pGLI (glioblastoma onc-prot)	C	-X4-	C	-X12,15-	H	-X3,4-	H	5	Kinzler et al. '88
		pSS-A/Ro (Sjögren antigen)	C	-X3-	C	-X10-	H	-X2-	H	1	Ben-Chetrit... '89
II	Soybean	Tgm5 (transposon)	C	-X3-	C	-X12-	C	-X2-	C	1	Vodkin... '89
	Maize	En-1 (transposon)	C	-X3-	C	-X12-	C	-X2-	C	1	Vodkin... '89
	Adenovirus	E1A (trans-activator protein)	C	-X2-	C	-X13-	C	-X2-	C	1	Culp et al. '88
	Drosophila	Notch gene product	C	-X4-	C	-X12,14-	C	-X1,4-	C	38	Kidd et al. '86
	Saccharomyces	GAL4 (TF galactose metab)	C	-X2-	C	-X13-	C	-X2-	C	1	Johnston '87
	cerevisciae	ARGRII (TF arginine metab)	C	-X2-	C	-X13-	C	-X2-	C	1	Messenguy... '86
		MAL63 (TF maltose metab)	C	-X2-	C	-X12-	C	-X2-	C	1	Kim/Michels '88
		CDC16 (cell division cycle)	C	-X3-	C	-X4-	C	-X3-	C	1	Icho/Wicker '87
		RPB2 (RNA polymerase II)	C	-X2	C	-X15-	C	-X2-	C	1	Sweetser et al. '87
	Kluyveromyces	LAC9 (TF lactose metab)	C	-X2-	C	-X13-	C	-X2-	C	1	Wray et al. '87
	Neurospora	qa-1F (TF quinate metab)	C	-X2-	C	-X13-	C	-X2-	C	1	Baum et al. '87
	Mouse	GF1 (erythroid growth factor)	C	-X2-	C	-X17-	C	-X2-	C	2	Tsai et al. '89
	Human	PKC (protein kinase C)	C	-X2-	C	-X13-	C	-X2-	C	2	Ono et al. '89
		UBI (ubiquitin)	C	-X4-	C	-X13,14-	C	-X2,4-	C	2	Salvesen et al. '87
		hTR (thyroid receptor)	C	-X2,5-	C	-X9,13-	C	-X2-	C	2	Weinberger... '86
		hGR (glucocorticoid receptor)	C	-X2,5-	C	-X9,13-	C	-X2-	C	2	Weinberger... '85
		hMR (mineralocorticoid rec)	C	-X2,5-	C	-X9,13-	C	-X2-	C	2	Arriza et al. '87
		hER (estrogen receptor)	C	-X2,5-	C	-X9,13-	C	-X2	C	2	Greene et al. '86
		hPR (progesterone receptor)	C	-X2,5-	C	-X9,13-	C	-X2-	C	2	Misrahi et al. '87
		hVDR (vitamin D receptor)	C	-X2,5-	C	-X9,13-	X	-X2-	C	2	Haussler et al. '88
		hRAR (retinoic acid receptor)	C	-X2,5-	C	-X9,13-	C	-X2-	C	2	Petkovich... '87
		COUP-TF (ovalbumin TF)	C	-X2,5-	C	-X9,13-	C	-X2-	C	2	Wang et al. '89
III	Tobacco streak	virus (coat protein)	C	-X2-	C	-X10-	C	-X2-	H	3	Sehnke et al. '89
	Tobacco rattle	virus (coat protein)	C	-X4-	C	-X18-	C	-X3-	H	2	Sehnke et al. '89
	Bov leuk virus	p12 (nucleic acid binding)	C	-X2-	C	-X4-	H	-X4-	C	2	Copeland et al. '83
	Hum T-cell leuk v	p15 (nucleic acid binding)	C	-X2-	C	-X4-	H	-X4-	C	2	Copeland et al. '83
	Hum HIV-1 virus	p7 (nucleic acid binding)	C	-X2-	C	-X4-	H	-X4-	C	2	South et al. '89
	Drosophila	Terminus gene product	C	-X2-	C	-X12-	H	-X4-	C	1	Baldarelli... '88
	melanogaster	Snail gene product	C	-X2-	C	-X12-	H	-X4-	C	1	Boulay et al. '87
	Xenopus laevis	Xfin (Krüppel pr homolog)	C	-X2-	C	-X12-	H	-X4-	C	1	Ruiz i Altaba.. '87
	Saccharomyces	Rad18 (error-prone repair)	C	-X2-	C	-X12-	H	-X3-	C	1	Chanet et al. '88
	Human	poly(ADP-ribose)polymerase	C	-X2-	C	-X28,30-	H	-X-2	C	2	Mazen et al. '89
	Saccharomyces	Rad18 (error-prone repair)	C	-X3-	H	-X6-	C	-X2-	C	1	Chanet et al. '88
	T4 phage	g32p (ss-DNA binding prot)	C	-X3-	H	-X5-	C	-X2-	C	1	Giedroc et al. '86
IV	Human	myc oncogene protein	H	-X4-	H	-X7-	C	-X4	H	1	Sunderman... '88

3. CLASSIFICATION OF FINGER-LOOP DOMAINS

Two major classes and some minor classes of finger-loop proteins have been delineated (Table 1). Class I proteins usually have 2 to 5 residues in the knuckles (*i.e.*, between the pairs of cys or his residues) and 9 to 15 residues in the central loop. The central loop typically contains conserved hydrophobic residues (phe, tyr, leu, ileu, val) that hypothetically stabilize the finger by forming hydrophobic bonds, and basic or polar residues (arg, asp, glu, his, thr) that may interact with DNA. The proteins of Class I generally contain multiple finger-loops. Class I includes proteins encoded by several morphogenic genes of *Drosophila*, such as the Krüppel and Hunchback genes that control initial segmentation of the embryo. Yeast regulatory proteins such as ADR1 and SWI5 are members of this class. Mammalian regulatory proteins of Class I include: Egr-1 and EGR-2, transcription factors that are induced by mitogenic stimulation of early growth response genes; Sp1, a protein that binds to GC-boxes and activates transcription of numerous viral and cellular promotors; and pGLI, a protein expressed in glioblastomas and embryonal carcinomas that is encoded by the GLI oncogene. Of particular note are pZFY, a zinc-finger protein on the Y chromosome ('testis determining factor'), and its homolog, pZFX, which is located on the X chromosome.

In finger proteins of Class II, the Zn-binding site comprises four cys residues. Some of these proteins contain a single finger, as in several transcription factors of yeast (*e.g.*, GAL4, LAC9), while other proteins of Class II have two fingers (including the steroid hormone receptors and related proteins, such as the thyroid-, glucocorticoid-, mineralocorticoid-, estrogen-, progesterone-, androgen-, and retinoic acid-receptors). In finger proteins of Class III, Zn^{2+} is coordinated to three cys residues and one his residue. The most thoroughly studied member of this class is gene 32 protein ('g32p') of bacteriophage T4. The existence of Class IV fingers, in which Zn^{2+} may be coordinated to three his residues and one cys residue, is speculative. The putative finger loop of the myc protein was suggested by Sunderman and Barber (1988) based on a computer search of the amino acid sequences of proteins encoded by oncogenes; Dang *et al.* (1989) subsequently reported that this segment of the human myc protein is responsible for DNA binding.

4. STUDIES OF METALS IN FINGER-LOOP PROTEINS

Investigations of metals in TFIIIA isolated from *Xenopus* oocytes are summarized in the upper section of Table 2. Hanas *et al.* (1983) showed that Zn can be removed from TFIIIA by chelation with EDTA or 1,10-phenanthroline, and that apoTFIIIA does not bind to the 5S RNA gene unless Zn is replaced. Additions of divalent Co, Ni, Mn, or Fe do not restore the binding. Miller *et al.* (1985) demonstrated that TFIIIA contains approximately 9 atoms of Zn per mole and that chelation of Zn in 7S ribonucleoprotein particles dissociates TFIIIA from 5S RNA, as does exposure of the particles to divalent Cd or Cu, which can evidently displace the native Zn. Diakun *et al.* (1986) showed by EXAFS analysis that Zn in TFIIIA is coordinated to 2 his and 2 cys residues, consistent with the predicted finger-loop structure. Studying a synthetic peptide corresponding to the second finger of TFIIIA, Frankel *et al.* (1987) showed that the peptide folds after Zn is added *in vitro* and that it also binds Co^{2+}, with a characteristic change of absorption spectrum. In further research, Berg and Merkle (1989) showed that Zn^{2+} or Mn^{2+} displace Co^{2+} bound to the synthetic peptide.

Studies of metals in microbial proteins with finger-loop domains are cited in the middle section of Table 2. Giedroc *et al.* (1986), Keating *et al.* (1988), and Pan and Coleman (1989) showed that metal-free apoproteins of g32p and GAL4 do not bind to DNA, but that the DNA binding of these finger-loop proteins can be partially restored by adding Cd^{2+} or Co^{2+}, instead of Zn^{2+}. South *et al.* (1989) performed NMR analyses upon [133]Cd-substituted nucleic acid-binding protein (NABP) from the HIV-1 virus. The NMR spectra indicate that cadmium is coordinated to 3 cys residues and 1 his residue, consistent with the predicted structure. The presence of a finger-loop domain in the virus that causes human AIDS raises the possibility of chemotherapy by chelation or displacement of the intrinsic Zn atom (Oxford *et al.* 1989).

Research on metals in mammalian finger-loop proteins is summarized in the bottom section of Table 2. The studies of Kadonaga *et al.* (1987) and Westin and Schaffner (1988) indicate that DNA-binding activity of transcription factor Sp1 is prevented by removal of Zn from the finger-loop domains of Sp1 and partially restored by Zn-replacement, but not by additions of divalent Cd, Co, Ni, Cu, or Mn. Freedman *et al.* (1988) showed by EXAFS analysis that the Zn atoms in human glucocorticoid receptor are coordinated to 4 cys groups, consistent with the predicted structure of the finger-loops. Surprisingly, the affinity of the glucocorticoid receptor for Cd is 200-fold that for Zn, suggesting that Cd might be substituted for Zn in the finger-domains of Cd-exposed cells or animals.

TABLE 2. STUDIES OF ZINC REMOVAL AND METAL SUBSTITUTIONS IN FINGER-LOOPS

System	Authors	Experiments	Observations
TFIIIA from Xenopus oocytes	Hanas et al. (1983)	Extraction of Zn^{2+} from TFIIIA with EDTA or 1,10-phenanthroline.	ApoTFIIIA does not bind to the 5S RNA gene unless Zn^{2+} is replaced. Additions of Co^{2+}, Ni^{2+}, Mn^{2+}, or Fe^{2+} do not restore the binding.
	Miller et al. (1985)	AAS analyses of TFIIIA; extraction of oocyte 7S particles with EDTA or 1,10-phenanthroline.	TFIIIA has ~9 Zn/mol; chelation of Zn in 7S particles dissociates TFIIIA from 5S RNA; exposure to Cd^{2+} or Cu^{2+} also dissociates the 7S particles, evidently by displacing Zn^{2+}.
	Diakun et al. (1986)	EXAFS study of TFIIIA.	Coordination sphere of the Zn^{2+} binding site has 2 his and 2 cys residues.
	Frankel et al. (1987)	Circular dichroism (CD) spectra of a synthetic peptide that corresponds to the second finger-loop of TFIIIA.	The peptide folds in the presence of Zn^{2+} with a shift of CD spectrum; the peptide also binds Co^{2+}, yielding a characteristic absorption spectrum.
	Berg & Merkle (1989)	Spectrophotometry of a peptide that corresponds to the second finger of TFIIIA, after metal substitutions.	Affinities of the peptide for divalent metal ions are ranked as follows: $Zn^{2+} > Mn^{2+} > Co^{2+}$.
Microbial finger-loops	Giedroc et al. (1986)	EDTA extraction of g32p from bacteriophage T4.	Cd^{2+} and Co^{2+} can substitute for Zn^{2+} in apo-g32p.
	Keating et al. (1988)	DNA-binding assay and thermal stability tests of metal-substituted g32p.	Cd- or Co-substituted g32p binds to poly(dT) in a manner similar to Zn-g32p; Cd-g32p has less thermal stability than Zn-g32p, whereas Co-g32p has greater stability than Zn-g32p.
	Culp et al. (1988)	E1A protein of adenovirus type 5.	ApoE1A protein binds 1 atom of Zn^{2+} per mole.
	Nagai et al. (1988)	Domain of 3 Zn-fingers of yeast SWI5 protein, expressed in E coli.	Zn^{2+} is required for specific DNA-binding; the Zn-peptide protects the binding site from DNAase.
	Pan & Coleman (1989)	Yeast GAL4 fragment expressed in E coli, treated with EDTA to extract Zn^{2+}.	Apoprotein does not bind to DNA; binding is restored by adding Zn^{2+}, Cd^{2+}, or Co^{2+}, which also protect the GAL4 protein from tryptic proteolysis.
	South et al. (1989)	18 residue finger peptide from HIV-1 virus nucleic acid binding protein p7.	NMR analysis of $^{113}Cd^{2+}$-substituted peptide shows Cd-coordination to 3 cys residues and 1 his residue
Mammalian finger-loop proteins	Kadonaga et al. (1987)	Extraction of Sp1 with EDTA to remove Zn^{2+} from the three finger-loop domains.	After EDTA treatment, Sp1 does not bind to the GC boxes of SV40; binding is partly restored by Zn^{2+} addition, but not by Co^{2+} or Ni^{2+}.
	Westin & Schaffner (1988)	Extraction of Sp1 with 1,10-phenanthroline or EDTA.	Chelators prevent Sp1 binding to a GC box in *Herpes simplex*; Sp1 binding is restored by adding Zn^{2+}; but not by additions of Cd^{2+}, Cu^{2+}, or Mg^{2+}.
	Freedman et al. (1988)	EDTA extraction and EXAFS analysis of finger-loop domains in the human glucocorticoid (hGR) receptor protein.	Zn^{2+} in hGR binds to 4 cys residues; apo-hGR does not bind to specific DNA, but binding is restored by Zn^{2+} or Cd^{2+}; affinity of hGR for Cd^{2+} is 200-fold that for Zn^{2+}.

5. Zn-FINGERS IN HORMONE RECEPTOR PROTEINS

The human steroid hormone receptors have similar structures: a variable immunogenic region located at the N-terminus; the DNA-binding 'C' region, which contains two zinc finger-loops ('CI' and 'CII'), located in the middle; and the hormone-binding site located near the COOH-terminus (Beato 1989). Binding of the appropriate hormone to the C-terminal site activates the receptor, possibly by releasing a heat-shock protein, 'HSP-90', that may block nuclear uptake of the receptor (Green and Chambon 1989). The receptor then binds to a palindromic hormone response element (HRE), causing hormone-specific transcriptional activation (Danielsen *et al.* 1987). Two molecules of the steroid receptor evidently bind on opposite sides of the double-stranded DNA segment that contains the palindromic HRE, forming a dimer that may be stabilized by complexation of an additional Zn^{2+} ion (Green and Chambon 1989).

Current understanding of the mechanisms whereby the finger-loops of steroid hormone receptors achieve site-specific recognition of their HREs is fragmentary; the available knowledge has been summarized by Beato (1989) and Berg (1989), based largely upon mutational analyses of Zn-fingers in several hormone receptor proteins, as epitomized in Table 3. Studies of Danielson *et al.* (1989) indicate that specificity of binding to the HRE is primarily governed by the two amino acids situated between the second pair of cys residues in finger-loop 'CI'. There are evidently two families of hormone receptors, the estrogen family, which contains glutamic acid and glycine at these positions, and the glucocorticoid family, which contains glycine and serine at these positions. Umesono and Evans (1989) found that the five amino acids between the first pair of cys residues in the second finger-loop, 'CII', discriminate between the thyroid and estrogen response elements within the estrogen family.

The clinical significance of Zn-fingers in hormone receptors was demonstrated by Hughes *et al.* (1989) who analyzed the amino acid sequences of the vitamin D receptors in two families of patients with vitamin D-resistant rickets. In one family, there was a single amino acid exchange at the tip of the CI finger-loop, and in the other family, there was a single amino acid exchange at the tip of the CII finger-loop. Evidently, both finger-loops are essential for correct binding of the vitamin D receptor to its HRE.

TABLE 3. MUTATIONAL ANALYSIS OF Zn-FINGERS IN HORMONE RECEPTOR PROTEINS

Authors	Experimental System	Observations
Severne *et al.*(1988)	DNA-binding Zn-fingers of rat glucocorticoid receptor (rGR).	Site-directed mutagenesis shows that, unlike the other cys residues, the fifth cys of the CII finger is nonessential.
Green & Chambon (1989)	Chimeric Zn-fingers of hGR and hER receptors and palindromic GRE and ERE response elements.	The CI finger specifies the GRE or ERE response element and the CII finger stabilizes DNA binding, possibly by forming a dimeric protein complex.
Mader *et al.* (1989)	DNA-binding Zn-finger domains of hGR and hER receptors; GRE and ERE.	Three amino acids located at the C-terminal side of CI finger discriminate between the GRE and ERE.
Danielson *et al.* (1989)	DNA-binding Zn-finger domains of mGR and mER; GRE and ERE response elements.	The 2 amino acids between the cys residues in the C-terminal knuckle of the CI finger discriminate between the GRE and ERE response elements.
Umesono & Evans (1989)	Zn-fingers of hGR, hTRβ, and hER; GRE, TRE, & ERE response elements.	The 5 amino acids between the cys residues in the N-terminal knuckle of the CII finger discriminate between TRE and ERE.

6. SPECULATIONS ABOUT THE PATHOPHYSIOLOGY OF METALS IN FINGER-LOOPS

The author proposes that substitutions of foreign metal ions for Zn^{2+} in finger-loops of DNA-binding proteins may induce genotoxicity, based upon the following rationale. Peptide complexes of Ni^{2+}, Co^{2+}, Cu^{2+}, and Cd^{2+} are known to generate oxygen free radicals (*e.g.*, $OH^{\bullet}$) in biological systems (Ueda *et al.* 1985, Sunderman 1986, Inoue and Kawanishi 1989). Zn^{2+} has an ionic radius of 0.74 Å, compared to the ionic radii of Ni^{2+} (0.69Å), Co^{2+} (0.74Å), Cu^{2+} (0.73Å), and Cd^{2+} (0.97 Å) (Weast 1984, Martin 1988). Furthermore, Ni^{2+}, Co^{2+}, Cu^{2+}, and Cd^{2+} can mimic Zn^{2+} by forming tetrahedral coordination complexes with thiol-sulfur and imidazole-nitrogen atoms, so the foreign metals might substitute *in vivo* for Zn^{2+} in finger-loop domains, affecting the conformation or stability of the DNA-binding structures. By generating oxygen free radicals close to specific gene loci, foreign metal ions in finger-loops could hypothetically cause

DNA cleavage or DNA-protein cross-links. In support of this concept, Cu-complexes are used as DNA-cleavage reagents in hydroxyl radical foot-printing assays (Spassky and Sigman 1985, Tullius 1987). If the DNA damage occurs in oncogenes or at other critical regulatory sites, carcinogenesis or teratogenesis could result (Sunderman 1989, Vainio 1989). Finally, the present author suggests that nutritional deficiency of zinc might cause depletion of Zn^{2+} from the finger-loop domains of the androgen receptor or the testis determining factor, resulting in the dwarfism and male hypogonadism that have been recognized as endocrine manifestations of dietary zinc deprivation (Prasad 1982).

7.   REFERENCES

Allison LA, Moyle M, Shales M, Ingles CJ (1985) Extensive homology among the largest subunits of eukaryotic and prokaryotic RNA polymerases. Cell 42:599-610

Arriza JL, Weinberger C, Cerelli G, Glaser TM, Handelin BL, Housman DE, Evans RM (1987) Cloning of human mineralocorticoid receptor complementary DNA: structural and functional kinship with the glucocorticoid receptor. Science 237:268-275

Baldarelli RM, Mahoney PA, Salas F, Gustavson E, Boyer PD, Chang MF, Roark M, Lengyel JA (1988) Transcripts of the *Drosophila* blastoderm-specific locus, *terminus*, are concentrated posteriorly and encode a potential DNA-binding finger. Dev Biol 125:85-95

Baum JA, Geever R, Giles NH (1987) Expression of qa-1F activator protein: identification of upstream binding sites in the *qa* gene cluster and localization of the DNA-binding domain. Mol Cell Biol 7:1256-1266

Beato M (1989) Gene regulation by steroid hormones. Cell 56:335-344

Ben-Chetrit E, Gandy BJ, Tan EM, Sullivan KF (1989) Isolation and characterization of a cDNA clone encoding the 60-kD component of the human SS-A/Ro ribonucleoprotein autoantigen. J Clin Invest 83:1284-1292

Berg JM (1986) Potential metal-binding domains in nucleic acid binding proteins. Science 232:485-487

Berg JM (1988) Proposed structure for the zinc-binding domains from transcription factor IIIA and related proteins. Proc Natl Acad Sci USA 85:99-102

Berg JM (1989) DNA binding specificity of steroid receptors. Cell 57:1065-1068

Berg JM, Merkle DL (1989) On the metal ion specificity of "zinc finger" proteins. J Am Chem Soc 111:3759-3761

Boulay JS, Dennefeld C, Alberga A (1987) The *Drosophila* developmental gene *snail* encodes a protein with nucleic acid binding fingers. Nature 330:395-398

Brown DD (1984) The role of stable complexes that repress and activate eukaryotic genes. Cell 37:359-365

Brown RS, Sander C, Argos P (1985) The primary structure of transcription factor TFIIIA has 12 consecutive repeats. FEBS Lett 186:271-274

Chanet R, Magana-Schwencke N, Fabre F (1988) Potential DNA-binding domains in the RAD18 gene product of *Saccharomyces cerevisiae*. Gene 74:543-547

Changelian PS, Feng P, King TC, Milbrandt J (1989) Structure of the NGFI-A gene and detection of upstream sequences responsible for its transcriptional induction by nerve growth factor. Proc Natl Acad Sci USA 86:377-381

Chavrier P, Lemaire P, Revelant O, Bravo, R, Charnay P (1988) Characterization of a mouse multigene family that encodes zinc finger structures. Mol Cell Biol 8:1319-1326

Chowdhury K, Deutsch U, Gruss P (1987) A multigene family encoding several "finger" structures is present and differentially active in mammalian genomes. Cell 48:771-778

Chowdhury K, Dressler G, Breier G, Deutsch U, Gruss P (1988) The primary structure of the murine multifinger gene mKr2 and its specific expression in developing and adult neurons. EMBO J 7:1345-1353

Copeland TD, Oroszlan S, Kalyanaraman VS, Sarngadharan MG, Gallo RC (1983) Complete amino acid sequence of human T-cell leukemia virus structural protein p15. FEBS Lett 162:390-395

Culp JS, Webster LC, Friedman DJ, Smith CL, Huang WJ, Wu FYH, Rosenberg M, Ricciardi RP (1988) The 289-amino acid E1A protein of adenovirus binds zinc in a region that is important for trans-activation. Proc Natl Acad Sci USA 85:6450-6454

Dang CV, van Dam H, Buckmire M, Lee WM (1989) DNA-binding domain of human c-myc produced in *Escherichia coli*. Mol Cell Biol 9:2477-2486

Danielsen M, Hinck L, Ringold GM (1989) Two amino acids within the knuckle of the first finger specify DNA response element activation by the glucocorticoid receptor. Cell 57:1131-1138

Danielsen M, Northrop JP, Konklaas J, Ringold GM (1987) Domains of the glucocorticoid receptor involved in specific and nonspecific deoxyribonucleic acid binding, hormone activation, and transcriptional enhancement. Mol Endocrinol 1:816-822

Diakun GP, Fairall L, Klug A (1986) EXAFS study of the zinc-binding sites in the protein transcription factor IIIA. Nature 324:698-699

Evans RM, Hollenberg SM (1988) Zinc fingers: guilt by association. Cell 52:1-3

Fairall L, Martin S, Rhodes D (1989) The DNA binding site of the *Xenopus* transcription factor IIIA has a non-B-form structure. EMBO J 8:1809-1817

Fairall L, Rhodes D, Klug A (1986) Mapping of the sites of protection on a 5 S RNA gene by the *Xenopus* transcription factor IIIA. J Mol Biol 192:566-591

Frankel AD, Berg JM, Pabo CO (1987) Metal-dependent folding of a single zinc finger from transcription factor IIIA. Proc Natl Acad Sci USA 84:4841-4845

Freedman LP, Luisi BF, Korszun ZR, Basavappa R, Sigler PB, Yamamoto KR (1988) The function and structure of the metal coordination sites within the glucocorticoid receptor DNA binding domain. Nature 334:543-546

Gibson TJ, Postma JPM, Brown RS, Argos P (1988) A model for the tertiary structure of the 28 residue DNA-binding motif ('zinc finger') common to many eukaryotic transcriptional regulatory proteins. Prot Engineering 2:209-218

Giedroc DP, Keating KM, Williams KR, Konigsberg WH, Coleman JE (1986) Gene 32 protein, the single-stranded DNA binding protein from bacteriophage T4, is a zinc metalloprotein. Proc Natl Acad Sci USA 83:8452-8456

Ginsberg AM, King BO, Roeder RG (1984) Xenopus 5S gene transcription factor, TFIIIA: characterization of a cDNA clone and measurement of RNA levels throughout development. Cell 39:479-489

Green S, Chambon P (1989) Chimeric receptors used to probe the DNA-binding domain of the estrogen and glucocorticoid receptors. Cancer Res 49S:2282-2285

Greene GL, Gilna P, Waterfield M, Baker A, Hort Y, Shine J (1986) Sequence and expression of human estrogen receptor complementary DNA. Science 231:1150-1154

Hanas JS, Hazuda DJ, Bogenhagen DF, Wu FYH, Wu CW. (1983) Xenopus transcription factor A requires zinc for binding to the 5 S RNA gene. J Biol Chem 258:14120-14125

Hartshorne, TA, Blumberg H, Young ET (1986) Sequence homology of the yeast regulatory protein ADR1 with Xenopus transcription factor TFIIIA. Nature 320:283-287

Haussler MR, Mangelsdorf DJ, Yamaoka K, Allegretto EA, Komm BS, Terpening CM, McDonnell DP, Pike JW, O'Malley BW (1988) Molecular characterization and actions of the vitamin D hormone receptor. Steroid Hormone Action 1:247-262

Hughes MR, Malloy PJ, Kieback DG, Kesterson RA, Pike JW, Feldman D, O'Malley BW (1988) Point mutations in the human vitamin D receptor gene associated with hypocalcemic rickets. Science 242:1702-1705

Icho T, Wickner RB (1987) Metal-binding, nucleic acid-binding finger sequences in the CDC16 gene of Saccharymyces cerevisiae. Nucl Acids Res 15:8439-8450

Inoue S, Kawanishi S (1989) ESR evidence for superoxide, hydroxyl radicals and singlet oxygen produced from hydrogen peroxide and nickel(II) complex of glycylglycyl-L-histidine. Biochem Biophys Res Comm 159:445-451

Johnston M (1097) Genetic evidence that zinc is an essential co-factor in the DNA binding domain of GAL4 protein. Nature 328:353-355

Joseph LJ, Le Beau MM, Jamieson GA Jr, Acharya S, Shows TB, Rowley JD, Sukhatme VP (1988) Molecular cloning, sequencing, and mapping of EGR2, a human early growth response gene encoding a protein with "zinc-binding finger" structure. Proc Natl Acad Sci USA 85:7164-7168

Kadonaga JT, Carner KR, Masiarz FR, Tjian R (1987) Isolation of cDNA encoding transcription factor Sp1 and functional analysis of the DNA binding domain. Cell 51:1079-1090

Kadonaga JT, Courey AJ, Ladika J, Tjian R (1988) Distinct regions of Sp1 modulate DNA binding and transcriptional activation. Science 242:1566-1570

Keating KM, Ghosaini LR, Giedroc DP, Williams KR, Coleman JC, Sturtevant JM (1988) Thermal denaturation of T4 gene 32 protein: effects of zinc removal and substitution. Biochemistry 27:5340-5245

Kidd S, Kelley MR, Young MW (1986) Sequence of the notch locus of Drosophila melanogaster: relationship of the encoded protein to mammalian clotting and growth factors. Mol Cell Biol 6:3094-3108

Kim J, Michels CA (1988) The MAL63 gene of Saccharomyces encodes a cysteine-zinc finger protein. Curr Genet 14:319-323

Kinzler KW, Ruppert JM, Bigner SH, Vogelstein B (1988) The GLI gene is a member of the Krüppel family of zinc finger proteins. Nature 332:371-374

Klug A, Rhodes A (1987) 'Zinc fingers': a novel protein fold for nucleic acid recognition. Cold Spring Harbor Symp Quant Biol 52:473-482

Lee MS, Gippert GP, Soman KV, Case DA, Wright PE (1989) Three-dimensional solution structure of a single zinc finger DNA-binding domain. Science 245:635-637

Mader S, Kumar V, de Verneuil H, Chambon P (1989) Three amino acids of the oestrogen receptor are essential to its ability to distinguish an oestrogen from a glucocorticoid-responsive element. Nature 338:271-274

Mardon G, Page DC (1989) The sex-determining region of the mouse Y chromosome encodes a protein with a highly acidic domain and 13 zinc fingers. Cell 56:765-770

Martin RB (1988) Nickel ion binding to amino acids and peptides. In: Sigel H (ed) Metal ions in biological systems. Dekker, New York, p 123-164

Mazen A, Menissier-de Murcia J, Molinete M, Simonin F, Gradwohl G, Poirier G, de Murcia G (1989) Poly(ADP-ribose)polymerase: a novel finger protein. Nucl Acids Res 17:4689-4698

Messenguy F, Dubois E, Deachamps F (1986) Nucleotide sequence of the ARGRII regulatory gene and amino acid sequence homologies between ARGRII, PPRI and GAL4 regulatory proteins. Eur J Biochem 157:77-81

Miller J, McLachlan AD, Klug A (1985) Repetitive zinc-binding domains in the protein transcription factor IIIA from Xenopus oocytes. EMBO J 4:1609-1614

Misrahi M, Atger M, d'Auriol L, Loosfelt H, Meriel C, Fridlansky F, Guiochon-Mantel A, Galibert F, and Milgrom (1987) Complete amino acid sequence of the human progesterone receptor deduced from cloned cDNA. Biochem Biophys Res Comm 143:740-748

Nagai K, Nakaseko Y, Nasmyth K, Rhodes D (1988) Zinc-finger motifs expressed in E. coli and folded in vitro direct specific binding to DNA. Nature 332:284-286

Ono Y, Fujii T, Ogita K, Kikkawa U, Igarashi K, Nishizuka Y (1989) Protein kinase C ξ subspecies from rat brain: its structure, expression, and properties. Proc Natl Acad Sci USA 86:3099-3103

Oxford JS, Coates ARM, Sia DY, Brown K, Asad S (1989) Potential target sites for antiviral inhibitors of human immunodeficiency virus (HIV). J Antimicrob Chemother 23A:9-27

Pan T, Coleman JE (1989) Structure and function of the Zn(II) binding site within the DNA-binding domain of the GAL4 transcription factor. Proc Natl Acad Sci USA 86:3145-3149

Pannuti A, Lanfrancone L, Pascucci A, Pelicci PG, La Mantia G, Lania L (1988) Isolation of cDNAs encoding finger proteins and measurement of the corresponding mRNA levels during myeloid terminal differentiation. Nucl Acids Res 16:4227-4237

Párraga G, Horvath SJ, Eisen A, Taylor WE, Young ET, Klevit RE (1988) Zinc-dependent structure of a single-finger domain of yeast ADR1. Science 241:1489-1492

Payre F, Vincent A (1988) Finger proteins and DNA-specific recognition: distinct patterns of conserved amino acids suggest different evolutionary modes. FEBS Lett 234:245-250

Pays E, Murphy NB (1987) DNA-binding fingers encoded by a trypanosome retroposon. J Mol Biol 197:147-148

Petkovich M, Brand NJ, Krust A, Chambon P (1987) A human retinoic acid receptor which belongs to the family of nuclear receptors. Nature 330:444-450

Prasad AS (1982) Clinical disorders of zinc deficiency. In: Prasad AS, Dreosti IE, Hetzel BS (eds) Clinical applications of recent advances in zinc metabolism. Liss, New York, p 89-120

Ruiz i Altaba A, Perry-O'Keefe H, Melton D.A (1987) *Xfin:* an embryonic gene encoding a multifingered protein in *Xenopus.* EMBO J 6:3065-3070

Salvesen G, Lloyd C, Farley D (1987) cDNA encoding a human homolog of yeast ubiquitin 1. Nucl. Acids Res 15:5485

Schiff LA, Nibert ML, Co MS, Brown EG, Fields BN (1988) Distinct binding sites for zinc and double-stranded RNA in the reovirus outer capsid protein σ3. Mol Cell Biol 8:273-283

Schneider-Gädicke A, Beer-Romero P, Brown LG, Nussbaum R, Page DC (1989) ZFX has a gene structure similar to ZFY, the putative human sex determinant, and escapes X inactivation. Cell 57:1247-1258

Schuh R, Aicher W, Gaul U, Côté S, Preiss A, Maier D, Seifert E, Nauber U, Schröder C, Kemler R, Jäckle H (1986) A conserved family of nuclear proteins containing structural elements of the finger protein encoded by *Krüppel*, a *Drosophila* segmentation gene. Cell 47:1025-1032,

Sehnke PC, Mason AM, Hood SJ, Lister RM, Johnson JE (1989) A "zinc-finger"-type binding domain in tobacco streak virus coat protein. Virology 168:48-56

Severne Y, Wieland S, Schaffner W, Rusconi S (1988) Metal binding 'finger'structures in the glucorticoid receptor defined by site-directed mutagenesis. EMBO J 7:2503-2508

Seyfert VL, Sukhatme VP, Monroe JG (1989) Differential expression of a zinc finger-encoding gene in response to positive versus negative signaling through receptor immunoglobulin in murine B lymphocytes. Mol Cell Biol 9:2083-2088

South L, Kim B, Summers MF (1989) [113]Cd NMR studies of a 1:1 Cd adduct with and 18-residue finger peptide from HIV-1 nucleic acid binding protein. J Am Chem Soc 111:395-396

Spassky A, Sigman DS (1985) Nuclease activity of 1,10-phenanthroline-copper ion: conformational analysis and footprinting of the lac operon. Biochemistry 24:8050-8056

Sunderman FW Jr (1986) Metals and lipid peroxidation. Acta Pharmacol Toxicol 59 (Suppl 7):243-255

Sunderman FW Jr (1989) Mechanisms of nickel carcinogenesis. Scand J Work Environ Health 15:1-12

Sunderman FW Jr, Barber AM (1988) Finger-loops, oncogenes, and metals. Ann Clin Lab Sci 18:267-288

Sweetser D, Nonet M, Young RA (1987) Prokaryotic and eukaryotic RNA polymerases have homologous core subunits. Proc Natl Acad Sci USA 84:1192-1196

Tautz D, Lehmann R, Schnürch H, Schuh R, Seifert E, Kienlin A, Jones K, Jäckle H (1987) Finger protein of novel structure encoded by *hunchback*, a second member of the gap class of *Drosophila* segmentation genes. Nature 327:383-389

Tsai SF, Martin DIK, Zon LI, D'Andrea AD, Wong GG, Orkin SH (1989) Cloning of cDNA for the major DNA-binding protein of the erythroid lineage through expression in mammalian cells. Nature 339:446-451

Tullius TD, Dombroski BA, Churchill MEA, Kam K (1987) Hydroxyl radical footprinting: a high resolution method for mapping protein-DNA contacts. Meth Enzymol 155:537-538

Ueda K, Kobayashi S, Morita J, Komano T (1985) Site-specific DNA damage caused by lipid peroxidation products. Biochim Biophys Acta 824:341-348

Umesono K, Evans RM (1989) Determinants of target gene specificity for steroid/thyroid hormone receptors. Cell 57:1139-1146

Umesono K, Giguere V, Glass C, Rosenfeld MG, Evans RM (1988) Retinoic acid and thyroid hormone induce gene expression through a common responsive element. Nature 336:262-265

Vainio H (1989) Carcinogenesis and teratogenesis may have common mechanisms. Scand J Work Environ Health 15:13-17

Vincent A, Colot HV, Rosbash M (1986) Sequence and structure of the *Serendipity* locus of *Drosophila melanogaster*. J Mol Biol 186:149-166

Vodkin MH, Vodkin LO (1989) A conserved zinc finger domain in higher plants. Plant Mol Biol 12:593-594

Wang LH, Tsai SY, Cook RG, Beattie WG, Tsai JM, O'Malley BW (1989) COUP transcription factor is a member of the steroid receptor superfamily. Nature 340:163-166

Weinberger C, Hollenberg SM, Rosenfeld MG, Evans RM (1985) Domain structure of human glucocorticoid receptor and its relationship to the v-*erb*-A oncogene product. Nature 318:670-672

Weinberger C, Thompson CC, Ong ES, Lebo R, Gruol DJ, Evans RM (1986) The c-*erb*-A gene encodes a thyroid hormone receptor. Nature 324:641-646,

Weast RC, Astle MJ, Beyer WH (1984) Handbook of chemistry and physics, Sixty-fifth ed., CRC Press, Palm Beach, 1984, p F165

Westin G, Schaffner W (1988) Heavy metal ions in transcription factors from HeLa cells: Sp1, but not octamer transcription factor requires zinc for DNA binding and for activation function. Nucl Acids Res 16:5771-5781

Wray LV Jr, Witte MM, Dickson RC, Riley MI (1987) Characterization of a positive regulatory gene, *LAC9*, that controls induction of the lactose-galactose regulon of *Kluyveromyces lactis*: structural and functional relationships to *GAL4* of *Saccharomyces cerevisiae*. Mol Cell Biol 7:1111-1121

Construction of Mouse Glutathione Peroxidase Gene and Its Expression

Kenji Soda[1], Manabu Sugimoto[1,2], Nobuyoshi Esaki[1], Hidehiko Tanaka[1,3], and Paul R. Harrison[4]

[1] Institute for Chemical Research, Kyoto University, Kyoto, 611 Japan
[2] Research Institute for Bioresources, Okayama University, Kurashiki, 710 Japan
[3] Faculty of Agriculture, Okayama University, Okayama, 700 Japan
[4] The Beatson Institute for Cancer Research, Garscibe Estate, Bearsden, UK

ABSTRACT

The essentiality of selenium can be ascribed mostly to the presence of enzymes containing selenocysteine (SeCys) as an integral moiety. Recently, the genes of mammalian glutathione peroxidase (GSHPx) and Escherichia coli formate dehydrogenase were cloned, and it was shown that a nonsense opal codon, TGA, encodes SeCys. However, the mechanism of the gene expression is still obscure. We here describe the construction of chimeric fused genes of mouse GSHPx and E. coli β-galactosidase (β-gal), and its expression in E. coli. DNA fragments of GSHPx gene which include the TGA codon encoding the SeCys residue, were inserted into the portion of lacZ' gene which encodes the amino terminal moiety of β-gal. When E. coli JM109 cells were transformed with the fused genes, β-gal activity appeared in the transformants under anaerobic conditions. These indicate that the UGA was readthrough to give the protein containing SeCys residue in-frame under anaerobic conditions. Moreover, selenium was incorporated into the fused proteins produced anaerobically.

KEY WORDS: Glutathione peroxidase, Gene, Selenocysteine

INTRODUCTION

Selenium is a homologue of sulfur and tellurium, and its toxicity is well-known (Diplock 1981). However, it is now recognized as an essential micronutrient for mammals, fish and several bacteria (Scott 1973; Stadtman 1979). Recently, it was shown that selenium deficiency causes Keshan Disease, an endemic cardiomyopathy in China, and also Kaschin-Beck Disease (Keshan Disease Research Group 1979).

The essentiality of selenium can be ascribed mostly to the presence of enzymes containing selenocysteine (SeCys)[1] as an integral moiety. Mammalian glutathione peroxidase (GSHPx), bacterial formate dehydrogenase (FDH) and some others contain a SeCys residue in their active centers (Stadtman 1980).

Recently, the genes of human and mouse GSHPx and E. coli FDH were cloned, and it was shown that a nonsense opal codon, UGA encodes SeCys (Chambers 1986; Zinoni 1986). However, the mechanism of the gene expression is still obscure.

We have reconstructed the chromosomal DNA of mouse GSHPx and fused it with the β-galactosidase (β-gal) gene to express the fused gene in E. coli cells. We have also attempted to introduce the nonsense codon UGA into a gene of a protein which has no SeCys inherently, to produce a seleno-protein.

---

[1]Abbreviations: SeCys, selenocysteine; GSHPx, glutathione peroxidase; β-gal, β-galactosidase; Amp, ampicillin; IPTG, isopropyl-β-D-thiogalactopyranoside; X-Gal, 5-bromo-4-chloro-3-indolyl-β-D-galactoside; MT, metallothionein.

EXPERIMENTAL PROCEDURES

Materials

Restriction endonucleases, T4 DNA ligase, and T4 polynucleotide kinase were purchased from Takara Shuzo, Japan, and New England Biolabs, U.S.A. Ampicillin (Amp), isopropyl-β-D-thiogalactopyranoside (IPTG) and 5-bromo-4-chloro-3-indolyl-β-D-galactoside (X-Gal) were from Nacalai Tesque, Japan.

Synthesis and Purification of Oligonucleotides

The Neurospora crassa metallothionein (MT) gene and primers used for mutagenesis were synthesized with an Applied Biosystems 381A automated DNA synthesizer. The oligonucleotides were purified by HPLC with a reverse-phase column under the conditions described (Sugimoto 1988).

Construction of Plasmids

Mouse genomic GSHPx gene was cloned by Chambers (1986). The structural gene of enzyme consists of about 250 bp segment encoding the peptide moiety between methionine and glutamine, and about 350 bp segment corresponding from glutamate to serine. They are linked with an intron with about 200 bp. We have reconstructed the gene as shown in Fig. 1.

We synthesized a 35 bp linker A and a 15 bp linker B, which have EcoRI and SacII sites, OxaNI and HindIII at both sites, respectively. SacII-OxaNI fragment was cut out from pUC12-GSHPx plasmid and subcloned into pUC19 treated with EcoRI and HindIII with linker A and linker B (pGSHPx-2).

We eliminated the intron from pGSHPx-2 as shown in Fig. 2. We substituted new four bases, ATCC for TATA on the 5'- end side of intron, and also three bases, TCC for GAA in the 5'- end side of exon by site-directed mutagenesis by means of primers, 5'-TGGTAGACGGGGTCCTAGGACCACAGGCTT-3'(30 mer), 5'-AAGTAA GAACGGTCCTAGGATCCCTTCTG-3'(29 mer), respectively. Consequently, a BamHI site occurred in the substituted DNA. The intron was removed by BamHI digestion and ligated both ends of the larger DNA segment (pGSHPx-4). Therefore, the GSHPx gene was constructed without frame-shift. The structural gene of GSHPx

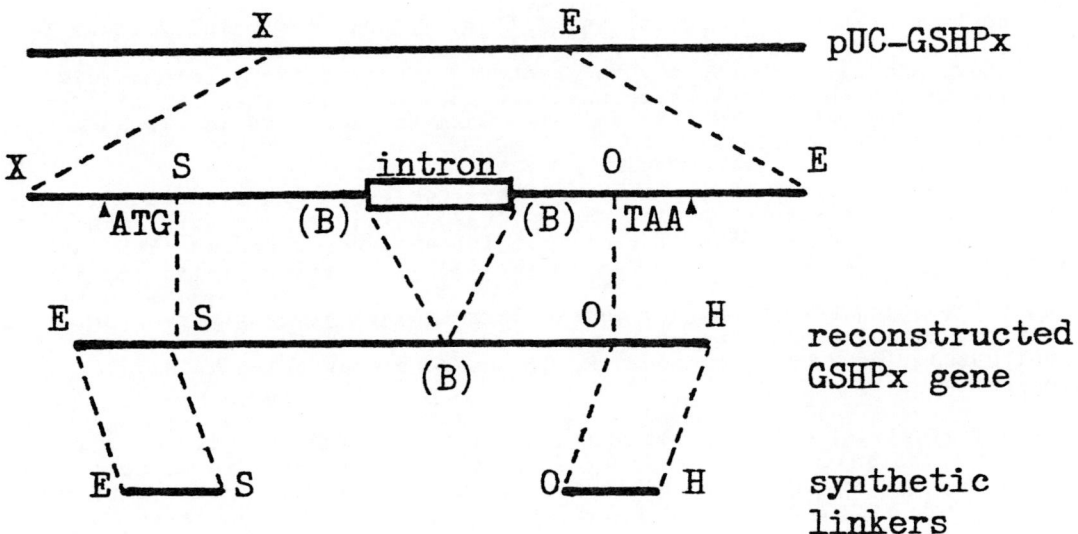

Fig. 1. Strategy for reconstruction of mouse GSHPx gene. Restriction sites are indicated by the following abbreviations: A, AccII; B, BamHI; E, EcoRI; H, HindIII; O, OxaNI; S, SacII; X, XmaI.

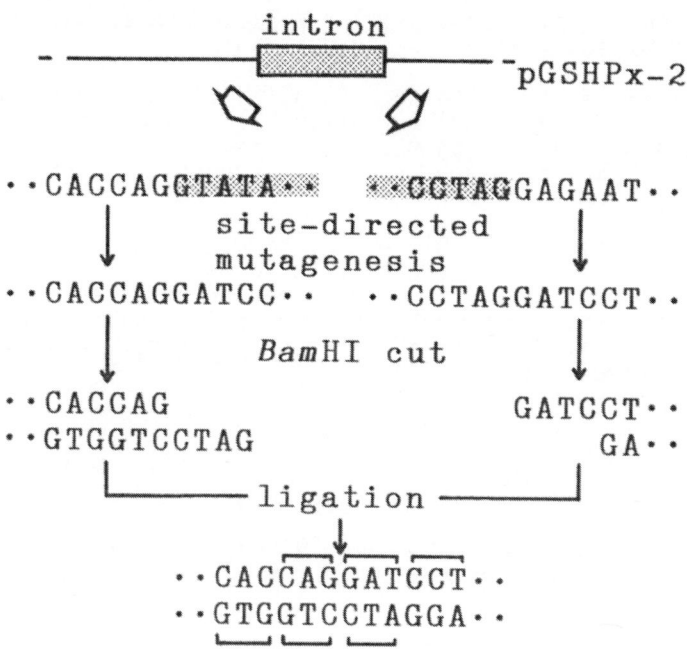

Fig. 2. Elimination of intron. BamHI site were introduced into both the ends of intron by the substitution of new four bases, ATCC for TATA on the 5' end side of intron, and also three bases TCC for GAA on the 3' end side of intron by site-directed mutagenesis.

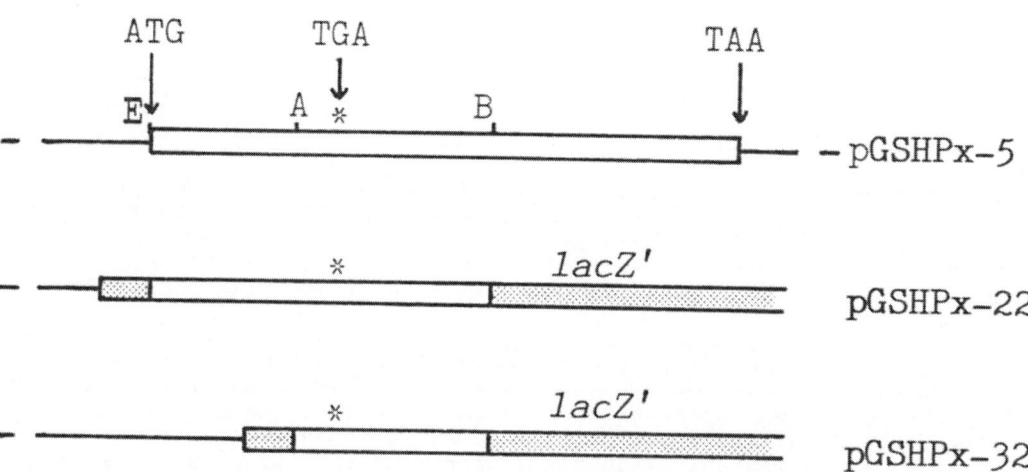

Fig. 3. Restriction map of subcloned mouse GSHPx gene and fused genes with lacZ'. The EcoRI-BamHI fragment (250 bp) and the AccII-BamHI fragment (110 bp) were prepared from pGSHPx-4, and fused with lacZ' gene of pUC118 and synthetic linker (pGSHPx-22, pGSHPx-32, respectively). The asterisk desig-nates the position of the TGA codon. Restriction sites are indicated by the following abbreviations: A, AccII; B, BamHI; E, EcoRI.

was obtained from pGSHPx-4 by <u>Eco</u>RI and <u>Hind</u>III digestion. We cloned it into
the E. <u>coli</u> expression vector, pKK223-3 to obtain the plasmid pGSHPx-5.

The <u>Eco</u>RI-<u>Bam</u>HI fragment of 250 bp and the <u>Acc</u>II-<u>Bam</u>HI fragment of 110 bp
from pGSHPx-5 were cut out, and fused with linkers and <u>lacZ</u>' gene of pUC118
plasmid digested with <u>Eco</u>RI and <u>Bam</u>HI, to give the right translational
reading frame as the latter gene (pGSHPx-22, pGSHPx-32, respectively) (Fig.3).

Oligonucleotides coding for the <u>N</u>. <u>crassa</u> MT gene (S1-S10) were
phosphorylated by ATP and T4 polynucleotide kinase. The mixture of
phosphorylated (S2-S9) and nonphosphorylated (S1, S10) oligonucleotides was
heated at 95°C for 3 min, and allowed to cool slowly to room temperature.
After ligation with T4 DNA ligase, the mixture was phosphorylated and cloned
into <u>Eco</u>RI/<u>Hind</u>III sites of pUC18. Thus, the plasmid, pFMT70, carrying the
MT gene was obtained (Fig. 4).

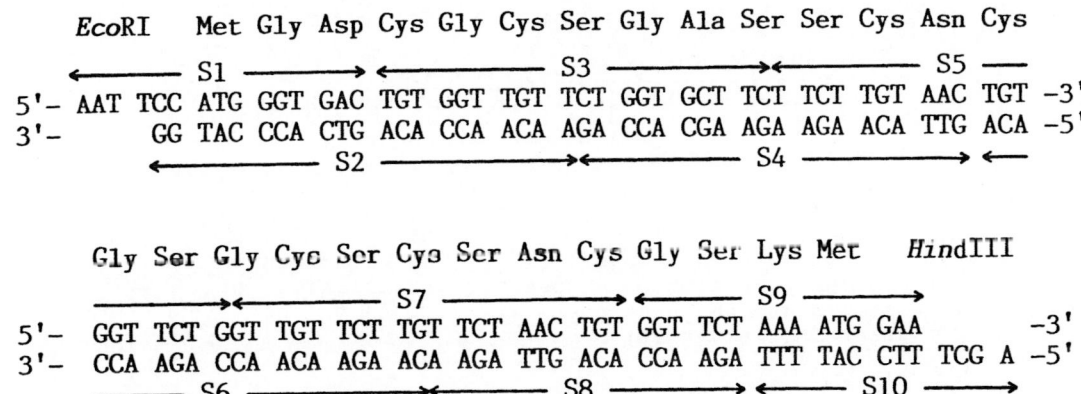

Fig. 4. Nucleotide and amino acid sequences of synthetic <u>N</u>. <u>crassa</u> MT gene.
Individual oligonucleotides (S1-S10) are shown by arrow.

RESULTS AND DISCUSSION

Expression of GSHPx Gene

E. <u>coli</u> expression vector, pKK223-3 has <u>tac</u> promoter upstream of the multi-
cloning site, and a terminater its downstream (Amann 1983). We inserted the
structural gene of GSHPx between <u>Eco</u>RI and <u>Hind</u>III site to obtain pGSHPx-5,
in which a distance between the initiation codon, ATG and Shine-Dalgarno
sequence is 10 bp. We transformed E. <u>coli</u> JM109 cells with pGSHPx-5 to
examine the expression.

We incubated the clone cells with ^{35}S-methionine and ^{75}Se-selenite to label
proteins, and analyzed them by SDS-polyacrylamide gel electrophoresis (SDS-
PAGE). The molecular weight of the enzyme subunit is about 22,000, but no
appreciable bands at the position corresponding to the value was detected,
namely, the enzyme was not synthesized under the conditions used (data not
shown).

We attempted to fuse the enzyme gene with β-galactosidase (β-gal) gene, and
express the fused gene. The GSHPx gene fragments in pGSHPx-22 and pGSHPx-32
encode the amino terminal moiety of β-gal, and contain the TGA codon encoding
the SeCys residue. Therefore, if the TGA codon is read-through, the fused
protein is produced, and easily detected by measurement of β-gal activity.
We transformed E. <u>coli</u> JM109 cells with pGSHPx-22 and pGSHPx-32, and then
determined β-gal of the transformants.

The cloned cells were cultivated on the agar medium containing Amp (50 μg/ml), IPTG (1 mM) and X-Gal (2 %). E. coli-pUC118 clone cells, a positive control turned blue. E. coli-pGSHPx-22 and pGSHPx-32 cells grown inside the agar medium only became blue gradually after incubation for two days. This suggests that β-gal may be produced only anaerobically. Therefore, the clone cells were incubated under anaerobic conditions, and β-gal was assayed.

The clone cells grown aerobically were then cultivated anaerobically for one day. IPTG was added and β-gal was determined periodically. The pUC118 cells began to produce β-gal immediately after the addition of IPTG, and pGSHPx-22 and pGSHPx-32 cells also produced β-gal after one day (Fig. 5). These show that the TGA codon of GSHPx gene is readthrough, and the fused gene is expressed under anaerobic conditions. The lag in β-gal production by pGSHPx-22 and pGSHPx-32 cells suggests that some factors to suppress the UGA codon probably are formed inducibly in the period. Then, we examined the incorporation of selenium into the fused protein.

We incubated clone cells anaerobically, and added IPTG. We added ^{75}Se-selenite to the medium after one day, and incubated for one more day. The cell extract was subjected to SDS-PAGE, and labeled proteins were visualized by autoradiography (Fig. 6). In pUC118 and JM109 cells selenium was incorporated into 80K, 110K products and tRNA. This shows that JM109 cells belong to the class 4 mutant described by Leinfelder (1988) and have selenium biosynthetic pathway. We found the occurrence of extra products about 125K and 120K in pGSHPx-22 and pGSHPx-32 cells, respectively. The molecular weights of these products coincide with those of the fused proteins. This shows that some factors suppressing the nonsense codon UGA are formed inducibly under anaerobic conditions, and consequently SeCys is incorporated into the fused protein.

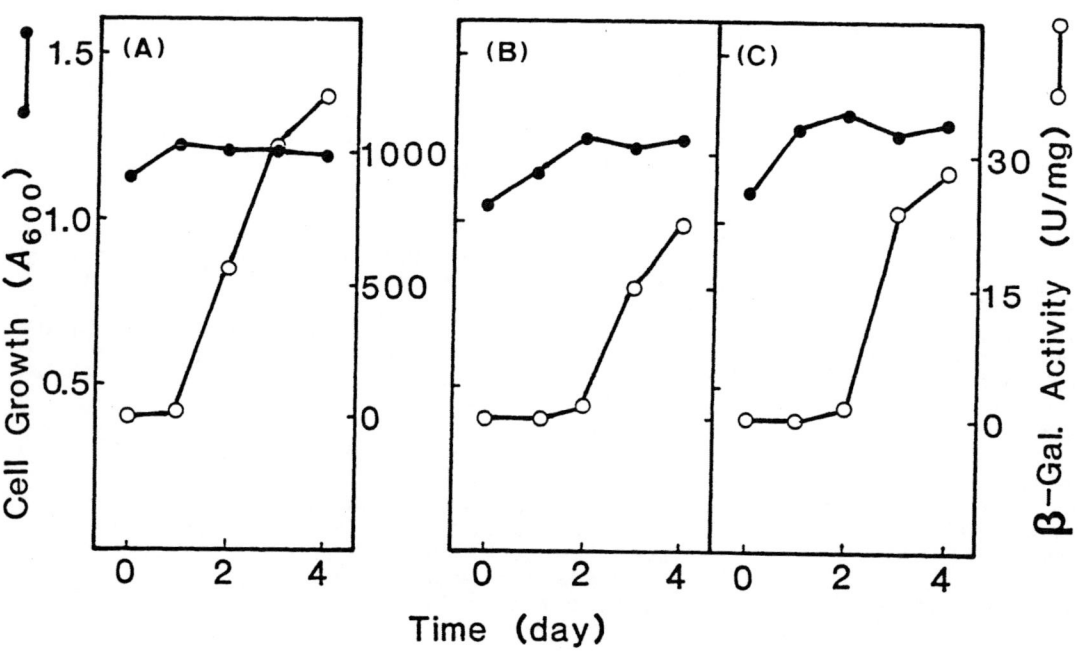

Fig. 5. β-galactosidase production under anaerobic conditions. Overnight cultures of E. coli JM109 cells harboring pUC118(A), pGSHPx-22(B) and pGSHPx-32(C) were inoculated to LB medium containing Amp, followed by the aerobic cultivation at 37°C to a cell density (A_{600}) of 1.0. Then cells were grown at anaerobically in GasPak Jar (BBL Mic. Sys.) without shaking at 37°C for one day, and IPTG was added to a final concentration of 1 mM.

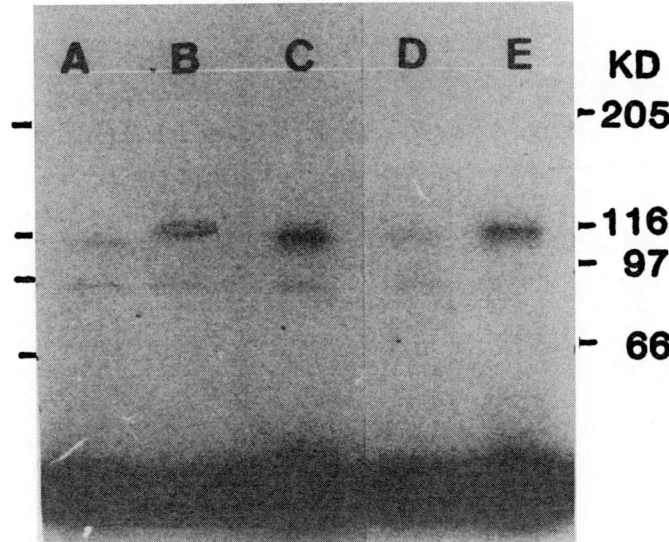

Fig. 6. Autoradiograph of ^{75}Se-labeled cell lysates separated by SDS-PAGE (7.5%). Clone cells were incubated anaerobically, and IPTG and K_2SO_3 were added to a final concentration of 1 mM and 2 mM, respectively. Then ^{75}Se-selenite was added to the medium at a final concentration of 1 μCi/ml after one day, and cells were incubated for one more day. A, pUC118; B, pGSHPx-22; C, pGSHPx-32; D, pFMT70; E, pFSeMT70 cells.

Incorporation of SeCys into Metallothionein

It is interesting to study whether SeCys is incorporated into other proteins by introduction of the nonsense codon UGA into genes and their expression under anaerobic conditions. Therefore, we studied biosynthesis of selenium analogue of metallothionein (MT) in which cysteine residues are replaced by SeCys residues.

Neurospora crassa produces a MT containing 7 cysteine residues in a single polypeptide consisting of 25 amino acid residues. We synthesized the MT gene chemically (Sugimoto 1988), and introduced it into the lacZ' gene (pFMT70). TGT encoding the first cysteine residue near the amino terminal was replaced by TGA by site-directed mutagenesis by means of a 21 mer primer (pFSeMT70) as shown in Fig. 7 (Sugimoto 1989). E. coli JM109 cells were transformed with pFSeMT70.

pFSeMT70 cells grown inside the agar medium containing Amp, IPTG and X-Gal, turned blue and produced β-gal under anaerobic conditions (data not shown). We also incubated pFSeMT70 cells anaerobically in the medium containing ^{75}Se-selenite to label proteins, and analyzed them by SDS-PAGE. pFSeMT70 cells incorporated selenium into the fused protein (Fig. 6). This suggests that the UGA suppressing factors formed inducibly under anaerobic conditions introduce SeCys not only into the inherently SeCys-containing proteins whose SeCys is encoded by the TGA codon, but also into other proteins that inherently have no SeCys.

Mechanism of SeCys Incorporation

Protein synthesis is known to be stopped by any of three termination codons, UAG (amber), UAA (ochre) and UGA (opal), in both prokaryotes and eukaryotes. However, these codons are sometimes read as though they encode an amino acid residue. Because of such alternation in the function, they are also called 'nonsense' codons.

```
        Asn Ser Met Gly Asp Cys Gly Cys Ser Gly      Coding sequence
5'-··AAT TCC ATG GGT GAC TGT GGT TGT TCT GGT···-3'    (pFMT70)
                             *
        3'- G TAC CCA CTG ACT CCA ACA A -5'           Mutant primer
```

⇩

```
        Asn Ser Met Gly Asp *** Gly Cys Ser Gly      Mutant coding
5'-··AAT TCC ATG GGT GAC TGA GGT TGT TCT GGT···-3'    sequence
                                                      (pFSeMT70)
```

Fig. 7. Nucleotide sequence of the synthetic primer used for site-directed mutagenesis. The primer, with asterisk designating the mismatch, introduces a stop codon TGA.

We have described that the nonsense codon UGA is suppressed in E. coli under anaerobic conditions, and SeCys is incorporated into proteins. Several protein genes whose translational termination is inherently determined by UGA occur in E. coli. However, SeCys is not incorporated into these proteins even when the cells are cultivated anaerobically. Then, we propose a mechanism of the anaerobic suppression of UGA in E. coli as follows.

(1) Termination codon: polypeptide chain termination involves a complex interaction among nucleotide signals on mRNA, the ribosome, GTP and the release factors (Tate 1984). There are specific nucleotide sequences around TGA that functions as an inherent termination codon in mRNA. These sequences are recognized by the 16S rRNA. The structure of this complex is interacted with release factors. Then, a polypeptide chain is released from the tRNA and the elongation of a polypeptide chain is stopped at the UGA codon.

(2) Nonsense codon: when the UGA codon occurs in frame within the open reading frame, and specific nucleotide sequences are present around the codon, UGA suppression factors which are produced inducibly anaerobic conditions function to SeCys into a protein (Fig. 8).

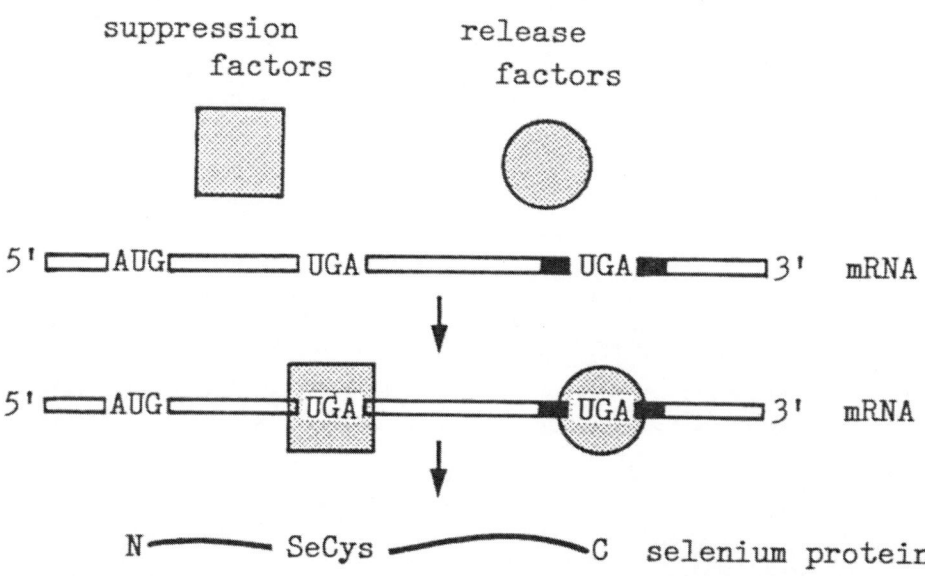

Fig. 8. Proposed scheme for regulation mechanism of SeCys incorporation.

However, at present we have no information about the UGA suppression factors induced only anaerobically, and also the mechanism of suppression in E. coli and also mammals. We must clarify those factors and the mechanism.

REFERENCES

Amann E, Brosius J, Ptashne, M (1983) Vectors bearing a hybrid trp-lac promoter useful for regulated expression of cloned genes in Escherichia coli. Gene 25:167-178

Chambers I, Frampton J, Goldfarb P, Affara N, McBain W, Harrison PR (1986) The structure of the mouse glutathione peroxidase gene: the selenocysteine in the active is encoded by the 'termination' codon, TGA. EMBO J 5:1221-1227

Diplock AT (1981) Metabolic and functional defects in selenium deficiency. Philosophical Transactions of the Royal Society of London B 294:105-117

Keshan Disease Research Group of the Chinese Academy of Medical Sciences (1979) Observations on effect of sodium selenite in prevention of keshan disease. Chinese Med J 92:471-476

Leinfelder W, Forchhammer K, Zinoni F, Sawers G, Mandrand-Berthelot M, Bock A (1988) Escherichia coli genes whose products are involved in selenium metabolism. J Bact 170:540-546

Scott ML (1973) Their chemistry and biology. Klayman DL, Gunther WHH (ed) John Wiley, New York, p 629

Stadtman TC (1979) Some selenium-dependent biochemical processes. Adv Enzymol 48:1-27

Stadtman TC (1980) Selenium-dependent enzymes. Annu Rev Biochem 49:93-110

Sugimoto M, Oikawa T, Esaki N, Tanaka H, Soda K (1988) Chemical synthesis and expression of copper metallothionein gene of Neurospora crassa. J Biochem 104:924-926

Sugimoto M, Esaki N, Tanaka H, Soda K (1989) A simple and efficient method for the oligonucleotide-directed mutagenesis using plasmid DNA template and phosphorothioate-modified nucleotide. Anal Biochem 179:309-311

Tate WP (1984) Peptide and Protein Reviews, Dekker, New York

Zinoni F, Birkmann A, Stadtman TC, Bock A (1986) Nucleotide sequence and expression of the selenocysteine-containing polypeptide of formate dehydrogenase (formate-hydrogen-lyse-linked) from Escherichia coli. Proc Natl Acad Sci USA 83:4650-4654

The Role of Natural Opal Suppressor tRNA in Incorporation of Selenium into Glutathione Peroxidase

Takaharu Mizutani[1], Teruaki Hitaka[1], Naosuke Maruyama[1], and Tsuyoshi Totsuka[2]

[1] Faculty of Pharmaceutical Sciences, Nagoya City University, Mizuho-ku, Nagoya, 467 Japan
[2] Department of Physiology, Institute for Developmental Research, Aichi Prefectural Colony, Aichi, 480-03 Japan

Abstract

This report contains three projects concerning Se incorporation into seleno-proteins. The first is the enzymatic conversion of bovine suppressor phosphoseryl(Ps)-tRNA to selenocysteyl(Sec)-tRNA which is co-translationally used for synthesis of glutathione peroxidase. The second shows the relationship between muscular dystrophy and suppressor tRNA. The ratio of suppressor to major tRNAs in dystrophic hindleg muscles was abnormally increased. At 3 months old, it was 1.7 times higher in dystrophic than in normal mice. The last shows the detection of natural opal suppressor tRNA in Escherichia coli. This suppressor seryl-tRNA was phosphorylated by a tRNA kinase in E. coli. We consider that the conversion of seryl-tRNA to Sec-tRNA through Ps-tRNA is universal in eukaryotes and prokaryotes.

Keywords
Selenium; Selenocysteine, tRNA; Muscular dystrophy; tRNA, suppressor; Glutathione peroxidase

Introduction

Selenium was found in 1817 by Berzelius and confirmed as an essential trace element (Schwarz 1957). For example, Se-deficiency causes Keshan disease, a chinese endemic disease (Keshan Disease Research Group 1979). There is a report showing a relation between the amount of Se in food and tumor of mammary gland (Schrauzer 1977). Selenium is demonstrated to be important for the production of sperm (Brown 1973), suggesting that Se may have a key role to maintain mammalian species. Thus Se is now widely accepted to be one of the most essential trace elements. Interestingly, most Se in our bodies is found in glutathione peroxidase (GSHPx), in which it is present in the active site as selenocysteine (Sec)(Forstrom 1978; Ladenstein 1979).

Glutathione peroxidase plays an important role in detoxification of hydrogen peroxide. The DNA sequence of the GSHPx gene has revealed that the codon for Sec is UGA, the opal termination codon (Sukenaga 1987). It is presumed that cellular cytosol contains a tRNA corresponding to UGA codon (natural opal suppressor tRNA) and that this tRNA plays to incorporate Sec into the GSHPx. However, since the UGA codon serves a stop codon having a role to terminate peptide elongation, UGA codon must be recognized by a releasing factor which releases peptides from ribosomes (Caskey 1980). Therefore, the in-frame UGA codon for Sec in GSHPx mRNA must escape from the reaction with a releasing factor. The fact that the concentration of natural

suppressor tRNA is as low as about one-fiftieth that of major tRNA suggests that there must be a very effective mechanism for suppressor tRNA to play this role. It is considered that 5'- and 3'-flanking regions of in-frame UGA codon take a hairpin structure (one of context mechanisms). This secondary structure of GSHPx mRNA must help the natural suppressor tRNA to behave as a normal tRNA to incorporate Sec (Mizutani 1988b). Meanhwile, it is suggested the presence of an elongation factor, a product of selB gene (Leinfelder 1988b), specfic for natural suppressor Sec-tRNA and Sec on the tRNA may be effectively incorporated into seleno-proteins with the elongation factor. This report contains three projects concerning Se incorporation into proteins. The first is the enzymatic conversion of bovine suppressor phosphoseryl(Ps)-tRNA to Sec-tRNA. The second shows the relationship between muscular dystrophy and suppressor tRNA. The last shows that the similar pathway of incorporation of Se as in vertebrates is present in bacteria.

The Enzymatic Conversion of Phosphoseryl-tRNA to Selenocysteyl-tRNA

Selenium is a trace and essential element for animals. Selenium is found as Sec in the active site of GSHPx. GSHPx plays an important role in the detoxification of hydrogen peroxides, organic hydroperoxides and lipid peroxides. Sec corresponds to the opal stop codon UGA on the murine and human mRNA of GSHPx (Chambers 1986; Sukenaga 1987). In order to understand the mechanisms of incorporation of Sec into GSHPx, the experiment with a perfused liver showed that the carbon source of the Sec in GSHPx came from serine by specific labelling of Sec with [^{14}C]Ser (Sunde 1987). This was also confirmed by the finding of Sec on tRNA prepared from cells cultured in the presence of [^{75}Se]Selenite and [^{3}H]Ser (Lee 1989). The synthesis of Sec from selenite has not been clarified .

Meanwhile, in mammals, Ps-tRNA and natural suppressor tRNA corresponding to UGA were found about two decades ago (Mäenpää 1970; Hatfield 1970; Hatfield 1985). This tRNA accepts serine and the primary sequence was determined (Diamond 1981). A human opal suppressor tRNA gene was also determined (O'Neill 1985). This suppressor seryl-tRNA was phosphorylated to Ps-tRNA by tRNA kinase (Mizutani 1984) and the incorporation of Ps into the read-through protein of β-globin was found (Mizutani 1986). The role of this tRNA is not clear. The possibility of participation of Ps-tRNA in phosphoserine aminotransferase catalysis was excluded (Mizutani 1988c). Natural suppressor tRNA did not participate in the regulation of the release reaction of RF, which has a stronger affinity to the stop codon UGA (K_a for Ps-tRNA:UGA, 8 x 10^3 M^{-1}; K_a for RF:UGA, 1.26 x 10^6 M^{-1})(Mizutani 1988a). The possibility remains that this natural suppressor tRNA participates in the incorporation of Sec into GSHPx. In the previous paper (Mizutani 1988b), we studied the mechanisms of incorporation of Sec into GSHPx through Ps-tRNA and suggested the conversion of Ps-tRNA to Sec-tRNA. Recently we showed some evidence of the in vitro conversion of Ps-tRNA to Sec-tRNA (Mizutani 1988d; Mizutani 1989b).

[^{3}H]Ps-tRNA was incubated with the H$_2$Se solution in the presence of Se-transferase (Mizutani 1989b). The results of analyses of the product are shown in Fig. 1, with use of an eluate from DEAE-cellulose as Se-transferase. Arrow 1 in Fig. 1 shows the position of serine. Arrow 2 shows the position of authentic cold Sec cochromatographed. The peak in front of arrow 2 corresponds to aliphatic amino acids, such as leucine and methionine, but we have not identified this spot. Some ^{3}H radioactivity was clearly found at the position coincident with Sec in Fig. 1. The amount of ^{3}H radioactivity in the spot of Sec was 0.2% of the total ^{3}H radioactivity applied on the TLC

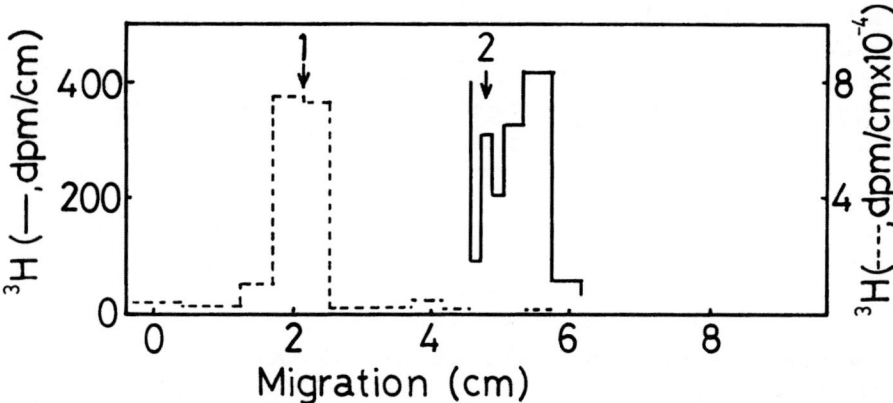

Fig. 1. Analyses of [³H] amino acids liberated from aminoacyl-tRNA by TLC on silica gel G in phenol:water (3:1). Authentic Sec and serine are indicated by arrows (1, serine;2, Sec). Silica on plates was scraped and the radioactivity was measured.

plate. This value indicates that 4% of Ps-tRNA was converted to Sec-tRNA, because Ps-tRNA was 5% of the total [³H]Ser-tRNA. This value, 4% of the product, is the standard level of many enzyme reactions and meaningful. It is difficult to get more Sec-tRNA in vitro, because Ps-tRNA is labile at neutral pH and hydrolyzed during long-time incubation. Therefore, the reaction was carried out by a restricted one set condition, such as at pH 6.5, 30°C and for 30 min. A more effective reaction from seryl-tRNA to Sec-tRNA through Ps-tRNA should be present in vivo, especially in the organelle synthesizing GSHPx such as reticulocytes. In our previous report (Mizutani 1988b), a small amount (0.016%) of Ps on casein was converted to Sec, but this value was much lower than the above value (4%) of Sec-tRNA. Therefore, we concluded that Sec binding on suppressor tRNA should be incorporated into GSHPx by co-translational mechanisms. In vertebrates, co-translational incorporation of Sec into seleno-proteins must be a major pathway or the only pathway.

When suppressor Ser-tRNA was used instead of Ps-tRNA as a substrate in Fig. 1, we could not find any conversion of Ser-tRNA to Sec-tRNA. The results showed that Sec-tRNA was not synthesized from Ser-tRNA but Ps-tRNA. Only Ps-tRNA is an effective substrate for Se-transferase. Thus, Sec-tRNA is synthesized in eukaryotes as follows:

Suppressor tRNA———→Ser-tRNA—————→Ps-tRNA——————→Sec-tRNA

     SerRS   tRNA kinase   Se-transferase

Muscular Dystrophy and Suppressor tRNA

GSHPx plays an important role in the detoxification of hydrogen peroxide and organic hydroperoxides, of which the major components are lipid peroxides of such polyunsaturated fatty acids as arachidonic acid (20:4) and docosahexaenoic acid (22:6). Meanwhile, this peroxidation is elevated in dystrophic muscles (Mechler 1984) and the level of malondialdehyde, a peroxidation product of polyunsaturated fatty acids, is also increased in the muscles (Burr 1987). Lipid peroxides are formed when oxygen radicals in solution reacted with polyunsaturated fatty acids in lipids (Murphy 1989). Polyunsaturated fatty acids in phospholipids of membranes are especially susceptible to such a damage by peroxidation. We have reported the decrease

of the 22:6 level in phospholipids of skeletal muscles in dystrophy mice (Futo 1989). It has been reported that the level of GSHPx relevant to decomposition of hydroperoxide increased in human dystrophic muscles (Burr 1987; Myllyla 1986). It seems likely that the high content of GSHPx may be caused by the high level of the above natural suppressor tRNA. In order to understand the role of natural suppressor tRNA in the pathogenesis of muscular dystrophy, we estimated the content of natural suppressor tRNA in mouse skeletal muscles by the dot blot hybridization with DNA probes (CGCCCGAAAGGTGGAA for suppressor tRNA and CGTAGTCGGCAGGAT for major serine tRNA)(Kumazawa 1988).

In order to study the relation between growth and the suppressor tRNA level, we prepared tRNA from skeletal muscles of dystrophic and normal mice of 3 weeks, 6 weeks, 3 months and 4 months old. Since the dystrophic symptom first appears at 2 weeks old, it is difficult to discriminate between dystrophic and normal littermate mice before 2 weeks. Therefore, in order to have information about a younger control level, we examined tRNA samples from 8-day-old normal mice of the mother strain C57BL.

By the analysis of the ^{32}P radioactivity of each dots with Bioimage-Analyzer, an arbitary unit for each dot was calculated. Table 1 shows age-related changes in the ratio of suppressor tRNA to major serine tRNA in normal and dystrophic muscles during the growth. As animals grew, the ratio of suppressor tRNA/major tRNA in normal muscles decreased but remained almost unchanged in dystrophic muscles. Significant differences between dystrophic and normal muscles were observed at the stages of 6 weeks and 3 months old. Noteworthy, an alteration in the dystrophic condition was already found at 3 weeks old (the symptom of dystrophy appears at 2 weeks old). Therefore, it seems highly possible that the high level of suppressor tRNA may be related to the pathogenesis of dystrophy.

Table 1 The Change of the Suppressor tRNA and Major tRNA Ratio from Normal and Dystrophic Mice during the Growth

| | The Ratio of Suppressor to Major Serine tRNA (Mean+S.E.) | | | | |
	1 week	3 weeks	6 weeks	3 months	4 months
Normal	0.29+0.02	0.26+0.02	0.20+0.02[*]	0.23+0.02[**]	0.26+0.03
Dystrophic	-	0.30+0.02	0.30+0.02	0.34+0.01	0.34+0.01

*, P<0.05; **, P<0.01; n=4-6.

It is well known that selenium relates to the growth of muscles. This report shows the high level of natural suppressor tRNA in murine muscular dystrophy. This tRNA, of which codon is UGA of opal stop codon, is considered to function in incorporating Sec into GSHPx. Sec-tRNA is derived from seryl-tRNA through Ps-tRNA which is a product of phosphorylation of seryl-tRNA by a tRNA kinase (Mizutani 1984). Thus far the findings indicate that the Sec at the active site of GSHPx must have been co-translationally incorporated into GSHPx. GSHPx is important for detoxification of peroxides and must thence be effectively synthesized in mammals. In dystrophic muscles which will contain the high level of peroxides, GSHPx is expected to be synthesized possibly resulting from the increase of suppressor tRNA. In fact, the level of GSHPx was significantly higher in dystrophic than in normal mice (data not shown).

Our previous report demonstrated that the content of docosahexaenoic acid (22:6) in phospholipid was low in skeletal muscle of dystrophic mice while there was no difference in its content between tongue muscles of dystrophic and normal mice (Futo 1989). This result supports the bone-muscle imbalance hypothesis proposed previously by Totsuka (1987). Meanwhile, it is well known that 22:6 in phospholipid is one of the biological substances most susceptible

to peroxidation.   The membranes are damaged by oxidation of 22:6 and then the oxidized fatty acid is removed by phospholipase A$_2$. Lysophospholipid is reacylated to become phospholipids by acyltransferase (van Kujik 1987).   The free oxidized fatty acid is further oxidized and hydrolyzed to malondialdehyde. Thus, membranes in dystrophic muscles are regarded to be exposed to such a toxic oxidative circumstance. In turn, the highly oxidative milieu will induce the synthesis of GSHPx which decomposes peroxides.   Thus, as a preceding phenomenon, the suppressor tRNA must increase because it will be necessary for the GSHPx synthesis. However, at present, it is difficult to explain the mechanisms of the specific increase of suppressor tRNA over many other species of tRNAs.

Selenocysteyl-tRNA in Bacteria

Selenium is essential also in prokaryotes (Pincent 1954) as well as in eukaryotes as demonstrated by the fact that selenium is found as Sec in active sites of formate dehydrogenase in Escherichia coli (Zinoni 1986) and (NiFeSe)hydrogenase in Desulfovibrio baculatus (Menon 1988).   In these enzymes, Sec corresponds to an in-frame UGA stop codon. In the case of formate dehydrogenase, incorporation occurs co-translationally (Zinoni 1987). In mammals, it has been showed that the natural opal suppressor seryl-tRNA should be converted to Sec-tRNA through Ps-tRNA (Mizutani 1989b).

In E. coli, it had been believed that there was no natural suppressor tRNA, however it has been shown that one of the genes (selC) relating to the synthesis of formate dehydrogenase corresponded to the opal suppressor tRNA (Leinfelder 1988a).   This tRNA is used in the co-translational incorporation of Sec into formate dehydrogenase (Zinoni 1987). It was shown that this tRNA did not accept Sec but serine. Despite these facts, the presence of suppressor tRNA in the cytosol of E. coli has not been shown.   Meanwhile, a method for purification of tRNA was developed: A dot blot hybridization method is used with a DNA probe corresponding to various tRNA parts (Kumazawa 1988). The principal difference between suppressor serine tRNA and the major serine tRNAs in E. coli (Sprinzl 1987) was found in the extra arm. We prepared a 12-mer DNA fragment (ACCGCTGGCGGC) corresponding to the extra arm in the tRNA-type structure of selC (Leinfelder 1988a) and used it as a probe in the dot blot hybridization method to detect the natural suppressor tRNA.   We show in this section the presence of natural suppressor tRNA in the E. coli tRNA preparation and the conversion of the seryl-tRNA to Ps-tRNA by a tRNA kinase in E. coli (Mizutani 1989a).

Fig. 2a shows the chromatographic pattern of the tRNA preparation from E. coli MC 4100 (cultured under anaerobic condition and in the presence of selenium).   The amount of ^{32}P by the dot blot hybridization is plotted in Fig.2a by closed circles and serine acceptor activity is shown by open circles.   Strong hybridization (specific hybridization, cpm/A$_{260}$ unit) is found in the last fraction of the tRNA peak (tubes 47-49) after the serine tRNA-3 of tube 44.   Weak specific hybridization is found in tubes 42-43 but the amount of tRNA in these fractions is more than that in tubes 47-49. Therefore, total amount of hybridization in tubes 42-43 is about twice as much as in tubes 47-49. Thus, there are two tRNAs which hybridize with the DNA probe to selC.   These two tRNAs are both active because the tRNAs are phosphorylated by a tRNA kinase.   We think that these two tRNAs come from one gene (selC), one being the hypo-modified tRNA of mature tRNA.   The same pattern with the tRNA preparation of E. coli wild-type strain was obtained.

Vertebrate opal suppressor seryl-tRNA was phosphorylated to become Ps-tRNA (Mizutani 1984).   This Ps-tRNA should be converted to Sec-tRNA by a Se-transferase (Mizutani 1989b). In order to confirm that, in E. coli, opal

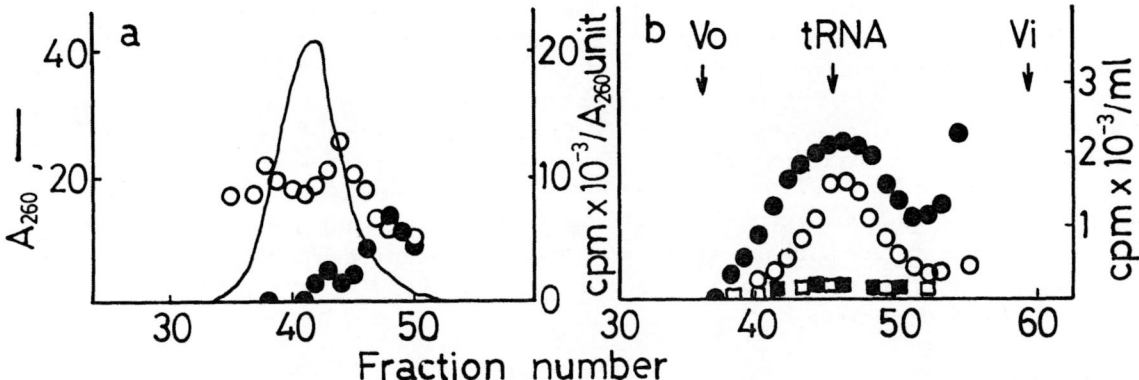

Fig. 2. Chromatographic patterns of tRNA. a, Chromatographic pattern of *Escherichia* *coli* MC 4100 tRNA on DEAE-Sephadex A50. Closed circles show the radioactivity of ^{32}P probe hybridized on tRNA. Open circles show the [^3H]serine acceptor activity. b, Chromatographic pattern of [^{32}P]phospho[^3H]seryl-tRNA on Sephacryl S-200 in 0.15M NaCl-0.01M acetate buffer at pH 4.6. Closed and open circles are ^{32}P and ^3H bound on tRNA after treatment in 10% cold TCA, respectively. Closed and open squares are radioactivities on tRNA after treatment in 10% hot TCA for 5 min, respectively. The peak of tubes 40-50 corresponds to the position of tRNA.

suppressor seryl-tRNA should also be converted to Sec-tRNA through Ps-tRNA. Therefore, as shown in Fig. 2b, we investigated the phosphorylation of suppressor seryl-tRNA. [^3H]Seryl-tRNA (tubes 47-49 in Fig. 2a) was phosphorylated with a tRNA kinase in *E.* *coli* B. The product was chromatographed on Sephacryl S-200 as shown in Fig. 2b. A [^{32}P]phospho[^{3}]seryl-tRNA peak can be seen in tubes 40-50 in Fig. 2b. The radioactivity (^{32}P and ^3H) of the peak disappeared following the hot TCA treatment, as indicated by the squares in Fig. 2b. These results show that the phosphate became bound to the serine on the tRNA. When non-acylated tRNA (serine free tRNA) was used as a phosphorylation substrate, no radioactivity of ^{32}P was found on tRNA. This result supports the fact that phosphate on the tRNA in Fig. 2b binds the OH residue of serine (seryl-tRNA). The presence of phosphoserine on phosphoseryl-tRNA was confirmed by analyses on an AG-1 column. Radioactivity of ^{32}P was found at the position of the authentic phosphoserine (color reaction of ninhydrin) with ^3H radioactivity. These results showed that natural suppressor seryl-tRNA in *E.* *coli* was phosphorylated by a tRNA kinase of *E.* *coli*.

This paper has demonstrated the presence of the tRNA kinase activity in *E.* *coli* and the phosphorylation of UGA suppressor seryl-tRNA. This tRNA kinase is an unexpected enzyme in *E.* *coli*. Our other results suggest that this tRNA kinase is a product of the *selA* gene (Leinfelder 1988b) and acts as dimer. Phosphoseryl-tRNA in *E.* *coli* should be converted to Sec-tRNA by Se-transferase (Se-transferase may be a product of *selD*), in the same manner as in vertebrates (Mizutani 1989b). Previously, it has been confirmed that phosphoseryl-tRNA is present in vertebrates (Mäenpää 1970). Thus far the findings suggest that universally in living beings there is a similar system to synthesize phosphoseryl-tRNA, which acts as an intermediate in the conversion of seryl-tRNA to Sec-tRNA and corresponds to the UGA stop codon as a natural opal suppressor tRNA. Also in *Proteus* *vulgaris*, the presence of natural suppressor tRNA has been shown (Heider 1989). As a conclusion, we consider that this pathway of biosynthesis of Sec-tRNA through Ps-tRNA is universal in eukaryotes and prokaryotes.

Note added in proof. The presence of phosphoseryl-tRNA in *E.* *coli* is unclear.

References

Brown DG, Burk RF (1973) Selenium retention in tissues and sperm of rats fed a torula yeast diet. J Nutr 102:102-108

Burr IM, Asayama K, Fenichel GM (1987) Superoxide dismutases, glutathione peroxidase and catalase in neuromuscular disease. Muscle & Nerve 10:150-154

Caskey CT (1980) Peptide chain termination. Trends in Biochem Sci 5:234-237

Chambers I, Frampton J, Goldfarb P, Affara N, McBain W, Harrison P (1986) The structure of the mouse glutathione peroxidase gene:the selenocysteine in the active site is encoded by the termination codon, TGA. EMBO J 5:1221-1227

Diamond A, Dudock B, Hatfield D (1981) Structure and properties of a bovine liver UGA suppressor serine tRNA with a tryptophan anticodon. Cell 25:497-506

Forstrom JW, Zakowski JJ, Tappel AL (1978) Identification of the catalytic site of rat liver glutathione peroxidase as selenocysteine. Biochemistry 17:2639-2644

Futo T, Hitaka T, Mizutani T, Okuyama H, Watanabe k, Totsuka T (1989) Fatty acid composition of lipids in tongue and hindleg muscles of muscular dystrophic mice. J Neurol Sci 91:337-344

Hatfield D, Portugal FH (1970) Seryl-tRNA in mammalian tissues:chromatographic differences in brain and liver and a specific response to the codon UGA. Proc Natl Acad Sci USA 67:1200-1206

Hatfield D (1985) Suppression of termination codons in higher eukaryotes. Trends in Biochem Sci 10:201-204

Heider J, Leinfelder W, Böck A (1989) Occurrence and functional compatibility within Enterobacteriaceae of a tRNA species which inserts selenocysteine into protein. Nucleic Acids Res 17:2529-2540

Keshan Disease Research Group of the Chinese Academy of Medical Sciences (1979) Observation on the effect of sodium selenite in prevention of Keshan disease. Chin Med J 92:471-476

Kumazawa Y, Yokogawa T, Miura K, Watanabe K (1988) Bovine mitochondrial phenylalanine tRNA, serine tRNA(AGY) and serine tRNA(UCN):preparation using a new detection method and their properties in aminoacylation. Nucleic Acids Res Symp Ser 19:97-100

Ladenstein R, Epp O, Bartels K, Jones A, Huber R, Wendel A (1979) Structure analysis and molecular model of the selenoenzyme glutathione peroxidase at 2.8Å resolution. J Mol Biol 134:199-218

Lee BJ, Worland PJ, Davis JN, Stadtman TC, Hatfield DL (1989) Identification of a selenocysteyl-tRNA[Ser] in mammalian cells that recognitizes the nonsense codon, UGA. J Biol Chem 264:9724-9727

Leinfelder W, Zehelein E, Mandrand-Berthelot MA, Böck A (1988a) Gene for a novel tRNA species that accepts L-serine and cotranslationally inserts selenocysteine. Nature 331:723-725

Leinfelder W, Forchhammer K, Zinoni F, Sawers G, Mandrand-Berthelot MA, Böck A (1988b) Escherichia coli genes whose products are involved in selenium metabolism. J Bacteriol 170:540-546

Mäenpää P, Bernfield MR (1970) A specific hepatic transfer RNA for phosphoserine. Proc Natl Acad Sci USA 67:688-695

Mechler F, Imre S, Dioszeghy P (1984) Lipid peroxidation and superoxide dismutase activity in muscle and erythrocytes in Duchenne dystrophy. J Neurol Sci 63:279-283

Menon NK, Peck HD, Gall JL, Przybyla AE (1988) Cloning and Sequencing of the gene encoding the large and small subunits of the periplasmic (NiFeSe)Hydrogenase of Desulfovibrio baculatus. J Bacteriol 170:4429

Mizutani T, Hashimoto A (1984) Purification and properties of ATP:suppressor seryl-tRNA phosphotransferase from bovine liver. FEBS Lett 169:319-322

Mizutani T, Tachibana Y (1986) Possible incorporation of phosphoserine into globin readthrough protein via bovine opal suppressor phosphoseryl-tRNA. FEBS Lett 207:162-166

Mizutani T, Hitaka T (1988a) Stronger affinity of reticulocyte release factor than natural suppressor tRNA for the opal termination codon. FEBS Lett 226:227-231

Mizutani T, Hitaka T (1988b) The conversion of phosphoserine residues to selenocysteine residues on an opal suppressor tRNA and casein. FEBS Lett 232:243-248

Mizutani T, Kanbe K, Kimura Y, Tachibana Y, Hitaka T (1988c) Non-participation of opal suppressor phosphoseryl-tRNA in phosphoserine amintransferase catalysis. Chem Pharm Bull 36:824-827

Mizutani T, Hitaka T, Maruyama N (1988d) Suppressor tRNA and its function to incorporate selenocysteine in a protein-synthesizing system. Seikagaku 60:632

Mizutani T, Maruyama N, Hitaka T, Sukenaga Y (1989a) The detection of natural opal suppressor seryl-tRNA in Escherichia coli by the dot blot hybridization and its phosphorylation by a tRNA kinase. FEBS Lett 247:345-348

Mizutani T (1989b) Some evidence of the enzymatic conversion of bovine suppressor phosphoseryl-tRNA to selenocysteyl-tRNA. FEBS Lett 250:142-146

Murphy ME, Kehrer JP (1989) Oxidative stress and muscular dystrophy. Chem-Biol Interactions 69:101-173

Myllyla V, Kihlstrom M, Takala TES, Tolonen U, Salminen A, Vinko V (1986) Activation of some antioxidative and hexose monophosphate shunt enzymes of skeletal muscle in neuromuscular diseases. Acta Neurol Scand 74:17-24

O'Neill VA, Eden FC, Pratt K, and Hatfield DL (1985) A human opal suppressor tRNA gene and pseudogene. J Biol Chem 260:2501-2508

Pincent J (1954) The need for selenite and molybdate in the formation of formate dehydrogenase by members of the Coli-aerogenes group of bacteria. Biochem J 57:10-16

Schrauzer GN, White DA, Schneider CJ (1977) Cancer mortality correlation studies. III. Statistical associations with dietary selenium intakes. Bioinorg Chem 7:23-34

Schwarz K, Foltz CM (1957) Selenium as an integral part of factor 3 against dietary necrotic liver degeneration. J Am Chem Soc 79:3292-3293

Sukenaga Y, Ishida K, Takeda T, Takagi K (1987) cDNA sequence coding for human glutathione peroxidase. Nucleic Acids Res 15:7178

Sprinzl M, Hartmann T, Meissner F, Moll J, Vorderwullbecke T (1987) Compilation of tRNA sequences and sequences of tRNA genes. Nucleic Acids Res 15:r53-r115

Sunde RA, Evenson JK (1987) Serine incorporation into the selenocysteine moiety of glutathione peroxidase. J Biol Chem 262:933-937

Totsuka T (1987) Normal diameter distribution of tongue muscle fibers in muscular dystrophic mice. Consistent with the bone-muscle imbalance hypothesis for the pathogenesis. Proc Japan Acad 63B:131-134

van Kujik FJGM, Sevanian A, Handelman GJ, Dratz EA (1987) A new role for phospholipase A_2:protection of membranes from lipid peroxidation damage. Trends in Biochem Sci 12:31-34

Zinoni F, Birkmann A, Stadtman TC, Böck A (1986) Nucleotide sequence and expression of the selenocysteine-containing polypeptide of formate dehydrogenase (formate-hydrogen-lyase-linked) from Escherichia coli. Proc Natl Acad Sci USA 83:4650-4654

Zinoni F, Birkmann A, Leinfelder W, Böck A (1987) Cotranslational insertion of selenocysteine into formate dehydrogenase from Escherichia coli directed by a UGA codon. Proc Natl Acad Sci USA 84:3156-3160

Transcriptional Activation of Human and Yeast Metallothionein Genes by Heavy Metal Ions

Petra Skroch, Carla Buchman, and Michael Karin

Department of Pharmacology, M-036 School of Medicine, University of California, San Diego, La Jolla, CA 92093, USA

ABSTRACT

Organisms from yeast to mammals express low molecular weight metal binding, cystein-rich proteins, the metallotheineins (MTs). The metal-dependent expression of MTs is controlled transcriptionally. Up to now no transacting factor conferring metal-inducibility upon mammalian MT-genes was isolated. Recently, however, the product of the yeast CUP2 gene was identified as a sequence specific Cu-dependent activator of the yeast MT, CUP1 gene. The N-terminal DNA-binding domain of CUP2 resembles CUP1 in its cystein-content and arrangement. Study of in vitro synthesized wild-type and mutant CUP2 proteins demonstrated the effect of metal-binding on specific DNA-affinity and the importance of single amino acid residues in this process.

Key Words: Metallothionein, Gene Regulation, Copper-Dependent Transcription Factors, DNA Binding Protein

INTRODUCTION

Organisms as diverse as yeast, drosophila and mammals express heavy metal binding proteins, metallothioneins. These cysteine-rich, low molecular weight proteins play a key role in trace metal ion homeostasis and detoxification. By regulating the levels and availability of intracellular Zn and Cu, they contribute indirectly to the regulation of various Zn requiring enzyme systems, some of which are involved in replication, transcription and DNA repair (Karin 1985a; Hamer 1986). The major form of regulation of MT activity occurs by increased synthesis of these proteins in response to elevated heavy metal ion concentrations and various steroid and polypeptide hormones. The activation of the metallothionein genes by heavy metal ions and hormones occurs on the transcriptional level, but whereas the highly conserved cis and trans-acting DNA elements involved in the response to hormone have been characterized and studied in detail, the trans-acting factors conferring metal inducibility could not be defined for a long time. Only recently a protein confering Cu inducibility upon the yeast metallothionein gene CUP1 was isolated by a genetic approach and its function and structure could be studied in detail.

HUMAN METALLOTHIONEIN GENES AND REGULATION

The human genome contains 12 distinct MT genes, out of which six or seven code for functional proteins, whereas the rest are nonfunctional pseudogenes (Karin and Richards, 1982; Hunziker and Kagi, 1983; Richards et al., 1984; Klauser and Kaegi, 1985; Schmidt et al., 1985; Heguy et al., 1986). The MTIIA gene is the major human MT gene, accounting for 50% of the MTs expresssed in cultured human cells and in liver (Karin and Herschman, 1980a). In addition to heavy metal ions the expression of the MTIIA gene is induced by glucocorticoid and progesterone hormones, interferon, interleukin I, serum factors, phorbol ester tumor promotors such as TPA and DNA damaged by UV irradiation (Karin et al., 1980b; Friedman and Stark, 1985; Karin 1985b; Imbra and Karin 1986; Angel et al., 1986).

The cis acting DNA elements involved in regulation of the MTIIA gene have been studied using in vitro mutagenesis and gene transfer experiments (Haslinger and Karin, 1985; Serfling et al., 1985; Karin et al., 1987a). The trans acting factors that bind these elements have been defined by the DNase I footprinting procedure and gel retardation assays. A GC-box located between nucleotides -57 to -68 relative to the start side of transcription is recognized by the specific transcription factor SP1. Interestingly SP1 is a zinc-finger DNA-binding protein (Kadonaga et al., 1987) which raises the possibility that MT might indirectly control its activity. A TPA responsive element (TRE), another common regulatory DNA element, spans the region from -96 to -105 (Lee et al., 1987). It is recognized by the transcription factor AP1 (Angel et al., 1987). AP1 was demonstrated to be composed of both homodimeric and heterodimeric complexes involving the cJun and cFos protooncoproteins (Chiu et al., 1988). Further upstream, transcription factor AP2 binds to three relatively divergent sequences in the MTIIA promoter (Imagawa et al., 1987). The element that confers glucocorticoid responsiveness upon the MTIIA gene is located between positions -240 and -270 and was shown to serve as a binding site for the glucocorticoid receptor (Karin et al., 1984a). The same element also confers a response to progesterone (Slater et al., 1988). Because of its unusual responsiveness to a large number of hormonal and environmental cues, the MTIIA gene is an excellent system for analysing the molecular mechanisms involved in signal transduction to the transcriptional machinery. The activity of multiple synthetic copies of the MTIIA high-affinity AP2 binding site increases after treatment of cells with phorbol ester or cAMP-elevating agents. In contrast, a synthetic enhancer specifically recognized by AP1 is activated only by phorbol ester. Therefore, in contrast to AP1, AP2 appears to mediate transcriptional activation in response to two different signal transduction pathways, one involving the phorbol ester- and diacylglycerol-activated protein kinase C, the other involving cAMP-dependent protein kinase A. Steroid hormones bind to and activate a specific cytoplasmic receptor and lead to its migration into the nucleus, where it binds to a specific site on steroid responsive genes (Evans, 1988).

Interspersed between the basal promoter elements are four metal response elements (MREs) (Karin et al., 1987a). MREs are short (12 to 17 basepairs long) repetitive sequences that are highly conserved within one species and also between different species. In spite of their high sequence homology different MREs differ in their abilities to confer metal inducibility when assayed individually. Culotta and Hamer (1989) showed, that one out of the five distinct MRE sequences present in the mouse MT-I gene, confers a stronger metal induction to a heterologous promoter then the others. Two copies of this individual MRE cloned in front of its authentic promoter led to a 10-20 fold induction, whereas a single copy enhanced the metal dependent transcription only 2-4 fold. This suggests a cooperative effect, which coincides with the occurrence of multiple MREs in natural MT gene promoters. For example each of the MREs of the MTIIA gene is adjacent to another cis element (Karin et al., 1987a) and competition experiments have demonstrated that several of these elements promote binding of metal-dependent transcription factors to the MREs. This is likely to occur by protein-protein interactions as originally suggested by Scholer et al (1986). Metal ion induced transcription of the MT genes is positively regulated, and the Zn induced transcription of MT mRNA in HeLa cells is insensitive to cycloheximide, which shows that the induction is not dependent on de novo protein synthesis (Karin et al., 1980c). These results are consistent with a model according to which a metal-responsive factor (MRF) exists in the cell prior to induction and is activated by metal ions. At elevated metal ion concentrations the factor might acquire an increased affinity for MREs as a prerequisite for transcriptional activation. As mentioned above, binding of the MRF may also depend on interactions with adjacent factors such as SP1 or AP1.

Generating a series of synthetic oligonucleotides containing point mutations Culotta and Hamer (1989) defined a core region in which single basepair exchanges destroyed the activity of the element and a flanking GC-rich region which proved to be less critical for function. This region exhibits significant homology to the core binding site for mammalian transcription factor SP1 , but a partially purified mouse SP1 preparation did not exhibit strong binding to this site. In addition purified human SP1 was never found to bind any of the MTIIA MREs (Lee et al., 1987). Using the exact same DNA

sequence Westin and Schaffner (1988), showed evidence that this individual mouse MT1 MRE is recognized by SP1 as well as by another protein, possibly the one confering metal inducibility. However the two factors were reported to compete for the same binding site rather than act synergistically. The basis for these descriptions is not clear, but it is certain that under normal conditions SP1 does not interact with the MRE. In the human MTIIA gene, one or two MREs are present adjacent to each of the basal level enhancer elements. Just as the MRE sequences appear to work in a cooperative manner when multimerized (Searle et al., 1987; Culotta and Hamer 1989), a possible cooperation between the basal level enhancer elements and the MREs was also demonstrated (Karin et al., 1987a). In vivo competition experiments demonstrated that, in the presence of heavy metal ions, the MTIIA enhancer is able to compete more efficiently with the SV40 enhancer for binding of cellular factors. It is likely that the positively acting factor that binds to the MRE in the presence of heavy metal ions is acting by stabilysing the binding of adjacent trans-acting factor via protein-protein interactions (Scholer et al., 1986). Thus the cooperative interaction between basal factors and the MRF appears to operate both ways.

Extracts were prepared from Cd-induced and uninduced Hela cells and partially purified, but failed to show any differences in their DNase I protection pattern on the MTIIA promoter. Actually the only upstream elements over which no protection was observed were the MREs (Karin et al., 1987b). However a DNA probe containing only a tandemly duplicated MRE sequence showed a footprint over the entire MRE sequence after incubation with either induced or uninduced HeLa cell extracts (Skroch, unpublished). Seguin and Hamer (1987) observed partial protection of some MRE sequences in footprinting experiments with the mouse MTI promoter using nuclear extracts from Cd treated mouse L cells. However, other groups including ours have failed to reproduce these experiments. Mueller et al. (1988) could demonstrate a change in protein-DNA contacts on the mouse MTI promoter by in vivo footprinting when cells are shifted from basal level expression to metal induced expression. The non-induced footprint pattern consists of interactions of factors such as SP1 and AP1 with the basal elements that are thought to be responsible for the moderate expression of this gene in the absence of added metal. By exposing the L-cells to Zn or Cd a new set of metal dependent footprints appeared over all five genetically defined MREs. These experiments provide further support for the existence of a positively regulated MRF interacting with the MREs.

Although several reports claiming detection of a protein with some of the features expected from the MRF have been published, none of them provided a conclusive proof that the detected factor was indeed the MRF. For example, the binding of some of these factors was affected only by Zn and not by Cd (Westin and Schaffner, 1988) or vice versa (Seguin and Hamer, 1987). The authentic MRF should respond to both of these metal ions. Furthermore, no purification and isolation of these factors has been reported. Our own attempts to detect an authentic MRF activity have failed (Skroch unpubl.). Since we could detect all of the other factors that bind to the MTIIA promoter we believe that the MRF is either extremely labile or present in very low amounts within the cell.

METALLOTHIONEIN GENE REGULATION IN YEAST

A convenient eukaryotic system for studying MT gene function and regulation is provided by the brewer's yeast Saccharomyces cerevisiae. Resistance to toxic effects of copper ions in Saccharomyces cerevisiae is mediated by the inducible synthesis of a low molecular weight, cystein-rich, metal-binding protein, known as copper chelatin or yeast metallothionein (Fogel and Welch, 1982; Butt et al., 1987). Various naturally occuring yeast strains differ over a 20fold range in their resistance to the inhibitory effects of copper on growth. Increased resistance correlates with over-production of yeast MT (Welch et al., 1983). Enhanced resistance reflects gene amplification as a consequence of sister chromatid exchange and gene conversion events. The 1.95kb repeat units containing the yeast MT gene CUP1 are tandemly arrayed and copper resistant strains contain as many as 15 reiterared repeats (Fogel and Welch, 1982). The entire nucleotide sequence of the CUP1 locus was determined. The predicted aminoacid composition derived from the DNA sequence

shows, that the yeast CUP1 protein bears only limited primary sequence homology to the mammalian MTs (Karin et al., 1984). However, the two types of protein are similar in their small size (60-61 amino acids) and unusual composition: CUP1 contains 20% cysteines while mammalian MTs contain 30% cysteines. Also it has been shown that overproduction of monkey MT can functionally substitute for the CUP1 gene, when transfected into yeast cells deleted of their CUP1 locus (Thiele et al., 1986a). The purified protein product of the CUP1 gene contains eight monovalent copper ions liganded by 12 cysteine residues, but it can bind in vitro also to eight silver ions or to four divalent zinc or cadmium ions (Winge et al., 1985). However, unlike the mammalian MT genes, transcription of the yeast gene is not affected by other ions such as cadmium, zinc or mercury (Karin et al., 1984b).

The regulatory region of the yeast CUP1 gene was studied in detail. By deletion analysis and construction of synthetic hybrid promoters the cis acting control elements that confer copper dependent transcription have been narrowed down to nucleotides -105 to -180. This region contains at least two functional elements and consists of two related sequences of 32 and 34 nucleotides, denoted UASp and UASd respectively (Butt et al., 1984; Thiele and Hamer, 1986b). A series of overlapping oligonucleotides that span the complete region was synthesized and introduced into yeast after fusion to a deleted CYC1 promoter driving expression of the E.coli galK reporter gene. Galaktokinase activity was tested in the presence and absence of copper. The region between -105 and -148 spanning the UASp region showed the best result, activating transcription 7.5 fold. The precise nucleotides that are responsible for copper induced transcription were identified by systematic point mutagenesis of the UASp. UASd alone was not able to confer copper inducibility under these conditions. (Fuerst et al., 1988). Transcription of the 0.5kb CUP1 gene itself is stimulated 10-20 fold in response to elevated copper concentrations (Karin et al., 1984b).

To identify putative trans-acting regulatory genes whose products may interact with the copper responsive elements of the CUP1 promoter to mediate copper induction and resistance, a collection of copper sensitive mutants, induced by ethyl methan sulfonate (EMS) treatment of a wildtype multicopy CUP1r strain, was screened for possible regulatory gene mutations. In one recessive isolate, a mutation designated cup2, the phenotypic copper resistance was dramatically reduced by two orders of magnitude. This cup2 strain thus exhibited an even more pronounced sensitivity than a cup1s strain, which contains only a single CUP1 gene. The results of of RNA blot analysis indicate that in cup2 cells, despite its presence in 11 copies, the CUP1 gene is expressed only at a very low basal level and is refractory to induction by exogenous copper ions. Consequently, cup2 was selected for extensive analysis based on the assumption that it is defective in production of a functional protein CUP2, which was supposed to act as a trans-acting factor for the metal dependent expression of the CUP1 locus. The mutant cup2 strain was transformed with a genomic wildtype DNA library cloned into a yeast expression shuttle vector. Transformants were isolated and tested for copper resistance. The wildtype allel of the CUP2 gene was rescued by retransformation of E.coli. by plasmids extracted from four independent Cu-resistent yeast colonies (Welch et al., 1989).

To determine whether the effect on CUP1 expression in the cup2 strain was attributable to an altered interaction of regulatory proteins with the CUP1 promoter, nuclear protein extracts were prepared from wildtype (CUP1r, X2180) and the otherwise isogenic cup2 strain. Incubation of a DNA probe spanning the copper responsive elements of the CUP1 promoter with the cup2 extract resulted in the formation of three different protein-DNA complexes. The same complexes formed after incubation with the wildtype extract, but now an additional distinct protein-DNA complex appeared. This result suggested, that mutant cup2 is defective in one of the DNA-binding activities which recognize the CUP1 promoter. Since introduction of a wildtype CUP2 allele into the cup2 strain restores copper resistance and CUP1 induction and leads to the appearence of the wildtype-specific protein-DNA complex, it was concluded that CUP2 encodes a regulatory protein required for Cu-dependent activation of CUP1 (Welch et al., 1989). Thiele (1988) used essentially the same approach to isolate a yeast gene named ACE1 that shows the identical features to CUP2: it restores copper resistance after reintroduction into a yeast strain that fails to grow in elevated copper concentrations. Sequencing of the cloned DNAs demonstrated the identity of the ACE1 and CUP2 genes.

The CUP2 coding sequences have the potential to form a 225 amino acid protein with a predicted molecular weight of 24kd. When expressed in E.coli., however, the protein migrates with an apparent molecular weight of 33 kd.

Examination of the distribution of charged amino acids encoded by the gene suggests, that CUP2 is a two domain protein. The amino terminal portion of the molecule, from amino acid 1 to 108, is unusually poor in hydrophobic and aromatic residues but is extremely rich in basic residues and has a predicted positive charge of +16. This part of the protein also contains 12 cysteine residues, ten of which are arranged in the CysXCys and CysXXCys configurations characteristic of the metallothionein gene CUP1 itself and mammalian MTs as well. This highly charged domain could be responsible for the abnormal migration of the CUP2 protein. In contrast, the carboxy-terminal portion of the molecule has a predicted negative charge of -14, and thus resembles several highly acidic sequences known to act as transcriptional activation domains in yeast (Fuerst et al., 1988; Welch et al., 1989). The bacterially expressed CUP2 protein appears to be partially cleaved by an endogenous protease to yield two fragments corresponding to the two predicted domains (Buchmann et al., 1989).

Although the primary sequences of the N-terminal domain of CUP2 and various MT sequences differ a lot, it is important to realize, that MTs themselves are highly divergent when comparing distantly related species, and that the primary sequence of the protein can vary considerably without excerting large effects on its metal binding properties and in vivo function. Based on the predicted structural similarities between CUP2/ACE1 and metallothionein, (Fuerst et al., 1988) devised a preliminary model for how its relatively small N-terminal domain could combine specific DNA binding and metal binding properties. They proposed that the core structure of the CUP2/ACE1 N-terminal domain is a cuboidal Cu8-S12 cluster in which paired cysteine residues coordinate common copper ions, as it has been suggested recently for the CUP1 protein.

To analyse the properties of the CUP2 protein we cloned it into a bacterial expression vector and first expressed it as a trpE fusion protein. Extracts prepared from transformed E.coli cells grown in the presence of 1mM $CuSO_4$ were assayed by gel retardation and were found to form one or two specific protein-DNA complexes with a CUP1 DNA probe containing the Cu responsive UASs. Expression of the N-terminal 122 amino acids of the protein followed by gel retardation analysis demonstrated that this domain alone is sufficient for binding to the CUP1 promoter. To examine the importance of copper for DNA binding of CUP2, E.coli cells expressing CUP2 were grown in the absence of copper and protein extracts were prepared. They were found to contain substancial amounts of CUP2 protein as shown by SDS polyacrylamide gel analysis, but hardly any DNA-binding activity was detected. Copper added to these extracts in the form of Cu^+ restored the DNA binding activity of the protein (Buchman et al., 1989). Cu^+ was chosen in these experiments instead of the Cu^{++} ion used for in vivo induction, as Cu^+ is the natural ligand of yeast MT (Winge et al., 1985). Indeed, while the DNA-binding activity of the trpE CUP2 fusion was also restored by Ag^+, which is electronically similar to Cu^+, the addition of Cu^{++} was much less efficient in restoring activity. Additionally, treatment of CUP2 protein with KCN, a chelator capable of removing Cu^+ ions from yeast MT led to loss of DNA binding activity, which could be restored upon addition of excess Cu^+ or Ag^+, whereas Zn^{++}, Cd^{++} or Hg^{++} were not effective (Buchman et al., 1989).

The two mutant yeast strains cup2 and ace1-1, that led to the identification of CUP2/ACE1 were isolated independently as ethyl-methansulfonate-induced copper sensitive mutants from their copper resistant parental strains. (Thiele 1988; Welch et al., 1989). To determine the molecular basis for the failure of these strains to induce CUP1 in response to Cu, we isolated molecular clones of the two mutant allels after their amplification by the polymerase chain reaction (Mullis and Faloona 1987). Sequence analysis revealed, that each of the mutants differed from the wild type allele only by a single point mutation. The cup2 allele contained a G to A transition converting Gly37 to a glutamic acid residue, while in the ace1-1 strain Cys11 was converted to a tyrosine residue also by a G to A transition. To examine the effect of these mutations on CUP2 protein function, both alleles were cloned into bacterial expression vectors. The mutant trpE fusion proteins appeared to be as stable as the wildtype protein and showed identical

mobility on SDS polyacrylamide gels. In a gel retardation analysis using the UASs as a DNA probe, trpE.ace1-1 had reduced DNA-binding activity, whereas trpE.cup2 did not exhibit any detectable binding to the CUP1 promoter. The activity of the N-terminal half of CUP2 is highly dependent on Cu^+ or Ag^+ ions suggesting that these specific ions interact with the cysteine residues to direct the folding of the DNA-binding domain. The reduced ability of the ace1-1 protein to bind to DNA, suggests, that Cys11 could be directly involved in formation of the Cu+ -coordinated DNA-binding domain. The reason for the dramatic effect of the cup2 mutation on DNA-binding is less obvious. One possibility is, that the Gly37 to Glu change introduced a negatively charged residue into a cluster of positively charged amino acids and thereby affected interaction with phosphate groups in the DNA-backbone. The other possibility is that the Gly to Glu change led to loss of conformational flexibility and interfered with proper folding of the DNA-binding domain. Unlike the Cys11 to Tyr mutation, the Gly37 to Glu mutation is not expected to affected Cu^+ binding per se (Buchman et al., 1989).

While the DNA-binding activity of the CUP2 protein is highly regulated by Cu, expression of the CUP2 gene is constitutive (Szczypka, et al., 1989). Taken collectively, these these findings suggest the following scheme for regulation of CUP1 expression. Cells grown with low levels of copper already contain a substancial amount of CUP2 , but because of the low intracellular levels of Cu^+, most of it is in the inactive apoprotein form and, hence, is incapable of binding to the UASs of CUP1. Once the intracellular level of Cu^+ is elevated by exposure to extracellular Cu, CUP2 becomes rapidly activated, binds to the UASs and leads to transcriptional activation of the CUP1 gene. Overexpression of CUP1 will lead to chelation of all available Cu^+ within the cell and thereby decrease the DNA-binding activity of the CUP2 protein. This will result in dissociation of CUP2 from the CUP1 promoter and inactivation of CUP1 transcription. This concerted mechanism would lead to very precise regulation of CUP1 synthesis and maintenance of Cu-homoeostasis.

CONCLUSIONS

Organisms, that are as far apart in evolution as yeast and mammals use similarly designed proteins to fulfill their need for metal ion homeostasis. The unusually abundant cysteines of these low molecular weight proteins are mainly arranged in a CysXCys or CysXXCys configuration and accomplish the association with the metal ions. The activation of MT genes in response to metal ions is regulated at the transcriptional level by preexisting trans-acting factors. The gene for transcription factor CUP2 that confers Cu-inducibility to its MT gene has been cloned from the brewer's yeast Saccharomyces cerevisiae. The primary sequence of the metal-and DNA-binding domain of this factor shows a stunning similarity to the yeast CUP1 gene itself. Cu ions are complexed by cysteine residues arranged in CysXCys and CysXXCys motifs, the resulting conformational change increases the affinity of the factor to its DNA binding site, thereby leading to transcriptional activation of the CUP1 gene. It is very likely that mammals use the same regulatory mechanism to activate their MT genes. The affinity of a trans-acting factor to its binding site on an MT promoter is largely increased by the interaction of this factor with metal ions, so its metal-binding domain might share the high content of cysteines and their specific arrangement with both the mammalian and the yeast MTs and the yeast MT-specific activator protein CUP2. It is very unlikely though, that the yeast and the mammalian factors share any primary sequence homology beside the cysteine-motifs. The difficulties in isolation of a mammalian metal responsive trans-acting factor or its coding gene can be taken as a hint for a relatively low DNA-binding affinity even in the activated form or for its very low abundance. In addition, the mammalian MRF may be regulated by a novel mechanism involving an interaction with another protein as a prerequisite for DNA-binding. Such a labile interaction, which is disrupted upon preparation of extracts of cellular proteins may have prevented its characterization up to now.

REFERENCES

Angel P, Poeting A, Mallick U, Rahmsdorf HJ, Schorpp M, Herrlich P (1986) Induction of metallothionein and other mRNA species by carcinogens and tumor promoters in primary human skin fibroblasts. Mol Cell Biol 6:1760-1766

Angel P, Imagawa M, Chiu R, Stein B, Imbra RJ, Rahmsdorf HJ, Jonat C, Herrlich P, Karin M (1987) Phorbol ester-inducible genes contain a common cis element recognized by a TPA-modulated TRans-acting factor. Cell 49:729-739

Buchman C, Skroch P, Welch J, Fogel S, Karin M (1989) The CUP2 gene product, regulator of yeast metallothionein expression, is a copper-activated DNA-binding protein. Mol Cell Biol 9:4091-4095

Butt TR, Sternberg EJ, Herd J, Crooke ST (1984) Copper metallothionein of yeast, structure of the gene, and regulation of expression. Gene 27:23-33

Chiu R, Boyle W, Meek J, Smeal T, Hunter T, Karin M (1988) The c-fos protein interacts with c-jun/AP1 to stimulate transcription of AP-1 responsive genes. Cell 54:541-552

Culotta V, Hamer D (1989) Fine mapping of a mouse metallothionein gene metal response element. Mol Cell Biol 9:1376-1380

Evans RM (1988) The steroid and thyroid hormone receptor superfamily. Science 240:889-895.

Fogel S, Welch JW (1982) Tandem gene amplification mediates copper resistance in yeast. Proc Natl Acad Sci USA 79:5342-5346

Friedman RL, Stark GR (1985) alpha-Interferon-induced transcription of HLA and metallothionein genes containing homologous upstream sequences. Nature 314:367-369

Fuerst P, Hu S, Hackett R, Hamer D (1988) Copper activates metallothionein gene transcription by altering the conformation of a specific DNA binding protein. Cell 55:705-717

Hamer DH (1986) Metallothionein. Ann Rev Biochem 55:913-951

Haslinger A, Karin M (1985) Upstream promoter element of the human metallothionein IIA gene can act like an enhancer element. Proc Natl Acad Sci USA 82:8572-8576

Heguy A, West A, Richards RI, Karin M (1986) Structure and tissue specific expression of the human metallotionein IB gene. Mol Cell Biol 6:2149-2157

Hunziker PE, Kaegi JHR (1983) Isolation of five human isometallothioneins. 15th meeting of the Federation of European Biochemical Societies, Brussels

Imagawa M, Chiu R, Karin M (1987) Transcription factor AP-2 mediates induction by two different signal-transduction pathways: protein kinase C and cAMP. Cell 51:251-260

Imbra RJ, Karin M (1986) Metallothionein gene expression is regulated by serum factors and activators of protein kinase C. Mol Cell Biol 7:1358-1363

Kadonaga JT, Carner KR, Masiarz FR, Tjian R (1987) Isolation of cDNA encoding transcription factor Sp1 and functional analysis of the DNA binding domain. Cell 51:1079-1090

Karin M, Herschman HR (1980a) Characterization of the metallothioneins induced in HeLa cells by dexamethasone and zinc. Eur J Biochem 107:395-401

Karin M, Herschman HR, Weinstein D (1980b) Primary induction of metallothionein by dexamethasone in cultured rat hepatocytes. Biochem Biophys Res. Commun 92:1052-1059

Karin M, Andersen RD, Slater E, Smith K, Herschman HR (1980c) Metallothionein mRNA induction in HeLa cells in response to zinc or dexamethasone is a primary induction response. Nature 286:295-297

Karin M, Richards RI (1982) Human metallothionein genes:Primary structure of the metallothionein II gene and a related processed pseudogene. Nature 299:797-802

Karin M, Haslinger A, Holtgreve H, Richards RI, Krautner P, Westphal HM, Beato M (1984a) Characterization of DNA sequences through which cadmium and glucocorticoid hormones induce human metallothionein-IIA gene. Nature 308:513-519

Karin M, Najarian R, Haslinger A, Valenzuela P, Welch J, Fogel S (1984b) Primary structure and transcription of an amplified genetic locus: The CUP 1 locus of yeast. Proc Natl Acad Sci USA 81:337-341

Karin M (1985a) Metallothioneins : Proteins in search of function. Cell 41:9-10

Karin M, Imbra RJ, Heguy A, Wong G (1985b) Interleukin I regulates human metallothionein expression. Mol Cell Biol 5:2866-2869

Karin M, Haslinger A, Heguy A, Dietlin T, Cooke T (1987a) Metal-responsive elements act as positive modulators of human metallothionein enhancer activity. Mol Cell Biol 7:606-613

Karin M, Imagawa M, Imbra RJ, Chiu R, Heguy A, Haslinger A, Cooke T, Sundramurthi S, Jonat C, Herrlich P (1987b) Hormonal and environmental control of metallothionein gene expression. In: Transcriptional control mechanisms; Alan R Liss Inc p:295-311

Klauser S, Kaegi JHR (1985) Characterization of isoprotein patterns in tissue extracts and isolated samples of metallothioneins by reversed-phase high pressure liquid chromatographie. Biochem J 209:71-80

Lee W, Haslinger A, Karin M, Tjian R (1987) Activation of transcription by two factors that bind promoter and enhancer sequences of the human metallothionein gene and SV40. Nature 352:368-372

Mueller PR, Salser S, Wold B (1988) Constitutive and metal-inducible protein:DNA interactions at the mouse metallothionein I promoter examined by in vivo and in vitro footprinting. Genes Dev 2:412-427

Mullis KB, Faloona FA (1987) Specific synthesis of DNA in vitro via a polymerase-catalysed chain reaction. Methods Enzymol 155:335-350

Richards RI, Heguy A, Karin M (1984) Structural and functional analysis of the human metallothionein IA gene: Differential induction by metal ions and glucocorticoids. Cell 37:263-272

Schmidt CJ, Jubier MF, Hamer DH (1985) Structure and expression of two human metallothionein I isoform genes and a related pseudogene. J Biol Chem 280:7731-7737

Scholer H, Haslinger A, Heguy A, Holtgreve H, Karin M (1986) In vivo competition between a metallothionein regulatory element and the SV40 enhancer. Science 232:76-80

Szczypka MS, Thiele DJ (1989) A cysteine rich nuclear protein activates metallothionein gene transcription. Mol. Cell. Biol. 9:421-429

Soarle FF, Stuart GW, Palmiter RD (1987) Metal regulatory elements of the mouse metallothionein-I gene. Exper Suppl 52: Metallothionein II; Birkhaeuser Verlag Basel p:407-414

Seguin C, Hamer D (1987) Regulation in vitro of metallothionein gene binding factors. Science 235:1383-1387

Serfling E, Luebbe A, Dorsch-Haesler K, Schaffner W (1985) Metal-dependent SV40 viruses containing inducible enhancers from the upstream region of metallothionein genes. EMBO J 4:3851-3859

Slater E, Cato A, Karin M, Baxter J, Beato M (1988) Progesterone induction of metallothionein-IIA expression. Mol Endo 2:485-491.

Thiele D, Walling MJ, Hamer D (1986) Mammalian metallothionein is functional in yeast. Science 231:854-856

Thiele D, Hamer D (1986) Tandemly duplicated upstream control sequences mediate copper-indiced transcription of the Sacharomyces cerevisiae copper-metallothionein gene. Mol Cell Biol 6:1158-1163

Thiele D (1988) Mol Cell Biol 8:2745-2752

Welch JW, Fogel S, Cathala G, Karin M (1983) Industrial yeast display tandem gene iteration at the CUP1 region. Mol Cell Biol 3:1353-1361

Welch J, Fogel S, Buchman C, Karin M (1989) The CUP2 gene product regulates the expression of the CUP1 gene, coding for yeast metallothionein. EMBO J 8:255-260

Westin G, Schaffner W (1988) A zinc-responsive factor interacts with a metal-regulated enhancer element (MRE) of the mouse metallothionein-I gene. EMBO J 7:3763-3770

Winge D, Nielson K, Gray W, Hamer D (1985) Yeast metallothionein. J Biol Chem 260:14464-14470

The Problems of
Nutritional Requirement of
Trace Elements

Recommended Dietary Intakes of Trace Elements: Some Observations on Their Definition and Interpretation in Comparison with Actual Levels of Dietary Intake

ROBERT M. PARR

International Atomic Energy Agency, Box 100, A-1400 Vienna, Austria

ABSTRACT

An overview is presented of current dietary guidelines for minor and trace elements based on recommendations of WHO, FAO and the United States National Academy of Sciences. Some difficulties are identified in applying these recommendations, and the newer concepts of "basal" and "normative storage" requirements are summarized. Data for various population groups around the world show that actual intakes are generally less than the currently recommended levels, indicating that basal requirements may be as low as, or lower than, the following daily intakes (calculated for a 60 kg person): Ca 420 mg; Cu 1 mg; Mg 170 mg; Mn 2 mg; Se 20 µg; Zn 6 mg.

Key words: recommended dietary intakes; basal requirement; storage requirement; dietary intakes of Ca, Cu, Mg, Mn, Se, Zn

1. INTRODUCTION

Particularly during the past two decades, public health authorities in many different countries have started to take an interest in defining desirable levels of nutrient intake for their populations (IUNS 1983). Some of these efforts have also been duplicated at the international level by bodies such as the World Health Organization (WHO) and the Food and Agriculture Organization (FAO) (e.g. FAO 1988, WHO 1973-89). For the most part, the guidelines and recommendations issued by the various expert committees that have considered the matter concentrate on energy and major components of the diet, such as protein, carbohydrate and fat. However, a significant literature has also been created on micronutrients such as the vitamins and trace elements. In conformity with the theme of this meeting, nutritional requirements for trace elements will be considered in this paper, along with the minor elements (calcium, magnesium, etc.) from which it is sometimes hardly meaningful to separate them. Table 1 gives an overview of current dietary guidelines for minor and trace elements based on the sources most often quoted in this connection, namely WHO, FAO and the National Academy of Sciences of the USA.

For various reasons outlined below, some of these guidelines may be criticized on the grounds that the underlying concepts are insufficiently clearly defined and that new evidence points to the need for revising some of the numerical values presently in use. Dietary guidelines for trace elements have apparently become a very topical issue judging by the fact that various countries have recently issued new guidelines or are in the process of doing so (e.g. Australia, Denmark, Finland, Norway, Sweden, UK, USA); also that this topic will be taken up by an expert group of WHO and FAO (together with IAEA) in 1990.

2. SOME PROBLEMS WITH THE DEFINITIONS AND USES OF "RDIs"

The recommended dietary intake (RDI) of an essential nutrient has traditionally been taken to be the estimated level of intake considered, on the basis of available scientific knowledge, to be adequate to meet the known

Table 1 Recommended daily dietary intakes of essential[1] and toxic[2] elements. (The values quoted refer to adults. For infants, children and adolescents, and in some cases for pregnant and lactating women, different intake values of essential elements are recommended.)

Element	Quantity	Intake value per day
Al	provisional tolerable intake	1 mg/kg Body Weight (BW)
As	provisional tolerable intake	2 µg/kg BW
Ca	recommended dietary allowance	800 mg
Cd	provisional tolerable intake	1 µg/kg BW
Cl	estimated safe and adequate intake	1700-5100 mg
Co	recommended dietary allowance	3 µg vitamin B_{12} [a]
Cr	estimated safe and adequate intake	50-200 µg
Cu	estimated safe and adequate intake maximum acceptable daily load	2-3 mg 0.5 mg/kg BW [b]
F	estimated safe and adequate intake	1.5-4.0 mg
Fe	recommended dietary allowance maximum acceptable daily load	10 or 18 mg (age/sex dependent) 0.8 mg/kg BW
Hg	provisional tolerable intake	0.7 µg/kg BW total Hg 0.5 µg/kg BW methyl-Hg [c]
I	recommended dietary allowance maximum tolerable intake	150 µg 17 µg/kg BW
K	estimated safe and adequate intake	1875-5625 mg
Mg	recommended dietary allowance	300 mg (female); 350 mg (male)
Mn	estimated safe and adequate intake	2.5-5.0 mg
Mo	estimated safe and adequate intake	0.15-0.5 mg
Na	estimated safe and adequate intake	1100-3300 mg
P	recommended dietary allowance	800 mg
Pb	provisional tolerable intake	7 µg/kg BW
Se	estimated safe and adequate intake	50-200 µg
Sn	maximum acceptable daily load	2 mg/kg BW [b]
Zn	recommended dietary allowance maximum acceptable daily load	15 mg 1.0 mg/kg BW [b]

a. expressed as weight of vitamin B_{12}, which contains about 4% by weight of cobalt
b. provisional value
c. expressed as weight of mercury

1. USA RDIs (USA 1980)
2. WHO/FAO recommendations (WHO 1982-89)

nutritional needs of practically all healthy persons in a group with similar specified characteristics (age, sex, type of diet, etc.). It is normally conceived to be an average requirement plus 2 standard deviations (generally +30%). For toxic elements, on the other hand, the level defined is based on the lowest exposure known to have a detectable toxic effect, reduced by an appropriate (and usually arbitrary) safety factor.

In some countries, particularly in the USA, the RDI is normally referred to as the RDA (recommended dietary allowance), but the present author shares the view, already expressed by others (IUNS 1983), that the term RDI is to be preferred since it is more easily understandable internationally.

Some of the possible uses of RDIs are:

- For assessing food intake of groups or individuals
- For planning therapeutic diets or institutional meals
- For planning national food supplies
- As the denominator for nutrition labelling
- To express nutritional quality of foods (e.g. nutrient density).

Most of the expert committees that have issued RDIs have been at pains to point out that their recommendations apply to groups and not to individuals. The US National Academy of Sciences report (USA 1980) puts it this way. "RDIs should not be confused with requirements for a specific individual. Differences in the nutritional requirements of individuals are ordinarily unknown. Intakes below the RDI for a nutrient are not necessarily inadequate, but the risk of having an inadequate intake increases to the extent that intake is less than the level recommended as safe."

Nevertheless, many individuals do indeed take the RDIs as being applicable to **them** - as indeed they are if that individual wishes to avoid **all** risk of possibly having an inadequate intake. Furthermore, the use of RDIs by individuals is implicitly sanctioned by their use as the denominator in food labelling, as suggested by FAO (1987).

Some of the presently used RDIs are specified in considerable detail, with values that vary according to age and sex. As an example, table 2 summarizes the values recommended for females in the US National Academy of Sciences report (USA 1980).

Table 2   RDIs for females [note: the actual RDIs are the values in mg; the other values (i.e. relative to body weight (BW) and energy intake (MJ)) are derived from these]

Age y	Kg	MJ	RDI (per day)		
			mg	mg/kg BW	mg/MJ
0-½	6	2.9	10	1.7	3.4
½-1	9	4.0	15	1.7	3.8
1-3	13	5.5	15	1.2	2.7
4-6	20	7.1	10	0.50	1.4
7-10	28	10.1	10	0.36	1.0
11-14	46	9.2	18	0.39	2.0
15-18	55	8.8	18	0.33	2.0
19-22	55	8.8	18	0.33	2.0
23-50	55	8.4	18	0.33	2.1
51-75	55	7.6	10	0.18	1.3

Some interesting consequences of these RDIs become apparent when one expresses them in terms of the body weight (BW) or the energy intake (in MJ per day). For example, the RDIs expressed in mg/kg BW for infants and young children actually exceed the maximum tolerable intake specified by WHO (1983), which is 0.8 mg/kg BW. Furthermore, looking at the recommended intakes normalized with respect to energy intake, one sees differences of almost a factor 4 between the lowest and the highest. For comparison, the energy density of iron in whole diets rarely exceeds 1.5 mg/MJ, and does not vary greatly with the age of the persons consuming them (except for babies). It may be questioned, therefore, whether the intakes quoted in Table 2, and the variations in intake as a function of age, are achievable in practice with the diets that most people actually consume.

Similar difficulties arise in respect of the **additional** intakes recommended for pregnant women. Calculations based on typical diets indicate that women would have to increase their energy intakes by around 4-8 MJ/day to give them the increase in intake recommended for elements such as Ca and Zn; however, the recommended increase in energy intake in pregnancy is only about 1.3 MJ/day (USA 1980).

Additional problems with RDIs as presently defined are that they are mostly based on scientific data for **healthy** populations eating a westernized diet with a relatively high proportion of dairy and meat products. It is somewhat unclear how these values can be applied in developing countries where the populations generally subsist on diets with much lower proportions of animal products (resulting in lower bioavailability of some trace elements), and where they have lower body weights, lower energy intakes, and high rates of endemic diseases, such as infestation with intestinal parasites and chronic diarrhoea, that might affect trace element absorption and excretion.

For these and other reasons, RDIs such as the ones illustrated in tables 1 and 2 are now regarded as being somewhat questionable. Difficulties have been recognized in applying these values, **new** data have come to light which – for some elements – call for a revision of the existing RDIs, and above all, there has been a change of philosophy – on the part of some scientists – in the way that RDIs should be defined (Beaton 1988).

Preparations are now underway for a **new** assessment of dietary intakes of trace elements, which is being coordinated by WHO and FAO (in collaboration with IAEA) and which is expected to lead to a new publication on this subject towards the end of 1990.

The new philosophy which will probably be introduced into this report will entail the definition not just of a **single** RDI for each element, but rather of at least 2 values, the so-called **basal requirement** and the **normative storage requirement**, defined as follows:

> BASAL REQUIREMENT: The amount needed to prevent a clinically demonstrable impairment of function. Individuals meeting this requirement will be well and will maintain normal growth and function. However, they will have very low or non-existent reserves of the nutrient in their tissues and hence may be susceptible to problems caused by short-term dietary inadequacies.

> NORMATIVE STORAGE REQUIREMENT: The amount required to maintain a reserve in body tissues. Conceptually this reserve is a store of nutrient that can be mobilized to meet essential needs without detectable impairment of function. The amount of reserve considered to be desirable is a normative judgement.

In addition, it is now proposed that the RDI as previously defined should henceforth be referred to as the **safe level of intake**. These new approaches to the definition of RDIs have already been adopted in the recent WHO/FAO report on Requirements for vitamin A, iron, folate and vitamin B_{12} (FAO 1988).

3. ACTUAL DIETARY INTAKES OF TRACE ELEMENTS

Whatever concepts are used to define RDIs, it is obviously necessary that they should bear some sort of reasonable relationship with the **observed** levels of dietary intake in healthy populations groups around the world; in other words there is a need to ensure that dietary recommendations have a secure basis in what healthy people are actually eating. This is not to argue that dietary recommendations should be based **only** on what apparently healthy people consume. Nevertheless, such data on actual intakes are certainly an important **component** of the scientific evidence that needs to be examined.

In this connection, the present author has recently been invited to coordinate the preparation of a report on dietary intakes of trace elements in various population groups around the world. This report will be one of the working papers for a group of experts that will meet in 1990 under the auspices of WHO and FAO to make **new** recommendations on human dietary intakes of trace elements. The preparation of this report is still in progress, but nevertheless a large database has already been created containing relevant data reported in the scientific literature during the past 20 years, together with data from an IAEA research programme (Parr 1987). In generating this database, emphasis has been placed on the inclusion of values obtained by the use of validated analytical techniques and adequately documented sample collection procedures.

For some elements, reliable data are available covering large areas of the world (though there are also important areas for which the data is either very scanty or completely missing, such as in parts of Asia, Africa, Latin America and eastern Europe). Figure 1 gives an illustration of the data so far collected for adult intakes of zinc. This data is presented for different geographical regions in the form of box plots (McGill 1978). These show the median of each set of data (the horizontal line in the middle of each box) as well as the 0, 25, 75 and 100 percentiles. [In other words the whole range is divided into four parts each containing one quarter of the values; half the values lie **within** the box and the other half lie outside it.] The values referred to are **literature** values. That is to say: each one of them is the mean or median of one study reported in the literature, including both measured and calculated values.

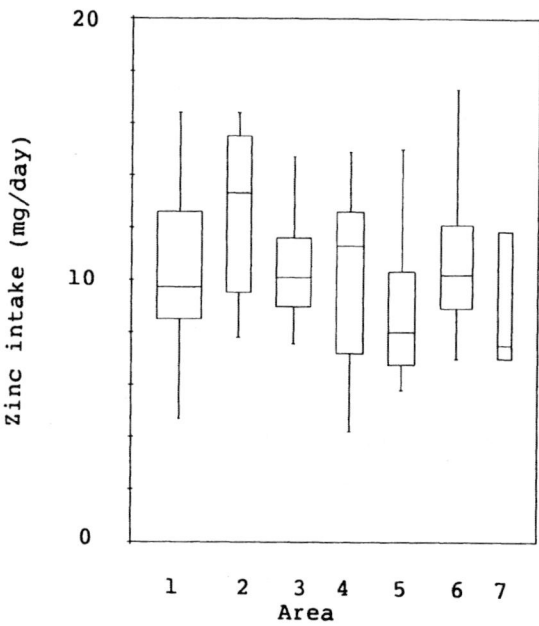

Fig. 1   Dietary intakes of zinc in various geographical areas (1 = north America; 2 = northern Europe; 3 = western and central Europe; 4 = southern Europe; 5 = Asia; 6 = Australia & New Zealand; 7 = Africa)

The present USA RDI for zinc, which dates from 1980, is 15 mg/day (table 1). Obviously, this level of intake is not being attained in most countries. However, actual intakes are in much better agreement with the more recent Canadian RDI (Canada 1983), which is only 9 mg/day for men. The conclusions that one can draw from this kind of comparison are (1) that the Canadian (but not the USA) RDI is reasonable in relation to what apparently healthy populations actually consume, and (2) that there are many population groups that have even lower levels of intake than the Canadian RDI without apparently manifesting any significant public health problems so far attributable to zinc deficiency. Perhaps, therefore, even the Canadian RDI for zinc may be too high to be regarded as a realistic estimate of the basal requirement.

All of the data just discussed refer to the dietary intakes of **adults**. For infants and young children, actual dietary intakes are much less well documented. However, just recently, the results of a study on dietary intakes of trace elements in babies through their mother's milk has been published by WHO and IAEA (WHO 1988). There were some interesting findings in this study which may have an important bearing on the RDIs for that age group.

Almost by definition one has to regard the dietary intakes of babies through their mother's milk as being adequate, at least for the first 3 months or so of life. However, for some elements there are startling differences between actual intakes and current values of the RDIs. This is illustrated in Table 3 for three elements - iron, manganese and molybdenum. Not only are the observed intakes in some cases **less than** the RDIs by a factor of **10 or more**, but also - as pointed out earlier - there is an interesting discrepancy between the US RDI for iron and the corresponding maximum tolerable daily intake proposed by WHO. [Bioavailability is an important issue that may partly justify some of these discrepancies. However, in this author's opinion, there are grounds for doubting that differences as large as a factor 10 could reasonably be accounted for in this way.]

Table 3. Intakes of trace elements in 3-months-old babies through their mother's milk; results for 6 countries (median values) in µg/day (WHO 1988)

Fe: IRON		Mn: MANGANESE		Mo: MOLYBDENUM	
Guatemala	220	Guatemala	2.4	Guatemala	1.3
Hungary	245	Hungary	2.7	Hungary	<0.3
Nigeria	336	Nigeria	10	Nigeria	1.7
Philippines	460	Philippines	25	Philippines	10
Sweden	357	Sweden	2.6	Sweden	0.3
Zaire	260	Zaire	5.2	Zaire	0.6
RDI (USA 1980)	10000	RDI (USA 1980)	500-700	RDI (WHO 1973)	12
				RDI (USA 1980)	30-60
MTI[a] (WHO)	4800				

a. Maximum tolerable intake for 6 kg body weight (WHO 1983)

How can data on actual dietary intakes contribute to the exercise of assigning numerical values to the RDIs? The present author takes the view that it is not unreasonable, as a first step, to examine the lowest levels of intake (e.g. the lowest 10% of literature values) in apparently healthy population groups. By definition, these values must represent upper limits on the basal requirements.

On this basis, using the IAEA database of literature values for dietary intakes in different population groups around the world, the following intakes can be proposed as being equal to, or above, the basal requirements for adults.

Table 4 Intakes of trace elements equal or above the basal requirements for adults

Element	Intake per kg body weight		Intake for 60 kg person	
Ca	7.0	mg	420	mg
Cr	0.5	µg	30	µg
Cu	0.016	mg	1.0	mg
Mg	2.8	mg	170	mg
Mn	0.033	mg	2.0	mg
Mo	0.0015	mg	0.1	mg
Se	0.33	µg	20	µg
Zn	0.1	mg	6	mg

It should be noted that nearly all these intakes are considerably lower than the RDIs listed in Table 1, which are presently the values most widely quoted and used by nutritional scientists around the world.

4. CONCLUSIONS

In this short review, the author has tried to demonstrate that some of the presently used RDIs for trace elements may be in need of revision. This is mainly because there have been some basic changes recently in the philosophy underlying the definition of an RDI. WHO and FAO have already given support to the proposition that there is a need not just for a single figure but for at least two levels of RDI (i.e. basal requirement and normative storage requirement). In addition (in the present author's view), RDIs may need to take account of what is reasonably achievable with normal foodstuffs, particularly the fact that the energy density (i.e. mg of element per MJ) is limited (unless one is willing to resort to fortification); also that changes in intake during life, apart from the very early years, are mainly a matter of changing the **amount** of food eaten, but not its energy density.

Viewed on the global scale RDIs also need to take account of basic differences in the kinds of diets eaten (e.g. meat-based or vegetarian), which can lead to large difference in the bioavailability of trace elements. Finally, there are large differences between countries in respect of the body weights, energy intakes and expenditure, and the frequency of occurrence of intestinal diseases, all of which may change the dietary requirements for trace elements.

5. REFERENCES

Beaton (1988) Nutrient requirements and population data. Proceedings Nutrition Society 47: 63-78

Canada (1983) Recommended Nutrient Intakes for Canadians. Department of National Health and Welfare, Canada

FAO (1987) Codex Standards and Guidelines for the Labelling of Foods and Food Additives. Report CAC/VOL. VI - Ed. 2, Food and Agriculture Organization, Rome

FAO (1988) Requirements of Vitamin A, Iron, Folate and Vitamin B_{12}. FAO Food and Nutrition Series No. 23, Food and Agriculture Organization, Rome

IUNS (1983) International Union of Nutritional Sciences, Report of Committee 1/5, Recommended Dietary Intakes Around The World. Nutrition Abstracts and Reviews in Clinical Nutrition - Series A 53/11: 939-1015

McGill R, Tukey JW, Larsen WA (1978) Variations of box plots. American Statistician 32: 12-16

Parr RM (1987) An international collaborative research programme on minor and trace elements in total diets. "Trace Element Analytical Chemistry in Medicine and Biology" (de Gruyter & Co., Berlin) 4: 157-164.

USA (1980) Recommended Dietary Allowances, Ninth Edition, National Academy of Sciences, Washington, DC

WHO (1973) Trace Elements in Human Nutrition. Technical Report Series No. 532, World Health Organization, Geneva

WHO (1974) Handbook of Human Nutritional Requirements, WHO Monograph Series No. 61, World Health Organization, Geneva

WHO (1982-89) Technical Report Series Nos. 683 (1982) [for Cu, Sn, Zn], 696 (1983) [for Fe], 751 (1987) [for Pb], and 776 (1989) [for Al, As, Cd, Hg, I], World Health Organization, Geneva

WHO (1988) Minor and Trace Elements in Breast Milk, Report of a Joint WHO/IAEA Collaborative Study. World Health Organization, Geneva

The Problems of Nutrition Requirements of Trace Elements: General Comments

ÅKE P.E. BRUCE

The National Food Administration, Box 622, Uppsala S751 26, Sweden

ABSTRACT

Trace elements in nutrition policy mostly concerns nutrient recommendations, fortification and supplementation, but also includes approved food composition tables and data bases for dietary surveys. In devising nutrient recommendations for affluent societies subclinical deficiencies must be taken into account and also chemopreventive and pharmacological effects. In the future, the chemopreventive effects may result in increased recommended intakes, which necessitate food fortification or supplementation of groups-at-risk. In the recently published Nordic Nutrition Recommendations, a table on the lower limits for the intake of some nutrients is presented and precautions have been taken for the recommendations on iron and selenium.

KEY WORDS

Nutrient recommendations, subclinical deficiency, chemoprevention, fortification,

INTRODUCTION

A governmental administration responsible for the national nutrition policy has to deal with trace elements and the various requirements in different population groups in various ways. It has to elaborate nutrient recommendations and to consider what actions have to be taken in order to ensure all individuals a sufficient intake of all essential nutrients including the trace elements. Another important task is to prepare the national food composition tables and to operate data bases to be used for calculating intakes of essential nutrients, including trace elements, for nutrition surveys and epidemiological studies.

This paper presents some general views and experiences that I have gained working as a nutrition scientist within the Swedish government. It reflects severals of the discussions and decisions which have been taken in Sweden and the other Nordic countries.

DOSES VS EFFECTS

Figure 1 gives a schematic version of the relationship between increasing intake or dosage and various physiological effects. In the lower left end of the curve are the insufficient intakes with the well-known clinical signs of deficiency: anaemia for iron; skin lesions for zinc; cardiomyopathy for selenium, etc. At a somewhat higher intake, the clinical signs are less obvious, or well known, but there are still significant biochemical markers: e.g., depletion of body stores; impairment of enzymatic functions. However, here is an area with no clinical signs but with discrete biochemical indicaters, the so-called subclinical deficiency. This area gradually merges into the minimal requirement level, which means that when intakes are below these levels for a long period, there is a risk that clinical or biochemical symptoms of deficiency might appear.

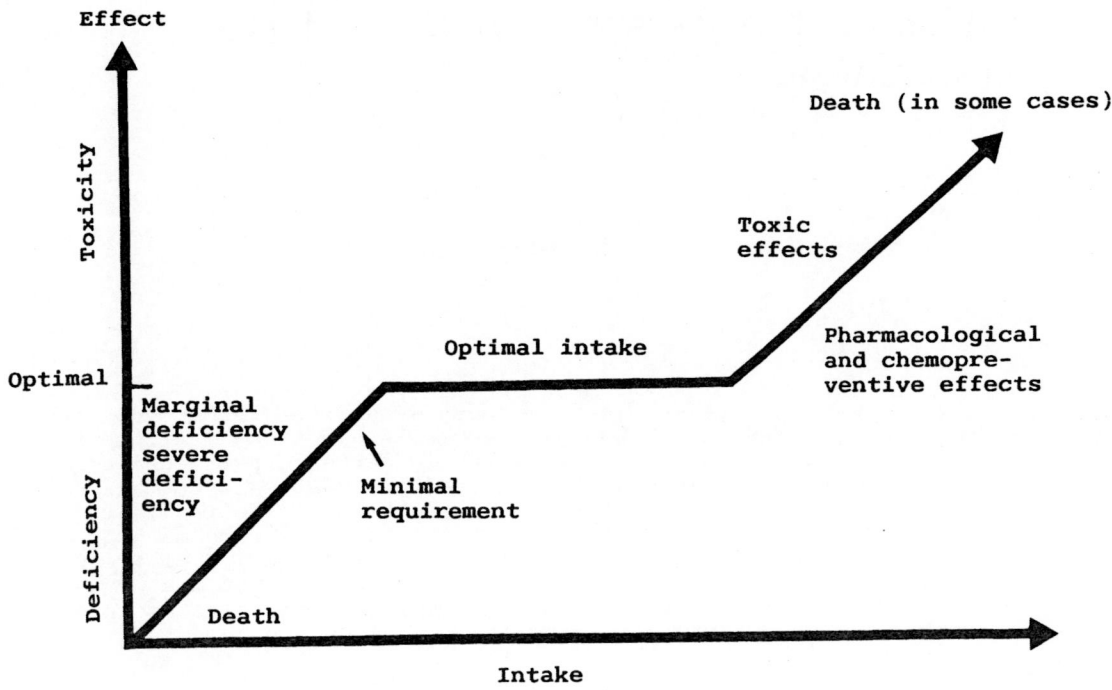

Figure 1.   Various nutrient effects.

At the upper right end of the curve, there are intake levels with a chemo-preventive effect, that is, the nutrients could prevent diseases or dis-orders which are not really nutritional but which could be modified by nutrients. Sometimes there are also pharmacological effects, and the nutrients could be used as drugs to prevent or treat various diseases that have no relation to nutrition whatsoever. Finally, there are the very high intake levels that could have toxic effects or, as is rather specific for the trace elements, could through interaction have negative effects on the status of other trace elements.

When it comes to the higher intake levels, it is not necessarily so that the dose relationships among chemopreventive, pharmacological and toxic effects are in the order indicated on the slide. Some of these effects can very well be at the same intake level as the optimum intake.

Subclinical Deficiency

This concept has mainly been elaborated by Brubacher (1979) and Pietrzik (1985) and originates from their studies on vitamins. However, these dis-cussions are also relevant for trace elements.

The sequence of events in the development of deficiency goes step by step with several physiological changes between the stages of optimum nutrient supply and manifest clinical nutrient deficiency. At first there is a lower-ing of body nutrient content to be found in different tissues. The next stage is characterized by reduced synthesis of various metabolites followed by a depressed activity of dependent enzymes and hormones. The next step is characterized by morphological or functional disturbances. After this third stage of deficiency extreme physiological situations or pathological condi-tions will rapidly lead to clinical symptoms and the nutrient deficiency will become manifest.

Chemopreventive Effects

Sometimes, a specific supplementation of nutrients and minerals is recommended as a means of protecting the population from the effects of chemical substances in the environment. A drastic example is the use of iodine preparations in order to prevent the uptake of radioactive iodine by the thyroid in the event of a nuclear accident.

Fluoride: The best current example of a chemopreventive effect, is fluoride for the prevention of dental caries. In the USA, the Food and Nutrition Board (1980) included fluoride in the RDA and, in my opinion, chemopreventive rather than nutritional data have been considered. Dental caries is almost always caused by an improper consumption of sucrose and perhaps also other sugars. However, it has been known for quite a long time that fluoride has a preventive effect against caries, particularly if it is taken continuously, for example through drinking water. Therefore, water fluoridation is usually very successful, although other means of employing fluorides, for example, in toothpaste or in dental procedures are also very effective. However, the effects of fluoride are not due to fluorine deficiency, but to the consumption of sugar in wrong amounts or at the wrong times.

Selenium is another relevant element in this discussion. There is overwhelming animal research data showing that high doses of selenium have cancer preventive effects, and these findings are supported by a number of epidemiological studies in the Nordic countries, the United States and elsewhere. However, the most reliable data have been presented from studies where serum selenium levels have been compared with the risk of developing cancer. Often a threefold increased risk has been found for the groups with the lowest selenium serum levels, but these data do not necessarily mean that their selenium intake has been decreased by the same magnitude. It is an urgent responsibility for those working with selenium research to establish what dietary intake levels are necessary for a significant risk reduction in human populations. When it comes to selenium and cardiovascular diseases, the data are far less convincing, and there are very few findings from animal research to support such a protective effect.

To what extent trace elements and other minerals can protect the individual against toxic minerals is not well known. However, calcium seems to have a positive effect against lead exposure either through environmental pollution or from paints. This should be considered in some areas where lead intoxication of children is one of the major causes of toxicity. Selenium has been discussed in relation to cadmium and mercury, but as far as I know has not been tested on populations.

Pharmacological Effects

In higher doses, certain vitamins and minerals can have pharmacological effects. This is best known for some vitamins, but fluoride is used with variable results for treating or secondary prevention of osteoporosis.

In Sweden, there was a surge of interest in selenium some years ago, based on the findings of veterinarians and others that various types of pain in muscles, bones or joints, particularly uncharacteristic low back pains were relieved by moderately high doses of selenium. Similar findings were also reported from other countries. Unfortunately, there are very few clinical studies supporting these popular beliefs, but it certainly increased the sale of selenium tablets in Sweden.

Mineral Interaction and Toxic Effects

The major side effect or toxic effect in relation to trace elements is the problem of mineral interaction. However, this phenomenon has very seldom been a true human nutritional problem, but rather has been seen in relation to high pharmacological doses of individual trace elements (particularly zinc and copper) for clinical reasons, for example, treating zinc or copper deficiencies, or overcoming metabolic disorders which impair the absorption

of individual trace elements. In the future, however, we might find minor
interaction phenomenon at intake levels far below those at which the
problems exist today.

FORTIFICATION AND SUPPLEMENTATION

If the government finds that groups of individuals or a whole population do
not consume a specific nutrient in amounts sufficient to avoid the risk of
developing a deficiency disorder, two different types of action could be
taken.

Food fortification is often carried out according to national standards
developed by a government agency. This is the deliberate addition of vitamins and/or trace elements to foods in order to increase their content to a
higher level than normal, as a means of providing the population with an increased level of intake. Food fortification is thus directed towards ensur-
ing population-wide dietary improvements, whereas supplementation is aimed
at increasing the individual's intake of the substances concerned. It
follows that supplementation offers the individual a choice, whereas fortification is for the whole population. Vehicles other than traditional food
may be used for fortification of the diet, for example, water is used as a
vehicle for fortification with fluoride and salt as a vehicle for providing
iodine.

Since the middle of the 1920s, table salt has been fortified with various
levels of potassium or sodium iodine in different countries. Goitre due to
iodine deficiency was once very common in certain regions, but has now
almost disappeared in countries such as New Zealand, Finland and Sweden due
to successful iodine fortification programmes.

Since 1985, Finland added selenium to all fertilizers, and this increased
the selenium content of a large number of its agricultural products. The
original intention was to increase selenium particularly in grains and plant
products, but there was also a considerable increase in the selenium content
of milk, egg, meat, etc., and the ultimate consequences for the Finnish diet
was more than a twofold increase in the selenium intake over a year.

Supplementation, i.e, individual intake of special preparations consisting
primarily of vitamins and/or essential minerals or trace elements to supplement the normal dietary intake of such substances, is particularly relevant
for iron and iron deficiency in fertile women.

THE NORDIC NUTRITION RECOMMENDATIONS (NNR)

One important principle with the Nordic recommendations is a clear distinction between nutrients providing energy and other nutrients (PNUN, 1989).
Diets should comply with the recommendations for protein, fat and carbohydrates including sugar. If the individual diets are varied and are kept to
such quantities that the requirements for energy are met, they should as a
rule provide a person with sufficient amounts of the different nutrients.
However, the recommendations include an extensive table for recommended
levels of daily intake of vitamins and minerals, including zinc, selenium
and iodine, and also a table on lower limits for intake of some nutrients.

Iron: During the work involved in preparing the NNR, some rather intensive
principle discussions took place about iron and selenium. Iron was discussed as to whether or not fortification is necessary. As background to the
iron discussion lies the fact that traditionally Sweden and Finland fortify
wheat flour as well as some other products with iron, while Norway and
Denmark fortify to a far lesser degree or not at all. Particularly between
Norway and Sweden, a number of discussions have taken place during the past f
decades as to whether or not fortification is necessary and subsequently
whether the recommendations for iron intake should be high or moderate for
fertile women. This refers to the well-known fact that there are rather
large groups of women who, due to their regular blood losses, need more iron
than others, and therefore the requirement distribution curve is skewed. The

discussions considered whether the recommendation should be 10 mg or 18 mg. Finally the Nordic recommendation for fertile women became 10 - 18 mg, thus letting the individual nations decide upon which value they should use.

Selenium: The Nordic countries, particularly Finland and Sweden, have for decades had a very low level of selenium intake due to the fact that the soils are low in selenium. No cases of obvious selenium deficiency have ever been identified in these countries. However, an increasing number of studies show that biochemical data indicating a low selenium status, perhaps not necessarily always due to low intakes, are related to various disorders, particularly cancer and cardiovascular diseases. The scientific data for a chemopreventive effect against cancer is rather good, while it is still very doubtful whether selenium has any preventive effects against cardiovascular diseases due to arteriosclerosis.

It was decided that only the nutritional aspects should be considered when selenium was discussed. Since there is no clinical selenium deficiency in the Nordic countries, this means that the traditional intake of around 30 - 40 µg per day is sufficient. The recommendation was set as an interval, 30 - 60 µg, and the individual countries could then make their own decisions.

Sweden decided upon 40 µg without any major discussions. Denmark ran into serious problems, however. A few scientists, with affiliations with pharmaceutical companies, seriously questioned this level and went public, including television, etc., telling the Danish people that their government and the Danish Food Administration deliberately increased the mortality in cancer and cardiovascular diseases because of this low recommendation. Finally, the responsible minister in the government asked representatives of the Food Administration and the complainants to discuss the matter with her, and after a few hours of more or less scientific discussions, she decided

that 50 µg was a scientifically well-founded recommendation. For a nutritionist, it was quite obvious that the pharmaceutical company felt threatened that their large sale of vitamin and mineral supplements containing selenium would decline seriously if the first official Danish recommendation were to be only 50 µg, which could rather easily be obtained through the normal diet.

REFERENCES

Brubacher G (1979) Borderline vitamin deficiency and the assessment of vitamin status in man. Ernährung/Nutrition 3: 4-9

Food and Nutrition Board (1980) Recommended Dietary Allowances, 9th ed. National Academy of Sciences, Washington, DC

Pietrzik K (1985) Concept of borderline vitamin deficiencies. In: Hanck A Hornig D (eds) Vitamins, Nutrients and Therapeutic Agents. Int. J Vit Nutr Res suppl 27, p 61

PNUN (1989) Nordic Nutrition Recommendations, 2nd ed. Standing Nordic Committee on Food, Nordic Council of Ministers, Copenhagen

Trace Element Requirements and Dietary Intake Recommendations: With Specific Reference to the Third World

Noel W. Solomons

Center for Studies of Sensory Impairment, Aging & Metabolism Hospital de Ojos Y Oidos "Dr. Rodolfo Robles V.", Diagonal 21 Y 19 Calle, Z 11 Guatemala City, Guatemala

ABSTRACT

Populations living in the world's developing countries are exposed to recurrent infections (respiratory; diarrheal), heat and humidity, prolonged lactation, recurrent pregnancies, all of which would tend to *increase* individual requirements for certain trace elements. The slower growth rates and smaller adult size of marginally malnourished populations, on the other hand, would tend to *decrease* individual nutrient requirements. As a consequence, nutritional requirements for trace elements, i.e. the amount that needs to be *absorbed* daily, may truly differ in the Third World from those for populations of industrialized countries in temperate climes. To determine the dietary intake recommendations, a factor for the average efficiency of absorption of the element from the customary regional diet must be added. Constituents such as phytic acid and dietary fiber are abundant in the unrefined cereal diets of most developing countries; these tend to reduce the biological availability of many trace elements. On balance, the predominance of factors common to Third World populations and developing country diets would suggest that nutritional requirements and dietary intake recommendations would be *different*, and generally *higher*, than those established for industrialized nations.

KEY WORDS: Trace elements -- Bioavailability -- Nutrient Requirements -- Dietary Intake Recommendations -- Developing Countries

INTRODUCTION

The "Third World" is constituted of the so-called less-developed countries, countries located primarily (but not exclusively) in between the Tropics of Cancer and Capricorn on the continents of Asia, Africa and South America, and in Oceania, the Caribbean, and Central America, including Mexico. They are commonly characterized as rural, largely agrarian, poor, with expanding populations and with limited economic, social, institutional and (occasionally) natural resources.

The Third World has been instrumental in the unravelling of our understanding of human trace element nutrition: human *zinc* deficiency was first described in Iran and Egypt (Prasad et al. 1961, 1963); human *copper* deficiency was recognized in Peruvian children (Cordano & Graham, 1964); human *chromium* deficiency was reported first from Jordan, Nigeria and Turkey (Saner G. et al. 1979); and *selenium* deficiency plays a conditioning role in Keshan disease in rural China (Keshan Disease Research Group, 1979). Despite those important and pioneering results in fundamental trace element research in developing countries, research has lagged behind that of industrialized countries in both quantity and quality. Several factors can be cited as causative for this situation. The most important is the generally low investment in biomedical research in less-developed societies. Additionally, nutrition research has concentrated on the four issues of: 1) protein-energy malnutrition; 2) nutritional anemias; 3) hypovitaminosis A; and 4) endemic goiter, almost to the exclusion of all others. Of course, iron (nutritional anemias) and iodine (endemic goiter) are two of the trace elements, but zinc, copper, selenium,

chromium, manganese and molybdenum issues in human nutrition have largely by-passed the Third World's scientific endeavor. Another factor is the lack of available analytical instrumentation for trace element assays in biological materials that are sophisticated, expensive, and not generally available in developing nations. Finally, despite the evidence produced above, the *myth* of the irrelevance of trace elements to the health and nutritional problems of the Third World has been promulgated by some, and accepted by many. I wish I had a funded research grant for every time a colleague -- from the First World -- had ridiculed the importance of trace elements in the nutrition of popula-tions in developing countries.

The present review will address the issues of trace elements' nutritional requirements in Third World countries in general, and in Guatemala, in par-ticular. We shall define and distinguish three entities: 1) individual nutritional requirements; 2) individual dietary intake requirements; and 3) dietary intake recommendations for populations (Table 1). The chief proposi-tions to be argued in this treatise is *that nutritional requirements, dietary requirements, and dietary intake recommendations for certain trace elements are specific to specific populations in different developing countries*. As a consequence, these parameters would be poorly predicted by research and obser-vations made in developed countries, and poorly expressed in the dietary recom-mendation documents published by international agencies. It is realized that this is both a controversial proposition and one that, if supported, has practical implications for nutritional planning and evaluation.

TABLE 1.   Key Definitions in the Discussion of Nutrient Requirements and Dietary Recommendations

Individual Nutritional Requirements: The amount of a nutrient (in this instance, a trace element) that has to be *absorbed* daily to replace endogenous losses and to provide for growth or milk secretion.

Individual Dietary Intake Requirement: The level of daily intake of a nutrient that, when consumed from the usual local diet, will satisfy the requirement for the given nutrient in a given individual.

Population Dietary Intake Recommendations: The level of intake of a nutrient that will cover the dietary intake requirements of the majority (usually construed somewhere over 90%) of the population in a specific age, gender, physiological-status group.

ISSUES OF INDIVIDUAL NUTRITIONAL REQUIREMENTS

Nutritional requirements may be influenced by a series of issues related to health, environmental circumstances, and growth and final adult size.

Health Factors

The assumptions concerning requirements and intake recommendations for nutri-ents are supposed to relate to "*healthy*" populations. The text of the 1980 Recommended Dietary Allowances (NAS 1980) of the Food and Nutrition Board of the U.S. National Academy of Sciences reads as follows: "Recommended Dietary Allowances (RDA) are the levels of essential nutrients considered ... to be adequate to meet the known nutritional needs of '*practically all healthy per-sons*". The reality of developing countries is one of a heavy burden of "chronic" and "recurrent" infectious conditions among large fractions, and at times the majorities, of the populations. They have high rates of intestinal parasitosis. In some locations of hyperendemicity, the majority of people have recurrent malaria. Acute infections -- respiratory and gastrointestinal -- are common in Third World populations, especially in children less than 5 years of age. Each child may spend 25 to 50% of his or her days ill over this period with some form of acute infection. Were we to restrict our considera-tions to the illness-free sector of the population, they would apply to so few individuals as to be irrelevant, or they would apply only to the elite so as to be a distortion of the overall national situation. For the Third World countries, the condition of "healthy" must be dropped as a restriction for the application of nutrient intake recommedations. We must substitute the phrase, "that apply to the vast majority of the free-living members of the population."

Several studies have demostrated the impact of diarrhea on zinc and copper losses. In Chile, Castillo-Durán et al. (1988) performed metabolic balance studies on children with acute diarrhea. While consuming an average of 51.4 ug/kg/day, the Chilean children were in negative balance of approximately minus 108 Ug/kg/day. A 10 kg child would lose a milligram of the metal daily from endogenous reserves. Studies in Guatemala by Ruz and Solomons (1988) focused on the losses of trace elements occuring during oral rehydration therapy for acute, secretory diarrhea. The rates of zinc averaged 8 ug/kg/h which adds up to 1.9 mg of zinc from endogenous stores lost during a day of acute gastroentertis by a 10 kg child if no foods were given. For copper, the excretion rates were 1.8 ug/kg/h in this study, projecting a daily loss of this metal of 0.4 mg for an average 10 kg child during 24 h of oral rehydration therapy (Ruz & Solomons: unpublished findings).

Environmental and Cultural Factors

Environmental and cultural factors common to some Third World societies could also influence nutritional requirements for certain trace elements, generally in a direction that would *increase* them. The hot, humid climatic conditions of tropical lowlands increase perspiration, and sweat losses of the more metallic trace elements can be enhanced under these conditions. Frequent and repeated pregnancies in countries with high birth rates will deplete the mothers of trace element reserves. This is well documented in the case of iron, in which up to 750 mg of iron can be transferred to the products of conception (fetus, placental). Lesser, but substantial, amounts of zinc, copper, manganese, and selenium, etc., are no doubt passed from maternal reserves to the placental tissue and to the offspring. Lactation follows each pregnancy, and milk production is another focus for the trace element nutrient drain (WHO/IAEA 1989). Children are often nursed up to age two years and beyond in developing countries, and the factor of frequency times the average duration of lactation is the true bottomline for estimating the losses of trace elements into maternal milk.

Growth Rates and Body Size

Genetic, nutritional, and health factors govern the rates of growth and final attainment of body size. In some regions of the developing world, such as rural Sudan (Ostroski et al., 1989), young children grow along the upper percentiles of the growth standards of the National Center for Health Statistics for both weight and height. By contrast, in Peru and Guatemala, the rates of growth are attenuated (Guzman et al., 1968; Trowbridge et al., 1987). Adults in most parts of the developing world are shorter and lighter than industrialized-country counterparts.

The issues of growth and body-size are not trivial for arguments about nutritional requirements. With respect to zinc in pediatric nutrition, Krebs and Hambidge (1986) have argued that, on days on which a child does not grow, he/she has no need to absorb zinc. This is a controversial position, and one would at least have to use the endogenous losses on a day a child failed to grow as his or her requirement for that day.

The average adult Guatemalan man weighs more like 50 kg than the 70 kg that has been the model for nutrient requirement assumptions in the United States. If total body trace element content is proportional to body weight, the amount of that nutrient to be accumulated over a life-time for a smaller adult man would be 28% less. If basal daily losses are related to body size (as in body surface) then another argument for a lower requirement in a more physically-small population emerges.

ISSUES OF DIETARY INTAKE RECOMMENDATIONS

Once the issues of individual and population nutritional requirements for trace elements have been established, there remain the issues of estimating the dietary intake levels necessary to allow for the absorption of the specified requirement amounts. Under what circumstances are dietary intake recommendations different from nutrient requirements? They are not equivalent when

the nutrients are not 100% absorbed. The dominant biology of trace elements is that dietary trace elements are usually not quantitatively absorbed. In soluble form, copper is efficiently absorbed (van Berge Henegouwen et al., 1977), and selenium is absorbed to a greater than 90% efficiency in certain chemical forms (Heinrich et al., 1977; Barbezat el al., 1984). At the other extreme is the case of inorganic iron, the absorption of which is less than 10% (Layrisse, 1975) and that of zinc, with absorptions from 17 to 41 % in the presence of meals (Solomons & Cousins, 1984). The domain of concern for the efficiency of nutrient aborption is called "biological availability."

Boyd O'Dell (1984) has defined bioavailability as the proportion of a nutrient in food that is absorbed and utilized. He further defined utilization as the process of transport, cellular assimilation and conversion to biologically active compounds following its uptake into the body. Although O'Dell's definition invokes the issue of utilization of the nutrients, in the present discussion we shall equate the term bioavailability strictly with *fractional absorption* of a nutrient.

Methods for assessing trace element bioavailability in humans are diverse (Rosenberg & Solomons, 1984). Some provide *relative* data, i.e. that one or another dietary situation produces a greater reduction in absorption of the trace element of interest. Intestinal perfusion studies, plasma uptake responses, single-isotope fecal monitoring studies represent these relative absorption approaches. Such methods do not provide quantitative numerical estimates of the fractional absorption. Single-isotope studies involving whole-body counting, and dual isotope studies using fecal monitoring can provide *precise* numbers for fractional absorption of the administered dose. With all of the aforementioned methods, the assumption that an extrinsic tag represents the behavior of the intrinsic metal in the food is necessary. Metabolic balance studies can be performed in which only the element intrinsic to the diet is involved, but any loss of endogenous element into the intestine will invalidate the equation of "apparent" absorption with "*true*" absorption.

A number of foods or specific chemical factors are recognized to affect the bioavailability of trace elements. Those constituents that *enhance* the intestinal absorption of trace elements are ascorbic acid, other organic acids and alcohols, and unknown factors in breast milk and meat. Chelating agents with a high affinity for cations such as oxalic acid and phytic acid tend to make transition metals insoluble in the intestinal lumen, and hence unavailable for absorption. The components of dietary fiber (cellulose, hemicellulose, pectins and lignins) have differential effects on trace element nutrients. Since the majority of traditional diets in developing countries depend on lightly milled and unrefined seeds of cereal, the intake of phytic acid and dietary fibers are high. On balance, this is detrimental to the bioavailability of most trace elements.

Factorial Approach to estimating Dietary Intake Recommendations

Trace elements did not really come onto the international scene until 1973, when the results of meeting of the World Health Organization sponsored expert committee was published in the form of the technical report monograph *Trace Elements in Human Nutrition* (WHO 1973). In this document, an approach to estimating trace element requirements, exemplified by the case of *zinc*, was presented (Table 2). It was based on a factorial approach in which an estimate of *requirements* (that to be absorbed) was made based on growth needs (in children and pregnant women), and losses of the metal from endogenous sources in the sweat, urine and stools. This provided an *average* requirement level for zinc on the basis of age, gender and physiological state (see the 4th column of Table 2). Then, three assumptions about the efficiency of zinc absorption: good (40%); moderate (20%); and poor (10%) were assigned to various foods. The factoral approach took into consideration how much needed to be absorbed (requirement) and how much needed to be *consumed*, given the bioavailability factor, to allow for the specified absorption.

This approach contains a conceptual error that can be recognized in the light of 17 years of further considerations of how to produce intake recommendations. This table provides *mean*, group-wise estimates of requirements. The development of recommendations is usually conceived of in terms of protecting the

TABLE 2. FACTORIAL BASIS OF PROVISIONAL REQUIREMENTS OF DIETARY ZINC IN RELATION TO ESTIMATES OF ASSUMED LOSSES AND AVAILABILITY

Age	Peak daily retention (mg)	Urinary excretion (mg)	Sweat excretion (mg)	Total required (mg)	Milligrams necessary in daily diet if content of available zinc is: 10%	20%	40%
Infants							
0 - 4 months	0.35	0.4	0.5	1.25	12.5	6.3	3.1
5 -12 months	0.2	0.4	0.5	1.1	11.0	5.5	2.8
Males							
0 -10 years	0.2	0.4	1.0	1.6	16.0	8.0	4.0
11 -17 years	0.8	0.5	1.5	2.8	28.0	14.0	7.0
18 years plus	0.2	0.5	1.5	2.2	22.0	11.0	5.5
Females							
1 - 9 years	0.15	0.4	1.0	1.55	15.5	7.8	3.9
10 -13 years	0.65	0.5	1.5	2.65	26.5	13.3	6.6
14 -16 years	0.2	0.5	1.5	2.2	22.0	11.0	5.5
17 years plus	0.2	0.5	1.5	2.2	22.0	11.0	5.5
Pregnant women							
0 -20 weeks	0.55	0.5	1.5	2.55	25.5	12.8	6.4
21 -30 weeks	0.9	0.5	1.5	2.9	29.0	14.5	7.3
31 -40 weeks	1.0	0.5	1.5	3.0	30.0	15.0	7.5
Lactating women	3.45	0.5	1.5	5.45	54.5	27.3	13.7

upper part of the requirement distribution. In other words, one could assume, on a conservative basis, that the variance around the mean of nutritional requirements is ± 15%. Using this *margin* of security, we would have to include a readjustment of the figures in Table 2. For lactating women, the mean requirement has been established as 5.4 mg to be absorbed. This led to the respective intake recommendations for this physiological status of: 13.7 mg/day (40% absorption); 25.4 mg/day (20% absorption); and 54.5 mg/day (40% absorption). However, bracketing these with a 15% margin for the expected variance around the mean requirements, we would re-estimate respectively: 15.5 mg/day (40%); 30.5 mg/day (20%); and 61 mg (10%).

Synthesis of Dietary Intake Recommendations

As we have seen, the recommendation takes into consideration the amount of a trace element nutrient that should be absorbed to cover losses and new tissue deposition as a function of the average absorbability from the traditional diet. Biological data is required to make refined estimates of both components. At least four factors common to developing societies tend to increase requirements, and most of the bioavailability issues in Third-World diets are felt to be inhibitory in nature.

One further consideration needs to be discussed in terms of a recommendation for a trace element. Should it accept as a reality that some of the remaining nutrients are not consumed in adequate amounts in a given population? Or should it be based on the assumption that all of the nutrients for *optimal growth* will be consumed by the population? As stated, growth is often a factor in the determination of requirements of trace elements. If some other nutrient(s) are *limiting* in the diet, and that is to be accepted as the population's reality, then the recommended intakes of a trace elment may have to be brought into balance with the prevailing limitation in dieatry quality. A lesser amount of the mineral in the diet will prevent its deficiency state and provide for an adequate body pool at a lower level, if demand is reduced by the limitation of another dietary nutrient. This consideration would prevail in processes in which committees deliberate dietary recommedations on a nutrient-by-nutrient or a nutrient-class basis. If the assumption upon publishing the recommendation for trace elements is that nutritional requirements will be met broadly across the gamut of dietary nutrients, then one would be more likely to project an intake level for the oligoelement that is based on the optimal rates of growth and metabolism This consideration is

more likely to prevail in processes in which the committee is dealing with the whole range of nutrients at the same time.

Finally, it is worth pointing out that the amount of food consumed by an individual is related to their energy metabolism and their caloric requirements. Thus, the *nutrient density* of a trace element, i.e. the quantity per 1000 kcal or per MJ of diet, comes into play. Referring again to the recommendations for zinc in Table 2, we would find out -- if we were to use the respective energy requirements for the distinct groups as the denominator -- we would find widely varying nutrient densities for zinc corresponding to the age, gender and physiological state brackets. In industrialized societies, in which dietary diversity is great, it is conceivable that individuals in different stages of the life-span can alter their diet to increase or decrease trace element density to maintain the recommended intake. Another general truism of the Third World is that the regional diets are "monotonous", that is, based on a limited range of food items. If this is true, then another strategy to develop dietary recommendations, is to specify a desirable nutrient density for the total diet. This also assumes, correctly, that most age groups consume the same limited number of foods and beverages. The recent deliberations in Caracas, Venezuela, that led to the publication of the *Guías de Alimentación: Bases para su Desarrollo en América Latina* (1988), an attempt at a universal set of recommendations for the Latin American region, took this approach. They specified a diet with the nutrient density for specified nutrients that would cover the group wtih the highest specific requirement, in relation to the expected energy intake of that age and sex. Once the density is established, the intake of any nutrient, in a person consuming his or her appropriate energy intake, would not be so much an independent recommendation but rather a *consequence*. Except for iron, no trace element nutrients have been considered in the Latin American Guide (UNU/Fundación Cavendes, 1988).

FACTORS AFFECTING ZINC BIOAVAILABILITY IN GUATEMALA

Studies investigating the influence of the foods and beverages consumed in Guatemala on zinc absorption have been conducted using the plasma appearance (zinc tolerance test) approach (Solomons et al., 1979a; 1979b). We found that a composite meal of black beans, tortilla, sweet roll and coffee, reduced to almost zero the rise in plasma zinc after a 25 mg oral dose administered as zinc sulfate. The work of Calloway and Kretsch (1979) indicated that the traditional Guatemalan food fare is high in dietary fiber, with 93 g consumed daily in their model diet. In our hands, coffee alone, as compared to water, reduced the excursion of the postdose zinc curve by 40%. When raw oysters were served as the source of zinc, one could still determine a plasma response in circulating zinc levels. With this model, it was clear that tortillas had a much more negative impact on zinc uptake into the plasma than did black beans. The constellation of findings in our studies suggested that individually and collectively, the major elements of the basic Guatemalan rural diet had negative effects on the absorption of zinc using this experimental approach.

How well zinc sulfate or oyster zinc represents the dietary pool of intrinsic zinc in Guatemalan foods cannot be assessed. Hence, although we feel that Guatemalan cuisine represents a "low bioavailability" diet with respect to zinc, we cannot as yet place a "fractional absorption" value for the factorial approach described above.

TRACE ELEMENT CONTENT OF REGIONAL FOODS

Food Composition Issues in the Third World

Tables of food composition offering estimates of the average content of trace elements in the foods and beverages of occidental diets are of variable availability. Several food composition tables exist for zinc (Gormican, 1970; Halsted et al., 1974; Murphy et al., 1975; Freeland & Cousins, 1976; Shils &

Young, 1988) and some for copper (Sandstead, 1982; Shils & Young, 1988). Those for selenium, manganese, chromium and molybdenum are in much scarcer supply. Thus, it should come as no surprise that the situation with respect to information on trace element content of foods and beverages consumed in regional diets throughout the Third World shows an even greater scarcity. In Guatemala, researchers from CeSSIAM conducted a field study in the peri-urban neighborhood of Guajitos in which up to 14 individual 24-h recall records were collected among 52 women in their third trimester of pregnancy. This produced 706 women-days of dietary information. Ms Sian Fitzgerald, graduate student from the University of Guelph, analyzed this data-set for the individual food and beverage items mentioned, and their frequency of mention. A total of 254 items were identified. (Fitzgerald, S: unpublished observations). Looking at the five reference sources available to us on zinc composition of foods (see above), we found that only 140 of the Guatemalan foods had any data reported for zinc. There was no zinc composition information in the literature for one-hundred fourteen items. The project involves analysis of some of the presuma-bly more contributory foods and beverages for their content of zinc, copper and manganese back in the laboratories in Canada.

Trace Element content of the Corn Tortilla

The traditional staple food of the Guatemalan Mayan, adopted by the entire contemporary population of the Republic is the corn tortilla, made from lime-soaked and milled corn (Tejada, 1979). In the highland village of Santa María Cauqué, this item constituted 70% of the dietary caloric intake (Mata, 1978). Ms Vivian Krause, an undergraduate student from Macdonald College, McGill University, in conjunction with Dr. Carmen Yolanda Lopez as a local counterpart from CeSSIAM, undertook a study of the trace element content of this food as prepared or consumed in Kekchi-speaking households in their native province of Alta Verapaz, and after migration to the nation's capital (Krause et al., 1989). This study revealed that the content of iron ranged from 1.0 to 3.4 mg/100 g of edible portion of tortillas with a mean of 1.4 mg. The amount of iron consumed by female heads of households *from tortillas* averaged 8.3 mg per day. For zinc, the food composition data was from 1.4 to 2.4 mg of the metal per 100 g of tortilla, with a mean of 1.8 mg. The contribution of tor-tilla zinc to the intake of the index women was 10.7 mg. For copper, the corresponding data were 0.10 to 0.22 mg/100 g with a mean of 0.15 mg, and an average daily intake of the mineral from this food source of 0.9 mg/day.

"Contamination" and Water Contribution to Trace Element Intake

Even when food composition data on trace elements become available for spe-cific foods of interest in regional diets of countries of the Third World, additional sources of intake must be included in the considerations of total intake. An obvious source is that of drinking water. In many cultures, and rural Guatemala is one of them, the water for household use is derived from some common community source, i.e. a well, a central faucet, a natural spring or waterway, and stored in a vessel during the day. The material of which the receptical is constructed may have a role in the net mineral intake. The major options in Guatemala are the traditional water jugs made of earthen clay and the "modern", light-weight variety composed of plastic. Clearly, more trace minerals could be leached into the water during standing in the former jugs as compared to the latter. How much of which trace elements are dis-solved into the drinking and cooking water in rural and peri-urban Guatemalan households has not been determined.

The aforementioned study of tortilla mineral content by Krause et al. (1989) also provided evidence for an influence of the processing sequence on final iron and zinc content of the tortilla. The dough produced exclusively on the basis of the traditional, stone grinding-implements, made tortillas with the lowest iron content, while the dough produced with a handmill and stone ele-ments combined in the grinding, gave the highest tortilla iron content. For zinc, differences were also found in this advential contamination of the dough. The complete reversal of the previously described relationship was observed for this nutrient, with the totally stone-ground corn having the greater zinc

content, with the handmill plus stone grinding process providing the lesser. For both iron and zinc, dough processed using electrical mills had intermediary levels of the metals in the final tortillas.

GUIDELINES FOR ESTABLISHING AND TESTING TRACE ELEMENT RECOMMENDATIONS FOR THIRD WORLD POPULATIONS

The most serviceable approach to developing recommendations for daily dietary intakes of trace elements for populations of developing countries is to begin with some estimate of the mean (average) and distribution (range and variance) of the nutritional requirement (amount to be absorbed daily) for the specific age, gender, and physiological-status group. This ideal, but unattainable level would have to be questioned. Some more information would come from direct studies in a representative subsample of the population of interest. The next step is to estimate the range of bioavailability (fractional absorption) from the usual diet consumed by the group of interest. From this, a common factor can be developed, based most logically on an average figure for absorbability. Introducing this bioavailability factor into the distribution of nutritional requirements for the trace element, one develops a new distribution curve of the estimates of individual dietary intake requirements for the population of interest. The *population* dietary intake *recommendation* is derived from this latter curve by determining the level that would provide the individual requirements for the selected percentage of the population group, e.g. 95%.

Even after such an exercise is performed using biological data and mathematical calculations, the resultant number can be looked at from two sequential perspectives: 1) reasonableness and 2) reliability. By "reasonableness" we mean the possibility that a large fraction of the population group could *actually consume such a level of the trace element* from a diet within the economic and cultural context of the region of interest. This again begs the question of having adequate data on the trace element content of the foods and composite diets used by the specific population. If the factorial analysis, for instance, indicates that an intake of 20 mg of zinc daily would cover 90% of the intake requirements with the usual diet of schoolchildren, while a survey showed that the 90th percentile for customary intake of that mineral in the regional diet of this age-group was 9 mg, the practical utility of setting the recommendation at a reasonable (lower) level would be set, with a footnote provision explaining the compromise that has been made.

Finally, whatever be the level established, some community based testing of the adequacy of this level to accomplish the goal, i.e. protect the target population from nutritional deficiency with respect to the trace element, can, in theory, be undertaken. It requires the existence of a reliable index of nutritional adequacy with respect to the nutrient, and preferably a *functional* index. Glutathione peroxidase levels in red cells or platelets might be considered such an indicator of selenium status. What one would then do is take a representative subsample of the population of interest and *assure a universal intake of the recommended level.* At the end of the feeding period, if 90% of the population manifested nutritional adequacy with respect to the trace element in question, one could conclude that the recommended level was an appropriate one.

SUMMARY AND CONCLUSIONS

Trace elements are relevant to the health and nutrition of the Third World. Issues of their nutritional requirements and dietary recommendations are unresolved. It is reasonable to question whether the current "universal" recommendations for dietary intake of trace elements apply to populations of developing countries. The contention offered in this review is that nutritional requirements for specific trace elements are different in Third World nations. Factors that could increase the relative nutritional requirements are: recurrent infections; environmental factors; successions of pregnancies; and prolonged lactation. Factors that could decrease the relative nutritional requirements are slower growth rates and smaller body mass. The issue of growth rates as they relate to trace element requirements needs further elucidation.

Dietary constituents of the regional cuisines of developing countries, such as dietary fiber, phytic acid, polyphenols (tannins), oxalates, ascorbic acid, etc., would be expected to influence the biological availability of trace elements, and, on balance, produce a lesser bioavailability. Within a population the distribution of individual dietary requirements for trace elements in Third World populations would be wider (more heterogeneous) than in an industrial nation. The level that would provide the individual intake requirements for over 90% of such populations would therefore be higher for a developing nation. The data-base for food composition tables of food and beverage contents of trace elements is meager, and this is an impediment to the practical application of a trace element requirement level even when one is developed. Moreover, trace element intakes may be supplemented by adventitial contamination of foods and beverages from metallic utensils and dishes used in the processing, preparation and service of foods in developing countries.

The "proof of the pudding" (practical verification) of a population dietary intake recommendation for a given trace element can come from prospective, community-level research in which the most sensitive indicator(s) of nutritional adequacy are applied to a population sample consuming the suggested level of the nutrient in question from the customary diet of the region.

REFERENCES

Barbezat GO, Casey CE, Reasbeck PG, Robinson MF, Thomson CD (1984) Selenium In: Solomons NW, Rosenberg IH (eds) Absorption and Malabsorption of Mineral Nutrients. Alan R Liss, New York, p 231

Calloway DH, Kretsch MJ (1978) Protein utilization in men given a rural Guatemalan diet and egg formulas with and without added oat-bran. Am J Clin Nutr 31:1118-1126

Castillo-Durán C, Vial P, Uauy R (1988) Trace mineral balance during acute diarrhea in infants. J Pediat 113:452-457

Cordano A, Graham GG (1964) Copper deficiency in infancy. Pediatrics 34:324-326

Freeland JH & Cousins RJ (1976) Zinc content of selected foods. J Am Dietet Assoc 68:526-529

Gormican A (1970) Inorganic elements in foods used in hospital menus. J Am Dietet Assoc 56:397-403

Guzmán MA (1968) Impaired physical growth and maturation in malnourished populations. In: Scrimshaw NS, Gordon JE (eds) Malnutrition, Learning and Behavior. M.I.T. Press, Cambridge MA p 42

Halsted JA, Smith JC Jr & Irwin MI (1974) A conspectus of research on zinc requirements of man. J Nutr 104:345-378

Heinrich HC, Gabbe EE, Bartels H, Oppitz KH, Bender-Gotze CH, Pfau AA (1977) Bioavailability of feed iron-(^{59}Fe), vitamin B_{12}(^{60}Co) and protein bound selenomethionine-(^{75}Se) in pancreatic exocrine insufficiency due to cystic fibrosis. Klin Wochenschr 55:587-593

Krebs N, Hambidge KM (1986) Zinc requirements and zinc intakes of breast-fed infants. Am J Clin Nutr 43:288-292

Keshan Disease Research Group of the Chinese Academy of Medical Sciences (1979) Observation on the effect of sodium selenite in prevention of Keshan disease. Chinese Med J 92:471-479

Krause VM, Kuhnlein HV, Lopez CY, Ruz M, Tucker K, Solomons NW (1989) Rural-urban variation in tortilla mineral content in Guatemala. XII International Conference on Preventive and Social Medicine, Montreal, Canada Abstracts p40

Layrisse M (1975) Dietary iron absorption. In: Kief H (ed) Iron Metabolism and its Disorders. Excerpta Medica, Amsterdam p 25

Mata LS (1978) The children of Santa María Cauqué. M.I.T. Press, Cambridge, MA

Murphy EW, Willis BW & Watts BK. Provisional tables on the zinc content of foods. J Am Dietet Assoc 66:345-355

National Academy of Sciences (1980) Recommended Dietary Allowances. 9th edn. Washington, D.C.

O'Dell BL (1984) Bioavailability of trace elements. Nutr Rev 42:301-308

Ostrowski ZL, Josse MCh, Ojaba E (1989) Nutritional status of infants in a remote area of Africa. XIV International Congress of Nutrition, Seoul, Korea, Abstracts p 607

Prasad AS, Halsted JA, Nadimi M (1961) Syndrome of iron deficiency anemia, hepatosplenomegaly, hypogonadism, dwarfism and geophagia. Am J Med 31:532-546

Prasad AS, Miale A, Farid Z, Sandstead HH, Darby WJ (1963) Biochemical studies on dwarfism, hypogonadism and anemia. Arch Intern Med 111:407-428

Rosenberg IH, Solomons NW (1984) Physiological and pathophysiological mechanisms in mineral absorption. In: Solomons NW, Rosenberg IH (eds) op cit Alan R Liss, New York p 1

Ruz M, Solomons NW (1990) Fecal excretion of endogenous zinc during oral rehydration therapy for acute diarrhea: Nutritional Implications. (Submitted for publication)

Sandstead HH (1982) Copper bioavailability and requirements. Am J Clin Nutr 35:809-814

Saner G (1979) Chromium and glucose metabolism in children. In: Shapcott D & Hubert J (eds) Chromium in Nutrition and Metabolism. p 129-144 Amsterdam, Elsevier/North Holland

Shils ME & Young YR (1988) Modern Nutrition in Health and Disease, 7th edn. Lea & Febiger, Philadelphia, p 1602

Solomons NW, Cousins RJ (1984) Zinc. In: Solomons NW, Rosenberg IH (eds) op cit p 125

Solomons NW, Jacob RA, Pineda O, Viteri FE (1979a) Studies on the bioavailability of zinc in man. I. Effect of the Guatemalan diet and of the iron-fortifying agent, NeFeEDTA. J Nutr 109:1519-1528

Solomons NW, Jacob RA, Pineda O, Viteri FE (1979b) Studies on the bioavailability of zinc in man. II. Absorption of zinc from organic and inorganic sources. J Lab Clin Med 94:335-343

Trowbridge FL, Marks JS, Lopez de Romana G, Madrid S, Boutton TW, Klein PD (1987) Body composition of Peruvian children with short stature and high weight-for-height. II. Implications for the interpretations of weight-for-height as an indicator of nutritional status. Am J Clin Nutr 46:411-418

United Nations University/Fundación Cavendes (1988) Guías de Alimentación: Bases para su Desarrollo en América Latina

van Berge Henegouwen GP, Tangedahl TN, Hofmann AF, Northfield TC, La Russo NF, McCall JT (1977) Biliary secretion of copper in healthy man. Quantitation by an intestinal perfusion technique. Gastroenterology 72:1228-1231

World Health Organization (1973) Trace Elements in Human Nutrition. World Health Organization Technical Report Series No. 532 Geneva, WHO

World Health Organization/International Atomic Energy Agency (1989) Minor and Trace Elements in Breast Milk. Geneva, WHO

Biochemical Characterization of Selenium Deficiency in China

Yiming Xia, Kristina E. Hill, and Raymond F. Burk

Division of Gastroenterology, C-2104 Medical Center N, Vanderbilt University, 1161 21st Avenue, South Nashville, TN 37232, USA

KEY WORDS: Keshan disease, Selenium, Glutathione peroxidase, Vitamin E

Dietary selenium intake varies greatly throughout the world for geographic, cultural, and economic reasons. The selenium content of food plants reflects the soil content and the availability of the element. Thus people at greatest risk for selenium deficiency are people living in areas where soil selenium availability is low who eat only food produced locally.

In the United States there are areas where plants have very low selenium concentrations, but the food distribution system ensures an adequate selenium intake. Thus, selenium deficiency does not occur among free-living people there. Similar situations occur in many other parts of the world.

However, some areas of China combine low soil selenium availability with local consumption of food, and this leads to a suboptimal intake of the element. A cardiomyopathy known as Keshan disease occurs in these areas. A controlled selenium supplementation study carried out in the 1970's by Chinese investigators showed that selenium supplementation prevented Keshan disease (Keshan Disease Research Group, 1979). Thus, it is now known that selenium deficiency is a prerequisite for the development of Keshan disease.

Recently we have investigated the selenium status of people living in a low-selenium area of China. The present manuscript presents some of our results. More details are available in a full paper (Xia et al., 1989). Our purpose was to define biochemically a selenium status associated with the development of Keshan disease. This would characterize a selenium deficiency which is deleterious to human beings.

The subjects for the study were male farmers or male children of farmers in Liangshan Prefecture of Sichuan Province. There were 19-20 subjects per study group. The selenium-deficient subjects were from Dechang County. Their daily selenium intake was estimated to be 11 µg per person based on results of a nutrition survey conducted in 1985. The incidence of Keshan disease in Dechang County is among the highest in China.

The control subjects were from Mianning County. Mianning County is included in a study of selenized salt being carried out by the Antiepidemic Station of Sichuan Province in Chengdu. Since mid-1983 all the salt sold in this county has contained 15 mg sodium selenite per kg. The 1985 nutrition survey estimated the daily per capita selenium intake in Mianning County to be 80 µg--11 µg from food and 69 µg from salt.

For comparison purposes, 10 men and 7 boys were recruited in Beijing for limited biochemical determinations in blood. Their selenium intake was approximately 80 µg per day in the form of natural foods.

All subjects were healthy and free of heart disease. After a blood sample was taken, selenium supplementation was started. Selenium was administered as sodium selenate in tablets containing 200 µg of selenium. Adults received 1 tablet daily. Children received half a tablet daily. Blood

samples were also taken 7 days and 14 days after the subjects began to receive the selenium supplements. Glutathione peroxidase activity, selenium concentration, and vitamin E were measured.

The study groups were chosen to be as similar as possible except for selenium status. The incidence of Keshan disease has been reported to vary from year to year and such variations were noted in Mianning and Dechang Counties. However, when selenium supplementation was begun in Mianning County, Keshan disease incidence fell strikingly. In 1987 ten cases of the disease were reported in Dechang County but only one was reported in Mianning County.

Table 1 presents plasma selenium contents and glutathione peroxidase activities in the initial blood sample. Unsupplemented values in the Dechang groups were 33 to 43% of Mianning values. Men had slightly higher values than boys. These results indicate the degree of selenium deficiency in these subjects. Plasma selenium values of Beijing subjects were severalfold higher than those of subjects from Mianning.

Table 1

Plasma Selenium Concentration and Glutathione Peroxidase Activity in Chinese Subjects [a]

	selenium		GSH-Px	
	μg/dl	% control	U/ml	% control
boys				
selenium-deficient	1.3 ± 0.5	33	29 ± 15	33
control	4.0 ± 1.1	-	87 ± 15	-
Beijing	11.9 ± 1.2	298	128 ± 35	147
men				
selenium-deficient	1.6 ± 0.4	38	51 ± 16	43
control	4.2 ± 1.1	-	118 ± 21	-
Beijing	11.1 ± 1.1	264	191 ± 16	162

[a]Values are means ± S.D.

In spite of the very low selenium concentrations, glutathione peroxidase activities in the Dechang groups were higher than we had expected them to be. The plasma activity of that enzyme in Dechang boys was 33% of the Mianning group values. If one compares Dechang values with the somewhat higher Beijing values, the figure still is 22%. In animal experiments, selenium deficiency routinely produces a plasma glutathione peroxidase activity below 5% of control. This suggests that the selenium deficiency in the people of Dechang County, while being as severe as any reported in humans, is not of the most severe nature when compared with animal experiments.

This study shows that children from a low selenium area have lower glutathione peroxidase activities than adults from the same population, and opens the possibility that children younger than 8 years of age will have even lower glutathione peroxidase activities than those reported here. Those younger children are the ones who develop Keshan disease and studies of them are needed. Plasma glutathione peroxidase activities were higher in the subjects from Beijing than in subjects from Mianning, suggesting that selenium supply in Mianning was not optimal to saturate the enzyme or that factors besides selenium affect glutathione peroxidase activity.

Plasma selenium concentration correlates well with glutathione peroxidase activity in the Dechang and Mianning subjects. The agreement of these two indices of selenium status suggests that these measurements provide good assessments of selenium nutriture under these conditions. The values from the Beijing subjects demonstrate a higher selenium value in proportion to glutathione peroxidase activity than the other groups. This indicates that some of the selenium in the group from Beijing is present in a form not represented by glutathione peroxidase.

We have employed two major approaches to determining the nutritional selenium status of human subjects. One is to measure selenium concentration in blood or blood fractions, and the other is to measure physiologically-active selenium such as glutathione peroxidase. Measurements of selenium require consideration of the different forms of the element present and their significance.

At least three forms of selenium are present in plasma. Physiologically-active selenium is present in glutathione peroxidase and in another protein which has been designated as selenoprotein P in the rat. Little is known about this protein in humans but it may contain more of the plasma selenium than does glutathione peroxidase (Avissar et al., 1989). The third known form of selenium in plasma is selenomethionine which is incorporated into plasma proteins in place of methionine. The presence of selenium in this latter form is dependent on its consumption in the diet. Selenomethionine is a major form of selenium found in plants, so it contributes to the plasma selenium in persons consuming plant foods with significant selenium content. The subjects studied in Beijing had a much higher dietary selenium intake from natural foods than the Sichuan groups. Much of their selenium intake was probably in the form of selenomethionine from plant sources. Thus their much higher plasma selenium concentrations might have been caused by intake of this form of dietary selenium. There is evidence that increased intake of selenium from natural foods leads to greater rises in blood selenium concentration than in blood glutathione peroxidase activity.

Plasma vitamin E levels in the subjects from Sichuan Province were below the normal range of 5 to 15 µg/ml. Selenium status had no effect on plasma vitamin E levels. Values from subjects in Beijing were higher and well within the normal range. Based on these values the subjects in Sichuan had marginal to deficient vitamin E status.

Table 2

Plasma Selenium Concentration after Supplementation[a]

	weeks of selenium supplementation		
	0	1 (μg/dl)	2
Boys			
Dechang	1.3 ± 0.5	2.7 ± 0.7	4.2 ± 0.7
Mianning	4.0 ± 1.1	5.2 ± 1.2	$6.3 \pm 1.5(16)$
Beijing	11.9 ± 1.2	-	-
Men			
Dechang	1.6 ± 0.4	3.4 ± 0.7	4.8 ± 0.8
Mianning	4.2 ± 1.1	6.1 ± 1.1	6.8 ± 1.0
Beijing	11.1 ± 1.1	-	-

[a]Values are means ± S.D.

After the initial assessment, the study groups received selenium supplementation for two weeks. Plasma selenium concentration increased by about the same amount in both groups (Table 2) but glutathione peroxidase activity (not shown) showed a plateau effect in the subjects from Mianning, suggesting that it was becoming saturated. This indicates that glutathione peroxidase activity increases along with plasma selenium in the Dechang groups, which have low plasma selenium, but does not increase as much as does selenium concentration in the Mianning groups, which have high plasma selenium. These results reflect the fact that plasma selenium is present in one or more forms in addition to glutathione peroxidase.

A complete assessment of selenium status will probably require measurement of physiologically-active selenium as well as selenium concentration. From these results, physiological adequacy of the element and the amount stored as selenomethionine can be assessed. While the present study allows this to be approximated, better understanding of the forms of selenium in plasma will be necessary for its refinement.

This study indicates that a population with a plasma glutathione peroxidase activity of 22% to 43% of control is at risk for developing Keshan disease. This corresponds to a plasma selenium less than 2 µg per deciliter. But, based on the complexity of plasma selenium, such a low physiological selenium status might occur with a higher selenium concentration. Other factors, such as vitamin E status, may be important in the causation of Keshan disease.

These results suggest that a selenium intake high enough to keep plasma glutathione peroxidase activity above 43% of control in adult men of a population is needed to prevent Keshan disease in its children.

References

1. Keshan Disease Research Group of the Chinese Academy of Medical Sciences. (1979) Epidemiologic studies on the etiologic relationship of selenium and Keshan disease. Chinese Med. J. 92:477-482.

2. Xia, Y., Hill, K.E., and Burk, R.F. (1989) Biochemical studies of a selenium-deficient population in China: Measurement of selenium, glutathione peroxidase, and some other oxidant defense indices in blood. J. Nutrition in press.

3. Avissar, N., Whitin, J.C., Allen, P.Z., Palmer, I.S., and Cohen, H.J. (1989) Antihuman plasma glutathione peroxidase antibodies: Immunologic investigations to determine plasma glutathione peroxidase protein and selenium content in plasma. Blood 73:318-323.

Ultratrace Elements: An Update

FORREST H. NIELSEN

Grand Forks Human Nutrition Research Center, 2420 2nd Avenue North, Box 7166, University Station, Grand Forks, ND 58202-7166, USA

ABSTRACT

Since the first ISTERH meeting in 1986, an increasing number of studies have examined the need for various ultratrace elements by animals and humans under some form of nutritional, metabolic, hormonal, or physiological stress. The findings indicate that under these conditions some of the ultratrace elements may be of nutritional significance. Under conditions in which calcium loss from the bone is quite likely, evidence has been obtained indicating that boron is important for optimal calcium and therefore bone metabolism. A diet low in taurine and cystine apparently makes the need for arsenic more obvious. The need for nickel may become most evident under conditions in which one of the vitamin B_{12}-dependent enzyme systems is functioning suboptimally. Recent findings suggest that, as humans age, the need for silicon increases, and this need is enhanced by the high consumption of aluminum. If vanadium is an essential nutrient, there is evidence suggesting that its need will be enhanced by a subnormal thyroid status. These observations suggest that some of the ultratrace elements are more important in human nutrition than is now generally acknowledged.

Key Words: boron, arsenic, nickel, silicon, vanadium

INTRODUCTION

Deficiencies of only four trace elements - cobalt as vitamin B-12, iodine, iron and zinc - occur with known sufficient frequency in humans so that they are acknowledged to be of practical importance in nutrition. Nonetheless, the trace elements are often suspected of being the missing link in some of the unexplained human diseases such as atherosclerosis, osteoporosis, osteoarthritis, hypertension, and ischemic heart disease. Efforts to demonstrate that trace element deficiencies are the missing links generally have been unsuccessful. Perhaps some of the failures have occurred because the experimental approaches were too limited or simple. Recent studies examining the need for various trace elements by animals under some form of nutritional, metabolic, hormonal or physiologic stress have indicated that these are situations in which some of the trace elements may be of nutritional significance.

Although trace elements play key roles in a variety of processes necessary for life, the occurrence of overt simple or uncomplicated deficiencies of any of the trace elements is probably relatively uncommon because of the powerful homeostatic mechanisms which the human body possesses. However, there are situations which may make a trace element nutritionally significant. These include 1) inborn errors of metabolism that affect absorption, retention, or excretion; 2) alterations in metabolism and/or biochemistry as a secondary consequence to malnutrition, disease, injury, or stress; 3) marginal deficiencies (slight deviation from an optimal intake of an essential nutrient) induced by various dietary manipulations or by direct or indirect interaction with another nutrient or drug; and 4) the enhanced requirement for a trace element caused by a sudden or severe change in the system requiring that element. The preceding probably can be summarized by the statement that the insufficient intake of a specific trace element probably becomes obvious only when the body is stressed in some way that enhances the need, or interferes with the utilization of that element.

Recently, Tapp and Natelson (1988) presented the formula:

Pathological Effects = Stress x Organic Vulnerability

This formula seems quite applicable to trace element nutrition. In other words, pathological effects are not likely to be seen if a trace element deficiency (organic vulnerability) is not multiplied by some significant stress. Likewise, pathological effects are not likely to be large if stress is not accompanied by an organic vulnerability, or a lack of a trace element. However, the multiplication of a suboptimal intake of a specific trace element times the presence of some nutritional, metabolic, hormonal, or physiologic stress affected by that element most likely would lead to serious pathological consequences.

The preceding concept is supported by the knowledge about the need for the established ultratrace element selenium. Selenium was first shown to be nutritionally important by using vitamin E-deficient animals (Schwarz and Foltz 1957). Close examination of the data indicates that a very limited number of deficiency signs are caused exclusively by selenium deficiency; most signs are enhanced by the lack of either vitamin E or by an elevated need for antioxidants (Shamberger 1984; Levander 1986). Human diseases involving selenium apparently are not simple selenium deficiencies. For example, it has been suggested that Keshan disease, which responds to selenium supplementation, also involves another factor. Suggested possibilities include various toxins, hypoxia, or infectious agents, particularly viruses (Yang et al 1984).

Examining the possibility that some of the ultratrace elements are of importance for humans under some form of stress has revealed several candidates of potential nutritional concern; foremost among these is boron. Other ultratrace elements which may be of concern include arsenic, nickel, silicon and vanadium; thus they will be discussed here.

BORON

At the first meeting of the International Society of Trace Element Research in Humans (ISTERH) in 1986, I reported that boron was emerging as a dynamic trace element that affects major mineral metabolism in animals, and suggested that persons exhibiting disorders of unknown etiology with disturbed mineral metabolism, e.g., osteoporosis, should be examined for possible abnormal boron status (Nielsen 1988). Animal studies subsequent to that meeting have reinforced those statements; generally they have shown that when the diet was manipulated to possibly cause changes in cellular membrane integrity (potassium or magnesium deficiency) or in hormone responsiveness (magnesium or cholecalciferol deficiency, aluminum toxicity), a large number of responses to dietary boron occurred, especially in variables associated with calcium and magnesium metabolism (Nielsen et al 1988bcd; Hunt 1988). However, when an animal was fed a diet apparently optimal in all respects, the response to dietary boron was not very marked. These findings suggested that the need for boron was not crucial, or was quite low, when the animal was not under any nutritional or metabolic stress, but that there was an enhanced need for boron when the animal needed to respond to a stressful situation that adversely altered hormonal or cellular membrane status.

The findings from the animal studies were a stimulus to study the possible nutritional importance of boron for humans under certain metabolic or nutritional stress, for example, with a low dietary intake of magnesium, or when hormonal changes occur (menopause) which causes an increased loss of calcium from bone. Thus far, I have overseen three studies performed with human subjects exposed to boron-low diets (about 0.25 mg/2000 kcal) for 42, 63 or 119 days. Although a 42 day depletion period induced a depression in serum ionized calcium similar to that when the depletion period was either 63 or 119 days, it did not induce most of the significant findings found with the longer boron depletions (unpublished observations). Thus, only the two experiments with a boron depletion period long enough to induce significant changes in several variables related to calcium metabolism will be reviewed here.

The first study with humans was designed to examine the possible effects of aluminum, magnesium and boron on mineral metabolism (Nielsen et al 1987, 1988a). Twelve postmenopausal women, who lived in a metabolic unit under close supervision for 167 days, were fed a 3-day menu rotation diet composed of conventional foods including beef, pork, rice, bread, and milk, but low in fruits and vegetables. At an intake of 2000 kcal, the diet provided per day: 600 mg calcium, 870 mg phosphorus, 116 mg magnesium and 0.25 mg boron. After an equilibration period of 23 days during which the basal low-boron diet supplemented with 200 mg of magnesium per day was fed, all women participated in four dietary periods of 24 days in which dietary aluminum and magnesium were varied. Completion of these four 24-day periods and the equilibration period meant that the volunteers were fed a diet low in boron for 119 days. After completing this phase of the study, the women participated in two additional 24-day dietary periods in which the basal diet was supplemented with 3 mg of boron as sodium borate per day in divided doses at mealtimes. Seven women were fed: 1) the boron basal diet only, and 2) the boron basal diet supplemented with 1000 mg of aluminum per day; thus these seven women were fed a diet low in magnesium for the full 48 days. The other five women were fed: 1) the boron basal diet supplemented with 200 mg of magnesium per day, and 2) the boron basal diet supplemented with 200 mg of magnesium and 1000 mg aluminum per day.

The data obtained indicated that dietary boron had a marked effect on major mineral metabolism in the postmenopausal women and this effect was not influenced by the changes in dietary aluminum. The boron supplementation markedly reduced the urinary excretion of calcium. Although the experimental design prevented a direct examination of the effect of magnesium status on the response to dietary boron, the differences in excretion seemed more marked in the low-magnesium than in the adequate-magnesium women. Boron supplementation elevated the serum concentration of 17β-estradiol; the elevation seemed more marked when dietary magnesium was low. Finally, boron supplementation elevated the serum concentration of ionized calcium, whether expressed as mg/dl or percent of total calcium; here also, the elevation was more significant in the magnesium-low women.

The findings from the first study were a stimulus to do a follow-up study in which additional variables related to calcium metabolism were determined (Nielsen 1989). Because magnesium deprivation seemed to enhance the effects of dietary boron, the diet was kept low in magnesium, 115 mg/2000 kcal, throughout the study. The subjects were five men over the age of 45, five postmenopausal women on estrogen therapy, four postmenopausal women not on estrogen therapy, and one woman who was thought to be postmenopausal but estrogen analysis during the study showed she was not. The subjects, who resided at their homes during the study, ate a conventional diet similar to that used in the first human experiment; it supplied 706 mg calcium and 0.23 mg boron per 2000 kcal. After an equilibration period of 14 days during which the basal low-boron diet supplemented with 3 mg boron per day as sodium borate was fed, there was a depletion period of 63 days when the basal diet was fed. This was followed by a 49 day repletion period when the basal diet was supplemented with 3 mg boron per day.

As shown in Table 1, after being fed a boron-low diet for 63 days, a boron supplement of 3 mg per day affected several variables associated with calcium metabolism in the 15 subjects. Plasma ionized calcium was higher during boron repletion than boron depletion. The difference was significant for each group separately or when all 15 subjects were combined. When all 15 subjects combined were used in the comparison, serum calcitonin and osteocalcin concentrations were lower during boron repletion than boron depletion. Similar changes were found when each group was examined separately. However, significance was achieved for calcitonin only with all the postmenopausal women, and marginal significance was achieved for osteocalcin only with the postmenopausal women on estrogen therapy. Subjects that were receiving estrogen therapy had the lowest serum calcitonin and osteocalcin concentrations. Boron supplementation changed, or tended to change these variables in the men and postmenopausal women not on estrogen in a manner that made them more like the women on estrogen therapy. This suggests that boron and estrogen were having similar effects. Thus, if estrogen is beneficial, boron must be beneficial to calcium metabolism and in the prevention of bone loss which occurs in postmenopausal women and older men.

Table 1. Effect of Boron on Plasma Ionized Calcium, Serum Calcitonin and Serum Osteocalcin in Subjects Fed a Low-Magnesium Diet[a]

Dietary Boron (mg/day)	Ionized Calcium[b] (mg/dl)	Serum Calcitonin[b] (pg/ml)	Serum Osteocalcin[b] (ng/ml)
Men Over Age 45 (n = 5)			
0.23	4.86 ± 0.04[c]	71 ± 14	3.7 ± 0.2
3.23	4.92 ± 0.04	60 ± 9	3.6 ± 0.6
P Value[d]	0.06	0.16	0.74
Postmenopausal Women (n = 4)			
0.23	4.90 ± 0.04	78 ± 8	3.8 ± 0.2
3.23	4.97 ± 0.03	52 ± 9	3.5 ± 0.4
P Value	0.008	0.02	0.60
Postmenopausal Women on Estrogen Therapy (n = 5)			
0.23	4.95 ± 0.04	61 ± 6	2.8 ± 0.5
3.23	5.04 ± 0.02	55 ± 7	1.8 ± 0.3
P Value	0.02	0.02	0.08
Above Combined Plus One Premenopausal Woman (n = 15)			
0.23	4.91 ± 0.02	74 ± 7	3.3 ± 0.2
3.23	4.98 ± 0.02	59 ± 5	2.8 ± 0.3
P Value	0.0001	0.001	0.06

[a]See text for details of experiment. [b]Values obtained during the last 42 days of depletion and last 35 days of repletion were compared. [c]SEM. [d]For each variable, a mean was computed for each dietary period for each volunteer. Paried t-tests were then used to test for dietary effects.

Moreover, findings by other investigators indicate that, depending upon one's point of view, boron deprivation causes, or boron supplementation prevents, changes in calcium-regulation hormones in a manner detrimental to maintaining bone mass in older people. Osteocalcin is regarded as an 1,25-hydroxycholecalciferol-regulated protein that inhibits osteoid mineralization (Price 1988). Thus, increased amounts of osteocalcin in serum or plasma may be an indicator of conditions enhancing the loss of bone mass; this is supported by the findings of an inverse relationship between plasma osteocalcin and bone mineral content in postmenopausal women (Johnson et al 1987). Thus the slightly elevated serum osteocalcin found during boron depletion can be construed as being detrimental to maintaining bone mass.

The finding that boron repletion depressed serum calcitonin seems contrary to the suggestion that boron helps maintain bone mass. Calcitonin is generally viewed as a hormone which prevents the loss of calcium from bone. However it has been found that calcitonin is elevated in women with postmenopausal osteoporosis (Tiegs et al 1985). In people with metastatic bone disease, the failure of the kidney to excrete the calcium from bone breakdown apparently is a major contributor to hypercalcemia in these patients (Benabe and Martinez-Maldonaldo 1978). Calcitonin administration decreases the hypercalcemia in these patients, perhaps through a calciuria effect (Hosking and Gilson 1984). If serum calcitonin is directly related to the loss of calcium from the kidney, then decreased serum calcitonin may help prevent the loss of calcium from the body. Once again, it is possible to construe that boron supplementation is beneficial to conserving calcium in the body.

In addition to the calcitonin results, the findings in Table 2 suggest that boron is active within the kidney. Blood urea nitrogen (BUN) and serum creatinine were significantly higher during boron depletion than boron repletion for each group separately (except for the serum creatinine in postmenopausal women) or for all 15 subjects combined. Also, when all 15 subjects were combined the urinary excretion of creatinine was significantly elevated during boron depletion; when each group was examined separately, the elevation achieved significance only in the postmenopausal women on estrogen therapy. The elevated BUN and serum creatinine levels were still in the normal range and therefore nowhere near those indicating renal failure; however, they may be indicators of detrimental changes in the kidney. Elevations in BUN and serum creatinine often are signs of a changed kidney function.

Table 2. Effect of Boron on Blood Urea Nitrogen (BUN), Serum Creatinine, and Urinary Excretion of Creatinine in Subjects Fed a Low-Magnesium Diet[a]

Dietary Boron (mg/day)	BUN[b] (mg/dl)	Serum Creatinine[b] (mg/dl)	Urinary Creatinine[b] (g/24 hr)
Men Over Age 45 (n=5)			
0.23	14.6 ± 0.3[c]	1.16 ± 0.04	2.23 ± 0.16
3.23	12.2 ± 0.5	1.04 ± 0.03	2.19 ± 0.13
P value[d]	0.02	0.0006	0.60
Postmenopausal Women (n=4)			
0.23	13.0 ± 0.5	0.94 ± 0.04	1.22 ± 0.08
3.23	11.4 ± 0.7	0.86 ± 0.06	1.18 ± 0.06
P value	0.07	0.25	0.96
Postmenopausal Women on Estrogen Therapy (n=5)			
0.23	13.1 ± 0.5	0.98 ± 0.05	1.40 ± 0.17
3.23	11.8 ± 0.3	0.89 ± 0.05	1.31 ± 0.17
P value	0.03	0.0008	0.03
Above Combined Plus One Premenopausal Woman (n=15)			
0.23	13.5 ± 0.3	1.03 ± 0.03	1.62 ± 0.14
3.23	11.6 ± 0.3	0.93 ± 0.03	1.56 ± 0.14
P value	0.0001	0.0001	0.04

[a]See text for details of experiment. [b]Values obtained during the last 42 days of depletion and last 35 days of repletion were compared. [c]SEM. [d]For each variable a mean was computed for each dietary period for each volunteer. Paired t-tests were then used to test for dietary effects.

In summary, many findings have been obtained which indicate that boron has a role in calcium metabolism, probably at the kidney level. Thus, boron most likely has an important role in the maintenance of normal bones. Moreover, this role seems to become more apparent under conditions in which increased calcium loss from the bone is quite likely.

ARSENIC

At the first ISTERH meeting, I described findings which suggested that arsenic has an essential role that is related to, or involved in, methionine or labile methyl metabolism (Nielsen 1988). Since that meeting, additional findings from animal studies have been obtained which support the suggestion that arsenic is involved in methionine metabolism. Among those studies was one which showed that arsenic deprivation depresses the concentration of putrescine, spermidine and spermine in liver of rats (Uthus et al 1989). The formation of these polyamines involve methionine.

Recently it was found that magnesium deficiency affects arsenic-supplemented rats more than arsenic-deficient rats (Uthus and Nielsen unpublished). As shown in Table 3, growth, plasma cholesterol and hemoglobin were lower in arsenic-deprived than arsenic-supplemented rats fed adequate magnesium. Magnesium deficiency depressed growth, plasma cholesterol and hemoglobin; however the depression was more marked in the arsenic-supplemented than the arsenic-deprived rats. Thus, arsenic deprivation did not markedly affect the magnesium-deficient rats. These findings also may be reflecting a relationship between arsenic and methionine. Casein-based diets are known to be marginal in methionine, low in cystine and devoid of taurine (Sturman 1983). Magnesium deprivation probably enhanced the need for methionine and cystine because it induces elevated levels of taurine in muscle and urine (Robeson et al 1979); taurine is formed from methionine via cystine. The finding that arsenic-supplemented rats become similar to arsenic-deprived rats when dietary magnesium is low, and thus when taurine metabolism is altered suggests that arsenic is an ultratrace element whose need by humans will be found to be influenced by nutritional stress involving the sulfur amino acids. Most likely arsenic is needed to form various metabolites from methionine (e.g. cystine, taurine). Thus a diet low in taurine and cystine may make the need for arsenic more obvious.

Table 3. Effect of Arsenic, Magnesium and Their Interaction on Growth, Hemoglobin and Plasma Cholesterol of Rats

Treatment		Growth	Hemoglobin	Cholesterol
As (µg/g)	Mg (µg/g)	(g)	(g/dl)	(mg/dl)
0	100	171	13.61	58
0	400	220	14.11	61
1	100	186	13.05	58
1	400	266	14.58	68
Analysis of Variance - P Values				
Arsenic		0.0001	NS	NS
Magnesium		0.0001	0.0001	0.01
As x Mg		0.003	0.003	0.10

NICKEL

At the first ISTERH meeting (Nielsen 1988), I reported that vitamin B_{12} status affects the growth response of methionine/methyl group-depleted rats to nickel deprivation. Further analyses of those rats revealed that an interaction between nickel and vitamin B_{12} also affected kidney weight/body weight ratio, plasma concentrations of copper, iron and molybdenum, liver concentrations of calcium, copper and molybdenum, and kidney concentrations of copper, manganese and nickel (Nielsen et al 1989). With almost all the variables affected by dietary nickel, the effects were influenced by vitamin B_{12} status. With many of the variables, vitamin B_{12} deprivation made, or tended to make, the nickel-supplemented rats essentially the same as the nickel-deprived rats. These results suggest that vitamin B_{12} is necessary for the optimal expression of the biological role of nickel. Thus, the need for nickel by humans may become evident under situations in which one of the vitamin B_{12}-dependent enzyme systems is functioning suboptimally.

SILICON

At the first ISTERH meeting, I mentioned that the nature of the signs of silicon deficiency in animals has resulted in the speculation that the lack of silicon is involved in several human disorders, including atherosclerosis, osteoarthritis, and hypertension, as well as the aging process (Nielsen 1988). Recently, findings have been obtained which suggest that silicon protects rats against aluminum-induced abnormal behavior (Carlisle and Curran 1987). The findings indicated that high dietary aluminum decreases the silicon content in selected regions of the brain, including those thought to be involved in Alzheimer's disease, and that brain aluminum content was elevated by aluminum supplementation of rats maintained on a low-silicon diet; no elevation occurred in rats maintained on a silicon-supplemented diet. Perhaps as humans age, the need for silicon increases, especially in the presence of a nutritional stress such as the high consumption of aluminum.

VANADIUM

Since the first ISTERH meeting, evidence has accumulated which suggests vanadium is an essential element. It was found that some haloperoxidases from red and brown algae (Vilter 1984, Krenn et al 1987) and from lichens (Plat et al 1987) require vanadium to be active. In rats, it was found that as dietary iodine increased from 0 to 0.33 µg/g diet, thyroid peroxidase activity decreased, with the decrease more marked in vanadium-supplemented (38.1 to 12.3 units/mg protein) than vanadium-deprived (18.7 to 10.3 units/mg protein) rats (Uthus and Nielsen 1988). If vanadium is an essential nutrient, perhaps there is an enhanced need for vanadium in animals and humans with subnormal thyroid status.

CONCLUDING STATEMENT

An increasing number of studies have been performed which have examined the importance of ultratrace element nurtiture in various forms of nutritional, metabolic, hormonal, or physiologic stress in animals and humans. These studies indicate that situations will be found in which an ultratrace element is of nutritional significance. Thus, it is likely that some of the ultratrace elements are more important in human nutrition than is now generally accepted.

REFERENCES

Benabe JE, Martinez-Maldonaldo M (1978) Hypercalcemic nephropathy. Arch Intern Med 138:777-779

Carlisle EM, Curran MJ (1987) Effect of dietary silicon and aluminum on silicon and aluminum levels in rat brain. Alzheimer Dis Associat Disord 1:83-89

Hosking DJ, Gilson D (1984) Comparison of the renal and skeletal actions of calcitonin in the treatment of severe hypercalcaemia of malignancy. Q J Med 53:359-369

Hunt CD (1988) Boron homeostasis in the cholecalciferol-deficient chick. Proc ND Acad Sci 42:60

Johnson PK, Hunt IF, Murphy N, Baylink DJ, Clemens RA (1987) Osteocalcin (OC), bone mineral content (BMC) and calcium intake in postmenopausal women. Fed Proc 46:902

Krenn BE, Plat H, Wever R (1987) The bromoperoxidase from the red alga Ceramicum rubrum also contains vanadium as a prosthetic group. Biochem Biophys Acta 912:287-291

Levander OA (1986) Selenium. In: Mertz W (ed) Trace elements in human and animal nutrition, 5th edn., vol 2. Academic Press, Orlando San Diego New York Austin London Montreal Sydney Tokyo Toronto, p 209

Nielsen FH (1988) Possible future implications of ultratrace elements in human health and disease. In: Prasad AS (ed) Essential and toxic trace elements in human health and disease. Alan R Liss, New York, p. 277

Nielsen FH (1989) Effect of boron depletion and repletion on calcium and copper status indices in humans fed a magnesium-low diet. FASEB J 3:A760

Nielsen FH, Hunt CD, Mullen LM, Hunt JR (1987) Effect of dietary boron on mineral, estrogen and testosterone metabolism in postmenopausal women. FASEB J 1:394-397

Nielsen FH, Mullen LM, Gallagher SK, Hunt JR, Hunt CD, Johnson LK (1988a) Effects of dietary boron, aluminum and magnesium on serum alkaline phosphatase, calcium and phosphorus, and plasma cholesterol in postmenopausal women. In: Hurley LS, Keen CL, Lonnerdal B, Rucker RB (eds) Trace elements in man and animals 6. Plenum, New York, London, p 187

Nielsen FH, Shuler TR, Zimmerman TJ, Uthus EO (1988b) Dietary magnesium, manganese and boron affect the response of rats to high dietary aluminum. Magnesium 7:133-147

Nielsen FH, Shuler TR, Zimmerman TJ, Uthus EO (1988c) Magnesium and methionine deprivation affect the response of rats to boron deprivation. Biol Trace Element Res 17:91-107

Nielsen FH, Zimmerman TJ, Shuler TJ (1988d) Dietary potassium affects the signs of boron and magnesium deficiency in the rat. Proc ND Acad Sci 41:61

Nielsen FH, Zimmerman TJ, Shuler TR, Brossart B, Uthus EO (1989) Evidence for a cooperative metabolic relationship between nickel and vitamin B_{12} in rats. J Trace Elements Exp Med 2:21-29

Plat H, Krenn BE, Wever R (1987) The bromoperoxidase from the lichen Xanthoria parietina is a novel vanadium enzyme. Biochem J 248:277-279

Price PA (1988) Role of vitamin-K-dependent proteins in bone metabolism. Annu Rev Nutr 8:565-583

Robeson BL, Maddox TL, Martin WG (1979) Muscle changes in rats fed magnesium and calcium deficient diets. J Nutr 109:1383-1389

Schwarz K, Foltz CM (1957) Selenium as an integral part of factor 3 against dietary necrotic liver degeneration. J Amer Chem Soc 79:3292-2393

Shamberger RJ (1984) Selenium. In: Frieden E (ed) Biochemistry of the essential ultratrace elements. Plenum, New York, London, p 201

Sturman JA (1983) Taurine in nutrition research. In: Kuriyama K, Huxtable RJ, Iwata H (eds) Sulfur amino acids: Biochemical and clinical aspects, Progress in clinical and biological research, vol 125. Alan R Liss, New York, p 281

Tapp WN, Natelson BH (1988) Consequences of stress: a multiplicative function of health status. FASEB J 2:2268-2271

Tiegs RD, Body JJ, Wahner HW, Barta J, Riggs BL, Heath H III (1985) Calcitonin secretion in postmenopausal osteoporosis. N Engl J Med 312:1097-1100

Yang G, Chen J, Wen Z, Ge K, Zhu L, Chen X, Chen X (1984) The role of selenium in Keshan disease. Adv Nutr Res 6:203-231

Uthus EO, Nielsen FH (1988) The effect of vanadium, iodine and their interaction on thyroid status indices. FASEB J 2:A841

Uthus EO, Poellot R, Brossart B, Nielsen FH (1989) Effect of arsenic deprivation on polyamine content in rat liver. FASEB J 3:A1072

Vilter H (1984) Peroxidases from Phaeophyceae: A vanadium(V)-dependent peroxidase from Ascophyllum nodosum. Phytochem 23:1387-1390

The Influence of the Ultra Trace Element Deficiency (Mo, Ni, As, Cd, V) on Growth, Reproduction Performance and Life Expectancy

M. ANKE, B. GROPPEL, W. ARNHOLD, M. LANGER, and U. KRAUSE

Sekt. Tierproduktion u. Veterinärmedizin Wissenschaftsber. Tierernährungschemie, Karl-Marx-Universität, Dornburger Str. 24, Jena DDR-6900, GDR

Abstract

Molybdenum, nickel, arsenic, cadmium, and vanadium are essential for animals. Primary deficiency of ultra trace elements need not be reckoned with in animals since feedstuffs contain sufficient amounts of them. The problems of detecting the essentiality of these inorganic components result from the necessity of composing highly nutritious semi-synthetic rations, of repeating the experiments at least five times, of taking the antagonistic effect of several elements into consideration and of registering the deficiency effects as comprehensively as possible. The application of reliable results from animal experiments to humans is possible.

Key words: Molybdenum, nickel, arsenic, cadmium, vanadium, essentiality.

During the passage of inorganic components of foodstuffs, water and air through the fauna, which lasted millions of years, the majority of the substances have probably become parts or activators of proteins, enzymes, hormones or other components of the body. Therefore, either a deficient or a toxic excess supply have to be reckoned with in a number of elements (Fig. 1).

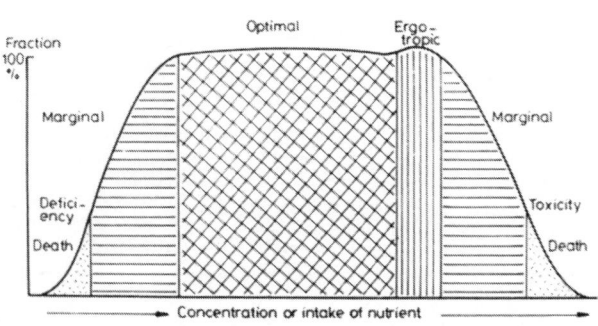

Fig. 1: Dependence of performance on trace element supply

In the transitional zones between deficient and sufficient supply as well as in the transition from normal to excess supply, adaptation reactions occur as described for Cu in sheep (Wiener and Field, 1970) and for Mn in goats and cattle (Anke et al., 1973). As a consequence, breeds of farm animals, which adapted themselves to the local trace element offer, came into being.

Dependent on the species, there is the pharmacodynamic or therapeutic range between the optimum and toxic trace element offer in some elements. Well-known examples for this effect of inorganic components of the diet are arsenic in poultry and humans and copper in pigs. The toxic effect of elements is species-specific. The results obtained in one species cannot be transferred to another without experimental testing. Cattle e.g. are extremely sensitive to an exposure of 10 mg Mo/kg dry matter of the ration and react with molybdenosis, whereas sheep tolerate the threefold and goats the thirtyfold molybdenum amount without difficulties (Falke and Anke, 1987).

1. Molybdenum

Richert and Westerfield (1953) and De Renzo et al. (1953) derived the essentiality of this element from its occurrence in the xanthine dehydrogenase. We know at present that the fauna disposes of 3 kinds of molybdenum-containing enzymes, the xanthine dehydrogenase, the aldehyde oxidase and the sulfite oxidase. These enzymes have a common cofactor termed molybdopterin. Molybdenum deficiency experiments with rats and chicks did not lead to retarded growth and reduced life expectancy (De Renzo et al., 1953; Richert and Westerfield, 1953; Higgins et al., 1956). Only a high supply of the molybdenum antagonist tungsten in the ratio of 1:1000 or 1:2000 resulted in a reduced growth of chicks (Higgins et al., 1956). Such tungsten amounts, however, are never found in feed- and foodstuffs. Furthermore, they strain the metabolism considerably as it was inpressively demonstrated in dairy cows (Graupe, 1965).

During ten years, we repeated experiments with growing, gravid and lactating goats ten times. They finally resulted in the first experimental data on the essentiality of Mo for the fauna. On an average, the molybdenum-deficient goats consumed a semi-synthetic ration with 24 μg Mo/kg dry matter, the control goats were fed a ration with 533 μg Mo/kg dry matter. Molybdenum-deficient kids weighed 6 % less than the controls (Table 1). The difference increased during the sucking period and was higly significant between the 14th and 91st day of life. The growth depression of molybdenum deficient kids during the sucking period might be due to 2 factors. There is no marked molybdenum storage in the body of the foetuses during gravidity (Table 2). The molybdenum deficient kids were already born with molybdenum depletion. The molybdenum content in their liver was 68 % lower than that of control

Table 1: The influence of molybdenum deficiency on the pre- and postnatal development of kids

day of life	control kids		Mo deficient kids		p	decrease %
	s	$\bar{x}$	$\bar{x}$	s		
1st (kg)	0.78	3.1	2.9	0.77	> 0.05	6
42nd (kg)	2.3	10.2	8.9	2.3	< 0.01	13
91st (kg)	4.2	19.6	15.3	3.9	< 0.001	22
100th – 268th female (g/day)	19	92	67	25	< 0.05	27
male (g/day)	41	131	97	13	< 0.05	26

Table 2: Effect of molybdenum deficiency and age on the molybdenum content of the liver of kids and their mothers (μg/kg dry matter)

age (n)	control animals		Mo deficient animals		p	decrease %
	s	$\bar{x}$	$\bar{x}$	s		
kids (4; 8)	297	1367	435	186	< 0.001	68
adult goats (35;16)	781	1211	432	408	< 0.001	64
p	> 0.05		> 0.05			
decrease %	11		1			–

kids. Adult molybdenum-deficient goats died with the same molybdenum content as one-day-old kids. The milk of molybdenum-deficient goats contained more molybdenum than the molybdenum-deficient ration with 24 μg/kg dry matter (Table 3). The molybdenum balance of molybdenum-deficient kids aggravated considerably with the increasing consumption of this ration. Live weight gain was reduced to the same extent. The 51 μg Mo/kg dry matter of the milk do apparently not meet the requirements whereas >100 do.
Under the conditions of the exclusive intake of the semisynthetic ration with 24 μg Mo/kg, the growth depression of molybdenum-deficient goats intensified to 27 and 26 %, resp. (Table 1) between the 100th and 268th day of life. Apart from our own investigations, the effect of molybdenum deficiency on reproduction has not yet been analyzed (Anke et al., 1977). Though the heat symptoms were normal, the success of the first insemination was significantly worse (Table 4). As a rule, repeated pairings led to gravidity in molybdenum-deficient goats as well. Their 12 % lower conception rate was insignificantly worse than that of control goats. The significantly higher abortion rate of molybdenum-deficient goats is most striking. 15 % of molybdenum-deficient goats aborted their foetuses. When the numbers of barren and aborting goats are added up, 44 % of molybdenum-deficient goats did not give birth to viable kids compared to 18 % of control goats. Molybdenum deficiency also had a significant effect on the mortality of kids. 28 % of molybdenum-deficient kids died till the 91st day of life.
The relatively high mortality of control goats (25 %) resulted from the weight of their kids. 12 % of them weighed more than 4.0 kg, whereas only 1 % of the kids of molybdenum-deficient goats was born with this weight. 61 % of molybdenum-deficient goats died.
The essentiality of molybdenum can be easily postulated in animals due to its occurrence in the molybdenum co-factor of several enzymes, growth depression, reproduction disorders and high mortality under the conditions of a molybdenum intake of 24 μg/kg ration dry matter. In humans, the essen-

Table 3: Effect of molybdenum on the molybdenum content of the milk (μg/kg dry matter)

lactation (n)	control animals		Mo deficient animals		p	decrease %
	s	$\bar{x}$	$\bar{x}$	s		
colostrum (32;23)	61	61	35	26	> 0.05	54
milk (76;48)	70	116	51	34	< 0.001	56
p	< 0.001		< 0.05			
increase %	190		146			-

Table 4: The influence of molybdenum deficiency on reproduction performance

parameter	control goats	Mo deficient goats	p
success of first insemination, %	69	57	< 0.05
conception rate, %	83	71	> 0.05
services per gravidity	1.5	1.9	< 0.05
abortion rate, %	1	15	< 0.01
kids per goat carrying to terms	1.5	1.7	> 0.05
sex ratio, ♀ ≙ 1, ♂ ≙ x	1 : 2.0	1 : 1.5	> 0.05
kids died between the 7th and 91st day of life, %	3	28	< 0.001
mortality of female goats, 1st year, %	25	61	< 0.001

tiality of molybdenum was demonstrated by Abumrad et al. (1984). After parenteral nutrition they observed deficiency symptoms which could be overcome by 300 μg Mo/day. Furthermore, Wadman et al. (1983) and Rajagopalan (1984) described a gene defect in babies which led to a reduced sulfite oxidase activity during intrauterine development and the sucking period. The molybdenum requirement of growing, gravid and lactating goats was calculated to be < 100 μg/kg ration dry matter (Anke et al., 1983). Due to the lacking molybdenum requirement of the rumen flora, monogastric animals may need even less molybdenum (Anke et al., 1983). The molybdenum requirement of adults is calculated to be 25 μg/day. In the case of a long parenteral nutrition, 10 to 12 μg/day might not meet the requirement (Anke and Groppel, 1988). According to the investigations and calculations of Anke et al. (1983) the molybdenum intake of farm animals in Central Europe amounts to >100 μg/kg dry matter. This is also true for areas where molybdenum deficiency occurs in legumes. In 1988, the molybdenum consumption of humans was systematically investigated in 4 test groups, each consisting of 7 men and women. The test persons came from different parts of the GDR and were subjected to a 7-day test period.
The evaluation of the results by means of the double variance analysis demonstrated that neither living area nor sex influenced the molybdenum content of the ration and beverage dry matter (Table 5). On the average of the 7-day test period, all persons consumed >100 μg Mo/kg dry matter. Only one woman took in significantly more molybdenum than the total population. She had a mean daily intake of 379 μg Mo/kg. A man from the district of Cottbus had the lowest daily molybdenum intake with 116 μg/kg dry matter. The molybdenum amount consumed per kg dry matter in the GDR is in accordance with the amount found by Parr (1987) (230 μg/kg dry matter) and with values from Switzerland (Wyttersbach et al., 1987) (206 μg/kg dry matter). The molybdenum amount (μg/day) taken in daily is determined by the dry matter con-

Table 5: Molybdenum content of consumed ration and beverage dry matter depending on living area and sex of consumers in μg/kg dry matter

| living area | women | | men | | influence |
	s	$\bar{x}$	$\bar{x}$	s	sex
district of Potsdam	136	201	224	231	
district of Cottbus	126	202	188	169	
district of Gera	109	203	191	87	>0.05
district of Erfurt	72	171	184	147	
influence site		>0.05			−

Table 6: Molybdenum intake of humans depending on site and sex in μg/kg dry matter

| living area | women | | men | | influence |
	s	$\bar{x}$	$\bar{x}$	s	sex
district of Potsdam	38	59	91	79	
district of Cottbus	45	65	68	74	
district of Gera	37	60	75	42	< 0.001
district of Erfurt	19	47	66	40	
influence living area		>0.05			−
all sites	36	58	74	62	< 0.01

sumption. On the average of the examined population, men took in 382 g dry matter, women 281 g/kg. The molybdenum consumption was in accordance with this intake (Table 6). The sex-specific difference was 28 %. No symptoms of disease were found in the 4 test groups.

The molybdenum intake registered in the GDR corresponds with that in New Zealand, England and Switzerland (48; 96; 128; 90 μg/day) (Robinson et al., 1973; Wyttersbach et al., 1987). The average daily molybdenum intake of US adults was calculated to be 350 μg by Schroeder et al. (1970) and 120 to 240 μg/day by Tsongas et al. (1980).

2. Nickel

When we began our nickel deficiency experiments with growing, gravid and lactating goats in 1972, nickel-dependent enzymes were unknown. At present we know that the hydrogenase, the carbon monoxide dehydrogenase and the methyl-coenzyme-M-reductase of several species of bacteria and the plant urease are nickel-dependent enzymes (Nielsen, 1987).

Our nickel deficiency experiments with growing, gravid and lactating goats, which were repeated 17 times, showed clearly that 100 μg Ni/kg ration dry matter do not meet the requirements of ruminants. On the average, control goats consumed a ration with 4000 μg Ni/kg dry matter. Deficiency experiments with selenium-, vanadium- and arsenic-poor rations showed that trace element deficiency can take a very different effect on feed consumption.

For that reason, the daily feed consumption of the goats was also calculated in the long-term experiments (Table 7). On the average of the 17 experimental years, the growing, gravid and lactating goats consumed 14 % less of the semisynthetic ration. In the following years of life, control and nickel-deficient goats consumed the same amounts of feedstuffs (Table 7). Intrauterine nickel deficiency did not take any significant effect on the mean birth weight of nickel-deficient kids. There were, however, significantly more kids with a live weight of < 1.6 kg among them at birth. These kids were usually not viable.

Table 7: Influence of nickel deficiency on the feed consumption of female goats (g/day)

age	control goats		Ni-deficient goats		p	decrease %
	s	$\bar{x}$	$\bar{x}$	s		
young goats	196	704	605	267	< 0.001	14
old goats	251	647	634	272	< 0.05	2

Table 8: Influence of nickel deficiency on the growth of control and nickel-deficient goats

day of life, unit of measurement	control goats		Ni-deficient goats		p	decrease %
	s	$\bar{x}$	$\bar{x}$	s		
1st (kg)	0.8	2.9	2.8	0.8	> 0.05	3
42nd (kg)	2.3	9.6	8.8	2.3	< 0.05	8
91st (kg)	4.5	18.1	15.8	4.0	< 0.001	13
100th - 268th (g/day)	32	96	73	52	< 0.001	24

The differences between control and nickel-deficient kids increase with age. At the beginning of lactation when the kids mainly take in the milk of their mothers the differences are not remarkable. This was due to the high nickel content of the milk of nickel-deficient goats (189 μg/kg dry matter) (Anke et al., 1983a),which contained more nickel than the semi-synthetic ration. The differences between the groups increased with the rising consumption of this ration and reached 24 % in growing and gravid young goats. A mean nickel offer of 100 μg/kg dry matter does not meet the requirements of goats.
In spite of normal heat symptoms, the success of the first insemination of the nickel-deficient goats was significantly worse than that of control animals (Table 9).

Table 9: Influence of nickel deficiency on the reproduction performance of female goats

parameter	control goats	Ni-deficient goats	p
success of first insemination, %	75	52	< 0.001
conception rate, %	83	76	> 0.05
abortion rate, %	1	21	< 0.001
services per gravidity	1.4	1.9	< 0.001
kids per goat carrying to terms	1.5	1.6	> 0.05
sex ratio, ♀ ≙ 1, ♂ ≙ x	1:1.8	1:1.6	> 0.05
died kids, 7th – 91st day, %	6	15	> 0.05
mothers died in the first year of life,%	10	50	< 0.001

Repeated matings resulted in reduced differences between the groups and led to insignificant differences in the conception rate between the groups. The high abortion rate (21 %) of nickel-deficient goats is extraordinarily important. When the proportions of barren and aborting goats are added as high a number as 45 % of the nickel-deficient goats did not give birth to viable kids. This was only true for 18 % of control goats. This difference between the groups is significant.
15 % of nickel-deficient kids died during the sucking period, compared to 6 % of control kids. The difference remained insignificant. This finding is certainly the result of the relatively high nickel content of the milk of nickel-deficient goats. In the goats with nickel-poor rations, the losses only occurred after the conversion to the semi-synthetic ration and the exhaustion of the nickel supplies which are mainly found in skeleton and liver (Anke et al., 1983a).As many as 50 % of nickel-deficient goats died within the first year of life.

Table 10: Influence of nickel deficiency on production and composition of milk during the first 56 days of lactation

parameter	control goats		Ni-deficient goats		p	decrease %
	s	$\bar{x}$	$\bar{x}$	s		
milk, ml day	566	1040	939	835	> 0.05	10
fat content, %	0.41	3.8	3.1	2.5	> 0.05	18
protein content, %	0.25	2.9	3.0	1.8	> 0.05	+ 3
fat, g/day	28	39	29	32	< 0.001	26
protein, g/day	17	30	28	25	> 0.05	7

The influence of nickel deficiency on the production and composition of milk is summarized in table 10. The nickel-deficient goats produced 10 % less milk with a reduced fat content. This resulted in a significantly decreased milk fat production of nickel-deficient goats.

The high mortality of nickel-deficient goats, which cover a certain extent of their protein requirement via 3 % urea in the semi-synthetic ration, has nothing to do with a too low conversion of N to ammonia which is used by bacteria for the microbiological protein synthesis (Table 11).

Table 11: Influence of nickel deficiency on the ammonia content and the urease activity of the rumen fluid of goats

parameter	control goats		Ni-deficient goats		p	decrease %
	s	$\bar{x}$	$\bar{x}$	s		
ammonia content, g/100 ml	7.0	14.2	11.0	5.9	> 0.05	23
urease activity, in U/ml	8.9	31.3	5.4	4.6	< 0.001	83

The urease activity in the rumen fluid of nickel-deficient goats was reduced by 83 %. Under the conditions of nickel deficiency the conversion of uric nitrogen to ammonia was carried out by ATP: urea amidolase. Thus, the ammonia content of the rumen fluid of the goats only differed insignificantly. An effect on the microbiological synthesis of volatile fatty acids in the rumen of ruminants might also be possible via nickel-depending hydrogenase, carbon monoxide dehydrogenase and methyl-coenzyme-M-reductase (Anke et al., 1985). The analysis of these fatty acids, however, did not show any differences between control and nickel-deficient goats (Langer, 1989).

Thus, the high mortality of nickel-deficient goats might be due to their reduced haemoglobin synthesis, their disturbed calcium metabolism which leads to a reduced zinc absorption. The damage registered at skin, hair and testicles underlines this assumption as much as the registered dwarfism and the reduced zinc incorporation in blood serum and milk (Anke, 1985). The nickel requirement of animals is calculated to be < 500 μg/kg ration dry matter. As demonstrated in the experiments with rats and mini-pigs, monogastric animals have a considerably lower nickel requirement than ruminants. It is, therefore, assumed that the nickel requirement of ruminants is covered by 300 - 350 μg Ni/kg and that of monogastric animals by < 200 μg/kg ration dry matter (Anke et al., 1983b). The increase of the urease activity after nickel supplementation is no symptom of an improved performance (Spears et al., 1986). The nickel requirement of humans is certainly not higher than that of monogastric animals. It might even be lower since humans grow more slowly. Nielsen (1987) calculated the nickel requirement of adults to be about 35 μg per day.

The nickel intake of humans in the GDR was investigated according to the same scheme as in the case of molybdenum (Table 12). It was detected that - in spite of cleaning, the removal of peels and grinding - the consumed foodstuffs and beverages contained 250 μg Ni/kg dry matter on the average of the 7-day periods and living areas. Furthermore, the double variance analysis demonstrated the significant influence of the living area on the nickel content of the consumed ration dry matter. It was significantly higher in the test group of Gera than in all other examined populations. Sex did not take a significant effect on the nickel proportion of the consumed food. The low nickel content of the ration dry matter in the districts of Potsdam and Cottbus might be due to the low nickel content of the flora of the moor sites in these living areas. Muschelkalk weathering soils and loess produce a flora with a higher nickel concentration

Table 12:Nickel content of consumed rations and beverages dry matter depending on living area and sex of consumers in µg/kg dry matter

living area	women s	$\bar{x}$	men $\bar{x}$	s	influence sex
district of Potsdam	124	358	397	112	
district of Cottbus	188	360	507	406	> 0.05
district of Erfurt	238	408	556	278	
district of Gera	741	657	607	481	
influence site	< 0.001				

Table 13: Nickel intake of humans depending on living area and sex in µg/day

living area	women s	$\bar{x}$	men $\bar{x}$	s	influence sex
district of Potsdam	131	157	140	61	
district of Cottbus	54	116	138	54	< 0.001
district of Erfurt	57	111	208	115	
district of Gera	152	183	256	284	
influence site	< 0.001				

(Anke et al., 1985). On the average of the test periods and the living areas the nickel consumption of adults exceeds 100 µg Ni/day (Table 13). Due to the higher dry matter consumption, men usually took in more nickel than women. The living area took the same effect on the nickel intake of both sexes.
Primary nickel deficiency symptoms in animals and humans need not be expected in Europe since foodstuffs and feedstuffs deliver much more nickel than required. An exception might be patients with long-term parenteral nutrition. They were given solutions with 6 to 39 µg Ni/l in the GDR (Anke et al., 1985b).Further investigations should solve the problem of nickel deficiency in parenteral nutrition.
The nickel intake of 111 to 256 µg Ni/day on the average of sex and living area or 358 to 657 µg/kg dry matter registered in adults in the GDR corresponds with that of 150 - 700 µg/day or 190 to 410 µg/kg dry matter (Nielsen, 1987; Anke et al., 1985; Parr, 1986) found in other countries.
Independent of nickel deficiency which need not be expected in humans and animals, this element is essential for both groups.

3. Arsenic

The arsenic deficiency trials began in 1973 with kids. They were repeated thirteen times. On the average of the applied charges, the semi-synthetic ration contained < 35 µg arsenic/kg dry matter, the control animals were given 350 µg arsenic/kg. The drinking water of the goats was distilled. In the first year of every trial, the young, growing, gravid and lactating goats of the control and the arsenic deficient group took in the same amount of feed (Table 14). On the average, the surviving adult arsenic-deficient goats ate 7 % more feedstuffs than control animals.

Table 14: Feed intake of young and adult control and arsenic-deficiency goats from July 1st to June 1st on the average of 12 years of trials (g/day)

age of goats	control goats		arsenic deficiency goats		p	%
	s	$\bar{x}$	$\bar{x}$	s		
young	234	678	680	237	> 0.05	99
adult	279	629	674	234	< 0.001	107
p	< 0.001		> 0.05			
%	92		99			

There was only a small significant influence of the arsenic-poor nutrition on the pre- and postnatal development of the kids (Anke et al., 1987). Arsenic deficiency significantly reduced the reproduction performance and especially the success of the first service (Table 15). The conception rate of arsenic-deficiency animals was also significantly lower, 29 % of the arsenic-deficiency goats remained barren. Arsenic deficiency increased the abortion rate significantly, whereas the number of born kids and the sex ratio of the offspring remained independent of the arsenic status.

Table 15: Influence of arsenic deficiency on the reproduction performance of female goats (n 131;113)

parameter	control goats	arsenic deficiency goats	p
success of the first insemination, %	75	57	< 0.01
conception rate, %	89	71	< 0.001
barren goats, %	11	29	< 0.001
services per gravidity	1.3	1.9	< 0.001
kids per goat carrying to terms	1.4	1.4	> 0.05
abortions, %	0.8	15	< 0.001
sex ratio, ♀ : ♂	1:1.6	1:1.7	> 0.05

Since the live weight gain of kids during the suckling period differed significantly and since this difference increased it had to be taken for granted that control kids dispose of other arsenic sources, which can only be intrauterine ones. The analysis of the organs of control and arsenic — deficiency kids which died at birth confirm this assumption (Table 16). Compared to control kids, arsenic-deficiency kids only had 2 or 12 % of the arsenic content in kidneys and cardiac muscle. The other parts of the body of control animals also stored more arsenic than those of deficiency kids. The arsenic content in the tissue of control kids showed considerable individual variations. This made it impossible to demonstrate the significance of these differences in the limited population. It must be assumed that a certain arsenic amount is accumulated in the body during the intrauterine development for the suckling period.

After birth, the arsenic offer via milk is important for the development of kids (Table 17). The maternal arsenic deficiency reduced the arsenic content of colostrum and mature milk. Compared to the normal arsenic content of feed- and foodstuffs, however, its arsenic content is low. The milk of control goats does not provide more arsenic than arsenic-deficiency diets.

Table 16: Arsenic content of several organs of kids of control and arsenic-deficiency goats (μg/kg dry matter)

Organ (n)		control kids		As-deficiency kids		p	%
		s	$\bar{x}$	$\bar{x}$	s		
kidneys	(5;4)	819	737	16	10	> 0.05	2
cardiac muscle	(7;4)	235	130	15	9.3	> 0.05	12
skeleton muscle	(6;5)	84	97	36	42	> 0.05	37
cerebrum	(7;5)	40	49	21	12	> 0.05	43
rib	(3;2)	100	90	39	46	> 0.05	43
lungs	(7;2)	34	44	24	1.9	> 0.05	55

Table 17: Arsenic content of the milk of control and arsenic deficiency goats (μg/kg dry matter)

kind of milk (n)	control goats		arsenic deficiency goats		p	%
	s	$\bar{x}$	$\bar{x}$	s		
colostrum (4;7)	3.4	10	7.8	3.6	>0.05	78
mature milk (22;37)	21	24	15	8.3	<0.05	62
p	> 0.05		< 0.05			
%	240		192			

The systematic investigation of skeleton and cardiac muscles or of the liver of arsenic deficiency goats shortly before exitus showed ultrastructural changes. Under the conditions of arsenic deficiency, electron-dense material was deposited in the mitochondrial membrane of skeleton muscles, cardiac muscle and liver. At an advanced stage, this electron-dense substrate is released from the mitochondrial membrane and is detectable in cytoplasm (Anke et al., 1987).

The arsenic requirement of goats, mini pigs, rats, and chicks was calculated to be < 50 μg/kg ration dry matter. It is influenced by several factors (Anke, 1986). Most arsenic compounds are well absorbed. A arsenic requirement of an adult man of 6 μg/1000 kcal or of 12 to 25 μg daily is derived from experiments with animals (Anke et al., 1984; Nielsen and Uthus, 1984). The arsenic requirement of animals and humans is met by feedstuffs, foodstuffs and water in the GDR (Krause, 1986).

4. Cadmium

On the average of ten experimental years the semi-synthetic control ration of goats and their kids contained 300 μg Cd/kg dry matter, the cadmium-deficiency ration less than 15 μg Cd/kg (Anke et al., 1987).

On an average, cadmium-poor nutrition did not influence the feed consumption of young goats in the first experimental year. The same is true for the feed intake of adult control and cadmium-deficiency goats though the latter took in 2 % more feedstuffs than control animals and though the difference is significant (Table 18).

Table 18: The feed consumption of young and adult control and cadmium - deficiency goats (g/day)

age of goats	control goats		cadmium-defi- ciency goats		p	%
	s	x̄	x̄	s		
first year of life	234	687	679	206	>0.05	99
adult goats	279	629	644	233	<0.05	102
p	<0.001		<0.001			-
%	92		95			

The cadmium-poor intrauterine development of kids did not influence their mean birth weight. The live weight development of control and cadmium-deficiency kids was identical (Figure 2). Insignificant differences only began to develop at the end of the lactation period. Summarizing all results it follows that the influence of a ration with 15 µg Cd/kg ration dry matter on the intrauterine and post-natal live weight gain of goats is small and insignificant.

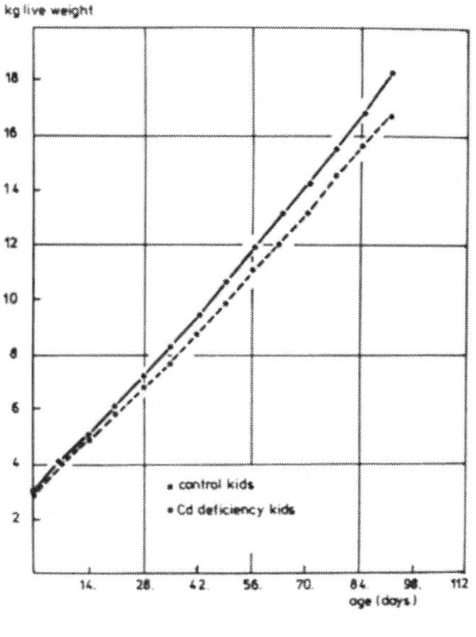

Fig. 2: Live weight development of control and Cd-deficiency kids within the first 13 weeks of life

The cadmium-poor nutrition influenced the reproduction performance (Table 19). All cadmium deficiency goats showed normal heat symptoms. In spite of that the success of the first insemination was significantly worse than in control animals. Repeated pairings improved their conception rate, which reached 72 % and thus differed only insignificantly from that of control animals. Generally, cadmium-deficiency goats needed one service more for conception than control animals. 12 % of the cadmium-deficient goats aborted their foetuses. Most abortions were registered in the 4th and 5th months of gravidity. The difference is statistically relevant. The number of kids per gravid goat was not affected by cadmium deficiency.

The intrauterinely cadmium-depleted kids of several test years were often very phlegmatic. They moved very little and were even too lazy to eat and drink. They had problems to hold their heads erect. The symptoms of this weakness of mobility were registered at different times of the lactation period.
In 1985, 9 kids of cadmium-deficient goats were at our disposal for further experiments for the first time. About 6 weeks after weaning, all animals showed clinical deficiency symptoms in the form of muscular weakness. First they moved very clumsily and stiffly. Later on they were no longer able to raise their heads and died. Up to the first symptoms of muscular weakness the kids were in a good nutritive state. Later on they took in less food and partly lost weight.

Table 19: The influence of cadmium deficiency on the reproduction performance of female goats (n 71;79)

data parameter	control goats	Cd-deficiency goats	p
success of first insemination (%)	73	46	< 0.001
conception rate (%)	85	72	> 0.05
services per gravidity	1.2	2.2	< 0.001
abortion rate (%)	0	12	< 0.01
kids per gravid goat	1.4	1.6	> 0.05
91-day-old kids per experimental goat	0.65	0.41	< 0.01

After 6 of the 9 cadmium-deficient kids had died, the three surviving animals (2 male ones and one female one) were given the ration of control animals with 300 µg Cd/kg dry matter. The animals regained their feed intake and mobility. They produced live weight again (Figure 3).

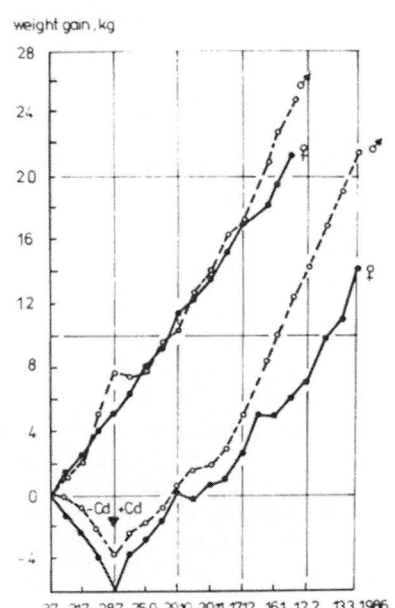

Later on they overcame their retarded development and had normal weight gains. In the following experiments the same muscle weakness occurred in lactating goats which led to death without cadmium supplementation.

Primary cadmium deficiency symptoms are not to be expected in farm animals, game and humans since, in Europe, the intake is considerable above this range. The cadmium requirement of goats is met by > 50 µg/kg ration dry matter (Anke et al., 1987).

Fig. 3: Live weight development of male and female control and Cd-deficient kids

5. Vanadium

In spite of a number of vanadium deficiency experiments with rats, chicks and goats, the essentiality of vanadium has not yet been reliably demonstrated. The contradictory data induced us to continue the vanadium deficiency experiments with goats started in 1980. The experiments were carried out with female goats. The vanadium content of the semi-synthetic ration amounted to < 10 µg/kg dry matter. It varied between 1 and 9 µg/kg. The control ration contained 0.5 mg/kg. The drinking water was distilled (Anke et al., 1989).

The vanadium-poor nutrition did not cause an intake depression, neither in young goats from the 100th day of life till the end of lactation nor in adult animals. On an average, the young goats even consumed 8 % more feed (Table 20).

Table 20: The influence of vanadium deficiency on the feed consumption of young and adult female goats (g/animal and day)

age	control goats		V-deficient goats		p	%
	s	$\bar{\bar{x}}$	$\bar{\bar{x}}$	s		
young goats	192	679	731	205	< 0.001	108
adult goats	274	647	656	247	> 0.05	101

Intrauterine vanadium deficiency did not at all influence weight gain development. This tendency continued systematically during the sucking period of kids up to the 91st day of life (Table 21). Independent of the vanadium supply, kids and goats had the same weight gain, vanadium-poor rations with < 10 ,ug V/kg dry matter do not cause growth depression.

Table 21: Effect of vanadium deficiency on the pre- and postnatal development of kids

parameter	control goats		V-deficiency goats		p	%
	s	$\bar{x}$	$\bar{\bar{x}}$	s		
birth weight (kg)	0.71	2.7	2.7	0.75	> 0.05	100
42nd day of life (kg)	1.8	8.6	8.6	2.0	> 0.05	100
91st day of life (kg)	4.1	16.4	15.6	4.4	> 0.05	95
168 days of life (g/day)	32	97	88	35	> 0.05	91

The vanadium-poor nutrition did also not take effect on the heat intensity of goats. Though the animals on heat were easy to identify, only 50 % of vanadium-deficient goats became gravid after the first mating (Table 22).

Table 22: The influence of vanadium deficiency on the reproduction performance of female goats

parameter	control goats	V-deficient goats	p
success of first insemination, %	74	50	< 0.01
conception rate, %	85	76	> 0.05
barren goats, %	15	24	> 0.05
abortion rate, %	0	18	< 0.001
services per gravidity	1.3	2.0	< 0.001
kids per goat carrying to terms	1.57	1.51	> 0.05
sex ratio, ♀ = 1, ♂ = x	1.72	0.71	< 0.001
died kids, 7th-91st day of life	8	40	< 0.001

Repeated pairings, it is true, increased the conception rate of vanadium-deficient goats too, but they did not reach that of control animals. The difference remained insignificant. As already demonstrated in previous publications, vanadium deficiency took a highly significant effect on the abortion rate.

About one fifth of vanadium-deficient goats aborted their foetuses, mainly during the last month of gravidity. When the proportions of barren and aborting vanadium-deficient goats are added up, 42 % of the vanadium-deficient goats did not give birth to viable kids in the 9 experiments. The difference to the control goats was significant.

Vanadium deficiency, however, surprisingly shifted the sex ratio significantly to the female sex (Table 22). Hornless goats usually give birth to significantly more male kids. This was also true for control goats. The perinatal mortality of male foetuses of vanadium-deficient goats was probably higher and resulted in this strikingly changed sex ratio in the offspring.

The high mortality of vanadium-deficient kids between the 7th and 91st day is most impressive. Previous publications already described pains at extremities, swollen joints and deformations at the fore extremities of vanadium-deficiency kids and growing goats (Anke et al., 1986). This vanadium deficiency syndrome already occurred at the birth of vanadium-deficient kids in the last years (Fig. 4). The skeletal damage did not disappear during the sucking period. In 1989, all kids of vanadium deficiency goats showed these symptoms. Their live weight gain was independent of these changes.

Vanadium deficiency reduced the life expectancy of depleted goats by almost 50 %. About 50 % of them died during the first year of life. The available data (reproduction, skeletal damage, life expectancy) demonstrate that vanadium is essential for goats and that the vanadium requirement is < 10 ug/kg ration dry matter.

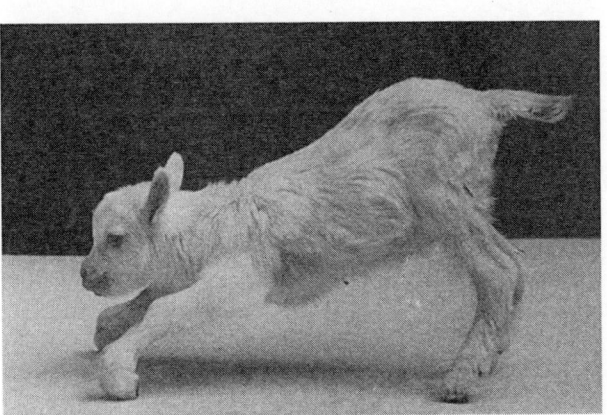

Fig. 4: Vanadium-deficient newborn kid

6. References

Abumrad NN, Schnieder AJ, Steel D, Rogers LS (1981) Amino acid intolerance during prolonged total parenteral nutrition reversed by molybdate therapy. J. Clin. Nutr. 34: 2551-2559

Anke M (1985) Nickel als essentielles Spurenelement. In: Gladke E, Heimann G, Lombeck I, Eckert I (eds) Spurenelemente. Georg Thieme Verlag Stuttgart, New York, p 106-125

Anke M (1986) Arsenic. In: Mertz W (ed) Trace elements in human and animal nutrition. Academic Press Inc. Orlando, Florida, USA, Vol. 2, p 347-372

Anke M, Groppel B (1988) Signifikanz der Essentialität von Fluor, Brom, Molybdän, Vanadium, Nickel, Arsen und Cadmium. Zent.bl. Pharm. Pharmakother. Lab. diagn. 127: 197-205

Anke M, Groppel B, Gruhn K, Košla T, Szilágyi M (1986) New research on vanadium deficiency in ruminants. In: Anke M, Baumann W, Bräunlich H, Brückner Chr, Groppel B (eds) 5. Spurenelementsymposium, Karl-Marx-Universität Leipzig, Friedrich-Schiller-Universität Jena, GDR, p 1266-1275

Anke M, Groppel B, Gruhn K, Langer M, Arnhold W (1989) The essentiality of vanadium for animals. In: Anke M, Baumann W, Bräunlich H, Brückner Chr, Groppel B, Grün M (eds) Trace element symposium - vanadium and molybdenum. Karl-Marx-Universität Leipzig and Friedrich-Schiller-Universität Jena, GDR, p 17-27

Anke M, Groppel B, Krause U (1989) Probleme beim Nachweis der Lebensnotwendigkeit von Ultraspurenelementen. Mengen- und Spurenelemente 9: 1-8

Anke M, Groppel B, Nordmann S, Kronemann H (1983a) Further evidence for the essentiality of nickel. In: Anke M, Baumann W, Bräunlich H, Brückner Chr (eds) 4. Spurenelementsymposium, Karl-Marx-Universität Leipzig, p 19-28

Anke M, Groppel B, Reissig W, Lüdke H, Grün M, Dittrich G (1973) Manganmangel beim Wiederkäuer. Arch. Tierernährung 23: 197-211

Anke M, Groppel B, Schmidt A (1987) New results on the essentiality of cadmium in ruminants. In: Hemphill DD (ed) Trace substances in environmental health - XXI, Univ. of Missouri, Columbia, USA, p 556-566

Anke M, Grün M, Groppel B, Kronemann H (1983b) Nutritional requirements of nickel. In: Sarkar B (ed) Biological aspects of metals and metal-related diseases. Raven Press, New York, p 89-105

Anke M, Grün M, Partschefeld M, Groppel B (1977) Molybdenum deficiency in ruminants. In: Kirchgessner M (ed) Trace element metabolism in man and animals 3: 230-233, Technische Universität München, FRG

Anke M, Krause U, Groppel B (1987) The effect of arsenic deficiency on growth, reproduction, life expectancy and disease symptoms in animals. In: Hemphill DD (ed) Trace substances in environmental health - XXI, Univ. of Missouri, Columbia, USA, p 533-550

Anke M, Risch M (1989) Importance of molybdenum in animal and man. In: Anke M, Baumann W, Bräunlich H, Brückner Chr, Groppel B, Grün M (eds) Sixth International Trace Element Symposium, Vanadium, Molybdenum. Karl-Marx-Universität Leipzig, Friedrich-Schiller-Universität Jena, GDR, p 216-232

Anke M, Schmidt A, Kronemann H, Krause U, Gruhn K (1985) New data on the essentiality of arsenic. In: Mills CF, Bremner I, Chesters JK (eds) Trace elements in man and animals - TEMA 5, Commonwealth Agricultural Bureaux, UK, p 151-154

Anke M, Szentmihályi S, Grün M, Groppel B (1983) Molybdängehalt und -versorgung der Flora und Fauna, Wiss. Z. Karl-Marx-Univ. Leipzig, Math.-Naturwiss. R. 33: 135-147

Anke M, Szentmihályi S, Regius A, Grün M (1985) Essentiality of nickel for flora and fauna. In: Pais I (ed) New results in the research of hardly known trace elements. Univ. Horticulture Budapest, p 15-60

De Renzo EC, Kaleita E, Heytler P, Hutchings BL, Williams JH (1953) Identification of the xanthin oxydase factor as molybdenum. Arch. Biochem. Biophys. 45: 247-256

Falke H, Anke M (1987) Die Reaktion der Ziege auf Molybdänbelastungen. In: Anke M, Brückner C, Gürtler H, Grün M (eds) Mengen- und Spurenelemente 7 Karl-Marx-Univ. Leipzig, GDR, p 448-452

Graupe B (1965) Molybdänstoffwechsel bei Wiederkäuern. Tagungsberichte, Dt. Akademie der Landwirtschaftswissenschaften zu Berlin, 85: 309-317

Higgins ES, Richert DA, Westerfield WW (1956) Molybdenum deficiency and tungstate inhibitions studies. J. Nutr. 59: 539-546

Krause U (1986) The site-spezific arsenic supply of ruminants in the GDR. In: Anke M, Baumann W, Bräunlich H, Brückner Chr (eds) Spurenelementsymposium - 5, Karl-Marx-Universität Leipzig and Friedrich-Schiller-Universität Jena, GDR, p 856-863

Langer M (1989) Die Bedeutung des Nickels für Tier und Mensch. Promotion A, Friedrich-Schiller-University Jena, GDR

Nielsen FH (1987) Nickel. In: Mertz W (ed) Trace elements in human and animal nutrition - Fifth Edition. Academic Press, INC; San Diego, p 245-273

Nielsen FH, Uthus EO (1984) Arsenic. In: Frieden E (ed) Biochemistry of the essential ultratrace elements. Plenum Press, New York and London, p 319-340

Parr RM (1987) An international collaborative research programme on minor and trace elements in total diets. In: Brätter P, Schramel P (eds) Trace element analytical chemistry in medicine and biology. W. de Gruyter Berlin, New York, p 157-164

Rajagopalan KV (1987) Molybdenum - An essential trace element. Nutr. Rev. 45: 321-328

Richert DA, Westerfield WW (1953) Isolation and identification of the xanthin oxydase factor as molybdenum. J. Biol. Chem. 203: 915-926

Robinson MF, McKenzie JU, Thomson CD, Van Rij AL (1973) Metabolic balance of zinc, copper, cadmium, iron, molybdenum and selenium in young New Zealand women. Br. J. Nutr. 30: 195-205

Schroeder HA, Balassa JJ, Tipton JH (1970) Essential trace metals in man: Molybdenum. J. Chron. Dis. 23: 481-499

Spears JW, Oscar TP, Starnes SR, Harvey RW (1986) Factors affecting biological responses to nickel supplementation in ruminants. In: Anke M, Baumann W, Bräunlich H, Brückner C, Groppel B (eds) 5. Internationales Spurenelementsymposium, Karl-Marx-Universität Leipzig, Friedrich-Schiller-Universität Jena, GDR, p 1134-1149

Tsongas TA, Meglen RR, Walravens PA, Chappell WR (1980) Molybdenum in the diet: An estimate of average daily intake in the United States. Am. J. Clin. Nutr. 33: 1103-1107

Wadman SK, Duran M, Beemer FA, Cats BP, Johnson JL, Rajagopalan KV, Drywawaych S (1983) Absence of hepatic molybdenum cofactor: An inborn error or metabolism leading to a combined deficiency of sulfite oxidase and xanthine dehydrogenase. J. Inher. Metab. Dis. 6 (suppl 1), 78-83

Wiener G, Field AC (1970) Genetic variation in copper metabolism of sheep. In: Mills CF (ed) Trace element metabolism in animals. Livingstone, Edinburgh and London, p 92-102

Wyttenbach A, Bajo S, Tobler L (1987) The concentration of 19 trace elements in the Swiss diet. In: Brätter P, Schramel P (eds) Trace element analytical chemistry in medicine and biology. W. de Gruyter Berlin, New York, p 169-178

Hormones and Vitamins

Manganese as a Potential Modulator of Second Messenger Pathways

MURRAY KORC

Division of Endocrinology, Departments of Medicine and Biochemistry, Med Sci I, D235, University of California, Irvine, CL 92717, USA

Abstract

Manganese deficiency and toxicity are not common in humans, and the mechanisms by which manganese exerts its effects are unknown. In the pancreatic acinar cell manganese regulates protein synthesis, modulates calcium fluxes, inhibits cholecystokinin-activated phosphatidylinositol hydrolysis, and participates in the regulation of protein phosphorylation. It is suggested that manganese may act, in part, by modulating hormonally activated second messenger pathways.

Key Words: Phosphorylation, calcium, pancreas, phosphatidylinositols, manganese.

Introduction

Manganese is the twelfth most abundant element in the earth's crust. It is an essential trace element that has been implicated in the regulation of several metabolic processes in humans. Although the total amount of manganese in the adult human is approximately 15 mg (Schroeder et al., 1966), relatively high concentrations of manganese are found in the liver and pancreas. Thus, the human liver and pancreas contain 0.170 mg and 0.076 mg of manganese per 100 gm of wet weight, respectively (Reiman and Minot, 1920). In the liver, manganese is especially abundant in nuclei and mitochondria, whereas in the pancreas it is high in the microsomal and supernatant fractions (Sakurai et al., 1985).

Biological Functions of Manganese

Manganese is an essential component of several proteins and enzymes, including arginase (Hirsch-Kolb et al., 1971), pyruvate carboxylase (Hurley, 1982), the manganese-containing superoxide dismutase (Weisiger and Fridovich, 1973), and the catalytic subunit of calmodulin-dependent protein phosphatase (Merat et al., 1984). It participates in the regulation of various pathways that control glucose homeostasis. Thus, pyruvate carboxylase catalyzes the carboxylation of pyruvate to oxaloacetate, an important step in the initiation of gluconeogenesis. Oxaloacetate is decarboxylated to phosphoenolpyruvate by manganese-activated carboxykinase (Bentle and Lardy, 1976). Manganese enhances hepatic gluconeogenesis, potentiates the effects of glucagon and epinephrine on gluconeogenesis, and induces hyperglycemia when given intraperitoneally to rats (Rognstad, 1981; Keen et al., 1984). Manganese can also activate glycogen synthase (Juang and Robinson, 1977) and phosphorylase kinase (Hallenbeck and Walsh, 1983), two enzymes that regulate glycogen metabolism. Studies with manganese deficient animals suggest that manganese may modulate insulin secretion, synthesis and degradation (Bell and Hurley, 1985), tissue responsiveness to insulin (Shani et al., 1972), and the levels of pancreatic amylase (Brannon et al., 1987).

Manganese Toxicity

Manganese toxicity can be the result of either chronic or acute exposure. Most cases of manganese poisoning are chronic, and related to occupational exposure. The inhalation of manganese oxide dust in manganese mines is the most frequent cause of chronic manganese toxicity (Mena, 1981). Signs of toxicity may appear within six months following the onset of exposure (Mena, 1981). In Chile, afflicted miners exhibit a psychotic period at the onset of the disease, followed by the appearance of Parkinson-like signs and symptoms (Mena, 1981). The victims sustain damage to the substantia nigra, striatum, and other brain structures that are also affected in Parkinson's disease (Mena, 1981). Both disorders respond to L-DOPA (Mena et al., 1970).

The absorption of manganese is a small but constant percentage of the amount ingested, and is independent of the total body stores of manganese (Mena, 1981). Therefore, excessive manganese ingestion may cause manganese toxicity. An example of this mode of poisoning was first reported in Japan in peple who drank well water that was contaminated with manganese (Kawamura et al., 1941). Some cases of acute manganese toxicity have been associated with pneumonitis (Cooper, 1984), hepatitis (Laitung and Mercer, 1983), and pancreatitis (Taylor and Price, 1982).

Manganese Deficiency

Manganese deficiency in adults is extremely rare, perhaps because of the widespread availability of manganese in the food supply. There is only one well-documented case of human manganese deficiency, in a volunteer who was fed a purified diet deficient in vitamin K that was inadvertently made deficient in manganese (Doisy, 1972). The subject lost weight, developed dermatitis, and exhibited growth retardation of the hair and nails and reddening of his black hair. In addition, his cholesterol and triglyceride levels declined.

A potential link between altered manganese homeostasis and certain medical disorders has been postulated, but remains to be proven. Thus, the successful treatment of a diabetic patient with manganese (Rubenstein et al., 1962) has given rise to the hypothesis that diabetes mellitus may be precipitated by manganese deficiency. Similarly, blood manganese levels are reportedly decreased in children with epilepsy (Papavasiliou et al., 1979; Dupont and Tanaka, 1985), raising the possibility that manganese deficiency may cause seizure disorders. However, the etiological relationship between manganese deficiency and these conditions is not generally accepted.

Mechanisms of Action of Manganese

The cellular and biochemical pathways whereby manganese causes its toxic effects, and the exact role of manganese in regulating eukaryotic cell function are not known. To study the mechanisms of action of manganese, we decided to use the pancreas as a model system. This tissue is mainly exocrine, but contains clusters of endocrine cells. The exocrine cells are either acinar or ductal, and synthesize digestive enzymes and bicarbonate rich fluid. The pancreatic model that we have used consists of a preparation of collagenase digested rat pancreatic acini (Williams et al., 1978). The advantage of this preparation is that it is made up of a relatively pure population of groups of acinar cells that have maintained their original polarity toward an open apical lumen. The acinar cells in this preparation have receptors for a variety of ligands (Figure 1), and are very sensitive to many different hormones, including vasoactive intestinal polypeptide (VIP) and cholecystokinin (CCK).

Our rationale for using the pancreatic acinar cell as a model for studying the mechanism of action of manganese was based on observations suggesting that manganese may regulate pancreatic exocrine function. Thus, the pancreas is relatively rich in manganese. Manganese is taken up from the systemic circulation by the pancreas and its concentration in the pancreatic duct is greater than in the systemic circulation (Burnett et al., 1952). Second

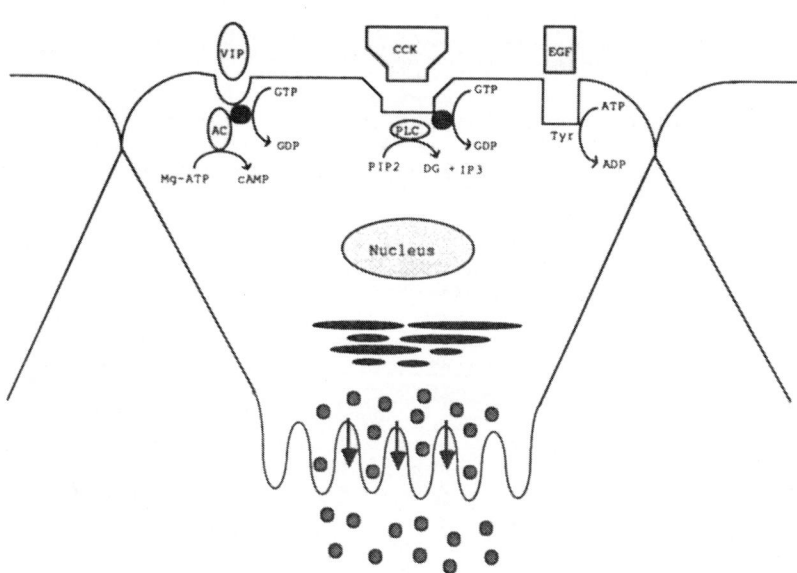

Fig. 1. Schematic representation of a pancreatic acinar cell. Zymogen
granules are being secreted (arrows) into an apical lumen. Receptors for
vasoactive intestinal polypeptide (VIP), cholecystokinin (CCK), and epidermal
growth factor (EGF) are shown on the basal membrane. VIP is coupled to
adenylate cyclase (AC). CCK is coupled to phospholipase C (PLC). EGF
contains tyrosine kinase activity (Tyr). Activation of PLC leads to
hydrolysis of membrane bound phosphatidylinositol-4,5-bisphosphate (PIP2), and
the generation of diacylglycerol (DG) and inositol-1,4,5-trisphosphate (IP3).

generation manganese-deficient animals may exhibit ultrastructural damage or
atrophy of the pancreatic acinar cell (Bell and Hurley, 1973). Manganese
stimulates pancreatic amylase secretion (Abdelmoumene and Gardner, 1981) and
hyperpolarizes the pancreatic acinar cell (Petersen and Ueda, 1976).

In our initial studies, we determined that manganese bypasses the
cell-surface receptors of the pancreatic acinar cell (Korc, 1983a, 1984), and
exerts a dual effect on protein synthesis, enhancing synthesis at low
concentrations (0.03 mM), and inhibiting synthesis at high concentrations.
The addition of calcium to the incubation medium abolishes the stimulatory
effect of manganese in normal rat acini, but enhances it in diabetic rat acini
(Korca, 1983). Further, in the diabetic acini a significant stimulatory
effect is observed at 0.01 mM manganese in the presence of calcium. These
observations suggest that manganese enhances protein synthesis in diabetic rat
acini via a mechanism that is dependent on calcium, and that cellular calcium
homeostasis is altered in these acini. In support of this hypothesis, we have
determined that by comparison to normal rat acini, diabetic acini exhibit an
enhanced sensitivity to the actions of manganese on the mobilization of calcium
from intracellular pools, a decrease in resting cytosolic free calcium levels,
and an attenuated rise in free calcium levels following the activation of the
CCK receptor (Korc and Schoni, 1988). Thus, insulin deficiency is associated
with multiple alterations in calcium homeostasis in the pancreatic acinar cell.

In addition to its direct effects, manganese rapidly antagonizes the
stimulatory effects of CCK on pancreatic enzyme secretion (Abdelmoumene and
Gardner, 1981) and protein synthesis (Korc, 1983b). Both of these biological
actions of CCK are mediated via calcium. Although manganese does not alter
the basal rates of calcium influx or efflux during short incubation periods,
it rapidly blocks CCK-stimulated calcium influx (Korc, 1983b). These
observations suggested that Mn may interfere with the actions of CCK by
attenuating the ability of CCK to enhance calcium influx. However, it is not
clear how manganese blocks CCK-mediated calcium iflux, and whether Mn

modulates the ability of CCK to mobilize intracellular calcium stores. In general, cellular calcium mobilization is mediated by inositol-1,4,5-tris-phosphate (IP3) which is generated following hydrolysis of phophatidylinositol-4,5- bisphosphate (PIP2) (Hokin, 1985; Whitaker and Irvine, 1984). PIP2 hydrolysis also leads to the formation of diacylgycerol (DG), a compound which activates protein kinase C (Takai et al., 1979). Further, phosphorylation of IP3 leads to the formation of inositol-1,3,4,5-tetrakisphosphate (IP4), a mediator of receptor activated calcium influx (Irvine et al., 1986; Morris et al., 1987). When we studied the effects of CCK on PIP2 hydrolysis in the AR42J pancreatic cancer cell line, we found that Mn rapidly diminishes CCK-stimulated IP3 and IP4 accumulation (Korc and Siwik, 1989). It is possible that Mn inhibits CCK-mediated calcium influx by attenuating the actions of CCK on PIP2 hydrolysis. Alternatively, Mn may inhibit calcium influx directly, thereby depleting an intracellular calcium pool which may be necessary for full activation of PIP2 hydrolysis by CCK.

In view of the importance of phosphorylation reactions in the mediation of many cellular processes that are activated following PIP2 hydrolysis, we also sought to determine whether Mn directly alters the state of phosphorylation of acinar cell proteins. We found that soluble extracts of rat pancreas contain a specific Mn-dependent kinase activity that induces the phosphorylation of numerous histones and a number of endogenous acinar cell proteins (unpublished observations). These findings suggest that Mn-mediated phosphorylation may exert important regulatory functions in certain tissues such as the pancreas. The ability of Mn to regulate CCK-mediated PIP2 hydrolysis and to directly alter the phosphorylation of acinar cell proteins suggests that Mn may act as a modulator of second messenger pathways in this cell type.

Acknowledgments

This study was supported by National Institutes of Health Grant DK-32561.

References

Abdelmoumene S, Gardner JD (1981) Effect of extracellular manganese on amylase release from dispersed pancreatic acini. Am J Physiol 241:G359-G364

Baly DL, Curry DL, Keen CL, Hurley LS (1985) Dynamics of insulin and glucagon release in rats: influence of dietary manganese. Endocrinology 116:1734-1740

Bell LT, Hurley LS (1973) Ultrastructural effects of manganese deficiency in liver, heart, kidney, and pancreas of mice. Lab Invest 29:723-736

Bentle LA, Lardy HA (1976) Interactions of anions and divalent metal ions with phosphoenolpyruvate carboxykinase. J Biol Chem 251:2916-2921

Brannon PM, Collins VP, Korc M (1987) Alterations of pancreatic digestive enzyme content in the manganese-deficient rat. J Nutr 117:305-311

Burnett WT, Bigelow, RR, Kimball AW, Sheppard CW (1952) Radiomanganese studies on the mouse, rat and pancreatic fistula dog. Am J Physiol 168:620-629

Cooper WC (1984) The health implications of increased manganese in the environment resulting from the combustion offuel additives: A review of the literature. J Toxicol Environ Health 14:23-46

Doisy EA Jr (1972) Micronutrient controls of biosynthesis of clotting proteins and cholesterol. In: Hemphill DD (ed) Trace Substances in Environmental Health - VI. University of Missouri Press, Columbia, p 193

Dupont CL, Tanaka Y (1985) Blood manganese levels in children with convulsive disorder. Biochem Med 33: 246-255

Hallenbeck PC, Walsh DA (1983) Autophosphorylation of phosphorylase kinase. J Biol Chem 258:13493-13501

Hirsch-Kolb H, Kolb HJ, Greenberg DM (1971) Nuclear magnetic resonance studies of manganese binding of rat liver arginase. J Biol Chem 246:395-401

Hokin LE (1985) Receptors and phosphoinositide-generated second messengers. Annu Rev Biochem 54:205-235

Hurley LS (1982) Clinical and experimental aspects of manganese in nutrition. In: Prasad AS (ed) Clinical, Biochemical, and Nutritional Aspects of Trace Elements. Alan R. Liss, New York, p 369

Irvine RF, Letcher AJ, Heslop JP, Beridge MJ (1986) The inositol tris/tetrakis-phosphate pathway-demonstration of Ins(1,4,5)P3 3-kinase activity in animal tissues. Nature 320:631-634

Juang K, Robinson JC (1977) Effect of manganese(ous) and sulfate on activity of human placental glucose-6-phosphate dependent form of glycogen synthase. J Biol Chem 252:3240-3244

Kawamura R, Ikuta H, Fukuzumi S, Yamada R, Tsubaki S, Kodama T, Kurata S (1941) Intoxication by manganese in well water. Kitasato Arch Exp Med 18:145-169

Keen CL, Lonnerdal B, Hurley LS (1984) Metabolism and biochemistry of manganese. In: Frieden E (ed) Biochemistry of the Essential Ultra-Trace Elements. Plenum Publishing Co, New York p 89

Korc M (1983a) Manganese action on pancreatic protein synthesis in normal and diabetic rats. Am J Physiol 245:G628-G634

Korc M (1983b) Effect of lanthanum on pancreatic protein synthesis in streptozotocin-diabetic rats. Am J Physiol 244:G321-G326

Korc M (1984) Manganese action on protein synthesis in diabetic rat pancreas: evidence for a possible physiological role. J Nutr 2119-2126

Korc M, Schoni MH (1988) Quin-2 and manganese define multiple alterations in cellular calcium homeostasis in diabetic rat pancreas. Diabetes 37:13-20

Laitung JK, Mercer DM (1983) Manganese absorption through a burn. Burns 10:145-146

Mena I (1981). Manganese. In: Bronner F, Coburn JW (eds) Disorders of Mineral Metabolism, Vol 1. Academic Press, New York, p 233

Mena I, Court J, Fuenzalida S, Papavasiliou PS, Cotzias GC (1970) Modification of chronic manganese poisoning. N Engl J Med 282:5-10

Merat DL, Hu ZY, Carter TE, Cheung WY (1984) Subunit A of calmodulin-dependent protein phosphatase requires Mn^{2+} for activity. Biochem Biophys Res Commun 122:1389-1396

Morris AP, Gallacher DV, Irvine RF, Petersen OH (1987) Synergism of inositol trisphosphate and tetrakisphosphate in activating Ca^{2+}-dependent K^+ channels. Nature 330:653-657

Papavasiliou PS, Kutt H, Miller ST, Rosal V, Wang YY, Aronson RB (1979) Seizure disorders and trace metals: Manganese tissue levels in treated epileptics. Neurology 29:1466-1473

Petersen OH, Ueda N (1976) Pancreatic acinar cells: the role of calcium in stimulus-secretion coupling. J Physiol 254:583-606

Reiman CK, Minot AS (1920) A method for manganese quantitation in biological material together with data on the manganese content of human blood and tissues. J Biol Chem 42:329-345

Rognstad R (1981) Manganese effects on gluconeogenesis. J Biol Chem 256:1608-1610

Rubenstein AH, Levin NW, Elliott, GA (1962) Hypoglycaemia induced by manganese. Nature 194:188-189

Sakurai H, Nishida M, Yoshimura T, Takada J, Koyama M (1985) Partition of divalent and total manganese in organs and subcellular organelles of $MnCl_2$-treated rats studied by ESR and neutron activation analysis. Biochim Biophys Acta 841:208-214

Schroeder HA, Balassa JJ, Tipton IH (1966) Essential trace metals in man: manganese. A study in homeostasis. J Chron Dis 19:545-571

Shani J, Ahronson Z, Sulman FG, Mertz W, Frenkel A, Kraicer PF (1972) Insulin-potentiating effect of salt bush (Atriplex halimus) ashes. Isr J Med Sci 8:757-758

Siwik SA, Korc M (1989) Manganese attenuates secretagogue-mediated phospholipid hydrolysis in AR42J cells. Life Sciences 45:1959-1965.

Takai Y, Kishimoto A, Iwasa Y, Kawahara Y, Mori T, Nishizuka Y (1979) Calcium-dependent activation of a multifunctional protein kinase by membrane phospholipids. J Biol Chem 254:3692-3695

Taylor PA, Price JDE (1982) Acute manganese intoxication and pancreatitis in a patient treated with a contaminated dialysate. Canad Med Ass J 126:503-505

Weisiger RA, Fridovich I (1973) Superoxide dismutase. J Biol Chem 248:3582-3592

Whitaker M, Irvine RF (1984) Inositol 1,4,5-trisphosphate microinjection activates sea urchin eggs. Nature 312:636-639

Williams JA, Korc M, Dormer RL (1978) Action of secretagogues on a new preparation of functionally intact isolated pancreatic acini. Am J Physiol 235:E517-E524

Metabolism of Trace Metals in Ascorbic Acid Deficiency: With Special Regard to Zinc Metabolism

SHIGEO TAKEUCHI

Department of Chemistry, Nihon University School of Medicine, 30-1 Oyaguchi, Itabashi-ku, Tokyo, 173 Japan

ABSTRACT

In order to see how metals are involved in delayed healing of wounds, fracture, and other symptoms observed in ascorbic acid (AsA) deficiency, experimentally prepared AsA-deficient guinea pigs and hemodialyzed patients with renal osteodystrophy, as well as severely handicapped children, were examined in terms of metal levels of blood and organs. It is suggested that symptoms of AsA deficiency observed in experimental animals might result not simply from AsA-related metabolic diseases, but from systemic nutritional disorders.

Key word: Ascorbic acid, Zn, metallothionein, skin burn, osteodystrophy

Ascorbic acid (AsA) is a vitamin which is known to serve as an antiscorbutic factor and an oxidizing-reducing agent in the body. AsA is thought to be important due to the biochemical as well as nutritional aspects of its functions. Concerning the biochemical roles of AsA, this substance has been found to act as a co-factor for the protocollagen proline hydroxylase involved in the synthesis of collagen and for other hydroxylases for various drugs (Povertson 1953; Hutton 1967). However, this alone cannot account biochemically for the typical AsA deficiency, scorbutus, and much remains to be resolved concerning AsA. Enzymes requiring various metals other than those requiring AsA as a co-factor are involved in the synthesis and metabolism of collagen. Of the metals, zinc plays an important role in the healing process of skin injury, and zinc deficiency is known to exert a suppressive effect on the synthesis of collagen, resulting in a reduction of the total amount and tension of collagen (Pories 1967; Oberleas 1971; McClain 1973). Thus, combined with the above findings, the relationship between the clinical symptoms of AsA deficiency and the metabolism of metals, especially of zinc, under AsA deficient conditions is of great interest for interpreting the symp-toms. Also, their kinetics warrant our attention from the biochemical standpoint.

DNA(RNA) polymerase	Mg^{2+}, Zn^{2+}
Proline (Lysine) hydroxylase	Fe^{2+}, AsA, α-Ketoglutarate
Glucosyl-galactosyl transferase	Mn^{2+}
Procollagen protease	Zn^{2+}
Monoamine oxidase	$Cu^{2+} \rightleftharpoons Zn^{2+}$
Collagenase	Ca^{2+}, Zn^{2+}

Fig.1 The major enzymes and their co-factors involved in the synthesis of collagen

The major enzymes and their co-factors involved in the synthesis of collagen are shown in Fig. 1. These metals as well as AsA play important roles in the synthesis of collagen in connective tissue. From this standpoint, the kinetics of metals in the serum and organs were investigated in guinea pigs with AsA deficiency, and the results for the relationship between AsA deficiency and the metabolism of metals, especially of zinc, are reported here.

1. Kinetics of metals in the serum and organs of guinea pigs with AsA deficiency

Male Hartley guinea pigs were assigned to three groups, a control group which received an AsA deficient diet for guinea pigs (Clea Japan Inc.) enriched with 1.3 g/kg AsA-Ca; an AsA deficient group which received the non-enriched diet; and an AsA excess group which received the diet enriched with 26 g/kg AsA-Ca.

They were fed on each diet for 3 or 4 weeks, and subjected to various assays.
Each sample was determined for various metals using an Hitachi 306 ICP-Emission Analysis System of the sequential type (Sasaki 1987).

a. Changes of serum metal levels in each group (Fig. 2)

The AsA deficient group revealed a
significantly higher serum Cu level
but a significantly lower serum Zn
level than the control group. The
serum levels of other metals in the
AsA group were not significantly
different from those in the control
group. The serum levels of all
metals in the AsA excess group were
similar to those in the control
group. These results suggest that
the lower serum Zn level in the AsA
deficient group was due to the AsA
deficiency.

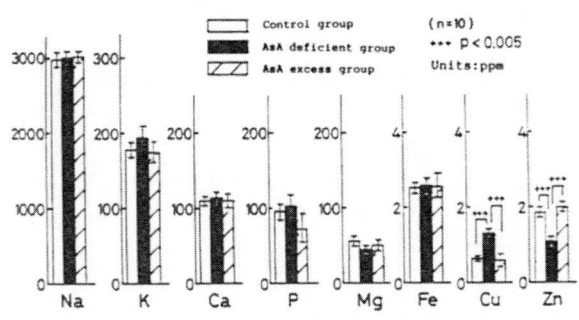

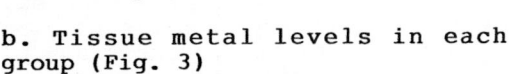

Fig.2 Metal Levels in the Serum of Each Group

b. Tissue metal levels in each group (Fig. 3)

No significant differences were
observed between the control group
and the AsA excess group except for the Fe levels
in the kidney. However, the AsA deficient group
showed a higher Ca level in the lung, lower Cu
and Fe levels in the liver, and a significantly
higher Ca level and higher Cu and Zn levels in
the pancreas. A significant decrease was also
noted in the Mg and Zn levels in the skin. It has
been known from ancient times that the absolute
amount of Zn contained in the skin is large.
Considering this together with the decrease in
the serum Zn level, the decrease of Zn levels in
the skin was presumed to be due to Zn supply from
the skin to more important tissues in which the
Zn levels were decreasing because of the AsA
deficiency.

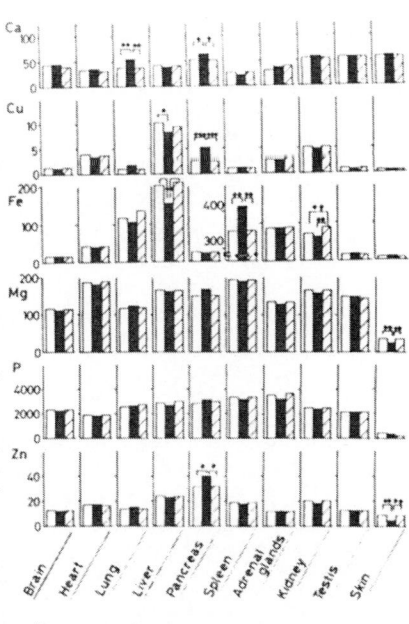

c. Time-course of uptake of orally administered ^{65}Zn into each organ and tissue in guinea pigs fed on the normal diet or AsA deficient diet for 4 weeks (Fig. 4)

After feeding for 4 weeks, 50 µCi of ^{65}Zn was
administered intragastrically via a catheter to
the guinea pigs in each group. At 12, 24 and 48
hrs after the ^{65}Zn administration and the animals
were then sacrificed. Each organ and tissue was
removed, and the Zn uptake was determined as the
radioactivity (cpm)/g wet tissue. In left slide,
A and B represent the control group and the AsA

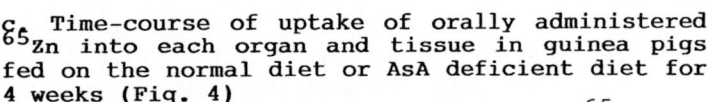

deficient group, respectively. The columns represent the Zn levels expressed
as the percentage of radioactivity in each tissue with respect to the total
radioactivity/g of all tissues observed after 12, 24 and 48 hrs. In the AsA
deficient group, the ^{65}Zn uptake was smaller in almost all tissues including
the serum compared with the control group, except that the uptake into the
liver was significantly greater after 12 hrs and still remained greater after
48 hrs than in the control group although some decrease was observed. In the
control group, the ^{65}Zn uptake into the pancreas as well as into the liver was
large after 12 hrs and decreased with the passage of time, whereas in the AsA
deficient group, the uptake into the pancreas increased with the passage of
time, suggesting a relationship between the secretory function of the pancreas
and ^{65}Zn uptake under AsA deficient conditions. A tendency towards increase
of the ^{65}Zn uptake into tissues such as the skin, brain, heart and testes was
observed with the passage of time in both the control and the AsA deficient
groups, but the increment in the latter was smaller than that in the former.

d. ^{65}Zn-binding substances in the liver cytoplasm at 48 hrs after administration of ^{65}Zn to guinea pigs fed on an AsA deficient diet for 3 and 4 weeks were analyzed by gel chromatography on a Sephadex G-75 column (Fig. 5)

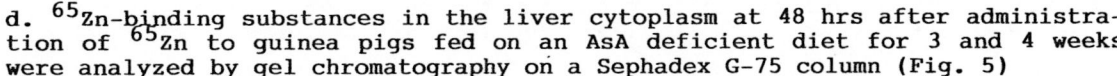

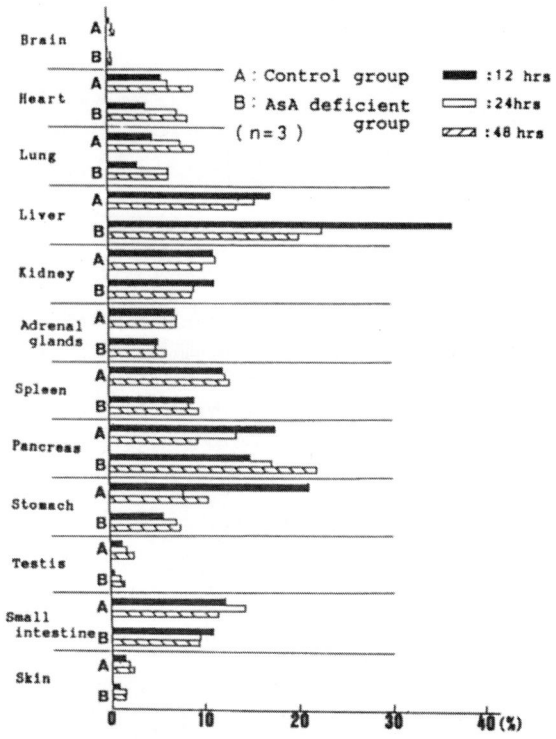

Fig.4   Changes in Uptake of Orally Administered ^{65}Zn
into Various Organs and Tissues of Guinea Pigs
in Each Group

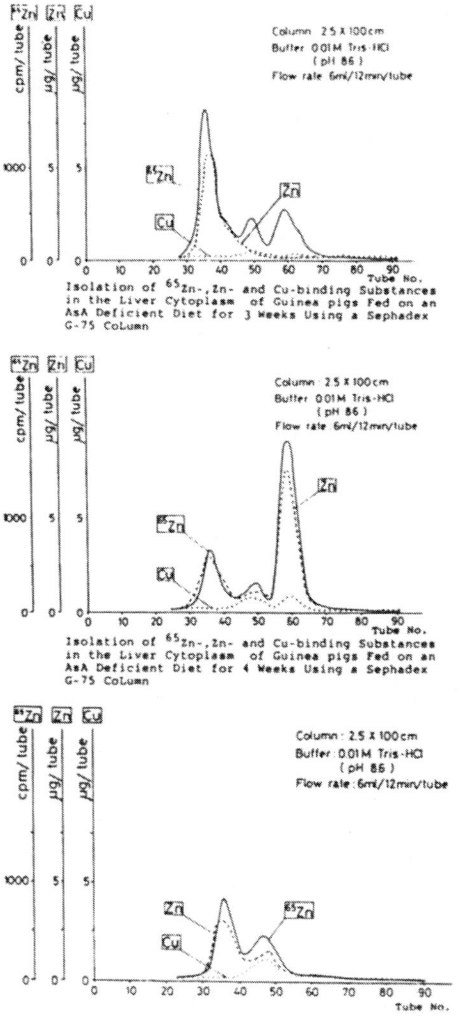

Isolation of ^{65}Zn-, Zn- and Cu-binding Substances
in the Liver Cytoplasm of Guinea pigs Fed on an
AsA Deficient Diet for 3 Weeks Using a Sephadex
G-75 Column

Isolation of ^{65}Zn-, Zn- and Cu-binding Substances
in the Liver Cytoplasm of Guinea pigs Fed on an
AsA Deficient Diet for 4 Weeks Using a Sephadex
G-75 Column

Fig.5   Isolation of ^{65}Zn-binding Fraction of Guinea Pig
Liver Using a Sephadex G-75 Column (Control,4W)

The ^{65}Zn-binding fraction in the samples
from animals fed for 3 weeks was
separated broadly into three fractions,
a macromolecule fraction near tube No.
36, a SOD fraction near tube No. 48, and
a metallothionein fraction near tube No.
59. Third faction indicated induction of
metallothionein. Chromatograms of the
fraction obtained from animals fed for 4
weeks showed a very large induction of
metallothionein.   They also revealed a large induction of non-labelled Zn,
which was considered to have been induced since the third week, when the AsA
deficiency exhibited its effect.   The isolation of the ^{65}Zn-binding fraction
obtained from control animals fed for 4 weeks is shown in Fig. 5 lower, where
induction of Zn-metallothinein was not observed.   Metallothionein has also
been reported to be induced in the liver by starvation and stress. In out
study,Cu-metallothionein was induced in starved control animals. No induction
of metallothionein was observed in the kidney or the small intestine under
similar conditions.

2. Metabolism of trace metals in skin burns of guinea pigs fed on an AsA deficient diet

Based on the results indicating that tissues such as the skin and liver play
an important role in the metabolism of zinc in AsA deficient guinea pigs with
low serum Zn levels, the kinetics of various metals in organs and tissues were
investigated using AsA deficient guinea pigs which bore second degree burns on
the skin in order to clarify the possible roles of Zn and AsA in the healing
process of such injuries. Male guinea pigs comprised the control and AsA deficient groups and were fed for 3 weeks before use.   The animals' haircoat on
the back was shaved, and under ether anesthesia, a copper plate which had been
heated to 90oC in hot water was placed on the naked skin of the back for 10
seconds to prepare second degree burns over about 3% of the total skin area.
The histological features of the burnt skin in the control group underwent a
favorable course of restoration, as represented by remarkable inflammatory
findings on the first day, more infiltrating inflammation but commencement of

alteration of the skin surface on the fourth day, and complete scarring of the surface and regeneration of internal cells on the seventh day. In contrast, the burnt skin in the AsA deficient group displayed few macroscopic and microscopic findings of inflammation on the first day, but appearance of inflammatory findings on the fourth day, and very severe inflammation on the seventh day, indicating severe impairment of the healing process of the injuries because of the AsA deficiency (Takeuchi 1989).

a. Kinetics of metals in the serum
As regards the time-course of the serum metal levels after induction of thermal burns, the serum levels of Zn in the control group underwent a marked change up to 24 hrs after formation of the thermal burns, but there was a tendency for gradual recovery to the level prior to burn formation by the seventh day. In the AsA deficient group, the serum Zn levels continued to increase on the seventh day. As for other metals, the serum Fe and Ca levels in the AsA deficient group showed a tendency to increase slightly, whereas those in the control group remained constant. No significant changes were observed in the serum levels of Cu in either the AsA deficient group or the control group.

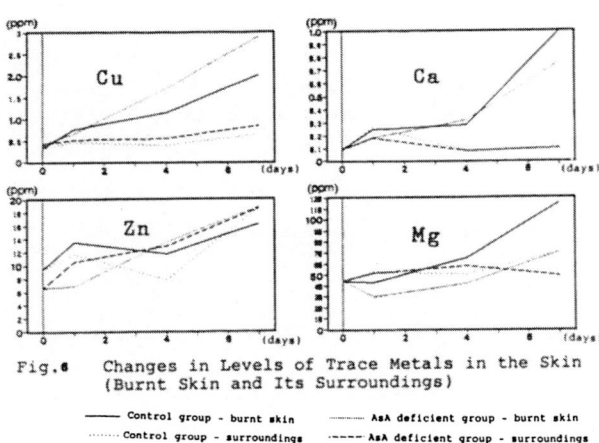

Fig.6   Changes in Levels of Trace Metals in the Skin (Burnt Skin and Its Surroundings)

——— Control group - burnt skin   ······· AsA deficient group - burnt skin
········ Control group - surroundings   ------ AsA deficient group - surroundings

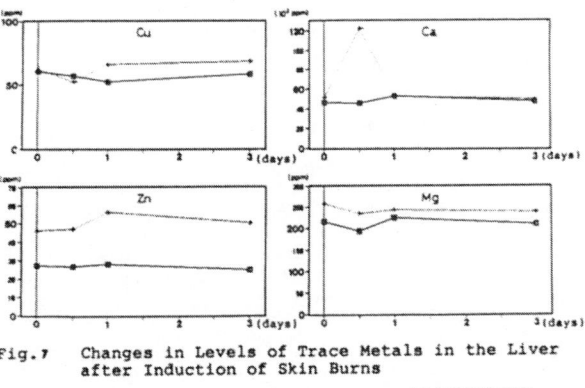

Fig.7   Changes in Levels of Trace Metals in the Liver after Induction of Skin Burns

—o—o— Control group   ···+···+··· AsA deficient group

b. Changes in levels of metals in burnt skin and its surroundings (Fig. 6)
The Zn levels in both the burnt skin and its surroundings in the control group increased after formation of the thermal burns. Before burn induction the Zn level in the skin was significantly lower in the AsA deficient group than in the control group. After the formation of thermal burns, the Zn levels in the AsA deficient group increased in both the burnt skin and its surroundings and the percentages of increase exceeded those in the control group from after 72 hrs onwards. The Cu and Ca levels were greatly increased in the burnt skin in both the control and AsA deficient groups. The Mg levels in both the burnt skin and its surroundings were significantly lower in the AsA deficient group than in the control group.

c. Changes in levels of metals in the liver   (Fig. 7)
In the AsA deficient group, the Zn levels in the liver remained high up to 72 hrs after, and the Ca levels did not change significantly up to 74 hrs after the formation of thermal burns except that the level was increased significantly after 12 hrs. There were no significant differences between the two groups in the Cu or Mg levels in the liver.

d. Time-course of uptake of orally administered ^{65}Zn into the skin and liver (Fig. 8-Left)
The ^{65}Zn uptake into the skin of guinea pigs in the AsA deficient group which had no thermal burns was smaller than that in the control animals with no burns. However, in the AsA deficient group after induction of thermal burns, a significantly large uptake was noted in both the burnt skin and its surroundings after 12 hrs, and the uptake declined towards the third day. In the control animals with thermal burns, there was a tendency for the uptake to decrease in the burnt skin and to increase slightly in its surroundings with the passage of time. Similarly in the liver, the uptake in the AsA deficient group was significantly greater than that in the control group during the

period from 12 to 24 hrs, and revealed a more gradual decline thereafter than in the skin.

e. Percentages of uptake of orally administered ^{65}Zn into the liver and skin in each group (Fig. 8-Right)

In general, the AsA deficient group showed lower percentages of ^{65}Zn uptake into both the burnt skin and its surroundings than the control group, except for a high percentage of uptake into the surroundings after 12 hrs. In contrast, the AsA deficient group revealed significantly higher percentages of uptake into the liver than the control group, although the percentage was slightly decreased in the early phase of restoration of the burns. These results indicate that the liver plays an important role in the response to Zn demand from the skin under AsA deficient conditions.

f. Separation of metal-binding substances in the liver cytoplasm by means of a Sephadex G-75 column (Fig.9)

The three graphs in the upper row of slide show the pattern of separation of ^{65}Zn-, Zn-, Cu- and Fe-binding substances in samples from the control group collected at 12 hrs after formation of the thermal burns. In the control group, ^{65}Zn-metallothionein was induced by the burns after 12 hrs, but decreased to a very low level after 72 hrs. In the AsA deficient group, as shown in the graphs in the lower row, a large amount of Zn- and Cu-metallothionein was induced at 12 hrs after formation of the thermal burns, and the induction

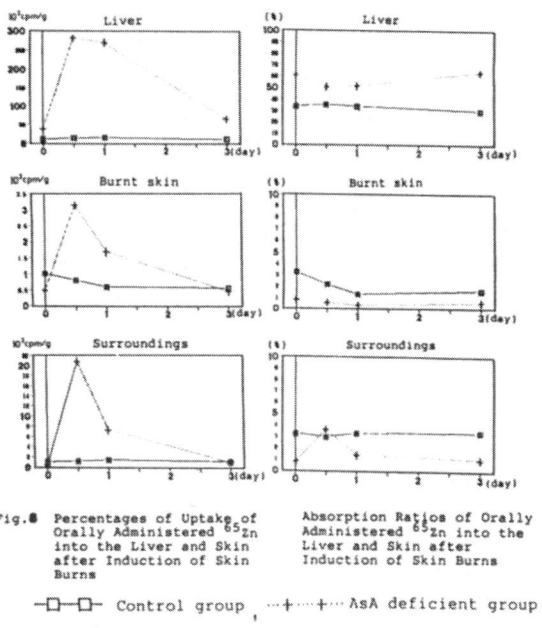

Fig.8 Percentages of Uptake of Orally Administered ^{65}Zn into the Liver and Skin after Induction of Skin Burns

Absorption Ratios of Orally Administered ^{65}Zn into the Liver and Skin after Induction of Skin Burns

—☐—☐— Control group , ····+····+··· AsA deficient group

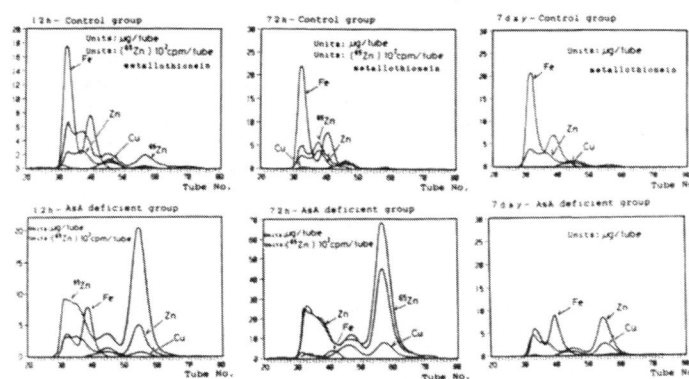

Fig.9 Separation of Metal-binding Substances in the Liver Cytoplasm of Guinea Pigs with Thermal Burns from Each Group Using a Sephadex-75 Column

was enhanced after 72 hrs. There was no ^{65}Zn-binding fraction in the graph for the 7 day-AsA deficient group, and its induction clearly declined as compared with that after 72 hrs. In the control group, a small amount of Zn-metallothionein was induced in the liver cytoplasm in the early phase of restoration of the burns, whereas in the AsA deficient group, a large amount was induced at the same phase. However, the amount decreased as the inflammation of the skin became severer, indicating a cross correlation between the delayed occurrence of ulcers and the induction of Zn-metallothionein. The principal role of metallothionein is thought to be the detoxication of heavy metals in the body. A moderate amount of Zn-metallothionein is also induced under AsA deficient conditions and a considerable amount is induced in the liver during the course of injury healing. It is presumed therefore that metallothionein acts as a Zn-pool and supplies Zn as a co-factor for enzymes involved in the synthesis of collagen in injured organs and tissues.

3. Relationship between AsA and trace metals in bones of AsA deficient guinea pigs

The relationship between the susceptibility to fracture and the levels of trace metals in the bones was investigated in guinea pigs under AsA deficient conditions.

a. Levels of Ca and Zn in the tibia and costal cartilage of the AsA deficient group and control group (Fig. 10)

In the costal cartilage, the Zn level was significantly lower but the Ca level was slightly higher in the AsA deficient group than in the control group. In the tibia, a significantly higher Cu level as well as slightly but not significantly higher Ca and P levels were observed in the AsA deficient group than in the control group, but the Zn levels were similar for the two groups.

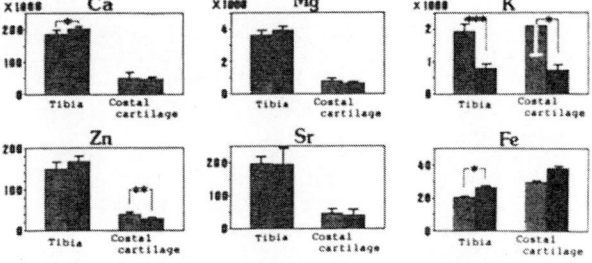

Fig.10　Levels of Trace Metals in Bones of Guinea Pigs

Control group ▆　　AsA deficient group ▆

Units:ppm　(n=8)　*:p<0.05　**:p<0.01　***:p<0.005

b. Percentages of ^{65}Zn uptake into the tibia and costal cartilage in the AsA deficient group and control group (Fig. 11)

In the AsA deficient group, the percentages of ^{65}Zn uptake into all organs except for the liver and the pancreas were significantly lower than those in the control group. The uptake into the tibia was also slightly lower, but the uptake into the costal cartilage was slightly higher than that in the control group. The Zn level in the costal cartilage was found to be lower in the AsA deficient group, but the levels of trace metals in the tibia were not significantly different from those in the control group, although there was some delay of bone growth. However, the ^{65}Zn uptake into the tibia of the AsA deficient animals was lower than that of the control animals, reflecting the gradual appearance of adverse effects of AsA deficiency on the bones. These results suggest that osteodystrophy in humans might be caused by long-term continuation of AsA deficiency.

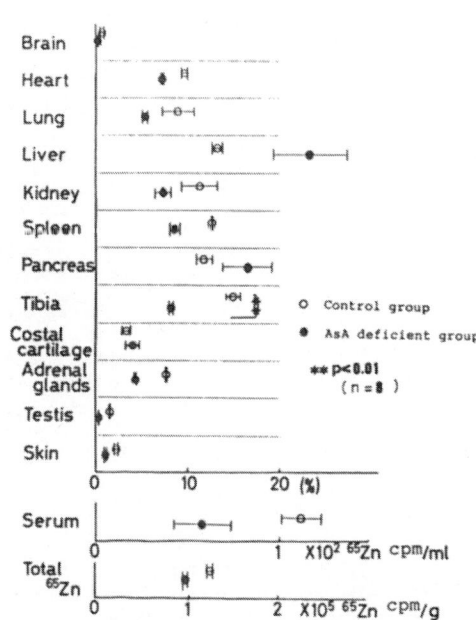

Fig.11　Percentages of ^{65}Zn Uptake into Various Organs and Tissues for Each Group (Including the Tibia and Costal Cartilage)

4. Relationship between latent AsA deficiency and renal osteodystrophy in dialyzed patients

PTH, calcitonin and Vitamin D_3 are known to be biochemical factors involved in the pathogenesis of osteodystrophy. These factors control the metabolism of metals such as Ca, P, Mg and Zn in the body. It is thought that inhibition of the synthesis of collagen, which is essential for osteogenesis, promotes the development of osteodystrophy. Since various enzymes are involved in the synthesis and metabolism of collagen and trace metals and AsA act as cofactors for the enzymes, the kinetics of these sudstaces are worthy of examination. Studies were therefore conducted to investigate the relationship between renal osteodystrophy in dialyzed patients with very low serum AsA levels and the susceptibility to fracture in severely-handicapped children on the one hand, and the levels of AsA and trace metals in the serum and bone-salt content on the other. The subjects comprised 80 dialyzed patients and 80 apparently healthy volunteers. The serum AsA level was 0.912±0.361 mg/100 ml in the normal volunteers, 0.273±0.066 mg/100 ml in patients before dialysis, and 0.193±0.189 mg/100 ml in patients after dialysis (Takeuchi 1987, 1988).

a. Levels of trace metals in the serum in normal volunteers and dialyzed patients before and after dialysis (Table 1)

The serum levels of Mg, Sr and K were significantly higher, and the serum levels of Cu, Fe, Zn and Na were significantly lower in the dialyzed patients before dialysis than in normal volunteers.

Table 1 Comparison of Serum Levels of Trace Metals
between Normal Subjects and Dialyzed Patients

	Ca	Cu	Fe	Mg	P	Zn	Sr
Normal subjects	84.2 ±3.3	0.96 ±0.17	1.37 ±0.46	18.52 ±1.18	114 ±13	0.79 ±0.10	0.025 ±0.009
Before dialysis	84.1 ±9.9	0.85 ±0.17**	0.79 ±0.51**	32.44 ±4.53**	146 ±27**	0.69 ±0.10**	0.083 ±0.009**

n=80   Mean±SD   Unit:ppm   **:p<0.01

b. Relationship between serum AsA levels and ΣGS/D and MCI in the second metacarpus in dialyzed patients (Table 2)

The total bone-salt content, ΣGS/D, and the metacarpal index (MCI) in the second metacarpus of the dialyzed patients were determined by micro densito-metry, and their correlation with various substances in the serum was examined. The serum AsA level was positively correlated with ΣGS/D and MCI, and the serum Sr level was negatively correlated with ΣGS/D. All these correlations were statistically significant. The serum Ca and Zn levels were negatively correlated with ΣGS/D, although these correlations were not significant. There were no statistically significant correlations between the serum levels of AsA and any of the trace metals in the normal volunteers, whereas the serum AsA level in the dialyzed patients before dialysis was negatively correlated with the serum Ca.

Table 2 Relationship between Serum Levels of Various
Substances and ΣGS/D and MCI, Determined in the
Metacarpus of Dialyzed Patients by the MD Method

	AsA	Ca	Cu	Mg	P	Zn	Sr
ΣGS/D	0.366*	-0.196	-0.130	-0.096	0.025	-0.184	-0.256*
MCI	0.379**	-0.003	-0.080	0.031	0.080	-0.138	-0.091

n=50   **:p<0.01 *:p<0.05

5. Relationship between susceptibility to fracture in severely-handicapped children and the serum AsA and Zn levels (Table 3)

The AsA and Zn levels in the serum and ΣGS/D were compared between 65 severely-handicapped children and 40 normal volunteers. The severely-handi-capped children were classified into three groups: Group A, patients confined to bed; Group B, patients who were able to sit up; and Group C, patients who were able to walk with some difficulty. Patients who had a history of fracture were picked up from these groups. The serum AsA and Zn levels and

Table **3**   Comparison of Serum AsA and Zn Levels and Bone-salt
Content between Several Groups of Severely Handicapped
Children and Normal Subjects

	No. of subjects	bone-salt content (ΣGS/D%score)		serum Zn (ppm)		serum AsA (μg/ml)	
		Range	Mean±SD	Range	Mean±SD	Range	Mean±SD
severely retarded patients Group A	23	0.43-0.80	0.59±0.11**	0.47-0.86	0.63±0.11**	0.18-12.43	4.54±3.50**
Group B	27	0.33-1.06	0.65±0.17**	0.47-0.82	0.68±0.09**	0.31-9.29	3.71±3.23**
Group C	15	0.56-0.92	0.70±0.11**	0.44-0.73	0.60±0.08**	0.26-18.18	4.33±4.11**
total	65	0.33-1.06	0.65±0.15**	0.44-0.86	0.63±0.09**	0.18-18.18	4.21±3.56**
patients with a history of fracture	7	0.33-0.80	0.57±0.15**	0.62-0.69	0.65±0.02**	0.92-7.09	3.02±2.26**
normal subjects	40	—	1.00	0.77-1.13	0.86±0.11	2.28-16.85	10.29±3.40

Group A: confined to bed

Group B: able to sit up   **p <0.01

Group C: able to walk with some difficulty

ΣGS/D were significantly lower in every group of severely-handicapped children compared with the control group. In particular, among those who had a history of fracture, 57% of the patients had AsA levels of 2 µg/ml or loss. Similar tendencies were observed for ΣGS/D and the serum Zn level. These results suggested the involvement of Zn metabolism under AsA deficient conditions in the development of osteodystrophy, which is commonly observed in dialyzed patients and severely-handicapped children. Under AsA deficient conditions, a different pathogenesis is postulated for the renal osteodystrophy observed in dialyzed patients who are generally elderly and for the osteodystrophy observed in younger severely-handicapped children. That is, AsA deficiency appears to cause the former by acting on the mechanism of release of minerals from the bones and the latter by acting on the mechanism of ossification.

In conclusion, the present studies on guinea pigs indicated that AsA deficiency affected the levels of various trace metals in the body, especially the Zn, Fe and Cu levels. The changes in levels of these metals appeared to be related mainly to changes in their absorption and excretion. Low serum levels of Zn were due to a decrease in its absorption and an increase in its excretion. Metallothionein was induced in the liver in response to the decrease of Zn in the body due to AsA deficiency. Zn-metallothionein was strongly induced in the early phase of the healing of thermal burns under AsA deficient conditions, but the induction declined as the inflammation in the skin lesions became severer, indicating a close relationship between the delayed occurrence of ulcers and the induction of Zn-metallothionein. These results suggested that Zn-metallothionein may supply Zn as a co-factor for enzymes involved in the synthesis of collagen in the injured organs. The synthesis of collagen containing fibers is necessary for the healing of skin injury, and AsA acts as a co-factor for the proline hydroxylase involved in collagen synthesis. However, this alone cannot provide a sufficient physiological explanation for the observations in AsA deficiency, and there is a strong possibility that clinical symptoms of AsA deficiency may be caused by changes in the metabolism of trace metals, especially depletion of Zn due to the AsA deficiency. Similarly, osteogenesis is mainly affected by the levels of metals such as Ca and P which make up the bones and by PTH, CT and vitamin D_3, which directly control the absorption and metabolism of these metals. However, since the Ca or P levels in hypoplastic bones or those susceptible to fracture due to AsA deficiency were not significantly different from those in normal bones, the hypoplasia and susceptibility to fracture appear to be due to impairment of the synthesis or metabolism of collagen-containing fibers, a matrix constituent of bones, and thus, to arise from depletion of trace metals including Zn, which act as co-factors for the enzymes involved in the synthesis of the fibers.

References

Hutton J J, Tappel Jr A L, Underfriend S (1967) Co-factor and substrate requirement of collagen proline hydroxylase. Arch Biochem Biophys 118:231-240

McClain P E, Wiley E R, Beecher G R, Anthony W L, Hsu J M (1973) Influence of zinc deficiency on synthesis and crosslinking of rat skin collagen. Biochim Biophys Acta 304:457-465

Oberleas D, Seymour J K, Prasad A S (1971) Effect of Zinc deficiency on wound-healing in rat. Amer J Surg 121:566-568

Pories W J, Henzel J H, Rob C R, Strain W H (1967) Acceleration of wound healing in man with zinc sulfate given by mouth. Lancet 1:121-124

Robertson W B, Schwartz B (1953) Ascorbic acid and the formation of collagen. J Biol Chem 201:689-696

Sasaki T, Takeuchi S, Kimoto I, Mano M, Tomioka E, Arakawa Y (1987) Technique for measurement of metals in serum by ICP spectrometry. In: Brown S S, Kodama Y (eds) Toxicology of metals. Fillis Horwood, London, p161-162

Takeuchi S, Kimoto I, Sasaki T, Mano M, Tomioka E, Arakawa Y, Sudo S, Sudo Y (1987) Significance of metals in serum of dialysis patients. In: Brown S S, Kodama Y (eds) Toxicology of metals. Fillis Horwood, London ,p53-54

Takeuchi S, Kimoto I, Mano M, Tomioka E, Sasaki T, Arakawa Y, Ikeda T (1988) Determination of plasma ascorbic acid levels by HPLC in patients on hemodialysis. In: The 4th asian-pacific congress of clinical biochemistry, Abstracts, Hongkong,p180

Takeuchi S, Kimoto I, Mano M, Tomioka E, Sasaki T, Arakawa Y, Tomita H (1989) Mineral metabolism in ascorbic acid deficient guinea pigs with scalded skin. Trace Metal Metabo. Tokyo, 17:1-9

Metal-Binding Proteins
and Metalloenzymes

Metallothionein: Biochemistry and Spatial Structure

JEREMIAS H.R. KÄGI, PETER HUNZIKER, and MILAN VAŠÁK

Institute of Biochemistry, University of Zürich, Winterthurerstrasse 190 CH-8057 Zürich, Switzerland

ABSTRACT

Metallothioneins (MTs) are polypeptides distinguished by an extremely high metal (Zn, Cd, Cu) and Cys content and by the arrangement of these components in unique metal-thiolate clusters. They feature importantly in the cellular pathways of essential (Zn, Cu) and nonessential (Cd, Hg, Pt, Bi, Ag, Au) metals and are also thought to play specific roles in stress response. The more than 60 MTs thus far sequenced have recently been subdivided into three structural classes. In the mammalian MTs which contain 20 Cys in a chain of 60 odd amino acid residues and normally bind seven bivalent metal ions (Me), the metals have been postulated to be located in two separate domains containing a $Me(II)_3(Cys)_9$ and a $Me(II)_4(Cys)_{11}$ cluster, respectively. This model has now been confirmed by the spatial structures of three mammalian MTs determined by two-dimensional NMR spectroscopy in aqueous solution.

Key words: Metallothionein, classification, amino acid sequences, spatial structure (2D NMR)

Of all the metal-containing biomacromolecules now known, metallothioneins are among the most unusual ones. They are of exceptional chemical composition, display an extreme metal-binding capacity, and have a unique spatial structure. The first metallothionein was discovered some 30 years ago by Margoshes and Vallee (1957) in their search for the compound responsible for the natural accumulation of the toxic element cadmium in the kidney of animals (Malyuga 1941). The material they isolated from equine kidney was a small protein (molecular weight 6000 - 7000) of an extremely high cadmium content. However, from the beginning it was evident that in different preparations the amount of cadmium in the protein was variable, the balance being made up of zinc and, occasionally, of copper (Kägi and Vallee 1960; Kägi and Vallee 1961). In fact, in most metallothioneins isolated from mammalian tissues other than kidney, zinc is the more plentiful and often the sole metallic constituent. Other exceptional features of these proteins are their extremely high sulfur content (>11%) in the form of cysteine residues, the unique distribution of the latter in the polypeptide chain, and the spectroscopic manifestations of oligonuclear metal-thiolate complexes, i.e., metal-thiolate clusters.

Definition and Biological Functions

Focussing on these unusual characteristics, the Committee on the Nomenclature of Metallothionein appointed at the Second International Meeting on Metallothionein and Other Low Molecular Weight Metal-binding Proteins in 1985 adopted the definition that any polypeptide resembling the initially described mammalian metallothionein in several of its features can be designated as "metallothionein" (Fowler et al. 1987). Polypeptides satisfying this phenomenological definition have now been found to occur throughout the animal kingdom as well as in higher plants, eukaryotic microorganisms, and some prokaryotes (cited in Hamer (1986) and Kägi and Kojima (1987).

In animals, metallothionein is most abundant in parenchymatous tissues, i.e., liver, kidney, pancreas, and intestines. There are wide variations in concentration in different species and tissues, reflecting effects of age, stage of development, dietary regimen, and other not yet fully identified factors. All metallothioneins thus far identified are inducible compounds. In experimental animals and in cultured cells, their concentration is raised sharply upon exposure to a wide variety of agents, among them metal ions, hormones and many cytotoxic substances (Table I). Such effects are also believed to underly the increased synthesis observed in various chemical and physical stress conditions also listed. This involvement of metallothionein in both specific and unspecific biological response mechanisms has engendered a strong interest in many

Table I: Factors reported to induce metallothionein synthesis in cultured cells or in vivo

Metal ions: Cd, Zn, Cu, Hg, Au, Ag, Co, Ni, Bi, Pb, Mn	Carageenan	Acetaminophen
	Dextran	Diethyldithiocarbamate
Glucocorticoids	Ethanol	2,3-dimercaptopropanol
Progesterone	Propanol	2,3-dimercaptosuccinic acid
Estrogen	Ascorbate	D,L-penicillamine
Catecholamines	Ethionine	EDTA
Interleukin I	Alkylating agents	
Interferon-α	5-azacytidine	Starvation
	5-aza-2'-deoxycytidine	Infection
Butyrate	6-mercaptopurine	Inflammation
Retinoate	Urethane	Laparatomy
Cyclic AMP	Chloroform	Physical stress
Phorbol esters	Carbon tetrachloride	X-Irradiation
Endotoxin		UV-Irradiation
		high O_2 tension

different fields. However, these circumstances have also made the assignment of a primary biological function difficult. Thus, more than three decades after the discovery of these compounds, this topic is still being disputed (Karin 1985; Bremner 1987). The main hypotheses considered are that (1) metallothionein serves as a rather unspecific metal-buffering ligand to either sequester or dispense metal ions within the cell, or that (2) it has a specialized metal-linked function in normal cellular metabolism or development (Cousins 1983; Bremner 1987). Pleiotropically, it may well serve several biological purposes. That metallothionein is the cellular component responsible for the intracellular sequestration of cadmium, bringing about the long biological half-life of this nonessential element, is undisputed and lends support to the view that the induction of the protein provides a means for attenuating temporarily the toxic effects of Cd and other nonessential metals such as Hg, Bi, Ag, Au, Pt (Webb 1987). In all likelihood, it also affords protection against the toxic consequences of an excessive influx of essential metals such as Zn and Cu. On the other hand, there is also good evidence from nutritional and biochemical studies that metallothionein may serve as a donator of metal ions, especially of Zn and Cu, in the biosynthesis of metalloenzymes and metalloproteins requiring these metals or in metal-dependent regulatory functions associated with cellular repair processes, growth and differentiation (Brady 1982; Beltramini and Lerch 1982). To what extent the functioning of metallothionein in such elementary processes of metal binding and dispensing is important to the overall metal economy and the biological state of cells and organisms is still unclear, however.

Primary Structure and Classification

In contrast to the persisting uncertainties concerning the functional significance of the metallothioneins, great advances have been made in recent years with respect to our knowledge of their chemical and structural features. Primary structure data obtained either by conventional protein-chemical sequencing or by nucleotide sequencing of cDNA or genomic DNA are now known either completely or in part for 65 metallothioneins. A representative selection of sequences is given in Fig. 1. The conspicuous common features of all forms is the multiple occurrence of the Cys-X-Cys tripeptide sequence, where X is an amino acid residue other than Cys. Its abundance has long been considered to be an essential conditioning factor for metal binding in these molecules (Kojima et al. 1976). Based on the existence or the lack of other structural relationships, the metallothioneins are now grouped into three different classes (Fowler ét al. 1987). Class I includes all forms obviously related to the mammalian metallothioneins. Class II subsumes a growing number of functionally analogous proteins which are neither related to one another nor bearing any clear relationship to the mammalian forms. Class III metallothioneins comprise a novel type of metal-binding polypeptides. Their structure, first determined by Kondo et al. (1984), shows them to be no proteins at all but constituting short atypical polypeptides built up of repetitive γ-glutamyl-cysteine units and thus being homologs of glutathione. They are of variable length and often oligomeric structures made up of two or more chains (Grill et

Class I
<pre>
                          1 b   20   40   60
Human (MT-2) c   MDP NCSCAAGDSCTCAGSCKCKECKCTSCKKSCCSCCPVGCAKCAQGCICKGASD KCSCCA
Chicken   MDPQDCTCAAGDSCSCAGSCKCKNCRCRSCRKSCCSCCPAGCNNCAKGCVCKEPASSKCSCCH
Trout (MTb)   MDP CECSKTGSCNCGGSCKCSNCACTSCKKSCCPCCPSDCSKCASGCVCKGKTC DTSCCQ
Crab (MT-2)   PDP C C NDKCDCKEGECKTGCKCTSCRCPPCEQCSSGC KCANKEDCRKTCSKPCSCCP
N. crassa   GDCGCSGASSCNCGSGCSCSNCGSK
</pre>

Class II
<pre>
                     1   20   40   60
Sea urchin (MTa)   MPDVKCVCCTEGKECACFGQDCCVTGECCKDGTCCGICTNAACKCANGCKCGSGCSCTEGNCAC
Yeast   QNEGHECQCQCGSCKNNEQCQKSCSCPTGCNSDDKCPCGNKSEETKKSCCSGK
Wheat germ (Ec protein) GCNDKCGCAVPCPGGTGCRCTSARSGAAAGEHTTCGCGEHCGGNPCACGGEGTPSGCAN d....
Cyanobacterium   TSTTLVKCACEPCLCNVDPSKAIDRNGLYYCCEACADGHTGGSKGCGHTGCNC
</pre>

Class III
<pre>
                                          1   10
S. pombe (cadystin A f)   eCeCeCG e
R. canina (phytochelatin, PC8)   eCeCeCeCeCeCeCeCG
P. vulgaris (homophytochelatin, h-PC7) eCeCeCeCeCeCeC-β-alanine
</pre>

Fig. 1: Classification of metallothionein; amino acid sequences of representative forms (see footnote a). For references see Kägi and Schäffer (1988). Supercripts: (a) open positions denote deletions introduced for optimal alignment of class I metallothioneins; (b) numeration refers to the sequence determined for human metallothionein-2; (c) specified isoform of metallothionein; (d) partial sequence; (e) "e" indicates glutamic acid residue linked by γ-glutamyl bond; (f) also designated "phytochelatin PC$_3$".

al. 1985; Bernhard and Kägi 1987). Their Cd-containing forms often contain inorganic sulfide (Murasugi et al. 1983). Thus far, they have been found in microorganisms and plants only.

All class I and II metallothioneins characterized thus far are single chain proteins. Mammalian forms contain 61 to 62 amino acid residues (Fig. 2). Chicken and sea urchin metallothionein (MTa) contain 63 and 64 residues, respectively. Shorter chains are formed in invertebrates and in certain fungi, the shortest one with 25 residues in Neurospora crassa. The hallmark of the class I metallothioneins is the striking correspondence in the alignment of the Cys along the chain.

Amino Acid Sequences of Mammalian Metallothioneins

A comparison of 30 mammalian metallothionein sequences illustrates the complete conservation of the 20 Cys (Fig. 2) and shows that nearly 60% of all residues are invariant. Noteworthy is also the preservation of seven of the eight basic residues (Lys or Arg), suggesting that they play a special structural or functional role in metallothionein (Kojima et al. 1976; Pande et al. 1985). Most of the amino acid substitutions are conservative both with respect to the chemical and the space-filling properties. Remarkably, there are no aromatic amino acids and only very few bulky aliphatic ones.

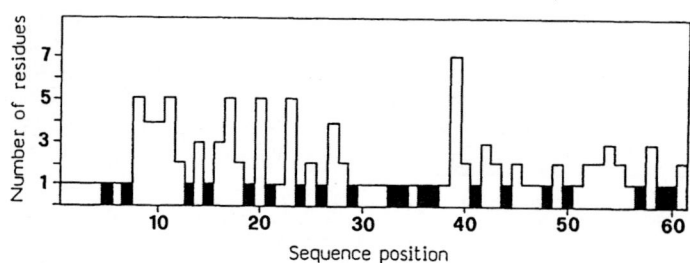

Fig. 2: Sequence variability of metallothionein in mammals. The number of different residues found at each sequence position in 30 mammalian metallothioneins is shown. Cys are marked by filled bars (adapted from Kägi and Kojima 1987).

Mammalian tissues usually contain two major metallothionein fractions designated MT-1 and MT-2 and carrying at neutral pH two and three negative charges,
respectively (Kägi and Kojima 1987). In the rabbit, in ungulates and in primates, there are additional isoforms within these charge-separable fractions.
In human liver tissue which contains up to 0.1% metallothionein (Onosaka et
al. 1986), we have identified by improved separating and amino acid sequencing
techniques six isometallothioneins designated MT-1e, MT-1f, MT-1h, MT-1i,
MT-1k, and MT-11 (Hunziker and Kägi 1988). Some of these forms differ from the
previously characterized and most abundant isoform MT-2 in as much as 15% of
all noncysteine residues. There is also good evidence that some of the multiple human metallothionein genes are expressed tissue-specifically, perhaps reflecting different functions of the respective gene products (Schmidt and
Hamer 1986; Varshney et al. 1986).

Metal-binding Sites

The abundance of Cys in metallothionein predisposes this protein toward binding of "soft" d^{10} metal ions through thiolate bonds. Mammalian metallothioneins
usually bind a total of seven equivs of bivalent metal. The details of metal
coordination have been elucidated by the application of a variety of spectroscopic methods enumerated in Table II. For many of these studies, homogenously

Table II: Spectroscopic methods employed in the study of the metal-binding sites of metallothionein.[a]

Ligand identification:	Absorption spectroscopy
	X-ray photoelectron spectroscopy
Coordination geometry:	Absorption spectroscopy
	Electron paramagnetic resonance (Co(II))
	Magnetic circular dichroism (Co(II))
	^{113}Cd nuclear magnetic resonance
	Perturbed angular correlation of gamma-ray (^{111m}Cd)
	Mössbauer (^{57}Fe)
	Extended X-ray absorption fine structure (EXAFS)
Metal-thiolate clusters:	^{113}Cd nuclear magnetic resonance
	Electron paramagnetic resonance (Co(II))
	Mössbauer (^{57}Fe)
	Absorption spectroscopy
	Circular dichroism
	Magnetic circular dichroism
	Luminescence (Cu(I))

[a]For citations, see Vašák and Kägi (1983) and Kägi and Kojima (1987)

substituted derivatives had to be prepared, in which the naturally bound metal
ions were replaced by metal ions or isotopes suitable for the chosen spectroscopic method. In particular, replacement studies with Co(II) have proven useful in establishing the geometry of metal binding (Vašák 1980; Vašák and Kägi
1981). The spectroscopic characterization of this derivative documented that
all metal ions are bound quite uniformly to four thiolate groups of Cys side
chains in tetrahedral geometry. Spectroscopic and magnetic studies showed,
moreover, that in fully reconstituted Co(II)-metallothionein these complexes
do not exist in separation but are joined to oligonuclear structures, in which
some of the thiolate ligands are shared by two adjacent metal ions (Fig. 3).
Such an arrangement or clustering of the metal ions is an obvious requirement
if the tetrahedral tetrathiolate coordination geometry of the metal complex is
to be reconciled with the measured stoichiometry of less than three thiolate
ligands per bivalent metal ion. In mammalian metallothioneins, the ratio of
20 Cys to seven metal ions demands that eight Cys serve as doubly coordinated
bridging thiolate ligands and twelve as singly coordinated terminal thiolate
ligands.

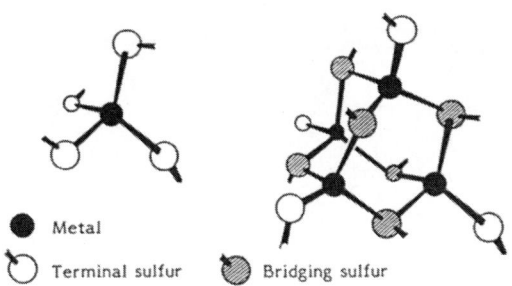

- ● Metal
- ○ Terminal sulfur
- ◐ Bridging sulfur

Fig. 3: Patterns of tetrahedral metal-thiolate coordination. Left: Mononuclear complex. Right: Oligonuclear complex (adamantane-like metal-thiolate cluster)

Metal-Thiolate Clusters

Of the many spectroscopic methods employed to establish the details of the metal complexation in metallothionein, ^{113}Cd NMR spectroscopy has proven to be the most powerful one (Sadler et al. 1978; Suzuki and Maitani 1978). The first direct evidence for the existence of discrete metal-thiolate clusters was provided nearly a decade ago by the elegant studies of Otvos and Armitage (1980) on ^{113}Cd-enriched rabbit liver metallothionein. As shown in Fig. 4, ^{113}Cd(II)$_7$-metallothionein displays in its ^{113}Cd NMR spectrum a set of seven resonances, each attributable to one of the seven metal sites. The spectrum shows also multiplet splitting of each of these resonances arising from ^{113}Cd-^{113}Cd scalar coupling and indicating a connection of neighbouring ^{113}Cd nuclei through bridging thiolate ligands. From crossirradiation experiments in which the resonances were selectively decoupled, Otvos and Armitage were able to deduce that the respective Cd sites are partitioned into two completely separate linkage groups, namely a metal-thiolate cluster made up of three metal ions and nine Cys, Me(II)$_3$(Cys)$_9$, having a cyclohexane-like structure, and a metal-thiolate cluster made up of four metal ions and eleven Cys, Me(II)$_4$(Cys)$_{11}$, having a bicyclo (3.3.1) nonane-like structure.

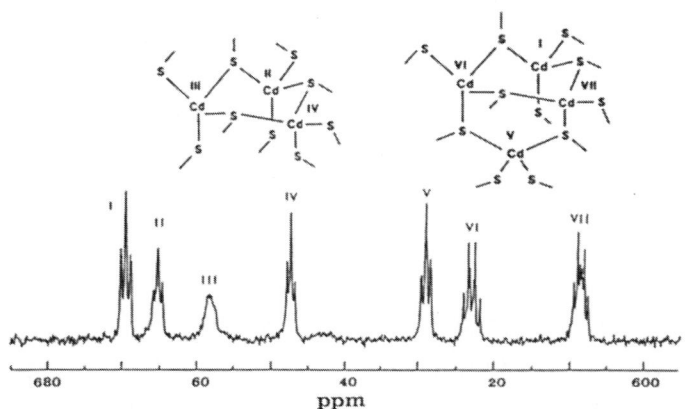

Fig. 4: ^{113}Cd NMR spectrum of rabbit liver ^{113}Cd$_7$-MT-2 and metal topology derived from homonuclear ^{113}Cd coupling (adapted from Otvos et al. 1985).

That the two clusters are not only separate entities but are also located in different regions of the protein was subsequently established by protein-chemical studies of Winge and Miklossy (1982) which revealed a two-domain structure of the molecule and which showed that the three-metal and the four-metal cluster are associated with the amino- and the carboxyl-terminal domains, respectively (Fig. 5). What these studies did not tell, of course, is the identity of the Cys. To get this information, one needs to distinguish between each of the 20 Cys in the polypeptide chain and to establish individually their topological relationships to the seven metal ions.

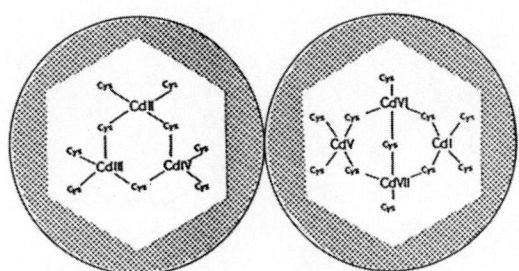

Fig. 5: Domain structure of mammalian Cd_7-metallothioneins. Left: Amino-terminal domain (β). Right: Carboxyl-terminal domain (α).

Cluster Topology

This aim has now been accomplished by the application of a newly developed heteronuclear two-dimensional ^1H ^{113}Cd NMR COSY technique (Frey et al. 1985) which allows monitoring of through-bond spin-spin coupling between the different ^{113}Cd nuclei and the three ^1H nuclei of the Cys bound to the metal (Fig. 6).

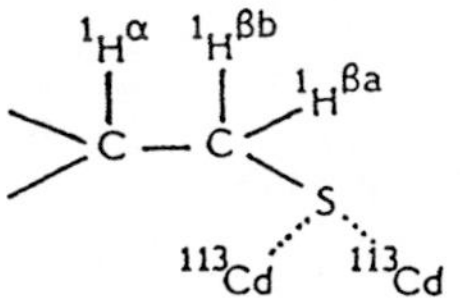

Fig. 6: Complexation of ^{113}Cd ions to a bridging Cys. Disposition of spin $1/2$ nuclei (^1H ^{113}Cd) showing selective heteronuclear through-bond coupling (adapted from Frey et al. 1985).

Such heteronuclear coupling is feasible since both ^{113}Cd and ^1H have a nuclear spin $I = 1/2$. Experimentally, these pair interactions manifest themselves by the generation of a cross-signal in the two-dimensional ^1H ^{113}Cd COSY spectrum. The representative ^1H NMR profiles shown in Fig. 7 are derived from such a two-dimensional spectrum. They indicate that each of the seven ^{113}Cd resonances couples specifically with a number of ^1H resonances attributable to particular

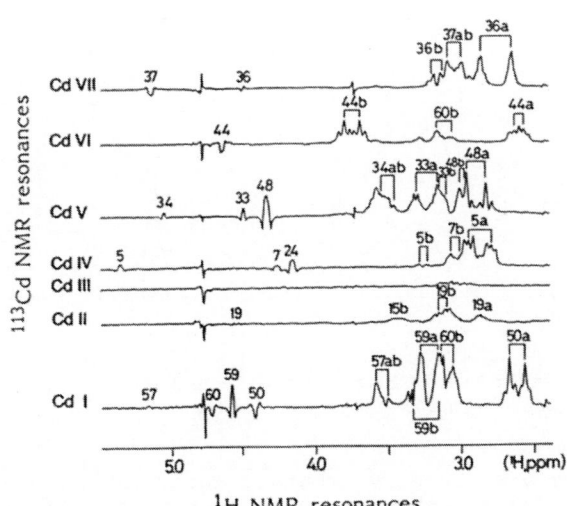

Fig. 7: Identification of Cd-Cys bonds in rabbit liver $^{113}Cd_7$-MT-2a by heteronuclear ^1H ^{113}Cd COSY spectroscopy. The spectra manifest cross-signals between the seven ^{113}Cd resonances (see Fig. 4) and sets of Cys ^1H resonances. The numbers placed over the Cys ^1H signals refer to the residue position in the sequence. The numbers without letters specify signals from Cys α-protons. a and b distinguish between the resonances from the pairs of Cys β-protons (see Fig. 6). The solid brackets indicate Cys ^1H resonance splittings due to (^1H ^{113}Cd) scalar coupling (adapted from Fig. 2 of Wagner et al. 1987).

sets of Cys β and α protons. The attribution of each of these ^{1}H signals to a given Cys in the chain, indicated by the numbers in the spectra, was made from homonuclear two-dimensional ^{1}H NMR COSY and NOESY data collected separately (Wagner et al. 1986).

From the [113]Cd ^{1}H coupling patterns all 28 Cys-Cd bonds in the Cd-thiolate clusters of rabbit liver MT-2a were identified, yielding the coordination scheme displayed in Fig. 8 (Frey et al. 1985). Recently, the same cluster topology was shown to pertain also to rat liver MT-2 (Schultze et al. 1988) and human liver MT-2 (Messerle et al., in preparation).

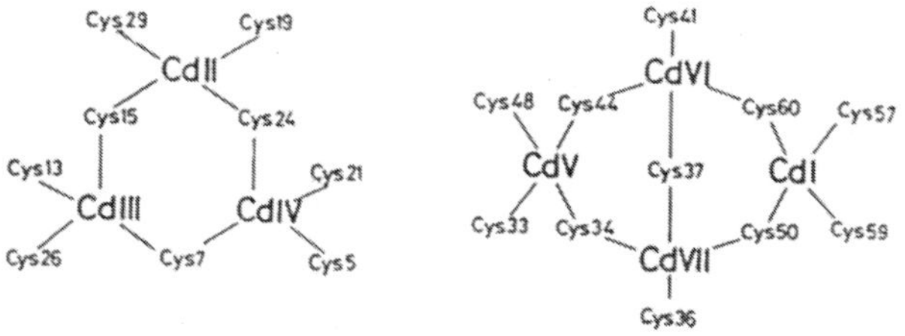

Fig. 8: Cd-Cys topology in mammalian metallothioneins. Left: 3-metal-thiolate cluster. Right: 4-metal thiolate cluster (from Frey et al. 1985). The arabic numerals refer to the residue position in the sequence. The roman numerals denote the Cd sites defined by Otvos and Armitage (see Fig. 4).

Spatial Structure

The most spectracular result of the spectroscopic studies of metallothionein is the recent determination of the steric organization of the metal-thiolate clusters and of the spatial folding of the polypeptide chain from two-dimensional NMR data using distance geometry calculations (Braun et al. 1986; Arseniev et al. 1988). A stereo view of the best structure computed for Cd_7-MT-2a from rabbit liver is shown in Fig. 9. For the sake of clarity, the model is simplified to show only the course of the polypeptide backbone, the orientation of the Cys side chains, and the positions of the seven [113]Cd ions. It documents that the protein is composed of two about equally sized globular domains. Their structures which in reality are connected through the Lys 30 - Lys 31 segment are drawn separately, since their mutual orientation is not sufficiently de-

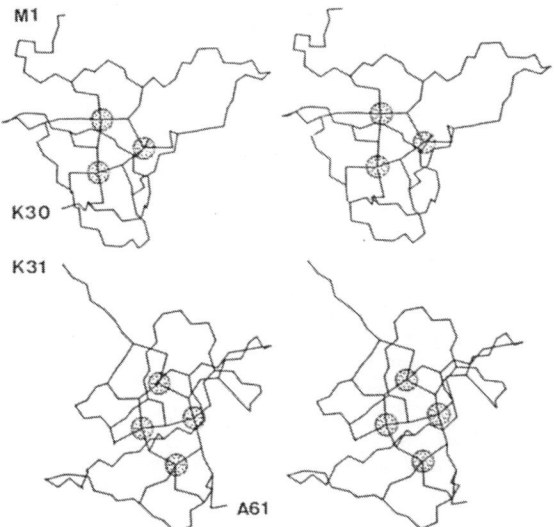

Fig. 9: NMR solution structure of amino-terminal domain (top) and carboxyl-terminal domain (bottom) of rabbit liver Cd_7-MT-2a. Stereo view of polypeptide backbone, Cys side chains, and metal positions (dotted spheres of radius 0.9 Å). Although the domains are connected by the Lys 30 - Lys 31 segment, they were calculated and drawn separately (see text). The figure is adapted from Arseniev et al., 1988.

fined by the available NMR data. Each domain contains in its interior the appropriate metal-thiolate cluster as a "mineral core" wrapped by two large helical turns of the polypeptide chain. In the amino-terminal domain, the spiral of the chain fold is right-handed, in the carboxyl-terminal, it is left-handed.

Very similar structures have also been found for the MT-2 isoforms from murine (Schultze et al. 1988) and human (Messerle et al., in preparation) liver. The close agreement between the rabbit and rat protein is especially remarkable since the two metallothioneins differ in more than 25% of their noncysteine amino acid residues. It attests that the organization of the metal-thiolate clusters and the conformation of the polypeptide chain enfolding them are dictated by the conserved arrangement of the Cys in the chains (see above).

With these results, the framework is now provided for addressing more specific questions concerning the many unusual physical and chemical features of these proteins and their functions in cellular processes. Of foremost importance are investigations of the structural basis of the mechanisms governing metal binding and release. Another pressing aim is the search for cellular partners interacting with the metallothioneins. Hopefully, in pursuing such questions, we may also learn why nature has chosen to endow mammalian metallothioneins with two separate domains, why the metal ions are bound in clusters, and why in evolution the incorporation of aromatic amino acids has not been tolerated.

Acknowledgment

This research was supported by Swiss National Science Foundation grant number 3.160-0.88.

References

Arseniev A, Schultze P, Wörgötter E, Braun W, Wagner G, Vašák M, Kägi JHR, Wüthrich K (1988) Three-dimensional structure of rabbit liver [Cd_7]metallo-thionein-2a in aqueous solution determined by nuclear magnetic resonance. J Mol Biol 201:637-657

Beltramini M, Lerch K (1982) Copper transfer between *Neurospora* copper metallo-thionein and type 3 copper apoproteins. FEBS Lett 142:219-222

Bernhard WR, Kägi, JHR (1987) Purification and characterization of atypical cadmium-binding polypeptides from Zea mays. Experientia, Suppl 52:309-315

Brady FO (1982) The physiological function of metallothionein. Trends Biochem Sci 7:143-145

Braun W, Wagner G, Wörgötter E, Vašák M, Kägi JHR, Wüthrich K (1986) Poly-peptide fold in the two metal clusters of metallothionein-2 by nuclear magnetic resonance in solution. J Mol Biol 187:125-129

Bremner I (1987) Nutritional and physiological significance of metallothionein. Experientia, Suppl 52:81-107

Cousins RJ (1983) Metallothionein - aspects related to copper and zinc metabo-lism. J Inher Metab Dis 6, Suppl 1:15-21

Fowler BA, Hildebrand CE, Kojima Y, Webb M (1987) Nomenclature of metallo-thionein. Experientia, Suppl 52:19-22

Frey MH, Wagner G, Vašák M, Sørensen OW, Neuhaus D, Wörgötter E, Kägi JHR, Ernst RR, Wüthrich K (1985) Polypeptide-metal cluster connectivities in metallothionein 2 by novel ^{1}H-^{113}Cd heteronuclear two-dimensional NMR experiments. J Am Chem Soc 107:6847-6851

Grill E, Winnacker E-L, Zenk MH (1985) Phytochelatins: The principal heavy-metal complexing peptides of higher plants. Science 230:674-676

Hamer DH (1986) Metallothionein. Ann Rev Biochem 55:913-951

Hunziker PE, Kägi JHR (1988) Metallothionein: A multigene protein. In: Prasad AS (ed) Essential and toxic trace elements in human health and disease. Alan R Liss, Inc, New York, p 349

Kägi JHR, Kojima Y (1987) Chemistry and biochemistry of metallothionein. Experientia, Suppl 52:25-61

Kägi JHR, Schäffer A (1988) Biochemistry of metallothionein. Biochemistry 27:8509-8515

Kägi JHR, Vallee BL (1960) Metallothionein: A cadmium- and zinc-containing protein from equine renal cortex. J Biol Chem 235:3460-3465

Kägi JHR, Vallee BL (1961) Metallothionein: A cadmium- and zinc-containing protein from equine renal cortex. II. Physicochemical properties. J Biol Chem 236:2435-2442

Karin M (1985) Metallothioneins: Proteins in search of function. Cell 41:9-10

Kojima Y, Berger C, Vallee BL, Kägi JHR (1976) Amino-acid sequence of equine renal metallothionein-1B. Proc Natl Acad Sci USA 73:3413-3417

Kondo N, Imai K, Isobe M, Goto T, Murasugi A, Wada-Nakagawa C, Hayashi Y. (1984) Cadystin A and B, major unit peptides comprising cadmium binding peptides induced in a fission yeast. Separation, revision of structures and synthesis. Tetrahedron Lett. 25:3869-3872

Malyuga DP (1941) On cadmium in organisms. Doklady Akad Nauk, USSR XXXI:145-147

Margoshes M, Vallee, BL (1957) A cadmium protein from equine kidney cortex. J Am Chem Soc 79:4813-4814

Murasugi A, Wada C, Hayashi Y (1983) Occurrence of acid-labile sulfide in cadmium-binding peptide 1 from fission yeast. J Biochem 93:661-664

Onosaka S, Min K-S, Fukuhara C, Tanaka K, Tashiro S-I, Shimizu I, Furuta M, Yasutomi T, Kobashi K, Yamamoto K-I (1986) Concentrations of metallothionein and metals in malignant and non-malignant tissues in human liver. Toxicology 38:261-268

Otvos JD, Armitage IM (1980) Structure of the metal clusters in rabbit liver metallothionein. Proc Natl Acad Sci USA 77:7094-7098

Otvos, JD, Engeseth HR, Wehrli S. (1985) Preparation and [113]Cd NMR studies of homogeneous reconstituted metallothionein: reaffirmation of the two-cluster arrangement of metals. Biochemistry 24:6735-6740

Pande J, Vašák M, Kägi JHR (1985) Interaction of lysine residues with the metal thiolate clusters in metallothionein. Biochemistry 24:6717-6722

Sadler PJ, Bakka A, Beynon PJ (1978) [113]Cd nuclear magnetic resonance of metallothionein. FEBS Lett 94:315-318

Schmidt CJ, Hamer DH (1986) Cell specificity and an effect of ras on human metallothionein gene expression. Proc Natl Acad Sci USA 83:3346-3350

Schultze P, Wörgötter E, Braun W, Wagner G, Vašák M, Kägi JHR, Wüthrich K (1988) The conformation of [Cd$_7$]-metallothionein-2 from rat liver in aqueous solution determined by nuclear magnetic resonance. J Mol Biol 203:251-268

Suzuki KT, Maitani T (1978) Cadmium-113 FT NMR-spectra of rabbit liver metallo-thioneins. Experientia 34:1449-1450

Varshney U, Jahroudi N, Foster R, Gedamu L (1986) Structure, organization, and regulation of human metallothionein I$_F$ gene: Differential and cell-type-specific expression in response to heavy metals and glucocorticoids. Mol Cell Biol 6:26-37

Vašák M (1980) Spectroscopic studies on cobalt(II) metallothionein: evidence for pseudotetrahedral metal coordination. J Am Chem Soc 102:3953-3955

Vašák M, Kägi JHR (1981) Metal thiolate clusters in cobalt(II)-metallothionein. Proc Natl Acad Sci USA 78:6709-6713

Vašák M, Kägi JHR (1983) Spectroscopic properties of metallothionein. Met Ions Biol Syst 15:213-273

Wagner G, Neuhaus D, Wörgötter E, Vašák M, Kägi JHR, Wüthrich K (1986) Sequence-specific ^{1}H-NMR assignments in rabbit-liver metallothionein-2. Eur J Biochem 157:275-289

Wagner G, Frey MH, Neuhaus D, Wörgötter E, Braun W, Vašák M, Kägi JHR, Wüthrich K (1987) Spatial structure of rabbit liver metallothionein-2 in solution by NMR. Experientia, Suppl 52:149-157

Webb M (1987) Toxicological significance of metallothionein. Experientia, Suppl 52:109-134

Winge DR, Miklossy K-A (1982) Domain nature of metallothionein. J Biol Chem 257:3471-3476

Metallothionein and Zinc Deficiency

IAN BREMNER

Rowett Research Institute, Bucksburn, Aberdeen, AB2 9SB, UK

ABSTRACT

There is a close relationship between the zinc status of animals and the concentrations of metallothionein in their tissues. Concentrations in urine and in blood cells and plasma are reduced on zinc deprivation, suggesting that assay of the protein can be used to diagnose zinc deficiency. An account is given of the effect of other physiological and nutritional factors on metallothionein concentrations in blood. The effects of experimental zinc deficiency in human volunteers are also described.

KEY WORDS

Zinc deficiency • Diagnosis • Blood • Metallothionein • Nutritional status

INTRODUCTION

Metallothionein is an ubiquitous metal-binding protein that occurs in most animal tissues (see Bremner 1987 for review). Even now, nearly 30 years after the discovery of the protein, there is considerable debate about its functions. However, one consensus view is that it plays major roles in the cellular detoxification of metals and homeostatic regulation of zinc metabolism. Synthesis of metallothionein is a frequent response to increases in tissue concentrations of metals such as cadmium, zinc and copper. A wide range of other physiological and pathological stimuli also induce its synthesis by hormone-mediated mechanisms involving increased gene transcription.

Although many metals, essential and non-essential alike, can effect the synthesis of metallothionein, it is apparent that zinc is of particular importance. For example when stress factors cause increased accumulation of metallothionein in liver, it is generally found that zinc is the metal associated with the protein (Sobocinski et al 1978). In addition, in our early work on the binding of copper and zinc to metallothionein in sheep and calf liver, we showed that the amount of the protein present was directly related to the liver zinc content (Bremner, Marshall 1974). This was the case even when copper was the only metal bound to the protein. Moreover, metallothionein was always absent from the livers of zinc-deficient animals. These results can be partially explained by the fact that zinc is a much better inducer of metallothionein synthesis than copper, at least in some species, although copper binds more avidly to the protein (Bremner 1987). There are also effects of zinc on the degradation of metallothionein, with its half-life in the liver of copper- or cadmium-injected rats being lower in zinc-deficient animals than in those of normal zinc status (Bremner et al 1978; Held, Hoekstra 1984).

Concentrations of metallothionein in the liver and other organs vary with nutritional status but it usually accounts for only a small proportion of the body burden of zinc. Because of the relatively short half-life of metallothionein, about 20 hours in rat liver, it is also a relatively labile pool (Feldman, Cousins 1976). Thus when animals are given a zinc-deficient diet, concentrations of metallothionein in the liver, pancreas and kidneys are rapidly reduced to almost non-detectable levels, even when measured by immunoassay (Bremner et al 1987)

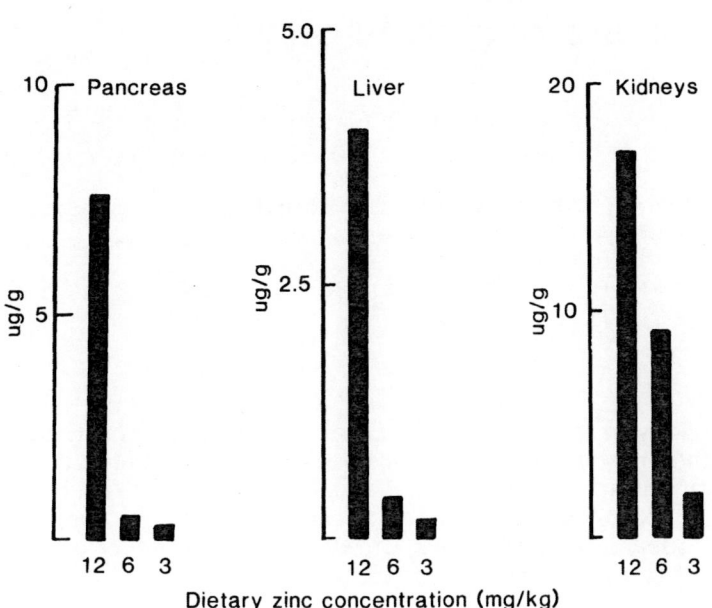

Fig. 1.   Tissue concentrations of metallothionein I in rats given diets of varying zinc contents for 7 days

(Fig. 1).   In addition, if pregnant or lactating rats are given a zinc-deficient diet, hepatic concentrations of metallothionein in their off-spring are greatly reduced (Morrison, Bremner 1987).   No such effects occur if the dams are given diets deficient in copper or iron (Gallant, Cherian, 1986).

DIAGNOSIS OF ZINC DEFICIENCY

It has long been recognised that improved tests are needed for the diagnosis of zinc deficiency.   Measurement of plasma zinc, for example, is commonly used but is unsatisfactory, mainly because hypozincaemia also occurs in other physiological or pathological states, such as during infection or in protein deficiency. Many research groups have sought another indicator which responds in different fashions in these conditions, so that changes in plasma zinc levels can be interpreted more reliably.   Since liver metallothionein concentrations are decreased in zinc deficiency but increased during stress or infection, and since there is a close correlation between liver and plasma metallothionein concentrations, it was suggested that assay of plasma metallothionein could be used to diagnose zinc deficiency (Sato *et al* 1984). Preliminary experiments with rats supported this view but problems were sometimes encountered in measuring the small amount of metallothionein present in the plasma of normal rats. Moreover intermediate concentrations were found in animals with concurrent zinc deficiency and stress.

More promising results are obtained when blood cells rather than plasma are analysed (Bremner *et al* 1987).   Concentrations are substantially greater in cells than in plasma and decrease rapidly in a dose-dependent manner when dietary zinc intakes are reduced from 12 to 6 or 3 mg/kg.   Stress factors have no major effect on blood cell metallothionein concentrations, with the result that concurrent zinc deficiency and stress still result in major decreases in metallothionein concentrations.

Analysis of urine is another possibility, since concentrations are again much higher than in plasma. They are reduced by about 50% in zinc-deficient rats and increased substantially in rats treated with endotoxin. Concurrent stress and zinc deficiency result in intermediate values which are difficult to interpret (Bremner *et al* 1987). Exposure to cadmium would also lead to increased urinary metallothionein levels (Nordberg *et al* 1982).

METALLOTHIONEIN IN BLOOD CELLS

Tanaka *et al* (1985, 1986) have reported that metallothionein is present in the erythrocytes of cadmium-dosed mice and that it remains there for the life-span of the cells. It is surprising therefore that injection of rats with zinc causes blood cell metallothionein concentrations to increase to a maximum level after 1 day but decrease again to control levels after 3 days (Morrison *et al* 1988). This is difficult to reconcile with the reported stability of the protein and with the inability of mature erythrocytes to synthesise proteins because of the absence of a nucleus. The rapid disappearance of metallothionein from blood cells of rats given a zinc-deficient diet also indicates that the protein is very labile.

Such responses are more typical of the occurrence of the protein in leucocytes rather than erythrocytes, but fractionation of rat blood on density gradients of Percoll did not support this view (Morrison *et al* 1988; Robertson *et al* 1989). Only trace quantities of metallothionein could be detected by radioimmunoassay in the white cells of normal or zinc-injected rats, although this may not be the case in human blood. There was not, however, a uniform distribution of metallothionein in rat red blood cells of different densities and therefore of different ages. Concentrations were considerably greater in the lightest and youngest cells, which contained most of the reticulocytes. Similar distributions occurred in blood from normal and zinc-injected rats. The implication is that metallothionein is synthesised in the reticulocytes or other precursor cells but degrades rapidly as these cells mature (Table 1).

Table 1. **Distribution of metallothionein-I in subfractions of rat blood cells obtained by density gradient centrifugation on Percoll**

Fraction	Density	% of Total MT-I	% of Total Hb	MT-I/Hb Ratio
1	1.085	3	1.6	1.9
2	1.093	41	1.7	23.6
3	1.099	35	19	1.8
4	1.105	21	80	0.3

Confirmation of this view is provided by the results of experiments on the maturation of reticulocytes *in vitro* (Noble, Bremner unpublished results). This technique involves the collection of reticulocytes of uniform age from rats which had previously been treated with chloramphenicol implants and subjected to phlebotomy (Noble *et al* 1989). The half-life of metallothionein when these reticulocytes were maintained in culture was only about 6 hours, which is considerably less than the rate of maturation of the reticulocytes, the loss of transferrin receptors, and the decline in activity of glycolytic enzymes (Fig. 2). These events are all characteristic features of the aging of red blood cells. The rapid disappearance of metallothionein in the cultured cells is consistent with the observation that it is present mainly in the youngest cells *in vivo*. Interestingly when reticulocytes were collected from rats dosed previously with cadmium, the half-life of metallothionein was extended to about 60 hours (Fig. 2). There are similar increases in the half-life of hepatic metallothionein both *in vivo* and *in vitro* when cadmium is bound to the protein. These results help explain how cadmium-metallothionein remains in erythrocytes for most of their life-span (Tanaka *et al* 1985, 1986), whereas the naturally-occurring protein, which is likely to contain zinc or copper as the bound metal, disappears quite rapidly.

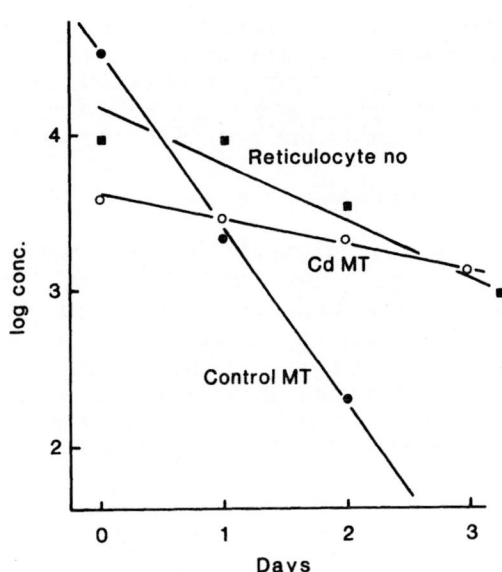

Fig. 2. Decrease in reticulocyte number (□) and in concentrations of metallothionein from control (●) and cadmium-treated (o) rats when reticulocytes were maintained in culture (Noble, Bremner, unpublished observations)

FACTORS INFLUENCING BLOOD CELL METALLOTHIONEIN LEVELS

An important implication of these results is that blood cell metallothionein concentrations are likely to be influenced not only by zinc status but also by factors that modify erythropoietic activity and population of blood cells. Thus concentrations are greatly increased in rats made anaemic by feeding an iron-deficient diet (Robertson *et al* 1989) or by administration of phenylhydrazine (Morrison *et al* 1988) (Table 2). The increases do not reflect specific induction of metallothionein synthesis but merely increased production of reticulocytes which have high basal metallothionein concentrations.

Conversely metallothionein concentrations are decreased in blood cells from rats subjected to treatments that inhibit erythropoiesis. Thus rats subjected to starvation, such as control rats pair-fed against anorexic iron-deficient animals (Robertson *et al* 1989), or severely protein-deficient rats (Morrison *et al* 1989) have greatly decreased metallothionein levels in the blood cells and also in the kidneys (Table 2). Protein deficiency inhibits erythropoietin production (Ito *et al* 1964) and reticulocyte counts were decreased in the severely protein-deficient rats given a diet with 5% protein. Metallothionein concentrations were however *increased* in blood cells and kidneys from rats given a diet with 12% protein instead of a normal 20% but this more moderate protein deficiency caused a slight increase in reticulocyte count.

If a similar situation occurs in man, assay of blood cell metallothionein would not be suitable for specific diagnosis of zinc deficiency although it could still have some use as a confirmatory test. However, there would be difficulties in the interpretation of results if there was concurrent iron or protein deficiency, which is not unlikely in some circumstances. Such extrapolation between species is not warranted and there is clearly a need for direct studies in man.

Table 2. **Concentrations of metallothionein-I in blood cells and kidneys of different nutritional state**

Treatment	Blood cells (ng/ml)	Kidneys (µg/g)
Iron-deficient	111 ± 12	34 ± 5
Control, *ad libitum*	15 ± 1	89 ± 7
Control, pair-fed	5 ± 1	42 ± 3
Phenylhydrazine-treated	40 ± 3	n.d.
Control, *ad libitum*	6 ± 1	n.d.
Protein-deficient (5% protein)	22 ± 1	23 ± 2
Protein-deficient (12% protein)	53 ± 5	75 ± 9
Control, *ad libitum* (20% protein)	43 ± 2	56 ± 3

ASSAY OF HUMAN METALLOTHIONEIN

The radioimmunoassay (Mehra, Bremner 1983) used to measure metallothionein in the above studies with rats is not suitable for assay of human metallothionein. It has therefore been necessary to develop an alternative enzyme-linked immunosorbent assay (ELISA), using the same antibody against a conjugate of rat liver metallothionein-I and rabbit IgG (Ghaffar *et al* 1989). The ELISA has the advantage that it is much quicker than the RIA and measures all isoproteins and metalloforms of metallothionein from most species. However, it has a lower limit of detection, has higher coefficients of variation and is more susceptible to interference and matrix effects.

In the ELISA, microwell modules are coated with purified human metallothionein whereupon additions are made of mixtures of primary antibody and either sample or metallothionein standards in appropriate buffer. After incubation the solutions are decanted, the wells washed and then treated with a secondary antibody-peroxidase conjugate. Following further incubation, the wells are again emptied and washed, whereupon substrate for the enzyme (ABTS) is added and the reaction allowed to proceed for 15 minutes. The reaction is then stopped with NaF and the absorbance at 405 nm measured on a microplate reader.

Using this procedure, there is a linear inverse relationship between abscrbance and log metallothionein concentration over the range 10 to 500 ng/ml. Linear standard curves are also obtained when standards are prepared in urine or lysates of erythrocytes; these are superimposable on the curves for standards prepared in buffer, indicating that for these samples at least there are no matrix effects. However, values for plasma are much higher than expected indicating that some component is affecting cross-reactivity. This was confirmed by analysis of plasma from zinc-deficient rats; zero levels of metallothionein were estimated by RIA but the ELISA gave concentrations of over 50 ng/ml, as were found in control rats. At the moment therefore we are unable to measure human metallothionein in plasma.

EXPERIMENTAL ZINC DEFICIENCY IN MAN

In order to establish whether metallothionein levels respond to changes in zinc status in humans, a metabolic study was carried out with a group of 5 male volunteers who were given a formula diet for a period of 55 days. During days 1-15 and 41-55 the diet provided about 85 µmoles zinc/day but during the depletion phase, days 16-40, this was reduced to about 13 µmoles zinc/day (Bosworth *et al* 1989). The adaptive response to variation in zinc supply was determined by oral administration of zinc-70. The efficiency of absorption of

zinc during the 3 phases, control, depletion and repletion, was 40 ± 4, 94 ± 1 and $63 \pm 8\%$ respectively. Endogenous losses of zinc at these times were estimated to be 29 ± 9, 13 ± 2 and 23 ± 8 µmol/day, confirming that there was effective homeostatic control over zinc metabolism as the dietary zinc supply was altered. One subject developed skin lesions characteristic of zinc deficiency and was severely hypozincaemic during the depletion phase. The other subjects showed only slight reductions in plasma zinc levels during depletion and were clinically normal.

Metallothionein concentrations in the blood cell lysates varied quite considerably for each individual but there was no characteristic pattern of decrease or increase during the depletion and repletion phases, even in the subject with signs of clinical zinc deficiency. Mean concentrations for the 5 subjects were about 120 ng/ml for most of the experiment (Fig. 3). Urinary metallothionein concentrations also fluctuated quite appreciably both between and within individuals. The results from one subject are shown in Fig. 3. Mean concentrations for the 5 subjects during the control, depletion and repletion phases were 35 ± 5, 36 ± 6 and 61 ± 11 ng/mg creatinine respectively. There was therefore no consistent decline in metallothionein excretion during the depletion phase, even in the subject with signs of zinc deficiency, but there was a 2-fold increase on zinc repletion.

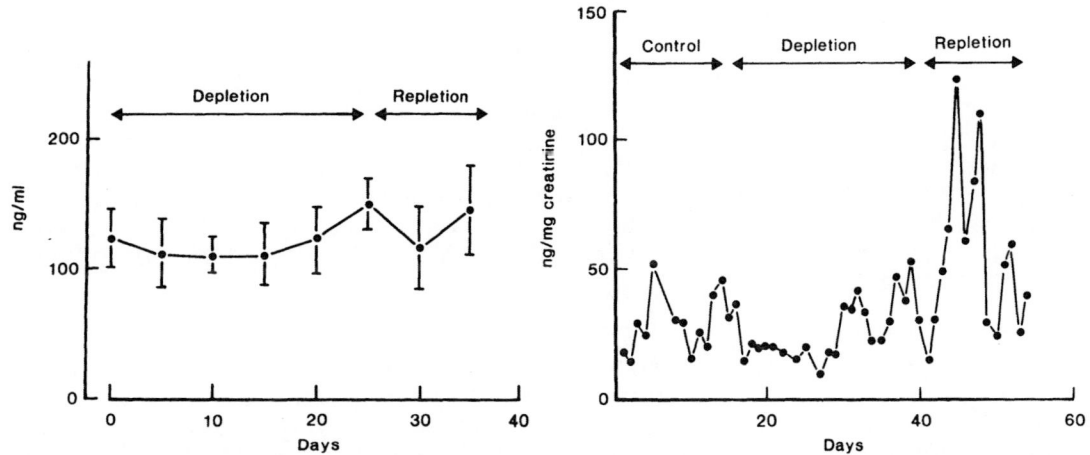

Fig. 3. Metallothionein concentrations in mixed blood cells and urine of human subjects during zinc depletion and repletion

CONCLUSIONS

Experimental zinc deficiency in animals reduces metallothionein concentrations in blood and urine. However, other nutritional and pathophysiological conditions may also affect the levels, either by induction of synthesis of the protein or indirectly by affecting blood cell populations. An ELISA method has been developed for human metallothionein but its application in preliminary studies with zinc deficient volunteers did not reveal major changes in levels of the protein in urine or blood cells.

These results do not indicate that assay of metallothionein in a population of mixed blood cells will be of particular value in the assessment of zinc status in man. However, account has to be taken of the variation in concentration of the protein in red blood cells of different ages. Moreover, unlike the situation in rat blood, the contribution of leucocytes may have to be considered. Thus concentrations of metallothionein in 'old' erythrocytes, 'young' erythrocytes (7% reticulocytes) and in monocytes were 49 ng/10^9 cells, 14 ng/10^9 cells and 15 ng/10^6 cells respectively (Branca, Bremner, unpublished observations). In addition 'negative' results were sometimes obtained from granulocytes, possibly because of their high endogenous peroxidase levels. It may therefore be essential to carry out sub-fractionation of human blood cells prior to assay if meaningful results are to be obtained.

REFERENCES

Bosworth CM, Bacon J, Bremner I, Aggett PJ (1989) The homeostatic regulation of zinc absorption and secretion in zinc depleted man. In: Southgate DAT, Johnson IT, Fenwick GR (eds) Nutrient Availability: Chemical and Biological Aspects. Royal Society of Chemistry, Cambridge (Special Publication No 72), p 213

Bremner I (1987) Nutritional and physiological significance of metallothionein. In: Kagi JHR, Kojima Y (eds) Metallothionein II. Birkhauser, Basel, p 81

Bremner I, Hoekstra WG, Davies NT, Young BW (1978) Effect of zinc status of rats on the synthesis and degradation of copper-induced metallothioneins. Biochem J 174:883-892

Bremner I, Marshall RB (1974) Hepatic copper- and zinc-binding proteins in ruminants. 2. Relationship between Cu and Zn concentrations and the occurrence of a metallothionein-like fraction. Br J Nutr 32:293-300

Bremner I, Morrison, JN, Wood AM, Arthur JR (1987) Effect of changes in dietary zinc, copper and selenium supply and of endotoxin administration on metallothionein I concentrations in blood cells and urine in the rat. J Nutr 117:1595-1602

Feldman, SL, Cousins, RJ (1976) Degradation of hepatic zinc-thionein following parenteral zinc administration. Biochem J 160:583-588

Gallant, KR, Cherian MG (1986) Influence of maternal mineral deficiency on the hepatic metallothionein and zinc in newborn rats. Biochem Cell Biol 64:8-12

Ghaffar A, Aggett PJ, Bremner I (1989) Development of an enzyme-linked immunosorbent assay for metallothionein. In: Southgate DAT, Johnson IT, Fenwick GR (eds) Nutrient Availability: Chemical and Biological Aspects. Royal Society of Chemistry, Cambridge (Special Publication No 72), p 74

Held DD, Hoekstra WG (1984) The effects of zinc deficiency on turnover of cadmium-metallothionein in rat liver. J Nutr 114:2274-2282

Ito K, Schmaus JW, Reissmann KR (1964) Protein metabolism and erythropoiesis. III. The erythroid marrow in protein-starved rats and its response to erythropoietin. Acta Hematol 32:257-264

Mehra RK, Bremner I (1983) Development of a radioimmunoassay for rat liver metallothionein I and its application to the analysis of rat plasma and kidneys. Biochem J 213:459-465

Morrison JN, Bremner I (1987) Effects of maternal zinc supply on blood and tissue metallothionein I concentrations in suckling rats. J Nutr 117:1588-1594

Morrison JN, Wood AM, Bremner I (1988) Concentrations and distribution of metallothionein-I in blood cells of rats injected with zinc or phenylhydrazine. J Trace Elements Exp Med 1:95-105

Morrison JN, Wood AM, Bremner I (1989) Effects of protein deficiency on blood cell and tissue metallothionein-I concentrations in rats. Proc Nutr Soc (in press)

Noble NA, Xu Q-P, Ward JH (1989) Reticulocytes I. Isolation and in vitro maturation of synchronised populations. Blood (in press)

Nordberg GF, Garvey JS, Chang CC (1982) Metallothionein in plasma and urine of cadmium workers. Envir Res 28:179-182

Robertson A, Morrison JN, Wood AM, Bremner I (1989) Effects of iron deficiency on metallothionein-I concentrations in blood and tissues of rats. J Nutr 119:439-445

Sato M, Mehra RK, Bremner I (1984) Measurement of plasma metallothionein in the assessment of the zinc status of zinc deficient and stressed rats. J Nutr 114:1683-1689

Sobocinski PZ, Canterbury WG, Mapes CA, Dinterman RE (1978) Involvement of hepatic metallothioneins in hypozincemia associated with bacterial infection. Am J Physiol 234:E399-E406

Tanaka K, Min K-S, Ohyanagi N, Onosaka S, Fukuhara C (1986) Fate of erythrocyte Cd-metallothionein in mice. Toxicol Appl Pharmacol 83:197-202

Tanaka K, Min K-S, Onosaka S, Fukuhara C, Ueda M (1985) The origin of metallothionein in red blood cells. Toxicol Appl Pharmacol 78:63-68

Synthesis and Properties of a Selenium Analogue of Metallothionein

Hidehiko Tanaka[1], Tadao Oikawa[2], Nobuyoshi Esaki[2], and Kenji Soda[2]

[1] Faculty of Agriculture, Okayama University, 1-1-1 Tsushima, Naka, Okayama, 700 Japan
[2] Institute for Chemical Research, Kyoto University, Uji, Kyoto, 611 Japan

ABSTRACT

The Neurospora crassa copper metallothionein (Cu-MT) consists of 25 amino acids including 7 cysteines to which 6 Cu(I) ions are bound. We studied the substitution of selenium for sulfur in the MT and the structure and function of the selenometallothionein (SeMT). SeMT was synthesized by the Merrifield method with Se-(p-methylbenzyl)-N-t-Boc-L-selenocysteine and purified by reversed phase HPLC. Cu(II) was bound with SeMT under anaerobic conditions to form Cu-SeMT. The spectrophotometric properties of Cu-SeMT were compared with those of Cu-MT prepared in a similar way. Both were characterized by a broad absorption band between 260 and 270 nm with a shoulder around 265 nm. In contrast to the native N. crassa Cu-MT, 3 Cu ions were bound with SeMT. ESR parameters of Cu-SeMT were similar to those of the oxidized form of Cu-MT. Copper EXAFS of Cu-SeMT and Cu-MT were measured by fluorescence mode at 18K. The XANES oscillation of Cu-SeMT showed a notable similarity to that of Cu-MT. These indicate that Cu-SeMT resembles Cu-MT in the coodination structure around CuS.

KEY WORDS
Synthesis of Selenometallothionein, Structure of Selenometallothionein, Copper EXAFS, Selenium EXAFS

INTRODUCTION

Selenium is located between sulfur and tellurium in the periodic table and has both metallic and non-metallic properties. It resembles sulfur in chemical properties, but generally the reactivity of selenium compounds with metals is higher than that of sulfur counterparts (Naganuma et al., 1983).

Metallothioneins (MTs) are a class of low molecular weight and cysteine-rich proteins, which bind with high amount of metal ions such as Zn, Cd and Cu (Margoshes and Vallee, 1957). All cysteinyl residues are involved in metal binding, The Neurospora crassa copper MT is the smallest MT isolated so far: it consists of only 25 amino acids including 7 cysteinyl residues to which 6 Cu(I) ions bind (Lerch, 1980). Therefore, the N. crassa MT is the simplest system to study the structure-function relationship of MTs. We are interested in examining whether its structure and properties are influenced by substituting selenium atoms for the sulfur ones. Thus, we have studied a chemical synthetic procedure for a selenium containing MT (selenometallothionein, SeMT), in which all the cysteinyl residues are replaced by selenium counterparts, and its properties to compare with those of MT synthesized in a similar manner. The results obtained here are described and discussed.

EXPERIMENTAL PROCEDURES

Materials

L-β-Chloroalanine was synthesized from L-serine by the method of Walsh et al (1971). Disodium diselenide was prepared by the method of Klayman and Griffin (1973). L-Selenocystine was prepared from disodium diselenide with

L-β-chloroalanine according to the procedure of Chocat et al (1985) with modification.

Synthesis of Se-(p-methylbenzyl)-L-selenocysteine   Se-(p-Methylbenzyl)-L-selenocysteine was prepared from L-selenocysteine prepared by borohydride reduction of L-selenocystine with α-bromo-p-xylene by a modification of procedure of Erickson and Merrifield (1973).   Triethylamine (58 ml, 410 mmol) was added to 200 ml of ethanol-water (1:1, v/v) in a N_2-flushed stoppered 1,000 ml three-necked flask and the resulting solution was stirred for 20 min.   Then L-selenocystine (2.84 g, 8.5 mmol) was dissolved in the solution, and NaBH$_4$ (2 g, 53 mmol) was added gradually.   After 1 hr, α-bromo-p-xylene (4.5 g, 24 mmol) was added, and the mixture was stirred at 40 °C for 14 hr.   The precipitated white crystals were collected and washed with water and ethanol thoroughly (yield 84%).

Synthesis of Se-(p-Methylbenzyl)-N-t-buthyloxycarbonyl(=Boc)-L-selenocysteine Se-(p-Methylbenzyl)-L-selenocysteine (4.3 g, 17 mmol) and triethylamine (15 ml, 108 mmol) were added to 11.3 ml of water.   To the solution, S-t-Boc-4,6-dimethyl-2-thiopyrimidine (4.81 g, 20 mmol) dissolved in 9.3 ml of dioxane was added, and then the mixture was incubated at room temperature for 65 hr under stirring.   After completion of the reaction, 50 ml of water was added to the reaction mixture.   After 5 min, unreacted carbonate was extracted twice with 50 ml of ethyl acetate.   The aqueous layer was cooled to 0°C and adjusted to pH 2.3 by addition of 5% KHSO$_4$ solution cooled to 0°C.   The precipitate appeared was extracted once with 100 ml and twice with 50 ml of ethyl acetate.   The combined ethyl acetate layer was dried over anhydrous sodium sulfate at 4°C.   After evaporation under vacuum, the resulting solid was dissolved in 3 ml of ethyl acetate, and 1 ml of petroleum ether was added.   The solvents were evaporated again, and the residue was cooled at 4 °C for 24 hr.   The resulting solid was broken into pieces and refluxed in 40 ml of petroleum ether for 1 hr.   Cream-colored, micro-powdered crystals were obtained (yield 94%).

Synthesis of SeMT and MT   Both the peptides were synthesized on an Applied Biosystems peptide synthesizer (430A).   The amino acid derivatives used for the synthesis were as follows: N-t-Boc-O-benzyl-L-serine, N-t-Boc-L-asparagine, N-t-Boc-glycine, N-t-Boc-L-alanine, N-t-Boc-L-aspartic acid benzyl ester, N-t-Boc-Se-(p-methylbenzyl)-L-selenocysteine, and N-t-Boc-S-(p-methylbenzyl)-L-cysteine. α-N-t-Boc-ϵ-benzyloxycarbonyl-L-lysine attached a polystyrane support was used.   The protecting groups for the side chains and solid support were removed from the peptide with anhydrous HF.   The dried peptide-resin (0.2-1.0 g) was transferred into a cleavage vessel containing 1.5 ml of anisol and 0.25 ml of ethyl methyl sulfide per g-resin as scavengers.   After addition of 10 ml of liquid HF condensed in -70°C, the mixture was stirred for 30 min at -20°C and then for 30 min at 0°C.   The HF was removed at 0°C under vacuum, and the residue was washed with anhydrous ether and chloroform, successively.

Amino acid composition

The peptide (about 7 nmol) was hydrolyzed with 200 μl of 6 N HCl at 108°C for 9 hr with an Waters PICO-TAG automatic acid hydrolysis system.   The hydrolyzates were dried, and the residue was dissolved in 0.2 M citrate buffer (pH 2.2) and analyzed with a Beckman 7300 high performance amino acid analyzer.   Cysteine was determined as cysteic acid after performate oxidation of the sample according to the procedure of Hirs (1967), and selenocysteine as Se-carboxyamidomethyl-selenocysteine after carboxyamido-methylation of the sample with monoiodoacetamide (Waxdall et al, 1968).

Preparation of Cu-SeMT and Cu-MT

All solutions used were degassed on a vacuum line, and the procedures were performed in a nitrogen-purged glove box (oxygen concentration, less than 10 ppm).   After apo Se-MT (800 μg, 312 nmol) in 20 mM Tris-HCl buffer (pH 8.6) was reduced with 2-mercaptoethanol, the mixture was incubated with CuSO$_4$ (3.74 μmol) at 37°C for 30 min.   The Cu-SeMT was separated from excess reagents with an Asahipak GS-310 HPLC column equilibrated with 20 mM Tris-HCL buffer (pH 8.6).   Cu-MT was prepared by the same way.

Other analytical methods

For preparation (10 X 250 mm) and analysis (4.5 X 150 mm) of the synthesized peptides, reverse-phase high performance liquid chromatography was performed with an ULTRON N-C18 column (Shinwa kako, Japan). The chromatography was carried out with a linear gradient between 0 and 50% acetonitrile in 0.1% trifluoroacetic acid at room temperature with a flow rate of 0.8 ml/min. Absorbance was recorded at 220 nm. Metal content was measured with a Shimadzu AA-670G atomic absorption spectrophotometer. The determination was repeated at least two times, and the content was calculated from the standard curves of each metal. Absorption spectra were measured with a Shimadzu MPS-2000 spectrophotometer. A sample was added into the black quartz cuvette with a mininert valve under anaerobic conditions, and other reagents were injected through the valve with a microsyringe. X-Ray absorption spectra for selenocystine were recorded over the energy range corresponding to the selenium K-edge. Data were collected at room temperature at an energy of 2.5 GeV with an avarage current of 193 mA. The X-ray absorption data were converted to EXAFS modulation spectra with a cubic spline fit background substraction. X-Ray spectra for Cu-SeMT and Cu-MT were measured by fluorescence mode with a Lytle detector at BL7C. Data were measured on the frozen solution, kept at 18°K, with a thermoelectric cooling module. Two data scans were averaged to give the reported spectra.

RESULTS AND DISCUSSION

Synthesis of SeMT and MT

We synthesized SeMT by the Merrifield method with Se-(p-methylbenzyl)-N-t-Boc-L-selenocysteine. The Se-p-methylbenzyl group was stable during peptide synthesis. The protecting groups for the side chains and the solid support were removed from the peptide with anhydrous HF. The crude SeMT and MT were reduced with NaBH₄ and purified by reversed phase HPLC (Fig. 1).

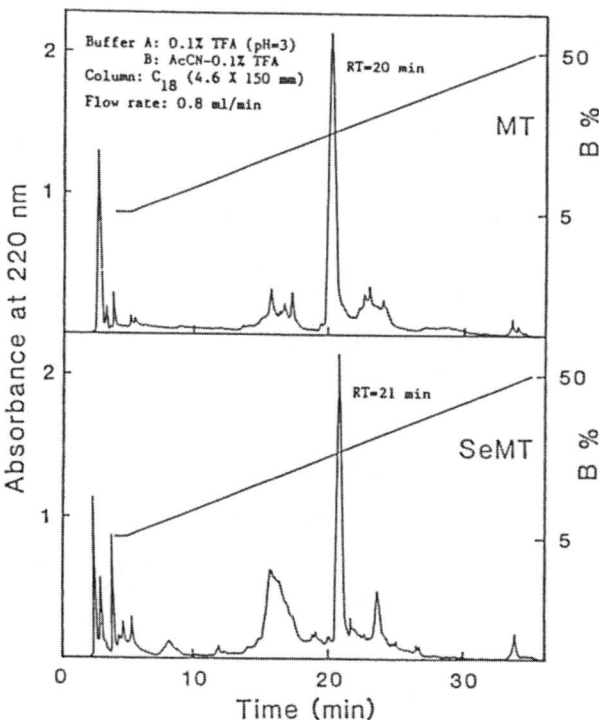

Fig. 1. Reversed phase HPLC of MT and SeMT

Table 1. Amino acid analysis of SeMT and MT

Amino acud	Number of residues		
	Syn. MT	Syn. SeMT	N. crassa MT
Asx	2.7	2.4	3
Ser	7.2	6.6	7
Gly	6.2	6.0	6
Ala	1.0	1.0	1
Cys	6.7	-	7
SeCys	-	7.1	-
Lys	0.8	0.7	1

Table 1 shows the amino acid composition of the purified SeMT and MT.   The
composition of MT was identical with that of N. crassa MT (Lerch, 1980).   In
the case of SeMT, the number of other amino acid residues than selenocysteine
accorded with that of N. crassa MT.   Seven residues of selenocysteine were
observed as Se-carboxymethyl selenocysteine in the reduced peptide alkylated
with iodoacetamide.   Unless alkylated, the selenium moiety was decomposed
during acid hydrolysis of the native peptide, and was detected neither in
hydrolysis of the native peptide nor in the peptide oxidized with performic
acid.   Thus, the amino acid composition of the purified SeMT and MT was
confirmed to agree with that of the expected value.

Copper content and spectrophotometric properties of Cu-SeMT and Cu-MT

We prepared copper bound SeMT with Cu (II) under anaerobic conditions.   The
Cu content of Cu-SeMT was determined by atomic absorption analysis.   In
contrast to the native N. crassa Cu-MT, 3 Cu ions were bound with SeMT.   The
Cu content of synthesized Cu-MT also was 3 g-atom per mol of peptide.   These
values agreed with that of the N. crassa Cu-MT substituted the copper ions
for the divalent ions such as Cd, Hg and Co (Beltramini et al., 1982).

The spectrophotometric properties of the Cu-SeMT were examined to compare
with those of synthesized Cu-MT.   Both the absorption spectra were rather
similar.   Both were characterized by a broad absorption band between 260 and
270 nm with a shoulder around 265 nm (Fig. 2).   No features were observed in
the visible region.   The broad band appears to result from Cu-Se bond
because of lacking the shoulder in the apo SeMT.

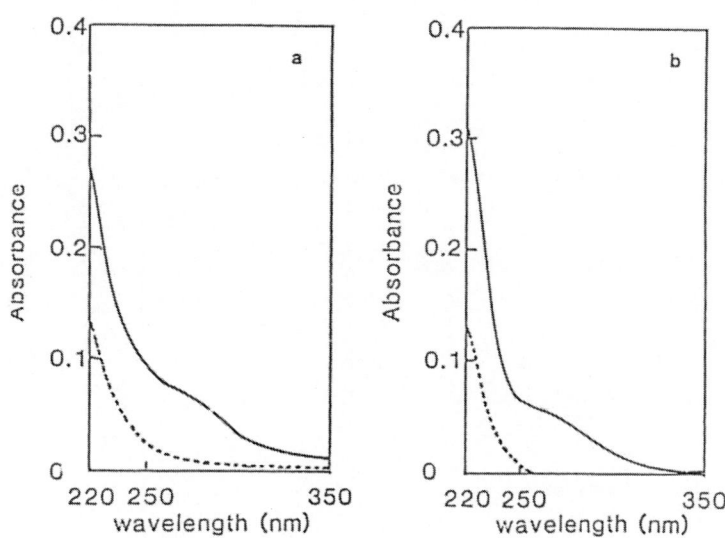

Fig. 2.   Absorption spectra of Cu-SeMT (a) and Cu-MT (b)
Solid lines, Cu bound form (in 20 mM Tris-HCl buffer, pH 8.6); dotted lines,
apo form (in 50 mM HCl)

ESR analysis of Cu-SeMT

Fig. 3 shows the ESR spectra of the Cu-SeMT and Cu-MT.   The derived g and A
values were obtained by scaling the spectrum with $MgCl_2$ as a standard.   The
ESR parameters of Cu-SeMT were similar to those of the oxidized form of N.
crassa Cu-MT (Table 2) (Beltramini and Lerch, 1982).   The notable change
resulting from the substitution of selenium for sulfur in copper-sulfur
cluster was not observed in the values of these parameters.   These results
indicate that the Cu-SeMT resembles the Cu-MT in the coordination mode around
Cu ions.

We added oxidizing, reducing and chelating reagents to Cu-SeMT and Cu-MT, and
examined the ESR spectral changes.   EDTA partially released copper ions from
Cu-SeMT and Cu-MT after 10 min. H_2O_2 did not affect on the ESR properties
of these MT.   The ESR signals of Cu-SeMT, however, disappeared on addition
of either cysteine or ascorbate, but those of Cu-MT were affected only by
cysteine.   This reflects the lower redox potential of Cu-SeMT than that of
Cu-MT.   These resuts indicate that the valece state of copper bound the SeMT
and MT is divalent.

Table 2.   ESR parameters of Cu-SeMT and Cu-MT

	g_o	$g_{//}$	$g_\perp$	A_o	$A_{//}$	$A_\perp$
			ESR parameters			
Cu-SeMT	2.112	2.234	2.052	81	189	27
Cu-MT	2.112	2.234	2.052	82	193	27
Cu-MT$_{ox}$	-	2.23	2.05	-	190	-

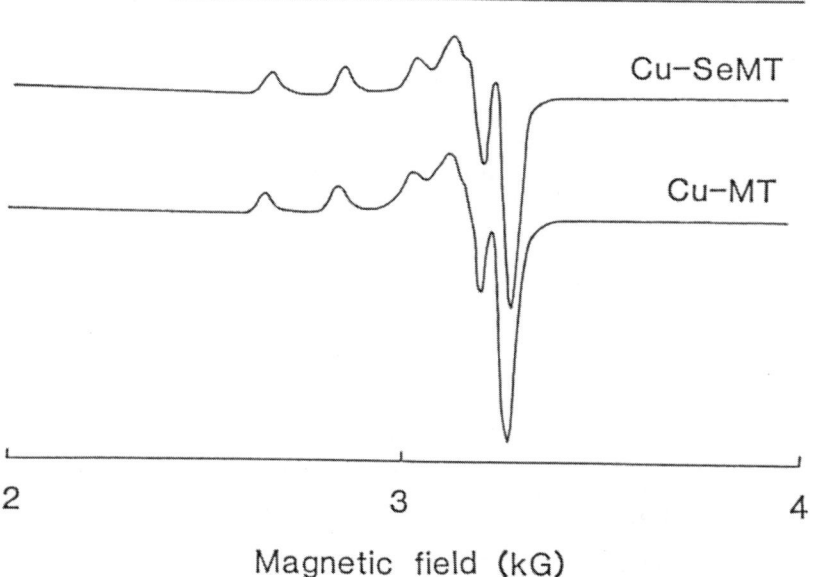

Fig. 3.   ESR spectra of Cu-SeMT and Cu-MT

418

EXAFS analysis of Cu-SeMT and Cu-MT

First, EXAFS measurements were carried out for aquaous solutions of
selenocystine and selenocysteine, and for selenocystine powder solidified
with polyethylene.   The X-ray absorption spectrum in the region of the Se K-
edge for aquaous solution of L-selenocystine was essentially identical to
that for L-selenocystine powder solidified with polyethylene.   The bond
lengths were determined as follows: C-SE, 1.88 A; and Se-Se, 2.32 A.   When
L-selenocystine was dissolved in 0.1 N HCl and analyzed, the bond lengths
were not influenced by dissociation of COOH and NH$_2$.

Based on these results, copper EXAFS of Cu-SeMT and Cu-MT were measured by
fluorescence mode at 18K.   The XANES oscillation of Cu-SeMT showed a notable
similarity to that of Cu-MT (Fig. 4).   This indicates that the structure of
Cu(II)-thiolate cluster in the Cu-MT is similar to that of Cu(II)-selenoate
cluster in the selenium counterpart.   CuS is identical with CuSe in the mode
of coordination; Cu is in a tetrahedral coordination to 3 S$_I$ (R= 2.09 Å) and
1 S$_{II}$ (R= 2.32 Å).   We have applied these values to the analysis of the
structure of Cu-MT complex.   Fig. 5 shows a proposed structure of the
Cu(II)-thiolate cluster.   We analyzed the structure of Cu-SeMT by the same
way: 3 Cu ions are coordinated to 7 selenoate anions to form a single
cluster.   Selenium EXAFS of Cu-SeMT also confirmed this structure (Table 3).

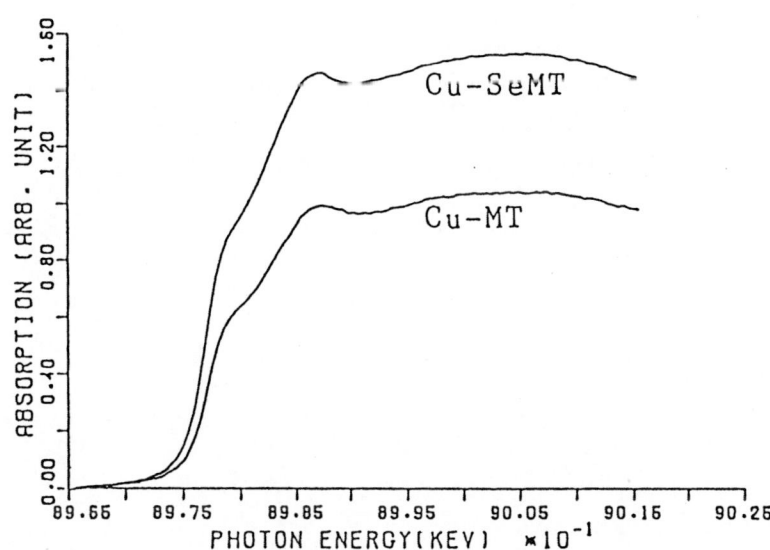

Fig. 4.   XANES spectra for Cu-SeMT and Cu-MT

Table 3.   Interatomic distance

	Cu EXAFS	Se EXAFS
Cu-Se$_I$	2.31	2.29
Cu-Se$_{II}$	2.36	2.29
Cu-Se$_{III}$	2.58	2.63
C-Se		1.99
Cu-Cu	3.14	
Se-Se		3.09
Cu-S$_I$, Cu-S$_{II}$	2.21	
Cu-S$_{III}$	2.32	

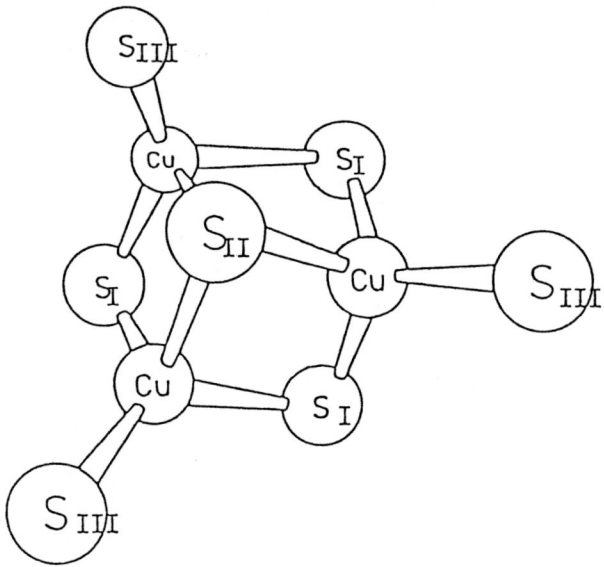

Fig. 5. Proposed arrangement of Cu-thiolate cluster in Cu-MT

REFERENCES

Beltramini M, Lerch K (1982) Copper transfer between Neurospora copper metallothionein and type 3 copper apo proteins. FEBS Lett 142:219-222

Beltramini M, Lerch K, Vasal M (1984) Metal substitution of Neurospora copper metallothionein. Biochemistry 23:3422-3427

Chocat P, Esaki N, Tanaka H, Soda K (1985) Synthetic of L-selenodjenkolate and its degradation with methionine γ-lyase. Anal Biochem 148:485-489

Erickson B W, Merrifield R B (1973) Acid stability of several benzylic protecting groups used in solid-phase peptide synthesis. Rearrangement of O-benzyltyrosine to 3-benzyltyrosine. J Am Chem Soc 95:3750-3756

Hirs C H W (1967) Performic acid oxidation. Methods Enzymol 11:197-199

Klayman D J, Griffin S T (1973) Reaction of selenium with sodium borohydride in protic solvents. A facile method for the introduction of selenium into organic molecules. J Am Chem Soc 95:197-199

Lerch K (1980) Copper metallothionein, a copper-binding protein. Nature 284:368-370

Margoshes M, Vallee B L (1957) A cadmium protein from equine kidney cortex. J Am Chem Soc 79:4813-4814

Naganuma A, Tanaka T, Maeda K, Matsuda B, Tabata J, Imura N (1983) The identification of selenium with various metals in vitro and in vivo. Toxycology 29:77-86

Walsh C, Schonbuemm A, Abels R T (1971) Studies on the mechanism of action of D-amino acid oxidase. Evidence for removal of substrate α-hydrogen as a proton. J Biol Chem 246:6855-6866

Waxdall M J, Konigsberg W H, Honey W H, Edelman G M (1968) The covalent structure of a human γ-G-immunoglobulin, II. Isolation and characterization of the cyanogen bromide fragments. Biochemistry 7:1959-1966

Transport of Copper

Edward D. Harris

Department of Biochemistry and Biophysics, Texas ASM University, College Station, TX 77842, USA

ABSTRACT:

Broadly interpreted, copper transport encompasses events and intermediates that move copper progressively from gut to cells and to copper requiring enzymes within. Ceruloplasmin, the major copper-binding protein in plasma, may be an important, but not exclusive source of copper for extrahepatic cells. We have attempted to learn the mechanism by which the six to seven tightly bound and catalytically functional copper atoms in ceruloplasmin are transferred to tissue enzymes. The data suggest that carriers for ionic copper may work in conjunction with a ceruloplasmin-mediated mechanism.

KEY WORDS:
Copper metabolism/ cellular copper uptake/ ceruloplasmin/ membrane transport

INTRODUCTION:

The transport of copper has been studied in plants, animals, bacteria, and fungi. Ultimately, the same questions are asked. How is copper conveyed to cells and through membranes. How do copper enzymes bind the metal. What is the mechanism. To answer these questions, workers have endeavored to identify and characterize plasma and membrane intermediates that bind copper and somehow mediate its passage through the membrane bilayer. All cells require copper. Yet, unlike iron which has a fairly broad and basic understanding, the mechanism of copper transport and uptake by cells is hardly known. Since copper is an essential nutrient, the void in understanding is a serious breach of science. In our work, we have focused on ceruloplasmin, the major serum copper source and are currently attempting to learn how the protein interchanges its tightly bound copper with cells. Efforts are under way to isolate and characterize membrane receptors that bind ceruloplasmin, to study serum factors that reduce the copper in the protein and facilitate its removal, and to study the ultimate target of copper utilization, intracellular copper enzymes.

WHY STUDY COPPER TRANSPORT:

We must not lose site of the importance of copper transport to more generalized understandings. Two disciplines, nutrition and medicine, benefit directly from a detailed knowledge of copper transport. Copper is recognized as an essential nutrient. Transport studies have helped us understand the mechanism of nutrient antagonism, nutrient interaction, and system adaptation to copper. In medicine, a knowledge of copper transport has given meaningful insight into genetic diseases such as Wilson's disease, Menkes' syndrome, where mismanagement of copper is the causative factor. Under abundance of copper to a living system can lead to metabolic and structural impairments, whereas over abundance of copper can cause toxicosis. Both conditions are devestating to normal physiological function. A deficiency in copper leads to a decrease in catalytic function of copper enzymes. Transport studies have shown that copper is a rate-limiting factor in enzyme expression; plasma levels and tissue enzyme levels show a positive correlation. Copper toxicosis shows us that our systems for handling and storing copper can be overwhelmed when copper is in excess. Both toxicosis and deficiency show us the importance of having the right amount of copper and ways the body adapts to mismatches of requirement with supply.

Copper transport in broader terms is but one phase of copper metabolism. To properly describe metabolism, we must also consider copper absorption from the gut, transport through the blood to cells, and uptake by the cells. For convenience, copper transport is divided into three phases: (1) an extracellular or plasma phase, (2) a membrane phase, and (3) an intracellular or cytosolic phase. We study each of these separate and apart at the same time realizing that the mechanism overall must involve coordinating the three phases into a coherent picture with potential for overlap.

HISTORICAL PERSPECTIVE:

Historically, copper transport was never given serious attention until the experiments of Berne and Kunkel (1954). These workers, while studying the metabolic defect in Wilson's disease, showed that newly absorbed copper transferred first from albumin to ceruloplasmin. Wilson's patients lacked this simple ability. As a consequence, Wilson patients typically had low levels of ceruloplasmin in plasma and showed other defects of interorgan copper homeostasis. Importantly, Berne and Kunkel showed that copper metabolism in both humans and animals is multifaceted. Owen followed with a remarkable series of studies in rats, showing as Berne and Kunkel had predicted, that absorbed copper passed first from albumin to ceruloplasmin before being taken up by extrahepatic cells (Owen 1965). Owen and Hazelrig (1966) later identified the liver as the site of ceruloplasmin biosynthesis and predicted that the liver was a major copper distribution center with ceruloplasmin the major mediator of copper to tissues. cDNA probes have now identified testes, choroid plexes and plancenta also as sites of ceruloplasmin synthesis (Aldred 1987).

Copper transport continues to be an important but controversial subject. Most attention continues to focus on blood transport agents, but more recently, studies have open inquiries on the plasma membrane and how copper ions are able to cross this formidable barrier to free copper movement. Ion channels for copper driven by ion Cu^{2+} currents and facilitated with sodium ions have been found in neurones of lower vertebrates (Weinreich 1987). In hepatocytes, HepG2 cells (Stockert 1986), and brain hypothalami (Hartter 1987), membrane proteins mediating passage of free copper ions or amino acid complexes have been described. Ettinger and colleagues (Schmitt 1983; Darwish 1983) have carried out an extensive characterization of mouse and rat hepatocyte copper transport. These workers have shown that the uptake of ^{64}Cu into suspended cells is rapid and is mediated by a saturable carrier. Half-saturation is reached at a medium copper concentration of 11 uM, a value considerably higher than the free copper level in plasma. Albumin in the medium tends to severely hamper rapid uptake of free copper (van den Berg 1984). This observation raises questions as to whether free copper transport occurs in vivo.

CERULOPLASMIN AND COPPER TRANSPORT:

Ceruloplasmin is the most readily available source of copper in serum, accounting for as much as 90-95% of the copper. A recent study by Weiss and Linder (1985) purport to give evidence for another high molecular weight protein other than albumin that transports copper. These workers have named the protein "transcuprein". The tightly-bound copper atoms in ceruloplasmin have been shown in time to appear in cytosolic fractions after injections of ceruloplasmin into animals (Campbell 1981; Linder 1977). More importantly, the transfer of ceruloplasmin copper into specific tissue enzymes has also been shown (Dameron 1987; Marceau 1973). Although tightly coordinated to the protein, the copper atoms in ceruloplasmin become labile when ascorbate (Scheinberg 1957), cysteine (Aisen 1965), or p-phenylenediamine (Owen 1975) is added to a solution containing the protein. Ceruloplasmin has been shown to bind to membranes of a number of blood cells (Barnes 1984; Kataoka 1985) and tissues (Stevens 1984) with binding constants estimated in the nanomolar range.

More recently, Dameron and Harris (1987) showed that copper from either ceruloplasmin or albumin was capable of entering aortic cells and activating aortic Cu,Zn superoxide dismutase (CuZnSOD). These workers stressed similarities rather than differences in the two copper transport proteins. Significantly, the transport of copper from ceruloplasmin occurred under conditions that matched the level of ceruloplasmin in vivo. Other studies (Hsieh 1975; Orena 1986; Terao 1976) have provided clear and compelling evidence that ceruloplasmin performs the function of copper transport in animal systems and is a source of copper for intracellular enzymes.

K562 CELLS AND COPPER TRANSPORT:

K562 cells, an erythroleukemic cell line, have been used extensively in iron transport studies (Klausner 1983; Bakkeren 1987). To date, these cells have been used only sparingly to study copper transport. K562 cells bind ceruloplasmin (Percival 1988) and when binding is carried out at $37^{0}C$, some of the copper remains bound to the cell (Percival 1989). The residual copper resists removal by mild acid (pH 2) washing, suggesting the copper is internal. Thus, K562 cells not only bind ceruloplasmin but can carry out a temperature-dependent transfer of the ceruloplasmin-bound copper to the cytosol. The rate at which copper accumulated in the cytosol was proportional to the ceruloplasmin concentration in the medium. These studies used nondenatured ^{67}Cu-labeled ceruloplasmin prepared by an ascorbate-catalyzed exchange of ceruloplasmin with $^{67}CuCl_2$ in vitro. The complex was stable under the conditions of the incubation (Percival 1989).

ASCORBIC ACID STIMULATES COPPER TRANSPORT FROM CERULOPLASMIN:

On a strictly chemical level, ascorbic acid has been shown to catalyze a pH-dependent decolorization of ceruloplasmin (Gunnarsson 1973). The reaction shows that ascorbate is a valid substrate for the oxidase activity of ceruloplasmin, although the physiological significance is not immediately apparent. In our studies (Percival 1989), we observed that L-ascorbic acid (100 uM) stimulated by four to ten fold the transfer of ^{67}Cu from ceruloplasmin to the cells. D-Isoascorbic acid worked as well as L-ascorbic acid, suggesting a common structural feature, perhaps the enediol group was the determining factor in the ascorbate effect. Ascorbate may be a cofactor or catalyst in the cross-membrane exchange of copper from ceruloplasmin.

IN SEARCH OF A MECHANISM:

The question of mechanism becomes paramount to these studies. Based on work with transferrin and iron transport, one is inclined at first to consider holo-ceruloplasmin the vehicle that not only brings copper to the cells, but mediates its passage into the cytosol. This would mean that receptors on the cell surface are part of an endocytic system for transferring and vesicularizing ceruloplasmin molecules. To test the mechanism, both ^{67}Cu-labeled ceruloplasmin and ^{125}I-labeled ceruloplasmin were prepared. Copper as ^{67}Cu taken up by the cells was analyzed on Percoll gradients. The same analysis was applied to cells that had been incubated with ^{125}I-ceruloplasmin. The data were definitive. ^{67}Cu was localized in buoyant fractions that sedimented at densities 1.03-1.05. 40%-80% of the acid-resistant ^{67}Cu was found in this region of the gradient. Hardly any radioactivity was found in the gradient when cells were incubated at 4°C. Cells that had been exposed to ^{125}I-ceruloplasmin for 1 hour failed to show any ^{125}I in the Percoll gradient fractions. The data suggested that the protein and copper had separated and only the copper had penetrated the cells. Thus, copper transport from ceruloplasmin shows a clear departure from iron transport via transferrin in these same cells.

ADVANTAGES OF A CERULOPLASMIN-MEDIATED MECHANISM:

The decided advantage of ceruloplasmin to other forms of copper is the extreme sensitivity of the reaction. In our studies, we observed that as little as 5 pmol of ^{67}Cu-ceruloplasmin produced measureable uptake of ^{67}Cu. Uptake was unimpeded by the presence of albumin (3%) in the medium. Ceruloplasmin is the major copper source in plasma. Only nanomolar amounts of ceruloplasmin are needed to bind to cell membranes.

A UNIFIED HYPOTHESIS OF COPPER TRANSPORT:

Plants, insects, bacteria, fungi all require copper, yet none of these species uses ceruloplasmin or albumin to deliver copper to the cells. Thus, there must be a commonality to all copper transport systems. If the mechanism shown by K562 cells applies to other cell types, then the role of ceruloplasmin in transport ends at the cell surface. Free copper carriers similar to those described by Ettinger and colleagues (Schmitt 1983) deliver copper to the interior. The role of ascorbate needs to be defined with certainty. Paradoxically, dietary ascorbate has been shown to interfer with copper absorption in a number of animal species (Hunt 1970; Van Campen 1968) and humans (Jacob 1987). The postabsorptive role of ascorbate on copper has not been studied extensively.

In seeking a unified hypothesis, one must consider that copper transport begins at the cell membrane. Ceruloplasmin can be used as a transport agent because cells have built-in mechanisms for dissociating copper from the protein. The mechanism may be linked to the redox properties of ceruloplasmin. Free copper may gain access to the cells through the same portals or using the same carriers that handle ceruloplasmin copper. It still is not clear whether copper taken in appears as free cytosolic copper or is trapped in vesicles. The latter offer the opportunity of having the copper "packaged" for movement through the cytosol. Endocytosis and pinocytosis have not been excluded from the uptake mechanism.

ACKNOWLEDGEMENTS: Funding for the work was obtained in part by NIH Grant DK35920 and Hatch Project H-6621 of the Texas Agricultural Experiment Station.

REFERENCES:

Aisen P, Morell AG (1965) Physical and chemical studies on ceruloplasmin. III. A stabilizing copper-copper interaction in ceruloplasmin. J Biol Chem.240:1974-1978
Aldred AR, Grimes A, Schreiber G, Mercer JFB (1987) Rat ceruloplasmin. Molecular cloning and gene expression in liver, choroid plexus, yolk sac, placenta, and testis. J Biol Chem 262: 2875-2878
Bakkeren DL, de Jeu-Jaspars CMH, Kroos MJ, van Eijk HG (1987) Release of iron from endosomes is an early step in the transferrin cycle. Int J Biochem 19:179-186

Barnes G, Frieden E (1984) Ceruloplasmin receptors of erythrocytes. Biochem Biophys Res Commun 125:157-162

Berne AG, Kunkel HG (1954) Localization of Cu^{64} in serum fractions following oral administration: an alteration in Wilson's disease. Proc Soc Exp Biol Med 85:44-48

Campbell CH, Grown R, Linder MC (1981) Circulating ceruloplasmin is an important source of copper for normal and malignant animal cells. Biochim Biophys Acta 678:27-38

Dameron CT, Harris ED (1987) Regulation of aortic CuZn-superoxide dismutase with copper. Ceruloplasmin and albumin transfer copper and reactivate the enzyme in culture. Biochem J 248:669-675

Darwish HM, Hoke JE, Ettinger MJ (1983) Kinetics of Cu(II) transport and accumulation by hepatocytes from copper-deficient mice and the brindled mouse model of Menkes' disease. J Biol Chem 258:13621-13626

Gunnarsson PO, Nylen U, Pettersson G (1973) Kinetics of the interaction between ceruloplasmin and reducing substrates. Eur J Biochem 37:41-46

Hartter DE, Barnea A (1987) Brain tissue accumulates ^{67}copper by two ligand-dependent saturable processes. A high affinity, low capacity and a low affinity, high capacity process. J Biol Chem 263:799-805

Hsieh HS, Frieden E (1975) Evidence for ceruloplasmin as a copper transport protein. Biochem Biophys Res Commun 67:1326-1331

Hunt CE, Landesman J, Newbern PM (1970) Copper deficiency in chicks: effects of ascorbic acid on iron, copper, cytochrome oxidase activity and aortic mucopolysaccharides. Br J Nutr 24:607-614

Hunt CE, Carlton WW (1965) Cardiovascular lesions associated with experimental copper deficiency in the rabbit. J Nutr 87:385-393

Jacob RA, Skala JH, Omaye ST, Turnland JR (1987) Effect of varying ascorbic acid intakes on copper absorption and ceruloplasmin levels of young men. J Nutr 117:2109-2115

Kataoka M, Tavassoli M (1985) Identification of ceruloplasmin receptors on the surface of human blood monocytes, granulocytes, and lymphocytes. Exp Hematol 13:806-810

Klausner RD, van Renswoude J, Ashwell G, Kempf C, Schecter AN, Dean A, Bridges KR (1983) Receptor-mediated endocytosis of transferrin in K562 cells. J Biol Chem 258:4715-4724

Linder MC, Moor JR (1977) Plasma ceruloplasmin. Evidence for its presence in and uptake by heart and other organs of the rat. Biochim Biophys Acta 499:329-336

Marceau N, Aspin N (1973) The intracellular distribution of the radiocopper derived from ceruloplasmin and albumin. Biochim Biophys Acta 293:338-350

Orena SJ, Goode CA, Linder MC (1986) Binding and uptake of copper from ceruloplasmin. Biochem Biophys Res Commun 139:822-829

Owen CA Jr (1965) Metabolism of radiocopper (^{64}Cu) in the rat. Am J Physiol 209:900-904

Owen CA Jr, Hazelrig JB (1966) Metabolism of Cu^{64} labeled copper by the isolated rat liver. Am J Physiol 210:1059-1064

Owen CA Jr (1975) Uptake of ^{67}Cu by ceruloplasmin in vitro. Proc Soc Exp Biol Med 149:681-682

Percival SS, Harris ED (1988) Specific binding of ceruloplasmin to hemin-induced K562 cells. J Trace Elements Exp Med 1:63-70

Percival SS, Harris ED (1989) Ascorbate enhances copper transport from ceruloplasmin into human K562 cells. J Nutr 119:779-784

Scheinberg IH, Morell AB (1957) Exchange of ceruloplasmin copper with ionic Cu^{64} with reference toWilson's disease. J Clin Invest 36:1193-1201

Schmitt RC, Darwish HM, Cheney JC, Ettinger MJ (1983) Copper transport kinetics by isolated rat hepatocytes. Am J Physiol 244:G183-G191

Stevens MD, DiSilvestro RA, Harris ED (1984) Specific receptor for ceruloplasmin in membrane fragments from aorta and heart tissues. Biochemistry 23:262-266

Stockert RJ, Grushoff PS, Morell AG, Bentley GE, O'Brien HA, Scheinberg IH, Sternlieb I (1986) Transport and intracellular distribution of copper in a human hepatoblastoma cell line, HepG2. Hepatology 6:60-64

Terao T, Owen CA Jr (1976) Effects of copper deficiency and copper loading on copper-67 in supernatants of rat organs. Tohuku J Exp Med 120:209-217

Van Campen D, Gross E (1968) Influence of ascorbic acid on the absorption of copper by rats. J Nutr 95:617-622

van den Berg GJ, van den Hamer CJA (1984) Trace metal uptake in liver cells. 1. Influence of albumin in the medium on the uptake of copper by hepatoma cells. J Inorgan Biochem 22:73-84

Weinreich D, Wonderlin WF (1987) Copper activates a unique inward current in molluscan neurones. J Physiol 394:429-443

Weiss KC, Linder MC (1985) Copper transport in rats involving a new plasma protein. Am J Physiol 249:E77-E88

New Analytical Techniques
for Trace Elements

ICP-MS-Application to Biological Samples

MASATOSHI MORITA

National Institute for Environmental Studies, 16-2 Onogawa, Tsukuba, Ibaraki, 305 Japan

Key Words: ICP-MS, ICP-AS, HPLC-ICP-MS, Biological Material

1.Introduction

Inductively coupled plasma mass spectrometry (ICP-MS) is a new method for elemental and isotopic analysis.[1)2)] After several years development, commercial instruments were announced in 1983 by SCIEX and by VG Instruments. At present there are four commercial manufacturers of ICP-MS instrumentation (the others are Yokogawa Electric and Seiko Electronic). The pace of improvements to instrumentation and of application studies are rapidly increasing as is shown in Table 1. Application of ICP-MS to biological samples including clinical samples is following the general trend of total publications and rapidly increasing.

Table 1   Number of Publications on ICP-MS[*]

Year Before	ICP-MS(Total)	Application to Biological Samples
1983	3	0
1983	7	0
1984	1	0
1985	9	0
1986	39	2
1987	79	4
1988	89	10

[*]CAS data base. Papers presented at conferences are not included.

ICP-MS allows the direct, rapid, and convenient recording of mass spectra of trace elements in solution. The most attractive features of the technique are low detection limits and a simultaneous multi-element capability. Its application to the biomedical field seems more promising than did inductively coupled emission spectrometry (ICP-AES) of ten years ago. At that time was some disillusionment among trace element scientists because of insufficient sensitivity. Simple spectra and the availability of isotope ratio information are other attractive features. Coupling with appropriate introduction methods such as laser ablation or high performance liquid chromatography can give further information on the state of elements such as localization or the chemical species present. The present paper discusses the application of the ICP-MS technique to trace element study in human nutrition and toxicology.

2.Elemental Analysis

In the few years after the first appearance of commercial instruments, users experienced difficulty in applying ICP-MS to the quantitative analysis of real samples. The instrument was highly sensitive but was rather unstable and not satisfactory in precision (and accuracy). However, improvements by

manufacturers in this period led to the successful establishment of useful instrumentation that is more sensitive, more stable, more precise and less expensive. Improvements have been achieved by developments in the size of the pin-holes of sampling and skimmer cones, in the shape and material of the cones, in the differential evacuation system, the nebulizer and chamber cooling, quaduru-poles and electronics, measurement to ultra-violet ray and so on.
For example, the detection limit of uranium in water was a few ppt in 1985 and improved to sub ppt in 1988. Effort to improve instrument are still continuing. Current studies on application are designed to minimize the analytical error in terms of precision, accuracy and spectrum interference by molecular ions.

Qualitative and Semi-quantitative Analysis

Most of the commercial ICP-MS instruments are designed to cover 4-250 amu. The full range spectrum is rather simple and easily assigned to individual elements present in solution. The fact that more than 40 elements can be found to be present in the sample is the result of extreme sensitivity of the method. Valuable information about the presence of possibly unexpected trace elements can be obtained in the scanning mode. Fig. 1 shows the mass spectra of human tissues. Kidney cortex ,for example, has a high concentration of cadmium. A small peak for palladium is observed near to cadmium. By using an appropriate calibration curve of signal intensities, either memorized before measurement or by comparison with a spiked internal standard, the recorded signal can be converted to a concentration.

Commercial instruments have an operational mode for semi-quantitative analysis which is designed to show the presence or absence of an element with a reliability within +100% error.

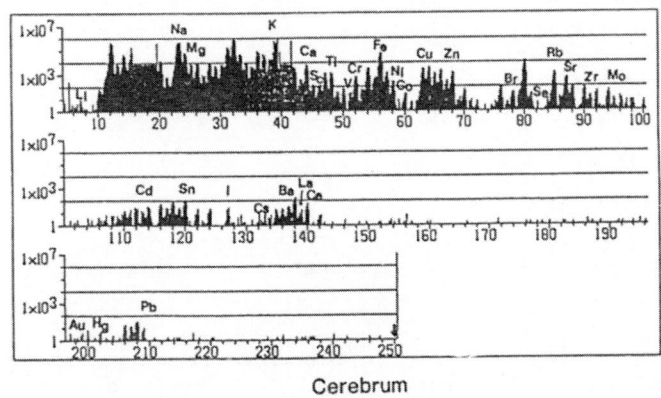

Cerebrum

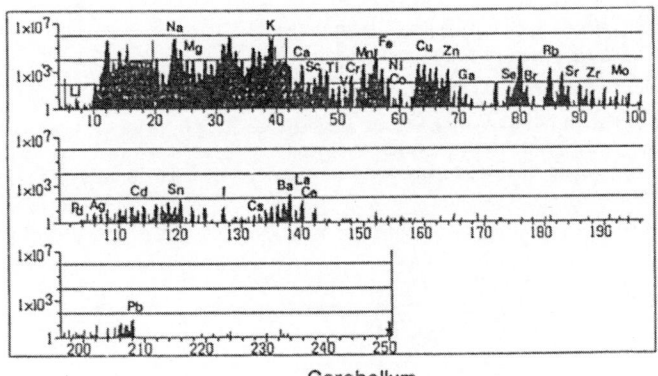

Cerebellum

Fig.1   ICP-MS Spectrum of Human Samples

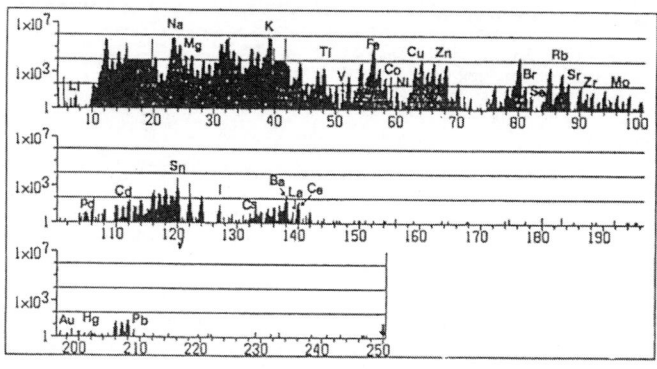

Heart

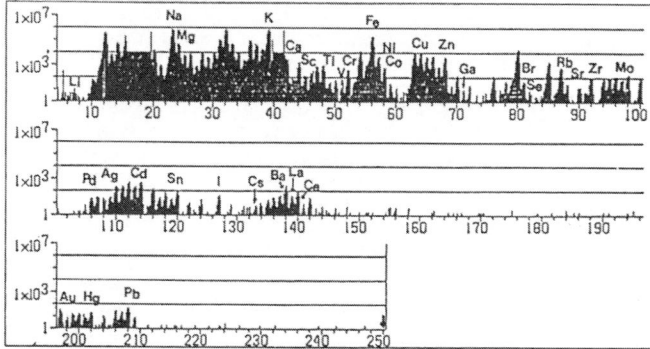

Liver

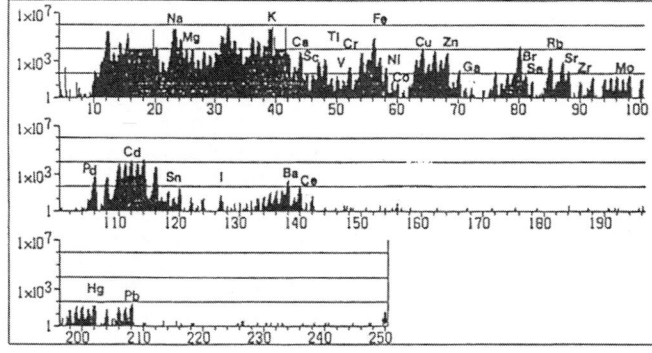

Kidney cortex

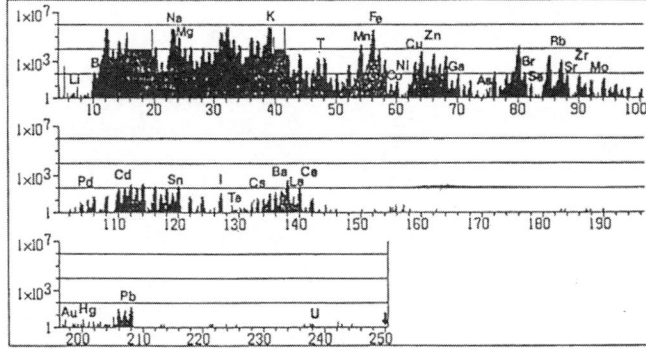

Spleen

Fig.1 (continued)

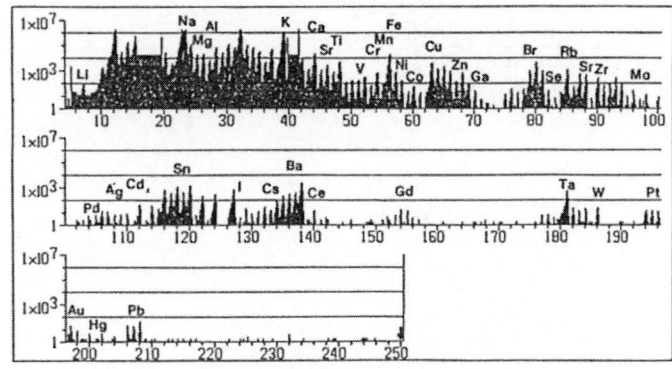

NIES NO.4   Serum

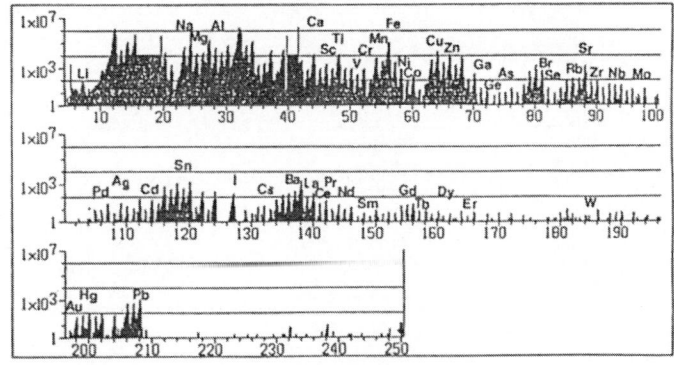

NIES NO.5   Hair

Fig.1 (continued)

Quantitative Analysis

Quantitative analysis is made by either scanning the full mass range or jump-scanning at masses of interest.   Scanning the full mass range requires time (and therefore sample volume).   Jump-scanning at selected mass numbers is often preferred because of the short measuring time and resulting lower signal drift.   Typical mass numbers used for analysis are shown in Table 2 along with isobasic interferences. Table 3 shows the precision of the method by measuring 61 elements in a mixed solution at concentrations of 100 ppb.   The different techniques used for measuring the concentrations of trace elements are subject to different sources of error: ICP-AES has problems with physical interference (nebulization and transport efficiency), chemical interference (matrix effect) and spectral interference(spectral overlapping).   In Graphite furnace atomic absorption spectrometry (GFAA),atomization efficiency, evaporation loss due to different chemical forms, and over- or under-correction of background can be misleading.   Spectra of ICP-MS are simple but considerable caution must be exercised to minimize errors.

Accuracy in real samples may be examined by analyzing certified reference materials. The most frequently analyzed sample by ICP-MS for this purpose may be NBS Bovine Liver Sample (NBS 1577a). MS spectra are shown in Fig. 2. The sample was acid-digested with nitric acid and perchloric acid and subjected to ICP-MS. Analysis was made by external standard and the result is shown in Table 4. When human blood serum was analyzed, the matrix effect by the major salt, sodium chloride, was the problem. If the sample was diluted, the effect is negligible but the concentration of trace elements becomes lower and sometimes falls below the detection limit. If the sample is not diluted, the

431

Table 2 Table of Isobaric and Polyatomic Interferences
Mass Elements Polyatomic Interferent

Mass	Elements	Polyatomic Interferent
23	23Na(100)	$46Ca2^+$
24	24Mg(78.8)	$48Ca2^+$
28	28Si(92.2)	$N2^+$, Co^+
29	30Si(4.7)	$N2H^+$
30	30Si(3.1)	No^+
31	31P (100)	NOH^+
32	32S (95.0)	$O2^+$
33	33S (0.8)	$O2H^+$
34	34S (4.2)	$18O16O^+$
35	35Cl(75.5)	$18O16OH^+$
37	37Cl(24.5)	$36ArH^+$
39	39K (93.1)	$38ArH^+$
40	40Ca(97.0)	$40Ar^+$
41	41K (6.9)	$40ArH^+$
42	42Ca(0.6)	$40Ar D^+$
44	44Ca(2.1)	$Co2^+$
45	45Sc(100)	
46	46Ti(8.0)	$No2^+$
47	47Ti(T3)	$31Po^+$
48	48Ti(74.0)	$32So^+$
49	49Ti(5.5)	
50	Ti50(5.3),50V(0.2),50Cr(4.3)	$36ArN^+$
51	51V (99.8)	$35ClO^+$,$34SOH^+$
52	52Cr(83.8)	$38ArN^+$, $40ArC^+$, 35ClOH
53	53Cr(9.6)	$37ClO^+$
54	54Cr(2.4),54Fe(5.8)	$40ArN^+$, $37ClOH^+$, 40ArMg
55	55Mn(100)	
56	56Fe(91.7)	$40ArO^+$, $40CaO^+$
57	57Fe(2.2)	
58	58Fe(0.3)	
59	59Co(100)	
60	60Ni(26.2)	$44CaO^+$
61	61Ni(1.25)	
62	62Ni(3.66)	
63	63Cn(69.1)	$PO2^+$, $40ArNa^+$
64	64Ni(1.2), 64Zn(48.9)	$32SO2^+$
65	65Cn(30.9)	$H32SO2^+$
66	66Zn(27.8)	$34SO2^+$
67	67Zn(4.1)	
68	68Zn(18.6)	
69	69Ga(60.2)	
70	70Zn(0.6), Ge70(20.6)	$35Cl2^+$
71	71Ga(39.8)	
72	72Ge(27.4)	$35Cl37Cl^+$
73	73Ge(7.7)	
74	74Ge(36.7)	$37Cl2^+$
75	75As(100)	$Ar35Cl^+$
76	76Ge(7.7)	$40Ar36Ar^+$
77	77Se(7.6)	$Ar37Cl^+$
78	78Se(23.5), 78Kr(0.4)	$40Ar38Ar^+$
79	79Br(50.5)	ArK
80	80Se(49.8), 80Kr(2.3)	$40Ar2^+$
81	81Br(49.5)	
82	82Se(9.2), 82Kr(11.6)	$C35Cl2^+$
83	83Kr(11.6)	
84	84Sr(0.6) 84Kr(56.9)	$C35Cl 37Cl^+$
85	85Rb(72.2)	
86	86Sr(9.9) 86Kr(17.4)	$C37Cl2^+$
87	87Rb(27.9)	
88	88Sr(7.0)	
89	89Y (100)	

Table 3 Precision in Repeated Measurement of 61 Elements

(100ppb) by ICP-MS

Element	m/z	x±	Element	m/z	x±
Li	7	102.5±11.6	Sb	121	97.8± 9.0
Be	9	101.4± 9.4	Te	130	98.4± 6.7
Na	23	98.8± 7.8	Cs	133	98.0± 6.4
Mg	24	99.0± 7.0	Ba	138	98.2± 6.6
Al	27	99.2± 6.8	La	139	98.3± 7.7
Sc	45	100.4± 9.2	Ce	140	98.0± 6.8
Ti	47	100.7±11.0	Pr	141	97.8± 6.7
V	51	100.4± 8.7	Nd	146	97.9± 7.7
Cr	53	98.0± 8.2	Sm	149	97.5± 6.7
Mn	55	100.3± 8.5	Eu	153	97.4± 5.8
Fe	57	97.9±12.1	Gd	157	98.8±10.0
Co	59	99.8± 8.8	Tb	159	97.6± 7.0
Ni	60	99.5± 8.5	Dy	163	98.1± 7.5
Cu	63	99.4± 7.5	Ho	165	97.8± 7.3
Zn	66	98.5± 6.4	Er	166	97.4± 6.9
Ga	69	99.2± 8.2	Tm	169	97.2± 6.5
Ge	72	99.5± 8.3	Yb	172	96.5± 5.0
As	75	99.5± 7.4	Lu	175	97.4± 6.6
Se	82	100.9±10.6	Hf	180	96.5± 5.7
Rb	85	99.1± 7.4	Ta	181	96.8± 6.3
Sr	88	99.1± 6.9	W	184	96.5± 5.5
Y	89	98.8± 6.8	Re	185	97.2± 7.4
Zr	90	99.1± 6.8	Pt	195	97.2± 8.3
Nb	93	98.7± 6.5	Au	197	96.8± 8.7
Mo	98	98.6± 7.2	Hg	202	96.0± 8.2
Rh	103	98.8± 6.8	Tl	205	96.7± 7.8
Pd	105	98.2± 6.9	Pb	208	96.7± 7.2
Ag	107	96.7± 9.6	Bi	209	97.2± 7.4
Cd	114	97.6± 5.4	Th	232	96.7± 7.3
In	115	98.1± 6.8	U	238	96.6± 6.9
Sn	118	98.2± 8.5			

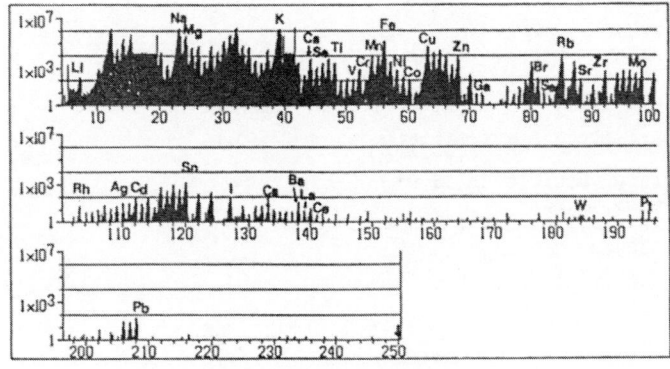

NBS 1577a Bovine Liver

Fig. 2 ICP-MS Spectrum of NBS 1577a Bovine Liver

matrix effect usually decreases the sensitivity (in special cases it may increase the sensitivity). As shown in Table 5, when blood serum is diluted by a factor of four after digestion, an apparent decrease of sensitivity is seen for copper and zinc. This is improved by diluting the digested solution by a factor of ten. Determination of trace levels of iron is not simple. The major iron isotope, iron-56, is interfered with by the molecular ion of Argon-oxide, and thus iron-57 is used instead, but its abundance is low and hence sensitivity is lost.

In Table 6, determined concentration of trace elements are compared with those of other analytical techniques. The values of Cd,Sr,Mo and Cs are in fairly good accordance with those determined by surface ionization mass spectrometry.

Table 4   Analytical Result of NBS 1577a

Element	m/e	Certified(ug/g)	Found(ug/g) (1)	(2)
Na	23	2430+130	2530+180	-----
Mg	24	600+15	953+53	603+45
Al	27	2	-----	3.3+0.3
V	51	0.099+0.008	-----	-----
Mn	55	9.9+0.8	10.8+1.2	12.2
Fe	57	194+20	144+20	186+13
Co	59	0.21+0.05	-----	0.28+0.01
Cu	63	158+7	175+15	170+4
Zn	66	123+8	143+19	135+2
As	75	0.047+0.006	0.078+0.011	(0.38)
Se	82	0.71+0.07	0.68+0.06	(1.05)
Rd	85	12.5+0.1	14.4+2.1	13.1+0.4
Sr	88	0.138+0.003	0.130+0.017	0.145
Mo	98	3.5+0.5	4.2+0.4	3.9
Ag	107	0.04+0.01	0.06+0.01	0.055
Cd	114	0.44+0.06	0.47+0.05	0.52+0.02
Sb	121	(0.003)	-----	(0.009)
Hg	202	0.004+0.002	-----	-----
Pb	208	0.135+0.015	-----	0.17
Th	232	(0.003)	-----	-----
U	238	0.00071+0.0003	-----	0.001

Table 5   Trace Elements in Serum Determined by ICP-MS (ppm)

	x40	x4	x4	Ref.V.
Mg	15.6	6.6	8.2	19.5
Fe	1.8	1.96	2.6	1.07
Cu	0.97	0.51	0.68	1.04
Zn	0.94	0.62	0.82	0.91

Table 6   Trace Elements in Serum Determined by ICP-MS (ppb)

	ICP-MS	Ref.
Ag	1.6	1.32(IDMS)
Cd	1.6	1.74(IDMS)
Tl	0.5	0.38(IDMS)
Sr	50	50 (ICP-AES)
Mo	4	3.6 (AA)
Rb	160	230 (INAA)
Cs	1	0.9 (INAA)

Analytical results for a human scalp hair sample (prepared as a certified reference material in our laboratory) is shown in Table 7. Analytical values are compared for ICP-MS and ICP-AES. 1g sample was acid digested and made up to 20 ml solution for ICP emission spectrometry. A quarter of the solution was further diluted and made up to 20 ml for ICP-MS. The data show good agreement for most elements except chromium. The chromium concentration in blood serum is supposed to be quite low. However when analyzed by means of the major Cr isotope of 52 (natural abundance 83.8%), ICP-MS gave the level of chromium in serum at levels of several ppb, which is unacceptably high. The mass number 52 may overlap with polyatomic ions such as $^{40}ArC^+$, $^{38}ArN^+$ or $^{35}ClOH^+$.

Table 7 Trace Elements in Hair Determined by ICP-MS and ICP-AES

(NIES No.5 HAIR:ppm)

	ICP-MS	ICP-AES		ICP-MS	ICP-AES
Na	25	26	Ni	1.6	1.8
Mg	154	162	Cu	16.2	16.3
Al	186	181	Zn	159	164
Cr	3.7	1.4	Sr	2.0	2.1
Mn	4.3	4.7	Cd	0.19	0.20
Fe	214	214	Pb	6.8	6.8

Interference by polyatomic ions may be removed when a high resolution mass spectrometer is used. The recently produced "VG Plasma Trace" may solve the problem if it has enough sensitivity. Our laboratory is also now building a high resolution mass spectrometer for the ICP ion source. In preliminary experiments, the peaks of samarium and barium oxide are completely resolved at a resolution of 10,000. ICP-MS can be used for routine analysis.

A typical routine analysis sequence in our laboratory is shown in Fig.3. Samples are sufficiently diluted and subjected to analysis. Firstly a high concentration standard and a zero standard solution are introduced to the plasma and then the sample solution follows.
After the analysis of five samples, standardization is again carried out and the analyses are repeated. By repeating these sequences, it is possible to analyze ten samples per hour or 80 samples in one day. Calculation and data-out put is carried out overnight.

Isotope Ratio Determination

ICP-MS provides a capability of the determination of isotopes ratios directly and rapidly for elements in solution. Sample throughout is about one sample every one to ten minutes which is approximately ten times as rapid as in surface ionization mass spectrometry. Precision for isotope ratios is in the range 0.1-3%, typically 0.2-1%, depending on measuring time and sample concentration. The sensitivity of the method can be lowered to the 10ng/ml range if 10 ml samples are available.

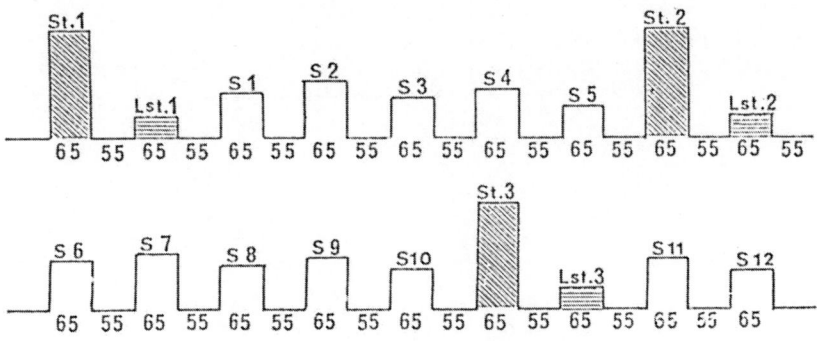

Fig. 3 Analytical Sequence

Precision on isotope ratios by ICP-MS is apparently inferior to surface ionization mass spectrometry which gives 0.01% precision or better. The uncertainty in isotope ratio determinations arises from plasma flicker, nebulizer noise, interface noise(?), detector noise and counting statistics. Two different measurement schemes have been used to overcome this. One is to use a multichannel detector to accumulate an averaged spectrum at high scanning speed.(ref.) The other is to measure each peak of interest with a comparatively long dwell time with subsequent averaging of the data.(ref.) Precision obtained with both methods is similar. Our attempt, with the cooperation of VG-isotope Co., to use variable scanning speeds was not markedly successful in improving precision.

It has been reported that mass discrimination effects of 1-2% can be seen. In our experiments using a Yokogawa PMS100 instrument, the results are shown in Table 8. The effects can be corrected by analysis of certified isotopic standards.
This rapid determination of isotope ratios in solution greatly facilitate the isotope dilution analysis and metabolism studies using a stable isotope as a tracer.

Table 8 Mass Discrimination

Pb-208	1.0141
Pb-207	1.0055
Pb-206	1
Pb-204	0.9919

Isotope Ratio-tracer

Studies of trace element bioavailability have been limited because such studies require the application of isotopic techniques. The accepted isotopic methods involve the use of radio-isotopes which cause internal exposure to radiation and,in some case, pose ethical problems. Much effort has been expended in developing suitable tracer methods based on non-radioactive stable isotopes. Determination of stable isotope ratios was carried out using techniques such as neutron activation analysis or thermal ionization mass spectrometry, where measurement was both complex and time consuming and sample preparation often difficult. ICP-MS is quite suitable for this problem, not only because measurement is easy and rapid, but also because the most critical measurement requirement, viz sufficient precision is achieved.

Precision in the determination of isotope ratio at quite low concentrations, can be met by this method. Application of ICP-MS to the bioavailability of essential elements has been reported by several groups.

Isotope Dilution Analysis

Isotope dilution analysis is a method which uses an artificially enriched isotope as an internal standard. Since the isotope ratio is precisely determined by ICP-MS, the concentration of elements in the original matrix is determined precisely when an appropriate amount of isotope is spiked. Our analytical results for lead in certified reference materials is given in Table 9.

Table 9 Lead Concentration in Certified Reference Materials
Determined by Isotope Dilution Mass Spectrometry Using ICP-MS

Samples	Found Value (ppm)	Certified Value (ppm)
NIES No.1 Pepper bush	5.5 ± 0.1	5.5 ± 0.8
NIES No.5 Hair	6.1 ± 0.1	(6.0)
NIES No.6 Mussel	0.91 ± 0.02	0.91 ± 0.04
NIES No.7 Tea Leaves	0.78	0.80 ± 0.03
NIES No.9 Sargasso	1.38	1.35 ± 0.04
NBS 1572 Citrus Leaves	13.7	13.3 ± 0.24
NBS 1568 Pine Needle	10.9 ± 0.1	10.8 ± 0.5
NBS 1571 Orchard Leaves	43.4 ± 0.5	44.5 ± 3

Flow Injection Analysis

Flow injection analysis seems well matched to ICP-MS. ICP-MS is rather weak in analyzing samples with high salt concentrations but the small volume of samples used for flow injection analysis can be acceptable. Signal intensities of trace elements are decreased less than 10% when the sodium chloride concentration is 0.1%. Matrix matching may give better result for real samples analysis.

HPLC-ICP-MS

By combining HPLC to ICP-MS, it is possible to get information on the chemical species of the element. Fig.4 is an example of arsenic species that were determined by HPLC-ICP-MS. The sample is a marine alga that Japanese eat. Arseno-sugars appear as major peaks. When we analyzed human urine, we could detect dimethylarsonic acid and arsenobetaine.

Laser Ablation ICP-MS

Samples for ICP-MS are usually in solution, but by using a laser ablation system, the analysis of solid samples becomes possible. Sensitivities of about 1 fg/sec. can be achieved. Using a Yag laser, the focused light produces a hole of 50-100 microns diameter, and hence the spatial resolution is approximately 100 microns.

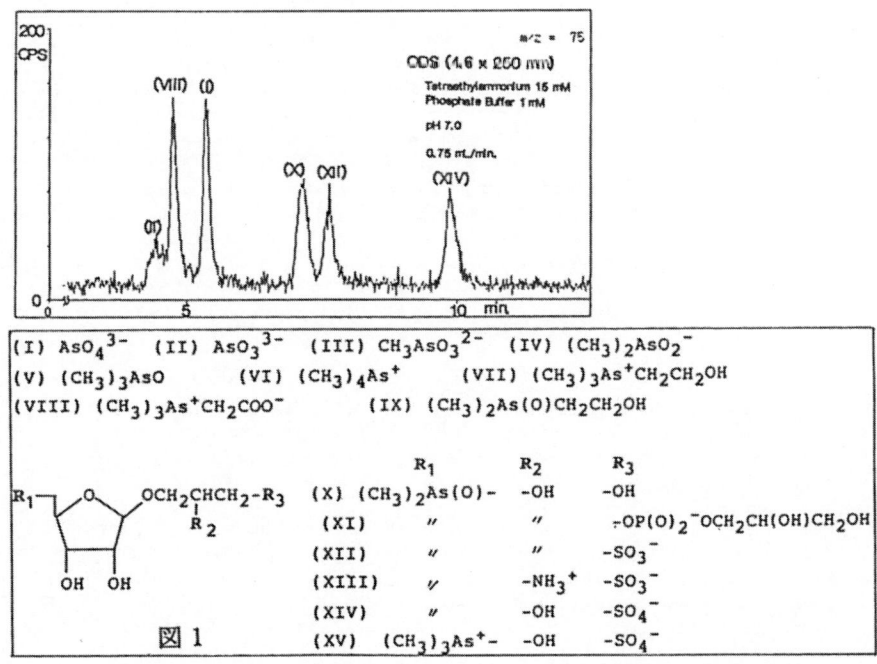

Fig. 4   Arsenic Species Determined by HPLC-ICP-MS

References
1) A.L.Grag:Analyst(London),100,289(1975)
2) R.S.Houk,V.A.Fassel,G.D.Flesch,H.J.Svec,
   A.L.Gray,C.E.Taylor :Anal. Chem.,52,2283(1980)

Role of Reference Materials for Validation of Analytical Data in Trace Element Analysis of Biological Materials

KENSAKU OKAMOTO

National Institute for Environmental Studies, 16-2 Onogawa, Tsukuba, Ibaraki, 305 Japan

ABSTRACT

Analytical quality assurance is essential to obtain meaningful results in trace element analysis of biological materials. The incorporation of appropriate reference materials in the scheme of analysis is the most convenient, cost-effective mechanism by which to assess and maintain analytical data quality. The National Institute for Environmental Studies (NIES) has issued a variety of biological and environmental reference materials to serve the needs of scientists engaged in trace element analysis of these types of samples. In this paper the preparation and trace element analysis of NIES "human" reference materials - Freeze-dried Human Serum and Human Hair - and NIES food reference material - Rice Flour-Unpolished - are described.

KEY WORDS: Certified reference material; Human serum; Human hair; Rice flour.

INTRODUCTION

Analytical quality assurance is essential to obtain meaningful results in trace element analysis of biological materials. For this purpose the use of appropriate reference materials is the most practical way to maintain accuracy and precision of analytical data. Reference material (RM) is " a material or substance one or more properties of which are sufficiently well established to be used for the calibration of an apparatus, the assessment of a measurement method, or for assigning values to materials" and certified reference material (CRM) is " a reference material one or more of whose property values are certified by a technically valid procedure, accompanied by or traceable to a certificate or other documentation which is issued by a certifying body" (ISO GUIDE 35-1985). In areas such as environmental or biological analysis where "real world" specimens may be exceedingly complex and variable, a wide variety of CRMs is urgently required to give assurance to analytical results. Since many modern, rapid, instrumental methods are subject to matrix interference effects, the careful analyst must choose a CRM that matches as closely as possible in, for example, matrix type and concentration of the element of interest, the real samples that are to be analyzed.

The National Institute for Environmental Studies (NIES), Environment Agency of Japan, has produced a variety of biological and environmental RMs to serve the needs of scientists engaged in trace element analysis of such matrices. Table 1 shows the present status of the NIES CRM program. The currently available CRMs from NIES include Pepperbush, Pond Sediment, Chlorella, Human Hair, Tea Leaves, Vehicle Exhaust Particulates, Sargasso and Rice Flour-Unpolished. NIES No.4 Freeze-dried Human Serum still remains as a candidate RM for the reasons described below.

In the CRM program special attention has been paid to prevent contamination during preparation so that the samples are representative of the natural material. Since CRMs must be homogeneous and stable, procedures such as drying, grinding, sieving, mixing, sterilizing, etc, are required for sample preparation. As already mentioned, the matrix of a CRM should match as close as possible to that of a real specimen being analyzed, so contamination from preparation apparatus and denaturation of the sample have to be reduced to an insignificant level. Assessment of the homogeneity of the prepared material

Table 1 Present status of NIES certified reference material program

	Material	Elements certified
No.1	Pepperbush	As,Ba,Ca,Cd,Co,Cu,Fe,K,Mg,Mn,Na,Ni,Pb,Rb,Sr,Zn
No.2	Pond Sediment	Al,As,Ca,Cd,Co,Cr,Cu,Fe,K,Na,Ni,Pb,Zn
No.3	Chlorella	Ca,Co,Cu,Fe,K,Mg,Mn,Sr,Zn
No.4	Freeze-dried Human Serum	candidate RM
No.5	Human Hair	Ca,Cd,Cr,Cu,Fe,Hg,K.Mg,Mn,Na,Ni,Sr,Zn
No.6	Mussel	Ag,As,Ca,Cd,Cr,Cu,Fe,K,Mg,Mn.Na,Ni,Zn(exhausted)
No.7	Tea Leaves	Al,Ca,Cd,Cu,K,Mg,Mn,Na,Ni,Pb,Zn
No.8	Vehicle Exhaust Particulates	Al,As,Ca,Cd,Co,Cr,Cu,K,Mg,Na,Ni,Pb,Sb,Sr,V, Zn
No.9	Sargasso	Ag,As,Ca,Cd,Co,Cu,Fe,K,Mg,Mn,Na,Pb,Rb,Sr,V,Zn
No.10	Rice Flour-Unpolished	Ca,Cd,Cu,Fe,K,Mg,Mn,Mo,Na,Ni,P,Rb,Zn
No.11	Fish Tissue	in certification for organotin compounds
No.12	Marine Sediment	in preparation

is essential before collaborative study can proceed and this has been carried out at NIES using atomic absorption and inductively coupled plasma emission analyses after acid dissolution of the samples. When homogeneity is guaranteed, collaborative studies on the material, using various analytical techniques, is organized and performed by 20-30 qualified participating laboratories. Thereafter, analytical data provided by cooperating laboratories are subject to technical consideration and statistical treatment to lead to certification.

For all of the NIES CRMs, certified values are provided for elements determined by at least three independent analytical techniques and whose values agreed within an acceptable range. Reference values for the materials are included in the certificates. In this paper NIES "human" RMs - NIES No.4 Freeze-dried Human Serum and NIES No.5 Human Hair - are described. NIES No.10 Rice Flour-Unpolished is also included as a new CRM of environmental and nutritional importance.

NIES FREEZE-DRIED HUMAN SERUM REFERENCE MATERIAL

It is highly desirable to establish normal ranges for concentrations of elements in human tissues and body fluids so that the nutritional and toxicological status of the individual can be better assessed. However,in practice this is difficult, and such data for trace elements often disagree. It is clear that there is a great demand for human serum RMs but availability of such RMs has been limited mainly because of difficulties in trace element determination in the complex matrix and in a large-scale preparation of contamination-free human serum.

NIES prepared Freeze-dried Human Serum RM for trace element analysis in 1983 but this RM still remains as a candidate RM, due to the following reasons: (1) difficulty in solubilizing the freeze-dried serum completely, (2) disagreement of the analytical values for trace elements, and (3) deviation from normal values for certain elements. It is preferable to use human serum RM of liquid form for daily analysis but from the standpoints of long-term stability and overseas distribution freeze-drying was applied to the serum. Prevention against possible diseases is also a very important consideration in preparing human serum RM and the NIES human serum was sterilized by Co-60 radiation (at 2 mega rad) for safe operation. These freeze-drying and sterilization procedures possibly caused the denaturation of serum proteins and, on reconstitution, it was not possible to obtain a clear serum solution. At present, total digestion of the freeze-dried serum in a vial with a mixture of mineral acids is required, or alternatively, centrifugation of the well-mixed serum and use of the supernatant for analysis.

Table 2 lists analytical results on the NIES freeze-dried human serum RM for three groups of elements at different concentrations: between 10 and 200 µg/ml for K, Ca, Mg; around 1 µg/ml for Fe, Cu, and Zn; less than 10 ng/ml for Mn, Cr and Co. Good agreement of the analytical values was achieved for each of the elements whose concentrations were greater than 1 µg/ml. For the elements lower than 10 ng/ml, however, the analytical values ranged over one order of

Table 2. Analytical results for NIES Freeze-dried Human Serum RM

	K	Ca	Mg	Fe	Cu	Zn	Mn	Co	Cr
	(µg/ml)			(µg/ml)			(ng/ml)		
169 FES	75.1 AAS	19.3 AAS	1.06 AAS	0.90 AAS	0.89 AAS	3.83	2.66	2.5
169 CPAA	76.6 CPAA	20.7 AAS	1.13 AAS	1.09 AAS	0.83 AAS	AAS	AAS	AAS
183 AAS	77.7 SP	19.9 AAS	1.08 AAS	1.02 AAS	0.83 AAS			
178 FES	88.2 AAS	19.6 AAS	1.1 AAS	1.02 AAS	0.89 INAA	6	1.0	4.5
178 AAS	75.8 ICP	18.0 ICP	1.0 INAA	1.06 AAS	0.93 AAS	INAA	INAA	INAA
179 AAS	86.8 AAS	19.6 AAS	1.10 ICP	0.98 ICP	0.88 AAS			
167 ICP	77.9 AAS	19.0 ICP	1.06 INAA	1.03 AAS	0.89 AAS	0.52	0.7	3
172 INAA	75.3 ICP	20.5 AAS	1.01 ICP	1.11 ICP	0.81 ICP	INAA	INAA	INAA
168 ICP	79 ICP	19.0 ICP	1.06 ICP	1.04 ICP	0.93 AAS	5 ICP		
172 FES	76 ICP	20.4 ICP	1.06 AAS	1.03 AAS	0.94 INAA		7.7	0.68
174 AAS	81.9 AAS			1.04 ICP	0.94 ICP		INAA	AAS
173 ICP	81.8 FES			0.99 ICP	0.92 AAS	1.4		
	79.5 ICP				0.91 ICP	AAS		

AAS:atomic absorption spectrometry, FES:flame emission spectrometry, ICP: inductively coupled plasma emission spectrometry, INAA:instrumental neutron activation analysis, CPAA:carged particle activation analysis, SP:spectro-photometry

Analytical value from each participating laboratory

magnitude. It is very difficult to accurately determine trace elements at ng/ml levels in human serum, thus development of analytical methods for these trace elements has been required.

NIES HUMAN HAIR REFERENCE MATERIAL

Trace element analysis of human hair has been carried out in laboratories throughout the world for the purpose of assessing the nutritional and toxicological status of the individual. However, the reliability of data provided by laboratories engaged in hair analysis is open to question, because there is a lack not only of reference methods of analysis but also of appropriate CRMs. An important requirement in such work is the application of suitable analytical quality control, for which purpose the availability of appropriate CRM is of great importance. The International Atomic Energy Agency prepared human hair reference material (HH-1) for interlaboratory study for trace and other elements (M'Baku, 1982) but, unfortunately, this reference material has been exhausted. Subsequently, the deveopment of a human hair RM was undertaken at NIES (Okamoto,et al., 1985).

In the preparation of a large quantity of homogeneous hair material, a most crucial point was the need to pulverize the material efficiently with insignificant contamination from grinding vessels. Therefore, we tested the following grinding apparatuses for grinding efficiency and contamination:(1) mortar grinder, (2) rolling ball mill, (3) planetary-type ball mill and (4) cryogenic grinder. Grinding with the mechanical mortar and rolling mill was not considered further, because of the low yield of powdered hair. The planetary-type ball mill was quite efficient. We found the agate ball mill to be most suitable for grinding hair samples. Use of the agate ball mill should diminish contamination to insignificant. Although it was unavoidable that the hair powder contained some agate debris because of the strenuous operating conditions, contamination with heavy metals in the agate debris is not considered serious for most analytical purposes. Cryogenic grinding showed the highest grinding capability but the disc mill used for cryogenic grinding was made of hardened steel, which made it inappropriate for trace element analysis. Considering all of these factors, we chose to use a planetary-type agate ball mill to pulverize the human hair RM on a large-scale.

Human scalp hair (black, from Japanese males) was collected from barber shops in Tsukuba and Tokyo. After removal of visible contaminants, the hair was washed by ultrasonic cleaning for 20 min in a 0.3 % solution of non-ionic surface-active agent. The detergent was removed by copious rinsing with

distilled water. The cleaned hair was dried in an air-oven at 80°C overnight, and about 2 kg was used in preparing this RM. About 60 g of the hair was ground initially in the agate ball mill (250 ml X 4) and this procedure was repeated for about 2 kg of the cleaned hair. The pulverized hair was sieved through a polyethylene net to remove undestroyed fibrous hair. The sieved powder was mixed in a V-blender for 2 h and packaged into acid-washed glass vials (1100 vials, 2 g each). The human hair RM in the packaged form was sterilized by Co-60 radiation at 2 mega rad at the Japan Atomic Energy Research Institute, Takasaki, Japan.

The use of scalp hair as a biopsy material has several advantages: (a)sampling being relatively noninvasive, samples are easily obtained and stored, (b)trace elements in hair are more concentrated than in blood, thus facilitating analysis, and (c) hair offers a good way of discerning long-term variations in trace element concentrations. However, there clearly is considerable variation according to age, sex, race and geographic location. Other factors also significantly affect trace element content; the extent and amount of exposure of hair to exogeneous materials, particularly shampoos, dyes and medications. Moreover, there may be variation across the scalp as a result of the choice of washing procedures. We believe that recent development of analytical techniques and improvement of sample-pretreatment procedures will soon make it possible to establish normal values for trace elements in human hair.

Table 3 compares our data on the elemental composition of NIES Human Hair CRM with some other published data on normal values for hair. We conclude that the elemental content of Ca, Zn, P, K, Na, Cu, Pb, Hg, Ni, Se, Cd and Co in NIES Human Hair can be considered typical for the Japanese male population and for U. S. males (except for Hg). Al, Fe, Ti, Mn (and probably Mg) concentrations in the CRM are unfortunately much higher than the actual values for hair, owing to contamination from the agate ball mill used to pulverize the material. This material is the only CRM of human hair currently available and, though its use for certain elements is limited, the NIES Human Hair CRM will be of practical use for analytical quality control, development of analytical methods, calibration of instruments, etc, in laboratories engaged in clinical and environmental analysis.

Table 3. Elemental composition of NIES Human Hair and reported normal concentrations (µg/g) (Okamoto, et al, 1985)

	NIES Human Hair	Japanese male	Japanese both sexes	U. S. male		NIES Human Hair	Japanese male	Japanese both sexes	U.S. male
Ca	728	482	1150	360-850	**Pb**	6	4.5		4.1
Al	240	12.3	12.8	4.5-5.5	**Mn**	5.2	0.8	1.1	0.14-0.45
Fe	225	17.0	35	26-32	**Hg**	4.4		4.2	1.7-1.9
Mg	208	84.2	119	53-135	**Sr**	2.3	2.0		
Zn	169	125	183	150-190	**Ni**	1.8			2.8-3.2
P	165	132			**Cr**	1.4		0.82	1.3-1.7
K	34	35	18.7		**Se**	1.4		1.18	0.58-0.76
Na	26	43.1	23		**Cd**	0.20	0.15	1.6	0.47
Ti	22			2.4-6.1	**Co**	0.10		0.08	0.03-0.045
Cu	16.3	14.2	13.1	15-17	**Sb**	0.07			0.073-0.2

NIES RICE FLOUR-UNPOLISHED REFERENCE MATERIAL

Food analysis is of great importance related to human health. There are several cereal RMs for use in trace element determination: NIST SRM 1567 Wheat Flour, NIST SRM 1568 Rice Flour, NIST 8412 Corn Stalk, NIST RM 8413 Corn Kernel, IAEA V-8 Rye Flour. In Japan, environmental pollution by Cd has become one of the major health issues. For example, "Itai-itai"(ouch-ouch) disease was caused mainly by consumption of rice and drinking water which was contaminated with Cd derived from a river basin. On the basis of investigation

on Cd in rice, which is the main Cd source for the Japanese, the following legal critical values has been set for Cd in rice.
(1)The surveillance level of Cd in unpolished rice is 0.4 µg/g.
(2)The maximum permissible level of Cd in unpolished rice is 1.0 µg/g.
In Japan,sale on the market of unpolished rice containing more than 1.0 µg/g of Cd has been prohibited by the Food Sanitation Law. At present, unpolished rice containing Cd between 0.4 and 1.0 µg/g is also prohibited to be sold on the market by an administrative measure of the Government. It has therefore been required to accurately determine a range of concentrations of Cd and other heavy metals in unpolished rice.

NIES Rice Flour-Unpolished RM consists of three samples, each containing different levels of Cd(low level, medium level, high level). The materials were each prepared in an identical manner from unpolished rice collected from three different locations in Japan. The unpolished rice containing an elevated level of Cd was produced in a Cd-contaminated paddy field. A batch (about 1.5 kg) of unpolished rice was pulverized with a rotor speed mill at 20,000 rpm, sieved to pass a 0.5 mm screen and oven-dried at 80°C for 4 h to reduce the moisture content to approximately 5 %. After repeating this procedure for the remaining batches, the unpolished rice flour (60 kg) was mixed in a V-blender for 2 h. The homogenized powder was packaged into 1,000 acid-washed glass bottles (60 g, each).

Table 4. Certified and reference values for NIES Rice Flour-Unpolished RM

	Certified values		
	Low level(Cd)	Medium level(Cd)	High level(Cd)
Minor constituents		Wt. %	
P	0.340±0.007	0.315±0.006	0.335±0.008
K	0.280±0.008	0.245±0.010	0.275±0.010
Mg	0.134±0.008	0.131±0.006	0.125±0.008
Trace constituents		µg/g	
Ca	93 ± 3	78 ± 3	95 ± 2
Mn	34.7±1.8	31.5±1.6	40.1±2.0
Zn	25.2±0.8	22.3±0.9	23.1±0.8
Fe	12.7±0.7	13.4±0.9	11.4±0.8
Na	10.2±0.3	17.8±0.4	14.0±0.4
Rb	4.5±0.3	3.3±0.3	5.7±0.3
Cu	3.5±0.3	3.3±0.2	4.1±0.3
Mo	0.35±0.05	0.42±0.05	1.6±0.1
Ni	0.19±0.03	0.39±0.04	0.30±0.03
Cd	0.023±0.003	0.32±0.02	1.82±0.06
	Reference values (µg/g)		
	Low level(Cd)	Medium level(Cd)	High level(Cd)
Cl	260	310	230
Al	3	2	1.5
Br	0.3	0.5	0.5
Sr	0.3	0.3	0.2
As	0.17	0.11	0.15
Cr	0.07	0.22	0.08
Se	0.06	0.02	0.07
Co	0.02	0.02	0.007
Hg	0.004	0.003	0.005

On a dry weight basis.

Table 4 shows the certified and reference values for NIES Rice Flour-Unpolish-ed RM. Certified values are provided for Ca, Cd, Cu, Fe, K, Mg, Mn, Mo, Na, Ni, P, Rb and Zn, while reference values are reported for Al, As, Br, Cl, Co, Cr, Hg, Se and Sr. The Cd content (0.023 µg/g) in the low level sample is close to those in NIST cereal SRMs (0.029 for Rice Flour and 0.032 µg/g for Wheat Flour) probably at background level. The medium level sample contains 0.32 µg/g of Cd, which is just below the surveillance level of Cd (0.4 µg/g) in unpolished rice. The Cd content in the high level sample is about the double of the maximum permissible level of Cd(1.0 µg/g). The elemental composition of NIES Rice Flour-Unpolished CRM is essentially the same as the average concentration in Japanese unpolished rice and also similar to those of NIST Wheat Flour and Rice Flour SRMs. The certified and reference values are provided for most "difficult" trace elements, which are of nutritional and environmental signi-ficance. NIES Rice Flour-Unpolished CRM will be of practical use for the vali-dation of analytical methods and for the calibration of analytical instruments, particularly for accurate determination of a wide range of Cd levels.

REFERENCES

M'Baku, Parr RM (1982) Interlaboratory study of trace and other elements in the IAEA powdered human hair reference material. J. Radioanal Chem 69: 171-180

Okamoto K, Morita M, Quan H, Uehiro T, Fuwa K (1985) Preparation and certifi-cation of human hair powder reference material. Clin Chem 31: 1592-1597

A Guideline for Application of Neutron Activation Analysis to Biological and Medical Samples

Shoji Hirai

Atomic Energy Research Laboratory, Musashi Institute of Technology, 971 Ozenji, Aso-ku, Kawasaki, Kanagawa, 215 Japan

ABSTRACTS

In the biological and medical field, a number of trace elements are known to play an important role in relation to health and disease, or are suspected of doing so. Many researchers have used neutron activation analysis of these samples to study the behavior of trace elements. And neutron activation analysis has become today a versatile, high sensitivity and accuracy analytical method in not only biology and medicine but also all fiels of science and technology. One of the reasons is that multi-elements in a little amount of sample can be determined simultaneously and nondestructively with high sensitivity and accuracy by means of this technique. Therefore, neutron activation analysis has born a part in decision of concentrations of many trace elements in most reference standard materials (prepared by NBS and NIES et al.) for a quality assuarance of analytical deta,too.

KEY WORDS

Neutron activation analysis, trace multi-elements, selenium, platinum.

1. INTRODUCTION

As the advantage of an neutron activation analysis shown in Table 1 has mainly three chracteristics, the analytical method is expensively used in many fields, including biological and medical science.   The chracteristics of the neutron activation analysis are high sensitivity, good accuracy, and nondestructive analytical method.

Table 1   Advantage of neutron activation analysis

---

① high sensitivity : determination in μg or ng range
(rich reactivity of neutron capture reaction)

② good accuracy : decision of certified values in standard reference
materials
(uniform irradiation)

③ nondestructive analysis : determination of simultaneous multielements
(high performance γ-spectrometry with computer)

Many atoms easily react with reactor neutron, and have radioactivity. Therefore, elements of amounts of μg or ng level are determined by this method measuring radioactivity. And, as the reactor neutron penetrate well into the whole sample in an irradiation using the nuclear reactor, the sample can be uniformly irradiated. As a result, the analytical values obtained represent the values of the whole sample analyzed. Also, since this nuclear reaction occur regardless of chemical forms of elements, the analytical values can be obtained with beter accuracy without a chemical interference beside a different in an chemical analysis. Therefore, the neutron activation analysis method can be expensively used for determination of analytical values of reference standard materials prepared by NBS (National Bureau of Standards) and NIES (National Institute for Environmental Studies of Japan). Moreover, as measuring radioactivity of the irradiated samples by use of a high resolution radiation detector and a high performance multichannel pulse hight analyser and computer, many elements can be simultaneously determined at once measuring time. Particularly, the neutron activation analysis which is nondestructive, and not procedure chemical separation is called as an instrumental neutron activation analysis. Today, the instrumental neutron activation analysis is one of the leading mainstreams of the neutron activation analysis. Then, I show a few example of the instrumental neutron activation analysis to biological samples, making a good use of these characteristics, and indicate a guideline for application of the neutron activation analysis to biological and medical samples.

2. DETERMINATION OF MULTI-ELEMENTS

The first example[1] is multi-elements of the determination by the instrumental neutron activation analysis. The analysed sample is the standard reference material, the Sagasso prepared by NIES. The about 100mg samples were irradiated at the Mussashi Institute of Technology Research Reactor, and after

Table 2 Irradiation and counting conditions for the analysis of the Sagasso

Counting method *	Thermal neutron flux(n·cm^{-2}·s^{-1})	Irradiation time	Cooling time	Counting time	Determined elements
	[Irradiation method : without Cd filter]				
G	1.5 x 10^{12}	30 sec	1-3 min	5 min	Mg, Al, S, Cl, V, Mn, I
	3.2 x 10^{12}	5 hr	4-20 d	2-15 hr	Na, K, As, Br, Sm, Au, U
			21-80 d	8-40 hr	Sc, Cr, Fe, Co, Zn, Rb, Sr, Ag, Sb, Te, Cs, Ba, Ce, Eu, Tb, Yb
A	1.5 x 10^{12}	10 sec	1-3 min	5 min	Mg, Al, Cl, Ca, V, Mn
	3.2 x 10^{12}	5 hr	4-20 d	2-15 hr	Na, K, As, Br, Mo, La, Sm, Au, U
			21-80 d	8-40 hr	Sc, Cr, Fe, Co, Zn, Rb, Sr, Zr, Ag, Sb, Te, Cs, Ba, Ce, Eu, Tb, Th
C	3.2 x 10^{12}	5 hr	44-75 d	9-50 hr	Se, Ba, Hf
	[Irradiation method : with Cd filter]				
G	1.8 x 10^{12} (Cd ratio = 6)	5 hr	5-12 d	1-15 hr	Na, As, Br, Sm, U
			14-71 d	8-50 hr	Sc, Fe, Co, Zn, Rb, Sr, Ag, Sb, Cs, Ba, Tb
A	1.8 x 10^{12} (Cd ratio = 6)	5 hr	50-73 d	15-50 hr	Sc, Fe, Co, Ni, Rb, Sr, Sb, Tb

* G;gamma-ray spectrometry, A;anticoincident counting, C;coincident counting

Table 3 Concentration of elements in the Sargasso[1] (μg/g: ppm)

E*	γ-spectrometry without filter	Anticoincidence without filter	γ-spectrometry with Cd filter	Anticoincidence with Cd filter	NIES[a]
Na	15000±1000	17000±2000	12000±1000		17000±800
Mg	6900±600	6100±500			6500±300
Al	210±20	220±30			(215)
S	53000±6000				(1200)
Cl	43000±2000	43000±4000			(5100)
K	59000±3000	60000±5000			61000±2000
Ca	12000±700	11000±1000			13400±500
Sc	0.090±0.004	0.100±0.003	0.082±0.008	0.078±0.012	(0.09)
Ti	<75				(9)
V	1.0±0.2	1.0±0.2			1.0±0.1
Cr	0.19±0.03	0.18±0.02			(0.2)
Mn	20±1	21±4			21.2±1.0
Fe	180±10	190±10	170±20	150±10	187±6
Co	0.11±0.01	0.11±0.01	0.12±0.01	0.12±0.01	0.12±0.01
Ni	<0.42	0.40±0.03	<0.38	0.43±0.03	
Cu	<51	<38			4.9±0.2
Zn	13±1	14±1	15±1		15.6±1.2
Ga			<28		
As	120±10	110±10	100±10		115±9
Se[b]	<0.049		<0.13		(0.05)
Br	270±10	290±10	210±10		(270)
Rb	22±1	23±1	21±1	21±2	24±2
Sr	990±40	1100±100	980±50	1000±100	1000±30
Zr	<4.7	4.3±0.5			
Mo	<0.58	0.56±0.20	<0.55		
Ag	0.30±0.01	0.30±0.02	0.28±0.02		0.31±0.02
Cd	<1.0		<1.4		0.15±0.02
In	<0.11				
Sn	<1.3		<3.2		
Sb	0.039±0.004	0.039±0.004	0.038±0.004	0.034±0.004	(0.04)
Te	0.46±0.07	0.43±0.02			
I	520±20				(520)
Cs	0.042±0.003	0.042±0.004	0.043±0.003		(0.04)
Ba[c]	9.4±1.4	11±1	8.7±0.9		
La	<0.088	0.094±0.010	<0.26		
Ce	0.18±0.01	0.21±0.01			
Nd	<0.52	<0.30	<3.1		
Sm	0.083±0.013	0.078±0.014	0.083±0.009		
Eu	0.0069±0.0005	0.0074±0.0012			
Tb	0.0049±0.0009	0.0046±0.0008	0.0045±0.0009	0.0046±0.0006	
Dy	<0.21	<0.23			
Yb	0.024±0.009	<0.059	<0.074		
Lu	<0.0032		<0.056		
Hf[d]	<0.011		<0.085		
Ta	<0.00085	<0.0030	<0.0027		
W	<0.57	<0.14	<0.40		
Au	0.0015±0.0003	0.0010±0.0001			
Hg	<0.061				(0.04)
Th	<0.0089	0.0055±0.0010	<0.017	<0.010	
U	0.37±0.02	0.38±0.03	0.36±0.02		(0.4)

* Element; a) data from NIES certified values(values in parentheses are reference values only.); Coincidence without filter: b) 0.044±0.003, c) 7.9± 1.0, d) 0.011±0.001

proper cooling time based on half-lives of the formed radionuclides, the irradiated samples were measured by a gamma-spectrometry. When measuring the short-lived nuclides, the samples are irradiated for a short time, and then immediately measured for a short time. On the other hand, when measuring the long-lived nuclides, the samples are irradiated for a long time, and after a long cooling time measured for a long time. As a result, the 31 elements were determined by this method under the three irradiating and counting conditions groups shown in Table 2. In addition, the 7 elements were determined by use of an anticoncidence counting method or a coincidence couting method using two gamma-ray detectors. Total 38 elements were determined by all together these methods. And also, the elements were determined with a good precission by use of the selective activation with a Cd filter in the irradiation. It is obvious that the determination of an element in agiven matrix is easer to perform, if the element of interest has a nuclide with a prominent resonance activation peak, whereas the matrix (interfering nuclides) has not.

Table 3 shows the analytical results for the Sagasso, reference standard material. The elements were determined over the range of concentration from minimum several ppb to maximun several % by these various methods. Each analytical value was good agreement with those obtained by various gamma-spectrometries, and with the certified and reference values of NIES. The 31 elements in total 38 elements were determined by the usual counting method, Ni, Zr, Mo, La and Th were determined only by the anticoincidence counting method and Se and Hf were determined only by the coincidence counting method. In this way, the neutron activation analysis is one of proper analytical techniques, which can detrmine multielements in a very small amounts of sample. However, a disadvantage of the neutron activation analysis takes a long analytical time from the irradiation to the gamma-ray counting, for example about one month to a complete analysis.

3. RAPID DETERMINATION OF SELENIUM

It is not always case that the analysis takes a long time. It can take a very short analytical time in the case of a special element. Its example[2] is the determination of Se. The trace Se in biological organs plays an important role in the fields of biology, physiology and medical scince and so on. It is known that Se is one of the essential elements in the biological tissues, but that an excess intaking causes a toxic symptoms, and that a deficient intaking causes a deficiency. And also it is known that Se has an action of competition with a mercury toxi or a cancer. Natural Se has six isotopes shown in Table 4. These isotopes produce the radioisotopes by the neutron capture reaction.

Table 4 Neutron capture reactions of selenium

Target Isotope	Natural Abundunce (%)	Isotope Formed	Half Life	Activation Cross Section (barns)	γ-ray Energy (keV) (Intensity) [%] (*:relative)
^{74}Se	0.87	^{75}Se	120 d	30	136(57),265(60),280(25)
^{76}Se	9.02	^{77m}Se	17.5 sec	22	162(50)
^{77}Se	7.58	^{78}Se	stable	42	
^{78}Se	23.52	^{79m}Se	3.9 min	0.36	96(9)
^{80}Se	49.82	^{81m}Se	56.8 min	0.1	103(8)
^{82}Se	9.19	^{83}Se	25 min	0.004	220(44),360(69),52?(59)
		^{83m}Se	70 sec	0.05	650(20*),1010(100*)

The shortest-lived ^{77m}Se (17.5 sec) in these radioisotopes was measured for a rapid determination of Se. The each sample was irradiated for 10 sec. After about 30 sec from the end of the irradiation, the gamma-rays were counted for 30 sec. And then the acquired gamma-spectrum was analysed for about 20 sec. In this way, the analysis from the irradiation to the output of the analytical result came to an end with very short time as about 2 min. The Bovine Liver of standard reference material of NBS was analysed by this method to examine the accuracy and precision. Table 5 shows the analytical resuls for various sample weights. The average concentration of Se was found to be 1.07 ± 0.11 ppm. This value is good agreement with NBS certified value 1.1 ± 0.1 ppm. Se of mouse organs and fish were determined as an example of this method. As shown in Table 6, Se in kidney was the highest concentration in various organs. The lower limit of determination of Se in these samples was about 10 ng.

Table 5 Determination of selenium in the Bovine Liver (SRM: 1577)[2]

Run No.	Sample weigt (mg/dry)	Concentration (μg/g:ppm)	Run No.	Sample weigt (mg/dry)	Concentration (μg/g:ppm)
1	239	1.02	7	179	1.06
2	249	0.973	8	114	1.04
3	469	1.05	9	154	1.10
4	218	1.10	10	164	0.951
5	114	1.27	11	231	1.26
6	217	0.94	[Average value		1.07± 0.11]
			[NBS certified value:		1.1 ± 0.1]

Table 6 Determination of selenium in mouse and fish[2]

Sample name	Sample weigt (mg/dry)	Concentration (μg/g:ppm)
Lung	61	Not Detected
Liver	449	0.30
Intestine	237	0.36
Stomach	62	0.59
Brain	85	0.45
Kidney	118	1.3
Aji(Horse Mackerel)	70	1.4
Iwashi(Sardine)	111	2.2
Kisu(Sillaginoid)	60	1.6
Saba(Mackerel)	198	1.4

In general, it is called that plant foodstuffs are better nutritious than animal foodstuffs. And it is known that soybean which is higher content of protein in many plant foodstuffs is higher content of Se, too. Se of soybean originating from Japan, China, Formasa, Burma, Brazil, Argentina and U.S.A. was determined by the neutron activation analysis[3]. The analytical results is shown in Fig. 1. Concentration of Se in foodstuffs can be devided into four nutritious levels, a deficiecy level, a sub-nutrition level, a nutrition level and an excessive level. The China soybean was classified into the deficiecy level or the sub-nutrition level, the Formasa, Burma, Brazil, Argentina and U.S.A. soybean were distributed over the wide level from the sub-nutrition level to the nutrition level and the Japan soybean was concentrated in the sub-neutrition level, which was lower content. Japanese people eat well the

soybean as "miso" bean paste, "toufu" bean curd and so on. The most raw
materials of these soybean products are imported from U.S.A., now. Therefore,
Japanese people have intaken the soybean with the nutrition level.

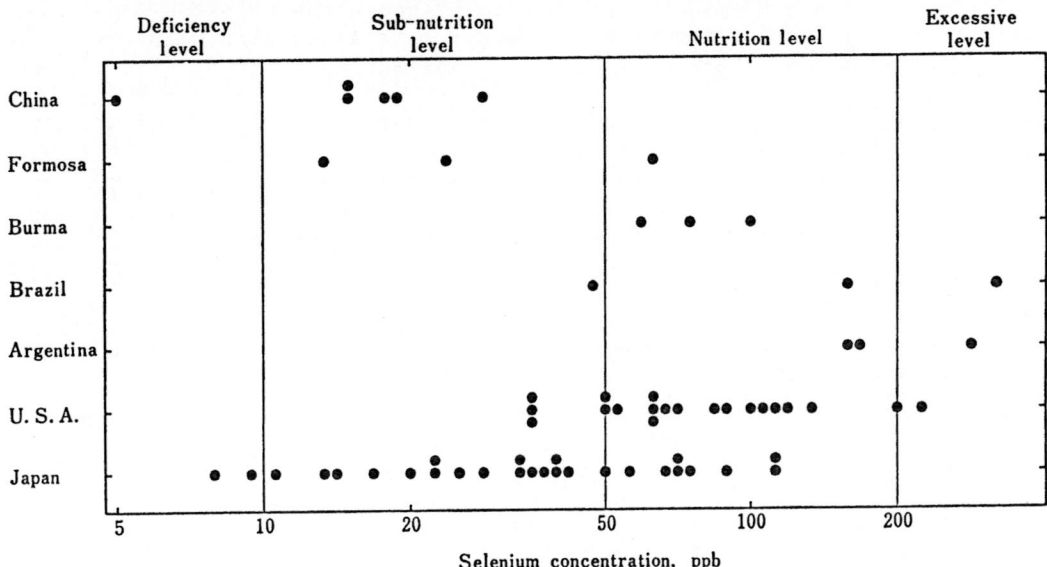

Fig. 1 Selenium concentration in various soybean-origins[3]

4. DETERMINATION OF PLATINUM FOR MEDICAL SCIENCE

Further, Se has carried out important roles in a medical science. Se has
an effect of suppression for a metalic poison as Hg and Cd poison and of
reduction for a cancer. A cisplatin being a platinum compound is an effective
drag as an anticancer, now. However, this drag produces a harmful aftereffect
to kidney. Se has been used for the reduction of the harmful aftereffect. This
metabolism of Se has not known well, and many studies have been carried on. Pt
and Se in biological organs can be determined by the neutron activation
analysis. The determination of Se was used by the same method above mentioned,
but the determination of Pt was used by a little different from the method. Pt
shown in Table 6 has six isotopes. The ^{199}Au nuclide produced by β decay from

Table 6 Neutron capture reactions of platinum

Target Isotope	Natural Abundunce (%)	Isotope Formed	Half Life	Activation Cross Section (barns)	γ-ray Energy (keV) (Intensity) [%]
^{190}Pt	0.0127	^{191}Pt	3 day	0.76	539(9), 360(4)
^{192}Pt	0.78	^{193m}Pt	4.3 day	2	136(small)
^{194}Pt	32.9	^{195m}Pt	4.1 day	87×10^{-3}	130(1), 99(11)
^{195}Pt	33.8	^{196}Pt	stable		
^{196}Pt	25.3	^{197m}Pt	80 min	60×10^{-3}	346(13), 279(3)
		^{197}Pt	18 hr	0.8	191(6), 77(20)
^{198}Pt	7.21	^{199}Pt	30 min	3.7	543(100), 494(38)
		$\downarrow \beta$			
		^{199}Au	3.15 day		208(8), 158(37)

[199]Pt was measured in the gamma-spectometry to determine Pt. When a content of Pt was high, a short-lived [197m]Pt could be used. In a case of using [199]Au, the time from the irradiation to the measurement took 3 or 4 days, but in a case of using [197m]Pt, the time took only 1 or 2 hours.

When long irradiating a liver of rat treated with the cisplatin, a small gamma-peak(158 keV) of [199]Au could be seen among many big gamma-peaks of [24]Na and [82]Br. A concentration of Pt was able to determine several ppm using a sample weight of about 100 mg. A concentration of Pt in organs of treated rat and a lower limit of determination are shown in Table 7. Each organ could be analysed with about 10% precision. In this case, Pt accumulated in the kidney as well as Se, too.

Table 7　Concentration of platinum in organs of rat treated with cisplatin

Organ	Concentration (μg/g:ppm)	Lower limit of determination (μg/g:ppm) at sample weight		Number of sample
Liver	5.3 ± 0.4 (8%)	0.4	(200mg)	11
Kidney	25 ± 2.2 (9%)	0.7	(100mg)	12
Whole blood	8.2 ± 0.9 (11%)	0.9	(100mg)	12

Several 10 mg of each organ of rats treated with the cisplatin, the sodium selenite or the cisplatin plus the sodium selenite were analysed by the neutron activation analysis method, and the distribution of Pt in the rats is shown in Fig. 2[4]. A concentration of Pt in the all organs except kidney was not almost various with either the treatment. However, a concentration of Pt in kidney was higher with the simultaneous treatment than with the individual treatment.

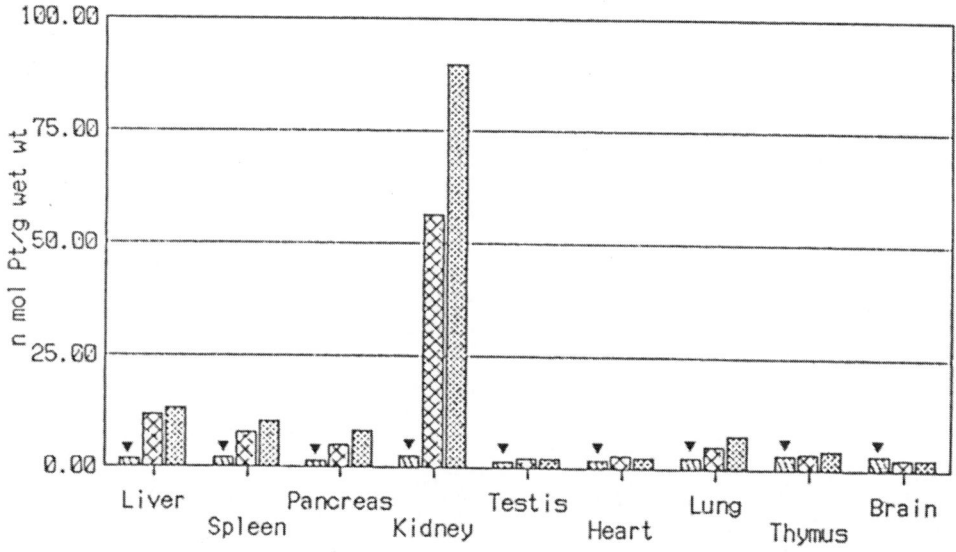

Fig. 2　Distribution of Pt after 24 hour in various organs of rats treated with cisplatin, sodium selenite or cisplatin plus sodium selenite[4].
▨ Pt level in rats with sodium selenite, ▧ Pt level in rats treated with cisplatin, ▨ Pt level in rats treated with cisplatin plus sodium selenite, ▼ ; not detected

As above mentioned, a μg or ng level of elements can be nondestructively determined by the neutron activation analysis using a small amounts of sample, and the behavior of trace elements in the organism can be known. Therefor, it is clear that the neutron activation analysis become an efficient analytical method to investigate many trace elements, which behave variously in the organism.

5. REFERENCES

1) Suzuki S, Saito K, Hirai S (1988) Determination of multielement in Sagasso reference material by instrumental neutron activation analysis. Bunseki Kagaku, 38:198-201
2) Hirai S, Suzuki S, Okamoto M (1978) Determination of selenium in biological materials by neutron activation analysis using ^{77m}Se. Bunseki Kagaku, 27: 435-440
3) Okada Y, Suzuki S, Hirai S (1987) Determination of selenium in soybean from different regions by instrumental neutron activation analysis. Bunseki Kagaku, 36:856-860
4) Sakurai H, Tsuchiya K, Hirai S, Okada Y, Haraguchi H (1988) Correlation of levels of Pt and Se in rats treated simultaneously with cisplatin and sodium selenite. Naturwissenschaften, 75:405-406

Summary of New Analytical Techniques and General Discussion

JORMA KUMPULAINEN

Central Laboratory, Agricultural Research Centre of Finland, 31600 Jokioinen, Finland

ABSTRACT

During the present decade, the most frequently employed analytical techniques in the field of trace element research, flame and electrothermal (ET) atomic absorption spectrometry (AAS) and inductively coupled plasma atomic emission spectrometry (ICP-AES) have reached full maturity, resulting in commercially available instruments capable of reliable routine determinations of low concentrations of trace elements in clinical and food samples. In particular, improvements in ETAAS have enabled routine determinations of ultratrace concentrations of elements in body fluids without preliminary digestion. Furthermore, the developments in ICP-mass spectrometry (MS) have facilitated simultaneous multielement determination of ultratrace concentrations of elements in biological samples without preconcentration. ICP-MS, when coupled with high performance liquid chromatography, allows speciation studies of trace elements. However, to meaningfully employ improvements in analytical instrumentation, rigorous methods for contamination and analytical quality control are needed.

Key words: trace elements, new analytical techniques

INTRODUCTION

Owing to the very rapid evolution in trace element analytical techniques during the present decade, it is impossible to deal with all of the interesting new developments. Thus, the present paper focuses on recent improvements in the major techniques presently in use, namely atomic spectroscopy, mass spectrometry as well as with the introduction of such novel and important hybrid systems as ICP-MS, HPLC-AAS and HPLC-ICP-MS. Also, the outlook for future trace element laboratories is presented.

ATOMIC SPECTROSCOPIC METHODS IN TRACE DETERMINATION OF ELEMENTS

Flame Atomic Absorption Spectrometry (FAAS)

Developments in microcomputer technology during the 1980s have resulted in highly automated FAAS capable of determining several elements in up to 140 samples completely automatically. Such previously manual operations as sample change, adjustment of burner height, optimizing fuel/oxidant ratio, change of lamp and lamp parameters, instrument calibration, calculation and reporting of results can be performed fully automatically thus shortening considerably the actual operator time required for the instrumental analysis.

Furthermore, instrumentation in which the flow injection analysis system (FIA) has been combined with FAAS or hydride generation has become commercially available this year. This system offers the advantages of lower sample volumes required, lower consumption of reagents, faster analyses, on-line addition of buffers or modifiers and on-line dilution of samples. The possiblity of using small sample volumes in the above system is very important for clinical samples, particularly in pediatric studies where only small-sized samples are available.

The above listed improvements in the automation, together with ease of operation, freedom from interferences, low operating and instrumentation costs and few sample pretreatments have made FAAS the most popular type of instrument in the field of trace element analytical chemistry at present. Although in biological materials the contents of Al, Cr, Ni, Mo and those of toxic elements are usually too low to allow analysis by FAAS, a recent report indicates that an on-line ion-exchange pre-concentration system with an enrichment factor of 30 may be applied to the FIA-FAAS thus greatly expanding the applicability of the method (Fang and Welz 1989).

Electrothermal Atomic Absorption Spectrometry (ETAAS)

Rapid developments in background correction techniques during the 1980s exemplified by the use of the Zeeman effect to improve efficiency in background correction (BC) have facilitated the determination of low concentrations of trace elements in difficult biological matrices with higher accuracy and less time-consuming sample pretreatment by ETAAS (Kumpulainen 1988; Paakki and Kumpulainen 1988). The "stabilized temperature platform furnace" (STPF) concept, which includes the use of matrix modification, platform or probe atomization and high rate of heat increase during atomization with accurate temperature control have made possible the determination of easily volatile elements such as lead, cadmium and selenium (Alfthan and Kumpulainen 1982) directly after dilution in biological fluids by ETAAS. The most recent developments feature determination of powdered dry biological samples using the so called slurry technique without sample ashing.

Furthermore, more sophisticated softwares and hardwares have made it possible to shorten the time of instrumental analysis from 3 minutes to less than 1 minute (Halls et al 1987), and possible improvements may further rerduce it down to 30 seconds (de Galan 1987). Fully automatic determination of several elements in up to 30 samples of clinical fluids using matrix modification and the method of additions under clean-room conditions is possible overnight without supervision. Thus, the relatively slow sample throughput in ETAAS is no longer that crucial. Moreover, there is a commercially available multichannel ETAAS with a Zeeman background correction system which considerably speeds up the time required for instrumental analysis. It is likely that ETAAS will continue for some time to be the most popular technique for the determination of ultratrace elements, particularly in view of the relatively low price of such instruments.

Atomic Emission Spectroscopy (AES)

During the present decade, the use of inductively coupled plasma (ICP) has prevailed over other plasma sources such as direct current or microwave in (AES). ICP-AES has the advantage of having fewer matrix interferences compared with the other plasma emission techniques. ICP-AES has a high dynamic range of 10^6, relatively high sensitivity and an adequate precision when an internal standard is employed. Improvements in sample introduction systems, particularly introduction of the ultrasonic nebulizer and in background correction techniques as well as rapid sample

throughput in instruments using simultaneous multichannel technique have made ICP-AES the method of choice for laboratories with a high sample load and a need to determine several elements per sample. As for FAAS, FIA may also be coupled to ICP-AES. However, ICP-AES is not sufficiently sensitive for the direct determination of such elements as Cr, Ni, Mo, Al, V, Se, Pb, Cd, Hg. As and Tl in most biological and clinical samples. Moreover, the relatively high cost of ICP-AES instruments in addition to high operation costs limit their use to large laboratories only.

RECENT DEVELOPMENTS IN MASS SPECTROMETRY OF TRACE ELEMENTS

Inductively Coupled Plasma-Mass Spectrometry (ICP-MS)

The most promising recent development in the mass spectrometry (MS) of trace elements is the use of ICP as the ionization source. ICP-MS is a highly sensitive method for all metallic elements in the range of 0.1-1 ng/ml, except for K, Ca and Fe for which the sensitivity is from 1-2 orders of magnitude lower. However, this is usually not a problem for biological materials whose Ca, K and Fe contents are relatively high. Developments in ICP-MS as well as interferences encountered have been excellently reviewed by Hieftje and Vickers (1989). ICP-MS combines the high sensitivity of ETAAS and the simultaneous multielement analysis capacity of ICP-AES. In addition, as the method yields information also on the isotopic ratios of the elements, it may be employed for metabolic studies using staple isotopes (Whittaker et al 1989; Ting et al 1989).

The practical adaptions of ICP-MS for the analysis of biological materials are still relatively few but the results are very promising (Munro et al 1986; Delves and Campbell 1988; Lyon et al 1988) Furthermore, there is a report indicating that FIA may be employed for the introduction of biological fluids with high salt content to ICP-MS (Kushida et al 1989). The high costs of operation and equipment limit, however, the use of this instrument to major laboratories only.

Other Mass Spectrometric Methods

Laser microprobe mass analyser (LAMMA) is another interesting development in MS. In this method, the sample has to be made into a thin film, and the elements are ionized using a laser beam followed by time-in-flight mass spectrometric determination. As correct isotopic abundances are obtained, use of altered isotopic abundances in intrinsically labelled samples is possible. So far, there are only preliminary results available on the quantitative applications of LAMMA, however, this method gives an important advantage in offering high spatial resolution allowing analysis of cellular and sub-cellular components. Similar high spatial resolution and relatively high sensitivity as for LAMMA has been described on the application of secondary ion mass spectrometry to trace element localization and quantitation in various biological tissues. Successful application of this method for the determination of Fe, I and Mn in thryoid gland, heart and kidney has been reported (Larras-Regard et al 1989)

X-RAY SPECTROMETRY

Use of a scanning electron microscope or an electron microprobe analyzer have the advantages of premitting analysis of certain major elements in cellular and even sub-cellular specimens. Unfortunately, the sensitivity is not sufficiently high to allow trace element determinations. However, particle induced ray emission (PIXE), particularly the microbeam technique often called μ-PIXE, has a sensitivity better by

a factor of approximately 100 thus permitting determination of several trace elements in biological and clinical samples (Johansson 1989). Recent developments in X-ray fluorescence feature *in vivo* determination of lead in bones and biodistribution of drugs used in chemotherapy, such as cisplatinum.

RECENT DEVELOPMENTS IN NEUTRON ACTIVATION ANALYSIS (NAA)

NAA is a very powerful multielement method for the quantitative trace analysis of biological materials requiring only very small samples. Recently, instrumental NAA has gained popularity due to the development of high resolution Ge(Li) gamma spectrometers and better computer software for automatic data processing. These improvements have made it possible to determine very short-lived isotopes of such trace elements as Se, Al, Mn and V thus considerably improving the analytical throughput (Hirai 1989). However, very high sensitivity may only be obtained employing radiochemical separation, which is very time-consuming. NAA is a powerful reference method, particularly for the certification of trace elements in biological reference materials (Hirai 1989)

HYBRID SYSTEMS

High performance liquid chromatography (HPLC) has been applied to the determination of trace elements in biological materials to a greater extent only within the past 5 years. This development reflects rapid improvements in sensitive on-line detectors as well as advances in column packing materials and instrumentation resulting in efficient high speed separations and quantitative determination of trace elements at relatively low concentrations (Nakagawa et al 1989). The advantage of HPLC over the spectroscopic methods is the possibility of determining different species of elements. However, its main problem still appears to be too low sensitivity of the conventional detectors available for the HPLC.

Therefore, HPLC will be increasingly used in hybrid systems for speciation studies. Particularly, hybrid systems are excellent for the study of trace elements associated with various proteins. Towards this end, hybrid systems have been developed by interfacing HPLC with AAS, ICP or MS. Thus far, the most successful applications have been made with an HPLC-AAS combination but examples of HPLC-ICP-AES systems are also available (Suzuki et al 1989). The high selectivity of AAS and relative insensitivity to mobile phase composition allows more scope for the design of elution systems to optimize the separation of desired peaks. In the field of speciation studies the most studied elements are the organo-metallic compounds of As, Hg, Pb and Sn as well as the various oxidation states of Cr, As and Se. However, the work in this important area is still in its infancy.

The most promising hybrid system consists of the triple hybrid HPLC-ICP-MS. This system combines the high separation power for macromolecules of HPLC with the ultra-high sensitivity and capability for isotopic ratio and simultaneous multielement analysis. Therefore, this system is the method of choice for future trace element speciation research and examples of successful applications have already been reported (Dean et al 1987 ; Shibata and Morita 1989)

FUTURE PROSPECTS

Except for trace element determinations in body fluids which can now be determined without preliminary sample digestion by AAS, the sample digestion has become a rate-limiting step in the trace analysis of biological samples. Recent innovations in

microwave digestion methods exemplified by a system comprising a microwave oven, a peristaltic pump, a closed flow system interfaced with an AAS instrument (Burguera et al. 1988) may offer a solution to this problem in the near future.

During the next decade we are probably going to see more sophisticated softwares for Laboratory Information Management Systems (LIMS) operating in microcomputer networks interfaced with laboratory robots and instruments including hybrid systems. Thus, the entire analytical process including sample grinding, weighing, digestion, analysis and reporting may be automatized (Hawk and Kingston 1988). However, obtaining meaningful results on contents of ultratrace elements in biological materials will require, also in the future, careful sample handling and rigorous methods in analytical quality control including the use of clean rooms and suitable certified reference materials.

REFERENCES

Alfthan G, Kumpulainen J (1982) Determination of selenium in small volumes of blood plasma and serum by electrothermal atomic absorption spectrometry. Anal Chim Acta 140:221-227.

Burguera M, Burguera JL, Alarcon OM (1988) Determination of zink and cadmium in small amounts of biological tissues by microwave-assisted digestion and flow-injection atomic-absorption spectrometry. Anal Chim Acta 214:421-427

Delves HT, Campbell JM (1988) Measurement of total lead concentrations and of lead isotope rations in whole blood by use of inductively coupled plasma source mass spectrometry. J. Anal At Spectrom 3:343-349.

Dean JR, Munro S, Ebdon L, Crews HM, Massey RC (1987) Studies of metalloprotein species by directly coupled HPLC-ICP/MS. J Anal At Spectrom 2:765.-772.

de Galan L (1987) A physicist's view on current questions in atomic spectrometry J Anal Atom Spectrom 2:89-95.

Fang Z, Welzt B (1989) High efficiency low sample consumption on-line ion-exchange pre-concentration system for flow injection flame atomic absorption spectrometry. JAAS 4: (in press)

Halls DJ, Mohl C, Stoeppler M (1987) Application of rapid furnace programmes in atomic absorption spectrometry to the determination of lead, chromium and copper in digests of plant materials. Analyst 112:185-189.

Hawk GL & Kingston HM (1988) Laboratory robotics for trace analysis In: McKenzie HA, Smythe LE, (eds) Quantative trace analysis of biological materials, Elsevier Science Publishers, Amsterdam, p. 285.

Hieftje GM, Vickers GH (1989) Developments in plasma source/mass spectrometry. Anal Chim Acta 216:1-24

Hirai S (1989) A guideline for application of neutron activation analysis to biological and medical samples In: Proceedings of the Second Meeting of the International Society for Trace Element Research in Humans, Springer-Verlag Tokyo, Inc. (in press).

Johansson S (1989) PIXE: A novel technique for elemental analysis. Endeavor, New Series, 13:48-53

Kumpulainen J (1988) Chromium In: McKenzie HA, Smythe LE (eds.) Quantitative trace analysis of biological materials, Elsevier Science Publishers, Amsterdam p 451.

Kushida K, Toshikazu T, Matsubayashi T, Matsuoka T (1989) Determination of trace elements in commercial IVH solutions by ICP-Mass Spectrometry using flow injection analysis J Trace Elem Exp Med 2:150

Larras-Regard E, Mony M-C, Olivo J-C, Aioun J, Kahn E. (1989) New advances in cellular and subcellular imaging and quantitation of trace elements. J Trace Elem Exp Med 2:183

Lyon TDB, Fell GS, Hutton RC, Eaton AN (1988) Evaluation of inductively coupled argon plasma mass spectrometry (ICP-MS) for simultaneous multi-element trace analysis in clinical chemistry. J Anal At Spectrom 3:265-270.

McLeod CW, Worsfold PJ, Cox AG (1984) Simultaneous multielements analysis of blood serum by flow injection-inductively coupled plasma atomic emission spectrometry. Analyst (London)109:327-332

Munro S, Ebdon L, McWeeny DJ (1986) Application of inductively coupled plasma mass spectrometry for trace metal determination in foods. J Anal At Spectrom 1:211-217.

Nakagawa T, Aoyama E, Hasegawa N, Kobayashi N & Tanaka H (1989) High-performance liquid chromatography-fluorometric determination of selenium based on selenotrisulfide formation reaction Anal Chem 61:233-236.

Paakki M, Kumpulainen J (1988) Molybdenum. In: Makenzie HA, Smythe LE (eds) Quantitative trace analysis of biological materials, Elsevier Science Publishers, Amsterdam p 463

Shibata Y, Morita M (1989) Speciation of arsenic in human urine by HPLC/ICP mass spectrometry. J Trace Elem Exp Med 2:184.

Slavin W, Manning DC, Carnrick GR (1989) Fast Analysis with Zeeman Graphite Furnace AAS. The Perkin-Elmer Corporation, Study No 245, CT, USA.

Suzuki K, Tamagawa H, Hirano S, Takahashi K and Shimojo N (1989) HPLC-ICP analysis of changes in distribution of Ca, Cu, Fe, Mg, P S and Zn in breast milk fractions of a healthy lactating mother. J Trace Elem Exp Med 2:167.

Ting BTG, Janghorbani M (1987) Application of ICP-MS to accurate isotopic analysis for human metabolic studies. Spectrochim Acta, Part B 42:21-30.

Whittaker PG, Lind T, Williams JG, Gray AL (1989) Inductively coupled plasma mass spectrometric determination of the absorption of iron in normal women. Analyst 114:675-678

Environment and Trace Elements

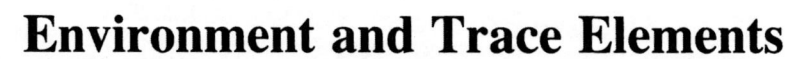

Aluminum Toxicity in Humans

ALLEN C. ALFREY

Renal Medicine, Veterans Administral Medical Center, University of Colorado Medical Center, 1055 Clermont Road, Denver, CO 80220, USA

ABSTRACT

Systemically administered aluminum can cause severe toxicity primarily effecting the neurological, skeletal and hematopoietic systems. Humans have largely been protected from toxicity because of the insoluble nature of most aluminum compounds. However, when administered with citrate aluminum compounds can be solubilized, and absorption markedly enhanced. Similarly the barriers to aluminum absorption (lungs, gastro-intestinal tract and skin) can be circumvented by the parenteral administration of aluminum and body burden of this element markedly increased. The kidneys are responsible for elimination of any systemic aluminum placing patients with renal failure at the greatest risk of developing toxicity.

Key Words: Aluminum, osteomalacia, citrate, Gastro-intestinal

INTRODUCTION

Although aluminum represents 8% of the earth's crust and man is continually exposed to this element it took unusual circumstances for aluminum toxicity to occur. This is largely a result of limited systemic exposure due to the insoluble nature of most aluminum compounds. In fresh water aluminum compounds undergo hydrolysis with formation of insoluble polynuclear complexes (Driscoll 1989). Of 175 natural water sources tested only 18 had aluminum levels greater than 18 μg/L (Miller 1984). Similarly in sea water aluminum levels are less than 1 μg/L. These low levels are felt to result from aluminum being adsorbed onto siliceous shells. Aluminum compounds present in soil and plants are also largely insoluble.

Not only man but other animals, aquabiotica and plants have been protected from exposure to this element and toxicity because of it's insolubility. However, rain-borne nitric and sulfuric acid from industrial pollution and acidifying fertilizers have caused aluminum compounds to become soluble in certain soils, and lakes having rock composition which do not buffer. This is especially prevalent in northern Europe and Canada. Under these conditions aluminum can exert considerable toxicity to plants and animals. In regards to plants soluble aluminum compounds have been shown to retard growth (Godbold 1988), in birds cause abnormal egg formation (Nyholm 1981) and produce gill necrosis in fish (Muniz 1980). In humans as well as experimentally in animals aluminum toxicity is manifested on the skeletal, hematological and neurological systems (Ward 1979; O'Hare 1982; Alfrey 1976).

NORMAL ALUMINUM METABOLISM:

Normally man ingests 10 to 15 mg/day of aluminum in food and an addition <0.01 to 2 mg/day from water. However, it would appear of the ingested aluminum only a very small fraction, 10 to 15 μg/day, is absorbed. Of the absorbed aluminum in health it is virtually completely eliminated by renal excretion. This is supported by the fact that tissue aluminum levels with the exception of lung are maintained at very low levels (<4 mg/kg dry weight) and do not increase with ageing. The total body aluminum burden has been estimated at 35 mg (Alfrey 1980). Although lung aluminum levels are considerably higher than other tissues, being up to 100 mg/kg or higher, the inhaled aluminum is trapped in the pulmonary tissue and largely if not entirely kept in the lung with little if any systemic transfer. Similarly the skin also is an impervious barrier to aluminum absorption.

There are factors however which can modulate aluminum absorption from the gastro-intestinal tract. These include increase aluminum ingestion (Kaehny 1977), the type of aluminum compound ingested (Froment 1989) and the uremic state (Ittel 1987). Kaehny et al (1977) found that urine aluminum increased from a normal value of 15 μg/day to over 400 μg/d following the ingestion of aluminum containing antacids by normal individuals. This documented for the first time that some aluminum was absorbed from these agents. Besides the amount of aluminum administered the type of aluminum compound is also an important determinate of the amount of this element absorbed. In rats follow gavage of very insoluble compounds such as aluminum hydroxide and sucralfate only about 0.016% of the administered dose is absorbed (Froment 1989). With slightly more soluble aluminum lactate 0.035% or twice as much aluminum is absorbed. With the most soluble compound studied, aluminum citrate, between 1 and 2% of the administered dose was absorbed. Froment et al (1989) showed in contrast to aluminum chloride the oral administration of aluminum citrate resulted in an early and marked enhancement of plasma aluminum levels. However, it would appear that solubility alone does not explain the marked enhancement of aluminum absorption by citrate. This is based on the fact that citrate has a variable effect on the solubility of a variety of aluminum compounds whereas under these conditions there is no relationship between solubility and absorbability of aluminum. To further characterize aluminum absorption aluminum citrate was administered with D (1-^{3}H) Glucose. The plasma peaks of glucose and aluminum following their gavage coincided supporting the fact that aluminum with citrate is absorbed in the proximal small bowel.

Studies were extended to the everted gut. Using this preparation it was found that at 37°C aluminum and citrate were taken up by the tissue and both transported into the serosal fluid. However, at 4°C tissue uptake of aluminum and citrate was markedly decreased as was serosal fluid transfer of citrate. In contrast serosal fluid transport of aluminum was unaffected by cold. Studies carried out with aluminum lactate demonstrated even greater tissue uptake of aluminum than found with citrate but no serosal fluid transport of aluminum. These studies suggested aluminum absorption from citrate was passive since it was not effected by cold, and through the paracellular pathway since there was no relationship between tissue uptake and serosal fluid transfer of aluminum. To further document that citrate opens the tight junction allowing aluminum to move through paracellular pathway two additional studies were performed using Ruthenium red and the Ussing chamber (Froment 1989).

Ruthenium red is an electron dense substance that does not permeate cells and can therefore be used as a marker of tight junction integrity. An intestinal loop was isolated and ruthenium red in association with normal saline, aluminum chloride and aluminum citrate was placed in the loop. After 1-hour the

intestinal loop was prepared for study by electron microscopy. The loops exposed to normal saline and aluminum chloride had only a small amount of Ruthenium red about the neck of a few goblet cells. In contrast in the intestine exposed to aluminum citrate there was extensive Ruthenium red around entire goblet cells.

The resistance across a segment of small intestine is maintained by the integrity of the tight junction which can be measured in the Ussing chamber using a short circuit current. Aluminum citrate, as found with sodium citrate, which is known to open the tight junctions markedly decreased the resistance across the small intestine. Since calcium is known to maintain the tight junction the effect of aluminum citrate on ionized calcium was studied. Aluminum citrate was found to totally bind the ionized calcium explaining how aluminum citrate was able to open tight junctions. Using more indirect methods Provan (1988) has also presented evidence that aluminum transport even without citrate, although much less, is also passive and through the tight junctions.

Another mechanism to enhance the body burden of aluminum is to circumvent the normal barriers to aluminum absorption, the skin, lung and gastro-intestinal tract by administering the aluminum parenterally. This was initially recognized in the dialyzed uremic patient who received parenteral aluminum as a result of aluminum contaminated dialysate. Since aluminum is largely bound in serum to transferrin the gradient from bath to patient is maintained throughout the dialysis procedure. In addition this protein binding of aluminum prevents its removal even when the dialysate is aluminum free (Kaehny 1977). Aluminum has also been given systemically as a result of aluminum contaminated albumin and total parenteral nutrition fluids (Klein 1982).

The kidney is the major excretory organ for any absorbed aluminum. Following an intravenously administered aluminum load to dogs over a subsequent eight hour period over 40% of the load was excreted in the urine whereas less than 1% was excreted in the bile (Kovalchick 1978). Even though the kidney is fairly efficient in excreting any systemic load of aluminum its capability is limited. Therefore large loads of systemically administered aluminum may exceed the ability of the kidneys to excrete the aluminum resulting in aluminum overload.

ALUMINUM OVERLOAD

Aluminum toxicity in humans has largely occurred in uremic patients, initially being recognized in patients dialyzed with aluminum contaminated dialysate (Alfrey 1976). However, it was subsequently recognized that aluminum overload could occur in dialyzed uremic patients being maintained with aluminum free dialysate and even non-dialyzed uremic patients.

As stated above aluminum absorption has been shown to be increased in animals with uremia. This might be a consequence of some compromise of the integrity of the tight junction in the uremic state. However, this has not been studied. Irrespective of the cause of the enhanced absorption of aluminum in uremia this may predispose the uremic patient to aluminum overload. However, there is no doubt that the loss of renal function plays the major role in enhancing the retention of any absorbed aluminum. In support of this is figure 4 which shows the tissue burden of aluminum measured in a large group of non-dialyzed uremic patients who were not receiving oral aluminum compounds (Alfrey 1980). It can be appreciated that over 50% of the patients had increased aluminum levels in bone and/or liver.

In addition patients who have received chronic total parenteral nutrition with aluminum contaminated fluids have also been found to have increased liver and bone aluminum loads frequently as high as found in uremic patients (Klein 1982). This also documents the fact that the ability of normal kidneys to excreted large parenteral loads of aluminum can be exceeded with retention of the aluminum.

Only two patients with normal renal function who have had no parenteral aluminum exposure have been reported to have an enhanced body burden of aluminum through only oral exposure (McLaughlin 1962; Lapresle 1975). At this time it is unknown whether patients with normal renal function ingesting aluminum compounds in association with citrate are at risk of developing aluminum overload.

ALUMINUM TOXICITY

Classical Neurotoxicity

The systemic toxicity of aluminum for man was first firmly established in the literature in 1976 (Alfrey 1976). This was based on the finding of increased brain gray matter aluminum in a group of dialyzed uremic patients who had died of a distinctive neurological syndrome. This disease was named dialysis encephalopathy or dialysis dementia and characterized by a distinctive speech disorder, myoclonus, seizures, hallucinations, personality changes and death within 6 to 8 months after onset of symptoms (Alfrey 1972). This syndrome is usually associated with other evidence of aluminum intoxication such as osteomalacia and a microcytic hypochromic anemia. Although this disease was initially seen in patients who had been maintained for several years using aluminum contaminated dialysate it also occurs in uremic patients ingesting large amounts of aluminum containing phosphate binding gels over extended periods of time. Diagnosis is dependent on the clinical picture in association with EEG alterations (normal background rhythm with multifocal bursts of slow or delta wave activity) and an elevated plasma aluminum level which is usually but not invariably between 100 and 200 $\mu g/L$.

Acute Neurotoxicity

An acute form of neurotoxicity has been recognized in both dialyzed uremic patients as well as non-dialyzed uremic patients. In dialyzed uremic patients this syndrome occurs in patients dialyzed with dialysate highly contaminated with aluminum (>500 $\mu g/L$) or patients receiving deferroxamine treatment for aluminum overload. More recently this syndrome has been recognized in non-dialyzed uremic patients in a matter of weeks after being placed on oral aluminum containing phosphate binding gels, for the control of serum phosphorus levels and sodium citrate for treatment of acidosis (Bakir 1986). The disease is characterized by an abrupt onset of seizures, obtundation, myoclonus, coma and frequently death in a matter of days. Laboratory reveals the classical EEG changes as described above and extremely high plasma aluminum levels (>500 $\mu g/L$).

Skeletal Toxicity

Clinically skeletal toxicity is manifested by bone pain, proximal myopathy and weakness and fractures which are not improved by vitamin D therapy (Hodsman 1981). Although very common when patients were receiving dialysis with aluminum contaminated dialysate recently it has also been reported to occur in

20 to 30% of patients whose only exposure to aluminum are the orally administered aluminum containing phosphate binding gels. Laboratory studies reveal mild hypercalcemia, relatively low parathyroid hormone levels and a modestly elevated plasma aluminum level (50-150 μg/L). The histological findings represent the hall mark of this toxicity. The bone shows absent bone formation frequently in association with classical osteomalacia. As demonstrated by aluminum staining the aluminum is present at the junction between the calcified and non-calcified bone (Maloney 1982).

Hematological Toxicity

Clinically hematological toxicity is manifested by a microcytic, hypochromic anemia in the presence of normal iron stores (O'Hare 1982). It frequently occurs in patients with skeletal and neurotoxicity resulting from aluminum.

Mechanism of Toxicity

It is suggested that the type of aluminum compound formed might be responsible for the toxicity noted. Free Aluminum would form hydroxides at the pH of the body. However, it forms tight complexes with transferrin and the majority of aluminum in blood is bound to transferrin (Martin 1986). The next tightest complexes are formed with citrate. However, there is so much more calcium available for binding that citrate would not be a major aluminum binder in blood although it could be in other tissues. The final form that aluminum would be
expected to exist in the body is as aluminum-phosphate hydroxide. This is probably the form aluminum is in when it exists in bone. Aluminum transferrin has been shown to compete with iron transferrin and inhibit erythropoiesis (Mladenovic 1988). Aluminum citrate has been shown to cause extremely high plasma levels of aluminum and induce acute neurotoxicity (Bakir 1986). This may result from the very high plasma aluminum level or the ability of aluminum citrate to cross the blood brain barrier. Finally aluminum-phosphate hydroxide is able to bind calcium promoting hypercalcemia and causing soft tissue calcification (Henry 1984). In addition it appears to cross link bone collagen rendering it unable to induce bone formation and inhibits calcification (Zhu 1989).

REFERENCES:

Alfrey AC, Legendre GR, Kaehny WD (1976) The dialysis encephalopathy syndrome. Possible aluminum intoxication. N Engl J Med 294:184-188

Alfrey AC, Mishell J, Burks J, Contiguglia SR, Rubolph H, Lewin E, Holmes JH (1972) Syndrome of dyspraxia and multifocal seizures associated with chronic hemodialysis. Trans Am Soc Artif Intern Organs 18:257-261

Alfrey AC, Hegg A, Craswell P (1980) Metabolism and toxicity of aluminum in renal failure. Am J Clin Nutr 33:1509-1516

Bakir AA, Hyrhorczuk DO, Berman E, Dunea G (1986) Acute fatal hyperaluminemic encephalopathy in undialyzed and recently dialyzed uremic patients. Trans Am Soc Artif Intern Organs 32:171-176

Driscoll CT, Baker JP, Bisogni JJ, Schofield CL (1980) Effect of aluminum speciation on fist in dilute acidified waters. Nature 284:161-163

Driscoll CT, Schecher WD (1989) Aqueous chemistry of aluminum . In: Aluminum and health. Marcel Dekker, Inc., New York, p 27

Froment DH, Buddington B, Miller NL, Alfrey AC (1989) Effect of solubility on gastrointestinal absorption of aluminum from various aluminum compounds in the rat. J. Lab and Clin Med (in press)

Froment DH, Molitoris BA, Buddington JB, Miller N, Alfrey AC (1989) Site and mechanism of enhanced gastrointestinal absorption of aluminum by citrate in the rat. Kidney Int (in press)

Godbold DL, Fritz E, Huttermann A (1988) Aluminum toxicity and forest decline. Proc Natl Acad Sci USA 85:3888-3892

Henry DA, JGoodman JWG, Nudelman RK,DiDomenico NC, Alfrey AC, Slatopolsky JE, Stanley TM, Coburn JW (1984) Parenteral aluminum administration in the dog. I. Plasma kinetics, tissue levels, calcium metabolism and parathyroid hormone. Kidney Int 25:362-369

Hodsman AB, Sherrard J, Wong EGC, Brickman AS, Lee DBN, Alfrey AC, Singer FR, Norman AW, Coburn JW (1981) Vitamin D resistant osteomalacia in hemodialysis patients lacking secondary hyperparathyroidism. Ann Intern Med 94:629-637

Ittel TH, Buddington B, Miller NL, Alfrey AC (1987) Enhanced gastrointestinal absorption of aluminum in uremic rats. Kidney Int 32:821-826

Kaehny WD, Alfrey AC, Holman R, Schorr W (1977) Dialyzability of aluminum. Kidney Int 12:361-366

Kaehny WD, Hegg A, Alfrey AC (1977) Gastrointestinal absorption of aluminum from aluminum-containing antacids. N Engl J Med 296:1389-1390

Klein GL, Ott SM, Alfrey AC, Sherrard DJ, Hazlet TK, Miller NL, Maloney NA, Berquist WE, Amenmt ME, Coburn J (1982) Aluminum as a factor in the bone disease of long-term parenteral nutrition. Trans Assoc Am Phys 95:155-164

Lapresle J, Duckett S, Galle P, Cartier L (1975) Documents cliniques anatomiques   et biophysiques dans une encephalopathie avec presence de depots d'aluminium.   CR Soc Biol 169:282-287

Maloney NA, Ott SM, Alfrey AC, Miller NL, Sherrard DJ (1982) Histological quantitation of aluminum in iliac bone from patients with renal failure. J Lab Clin Med 99:206-216

Martin RB (1986) The chemistry of aluminum as related to biology and medicine. Clin Chem 32:1797-1806

McLaughlin AIG, Kazantzis G, King E, Teare D, Porter RJ, Owen R (1962) Pulmonary fibrosis and encephalopathy associated with the inhalation of aluminum dust. Brit J Industr Med 19:253-263

Miller RG, Kopfler FC, Kelty KC, Stober JA, Ulmer NS (1984) The occurrence of aluminum in drinking water. J Am Water Works Assoc 76:84-91

Mladenovic J (1988) Aluminum inhibits erythropoiesis in vitro. J Clin Invest 81:1661-1665

Nyholm NEI (1981) Evidence of involvement of aluminum in causation of defective formation of eggshells and of impaired breeding in wild passerine birds. Environmental Research 26:363-371

O'Hare JA, Murnaghan DJ (1982) Reversal of aluminum induced hemodialysis anemia by low-aluminum dialysate. N Engl J Med 306:654-656

Provan SD, Yokel RA (1988) Aluminum uptake by the in situ rat gut preparation. J Pharm Exp Therapeutics 245:928-931

Ward MK, Feest TG, Ellis HA, Parkinson IS, Kerr DNMS, Herrington J, Goode GL (1978) Osteomalacic dialysis osteodystrophy: evidence for a water-bourne etiological agent, probably aluminum. Lancet i:841-845

Zhu JH, Alfrey AC (1989) Mechanism of aluminum induced bone disease. Am Soc Nephrology Abstracts (in press)

Methylmercury Toxicity

THOMAS W. CLARKSON

Environmental Health Sciences Center, University of Rochester School of Medicine, Rochester, NY 14642, USA

ABSTRACT

Mercury vapor, emitted from both natural and man-made sources is distributed globally by the atmosphere and returns to the earth's surface in a water soluble form. Biomethylation converts mercury into a form avidly accumulated by fish. Human toxicology is characterized by damage mainly to the central nervous system. The in utero period is the most susceptible stage of the life cycle. Neuronal migration and cell division appear to be two key processes damaged in the developing nervous system. Biochemical lesions include inhibited protein synthesis and the assembly of microtubules.

KEY WORDS

Acid Rain, Neuropathology, Blood-Brain Barrier

INTRODUCTION

Two expert groups have reviewed the environmental toxicology of methylmercury compounds in the past year (WHO 1988, 1989; NIEHS 1989). This paper will summarize recent findings.

Environmental Fate of Methylmercury and Human Exposure

Since the discovery in the late 1960's of the natural methylation of mercury in the environment (Jensen and Jernelov 1969; Wood et al. 1968) and its bio-accumulation up aquatic food chains (Miller and Berg 1969), several new environmental factors have been found that affect methylmercury levels in fish. Models of the global cycle of mercury (for review, see Fitzgerald 1989) indicate that the atmosphere is the vehicle for transport and dispersion of mercury on the earth's surface. Most mercury (99% of the total concentration) is in a gaseous form and most of this (95-100%) is elemental mercury vapor. Its residence time in the troposphere is, on the average, about one year so that transport takes place over global distances. It is slowly converted to a soluble form and desposited on the earth's surface in rain water. It has been estimated that human activities account for about one-third of the total input to the atmosphere with the burning of fossil fuels, one of the major anthropogenic sources. In a recent study of a lake in north central Wisconsin that receives its water by seepage, it was found that atmospheric deposition can account for the total mass of mercury in fish (Fitzgerald and Watra 1989; Wiener 1989). It is of interest that the small fraction of monomethylmercury compounds recently detected in the general atmosphere may account for up to 50% of the methylmercury in the fish in this lake. Thus areas remote from a source of mercury may be contaminated and modest increases in atmospheric levels may be important in raising methylmercury in fish to levels exceeding human health advisories.

Global models (Fitzgerald 1989) estimate that the mean residence time of mercury in the oceans is about 350 years. Atmospheric deposition is the dominant input for ocean water. Given anthropogenic input at about one-third of the total, increased release of mercury over the industrial era should have contributed to ocean concentrations and therefore oceanic fish levels of mercury. The finding that levels in tuna fish caught at the turn of this century are on the average similar to current values has been used to argue against any increase of methylmercury in ocean fish. However, these values cover such a range that an increase in levels over recent times cannot be excluded.

Any discussion of the environmental fate of mercury is not complete without reference to the considerable challenge presented to the analytical chemist to measure and speciate the extremely low concentrations in environmental samples (for review, see Bloom 1989). The average concentrations in the ocean and freshwater are now reported in the range of 1ng Hg/liter and in the troposphere in the range of 1ng Hg/cubic meter. These values are for total mercury. Measurement of different physical and chemical species is an even greater challenge. Thus methylmercury compounds may account for no more than 5% of the total mercury but are very important in the aquatic bioaccumulation process. Such low levels require considerable care in the collection, transport and storage of samples. Analytical laboratories must undertake heroic efforts to decontaminate the working environment. In fact it has been stated recently that practically all environmental measurements on air and water samples made before 1980 cannot be trusted. This point was illustrated by Bloom (1989) who reviewed the literature to show that "accepted" values for mercury in uncontaminated surface waters had fallen from near 100 ng/l in the early 1960s to about 1 ng/l in the mid 1980s.

In addition to long distance atmospheric transport, two other factors are believed to affect methylmercury levels in freshwater fish. The acidification of lakes has now been established as an important cause of elevated levels of methylmercury in fish. The first observations described correlations between reduced pH and concentrations in freshwater fish. More recently, experimental studies have eliminated other variables by deliberately acidifying one of two otherwise identical lakes and demonstrating the increase of methylmercury in fish living in the acidified lake (Wiener 1989). Other experimental studies have shown that the pH effect is probably on the microorganisms responsible for methylation and demethylation, tipping the balance in favor of the methylation process (Rudd et al. 1980).

The impoundment of bodies of freshwater for hydroelectric purposes has also been associated with elevated levels of methylmercury in freshwater fish. This impoundment involves the damming and redirection of rivers to form hydroelectric reservoirs. This is followed by an increase of methylmercury in fish and in one case with an associated increase in levels in humans in communities dependent on fish from the impounded water (Manitoba Report, 1987). The long-term persistence of elevated levels is not known and may well vary from one hydroelectric system to another. The mechanisms responsible for this effect on methylmercury levels are under intense study at this time. It is believed that the raising and lowering of water levels according to electric power demands may be an important causal link. Thus more vegetation falls from the banks, adding to vegetation already inundated by the original impoundment, serving as a substrate for methylating microbes. This has the effect of eroding banks, inundating more vegetation and exposing mercury in sediments to atmospheric oxygen possibly resulting in oxidation to more soluble and biologically available compounds of mercury.

HUMAN TOXICOLOGY

The differences between adult and prenatal human poisoning is sufficiently great to justify separate treatment of each.

Adult

The main features of adult poisoning have been well described in many reviews e.g., WHO 1976. The effects of methylmercury appear to be restricted mainly to the nervous system and predominantly to the central nervous system. Severe effects include ataxia, constriction of the visual fields and loss of hearing. The first effect to appear is usually the symptom of paresthesia and may be the predominant effect in milder cases. These signs and symptoms appear only after a latent period usually lasting for weeks or months. The pathological changes in the brain, described in detail by Takeuchi (1968) in the Minamata victims, are characterized by highly focal areas of damage to the cerebellum, the calcarine fissure and the precentral gyrus. Neuronal cells are destroyed in these areas thus resulting in permanent damage.

The earliest biochemical lesion is inhibition of protein synthesis in the brain. This was first shown by Yoshino and coworkers (1966) in Japan and has been amply confirmed by other workers (for review see Verity et al. 1989). Dr. Verity summarized the studies from his laboratory indicating that inhibition of protein synthesis preceded other effects on the cell such as changes in mitochondrial function, ion flux, ATP content or RNA synthesis. The first biochemical lesion is probably at the initiation stage of protein synthesis and is likely due primarily to perturbation in the aminoacylation of select amino acid tRNAs (Haregawa et al. 1988).

A major question is whether this early biochemical lesion is actually causally related to the pathogenesis of selective neuronal damage seen in methylmercury intoxication in the adult brain. Thus Dr. Verity noted that inhibition of protein synthesis by other agents such as puromycin or cycloheximide are not associated with equivalent degrees of cell death as that associated with methylmercury. It is of interest that agents that reduce lipid peroxidation also protect against methylmercury-induced neuronal cytotoxicity raising the possibility that free radical formation may be the key biochemical event in the pathogenesis of adult brain damage by methylmercury.

Nevertheless, important questions remain for future research, for example, the long latent period and the selective susceptibility of different regions of the brain and of different populations of neuronal cells. With respect to the latter, Syversen (1977) and Jacobs et al. (1977) have proposed the neuronal cells differ in their ability to repair the initial damage induced by methylmercury.

Prenatal Damage

The sensitivity of the fetal brain to damage from methylmercury has been amply demonstrated by clinical and epidemiological studies in the Minamata outbreak in Japan and in the 1971-72 outbreak in Iraq, in studies of exposed populations in Canada and New Zealand and in animal experiments (for review, see Choi 1989). Takeuchi (1968) noted important distinctions between the neuropathology of adult and fetal brain damage. In contrast to the focal damage to specific areas of the adult brain, damage to the developing brain is more diffuse and general. Choi (1989) in his review of both the Japanese reports and his own observations on autopsies of prenatally exposed Iraqi infants concluded that "the characteristic neuropathological alterations... are attributable to disturbed neuronal migration, deranged cortical organization and differentiation and exuberant white matter astrocytic reaction."

Miura (1989) and Reuhl (1989) have recently reviewed studies to elucidate the cellular and biochemical lesions underlying this neuropathology. While it is recognized that many mechanisms could operate in the complex structural and functional changes taking place during the in utero development of the brain, there is now mounting evidence that the microtubular structures of the cytoskeleton are affected at the lowest doses and damage to these structures provides a plausible explanation for at least some of the main features of the neuropathology (for review, see Miura and Imura 1987). For example, microtubules disappear from cells in culture at concentrations of methylmercury about ten times lower than concentrations affecting other cellular elements. This effect is specific to methylmercury as compared to inorganic mercury. The mitotic arrest seen in both cell culture (Miura et al. 1978) and in granule cells in the developing mouse brain (Sager et al. 1982; 1984) is explained by the destruction of the microtubules of the spindle apparatus. According to studies using cultures of human fetal brain cells, disruption of microtubules may be the reason for inhibition of neuronal migration seen in the autopsy cases.

The reason why microtubular structures are specially sensitive to methylmercury is still a major research question. Microtubules are formed by a treadmilling process (Margolis and Wilson 1981). Formation of the tubule takes place at one end (polymerization) from dimers of alpha and beta tubulin and depolymerization occurs at the opposite end. This treadmilling process is responsible for a rapid turnover of microtubules, or at least for certain classes of these tubules. The tubulin monomers contain many SH groups and therefore avidly bind methylmercury (Vogel et al. 1985). Thus it is possible that even sub-stoichiometric amounts of methylmercury could disrupt the assembly process whereas disassembly continued uninterrrupted resulting in rapid depolymerization of the microtubules.

PHYSIOLOGICAL DISPOSITION AND TRANSPORT OF METHYLMERCURY

A. Kinetic Model of Disposition and Elimination

Observations on both animals and humans have established that elimination of mercury from the body is a first order process, that the daily rate of excretion is proportional to the simultaneous body burden (for review, see Clarkson et al. 1988). Berglund and Berlin (1969) interpreted this kinetic behavior to indicate that the elimination process was the rate limiting or slowest process as compared to the relatively more rapid processes of distribution and transport between organs and tissues within the body.

Measurements of methylmercury concentrations in samples of blood or hair during intake and elimination of mercury support this "single compartment" model. The head hair is a convenient sample for biological monitoring of methylmercury in humans (for review, see Suzuki 1988). The concentration of methylmercury in newly formed hair appears to be directly proportional to the simultaneous concentration in blood. Once incorporated into the formed hair strand, the concentration of methylmercury remains unchanged. Since hair grows at about 1 cm per month, each 1 cm segment measured from the scalp provides a recapitulation of prior blood levels. The parallel between blood and hair levels is remarkable.

To what extent this model applies to concentrations in other tissues particularly the target organ, the brain, is not so well established. Dr. Friberg (this conference) has reviewed evidence that methylmercury breaks down to inorganic Hg in the brain suggesting at least two "compartments" should be considered.

B. Mechanisms of Transport

Our recent studies have foccused upon mechanisms of transport of methylmercury within the mammalian body. In particular, we have examined the mechanisms of elimination from the body and of transport across the blood-brain barrier. In brief, we find no evidence to support the view that methylmercury acts as a lipid soluble species. Nor do we find evidence that the methylmercury cation jumps from one -SH group to another to cross the membrane by a "ligand exchange" medium as proposed for alkyl gold cations. Instead, our evidence indicates that methylmercury is transported as water soluble compounds on carrier mediated mechanisms intended for structurally related endogenous substrates. It is our contention that "Molecular Mimicry" will be important for other not-yet-studied transport processes and may be important in the biochemical mechanisms of damage.

The main steps in the elimination have been established for some time (for review, see Clarkson et al. 1988). Methylmercury is excreted mainly via the feces. The process starts with secretion from liver to bile. Part of the secreted methylmercury is reabsorbed in the small intestine. The remainder is converted to inorganic mercury by intestinal microflora and excreted in the feces.

The secretion from liver to bile appears to involve the formation of a methylmercury conjugate with reduced glutathione and the carrier mediated transport across the cannicular membrane of the liver cell into bile. The glutathione carrier on the cannicular membrane of the liver cell is known to accept conjugates of glutathione. Thus the benzyl conjugate is accepted. This is a much less bulky molecule than the methylmercury cation so that the latter should also be accepted by the same carrier.

Following secretion into bile, glutathione is subject to hydrolytic cleavage by gamma glutamyltranspeptidase (gamma-GT), and further degraded by peptidases to its constituent amino acids. Thus a number of methylmercury-thiol complexes have been found in bile in addition to the glutathione complex and may even predominate over the original glutathione complex.

Transport across the blood-brain barrier may involve a methylmercury-cysteine complex that travels on the large neutral amino acid in view of its structural similarity to L-methionine. This was first studied in detail by Dr. Hirayama (1980; 1985). She found that water soluble amino acids and peptides, all containing at least on free SH group, stimulated brain uptake

of methylmercury. Large neutral amino acids not containing SH groups blocked the stimulatory effects of the thiol compounds. Thus, she proposed that the large neutral amino acid carrier, known to be present on both membranes of the endothelial cells of brain capillaries, were involved in the transport of amino acid complexes of methylmercury. Our studies support Hirayama's proposal that the large neutral amino acid carrier is involved in the transport of water soluble thiol complexes of methylmercury (Aschner and Clarkson 1988a; 1988b).

SUMMARY

Recent advances in analytical techniques and sample collection procedures have led to major revisions in the global model for mercury. The ocean is now seen to have a more rapid turnover time with correspondingly smaller amount of total mercury. Anthropogenic input can now be regarded as important over the industrial era. Long distance transport in the atmosphere is seen to account for pollution of lakes remote from known sources. Acid rain and impoundment of water for hydroelectric purposes are factors leading to increase of methylmercury levels in freshwater fish.

The human toxicology of methylmercury is characterized by the important distinction between adult and prenatal poisoning. Both are due almost exclusively to selective damage to the brain. Damage to the adult brain is selective to certain areas whereas it is diffuse and generalized in the developing brain. The earliest biochemical lesion in the adult brain is inhibition of protein synthesis although it is not establsihed that this is the cause of the observed neuropathology. In the developing brain, the microtubules of the cytoskeleton appear to be one of the earliest targets and damage to these structures could explain much of the neuropathology.

The transport of methylmercury across key barriers in the body may depend upon mechanisms of molecular mimicry. The first step in elimination from the body--the secretion from liver to bile--probably involves a complex of methylmercury with glutathione that mimics other glutathione conjugates to secure "free passage" on the glutathion carrier at the cannicular membrane of the liver cell. Entry into the target tissue is achieved by forming a complex in blood plasma with L-cysteine that structurally mimics L-methionine that travels on the large neutral amino acid carrier.

REFERENCES

Aschner M, Clarkson TW (1988a) Distribution of 203 mercury in pregnant rats and their fetuses following systemic infusions with thiol-containing amino acids and glutathione during late gestation. Teratology 38:145-155
Aschner M, Clarkson TW (1988b) Uptake of methylmercury in the rat brain: effects of amino acids. Brain Res 462:31-39
Berglund F, Berlin M. (1969) Human risk evaluation for various populations in Sweden due to methylmercury in fish. In: Miller MW, Berg GG (eds)Chemical Fallout. Charles C. Thomas, Springfield, Illinois, p 423-432
Bloom N (1989) Analysis of low levels of mercury in environmental samples. Presented at symposium on Methyl Mercury Exposure: Effects on Human Neurotoxicity, July 6-7, 1989, NIEHS Research Triangle Park, North Carolina
Choi BH (1989) The effects of methylmercury on the developing brain. Prog Neurobiol 32:447-470
Clarkson TW, Hursh JB, Sager PR, Syversen TLM (1988) Mercury . In: Clarkson TW, Friberg L, Nordberg GF, Sager PR (eds) Biological Monitoring of Toxic Metals. Plenum Press, New York, p 199-246
Fitzgerald WF, Watras CJ (1989) Mercury in surficial waters of rural Wisconsin lakes. Sci Total Environ, in press.
Fitzgerald WF (1989) Global cycling of mercury. Presented at symposium on Methyl Mercury Exposure: Effects on Human Neurotoxicity, July 6-7, 1989 NIEHS, Research Triangle Park, North Carolina
Friberg L (1989) This conference
Haregawa K, Omata S, Sugano H (1988) In vivo and in vitro effects of methylmercury on the activities of amino-tRNA synthetase in rat brain. Arch Toxicol 62(2):470-472
Hirayama K (1980) Effect of amino acids on brain uptake of methylmercury. Toxicol Appl Pharmacol 55:318-323

Hirayama K (1985) Effects of combined administration of thiol compounds and methylmercury chloride on mercury distribution in rats. Biochem Pharmacol 34:3030-3032

Jacobs JM, Carmichael N, Cavanagh JB (1977) Ultrastructural changes in the nervous system of rabbits poisoned with methyl mercury. Toxicol Appl Pharmacol 39:249-261

Jensen S, Jernelov A (1969) Biological methylation of mercury in aquatic organisms. Nature 23:753-754

Manitoba Report (1987) Canada-Manitoba agreement on the study and monitoring of mercury in the Churchill River Division. Environment Canada, Hull, Quebec

Margolis RL, Wilson L (1981) Microtubule treadmills - possible molecular machinery. Nature 293:705-711

Marsh DO, Clarkson TW, Cox C, Myers GJ, Amin-Zaki L, Al-Tikriti S (1987) Fetal methylmercury poisoning. Arch Neurol 44: 1017-1022

Miller M, Berg GG (1969) Chemical FAllout. Charles C. Thomas, Springfield, Illinois

Miura K (1989) The role of the microtubule disruption by methylmercury in the expression of methylmercury cytotoxicity. Presented at symposium on Methyl Mercury Exposure: Effects on Human Neurotoxicity, July 5-7, 1989, NIEHS, Research Triangle Park, North Carolina

Miura K, Imura N (1987) Mechanisms of methylmercury cytotoxicity. CRC Crt Rev Toxicol 18:184-189

Miura K, Suzuki K, Imura N (1978) Effects of methylmercury on mitotic mouse gloma cells. Environ Res 17:453-471

NIEHS (National Institute of Environmental Health Sciences) (1989) Symposium on Methyl Mercury Exposure: Effects on Human Neurotoxicity, July 6-7, 1989, Research Triangle Park, North Carolina

Reuhl K (1989) Mechanisms of methyl mercury-induced brain dysgenesis. Presented at symposium on Methyl Mercury Exposure: Effects on Human Neurotoxicity on July 6-7, 1989, NIEHS, Research Triangle Park, North Carolina

Rudd JWM, Furutani A, Turner MA (1980) Mercury methylation by fish intestinal contents. Appl Environ Microbiol 40:777-782

Sager PR, Aschner M, Rodier PM (1984) Persistent differential alterations in developing cerebellar cortex of male and female mice after methylmercury exposure. Brain Res 12:1-11

Sager PR, Doherty RA, Rodier PM (1982) Morphometric analysis of the effect of methylmercury on developing mouse cerebellar cortex. Toxicologist 2:116

Suzuki T (1988) Hair and nails: advantages and pitfall when used in biological monitoring. In: Clarkson TW, Friberg L, Nordberg GF, Sager PR (eds) Biological Monitoring of Toxic Metals, Plenum Press, New York p 623-640

Syversen TLM (1977) Effects of methylmercury on in vivo protein synthesis in isolated cerebral and cerebellar neurons. Neuropath Appl Neurobiol 3: 225-236

Takeuchi T (1968) Pathology of Minamata Disease. In: Minamata Disease. Study group of Minamata disease, Kumamot Univ Press, Japan p 178-194

Verity MA, Cheung M, Sarafian T (1989) Biochemical mechanisms of methyl mercury neurotoxicity. Presented at symposium on Methyl Mercury Exposure: Effects on Human Neurotoxicity on July 6-7, 1989, NIEHS, Research Triangle Park, North Carolina

Vogel DG, Margolis RL, Mottet NK (1985) The effects of methylmercury binding to microtubules. Toxicol Appl Pharmacol 80:473-486

Wiener J (1989) Lake acidification: effects on concentration of methyl mercury in fish. Presented at symposium on Methyl Mercury Exposure: Effects on Human Neurotoxicity on July 6-7, 1989, NIEHS, Research Triangle Park, North Carolina

Wood JM, Scott KF, Roser CE (1968) Synthesis of methylmercury compounds by extracts of a methanogenic bacterium. Nature 200:173-174

WHO (1976) Environmental Health Criteria 1: Mercury. World Health Organization, Geneva, Switzerland

WHO (1988) Task group to update criteria document on mercury. December 6-12, 1988, Rochester, New York (unpublished)

WHO (1989) Update of criteria document on mercury. June 5-9, 1989, Bologna, Italy (Unpublished)

Yoshino Y, Mozai T, Nakao K (1966) Biochemical changes in the brain in rats poisoned with an alkyl mercury compound, with special reference to the inhibition of protein synthesis in brain cortex slices. J Neurol 13:1233-1230

Advances in the Toxicology of Inorganic Mercury

LARS FRIBERG and MAGNUS NYLANDER

Department of Environmental Hygiene, Karolinska Institute, S-104 01 Stockholm, Sweden

ABSTRACT

The accumulation of inorganic mercury from different sources (occupational, dental amalgams, methylmercury in fish) are discussed against new data on exposure, accumulation and biokinetics, including demethylation of methyl-mercury, interaction with selenium and long biological half-time of part of the inorganic mercury.

The central nervous system is still the critical organ in most cases, although emphazis is now on minor effects at concentrations previously considered unharmful. Animal and human studies show that the immune system and the kidneys may become critical organ in sensitive subjects. Differences in sensitivity is partly regulated by genetic factors.

The nature and occurence of possible effects associated with dental amalgams within small fractions of the general population are discussed. The possibilities of foetal effects due to exposure from dental amalgams are discussed.

KEY WORDS: brain, demethylation, dental-amalgam, immuno-toxicology, methylmercury

It has been known for more than 50 years (Stock 1939; Frykholm 1957) that amalgam fillings can release mercury vapor. During the last decade a number of studies have shown in more detail how mercury vapor is released from amalgam fillings and that the release rate of mercury vapor increases dramatically when the amalgam is stimulated by chewing (Gay et al. 1979; Svare et al. 1981; Abraham et al. 1934; Patterson et al. 1935; Vimy and Lorscheider 1985a,b; Nylander et al. 1988; Berglund et al. 1939; Aronsson et al. 1989). It may take an hour or longer until the release rate of mercury has declined to the basal value preceding the stimulation (Vimy and Lorscheider 1985a,b).

Friberg and Nylander (1987) and Clarkson et al. (1988a) have reevaluated and estimated the average uptake of mercury vapor and steady state contribution to blood, urine, brain and kidney from amalgam fillings in four different studies. For these estimations data on metabolic models for mercury were used. The empirical data on uptake thus achieved varies between 2.5-17.5 µg Hg/day with a mean of 8 ug/day. The predicted average contribution to brain levels is 12.5 µg Hg/wet weight, which can be compared with empirical data of an average increase of approximately 10 µg Hg/kg from autopsy data on subjects with varying numbers of dental amalgams as reported by Nylander et al. (1987). The predicted average increase in urinary concentration of 2.3 µg Hg/l is in reasonable agreement with the average increase in urinary excretion related to amalgam as reported by Langworth et al. (1988). The average kidney mercury content, generated from amalgam, was estimated to 690 ug Hg/kg, which may be compared to an average of about 400 µg Hg/kg derived from amalgam, as reported by Nylander et al. (1987) among amalgam-bearing subjects.

Part of the uptake of mercury in the body could be a result of absorption via the gastrointestinal route. We know that mercury is released from fillings and swallowed. The absorption from the gut is only 5-10%, however. Furthermore, the fraction of the absorbed mercury that penetrates the blood barrier

is very small. Already in 1956, Friberg (1956) showed that in rats less than a tenth of a percent of subcutaneously administered mercuric chloride penetrated the blood brain barrier. We have now similar results from studies on two Macaca Fascicularies monkeys exposed during 2-3 months intravenously to mercuric chloride (0.5 mg Hg/day), concentrations of mercury in the kidneys at the end of exposure was about 250 000 µg Hg/kg wet weight, but only about 50 µg Hg/kg in the brain (Friberg and Mottet 1989).

Both the predicted uptake of mercury vapor from amalgam and the observed accumulation of mercury in the body are average values. It is obvious that there is a substantial individual variation. Thus "worst case" situations, where the uptake will be several times higher than the average, can be expected.

The biokinetics of methylmercury is considered to follow a one-compartment model with a biological half-time of about 70 days on an average. It is on the other hand obvious from studies on both animals and humans (Berlin et al. 1975; Clarkson et al. 1988b) that a substantial demethylation takes place in several organs, e.g. kidneys and liver. The general opinion is that any bio-transformation in the Central Nervous System, the critical organ, is negligible.

Already Kitamura et al. (1976), examining 20 Japanese subjects without known excessive exposure to methylmercury, found a median concentration of total mercury in cerebrum of 97 µg Hg/kg wet weight of methylmercury only 12 µg Hg/kg, which means that almost 90% was inorganic mercury. Similar results have recently been reported by Takizawa (1986) when examining 16 individuals from a non-polluted district. Total mercury in cerebrum was on an average 78 µg Hg/kg wet weight compared to only 9 µg as methylmercury, which means that 88% of the mercury was in an inorganic form. The levels of inorganic mercury in the Japanese studies are about 10 times higher than can be explained by the exposure from amalgam, as judged by results from our Swedish study (Ny-lander et al. 1987) and one US study by Eggleston and Nylander (1987) from the Los Angeles area, where total mercury was on an average only about 10 µg Hg/kg wet weight. In 6 of the Swedish cases speciation was carried out and about 80% of the mercury was inorganic mercury. The difference between the Japanese and the Swedish and US populations studied could instead easily be explained by differences in exposure to methylmercury from fish consumption.

Takizawa (1986) also reports a very high accumulation of inorganic mercury in the brain of victims from the Minamata incident. In 30 autopsy cases who had died from 20-100 days up to 18 years after onset of symptoms of methylmercury poisoning on an average 70 to 80% of the mercury was present as inorganic mercury. The total mercury concentration varied from 350 to 21 400 µg Hg/kg.

In the kidneys the fraction of inorganic mercury in the Japanese studies was still higher, between 90-100%.

The Japanese results are very important, and taken at their face value, demonstrate that with time the fraction of inorganic mercury in both brain and kidneys will increase relative to organic mercury due to demethylation. Total mercury was determined by flameless atomic absorption and methylmercury with gas chromatography. Unfortunately, quality control data are not presented. Recent studies we have carried out in collaboration with Mottet et al., University of Washington, Seattle, on monkeys (Macaca Fascicularis) support the Japanese findings. The monkeys were exposed to high levels of methylmercury for a period of years and sacrificed during ongoing exposure or half a year up to two years after end of exposure (Mottet and Burbacher, 1988; Lind et al. 1988). At the end of the exposure period 10-33% of the mercury in the brain was in inorganic form. In monkeys who had been without mercury exposure for long periods, the relative concentration of inorganic mercury was still higher, about 90%. In three of the monkeys sacrificed during the exposure, selenium analyses were made. In all three monkeys a molar ratio of 1:1 for inorganic mercury and selenium was found (Nylander et al., reported during this conference, to be published). We used another analytical method for the speciation of the different mercury compounds, the method of Magos.

Theoretically, the inorganic mercury in the brain could be a result of a demethylation in other organs and a transport to the brain via the blood. As mentioned earlier only a fraction of a percent of absorbed inorganic mercury penetrates the blood brain barrier.

The kinetics of the accumulation of inorganic mercury will depend on the magnitude of the fraction of methylmercury that is demethylated, and the relation between the biological half-times for methylmercury and inorganic mercury. Tracer studies have shown (Hursh et al. 1976, 1980; Clarkson et al. 1988b; Newton and Fry, 1978) that the half-time in the brain for most of the inorganic mercury during the first month after exposure to metallic mercury vapor is only a few weeks. This is much shorter than that for methylmercury, which is approximately 70 days. On the other hand our studies on monkeys have clearly indicated that one fraction of the inorganic mercury has a very long half-time.

Miners exposed to inorganic mercury in Yugoslavian mercury mines had very high levels of mercury in several organs, including the brain, several years after end of exposure (Kosta et al., 1975). A recent Swedish study of deceased dentists (Nylander et al. 1989) has similarly shown high concentrations of mercury in several organs, including the pituitary gland, brain, and thyroid gland. At least two of the cases had been retired for several years.

The toxicological importance of the formation and accumulation of inorganic mercury from methylmercury is not known. We don't know in what form the mercury exists more than it behaves like inorganic mercury during analysis. Nor do we know where the inorganic mercury accumulates in the brain. As mentioned earlier, some data from our studies on monkeys, indicate that the inorganic mercury formed from methylmercury seems to accumulate bound to selenium. The mercury that accumulates in the brain after exposure to metallic mercury vapor is oxidized to mercuric mercury. There is reason to believe that this mercury, after exposure to high concentrations of mercury vapor, is accumulated together with selenium in a molar ratio of 1:1 or close to 1:1 as judged by result of analyses of brain, pituitary gland and thyroid gland of Yugoslavian miners and Swedish dentists (Kosta et al. 1975; Nylander and Weiner 1989).

The exact accumulation might well differ depending on the exposure type, as the ultimate localization will depend on where the original mercury species, methylmercury or metallic mercury vapor, was deposited.

The ongoing exposure to metallic mercury vapor can be estimated by analyses of mercury in urine or, if methylmercury exposure can be excluded, by analyses of blood. Urine or blood levels give no information of the actual accumulation of inorganic mercury in the brain or kidneys. For an evaluation of the concentrations of inorganic mercury accumulated as a consequence of a demethylation of methylmercury, there are at present no indicator media. The formation of inorganic mercury through demethylation is so small per time unit that it will be impossible to detect in e.g. urine.

We (Nylander et al., to be published) have found a close correlation between mercury in abdominal muscle and occipital cortex among those individuals where a correlation between amalgam load and levels in the brain were studied. The results are so far preliminary and have only been carried out for total mercury. It is not known to what extent the correlation reflects exposure to methylmercury or inorganic mercury.

The critical organ after long-term exposure to metallic mercury vapor is for the majority of people the central nervous system. At high exposure levels, tremor and/or severe behavioral changes dominate. The critical effects at lower exposure levels are unspecific psychasthenic and vegetative symptoms, often referred to as micromercurialism.

An increased occurrence of proteinuria in workers exposed to mercury vapor has been observed. There are two types of kidney damage, involving the glomeruli or the tubuli, that can occur. A nephrotic syndrome with massive albuminuria was first described in workers in an ammunition factory. The symptoms were explained as manifestations of hypersensitivity to mercury.

The nephrotic syndrome seems to occur only occasionally and may have an immunological background. It has been observed also among dark skinned subjects from the general population who have used so called skin lightening cream or soap containing a few percent mercury iodine or ammoniated mercury (for review, see Kazantzis 1978; WHO, under publication). The products which are manufactured by different European companies are sold as germicidal cream or soap. As a result of the use of skin lightening cream mercury concentrations of 200-300 μg Hg/l urine are often seen, which is well above any acceptable level in industry.

The other form of renal injury is tubular damage giving rise to an increased urinary excretion of low molecular proteins or certain tubular enzymes, N-acetyl beta glucosaminidase (NAG). There are reports indicating dose-related effects down to 50 μg Hg/g creatinine or even lower levels (Rosenman et al. 1986; Langworth et al. 1987).

During recent years a great interest has been shown in immunologic effects of inorganic mercury. The development of a nephrotic syndrome may, as mentioned above, have an immunologic background. There are also reports that workers exposed to mercury vapor in concentrations corresponding to only about 50 μg Hg/g creatinine have developed antibodies against a component in basal membranes (laminine) or developed circulating immune complexes (Lauwerys et al. 1983; Stonard et al. 1983).

Certain strains of mercury exposed rats develop immune mediated glomerulo-nephritis with an initial phase of antiglomerular basement reactivity followed by mesangial deposits (Bellon et al. 1982; Bowman et al. 1983; Michaelson et al. 1985). The animals can demonstrate a proteinuria sometimes including a nephrotic syndrome. Transiently raised concentrations of circulating immune complexes may appear in the animals, involving an unknown antigen or antigens (Hirsch et al. 1982; Houssin et al. 1983; Henry et al. 1988).

Certain strains of mice develop circulating immune complexes about 10 days after mercury treatment which form deposits in the glomerular mesangial areas as well as in vessel walls (Hultman et al. 1989). This runs parallel to an hyperimmunoglobulinemia (IgG1, less concentrations of IgG2a and 2b) with the occurrence of anti nucleolar reactivity (Hultman and Eneström 1988).

Details of the mechanisms behind the development of the mercury induced auto-immune disease are not known. It seems clear, however, that the mercury model represents a unique tool to evaluate the relationship between genetics and chemical induced immune dysregulation (Pelletier et al. 1987a,b).

During recent years there has been an intense debate in some countries (e.g. Sweden and USA) on possible health hazards from dental amalgams (SOS 1937; Ziff 1984), partly because it has been shown that mercury is released from amalgam and taken up by the body, and partly because there is a large number of individuals who claim that they suffer from a variety of invalidating symptoms which they associate with exposure to mercury from amalgam. No doubt these persons are invalidated by their symptoms, which include pains in joints and muscles, fatigue, dissiness, headache, gastrointestinal symptoms and tremor. A Swedish Task group (SOS 1987) recently concluded, however, that there are no studies admitting conclusions about the cause of the reported symptoms but that research in the field is very important.

Metallic mercury vapor penetrates the placental barrier as demonstrated in humans as well as in animals. Some recent studies of guinea pigs by Yoshida et al. (1986, 1987) showed considerably lower concentrations in blood, brain and kidneys but higher in liver of the fetus than in the mothers. Vimy et al. (reported during this conference) on the other hand, when studying the distribution of mercury after insertion in sheep of radioactive mercury in dental amalgams, found elevated concentrations both in blood and several tissues, including certain brain structures of the fetuses.

In some studies of animals as well as on humans, reproductive effects in the form of increased incidence of abortions and even malformations have been observed but there are also studies where such effects have not been demonstrated. In one Polish study (Sikorski et al. 1987) where dental

personnel had been exposed to high concentrations of mercury, a very high prevalence of spina bifida was observed. In a much larger Swedish study (Ericson and Källén 1989) where the exposure to mercury probably had been lower, no increased prevalence of malformations was reported. The upper 95% confidence limit for the risk ratio for spina bifida was high, however, 2.1, and the study is therefore not conclusive even if the results are not consistent with the results of the Polish study.

A riskestimation for inorganic mercury can be made. It is well documented that high exposure to mercury can give rise to toxic effects in the form of tremor and psychological changes. Kidney damage is also observed. A number of recent studies have shown subtle but clear effects from the central nervous system and the kidneys after exposure to elemental mercury vapor corresponding to average urinary concentrations of about 50 µg Hg/g creatinine. In some studies effects have been noted already at an exposure corresponding to average urinary levels of 20-35 µg/g creatinine. Effects at still lower mercury exposure are unknown. It is reasonable that effects may occur also at lower levels if larger groups of exposed people are examined, but the association between the degree of exposure and effects and response is not known.

The exposure due to amalgam constitutes a special problem as the number of exposed is so large. As judged from urinary concentrations of mercury, average exposure levels from dental amalgam fillings among the general population are below average exposure levels associated with effects among occupationally exposed individuals. The individual spread is considerable, however, and it is furthermore difficult to extrapolate from occupational exposure to the general population. The safety margin is therefore small, if at all existent.

A special problem is the fact that mercury can give rise to allergic and immunotoxic reactions, which are partly genetically regulated. There may well be a small fraction of the population which is particularly sensitive as has been observed in animal studies. Some of the kidney changes reported after exposure to inorganic mercury can have an immunological background. One may therefore suspect that kidney disease of clinical significance may occur occasionally also after low exposure. The consequences of such, and other immunological changes, can be important if the exposure is very prevalent in the population, as is the case for dental amalgam. Another consequence of an immunological etiology is that, based on present knowledge, it is not scientifically possible to set a limit for mercury, in e.g. blood or urine, below which in the individual case mercury related symptoms could not occur, as long as dose-response studies for groups of immunologically sensitive individuals are not available.

It is not possible to make a riskestimation for the inorganic mercury that accumulates in the central nervous system and kidneys as a result of demethylation of methylmercury. Dose-effect and dose-response relationships must not necessarily be the same as after exposure to metallic mercury vapor.

Metallic mercury vapor penetrates the placental barrier, and will accumulate in the fetus. In some studies different reproductive effects have been reported, in other not. Till we know more we think that the statement of a WHO Study Group in 1980 is prudent where it is said: The exposure of women of childbearing age to mercury vapor should be as low as possible. The Group was not in a position to recommend a specific value.

REFERENCES

Abraham JE, Svare CW, Frank CW (1984) The effect of dental amalgam restorations on blood mercury levels. J Dent Res 63:71-73.
Aronsson AM, Lind B, Nylander M, Nordberg M (1988) Dental amalgam and mercury. Biol Metals 2:25-30.
Bellon B, Capron M, Druet E, Verroust P, Vial MC, Sapin C, Girard JF, Foidart JM, Mahieu P, Druet P (1982) Mercuric chloride induced auto-immune disease in Brown-Norway rats: Sequential search for antibasement membrane anti-bodies and circulating immune complexes. Eur J Clin Invest 12:127-133.

Berglund A, Pohl L, Olsson S, Bergman M (1988) Determination of the rate of release of intra-oral mercury vapour from amalgam. J Dent Res 9:1235-1242.

Berlin M (1989) Memo concerning the use of mercury containing remedies. Report to WHO.

Berlin M, Carlson J, Norseth T (1975) Dose-dependence of methylmercury metabolism. Arch Environ Health 30:307-313.

Bowman C, Peters DK, Lockwood CM (1983) Anti-glomerular basement membrane autoantibodies in the Brown Norway rat: detection by a solide-phase radio-immunoassay. J Immunol Methods 61:325-333.

Clarkson TW, Friberg L, Hursh JB, Nylander M (1988a) The prediction of intake of mercury vapor from amalgams. In: Clarkson TW, Friberg L, Nordberg GF, Sager P (eds) Biological monitoring of metals. Plenum Press, New York, 247-264.

Clarkson TW, Hursh JB, Sager PR, Syversen TLM (1988b) Mercury. In: Clarkson TV, Friberg L, Nordberg GF, Sager P (eds) Biological monitoring of metals. Plenum Press, New York, 199-246.

Eggleston DW, Nylander M (1987) Correlation of dental amalgam with mercury in brain tissue. J Prosthet Dent 58:704-707.

Ericson A, Källen B (1989) Pregnancy outcome in women working as dentists, dental assistants or dental technicians. Arch Occup Environ Health (in press).

Friberg L (1956) Studies on the accumulation, metabolism and excretion of inorganic mercury (Hg^{203}) after prolonged subcutaneous administration to rats. Acta Pharmacol Toxicol 12:411-427.

Friberg L, Mottet NK (1989) Accumulation of methylmercury and inorganic mercury in the brain. Biol Trace Elem Res (in press).

Friberg L, Nylander M (1987) The release and uptake of metallic mercury vapour from amalgam. In: Mercury/amalgam - health risks. Report by an expert group. Stockholm, National Board of Health Welfare, 65-79 (in Swedish with English summary).

Frykholm KO (1957) Mercury from dental amalgam. Its toxic and allergic effects and some comments on occupational hygiene. Acta Odont Scand 22:1-108.

Gay DD, Cox RD, Reinhardt JW (1979) Chewing releases mercury from fillings. Lancet 8123:985-986.

Henry GA, Jarnot BM, Steinhoff MM, Bigazzi PE (1988) Mercury-induced auto-immunity in the MAXX rat. Clin Immunol Immunopathol 49:187-203.

Hirsch F, Couderc J, Sapin C, Fournie G, Druet P (1982) Polyclonal effect of $HgCl_2$ in the rat, its possible role in an experimental autoimmune disease. Eur J Immunol 12:620-625.

Houssin D, Druet E, Hinglais N, Verroust P, Grossetete J, Bariety J, Druet P (1983) Glomerular and vascular lgG deposits in $HgCl_2$ nephritis: role of circulating antibodies and of immune complexes. Clin Immunol Immunopathol 29:167-180.

Hultman P, Eneström S (1988) Mercury induced antinuclear antibodies in mice: characterization and correlation with renal immune complex deposits. Clin Exp. Immunol 71:269-274.

Hultman P, Skogh T, Eneström S (1989) Circulating and tissue immunecomplexes in mercury treated mice. J Clin Lab Immunol (in press).

Hursh JB, Clarkson TW, Cherian MG, Vostal JV, Mallie RV (1976) Clearance of mercury (Hg-197, Hg-203) vapor inhaled by human subjects. Arch Environ Health 31:302-309.

Hursh JB, Greenwood MR, Clarkson TW, Allen J, Demuth S (1980) The effect of ethanol on the fate of mercury vapor inhaled by man. J Pharmacol Exp Ther 214:520-527.

Kazantzis G (1978) The role of hypersensitivity and the immune response in influencing susceptibility to metal toxicity. Environ Health Perspect 25:111-118.

Kitamura S, Sumino K, Hayakawa K, Shibata T (1976) Mercury content in human tissues from Japan. In: Nordberg GF (ed) Effects and dose-response relationships of toxic metals. Elsevier, Amsterdam, 290-298.

Kosta L, Byrne AR, Zelenko V (1975) Correlation between selenium and mercury in man following exposure to inorganic mercury. Nature 254:238-239.

Langworth S (1987) Renal function in workers exposed to inorganic mercury. In: International Congress on Occupational Health, Sydney, Australia, 27 September - 2 October (abstract p 237).

Langworth S, Elinder CG, Åkesson A (1988) Mercury exposure from dental fillings. Swed Dent J 12:69-70.

Lauwerys R, Bernard A, Roels H, Buchet JP, Gennart JP, Mahieu P, Foidart JM (1983) Anti-laminin antibodies in workers exposed to mercury vapour. Toxicol Lett 17:113-116.

Lind B, Friberg L, Nylander M (1988) Preliminary studies on methylmercury biotransformation and clearance in the brain of primates: II. Demethylation of mercury in brain. J Trace Elem Exp Med 1:49-56.

Michaelson JH, McCoy JP, Hirszel P, Bigazzi PE (1985) Mercury-induced autoimmune glomerulonephritis in inbred rats. Surv Synth Path Res 4:401-411.

Mottet NK, Burbacher TM (1988) Preliminary studies on methylmercury biotransformation and clearance in the brain of primates: I. Experimental design and general observations. J Trace Elem Exp Med 1:41-47.

Newton D, Fry FA (1978) The retention and distribution of radioactive mercuric oxide following accidental inhalation. Ann Occup Hyg 21:21-32.

Nylander M, Friberg L, Eggleston DW, Björkman L (1989) Mercury accumulation in tissues from dental staff and controls in relation to exposure. Swed Dent J (in press).

Nylander M, Friberg L, Lind B (1987) Mercury concentrations in the human brain and kidneys in relation to exposure from dental amalgam fillings. Swed Dent J 11:179-187.

Nylander M, Friberg L, Lind B, Åkerberg S (1988) Effects of breathing patterns on mercury release. Appendix II to Clarkson TW, Friberg L, Hursh JB, Nylander M: The prediction of intake of mercury vapor from amalgams. In: Clarkson TW, Friberg L, Nordberg GF, Sager PR (eds) Biological monitoring of toxic metals, Plenum Press, New York, 247-264.

Nylander M, Weiner J (1989) Relationship between mercury and selenium in pituitary glands of dental staff. Br J Ind Med (in press).

Patterson JE, Weissberg BG, Dennison PJ (1985) Mercury in human breath from dental amalgams. Bull Environ Contam Toxicol 34:459-468.

Pelletier L, Pasquier R, Vial MC, Mandet C, Moutier R, Salomon JC, Druet P (1987a) Mercury-induced autoimmune glomerulonephritis: Requirement for T-cells. Nephrol Dial Transplant 1:211-218.

Pelletier L, Galceran M, Pasquier R, Ronco P, Verroust P, Bariety J, Druet P (1987b) Down modulation of Heymann's nephritis by mercuric chloride. Kidney Int 32:227-232.

Rosenman KD, Valciukas JA, Glickman L, Meyers BR, Cinotti A (1986) Sensitive indicators of inorganic mercury toxicity. Arch Environ Health 41:208-215.

Sikorski R, Juszkiewicz T, Paszkowski T, Szprengier-Juszkiewicz T (1987) Women in dental surgeries: Reproductive hazards in occupational exposure to metallic mercury. Int Arch Occup Environ Health 59:551-557.

SOS (1987) Mercury/amalgam - Health risks. Report by the LEK-Committee, Stockholm, National Board of Health and Welfare (in Swedish with English summary).

Stock A (1939) Die chronische Quesksilber- und Amalgamvergiftung. Zahnärztl Rundschau 10:371-377, 403-407.

Stonard MD, Chater BV, Duffield DP, Nevitt AL, O'Sullivan JJ, Steel GT (1983) An evaluation of renal function in workers occupationally exposed to mercury vapour. Int Arch Occup Environ Health 52:177-189.

Svare CW, Peterson LC, Reinhardt JW, Boyer DB, Frank CW, Gay DD, Cox RD (1981) The effect of dental amalgams on mercury levels in expired air. J Dent Res 60:1668-1671.

Takizawa Y (1986) Mercury content in recognized patiens and non-recognized patients exposed to methylmercury from Minamata Bay in the last ten years. In: Tsubaki T, Takahashi H (eds) Recent advances in Minamata disease studies, Kodansha Ltd, Tokyo, 24-39.

Vimy MJ, Lorscheider FL (1985a) Intra-oral air mercury released from dental amalgam. J Dent Res 64:1069-1071.

Vimy MJ, Lorscheider FL (1985b) Serial measurements of intra-oral air mercury: Estimation of daily dose from dental amalgam. J Dent Res 6:1073-1085.

WHO (1980) WHO Technical Report Series No 647 (Recommended Health-Based Limits in Occupational Exposure to Heavy Metals).

WHO. Environmental Health Criteria for Inorganic Mercury (under publication).

Yoshida M, Yamamura Y, Satoh H (1986) Distribution of mercury in guinea pig offspring after in utero exposure to mercury vapor during late ge gestation. Arch Toxicol 58:225-228.

Yoshida M, Aoyama H, Satoh H, Yamamura Y (1987) Binding of mercury to metallothionein-like protein in fetal liver of the guinea pig following in utero exposure to mercury vapor. Toxicol Lett 37:1-6.

Current Topics on the Toxicology of Inorganic Lead

STAFFAN SKERFVING

Department of Occupational and Environmental Medicine, University Hospital,
S-221 85 Lund, Sweden

ABSTRACT

Some recent information on the exposure, metabolism and effects of lead is summarized. *Exposure*: The lead exposure is decreasing rapidly in several countries, *partly* as a result of reduction of lead in petrol. *Metabolism: In vivo* determination of lead level in bone is a time-integrated lead-uptake index. Chelatable lead does *not* indicate the lead level in compact bone, and thus not the total body burden. *Effects:* Lead is an important cause of chronic renal disease with gout in some areas. Slight renal tubular effects occur at low uptakes (average blood-lead levels of about 300 µg/l). Possibly, in adults, effects on the blood pressure occur in the same range. Effects of lead on the CNS in fetuses/infants may occur at low uptakes (possibly already at B-Pbs in the range 100-150 µg/l).
Key words: Exposure, metabolism, skeleton, effect, kidney.

BACKGROUND

The literature on the toxicology of lead is enormous. A large number of reviews have been published, some recently (Skerfving 1988a and 1988b; U.S. ATSDR 1988; WHO 1989). Particularly comprahensive is the review by the U.S. EPA (1986). In the present paper, some current topics on the exposure, metabolism, and effects of lead will be discussed.

EXPOSURE

Lead has had a widespread use as pigment in house *paint* in some countries, and weathering, chalking, and peeling paint may cause exposure (U.S. ATSDR 1988). Organolead compounds are added to *petrol*. During combustion in the engine, organic lead is transformed into inorganic lead, and is emitted almost entirely as such. This causes exposure to inorganic lead, in particular in subjects living in areas with heavy traffic (Schütz et al 1984 and in press; Skerfving et al 1986). Also, *industrial emissions* may cause exposure in neighborhood populations (Schütz et al 1984 and in press; Skerfving et al 1986).

These, and other sources, may cause exposure *via* inhalation and ingestion (of food, drinking water, alcohol beverages, and soil/dust). In the general environment, ingestion is the major route.

We have studied the development in time of blood-lead levels (B-PB; as an index of exposure) in Southern Sweden. The geometric mean of 1,781 samples obtained during the period 1978-88 was 46.9 µg/l (Schütz et al, in press). There was a significant (p<0.001) decrease by an estimated 7% per year, both in rural and urban areas. In 134 children, who were sampled twice, the average decrease over a two-year period was 18% (p<0.001).

There has been a major deacrease of lead in petrol sold in Sweden; it amounted to 1,432 tons in 1978, but by 1988 had decreased to 542 tons. In a regression analysis, adjusting for sex, age, site of living (near smelter/urban/rural) and potential lead-exposing hobbies, the percentage of total variance explained was 16.4%. Adding petrol lead values in each year increased the this to 27.5%. It might be, that it takes some time before the lead emissions from petrol are reflected in the B-Pbs. Thus, a lag period was applied. The

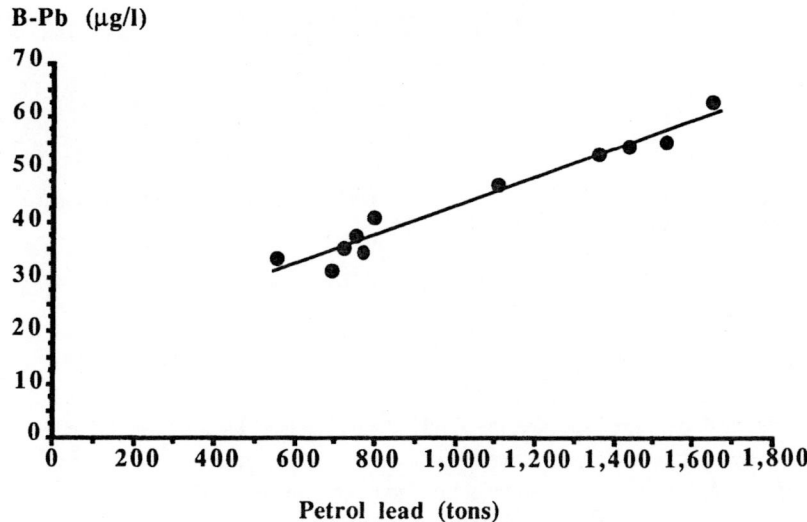

Fig.1. Relationship between lead in petrol sold in Sweden 1976-86 and average blood-lead levels (B-Pb; adjusted for sex, age, site of living, and potentially lead-exposing hobbies) 2 calendar years later (1978-88) in 1,773 children from the south of Sweden. Y=0.027X+16. r=0.96.

explained variance then increased to 30.5 and 32.7% (**Fig. 1**) for 1 and 2 years´ lag time, respectively; for 3 years it fell back to 30.9%.

However, other sources of lead exposure have decreased simultaneously, e.g. the use of lead-soldered cans and the lead emissions from industries in Sweden and from industries and traffic in other parts of Europe; lead in paint and drinking water is no problem in Sweden.In the regression analysis, the addtion of a term in calendar year to the above model containing petrol lead, increased the explained variance to 34.3%. Both the petrol term and the year term were highly significant (p<0.0001). The average estimated yearly decrease accounted for by petrol lead was 2.3%. The additional decrease was 4.2% per year. Thus, it seems that the reduction of the other sources has had an even greater effect than the reduction of petrollead. Thus, the slope of the regression line in **Fig. 1** overestimates the effect of a reduction of petrol lead.

Decreasing B-Pbs have been reported from several other industrialized countries (cf. Skerfving 1988a and 1988b).

A surprising finding in the studies of B-Pbs in children was an association between involuntary exposure to tobacco smoke (passive smoking) and B-Pb (Andrén et al 1988; Willers et al 1988). The mechanism is not known; it might be an effect on the clearance of lead particles from the airways.

METABOLISM

Traditionally, lead exposure has been estimated by analysis of lead levels in blood or urine, or by analysis of metabolites, accumulated and excreted as a result of the effects of lead on the heme metabolism (Skerfving et al 1985; cf. Skerfving 1988a).

One problem with these indices is that they mainly reflect the exposure during the last weeks/months (Schütz and Skerfving 1976; Schütz et al 1987c). However, some of the effects of lead are supposed to be due to chronic exposure. Thus, an exposure index covering a longer time period is valuable. Lead levels in calcified tissues have then been employed, as they accumulate calcium and have a slow turnover. Lead in

teeth has been used extensively in children (cf. Skerfving 1988a; Nørby Hansen et al 1989). Another possibility is lead levels in bone. This may be determined *post mortem* (Skerfving et al 1983) and in bone biopsies (Schütz et al 1987a).

In the last years, methods for *in vivo* determination of lead in bone by X-ray fluorescence have been developed. Several different methods have been applied. Originally, K X-rays from a [57]Co source were used for determinations in finger bone (Ahlgren et al 1976 and 1980) or tibia (Ahlgren et al 1976). Later, K X-rays from [109]Cd sources have been used for determination of lead in tibia and calcaneus (Somervaille et al, in press). Also, L X-rays from ^{125}I have been used for determination of lead in tibia (Wielopolski et al 1986). The K X-rays methods estimate the lead content in a larger volume of bone than do the L methods, which mainly measure superficial bone. This may be of importance, as there seems to be variations in lead content even within a small quantity of bone.

In all the methods, the time of measurement is about half an hour and the radiation dose is low, comparable to ordinary simple, diagnostic X-ray examinations.

By use of these methods, the metabolism of lead has been studied. In lead workers, there is an increase of lead levels in finger bone (Christoffersson et al 1984; Skerfving et al 1987), tibia, and calcaneus (Somervaille et al, in press) with rising time of exposure. The levels may thus serve as time-integrated indices of lead uptake, e.g. in epidemiological studies of effects.

After end of exposure, there is a slow decrease, in finger bone corresponding to a biological half life of about a decade (Christoffersson et al 1987a). Thus, even in lead workers, who had been retired for a long time, there may be considerable lead levels in finger bone (Christoffersson et al 1984 and 1987a; Schütz et al 1987c), tibia, and calcaneus (Somervaille et al, in press). However, the bones differ. Vertebral bone, has a much faster turnover, with a half life of a couple of years (Schütz et al 1987a).

When the accumulation of lead in the skeleton is large, there is a considerable "endogenous" lead exposure (Christoffersson et al 1984; Christoffersson et al 1987a and 1987b; Schütz et al 1987a and 1987c). In lead workers, this exposure may even make up for more than half of the B-Pb. The release of lead may increase in certain occasions. Thus, it has long been suspected, though never adequately shown, that bone tumours may cause release (cf. Schütz 1986). Recently, data have been presented to indicate that pregnancy and lactation (Thompson et al 1985), and postmenopausal osteoporosis (Silbergeld et al 1988) may cause mobilisation of the skeletal lead pool.

On the basis of this information, a metabolic model, containing two bone compartments (trabecular and compact/cortical bone) was established (Christoffersson et al 1987b). The model is a useful mean to predict the impact on the body burden of changes in exposure.

Lead excreted in urine ("chelated lead") after administration of a chelating agent (calcium disodium edetate, EDTA; or penicillamine, PCA) has often been used as an index of the total body burden of lead, including the skeleton.

However, the anatomical position of the chelatable lead has not been fully elucidated. Thus, we have studied the relationship between PCA-chelated lead and bone lead levels in lead workers (Schütz et al 1987b). There was a good (non-linear) correlation between chelated lead and B-Pb. However, the correlation with levels in finger bone, which is mainly cortical, was poor. However, the trabecular vertebral bone did correlate fairly well with chelated lead.

EDTA is a more potent chelating agent than is PCA. In 20 lead workers, there was a close correlation ($r=0.86$, $p<0.0001$) between urinary excretion of lead during 24 h after intravenous infusion of 1 g of the

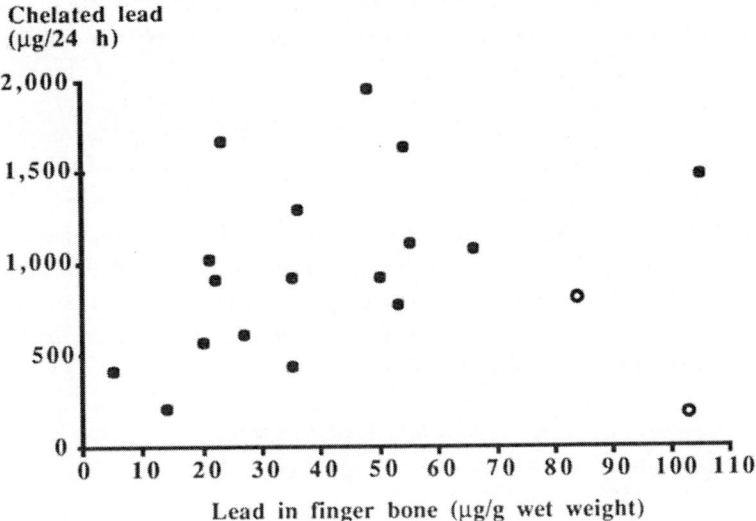

Fig. 2. Lead levels in finger bone and lead excretion in urine 24 h following EDTA chelation (Chelated lead) in 19 lead workers. A retired worker and one removed from active work are denoted by open symbols. Data from Tell et al (1988).

EDTA and the B-Pb (Tell et al, to be published). Chelation produced no significant change of lead level in either tibia or calcaneus. There was a significant correlation between chelated lead and bone lead (e.g. for calcaneus r=0.62) in currently exposed workers. However, there was no significant relationship if a retired worker and an inactive worker were included (r=0.14; **Fig. 2**). Chelatable lead thus mainly reflects blood and the soft tissue lead pool, which is only partly dependent upon the skeletal lead content, the major part of the total body burden (**Fig. 3**).

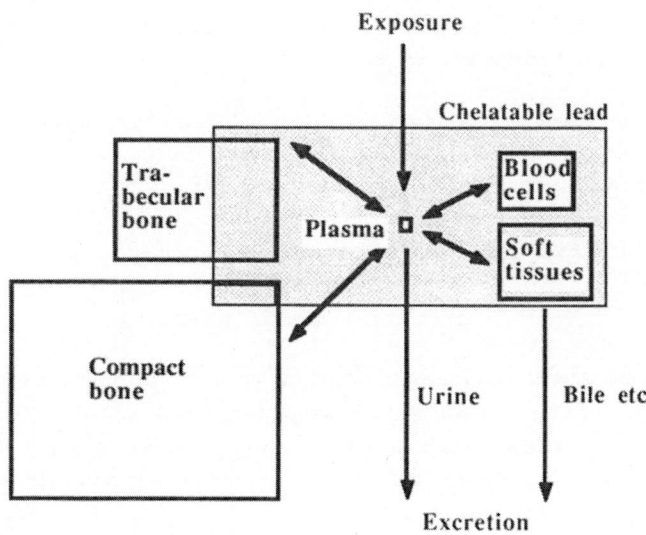

Fig. 3. Metabolic model for the turnover of lead in the body. The approximate anatomical position of the chelatable lead is indicated (dotted area).

EFFECTS

Lead exposure may, if sufficiently intensive and prolonged, adversely affect several organs (cf. Skerfving 1988a and 1988b).

Kidneys and blood pressure

It has been known for a century that lead may cause *kidney* damage (proximal tubular damage and chronic interstitial nephritis; cf. Goyer 1989). Also, lead may cause saturnine gout. It seems that, in some areas of the world, a considerable fraction of cases with uremia have a history of lead exposure, and increased body burden of lead.

Recently, it has become clear, that lead, even at a rather low intensity of exposure, may affect the kidney. We studied the kidney function in 80 smelter workers and 20 matched controls. In spite of rather low B-Pbs, there were slightly, but significantly increased excretions in urine of albumin and N-acetyl-ß-D-glucosaminidase (NAG), an enzyme present in the cells lining the proximal tubuli (Tell et al 1988; **Table 1**). In contrast, the classical indicator of an effect on the proximal tubuli, β_2-microglobulin, was not affected; an effect on the excretion of this protein has earlier been noted to be a rather late phenomenon in lead exposure. There were no associations between either albumin or NAG excretion in urine and finger bone lead levels, as measured by X-ray fluorescence. This may indicate that the tubular effects are not due to a longterm accumulation of lead, but rather dependent upon recent exposure. Accordingly, limited data indicate that an effect on NAG is reversible, if the exposure is interrupted (Coratelli et al 1988). The health impact of the slight changes are not known.

An association between lead exposure and *blood pressure* was reported already a century ago. This was long supposed to be the result of excessive exposure. However, several large studies of samples of the general population in the U.S., the U.K, the Netherlands, Canada, France, and Denmark (cf. Pocock et al 1988; Grandjean et al 1989) provides reasonably consistent evidence, that an association between B-Pb and blood pressure may occur already within the B-Pb range of the general population.

However, a causal relationship has not been established. There are possibilities of reversed causality (blood pressure affects the glomerular filtration rate, which, in turn, causes a decrease of lead excretion) and confounding (blood hemoglobin and alcohol intake are both associated both with B-Pb and blood pressure; Grandjean et al 1989).

Central Nervous System

Several recent crossectional studies have indicated effects on the central nervous system (CNS) of children (cognitive abilities and behavioural functions) at extremely low-level lead exposure (cf. U.K. MRC 1988; Nørby Hansen et al 1989).

Table. Blood lead and some urinary parameters (means) in 80 smelter workers and 20 referents.

	Exposed subjects	Referents	P
B-Pb (µg/l)	331	64	<0.0001
U-Albumin (mg/mmol crea)	2.1	1.2	=0.01
U-β_2-Microglobulin (µg/mmol crea)	11	9.3	NS
U-N-Acetyl-ß-glucose-aminidase (U/mmol crea)	0.22	0.12	<0.001

Further, three large prospective studies with follow-up from pregancy up to the age of four have indicated that CNS effects were associated with B-Pbs in the pregnant woman and/or the infant in the range of 100-150 µg/l (cf. U.K. MRC 1988), which is a concentration range generally found in many countries. However, there are enormous problems in controlling (and not over-controlling) confounding factors, and to avoid bias from reversed causality. Other, similar prospective studies have not shown clearcut effects.

Another effect on the CNS at rather low exposure, is the endocrine disturbances of the hypothalamic-pituitary-testis axis, reported in lead-exposed smelter workers (Gustafsson et al 1989), possibly mediated over an effect of lead on the turnover of selenium (Gustafsson et al 1987).

ACKNOWLEDGEMENTS

Some of the work quoted in this review was supported by the Swedish Work Environment Fund (grants 79-72, 82-0026, 82-160, and 86-0049) and the Swedish National Environmental Protection Agency (grant 611-644-1985-UF).

REFERENCES

Ahlgren L, Haeger-Aronsen B, Mattsson S, Schütz A (1980) In vivo determination of lead in the skeleton following occupational exposure. Brit J Ind Med 37:109-113.

Ahlgren L, Lidén K, Mattsson S, Tejning S (1976) X-ray fluorescence analysis of lead in human skeleton in vivo. Scand J Work Environ Hlth 2:82-6.

Andrén P, Schutz A, Vahter M, Attewell R, Johansson L, Willers S, Skerfving S (1988) Environmental exposure to lead and arsenic among children living near a glassworks. Sci Tot Environ 77:25-34.

Christoffersson JO, Schütz A, Ahlgren L, Haeger-Aronsen B, Mattsson S, Skerfving S (1984) Lead in finger-bone analysed in vivo in active and retired lead workers. Am J Ind Med 6:447-457.

Christoffersson J O, Schütz A, Skerfving S, Ahlgren L, Mattson S (1987a) Decrease of skeletal lead after end of occupational exposure. Arch Environ Hlth 41:312-8.

Christoffersson JO, Schütz A, Skerfving S, Ahlgren L, Mattsson S (1987b) A model describing the kinetics of lead in occupationally exposed workers. In: Ellis KJ, Yasumura S, Morgan WD (eds) In vivo body composition studies. The Institute of Physical Sciences in Medicine, London, IPSM 3, Bocardo Press Ltd, Oxford 1987b. ISBN 0 904181 50 2, pp 334-347.

Coratelli P, Giannattsio M, Lomonte C, Marzolla R, Rana F, L'Abbate N (1988) Enzymuria to detect tubular injury in workers exposed to lead: a 12-month follow-up. Contr Nephrol 68:207-11

Goyer RA (1989) Mechanism of lead and cadmium nephrotoxicity. Toxicol Lett, 46:153-62.

Grandjean P, Hollnagel H, Hedegaard L, Christensen JM, Larsen S (1989) Blood lead-blood pressure relations: Alcohol intake and hemoglobin as confounders. Am J Epidemiol 129:732-739.

Gustafson Å, Schütz A, Andersson P, Skerfving S (1987) Small effect on plasma selenium level by occupational lead exposure. Sci Tot Environ 66:39-43.

Gustafson Å, Hedner P, Schütz A, Skerfving S (1989) Occupational lead exposure and pituitary function. Int Arch Occup Environ Health 61:277-281.

Nørby hansen O, Trillingsgaard A, Beese I, Lyngbye T, Grandjean P (1989) A neuropsychological study of children with elevated dentine lead level: assessment of the effect of lead in diffrent socio-economic groups. Neurotoxicol Teratol 11:205-213.

Pocock SJ, Shaper AG, Ashby D, Delves HT, Clayton BE (1988) The relationship between blood lead, blood pressure, stroke, and heart attacks in middle-aged Britsih men. Environ Res 78:23-30.

Schütz A (1986) Metabolism of inorganic lead at occupational exposure. Bloms Boktryckeri, Lund, 45 pp. ISBN 91-7900-165-3. Thesis

Schütz A, Attewell R, Skerfving S (In press) Decreasing blood lead levels in Swedish children, 1978-88. Arch Environ Health.

Schütz A, Ranstam J, Skerfving S, Tejning S (1984) Blood-lead levels in schoolchildren in relation to industrial emission and automobile exhausts. Ambio, 13:115-117.

Schütz A, Skerfving S (1976) Effect of a short, heavy exposure to lead dust upon blood lead level, erythrocyte ∂-aminolevulinic acid dehydratase activity and urinary excretion of lead, ∂-aminolevulic acid, and coproporphyrine. Scand J Work Environ Hlth 3:176-184.

Schütz A, Skerfving S, Christoffersson JO, Ahlgren L, Mattson S (1987a) Lead in vertebral bone biopsies from active and retired lead workers. Arch Environ Hlth 42:340-360.

Schütz A, Skerfving S, Christoffersson JO, Tell I (1987b) Chelatable lead vs. lead in human trabecular and compact bone. Sci Tot Environ 61:201-9.

Schütz A, Skerfving S, Ranstam J, Gullberg B, Christoffersson JO (1987c) Kinetics of lead in blood after end of occupational exposure. Scand J Work Environ Hlth, 13:221-31.

Silbergeld EK, Schwartz J, Mahaffey K (1988) Lead and osteoporosis: Mobilization of lead from bone in postmenopausal women. Environ Res 47:79-94.

Skerfving S (1988a) Biological monitoring of exposure to inorganic lead. In: Clarkson TW, Friberg L, Nordberg GF, Sager PR (eds) Biological monitoring of toxic metals. Plenum Press, New York, pp 169-198

Skerfving S (1988b) Toxicology of inorganic lead. In Prasad A (ed) Essential and Toxic Elements in Human Health and Disease. Alan R Liss, New York, pp 611-630.

Skerfving S, Ahlgren L, Christoffersson JO, Haeger-Aronsen B, Mattsson S, Schütz A (1983) Metabolism of inorganic lead in occupationally exposed humans. Arch Hig Rada Toksikol 34:277-286.

Skerfving S, Ahlgren L, Christoffersson JO, Haeger-Aronsen B, Mattson S., Schütz A, Lindberg G (1985) Metabolism of inorganic lead in man. Nutr Res, Suppl 1:601-607.

Skerfving S, Christoffersson JO, Schütz A, Welinder H, Spång G, Ahlgren, L, Mattsson S (1987) Biological monitoring, by in vivo XRF measurements, of occupational exposure to lead, cadmium, and mercury. Biol Trace Elem Res 13:241-251.

Skerfving S, Schütz A, Ranstam J (1986) Decreasing lead exposure in Swedish children 1978-84. Sci Tot Environ 58:225-229.

Somervaille LJ, Nilsson U, Chettle DR, Tell I, Scott MC, Scütz A, Mattsson S, Skerfving S (In press) In vivo measurements of bone lead - a comparison of two X-ray fluorescence techniques used at three different bone sites. Phys Med Biol.

Tell I, Schütz A, Bensryd I, Nilsson U, Skerfving S (1988) Lead in the skeleton - metabolism and relation to effects. Report in Swedish from Department of occupational and environmental medicine, Lund university.

Tell I, Somervaille LJ, Nilsson U, Bensryd I, Schütz A, Chettle DR, Scott MC, Skerfving S (To be published) Chelated lead and bone lead.

Thompson GN, Robertson EF, Fitzgerald F (1985) Lead mobilization during pregnancy. Med J Austr 143:131.

U.K. Medical Research Council (1988) The neuropsychological effects of lead in children. A review of recent research 1984-1988. Medical Research Council, London.

WHO. Lead - Environmental aspects. Environmental Health Criteria. 1989. 106 pp.

U.S. ATSDR (1988) U.S. Agency for Toxic Substances and Disease Registry. The nature and extent of lead poisoning in children in the United States: A report to the congress. U.S. Department of Health and Human Services, Atlanta 1988.

U.S. EPA (1986) Environmental Protection Agency. Air quality criteria for lead. EPA-600/8-83/028aF, Vol I-IV. Environmental Protection Agency, Environmental Criteria and Assessment Office, Research Triangle Park, N.C.

Wielopolski L, Ellis KJ, Vaswani AN, Cohn SH, Greenberg A, Puschett JB, Parkinson DK, Fetterolf DE, Landrigan PJ (1986) In vivo bone lead measurements: A rapid monitoring method for cumulative lead exposure. Am J Ind Med 9:221-226.

Willers S, Schütz A, Attewell R, Skerfving S (1988) Relation between lead and cadmium in blood and involuntary smoking of children. Scand J Work Environ Hlth 14:385-389.

Cadmium Accumulation in Human Tissues: Relationship to Development of Toxic Effects

G. Nordberg, L. Gerhardsson, N.-G. Lundström, D. Brune, and P.O. Wester

Department of Environmental Medicine, University of Umeå, S-901 87 Umeå, Sweden

ABSTRACT

Toxicity of cadmium to the kidney is regarded as the critical effect in long term exposures. Another effect presently under consideration as a possible critical effect is carcinogenicity to the lung. Cadmium concentrations were analysed in autopsy specimens from smelter workers. Higher concentrations of cadmium were found in lung tissue from workers who died from lung cancer compared to those with other diagnoses. It is however not possible at present to relate lung cancer risk to elemental accumulation in lung tissue. Cadmium concentrations in the kidney of the workers were lower than those giving rise to renal damage. Evidence from the literature is reviewed concerning mechanisms of cadmium damage to the kidney and the critical concentration of cadmium in kidney cortex giving rise to proteinuria. After long term exposure approximately 200 µg/g has been reported in humans to be related to approximately a 10% risk of development of proteinuria. In other exposure situations other values can be expected.

KEY WORDS

Cadmium, critical concentration, lung cancer, kidney dysfunction

INTRODUCTION

Human cadmium exposure in the general environment is influenced by long range transport of cadmium via the atmosphere. By acidification of the environment cadmium can be mobilized from soils with subsequent increases of human exposure via the food chain. Environmental and industrial cadmium exposure to humans has attracted considerable interest during recent years and there is an ongoing debate concerning effects of cadmium that should be regarded as "critical" effects (Nordberg 1988). Usually the effect of cadmium on the kidney giving rise to proteinuria, is considered to be the critical effect. However, other effects of possible importance in discussions of critical effects are carcinogenicity, particularly to the lung and prostate, and reproductive toxicity mediated through toxicity to the placenta. This presentation will discus cadmium accumulation observed in human lung tissue and kidney cortex. The possibility that carcinogenicity to the lung by inhaled cadmium should be regarded as a critical effect will be considered. For the kidney, the critical concentration for development of proteinuria will be discussed.

Carcinogenicity Of Cadmium To The Lung

Available evidence on mutagenicity of cadmium has been summarized by IARC (1987a) and show both positive and negative results. Animal data on carcinogenicity of cadmium are sufficient according to an

evaluation by IARC (1987b). Of particular interest are animal studies (Takenaka et al 1983) showing in Wistar rats that inhalation of
cadmium chloride aerosol for 18 months gave rise to 15, 33 and 71%
lung cancers in the groups exposed to 12.5, 25 and 50 µg/m^3 respectively. Adenocarcinomas, epidermoid carcinomas and microepidermoid
carcinomas and combined epidermoid and adenocarcinomas were found.
Other cadmium compounds such as CdO, CdSO$_4$ and CdS may also give
rise to lung cancer in rats. It is not clear whether this effect can
also be reproduced in other animal species.

Epidemiologic data relating industrial cadmium exposure to occurrence of cancer in humans (IARC 1987) provide limited evidence for
carcinogenicity. Particularly lung cancer and prostatic cancer is
considered. Concerning lung cancer a study of 6995 cadmium exposed
workers by Kazantzis and Armstrong (1983) is of interest. For workers with more than 10 years of exposure there was an excess risk of
lung cancer.

As a basis for a discussion of a possible relationship between tissue concentrations of cadmium and some other elements in lung tissue
and an increased risk of lung cancer in humans, our data will be
presented concerning such tissue levels in smelter workers analyzed
at autopsy.

MATERIAL AND METHODS

Autopsy samples were obtained from 86 male workers at a non ferrous
metal smeltery in northern Sweden who died in the period 1975 to
1985. (Detailed description by Gerhardsson 1986.) The workers were
subdivided into groups according to diagnoses as shown in Table 1.
Mean length of retirement of the workers before death was 7.4 years.
There was a markedly increased incidence of lung cancer among workers in this smelter during the period that these samples were collected (cf Gerhardsson et al 1988).

Analysis of cadmium concentrations was made by neutron activation
analysis. Cadmium levels were also checked by atomic absorption
spectrophotometry and lead and zinc were analysed by atomic absorption spectrophotometry. Analytical quality assurance was certained
by analysis of certified reference materials and by intercomparison
of two different methods (AAS and NAA). Statistical comparisons were
made by use of non-parametic statistical methods (Kruskal-Wallis one
way analysis of variance, Mann-Whitney's U-test and Spearman's rank
order correlation).

RESULTS AND DISCUSSION OF TISSUE CONCENTRATIONS IN SMELTER WORKERS

Median values of cadmium, lead and zinc in lung, liver and kidney
tissue in relation to diagnosis are shown in Table 1. In lung tissue
relatively high concentrations of cadmium were found in the lung
cancer group (1A) in comparison to most other diagnoses, e.g. 1B (GI
cancer), 2B (cerebrovascular disease), 1C (other cancers), 2C (other
causes) (p<0.04). Concentrations of cadmium in the lung of all workers were higher than for both rural and urban controls. In the lung
cancer group there were 4 smokers, 2 ex-smokers and 1 non smoker.
There were 3 squamous cell carcinomas (1 was the non smoker), 2
anaplastic carcinomas, 1 adenocarcinoma and 1 unspecified. The distribution of smokers among the 86 smelter workers was 23 smokers, 26
non smokers 27 ex-smokers and 10 with unknown smoking habits. The
rural control group included 2 smokers, 2 ex-smokers, 10 non smokers
and 2 with unknown smoking habits. Cadmium in the lungs of exposed
workers was significantly raised (p<0.001) among the smokers compared with both ex-smokers and non smokers. The smoking pattern thus
can be an important factor as a partial explanation of the differences in cadmium values in lung tissue. However, occupational exposure
is also present since non smoking workers had a median cadmium value
in lung twice the value in rural controls, the latter group including also some smokers.

Table 1.

MEDIAN VALUES OF CADMIUM, LEAD AND ZINC IN LUNG, LIVER AND KIDNEY
TISSUES IN SMELTER WORKERS (groups 1A—2C) AND CONTROLS (groups 3A—3B)
(From Gerhardsson et al, Sci Total Env 50, 65, 1986)

Group	Cadmium (ppb, $\mu g\,kg^{-1}$ wet weight)			Lead (ppb, $\mu g\,kg^{-1}$ wet weight)			Zinc (ppm, $mg\,kg^{-1}$ wet weight)		
	Lung	Liver	Kidney	Lung	Liver	Kidney	Lung	Liver	Kidney
1A	390	2005	13950	320	425	280	13.7	80.5	24.2
1B	76	310	14000	140	380	150	11.6	66.0	31.0
1C	65	1025	10185	198	475	295	10.6	62.6	25.1
2A	246	660	17300	140	620	390	10.9	45.0	29.1
2B	114	1605	19725	151	665	440	12.0	64.4	32.0
2C	108		27600	130		400	10.0		39.0
All workers	166	965	17300	140	470	348	11.5	55.7	31.0
3A	40	350	3025	54	285	235	10.2	49.8	26.3
3B	79			39			12.2		

1A: Lung Cancer (n= 7), 1B: GI Cancer (n= 11), 1C: Other Cancers (n= 8),
2A: Cardiovascular Disease (n= 46), 2B: Cerebrovascular Disease (n= 8),
2C: Other Causes (n= 6), All Workers (n= 86), 3A: Rural Controls (n= 16),
3B: Urban Controls (n= 10).

Cadmium concentrations in lung, liver and kidney tissue were higher
in workers than in controls. Even if the kidney levels were considerably higher than in the lung tissue they did not reach concentrations which usually are associated with adverse effects on the
kidney as will be discussed later.

Lead concentrations were higher in lung tissue of smelter workers
than in rural and urban controls and the liver values for the workers were significantly higher than for the rural controls.

In addition to the observations on lead, cadmium and zinc, displayed
in detail in Table 1, concentrations of a number of other elements
were also determined in studies made parallel to the present one
(Gerhardsson et al 1985, 1988). It is of special interest that elevated concentrations of the following metals were recorded (concentrations in rural controls in parenthesis) antimony 260 ppb (32), arsenic 35 (7), chromium 450 (110), cobalt 17 (7), lanthanum 9.6 (5),
selenium 152 (110). When quotients were calculated between concentrations of the mentioned elements and selenium concentration in individual lung specimens, significantly higher quotients were found
in the lung cancer group. The quotients arsenic/selenium, cadmium/-
selenium, lanthanum/selenium, lead/selenium and antimony/selenium
were significantly higher in the lung cancer group than in other
groups of smelter workers and controls in 28 of 35 comparisons, indicating that low selenium concentration in combination with tissue
accumulation of one or several of the "toxic" elements may be of
possible importance in relation to the development of lung cancer
(Gerhardsson et al 1988). It is, however, not possible at present to
relate tissue concentrations to cancer risk.

CRITICAL CONCENTRATION OF CADMIUM IN THE KIDNEY

There is much more information available concerning kidney damage by
cadmium than on carcinogenicity. Kidney damage is an effect generally recognized to be caused by long term cadmium exposure both in
animals and in man. This effect is believed to be of the non stoch-
astic or deterministic type and a critical organ (or tissue) concentration can thus be discussed. Our own observations on cadmium
accumulation in the kidney of workers in a smelter do not add much
to the discussion about critical concentrations and the following
discussion therefore will be based almost entirely on data from the
literature.

Mechanism Of Damage

Ideally, for arriving at a quantitative estimate of relationships
between exposure and risk of adverse effects, the mechanism of damage should be known. Concerning cadmium damage to the kidney only
partial understanding is available and there is still considerable
uncertainty involved in the quantitative estimates of risk calculated on the basis of metabolic models and critical concentration estimates. Cadmium damage to the kidney could be of tubular type or
mixed glomerular/tubular type (Friberg 1950). Sometimes a glomerular
type lesion can occur without preceding tubular lesion (Bernard
1988). The mechanisms to be discussed and the quantitative estimates
of risk relate to the tubular type effect which is usually the
earliest damage.

Available knowledge concerning kinetics and metabolism of cadmium in
animals and man including a mechanistic model of the tubular damage
to the kidney has been presented by Nordberg et al 1985 and Nordberg
1989. After uptake from the g.i. tract or the lung, cadmium is initially bound to albumin in plasma. This form of cadmium is predominantly distributed to the liver and other tissues and only to a
limited extent to the kidney. Cadmium uptake in the liver induces
the synthesis of metallothionein, which binds to cadmium intracellularly. A small proportion of Cd-metallothionein (CdMT) from the
liver and other tissues will be released into plasma. Because of its
small size, Cd-MT in plasma will be efficiently filtered through the
glomerular membrane and subsequently reabsorbed in the proximal
renal tubule. In the kidney Cd-MT is taken up by lysosomes and the
metallothionein degraded so as to release Cd-ions which will quickly
bind to other ligands in the cell. It is believed that this non-MT
bound fraction of Cd in the cell produces toxicity. Since renal
cells can also synthesize metallothionein such metallothionein synthesis in renal cells can bind Cd and protect from renal damage (Jin
et al 1987). It is believed that when non-metallothionein Cd in the
kidney cells exceeds a certain concentration cell damage will occur.
Although experiments have been reported where both CdMT and non-MT
bound Cd in kidney tissue have been measured (Nomiyama and Nomiyama
1982), a clear relationship between non-MT Cd and the development of
damage has not yet been demonstrated. Also, the intracellular biochemical target responsible for cell damage in Cd-toxicity to the
kidney has not yet been clearly identified. Interference with membrane stability, possibly mediated via interference with cellular
calcium balance is one possibility (Jin et al 1987).

Estimates Of Critical Concentrations Of Cd In Renal Cortex

Although it would be desirable in the future to be able to discuss
concentrations of non-metallothionein cadmium in relation to development of renal toxicity, this is not yet possible. Concentration
of total Cd in kidney cortex is the best measurement to be related
to renal damage. Animal experiments in different species with
various routes of exposure have been summarized by Kjellström
(1986). It can be concluded from these data that tubular proteinuria
and/or morphological changes in the renal tubules were observed with
long term exposures (months to years) at kidney cortex concentrations around 200 µg/g wet weight. Histological changes of less
severe type (horse, rabbit, rat) were sometimes observed at lower
concentrations (about 50 ppm) whereas total proteinuria and/or β-2-
microglobulinuria was in one study not observed until 450 or 800 ppm
was reached (monkey).

It has been shown in animal experiments (e.g. Bernard et al 1981a;
Nomiyama and Nomiyama 1982) that in continuous exposures the cadmium
concentration in kidney cortex increases continuously until renal
tubular damage is manifest. At this point urinary cadmium excretion
increases and cadmium concentrations in renal cortex level off or
even decrease. A similar course of events is likely to be present in
humans exposed to cadmium and this makes it difficult to interpret
autopsy data from persons who suffered more severe renal disease induced by cadmium. In such cases comparatively low concentrations of
cadmium are found in the kidney cortex.

During the last 10 years in vivo neutron activation analysis of cadmium in kidney and liver has been used for studies of relationships
between cadmium concentrations and occurrence of effects in Cd exposed persons with only slight renal tubular damage. Since liver
concentrations of cadmium do not change when tubular damage develops
in the kidney, liver concentrations of cadmium can be used to evaluate the stage of exposure as shown by Roels et al (1981b) and Ellis
et al (1981). In vivo neutron activation data from those two studies
constitute the best information available to make an estimate of
the critical concentration of cadmium in the renal cortex of humans.
Based on observations of kidney cadmium, liver cadmium and the occurrence of proteinuria, β-2-microblobulinuria and albuminuria,
Roels et al (1981b) concluded that the critical concentration of Cd
in kidney cortex is found in the range 160-285 mg/kg. The authors
used a factor of 1.5 between renal cortex concentration and total
concentration in the kidney, whereas a more accurate number would be
1.25 (Svartengren et al 1986). If 1.25 is used the mentioned range
would correspond to 138-238 mg/kg. In another paper Roels et al
(1983) concluded that 10% of workers would have renal dysfunction at
216 ppm in cortex (if recalculated with factor 1.25, 180 ppm).

Ellis et al (1984), on the basis of data from 30 controls and 31
active workers calculated a dose response relationship between pro-
bability of kidney dysfunction and kidney Cd or liver Cd. From this
paper it can be seen that a 10% response rate occurs at about 22 mg
Cd in the whole kidney, correspoding to 190 ppm in kidney cortex
(factor 1.25).

There is thus an apparent agreement between estimates by different
authors and a reasonable similarity between human and animal data
concerning the critical concentration of cadmium in kidney cortex.
The estimate implies a 10% risk of development of kidney dysfunction
at approximately 200 ppm wet weight in kidney cortex. It should be
kept in mind, however, that there are several uncertainties involved
in arriving at these estimates of critical concentrations and the
apparent concordance between two independent studies does not necessarily consitute evidence of a high precision. Based on considerations from evidence concerning the mechanism of damage several factors including type and intensity of exposure might influence the
critical concentration.

It seems reasonable, as a contribution to risk assessment, to use
the estimated critical concentrations in combination with information about the metabolic model for cadmium in calculations of dose-
response relationships for long term exposure to Cd. Such calculations have been presented e.g. by Hutton et al (reviewed by Kjellström 1986) and provide estimates of dose response relationships
which agree fairly well with direct estimates of exposure versus
response (Friberg et al 1986).

REFERENCES

Bernard A, Lauwerys R, Gengoux P (1981a) Characterisation of the
    proteinuria induced by prolonged oral administration of cadmium in
    female rats. Toxicology 20:345-357
Bernard AM, Lauwerys R (1988) Decrease of erythrocyte and glomerular
    membrane negative charges in chronic cadmium poisoning. Brit J In-
    dust Med 45:112-115
Ellis KJ, Morgan WD, Zanzi I, Yasamura S, Vartsky D, Cohn SH (1981)
    Critical concentrations of cadmium in human renal cortex: dose ef-
    fect studies in cadmium smelter workers. J Toxicol Environ Health
    7:691-703
Ellis KJ, Yuen K, Yasumura S, Cohn SH (1984) Dose-response analysis
    of cadmium in man: body burden vs kidney dysfunction. Environ Res
    33:216-226
Friberg L (1950) Health hazards in the manufacture of alkaline accu-
    mulators with special reference to chronic cadmium poisoning. Acta
    Med Scand 138 (supp 240):1-124
Friberg L, Elinder CG, Kjellström T, Nordberg GF (1986) Cadmium and
    health, a toxicological and epidemiological appraisal, vol II, Ef-
    fects and response, CRC Press, Cleveland, pp 303

Gerhardsson L, Brune D, Nordberg GF, Wester PO (1985) Protective effect of selenium on lung cancer in smelter workers. Br J Ind Med 42:617–626

Gerhardsson L (1986) Trace elements in tissues of deceased smelter workers – relationship to mortality. Umeå university medical dissertations

Gerhardsson L, Brune D, Nordberg GF, Wester PO (1986) Distribution of cadmium, lead and zinc in lung, liver and kidney in long-term exposed smelter workers. Sci Tot Environ 50:65–85

Gerhardsson L, Brune D, Nordberg GF, Wester PO (1988) Multielemental assay of tissues of deceased smelter workers and controls. Sci Tot Environ 74:97–110

Hutton M (1983) Evaluation of the relationships between cadmium exposure and indicators of kidney function. London Chelsea College, University of London, 46 pp (MARC Report No 29)

IARC (1987a) IARC Monographs. Genetic and related effects, suppl 6, IARC/WHO, Lyon pp 729

IARC (1987b) IARC Monographs. Overall evaluations of carcinogenicity, suppl 7, IARC/WHO, Lyon, pp 440

Jin T, Leffler P, Nordberg GF (1987) Cadmium-Metallothionein nephrotoxicity in the rat – Transient calcuria and proteinuria. Toxicology 45:307–317.

Kazantzis G, Armstrong BG (1983) A mortality study of cadmium workers in seventeen plants in England. In: Wilson D, Volpe RA (eds) Proceedings of the Fourth International Cadmium Conference, Munich 1982, London, Cadmium Association, pp 139–142

Kjellström T (1986) Renal effects. In: Friberg L, Elinder C, Kjellström T, Nordberg G (eds) Cadmium and health: a toxicological and epidemiological appraisal, vol II. CRC Press, Boca Raton, FL, 22–109

Nomiyama K, Nomiyama H (1982) Tissue metallothionein in rabbits chronically exposed to cadmium with special reference to the critical concentration of cadmium in the renal cortex. In: Foulkes (ed) Biological roles of metallothionein. Elsevier/North Holland, Amsterdam, pp 47–67

Nordberg GF, Kjellström T, Nordberg M (1985) Kinetic model of cadmium metabolism. In: Friberg, Elinder, Kjellström, Nordberg (eds) Cadmium and Health, volume 1: Exposure, dose and metabolism. CRC Press, Boca Raton, Fla, 179–197

Nordberg GF (1988) Current concepts in the assessment of effects of metals in chronic low-level exposures – considerations of experimental and epidemiological evidence. Sci Tot Environ 71:243–252.

Nordberg GF (1989) Modulation of metal toxicity by metallothionein. Biol Trace Element Res (in press)

Roels HA, Lauwerys RR, Buchet JP, Bernard A, Chettle DR, Harvey TC, Al-Haddad IK (1981b) In vivo measurement of liver and kidney cadmium in workers exposed to this metal: its significance with respect to cadmium in blood and urine. Environ Res 26:217–240

Roels H, Lauwerys R, Dardenne AN (1983) The critical level of cadmium in human renal cortex: a reevaluation. Toxicol Lett 15:357–360

Svartengren M, Elinder CG, Friberg L, Lind B (1986). Distribution and concentration of cadmium in human kidney. Environ Res 26:217–240.

Takenaka S, Oldiges H, Koenig H, Hochrainer D, Oberdörster G (1983) Carcinogenicity of cadmium aerosols in Wistar rats. J Natl Cancer Inst 70:367–371.

Factors Aggravating Cadmium Health Effects in Old Animals

KAZUO NOMIYAMA, HIROKO NOMIYAMA, ABDEL-AZIZ M. KAMAL,
OSAMU YAMAZAKI, YASUTAKA ISHIMARU, MADHUSUDAN G. SONI, and
HITOSHI OHSHIRO

Department of Environmental Health, Jichi Medical School, Minamikawachi-machi, Tochigi,
329-04 Japan

ABSTRACT

Critical daily dose of cadmium in residents in cadmium-polluted areas in
Japan, were quite diverse by area. Some elucidations for the diversion shall
be discussed based on our recent experimental data.
1) Aging aggravated cadmium health effects in rabbits.
2) Excessive dietary zinc and/or copper aggravated cadmium health effects in
old animals, but did not in young animals. The effects were quite diverse by
age.
3) Low-protein diet was a potent factors aggravating cadmium health effects
especially in old rats.
4) It may be difficult, therefore, to estimate a single and unanimous
critical daily dose of cadmium, but it is highly recommended for old
residents in cadmium-polluted areas, and probably also for old cadmium
workers, to avoid excessive exposure to zinc and/or copper, and to take
enough protein to avoid aggravating cadmium health effects.

KEY WORDS: Cadmium Toxicity, Aging, Zinc, Copper, Protein

INTRODUCTION

Health surveys were performed on residents in cadmium-polluted areas in
Japan (Shigematsu, 1989). Cadmium health effects were observed in 5
districts with minimal daily intake of 139-177 ug in Akita, while no health
effects were observed in other areas with maximal daily intake of 300 ug in
Fukuoka. We may wonder why so diverse the critical daily dose of cadmium to
induce health effects by area was, and what factors induced such a large
diversion by area in cadmium health effects.

EPIDEMIOLOGICAL STUDIES SUGGESTED POSSIBLE FACTORS AGGRAVATING CADMIUM HEALTH
EFFECTS

Cadmium-induced health effects were compared among 236 residents in 3
cadmium-polluted areas in Toyama, Gunma and Ibaragi prefectures, relating
urine cadmium, which represented for the dose level of cadmium (Nomiyama et
al, 1983). Multivariate statistical analysis for elevated urinary β_2-
microglobulin indicated that the change of square of multiple correlation
coefficient, that is, the contribution ratio, of area was 48 %, age 7%, and
urine cadmium only 1%.

It is also peculiar that urinary β_2-microglobulin increased in parallel with
urine copper, but not with urine cadmium (Shigematsu et al, 1979). We may
even wonder so-called cadmium health effects being caused by copper exposure.

We may presume that cadmium health effects are diverse by several aggravating
factors of the area, such as simultaneous contamination with other metals
such as zinc and copper, chemical form and intestinal absorption of cadmium,
and daily intake of protein, by depressing detoxication of tissue cadmium by
suppressed metallothionein induction.

We then performed a series of experiments to clarify factors aggravating cadmium health effects.

EXPERIMENT 1: EFFECTS OF AGING ON CADMIUM HEALTH EFFECTS (Nomiyama et al, 1984)

First of all, we examined whether cadmium health effects was aggravated by aging or not.

Experimental animals were 63 male rabbits of 1 week, 4 months or 17 months old, and were given subcutaneous injections of cadmium at a dose level of 0, 0.1 or 0.25 mg/kg, 5 times a week, for 24 weeks. Every other week, urine and blood were subjected to determine 50 items for assessing cadmium health effects.

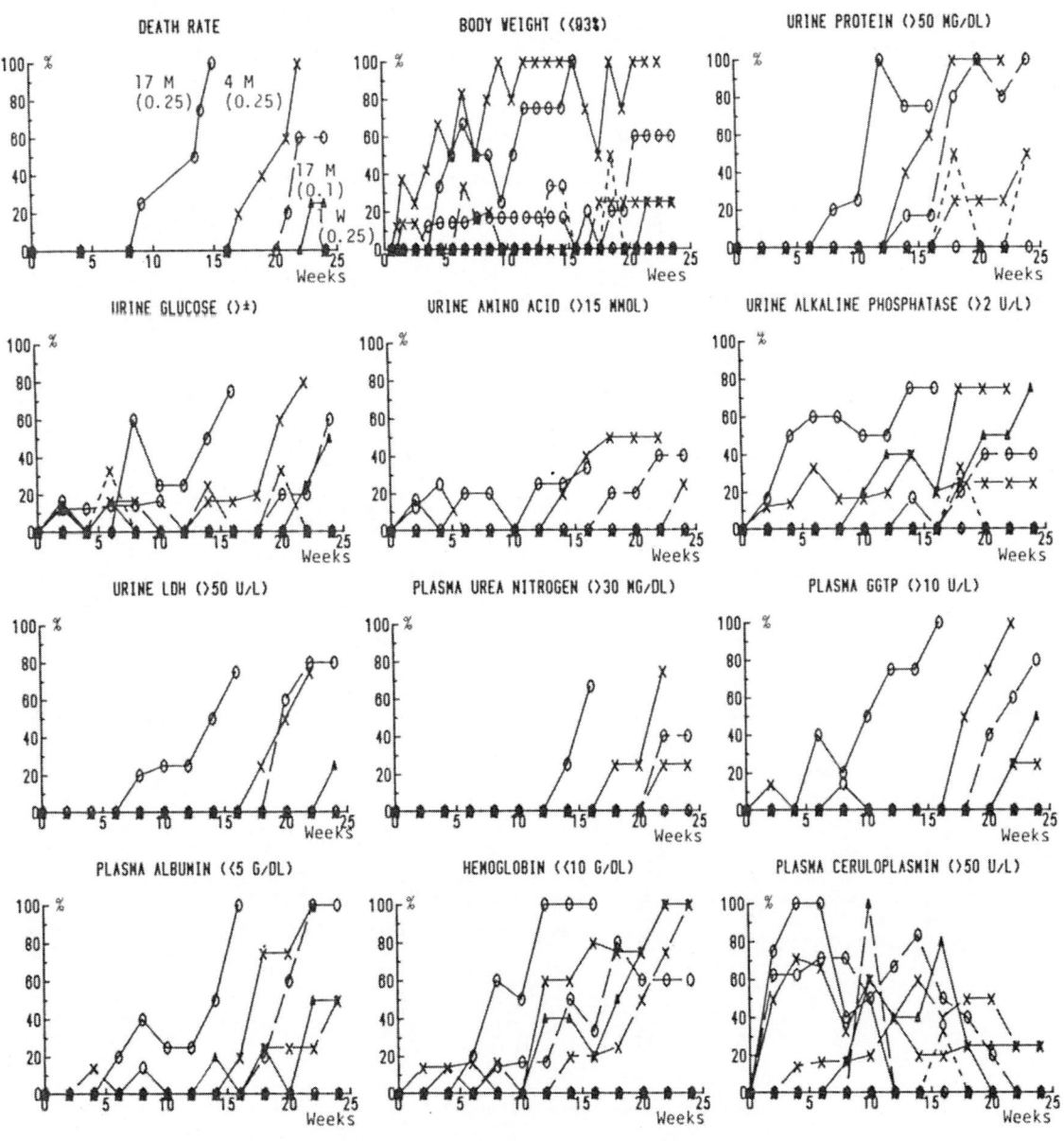

Fig. 1 Aging, a factor aggravating chronic toxicity of cadmium in rabbits (Nomiyama et al, 1984)

Fig. 1 indicate that accumulated rate of death was highest in 17 months old rabbits, followed by 4 months old and then 1 week old rabbits. Incidences of proteinuria, glycosuria, aminoaciduria and enzymuria were also highest in 17 months old rabbits.

Incidences of other health effects, such as elevated plasma urea nitrogen, γ-glutamyl transpeptidase, plasma albumin, hemoglobin and plasma ceruloplasmin were also aggravated by aging.

Following experiments on the effects of metals other than cadmium or of low-protein diet on cadmium health effects, were performed, when applicable, in combination with aging, a potent factor aggravating cadmium toxicity, because most residents suffering from cadmium health effects were the aged above 50 years old.

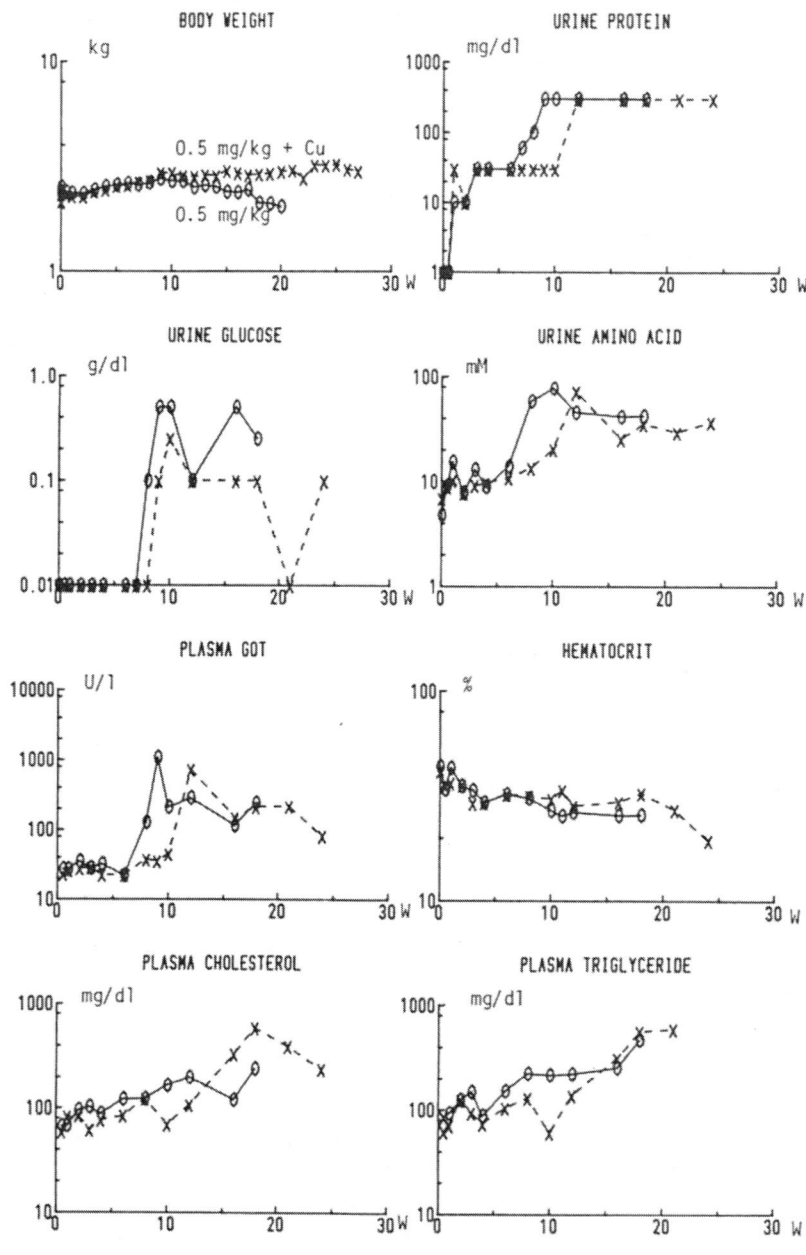

Fig. 2 Preventive effects of intravenous copper on cadmium health effects in rabbits (Nomiyama and Nomiyama, 1983)

EXPERIMENT 2: EFFECTS OF INTRAVENOUS COPPER SUPPLEMENTATION ON CADMIUM HEALTH EFFECTS (Nomiyama and Nomiyama, 1983)

We then examined the effects of intravenous copper injections on cadmium health effects.

Thirty-three male rabbits of 2 months old were given subcutaneous injections of cadmium at a dose level of 0 or 0.25 mg/kg, 7 times a week, for 24 weeks, simultaneously with intravenous copper at a dose level of 0 or 50 μg/kg, once a week.

As seen in Fig. 2, body weight loss was not so serious in the copper-supplemented group. The increases in urinary excretion of protein, glucose and amino acids delayed in the copper supplemented group by 1-4 weeks compared with the not-supplemented group. In the copper-supplemented group, plasma aspartate aminotransferase (GOT), alanine aminotransferase (GPT) and γ-glutamyl transpeptidase (γ-GTP) were elevated, and hemoglobin and hematocrit decreased later than those in the not-supplemented group. Cadmium, copper and zinc concentrations in urine increased in the copper supplemented group later than those in the not-supplemented group as well. Copper supplementation, however, could not prevent the decrease in hepatic and renal copper.

The above results may indicate that copper supplementation do not completely prevent nor treat cadmium health effects (Nomiyama and Nomiyama, 1982; 1986).

EXPERIMENT 3: EFFECTS OF EXCESSIVE DIETARY ZINC ON CADMIUM HEALTH EFFECTS (Nomiyama et al, 1984)

It was well known that zinc treatment alleviated cadmium-induced testicular necrosis and hypertension (Parizek, 1957; Perry et al, 1977). On the other hand, toxic dose of zinc was reported 100 times as much as required dose in animals (NRC, 1979).

We then examined whether dietary zinc supplementation alleviated cadmium health effects in young animals as well as in old animals.

Totally 81 male rats aged 2 and 18 months were fed with pelleted food (15 g/day) containing 12 (nutritional requirement) (NAS, 1972) or 120 mg/kg zinc, and given subcutaneous injections of cadmium at a dose level of 0, 0.1 or 0.25 mg/kg, 6 times a week, for 14 weeks. Totally 70 items were determined to assess effects of excessive dietary zinc on cadmium health effects, at the end of the experiment.

As seen in Fig. 3, excessive dietary zinc, 10 times as much as required dose, did not induce adverse effects in young rats, but even indicate a tendency to alleviate cadmium health effects such as loss of body weight in young rats.

On the contrary, a large number of old rats, fed with pelleted food containing 120 mg/kg zinc, died of cadmium, whereas no young rats died of cadmium. Excessive dietary zinc also increased plasma urea nitrogen and decrease hematocrit as well as plasma ceruloplasmin. Results were quite diverse in between old and young rats.

EXPERIMENT 4: EFFECTS OF EXCESSIVE DIETARY ZINC PLUS COPPER ON CADMIUM HEALTH EFFECTS (Nomiyama et al, 1989b)

We then studied the modified cadmium health effects by excessive dietary zinc plus copper.

Sixty-seven male mice of 2 or 18 months old were given pelleted food (5 g/day) containing 0 or 300 mg/kg cadmium, 12 (nutritional requirement) or 120 mg/kg zinc, and 50 (nutritional requirement) (NAS, 1978) or 1,000 mg/kg copper, for 12 weeks. Seventy items were determined to assess effects of

excessive dietary zinc plus copper on cadmium health effects at the end of the experiment.

In old mice, as indicated in Fig. 4, excessive copper diet aggravated cadmium health effects: body weight loss became remarkable either by excessive copper or by excessive copper plus zinc diet. Increased renal and hepatic weights by cadmium showed a tendency to be intensified both in the excessive copper group and in the excessive copper plus zinc group of old mice. Hemoglobin and hematocrit was depressed significantly either by excessive zinc, copper, or copper plus zinc in old mice.

No beneficial nor adverse effects either of excessive zinc, copper, and copper plus zinc diet were observed in young mice, except plasma ceruloplasmin recovering to the normal level. Effects were quite diverse in between old and young mice.

Intravenous copper supplementation alleviated cadmium health effects in young animals, as was seen in Experiment 3. However, cadmium health effects were

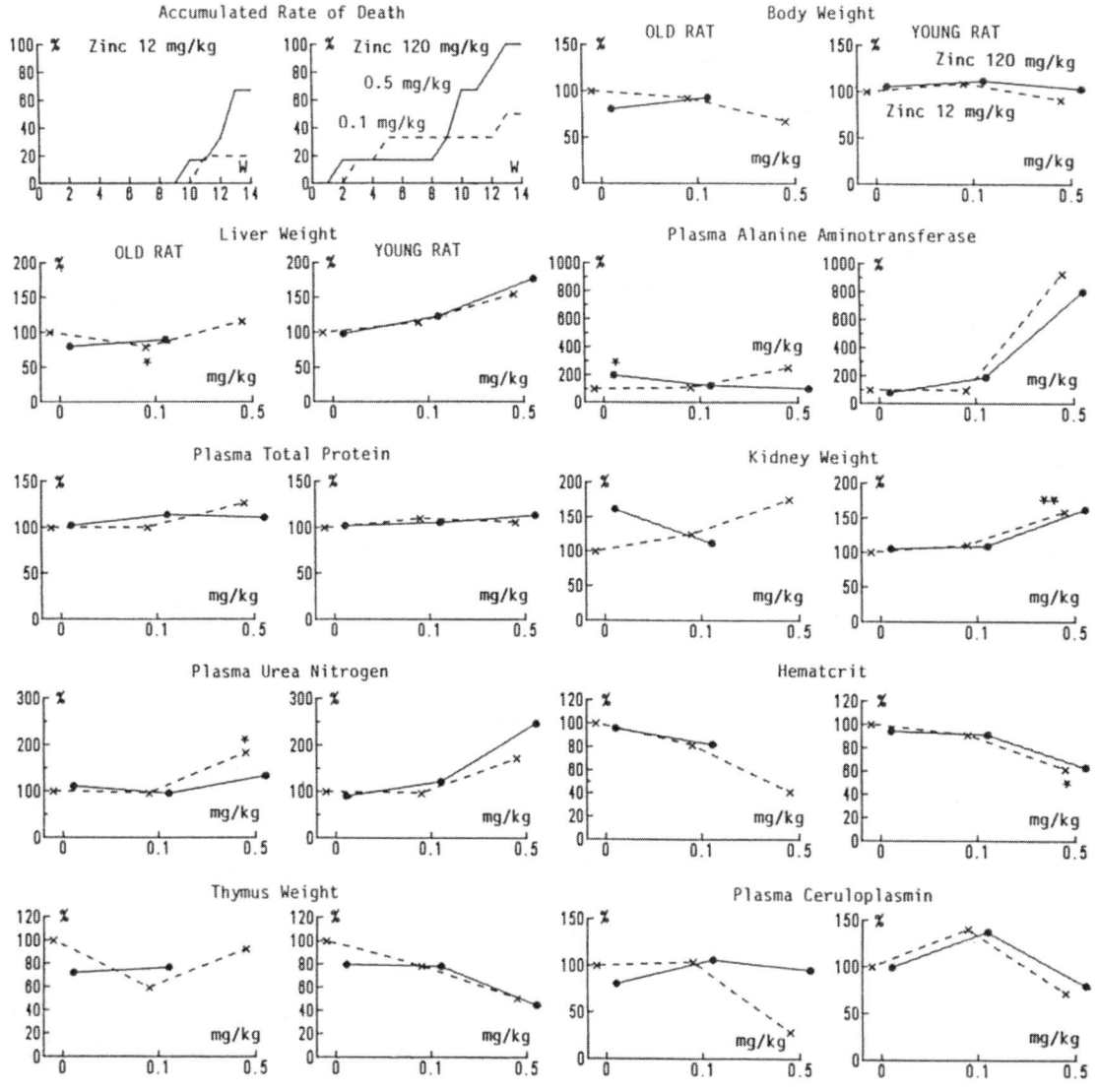

Fig. 3 Excessive dietary zinc, a factor modifying toxicity of cadmium in young and old rats given subcutaneous injections of cadmium (Nomiyama et al, 1989a)

unchanged or not alleviated by oral administration of excessive copper as was found in the present experiment. Plasma ceruloplasmin was elevated either in the excess copper group or the excess copper plus zinc group in young mice, but it is possible for copper not to improve cadmium health effects because of depressed intestinal absorption of copper by its binding to intestinal metallothionein (Davies and Campbell, 1977) or of different chemical forms in the target organs in between oral and intravenous administration.

Only beneficial effects of zinc on cadmium health effects were reported so far (Parizek, 1957; Perry et al, 1977), but aging differed the effects of excessive zinc and copper on cadmium health effects: aggravating effects in old animals, and alleviating or no effects in young animals. It is, therefore, strongly recommended to study interactions of copper and/or zinc with cadmium in old animals to innovate countermeasures to prevent or cure cadmium health effects prevailed among residents in cadmium polluted areas, because elderly residents are suffering from cadmium health effects.

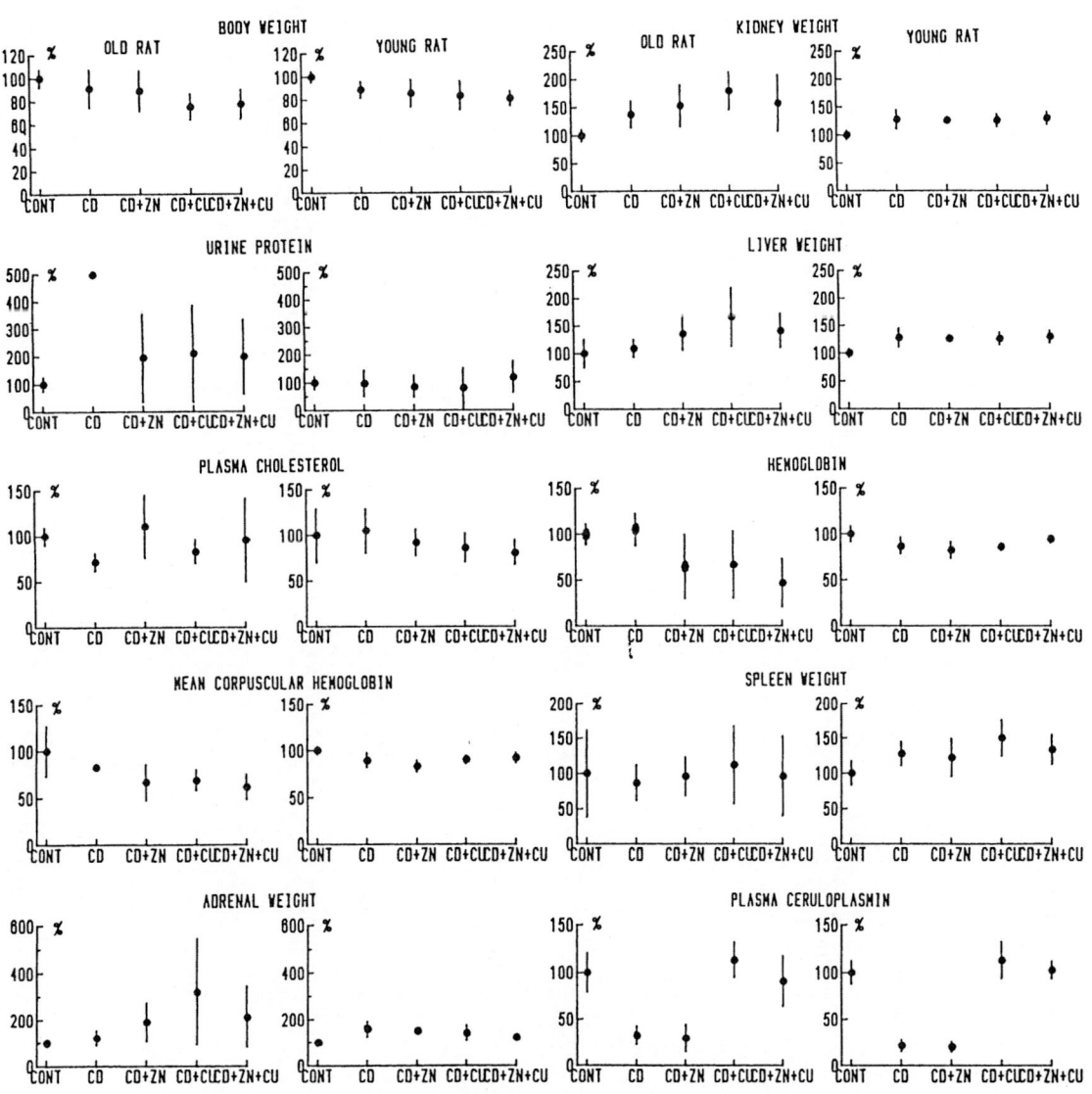

Fig. 4 Excessive dietary copper and zinc, a factor modifying toxicity of cadmium in young and old mice given oral cadmium (Nomiyama et al, 1989b)

EXPERIMENT 5: EFFECTS OF LOW-PROTEIN DIET ON CADMIUM HEALTH EFFECTS (Nomiyama et al, 1989c)

We turned to examine effects of low-protein diet on cadmium health effects.

Fifty-six male mice of 2 or 18 months old were given pelleted food (5 g/day) containing 0 or 300 mg/kg cadmium, and 20 (nutritional requirement) (NAS, 1972) or 7 % protein (marginal level) (NAS, 1972) for 12 weeks.

Cadmium elevated significantly the accumulated rate of death in 18 months old mice fed with low-protein diet containing 7% protein. Accumulated rate of death was 50% in 6 weeks, while no young animal fed low-protein diet of cadmium.

Plasma ceruloplasmin was also significantly depressed in the old mice fed low-protein diet.

Low-protein diet aggravated cadmium health effects, such as body weight loss, increased renal weight, decreased plasma ceruloplasmin and anemia both in old and young mice.

Cadmium health effects seemed more serious in old mice than in young mice on low-protein diet, as a conclusion.

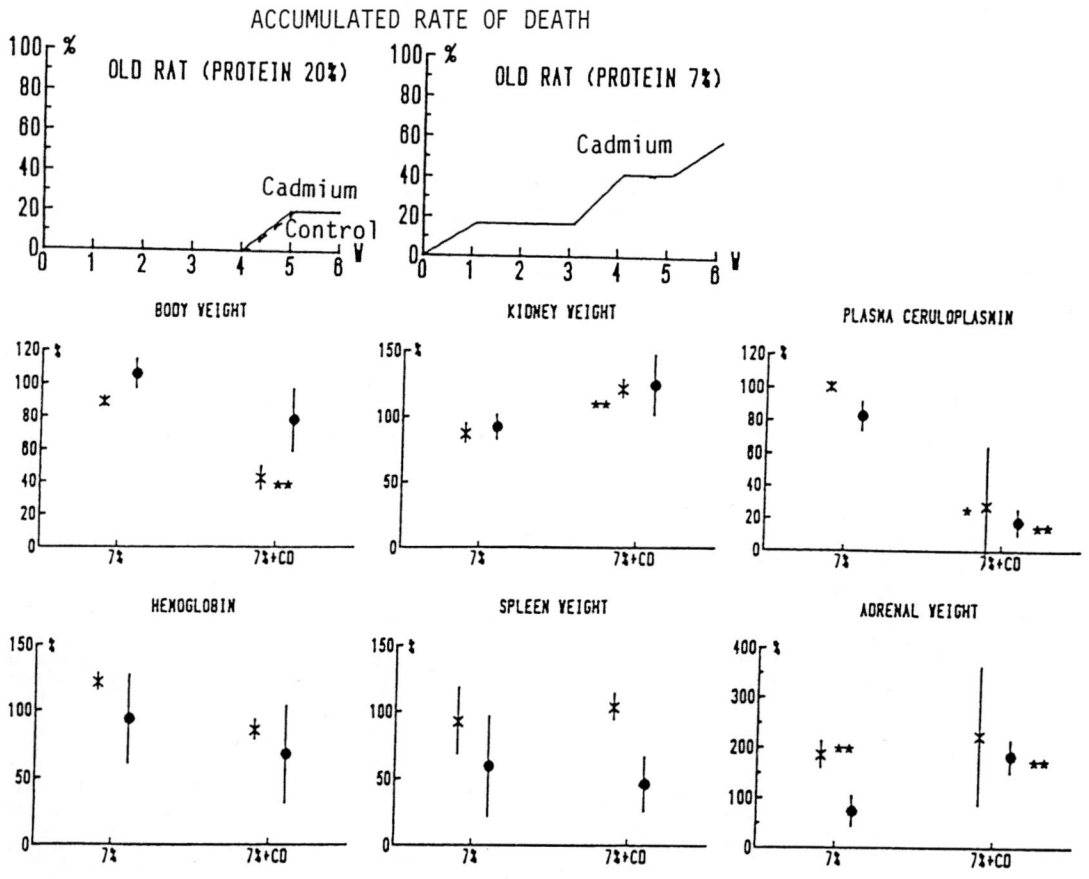

Fig. 5 Low-protein diet, a factor aggravating toxicity of cadmium in young and old rats given oral cadmium (Nomiyama et al, 1989c)
Upper 2 figures show accumulated rate of death in old rats fed diet containing 20 and 7 %. Dots and crosses in lower 6 figures represent for means of old and young rats. Single and double asterics indicated the statistical significance at 5 and 1 % level.

CONCLUSION

1) Remarkable diversion by area in the critical daily dose of cadmium in residents in cadmium-polluted areas in Japan, may be mainly caused by aging, metal pollution (zinc and copper) other than cadmium, and low-protein diet.

2) It may be difficult, therefore, to establish a single, conclusive and unanimous critical dose of cadmium for residents in cadmium-polluted areas.

3) It should be stressed to avoid excessive exposure to cadmium, zinc and copper in old residents in cadmium-polluted areas not to make cadmium health effects more serious.

4) Enough daily intake of protein are recommended for residents in cadmium-polluted areas to prevent aggravating cadmium health effects.

ACKNOWLEDGEMENT

We appreciate Mr. S. Watarai of Funabashi Pharm Co. Ltd. for his adequate advice in manufacturing synthetic diet used in the present study. The present study was supported in part by Japan Ministry of Education, Science and Culture (Grant-in-Aid for Scientific Research A-58440040).

REFERENCES

Davies NT and Campbell JK (1977) The effect of cadmium on intestinal copper absorption and binding in the rat. Life Sci 20:955-960

National Academy of Science (1978) Nutritional requirements of laboratory animals No. 10 Rat, mouse, gerbil, guinea pig, hamster, vole, fish. National Research Council, Washington

National Research Council (1979) Zinc, National Academy of Science, Washington

Nomiyama K and Nomiyama H (1982) Tissue metallothioneins in rabbits chronically exposed to cadmium, with special reference to the critical concentration of cadmium in the renal cortex. In: Foulkes EC (ed) Biological roles of metallothionein. Elsevier, N. Y., p 47-67

Nomiyama K and Nomiyama H (1983) Prevention and treatment of cadmium health effects by intravenous copper administration in rabbits. Proc 56th Ann Meeting Japan Assoc Ind Health 524-525

Nomiyama K and Nomiyama H (1986) Modified trace element metabolism in cadmium-induced renal dysfunctions. Acta Pharmacol Toxicol 59 (Suppl VII):427-430

Nomiyama K, Nomiyama H, Kamal AM and Ohshiro H (1984) Aging, a factor aggravating chronic toxicity of cadmium in rabbits. Jpn J Hyg 39:466

Nomiyama K, Nomiyama H, and Soni MG (1989a) Factors, modifying toxicity of cadmium in old animals (1) Excess dietary zinc in rats given subcutaneous cadmium. Jpn J Hyg 44:210

Nomiyama K, Nomiyama H, and Ishimaru Y (1989b) Factors, modifying toxicity of cadmium in old animals (3) Excess dietary cooper in mice given oral cadmium. Jpn J Hyg 44:212

Nomiyama K, Nomiyama H, and Ishimaru Y (1989c) Factors, modifying toxicity of cadmium in old animals (4) Low-protein diet in mice given oral cadmium. Jpn J Hyg 44:210

Nomiyama K, Yotoriyama M and Nomiyama H (1983) Dose-effect relationship between cadmium and β_2-microglobulin in the urine of inhabitants of cadmium-polluted areas (Japan). Arch Environ Contam Toxicol 12:147-150

Parizek J (1957) The destructive effect of cadmium ion on testicular tissue and its prevention by zinc. J Endocrin 15:56-63

Perry HM, Erlanger M and Perry EF (1977) Elevated systolic pressure following chronic low level cadmium feeding. Am J Physiol 232:H114

Shigematsu I (1989) The health survey of residents in areas with environmental cadmium pollution. Kankyo Hoken Report 56:69-345

Shigematsu I, Minowa M, Yoshida I and Miyamoto K (1979) Recent results of health examinations on the general population in cadmium-polluted and control areas in Japan. Environ Health Perspect 28:205-210

Author Index

Key Word Index